Guide to standard floras
of the world

Guide to standard floras of the world

An annotated, geographically arranged systematic bibliography of the principal floras, enumerations, checklists, and chorological atlases of different areas

D. G. FRODIN

Department of Biology
University of Papua New Guinea

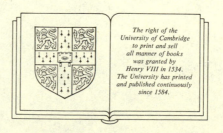

The right of the
University of Cambridge
to print and sell
all manner of books
was granted by
Henry VIII in 1534.
The University has printed
and published continuously
since 1584.

Cambridge University Press
Cambridge
London New York New Rochelle
Melbourne Sydney

Published by the Press Syndicate of the University of Cambridge
The Pitt Building, Trumpington Street, Cambridge CB2 1RP
32 East 57th Street, New York, NY 10022, USA
296 Beaconsfield Parade, Middle Park, Melbourne 3206, Australia

First published 1984

Printed in Great Britain by the University Press, Cambridge

Library of Congress catalogue card number: 82–4501

British Library Cataloguing in Publication Data
Frodin, D. G.
Guide to standard floras of the world.
1. Botany
I. Title
581   QK45.2

ISBN 0 521 23688 6

UP

To Sidney Fay Blake and Alice Cary Atwood
authors of *Geographical Guide to Floras of the World*
and
E. J. H. Corner
who has eloquently reminded us of what Floras are *for*.

Contents

Prologue

No branch of botanical literature is more useful, and at the same time more neglected than [floras]...For a beginner [the Flora] is the first, and one of the most important aids for obtaining botanical knowledge.

> de Candolle and Sprengel, *Elements of the philosophy of plants* (Edinburgh, 1821).

Quatenus bibliotheca in omni scientia primum à studioso evolvi debeat, ita etiam est Botanico *maxime necessaria*, quum multiplex usus inde deducitur...

> Linnaeus, *Bibliotheca botanica* (Amsterdam, 1736; reprinted Halle 1747).

Now, there are two different attitudes towards learning from others. One is the dogmatic attitude of transplanting everything, whether or not it is suited to our conditions. This is no good. The other attitude is to use our heads and learn those things which suit our conditions, that is, to absorb whatever experience is useful to us. That is the attitude we should adopt.

> Mao Tse-tung, 27 February 1957, in *Quotations from Chairman Mao Tse-tung* (New York, 1967).

Of all forms of human activity related to plants, that of knowing the kinds, properties and uses of such plants as grow in one's *Landschaft*, or 'parish', is perhaps the longest-established. Most, if not all, 'traditional' cultures centered on the land possess, or once possessed, a comparatively detailed knowledge of the local flora, in many cases recognizing the same species (and sometimes genera) as would a modern professional botanist; and, in like manner to many 'advanced' societies, this knowledge is best developed amongst a comparatively small circle of *savants*.

It is thus not surprising that, in those civilizations which achieved literacy in the pre-Columbian era, this kind of botanical knowledge should have come to be recorded at an early date. However, such works as are now known were largely compilations of what was common knowledge, considerable though this might have been, and for long were conceptually pragmatic. This obtains, for instance, in the oldest known extant geographically oriented botanical works, the treatises of Theophrastus reporting discoveries on Alexander's campaigns in the fourth century B.C.E. and *Nan-fang ts'ao mu chuang* of the fourth century C.E. on south China and Indochina, whose post-Columbian semantic descendants include the Swedish surveys of Linnaeus, the many geographical accounts of distant lands of the nineteenth century containing substantial botanical information, and, with greater impersonality, the Australian and Pacific-area terrain studies by various military and civilian agencies in the mid-twentieth century. Not, however, until the rise of the Western European tradition of independent scientific enquiry and the consequent development of a systematics based on the *nature* of plants *in themselves* rather than on traditionally pragmatic values did the compilation of organized area floristic accounts and plant lists become a distinct activity, which after the Linnean revolution and during the century of European colonial expansion came to constitute a significant proportion of what in some lands was to be called 'special botany' (*spezieller Pflanzenkunde, bijzondre plantkunde*) or, in more modern parlance, systematic botany in a broad sense. Floristic studies and flora- and checklist-writing have ever since constituted an important part of the work of this subdiscipline and the published (and, increasingly, semi-published) results cumulatively contain an immense amount of botanical information. To the non-specialist, these works, along with the provision 'on demand' of identification and information services, perhaps represent the most easily comprehensible aspect of the systematist's work.

The relative importance of area floras and checklists in the world of systematic botany has varied over nearly four centuries, but since the 1930s, and especially since 1945 and the advent of liberal but often short-term state support, floras, checklists, and related area works and contributions thereto have come to predominate. Stafleu[1] has termed the present cycle as 'an age of floras and floristic work', but at the same time notes that this has partly come about at the expense of serious monographic and revisionary work, a trend strongly aggravated by the virtual destruction of the Berlin Herbarium and the German systematics profession by the end of World War II. Current indications are, as Jäger[2] has noted, that this pattern will continue, causing the collective mass of floristic works, especially the more significant ones, to become the single most important source of modern taxonomic knowledge, and thus by default supplanting the great synthetic works of the mid-nineteenth to early-twentieth centuries such as the *Prodromus systematis naturalis regni vegetabilis, Monographiae Phanerogamarum, Die natürlichen Pflanzenfamilien,* and *Das Pflanzenreich.* Of a verity have floras expanded in scope far beyond what was originally conceived: from simple inventory (and diagnosis) they have successively assumed the roles of identification manual and taxonomic encyclopedia, in the latter case now often also accounting for current notions on the classification of a given group above, as well as at, species level. In too many instances, however, their *effective* role has been lost sight of.

That floras and checklists had a distinctive place in botanical literature was already recognized from late in the eighteenth century and by 1820 had become canonical. Classified lists of those which were independently published appear in general bibliographies from Linnaeus onwards, but by 1879 their importance had become so recognized that a separate list was deemed necessary. This first list, a slender but closely printed pamphlet of twelve pages, was *The floras of different countries* by G. L. Goodale of Harvard University. Two others have followed since: the Lloyd Library's *Bibliography of the floras* by W. Holden and E. Wycoff (1911–14), rather more comprehensive than Goodale's list but like it compiled 'in-house' and (following traditional practice) limited to independently published works, and the original and critical *Geographical guide to floras of the world* by S. F. Blake and A. C. Atwood (1942–61, not completed). Several regional and national lists have also been produced.

These earlier general guides to floras, however, were produced when the totality of botanical literature, even accounting for that published in periodicals and serials, was far less and overall bibliographical control more satisfactory (particularly before World War I). These conditions no longer existed by the 1960s. Twenty years alone were required by Blake for distillation of volume 2 of the *Geographical Guide* from the vast mass of Western European floristic literature, and by the end of that period, volume 1 was already in need of revision, although the flow of new literature had not yet taken on the proportions of the 1970's flood. Developments since 1960 have been such that, although Blake was said to be well aware of the magnitude of his task,[3] it is likely that in the years after his death, even had the will and the means existed, completion of the work on the original plan would, for a variety of reasons, have been very difficult if not impossible.

At the present time the climate for a revised and completed version of the *Geographical Guide* along its original lines is even less favorable, however much it may be desired in some quarters. The exponential growth of biological literature in the 1980s, and its control, is but one factor: others include the effects of the disruption and fragmentation of the world botanical information system due to two World Wars, trans-Atlantic isolationism in the inter-war period, additional centers of botanical activity and publication, changes in scientific fashions, and political and social developments, of recent decades including the currently changing relationship between science and society in a more austere economic climate.[4]

Moreover, much current retrospective bibliographical work has been directed elsewhere: *Index nominum genericorum*, *Taxonomic literature* (and its 2nd edition referred to as *TL-2*), *Bibliographia Huntiana*, etc. With respect to floristic bibliography, the fragmentation and partial disintegration of the botanical *referat* system alone has posed significant obstacles which only sophisticated organization and large financial expenditure can overcome. A number of the principal sources utilized by Blake and Atwood no longer exist; these include their primary source, the USDA Botany Subject Catalogue, terminated in mid-1952 (fortunately for others, it appeared in book form in 1958) and current literature coverage, especially of independently published items which comprise the majority of significant contributions, is more diffuse and uneven and less complete than in the past, although since 1950 two indexing journals specifically dealing with systematic botany have come into being: *Excerpta Botanica*, A (from 1959) and *Kew Record of Taxonomic Literature* (from 1971). Lately, 'semi-publication' has presented an increasing problem to bibliographers as inexpensive, comparatively permanent modes of offset printing have become widely diffused. Another approach was needed if the heterogeneous flood of floristic literature, which had increased greatly during the 1950s and later came to be considered a key contributor to what Heywood[5] has termed the contemporary 'crisis' in taxonomy, was ever to be mastered and meaningful world-wide coverage once more provided.

The actual stimulus for the present book, from which grew its basic idea, arose from a conversation in the summer of 1962 with a fellow student at the University of Michigan Biological Station in the northernmost part of the state's Lower Peninsula. As an invertebrate zoologist planning to participate in the 1963 International Indian Ocean Expedition, he desired to obtain some basic references on the vascular plants of the islands in the region. A search through the first volume of the *Geographical Guide* revealed a goodly number of titles, but upon reflection it became apparent that many were too specialized or restricted in scope for the kinds of information sought. Ultimately it was found that a comparatively limited selection of floras and enumerations would provide, within a reasonable compass, a proportionately high degree of useful information about the region's vascular plants; in other words, these works could be viewed as 'standard' floras.

From this beginning there developed the idea that such a selective process could, with variations, be applicable world-wide, and that this would in due time enable the preparation of a one-volume annotated general bibliography of 'standard' floristic works on vascular plants which would cover the entire world, region by region. I also came to believe that such a work would be of particular interest to non-botanists as well as to botanists without a detailed knowledge of regional floristic literature outside their own sphere. Other factors contributing to a decision to prepare such a bibliography were the limited nature of lists of 'useful' floras provided in systematics textbooks as well as the unlikely prospect, noted above, that the *Geographical Guide* would ever be completed, especially considering the death of its senior author in 1959 (as it stands, it does not cover central or eastern Europe or the continent of Asia). Furthermore, in addition to Part I becoming increasingly out of date, the size of Part II appeared likely to daunt all those not having some familiarity with the vast corpus of western European floristic literature.

During 1962–3, various experiments in relation to depth of coverage were attempted, but the main catalyst proved to be in the pair of 'Green Books'[6] published by the Flora Europaea Organization which came to my attention in March 1963 on a visit to the University of Michigan Herbarium. Therein was given a list, with supplement, of 'standard' floras of Europe deemed most significant for the preparation of *Flora Europaea*. (The 'standard' flora concept had itself evidently been formulated by the Organization in the mid-1950s.)[7]

The final result, for which work originally began in a substantial way during the summer of 1963 at the Field Museum of Natural History, Chicago, is represented by the present book. However, lack of experience as well as time suggested that the *Guide* be first written up and distributed in short-title form without annotations or commentary. That effort materialized as the mimeographed booklet written largely at the University of Tennessee, Knoxville, and issued from its Department of Botany in 1964.

The consequent strong and continuing demand for that booklet, even to the time of final revision of this preface, ultimately led me to consider an expanded, more definitive edition. For various reasons, however, no serious research was begun until the end of 1967 when, encouraged by representations from colleagues all over the world as well as a publication proposal from the University of Tennessee Press, I felt compelled to undertake the task, one which would be greatly facilitated by being at the time at the University of Cambridge. Primary compilation of the necessary material was undertaken largely in Cambridge and London, with additions from Australian libraries in 1971 following my move to an academic position at Port Moresby (Papua New Guinea), but short visits were made to libraries in several other centers.

It was a basic tenet of both the preliminary and the present versions of this book that as far as possible all titles selected for inclusion should be examined and annotated at first hand. To a very large degree, this has been achieved, in a few cases with the aid of photocopied extracts. Where an entry has had to be based upon a secondary source, that source has been indicated.

The original selection of titles was made by systematic browsing along the shelves of the Botany Library in the Field Museum. Additions were made through work in the University of Tennessee (Knoxville) Libraries and short visits to some special botanical libraries in the central and eastern United States. Guidelines for the selection process were also provided by a number of secondary sources as well as advice from colleagues. For the present version, the botanical libraries at Cambridge (England), the Royal Botanical Gardens, Kew, and the British Museum (Natural History) were extensively utilized, along with the working library of the Flora Europaea Secretariat (at Liverpool, later at Reading), the library of the Komarov Botanical Institute, Leningrad, and the libraries of the New York Botanical Garden and the Arnold Arboretum/Gray Herbarium at Harvard University. Small amounts of work were done at additional special botanical libraries as opportunities arose. Advice was also sought from a great number of other botanists, both in person and in writing. It may here be noted that the number of botanical libraries in which a substantial primary search for floras and related works may be carried out efficiently is comparatively small: five in the United States (in four centers) and three in Europe (in two centers). It is in London that the most substantial collections of these works exist, and it has been my good fortune to have been able to make extensive use of them over the years.

As might be expected, the coverage of material in the periodical literature has presented the greatest problems, both in ferreting out references and in seeing the articles concerned. No good cumulative classified index is currently available and extensive searches of the various abstracting and indexing journals would have been tedious and very time-consuming. Furthermore, floristic material is found in a wide and scattered range of biological, zoological, general scientific, and other periodicals as well as in those more specifically concerned with botany. In more recent years, material published or 'semi-published' in various kinds of technical series or runs of 'occasional papers' emanating from a plethora of university departments, institutes, and other organizations has proliferated to an inordinate degree. A great misfortune has been the above-mentioned discontinuance of the botany subject catalogue in what is now the United States National Agricultural Library; this provides the best classified source for the first half of the twentieth century. Its suspension without an adequate replacement can only be deplored. Fortunately, in more recent decades there has been a marked rise in the number of regional compilations of botanical literature (both bibliographies and indices), and much use was made of them; they are now available over many parts of the world though variable in scope and quality. Some of these provided their own selections of key floristic works. Lists of references in major floras themselves were searched for periodical material. It must be confessed, however, that a goodly number of items were yet found 'by chance'. In all respects, having made a systematic study of a world-wide tropical and subtropical genus, *Schefflera* (Araliaceae), which followed earlier work on *Cytisus* and allied genera (Leguminosae – Genisteae), proved a considerable asset.

Principal secondary sources utilized included, above all, the two volumes of Blake and Atwood's *Geographical Guide*. Another useful but older general source was *Bibliography relating to the Floras* (1911–14) in the *Bibliographical Contributions* series of the Lloyd Library. Other key works were, in the main, regional: among them were the bibliographies in Hultén's *The amphi-Atlantic plants* (1958) and *The circumpolar plants*, II (1971); *Bibliography of eastern Asiatic botany* by Merrill and Walker (1938) and its *Supplement* by Walker (1960); the two volumes of *Island bibliographies* by Sachet and Fosberg (1955, 1971); *Botanical bibliography of the islands of the Pacific* by Merrill, with subject index by Walker (1947); *Vvedenie v botaničeskuju literaturu SSSR* by Lebedev (1956) and *Literaturnye istočniki po flore SSSR* by Lipschitz (1975); the *Guide for contributors to 'Flora Europaea'* and its *Supplement*, both by Heywood (1958, 1960), otherwise known as the 'Green Books'; *History of botanical researches in India, Burma, and Ceylon*, II: *Systematic botany of angiosperms* by Santapau (1958), and *A guide to selected current literature on vascular plant floristics for the contiguous United States, Alaska, Canada, Greenland, and the US Caribbean and Pacific Islands* by Lawyer, Miller, Morse, and Kartesz (in press). Some individual library or union catalogues were useful, particularly the *Botany Subject Index* (1958), which constitutes the above-mentioned former USDA botany subject catalogue of 1906–52 in book form, the *Catalogue of the Library, Royal Botanic Gardens, Kew* (1974) with both author and classified divisions, the *Catalogue of the Library of the Arnold Arboretum* (1914–33), and, for bibliographic control, the *National Union Catalog* [USA]: *Pre-1956 imprints* and its retrospective and post-1956 supplements together with the *Botany Subject Index* and *Biological Abstracts*. Major indices used from time to time included *Excerpta Botanica* and *Kew Record of Taxonomic Literature* and, at regional level, *Index to American Botanical Literature* (and the former *Taxonomic Index*), *AETFAT Index*, *Flora Malesiana Bulletin*, and the European and Australasian indices published through the International Association for Plant Taxonomy in the 1960s. For search purposes, however, only occasional use was made of *Biological Abstracts*, however, and with the advent of the many regional botanical bibliographies now in existence there proved relatively little need to consult the older general indices, even had they been readily available. Of general current awareness lists, those extensively utilized included the *referat* sections in *Taxon* and *Progress in Botany* [Fortschritte der Botanik] as well as the 'semi-published' accession lists from the New York Botanical Garden (now defunct) and Kew libraries (the latter classified); these were supplemented by a range of dealers' catalogues (mainly Antiquariat Junk, Koeltz, Krypto, Scientia, Stechert-Hafner, and Wheldon and Wesley) and trade announcements (the latter sometimes providing descriptions). None of these,

however, acted as substitutes for examination of the originals save when no other opportunity was available, but nevertheless they prove especially valuable whilst working in a relatively remote country such as Papua New Guinea.

The actual preparation of the *Guide*, although undertaken in 1970, was unfortunately considerably prolonged on account of my many university responsibilities as well as the attractions of a tropical flora, and only in late 1975 could it be terminated. The remoteness of Port Moresby was also a handicap, but on account of circumstances perhaps less so than might be imagined. More importantly, it enabled the work to be written from the point of view of a botanist attempting to cope with an imperfectly known tropical flora and actively involved in teaching. Much of the writing was accomplished during spells in remote outstations and camps while 'on patrol', often when waiting for airplanes or sitting out the rain. Following submission of the manuscript, a variety of technical difficulties led to a long delay in publication and in January 1979 it was formally transferred to Cambridge University Press. Accumulating additions and other changes as well as ideological refinements necessitated complete revision of the manuscript and this was largely carried out in Papua New Guinea and Australia during study leave from July 1979 to February 1980. Overseas visits in 1973, 1975–6, 1976–7, and 1978–9 enabled coverage of new or overlooked works. As far as possible 1980 is taken as the 'cut-off' year, with some indication of likely future developments and publications given in the various regional commentaries.

It is hoped that the *Guide* as now presented will meet the needs of a wide range of users, both botanical and non-botanical. It has been written in the belief that, since a thorough revision and completion of Blake and Atwood's *Geographical Guide* is not likely in the foreseeable future (and in any case would have to be an institutional project), a simpler one-volume analytical work would serve as a practical and more easily realizable alternative which would yet suffice for a majority of interested persons.

The work as it stands, though, is also intended to draw attention to the need for developments in floristic and other botanical bibliography comparable with what Heywood[8] and some other authors have called for with regard to floras generally. Although the necessity for various kinds of functional articulation and resource redeployment was long ago recognized in

bibliographic science through sheer force of circumstances, it has been slow to come to systematic botany:
the dream of the definitive, hard-cover, omnibus work
has been long-persistent. Yet two (or more) functions
are served in both floras and botanical bibliography –
chiefly the archival and the practical – which in most
cases can no longer usefully be combined within a single
work and now require separation in publication. It is
here suggested, for instance, that comparable selectivity
with articulation is as necessary for flora-bibliography
as for floras themselves, and that this is but part of a
continuing process in information handling with implications for all fields of knowledge.[9] A work such as the
Geographical Guide – considered in its day as 'selective'
in relation to the general corpus of systematic botanical
literature, and representative of the 'new trend' of
scholarly bibliography which arose out of World War
I[10] – is seen here as to a marked extent now archival,
whereas the present *Guide*, though less extensive in its
coverage, should prove useful at a more practical level
whilst still remaining a meaningful indicator. It thus
continues to stand for the 'state of the art' in Malclès'
sense. Put in another way, it represents a level of
selection twice removed from the coverage spread
represented by the last of the great retrospective subject
bibliographies (Pritzel, 1871–7; Rehder, 1911–18).
'Standard' floras may be viewed as having a place in
floristic literature comparable to the head of a comet;
the rest forms the gradually thinning tail.[11] As with
floras themselves, the overall view is that there is room
for both kinds of bibliographical works.

Any deliberate abridgement of the kind represented by the present work, though, always involves
subjective decisions over inclusion or exclusion of
particular titles, even though they be based upon
heuristic criteria. Many items inevitably will have a
'borderline' status, even given the intuitively recognized
'point of balance' which limits this work. Such items
may show a 'shift' in that they possess rather more
importance in a local as opposed to a global context.
All that can be said is that all care has been taken in
such decisions, using the only computer available.
Nonetheless, I shall always welcome any reasonable
suggestions for addition (or deletion) of titles (within
limits) with appropriate arguments. It should also be
noted that the actual preparation of this work has to a
considerable extent been carried out at locations remote
from large botanical libraries, making quick rechecking
or reinterpretation of sources difficult or impossible;
unintentional errors may, therefore, have crept in. Any
technical omissions or errors or misleading statements
should, if possible, be brought to my attention. All
changes accepted would be incorporated in a supplement
contemplated for publication in the late 1980s.

Finally, it should be noted that whereas earlier
bibliographies of floras have been largely empirical or
descriptive, the present work attempts as well to be
analytical and interpretative, essaying also some
integration on historical principles. The belief has,
latterly, grown in my mind that a classified subject
bibliography should not only present and describe titles
but also reach outwards: to act as a *Spiegelbild der
Forschungsergebnisse*, a mirror on the progress of the
subject,[12] as well as to guide – in the words of an earlier
promotor of bibliographic science – 'a young man
[who], instead of wasting months getting lost in unimportant reading...would be [thus] directed toward
the best works and more easily and quickly attain a
better education'.[13] It is hoped that this *Guide*, at least
to some extent, fulfills these ideals, which with
variations are of long standing in bibliography. Modern
methods of bibliographical analysis moreover, indicate
that a literature cross-section of the kind presented here
can be about as meaningful as a comprehensive
bibliography in revealing patterns of development in
the subject, in this case floristic botany. Further
research on the themes embodied here might (1) utilize
citation analysis of a wide range of floristic articles as
a means of quantifying the selection criteria and the
'point of balance', and (2) estimate patterns of usage
through time by analysis along similar lines of a series
of historical cross-sections of the literature. Both could
serve as contributions to the history of systematic
botany; and other insights might also be obtained in
ways not yet suspected.

D. G. Frodin
Port Moresby, Papua New Guinea
August 1980/January 1982

Notes

1 Stafleu, F. A., 1959. The present status of plant taxonomy. *Syst. Zool.* **8**: 59–68.

2 Jäger, E. J., 1978. Areal- und Florenkunde (Floristische Geobotanik). *Prog. Bot.* **40**: 413–28.

3 Schubert, B. G., 1960. Sidney Fay Blake. *Rhodora*, **62**: 325–38.

4 Drucker, P. F., 1979. Science and industry, challenges of antagonistic interdependence. *Science*, **204**: 806–10.

5 Heywood, V. H., 1973. Taxonomy in crisis? or taxonomy is the digestive system of biology. *Acta Bot. Acad. Sci. Hung.* **19**: 139–46.

6 Heywood, V. H., 1958. *The presentation of taxonomic information: a short guide for contributors to Flora Europaea.* 24 pp. Leicester: Leicester University Press; *idem*, 1960. *Supplement.* 20 pp. Coimbra, Portugal.

7 Heywood, V. H., 1957. A proposed flora of Europe. *Taxon*, **6**: 33–42.

8 Heywood, V. H., 1973. Ecological data in practical taxonomy. In *Taxonomy and ecology* (ed. V. H. Heywood), pp. 329–47. London: Academic Press.

9 Garfield, E., 1979. *Citation indexing.* New York: Wiley.

10 Malclès, L. N., 1961. *Bibliography* (Trans. T. C. Hines). New York: Scarecrow (reprinted 1973). [Originally publ. 1956, Paris, as *La bibliographie.*]

11 Garfield, E., 1980. Bradford's Law and related statistical patterns. *Current Contents/Life Sciences* **23**(19): 5–12.

12 Simon, H.-R., 1977. *Die Bibliographie der Biologie*, p. 75. Stuttgart: Hiersemann.

13 Napoléon I to Finkestein, 19 April 1807; quoted in Maclès, 1961, p. 75 (see n. 10).

Acknowledgments

The preparation of both the preliminary version and this present edition of the *Guide to standard floras of the world*, especially the latter, has necessitated the consultation over several years of a great many sources, as noted in the Preface, and furthermore has involved the assistance in various ways of numerous individuals and institutions. These latter must now be acknowledged formally, for without their aid this book could not have appeared in its present form, or indeed at all. The author wishes here to express his deep appreciation to all the support and assistance given him over the nearly two decades required for gestation of the work in its present form.

For the original version (1963–4), the author wishes to record his sincere gratitude to all those in charge of library collections for granting him access to them, particularly the late John Millar, then Chief Curator of Botany at the Field Museum of Natural History, Chicago, and those in charge of the University of Tennessee (Knoxville) libraries. Thanks are also due to the authorities of the Biology Library of The University of Chicago; the Lloyd Library, Cincinnati, Ohio; the Missouri Botanical Garden Library, St Louis; the library of the Department of Botany, Smithsonian Institution; the New York Botanical Garden Library; and the libraries of the Arnold Arboretum and Gray Herbarium of Harvard University, Cambridge, Massachusetts.

Advice and assistance was also given by many

individuals, but the author is particularly indebted to the following: E. G. Voss, University Herbarium, University of Michigan, for an introduction to the *Flora Europaea* 'Green Books', L. B. Smith, Washington, for assistance with South American references; and above all to A. J. Sharp and other staff and students at the Department of Botany at the University of Tennessee, Knoxville, for their continuing interest in and support of the project. It was Prof. Sharp who made it possible for the preliminary edition to be reproduced and circulated around the world.

The preparation for and writing of the present version has unfortunately extended over a much longer period (late 1967 to mid-1980), owing to the considerably expanded format, changes in the philosophy of the work, publication difficulties, and the author's many other responsibilities while at Cambridge and in Port Moresby. A major contributing factor to the time span was naturally the decision to annotate, as far as possible, all floristic works included in the *Guide*. This made it necessary to examine personally, or obtain full notes upon, the contents, style, and philosophy of each title, and the author considers himself fortunate to have been able to carry out much of the work of compilation in Europe and especially in London. For completeness of world-wide coverage and for convenience of access and usage, the libraries of the Royal Botanic Gardens, Kew, and the Department of Botany, British Museum (Natural History), are perhaps without peer for research on a work of this kind; and it was, as noted in the Prologue, at these two libraries that the greater part of the materials for the present edition was compiled during 1968–70 and in short intervals in the succeeding ten years. Special thanks are therefore due to R. G. C. Desmond and V. T. H. Parry, successively Librarians at the Royal Botanic Gardens, Kew, and their assistants, and to Miss P. I. Edwards, formerly Botany Librarian, Department of Botany, British Museum (Natural History), and her successors, for their help (and patience!) during my extended visits to their libraries.

A significant amount of compilation was also carried out in 1971 and again in 1979–80 at the library of the Royal Botanic Gardens and National Herbarium of Victoria in Melbourne. This resource is perhaps the most extensive of its kind in Australasia, despite past neglect, and proved of great value at a time when substantial work on the general chapters and area commentaries was necessary but owing to circumstances beyond my control could not be done in Europe or the United States. My thanks are due to the director and staff of that institution, but especially to J. H. Ross, Senior Botanist, and Miss Olwyn Evans, assistant in the library. The help of J. Ashworth, Assistant Secretary, Department of Lands and Environment of Victoria, in resolving an unforeseen crisis over access to the facilities is also hereby acknowledged. During the second period in Melbourne, much use was also made of the Baillieu Library and of the branch library in the Department of Botany in the University of Melbourne, and the opportunity to make use of these well-endowed resources is much appreciated.

Extensive use was naturally made of the University Library, the Scientific Periodicals Library, and the Libraries of the Department of Botany and of the Botanic Garden in the University of Cambridge whilst the author was in residence as a Research Student from 1967 to 1970. As one of the centers for preparation of *Flora Europaea*, the Department of Botany housed a fine collection of major European floras, ably cared for in the Herbarium by P. D. Sell and (at the time) S. M. Walters. It was under the guidance of Prof. E. J. H. Corner, however, that the varying worth of tropical floras came to be appreciated through research into the large genus *Schefflera* (Araliaceae), a stimulus enriched by subsequent personal experience. This augmented earlier experience at Liverpool in 1964–5 when a study was made of *Cytisus* and its allies (including preparation of an account for *Flora Europaea*) under the direction of Prof. V. H. Heywood.

Other resources substantively utilized include the libraries of the Royal Botanic Gardens, Sydney, and the Commonwealth Scientific and Industrial Research Organization, Black Mountain, Canberra; the library of the Komarov Botanical Institute, Academy of Sciences of the USSR, Leningrad; the library of the Conservatoire et Jardin Botaniques, City of Geneva; the libraries of the Institut für systematische Botanik, Universität Zürich, the Botanische Staatssammlung, München, the Rijksherbarium, Leiden University, and the Botaniska Avdeling, Naturhistoriska Riksmuseet, Stockholm; the library of the New York Botanical Garden; the libraries of the Arnold Arboretum and Gray Herbarium of Harvard University, Cambridge, Massachusetts; and the libraries of the Linnean Society of London and the Commonwealth Forestry Institute, Oxford. Use was also made of the library of the Flora

Europeaea Secretariat, both in Liverpool and in Reading, and of a number of private collections. The author is much indebted to all those persons in charge of institutional libraries as well as private owners for permission to consult the collections in their care and for their assistance in locating needed references.

As with the earlier version of this work, the author is indebted to all those who freely gave assistance during the various stages of preparation and writing of the present edition. The difficult task of searching out, selecting, and locating the various European floras and manuals occupied a goodly amount of attention in the early stages; in this connection, particular thanks are due to the Flora Europaea Organization (and especially to S. M. Walters) for arranging to have a draft of the bibliographic text of Division 6 (Europe) typed, mimeographed, and sent from Reading to all regional advisers for comment. To all those who replied, many thanks. Thanks are also due to Prof. V. H. Heywood, now at Reading, for advice on European floras generally, and especially A. O. Chater, London (formerly Leicester), for assistance over several years (mainly before 1977) in locating and annotating obscure works and for arranging contacts with Soviet botanists.

The very exacting and time-consuming task of selecting titles and preparing text for those sections of the book covering the Soviet Union was considerably eased through the generous assistance of M. E. Kirpicznikov of the Komarov Botanical Institute, Leningrad. Not only did he prepare extracts and sample pages from a goodly number of works scarcely available outside the Soviet Union but he also sent a copy of S. J. Lipschitz' *Literaturnye istočniki po flore SSSR*, mentioned above, by air post to New Guinea immediately upon its publication. Moreover, during my visit to Leningrad in the summer of 1975, he graciously read through the completed manuscript for those portions covering the USSR and made many valuable suggestions. In addition, V. I. Grubov of the same Institute gave advice on his special region, central Asia (i.e., from Tibet to Mongolia). The author is also indebted to Prof. Al. A. Fëdorov, Director of the Institute, for permission to make use of the Institute library as well as the collections for a period of several days following the International Botanical Congress, as well as during the Congress itself.

Other botanists in Europe who gave assistance in various forms and to whom the author is likewise indebted include K. Browicz, Zakład Dendrologii, PAN, Kórnik, Poland (Poland and adjacent countries); H. M. Burdet, Geneva (Corsica and other parts of the Mediterranean, as well as general advice); Prof. E. Hultén, Stockholm (Eurasia in general); L. A. Lauener, Edinburgh (China); Prof. C. G. G. J. van Steenis and other staff members of the Rijksherbarium, Leiden (Malesia and adjacent regions); F. White, Oxford, and Prof. J. Léonard, Brussels (Africa); and especially Prof. F. A. Stafleu, Utrecht, for his general advice, criticism, and support.

Botanists in Asia who rendered assistance include R. I. Patel, Baroda, Gujarat State, India (India); Profs. H. Hara and H. Inoue, Tokyo (works published in Japan; Korea); K.-C. Oh (Korea); Prof. Pham Hoang Hô, Ho Chi Minh City (Indo-China); Stella Thrower and Y. S. Lau, Hong Kong (China, Hong Kong); H. Keng, Singapore (China and other areas); and B. C. Stone, Kuala Lumpur (miscellaneous).

In Australia, advice was received from J. H. Ross, Melbourne (Africa and Australia) and Hj. Eichler and A. Kanis as well as the late Nancy Burbidge, Canberra (Australia, Europe, and in general). Casual comments were advanced by many other colleagues on that continent in the course of a number of visits over the past decade while pursuing this and other, perhaps too ambitious, projects.

Assistance from Africa was received from E. A. C. L. E. Schelpe, Cape Town, and Prof. H. P. van der Schijff, Pretoria (southern Africa). Some pointers regarding the Tethyan side of that continent were suggested by Marie-Thérèse Misset, Oran, Algeria.

As with Europe and the Soviet Union, sifting through the mass of more recent floristic literature on North America was a not inconsiderable task. Special thanks accrue to E. L. Little, Jr., Washington, DC, for his general advice and for information on woody floras and on the Americas in general, and to L. E. Morse, New York, for general advice and for sending a copy, in advance of publication, of the typewritten manuscript of *A guide to selected current literature* by Lawyer and others, mentioned in the Prologue. Other advice was received from A. Cronquist and N. Holmgren, New York; P. F. Stevens, Cambridge, Mass.; and S. G. Shetler and C. R. Gunn, Washington.

Botanists in the United States who gave advice for other parts of the world included L. B. Smith, Washington, DC (Brazil); In-cho Chung, Mansfield,

Pa. (Korea); Prof. A. Löve, formerly of Boulder, Colo. (miscellaneous, but especially the Arctic regions); E. H. Walker, formerly of Washington (China); and especially F. R. Fosberg, Washington (general advice and support, and for steering me through the scattered shoals of the insular literature).

During the years the author was resident in England and on subsequent visits, many staff members of both the Royal Botanic Gardens, Kew, and the Department of Botany, British Museum (Natural History), gave freely of their time and knowledge to answer my questions regarding the floristic literature of many different parts of the world, thus enabling a kind of collective picture to be formed. My 1970 sojourn at Kew furthermore coincided with a series of staff briefings related to an internal reorganization of responsibilities in the Herbarium; I am indebted to Prof. J. P. M. Brenan, then Keeper, for copies of the area circulars produced for these briefings.

Special assistance on China and Korea was received from J. Needham, Cambridge, and E. Wu, Harvard-Yenching Institute, Cambridge, Mass. Advice on bibliographic matters was given by W. T. Stearn, London, and by the staffs of the University of Tennessee Press and Cambridge University Press.

Members of the botanical sodality in New Guinea, past and present, as well as professional and personal visitors, assisted in various ways and gave moral support. Among them were J. Croft, J. Dodd, V. Demoulin, Elizabeth Gagné, Lord Alistair Hay, M. Heads, Camilla Huxley, R. J. Johns, I. M. Johnstone, P. Kores, G. Leach, P. van Royen, B. Verdcourt, Estelle van der Watt, and A. Wheeler. Margaret O'Grady, formerly on the staff of the University of Papua New Guinea Library, assisted in tracking down some obscure items, one of which proved not to exist as cited.

The author also owes much to the late J. L. Gressitt, and to his associates H. Sakulas and A. Allison for enabling accommodation to be made available at Wau Ecology Institute, in the mountains south of Lae in Papua New Guinea, for an extended period in 1979. This meant that much of the final revision of this book could be accomplished in comparative comfort. Some chapters were rewritten at their nearby Mt Kaindi branch house, where near-total isolation at 2362 m in a diurnal temperature range of 9 to 23 degrees C acted as strong incentives! All associated with the Institute gave support and encouragement to the work. In Melbourne over a two-month period from December 1979 to February 1980, residence at Graduate House, University of Melbourne, was kindly granted by the Warden, W. E. F. Berry. Several residents, botanists and otherwise, provided conversation and moral support over this period; particular thanks are due to P. Bernhardt and D. Fleming. A further short visit to Melbourne and Canberra was made just before completion of the manuscript, which had been unavoidably delayed; at this time A. Kanis, Hj. Eichler and P. Bernhardt kindly undertook to go through the general chapters. A debt of gratitude is also owed to the University of Papua New Guinea and to its Department of Biology for the grant of six months' study leave to carry out the task of thorough revision of the manuscript which, as noted in the Preface, had originally been completed in 1975 but owing to technical and other difficulties was not published as intended.

Finally, I wish to thank my father, Reuben Frodin, for advice and encouragement at all times and for assistance in locating some obscure references and arranging for notes and copies of sample pages to be sent; to the late L. T. Iglehart, of The University of Tennessee Press, for early financial assistance, much advice and encouragement, and above all patience with the long drawn-out initial period of preparation of this book and sympathy when publication arrangements had to be terminated; to A. Winter and M. Walters at Cambridge University Press for advice, encouragement, and gentle nagging; to D. J. Mabberley, Oxford, for assistance at a critical stage in 1978, at a time when the author also suffered severe losses in an office and herbarium fire; to the Society of the Sigma Xi, USA, for a grant-in-aid in 1970 to enable visits to botanical libraries in Australia and elsewhere; to the Research Committee, the University of Papua New Guinea, for a grant-in-aid towards expenses associated with replication of the manuscript and carriage of two copies by air to the United Kingdom; to A. Butler, Librarian, and his staff in the University of Papua New Guinea Library, for the opportunity to utilize their extensive general bibliographical resources during final corrections to the manuscript in November and December 1981; to Prof. E. J. H. Corner, Cambridge and Great Shelford, for general encouragement over many years; to C. J. Humphries, British Museum (Natural History), London, and to R. Wetherbee in Melbourne, G. J.

Leach in Port Moresby, and M. Heads, formerly in Bulolo, for real support during the final stages of the project; and lastly (but no less importantly) to the staff of CUP for a thorough editing of a manuscript written under unconventional circumstances to say the least. Full responsibility for the text, including the onerous task of typing and retyping some 1800 pages of manuscript is, nevertheless, mine and mine alone.

Chapter 3 of the general introduction to this book is based upon the author's essay of the same title which appeared in *Gardens' Bulletin, Singapore* **29**: 239–50 (1976 (1977)). Acknowledgment is hereby made to the Government of Singapore for permission to reuse this material.

The map on p. 20 in Chapter 2 of the General introduction, depicting the relative state of present floristic knowledge for different parts of the world, was kindly supplied by E. J. Jäger, Halle/Saale, German Democratic Republic. It is a revised version of that which appeared in *Progress in Botany*, **38**: 317 (1976).

The not inconsiderable task of proofreading, carried out at Port Moresby, was assisted by Nancy Birge, G. J. and Amanda Leach, P. Osborne, and N. V. C. Polunin. Preparation of the indices was much facilitated by the use of 'Profile II', a Radio Shack (Tandy Corporation) proprietary file management package run on a Tandy TRS-80 Model II microcomputer; access to this machine was kindly granted by E. D'Sa and J. C. Renaud of the Mathematics Department, University of Papua New Guinea.

PART I

General Introduction

1

An analytical–synthetic systematic bibliography of 'standard' floras: scope, sources and structure

Die Bibliographie ist in ihrem weiteren Umfange der Codex diplomaticus der Literar-Geschichte, der sicherste Grad- und Höhenmesser der literarischen Kultur und Tätigkeit.

> Ebert, *Allgemeines bibliographisches Lexikon* (1821); quoted from Simon, *Die Bibliographie der Biologie* (1977).

Primarius noster scopus hic est ad redigendos auctores in ordinem, seu libros botanicos in *methodum naturalem*, ut tyrones sciant quos libros eligere debeant, auctoresque noscant, qui in hac vel illa scientiae nostrae partae scripserint.

> Linnaeus, *Bibliotheca botanica* (1736).

The difficulty in publishing an extended list of floras is to know where to stop.

> Turrill, Floras; in *Vistas in Botany* (ed. Turrill), vol. 4 (1964).

Definition and scope of the work

The aim of the present publication, as indicated in the Preface, is essentially the same as envisaged for the preliminary edition of 1964: to furnish, in bibliographic form, a one-volume guide to the most useful nominally complete floras and checklists dealing with the vascular plants of the world and its component parts. Its contents, however, have been entirely revised in the intervening decade and a half, and the scope of the work increased by the inclusion of annotations for each work along with references to general and local bibliographies and indices. Additional features include a map, geographical conspectuses, and short reviews of the state of floristic knowledge in different parts of the world. All titles up to and including those published in 1979 (and, as far as possible, those of 1980) of which the author has been aware and which fall within this work's scope have been accounted for.

Like its predecessor, the present work differs sharply from *Geographical guide to floras of the world* by S. F. Blake and A. C. Atwood (1942; 2nd volume by S. F. Blake, 1961) in that a much more rigorous selection from the available literature has been made, with only one to a few 'standard' works listed for each geographical unit, and that these units are arranged according to a geographical scheme developed especially for this book. There is, by and large, no detailed coverage of florulas and lists of comparatively local scope, nor has any but passing mention been made of works on weeds, poisonous plants, or plants of economic importance. Such limitations, admittedly severe, have made it possible to cover all of the most generally useful and/or comprehensive floras, enumerations, lists, and related works for different parts of the world in an approximately uniform fashion within a single, not too bulky volume. References to more specialized or extensive bibliographies, guides, and indices are provided throughout the work for the use of persons with interests in a particular geographical unit or units, thus enabling it also to act as a bibliography of bibliographies. Indeed, as with Linnaeus' *Bibliotheca botanica* of 1736, the work is seen not merely as a bibliography, but also as an introductory digest for the benefit of students.

Sources and the historical background
General

The present *Guide* is by no means the first list of floras to be published, nor is it the most extensive of its kind, although it is the first of world-wide scope to be written since 1914. Since early in the eighteenth century, various general world lists of greater or lesser scope have been produced, supplemented in more recent times by numerous local, regional, or supra-regional bibliographies. All have appeared as independent books or pamphlets, periodical or serial articles, or as parts of more general botanical or scientific bibliographies. The following paragraphs are devoted to a review of the most significant of these works, beginning with the general botanical bibliographies and followed by the specific flora bibliographies.

The first world-wide guide to regional and local floristic literature arranged on a geographical basis appears as Class 8 of *Bibliotheca botanica* by Linnaeus (1736; 2nd edn., 1751).[1] In this exuberant, highly didactic work, written in Holland and put forward as part of its author's comprehensive botanical reform campaign,[2] the listings, with sometimes pointed commentaries, are arranged geographically in a hierarchical fashion not unlike the present work. A remarkable feature of the *Bibliotheca* is the treatment of the books in an allegorical fashion as 'species', with country subdivisions as 'genera' – corresponding to the basic geographical unit in this book – and countries as 'orders' (with all extra-European works grouped together in a single 'order', *Extranei*). It was a sign of Linnaeus' scholastic view of the world that he could thus classify books (and people) in the same way

as organisms according to his so-called *methodus naturalis*, which, as Cain[3] and Stearn[4] have pointed out, was essentially based upon Aristotelean logic but contained elements of empiricism. Later classifications of literature, including arrangements of geographical entities, were by and large rather more empirically based (except within the mnemotechnic Dewey Decimal System (DDC), where a uniform numerical sequence of geographical entities, established in the first instance for human history, has been applied for regional literature in all other fields, with appropriate positioning of the decimal point). Through the centuries, however, there has always been a basic conflict between essentialism and empiricism in the theory and practice of any kind of classification,[5] with in more recent times the addition, at least in biological classification, of nominalism and Popperian objectivism as further doctrines. Most solutions adopted represent a balance of one kind or another between these values.

Although the literal application to bibliography of the Linnean *methodus naturalis* was later to be viewed as too extreme and impractical, the concept of a didactic subject classification as put forward in the *Bibliotheca botanica* appears to have gained more general currency by the end of the century, albeit in a rather more empirical and rational form. This development is demonstrated by the next significant listing of floras and related works within the third volume (Botany) of *Catalogus bibliothecae historico-naturalis Josephi Banks* by J. Dryander (1798), one of a five-volume set published between 1797 and 1800. Although based upon a single book collection,[6] this dry but very scholarly catalogue was of such a quality and completeness as to attract the sobriquet of an *opus aureum* as well as cause it to become the standard natural history bibliography for half a century. Only in the mid-nineteenth century was it for practical purposes superseded by the zoological bibliographies of Agassiz (1848–54) and Carus and Engelmann (1861) and the botanical bibliography of Pritzel (1847–52). In Dryander's work, which was limited to independently published books and papers, floras comprise classes 126 through 163 on 63 pages, arranged according to an empirical geographical scheme lacking any hierarchy of categories, a method followed in most succeeding botanical bibliographies. Nevertheless, the Banks library catalogue is in the first instance a *classified* work, giving the user a quick impression of the kinds of botanical work being accomplished and, more specifically, providing an insight into the state of floristic knowledge of the day. It is considered by Simon[7] to mark the beginning of the more modern tradition of monographic subject bibliographies which, although marked by increasing specialization, reached its fullest development in the century after 1815.

At the apogee of this nineteenth-century tradition stands the famous *Thesaurus literaturae botanicae* by G. A. Pritzel (1847–52; 2nd edn., 1871–7),[8] followed by its partial derivative, prepared for the English-speaking world by B. Daydon Jackson, *Guide to the literature of botany* (1881). Pritzel's *Thesaurus*, still much used today, is a highly critical work featuring a primary arrangement by authors but showing historical sensibility in its chronological arrangement of works by a given author, along with (in many cases) concise biographical notes. However, all entries are treated in short-title form in the elaborate, partly hierarchical classified index, which in both editions (but at much greater length in the second) includes several classes dealing with regional and local floristic literature; with the large format, this provides a good overview of the state of progress in description and analysis of the world's flora. Pritzel's *Thesaurus* has long been considered a landmark in botanical bibliography; in one work it fuses various traditions of earlier bibliographers but at the same time it is the last of the general botanical bibliographies. A successor work, to cover the period from 1870 through 1899, was projected in the early-twentieth century by the American bibliographer J. Christian Bay but apart from the issue of a list of bibliographies in 1909 (see under General bibliographies in Appendix I) nothing was realized and from 1881 coverage of botanical literature, even given the areas treated by Pritzel and by Jackson, has been fragmented and incomplete.

Jackson's *Guide*, although offered as a companion to the *Thesaurus*, is an independent work in its own right. However, in contrast to the *Thesaurus*, it lacks an alphabetical author section; the entire work is organized by empirically derived subject classes, continuing the pragmatic tradition established by Dryander. Unlike the Banks Catalogue, though, Jackson's work is a short-title catalogue, characteristic-ally an English bibliographical form. A substantial portion of this work (over 180 pages) is devoted to geographically arranged classes of regional and local floras, enumerations, and lists. One refinement

introduced by Jackson is that geographical subdivisions, especially for regions outside Europe, have been more finely drawn than in Pritzel's work – thus ac-knowledging the presence of a rapidly increasing body of 'overseas' literature, notably in North America.

In neither of these general compilations, however, is there extensive commentary; annotations are but few and for the most part strictly bibliographic, although brief critical notes do appear in the *Thesaurus*. Moreover, in both works only independently published work is covered; for reasons of economy in time and space the already significant body of relevant periodical literature was bypassed (save for articles which had also been independently published). A further important justification for this omission was the existence of the great *Catalogue of Scientific Papers* project of the Royal Society of London (indeed acknowledged by Pritzel in the preface to the second edition of his *Thesaurus*), of which the first volume had appeared in 1867. References to each author's papers as given in the *Catalogue* appear throughout the author section of the *Thesaurus*. It was indeed fortunate for Pritzel that this critical reference existed, as for some years before his death in 1874 he was debilitated by severe illness; indeed, the completion of the second edition of the *Thesaurus* itself had to be superintended by his associate K. F. W. Jessen, himself the author of an important culturally oriented history of botany, *Botanik der Gegenwart und Vorzeit* (1864). There was, however, also a then-prevalent belief that periodical papers were somehow 'ephemeral', being produced essentially as a form of interim communication and as precursors to monographic works, and consequently lacked the 'status' of the latter.[9] Nowadays, it is the monographic work which is a comparative rarity, especially one by a single author.

World guides to floras

Almost contemporaneously with the works of Pritzel and Jackson, and inspired by the former, there appeared, just over 100 years ago, the first – albeit slender at only 12 pages – annotated selected guide to the more important floras of different parts of the world, namely G. L. Goodale's *The floras of different countries* (1879). This compilation, published under the imprint of the Harvard University Library and, by and large, limited to independently published works available within the Harvard library system, is, however, of a plan and scope which clearly anticipate the present work, although, as in the *Thesaurus*, the brief annotations are mainly bibliographical. A notable feature is the omission of the great majority of the smaller local floras, by then very numerous in Europe and increasing in number in North America and some other parts of the world. The geographical arrangement follows that used in the *Thesaurus*. At the end of the list is an interesting appendix entitled 'Botanical Handbooks for Tourists'. In his brief foreword, Goodale indicated that his list was put together as 'simply an attempt to answer questions frequently asked respecting the systematic treatises upon the vegetation of different countries'. This little work, although largely derivative and nowadays of scarcely more than historical value, was an early example of the life-long interest in public relations and popular education on the part of the creator of the Harvard Botanical Museum and its famous 'glass flowers'.[10]

The next significant guide to floras of the world to appear, also under the aegis of a library, was the rather larger but mostly unannotated series of contributions by the Lloyd librarians W. Holden and E. Wycoff entitled '*Bibliography relating to the Floras*', which appeared as volume 1 of *Bibliographical Contributions from the Lloyd Library* (1911–14). Like Goodale's list, this series was produced in the interest of service to the public and in fact is an attempt to list all known independently published floras – with those present in the Lloyd Library[11] especially indicated – for all parts of the world. The work is divided into major geographical units which correspond to those in use in the Lloyd Library and are loosely based upon those in Jackson's *Guide*; however, within each major unit the arrangement of titles is alphabetical by author. Although by now long outdated, this set of lists remains the most recent substantial guide to floras completely covering the earth's surface. However, it is characterized by certain omissions, since, as acknowledged on page 2 of that work, its sources (apart from holdings within the library) were largely secondary; no special trips outside Cincinnati (Ohio) were essayed and great reliance thus had to be placed on such works as the *Thesaurus* and Jackson's *Guide* as well as the then-available volumes of the catalogue of the library of the British Museum (Natural History), *Botanisches Centralblatt* (founded in 1880 and then widely used in North America) and the *Index to American Botanical Literature* (begun in 1886).

Both of these early works, however, were essen-tially practical and predate or lie outside the mainstream of critical American bibliographical scholarship, which developed rapidly from the end of the nineteenth century and in botany by 1920 had resulted in two key works: *Bibliographies of Botany* by J. C. Bay (1909), as already noted a precursor to a projected retrospective supplement to the *Thesaurus* for the years 1870–99, and, far more substantially, the *Bradley Bibliography* by A. Rehder (5 vols., 1911–18). The former comprises a list of available botanical bibliographies, both general and regional, while the latter contains in its first volume (Dendrology, 1) a classified list of woody floras and 'tree-books' published through 1900. An innovation in the 'Bradley' is the inclusion of papers in periodicals. Despite its imposed limits, Rehder's work is of a scope and standard such that it can be considered the only real successor to the *Thesaurus*, save perhaps for *die moderne 'Pritzel'*, the monumental new second edition of *Taxonomic Literature* by Stafleu and Cowan (1976–   ; known also for short as *TL-2*).

Over the subsequent generation critical biblio-graphic scholarship filtered through to more specialized biological fields in their own right, including vascular plant floristics, and both in Europe and in North America several key monographic bibliographies were produced.[12] It was once more in the United States, though, that the next bibliography of floras appeared: *Geographical guide to floras of the world* by S. F. Blake and A. C. Atwood (volume 1, 1942; volume 2, 1961). The first volume, completed by 1940, covers Africa, the Americas, Australasia, and the islands of the Atlantic, Indian, and Pacific Oceans; the second volume, completed by Blake alone in 1959, provides detailed coverage for most of western Europe (save the German states). Yet this represents the most comprehensive and original contribution to the bibliography of floras to be published; it was based upon a wide range of primary and secondary sources and many years of critical research and experience on the part of its authors, the senior a botanist with the Crops Division of the United States Department of Agriculture, and the junior a librarian with the outstanding National Agricultural Library in the same Department. Unfortunately, the work, left incomplete upon the death of Blake in 1959, fails to cover eastern Europe and the continent of Asia, and no official plans have ever been made to complete it,[13] although in one of his last contributions Turrill[14] considered this to be a task of high priority.

The arrangement of the *Geographical Guide* is

fairly simple, with continents and their subdivisions arranged alphabetically in volume 1 and the countries and their administrative subdivisions similarly arranged in volume 2. Coverage extends to local florulas and checklists as well as encompassing the more important larger works, and many works on applied botany (medicinal and poisonous plants, economic plants, and weeds) are included as befitting the circumstances of its preparation. A major innovation over its two predecessors, paralleling the *Bradley Bibliography*, is its detailed coverage of floristic contributions in periodical and serial literature appropriate to the scope of the bibliography; this represented a major search effort which to some extent was eased by the availability of the detailed National Agricultural Library subject card catalogue, an asset with few rivals elsewhere.[15] Each primary citation moreover contains extensive bibliographic details and is briefly annotated; associated with these are many secondary citations (supplements, reviews, related or superseded works, etc.). Geographical and author indices are also provided. Like the works of Dryander and Pritzel, the *Geographical Guide* may also be considered an *opus aureum*. As far as it has been published, it has been a primary source for the present work.

Other lists of floras, usually much abridged and not representing an exhaustive search of the literature, have appeared in various textbooks of systematic botany, notably in *Taxonomy of Vascular Plants* by G. H. M. Lawrence (1951), *Taxonomy of Flowering Plants* by C. L. Porter (1959; 2nd edn., 1967), and *Vascular Plant Systematics* by A. E. Radford and others (1974) (see also Appendix I).

Regional guides

A fair number of guides to floras possessing a regional or local scope have appeared since the late-nineteenth century and the start of 'fragmentation' of general botanical bibliography, either independently or within more general regional botanical (or biological) bibliographies or, in some cases, 'national' bibliographies. Only the more salient aspects of this literature, now fairly extensive, will be dealt with here.

The earliest purely regional bibliography devoted exclusively to floras appears to be *A list of state and local floras of the United States and British America* by N. L. Britton (1890),[16] in which a geographically arranged listing of 791 works was given. It was partly superseded in 1930 by *State and local floras* by A. C. Atwood and S. F. Blake[17] and more fully in 1942 by the North American section of the *Geographical Guide*. The work parallels Linnaeus' *Bibliotheca botanica* in that it was part of an overall effort towards reform in North American taxonomy and floristics and the creation of a nominalistic but ultimately influential 'American' school of taxonomy which would be more 'scientific' and at the same time less reliant on 'tradition'.

Most other lists of North American (or United States) floras are of much more recent date and as often as not are essentially limited to the most significant works. Short lists for the United States have been published by Gunn (1956) and for North America north of Mexico by Shetler (1966). More substantial is to be the new list by Lawyer *et al.* (1980). Popular floras of the United States, including 'wild-flower books', have been covered in some detail by Blake (1954) and later, but less thoroughly, by E. R. Shetler (1967). United States tree books have similarly been rather fully covered by Little and Honkala (1976) in succession to Dayton (1952). However, apart from the existence of the non-computerized *Index to American Botanical Literature*, there is no current floristic (and taxonomic) information system of the kind now under development in Europe.

In the Old Continent, where until recently the botanical community was rather fragmented and to a large extent had become nationalistically or externally orientated subsequent to the waning of the universalism of the Linnean and post-Linnean periods, it is perhaps of interest to note that the most comprehensive work on its floristic literature is by an outsider – namely the second volume of the *Geographical Guide*, which represented an effort of some 20 years, largely by Blake alone – and even this work does not cover the German states, the remainder of central Europe, the Balkans or Greece, or the European part of the Soviet Union. For information on these works, European botanists have largely depended on general works like Pritzel and Jackson and upon the many national botanical bibliographies. The first European lists of floras dealing with the whole of that continent did not make their appearance until after the initiation of the *Flora Europaea* project in the 1950s (Heywood 1958, 1960; Lawalrée, 1960), and were purposely limited to what their authors considered to be the most significant and/or generally useful works, thus obtaining a depth of coverage comparable to that in the present *Guide*. Heywood's list has appeared, with successive revisions, in every volume of *Flora Europaea* (1964–80). An additional survey of significant floras in Europe – and, less thoroughly, other parts of the Holarctic zone – has appeared as part of a recent botanical bibliography of central Europe (Hamann and Wagenitz 1970; 2nd edn., 1977). The literature on the countries surrounding the Mediterranean was listed in a series of country reports edited by Heywood (1975), in continuation of a tradition established in 1963 with the publication of a similar set of surveys covering Europe as a whole (itself partially updated in 1974–5).

Soviet floras have been partially surveyed in a critical review by M. E. Kirpicznikov (1968–9), but at this writing (late 1980) this has covered only the European part of the USSR; alternatively, good listings can be found (up to 1955) in Lebedev's historico–didactic but selective *Vvedenie v botaničeskuju literaturu SSSR* (1956) as well as in Lipschitz' empirical but more complete *Literaturnye istočniki po flore SSSR* (1975).

A current development in Europe, first proposed in 1977, is a project for a 'European Floristic, Taxonomic and Biosystematic Documentation System' which would continue the integrating processes in European botany initiated by the *Flora Europaea* project. While no details have been seen, there will presumably be some mechanism for collection and processing of floristic bibliographic information. The new project is currently under the aegis of the Committee of the European Science Research Councils, since 1975 part of the European Science Foundation.[18]

Few separate surveys of floras and related floristic works have appeared for other parts of the world. A number of regional (and local) botanical bibliographies contain among other areas classified lists of indices of such works: examples of these are Merrill and Walker (1938; supplement by Walker, 1960) for eastern Asia, van Steenis (1955) for Malesia and adjacent areas, the Field Research Projects' bibliography for southwestern Asia (1953–72), and Hultén's excellent source bibliographies (1958, 1971) covering the whole of the north temperate and polar zones. A number of national botanical bibliographies (or botanical parts of general bibliographies) have also appeared in various parts of the world, e.g., for Turkey, Pakistan, Sri Lanka, India, the countries of Indochina, Mexico, Argentina, South Africa, and some Pacific countries and territories. Some brief continental or subcontinental literature surveys have also been

written, among them those by Léonard (1965) for Africa and the islands of the soutwestern Indian Ocean and Zohary (1966) for southwestern Asia and adjacent areas; like Heywood's lists for Europe, they cover 'standard' works and thus provide a level of coverage comparable to this *Guide*. The fact that more lists of floras have not appeared in the past may perhaps be explained by the existence of the Blake and Atwood *Geographical Guide* and/or the lack of felt need for such lists as distinct from local and regional botanical bibliographies of wider scope.

Periodical indices

The current situation respecting the coverage of floristic (and taxonomic) botany in periodical indices and abstracting journals is not entirely satisfactory. The long absence of a botanical equivalent of *Zoological Record* and the demise during World War II of *Botanisches Centralblatt*, *Just's Botanischer Jahresbericht*, and *Naturae novitates* (as well as the *International Catalogue of Scientific Literature* after World War I) have been but partly made up for by two works: the recently established *Kew Record*, covering literature from 1971 and absorbing two regional indices begun in the 1960s, and *Excerpta Botanica*, begun in 1959 by the publishers of the defunct *Centralblatt* under an agreement with the International Association for Plant Taxonomy and currently edited from Kassel (W. Germany). The latter includes short summaries for each title, but is less thoroughly classified than *Kew Record*. In each case, however, published coverage ranges from 2–3 years behind, and *Kew Record* has reportedly encountered technical difficulties. Apart from these, reliance, especially for more up-to-date coverage, has customarily had to be placed upon more general botanical and biological abstracting and indexing journals, worldwide and regional newsletters with literature lists, booksellers' catalogues, advertising leaflets, and announcements in *Taxon* and other journals. Summary lists of new floras and related works have also appeared in most years as part of the appropriate reviews in the annual *Progress in Botany* [Fortschritte der Botanik], begun in 1932. On the other hand, *Current Contents* (*Agriculture, Biology, and Environmental Sciences*), a commercial publication begun in 1970, is, because of its emphasis upon more widely used journals (as measured through citation analysis)[19] and more prominent symposium reports, is of comparatively little value for floristic botany on an international scale. [The relative strengths and weaknesses of the various periodical indices are considered along with other general sources in Appendix A.]

Progress reports and reviews

In recent decades, the publication of review articles and reports has extended to include reports on the state of floristic knowledge for different parts of the world. This is, in part, related to the growth of the conservation movement as well as increased general awareness of the tropical biota. Such reports vary considerably in scope and quality, and range from isolated articles to sometimes elaborate surveys covering large areas; more or less extensive bibliographies may be included. Examples of such reports include the surveys of European and Mediterranean floristics referred to above; the 1978 review of the state of tropical floristic inventory by Prance;[20] the many articles in *Plants and Plant Science in Latin America* edited by F. Verdoorn (1945), and the reviews presented at some of the congresses of AETFAT (Association pour l'Étude Taxinomique de la Flore d'Afrique Tropicale) and the Pacific Science Association. Collectively, they constitute a valuable source of information on the literature in conjunction with the progress of floristic research and the institutional background, but regrettably are scattered far and wide through the literature, and sometimes may be overlooked.[21] In some cases, they have been intertwined with historical surveys of botanical exploration. Valuable also are the introductory portions of many floras and checklists.[22] On the other hand, as Jonsell[23] has warned, the user should take note of the standard of these reviews and surveys; many are not well documented and in addition may be unreliable, although they often give the kind of statistical information beloved of bureaucrats, parliamentarians, and laymen.[24] Reference is made to the more important of these 'floristic reports' in the various divisions of this book.

The latest world-wide survey of progress, with literature references, is by E. Jäger in volumes 40 and 41 of *Progress in Botany*.[25] This follows the publication by the same author of a world map depicting floristic progress based upon four criteria.[26] A revised version of this map is presented here as Map II (see p. 20).

Plan and philosophy of the present work
Definition of a 'standard' flora

For the purposes of this *Guide*, a 'standard' flora (or corresponding manual, manual-key, enumeration, or list) is considered to be a current scientific work which yields the maximum information about the vascular plants of a given geographical unit within the parameters set by the nature and style of the work, and thus saves the inquirer an extensive (and often time-consuming) search in the more detailed taxonomic and floristic literature (which often may be very scattered, and whose retrieval has been made more difficult by the disruptions and changing styles which have characterized the biological information network in the present century). Put in another way, standard floras are generally those which one turns to first for information about the plants of a given region, state, or country; in many instances they may be the only ones consulted, as they are likely to suffice for the query in hand. Such works are regarded by specialists as providing the most representative coverage for the region (or regions) concerned. They represent among floristic literature an optimum ratio of information to effort, with descriptive floras having a higher information coefficient than enumerations or checklists.

Standard floras may in such wise be contrasted with those which are less comprehensive, such as county or provincial floras or checklists. These latter normally deal only with areas of relatively limited extent and are, comparatively speaking, of more interest to specialists on local floristics, local amateurs, and persons engaged on detailed monographic, revisionary, or chorological work. For some parts of the world, above all Europe, this literature is very large indeed and the extraction of desired information can become a very lengthy process.

No originality is claimed for the concept of a standard flora as expressed herein, and in fact a number of regional lists of floras now exist in which the depth of coverage and the kinds of works selected are clearly comparable to the present work. Even French and German equivalents for the term now exist, respectively 'flore de base' and 'Standardflora'. The first usage of the standard flora concept appears to have been by Heywood[27] in his 1957 report on the organization of the *Flora Europaea* project, when it was suggested that a list of about 100 titles had to be considered in obtaining a general overview on any given European taxonomic or floristic problem. The value of the concept was shortly afterwards reiterated by T. G. Tutin in his foreword to the *Flora Europaea* 'Green Book' of 1958: 'It is our belief that the list of Standard Floras...will be generally welcome. These floras, as far as we can ascertain, are the ones most generally acknowledged by botanists in the countries concerned.'[28] Although originally developed in a European context, I hold the belief that the standard flora concept is, with variations, applicable worldwide – a belief for which there is now considerable evidence. Indeed, a very satisfactory paraphrase of Tutin's words might read as follows: 'Standard floras, as far as can be ascertained, are the ones most generally acknowledged by botanists in, or working upon, the countries or other regions concerned.'

Selection and coverage of standard floras

The preparation of a comprehensive list of standard floras, no matter what definitions or guidelines are available or may be evolved, necessarily entails a difficult process of evaluation and selection. It is also essential that a reasonably uniform standard of coverage be adhered to throughout the bibliography, and every effort has been made to achieve this goal as far as possible. However, the total corpus of regional literature varies greatly from one part of the world to another with respect to its nature, quantity, and quality. For instance, many tropical areas, such as the island of New Guinea, have no general floras or enumerations of relatively recent date, and the student or nonspecialist is faced with an ill-digested mass of florulas, expedition reports, and scattered 'contributions', revisions, notes, and the occasional monograph of varying scope. By contrast, the bulk of Europe is covered for the most part by a plethora of local, national, and regional floras and lists of varying dates from which it was necessary to make a careful and limited choice for the present compilation. Less advantaged areas, however, may also be more or less well covered by supplementary bibliographies; in this respect the greater Malesian region is exceptionally well documented, a state of affairs also true of the Pacific islands, southeast Asia, and (to a lesser extent) tropical Africa.

Fortunately, the exacting tasks of selection and establishment of an approximately uniform standard of coverage have been greatly facilitated by the existence of several useful guidelines: (*a*) the regional lists of floras already referred to in passing (including the 'Green Books' and the lists of Léonard, van Steenis, and Zohary) (*b*) the selected lists in the standard textbooks referred to on p. 536, and (*c*) two lists of works considered to be of 'greatest general utility' in the *Geographical Guide* of Blake and Atwood (vol. 1, pp. 15–16; vol. 2, pp. 27–8). Other reference points have included a series of mimeographed memoranda on various regions prepared in 1970 for internal use in the Kew Herbarium as part of a major institutional reorganization (related in part to imperial dissolution); a 1979 list prepared at Geneva for the projected 'Med-Check List'; Shetler's 1966 list of principal works prepared for the Flora North America Program; published 'state of knowledge' reports for a wide variety of countries and geographical areas; and verbal and written advice from a number of specialists and others with local knowledge. The final responsiblity for the choice of titles, however, is mine.

As far as possible, every primary entry in this book has been provided with an annotation describing its style and contents. These commentaries have been based upon personal examination of the works concerned, except for a small number which the author was unable to consult; with regard to these latter the reviews have been consequently based on notes and/or extracts supplied by correspondents, who have been acknowledged in the text, or published or circulated secondary sources. Any material not seen at first hand has been so indicated. Subsidiary titles – i.e., those without separate headings – are provided with briefer commentaries, which may be of an adjectival nature, coming before the actual citation. In general, the style of the present work is modeled on that of the *Geographical Guide*, but with a greater amount of analysis and commentary and less detailed in its formal citations.

Some works of a more local scope, covering parts of basic geographical units as delineated in this work, have been included where warranted. This involves cases in which sizeable partial or local works, to a greater or lesser extent, bridge a gap left by the absence, relative antiquity, or inadequacy of a general work or works, or in which the local work concerned is of an exceptionally high standard or of acknowledged value well beyond its nominal circumscription. Amelioration of the limitations on coverage has also been applied with respect to sets of 'contributions' and/or expedition reports covering imperfectly known areas where these appear to be of exceptional importance or are otherwise often routinely consulted.

Provision has also been made for certain kinds of ancillary works. Atlases of illustrations, if of major importance, have usually been accorded the status of primary entries, unless they are clearly companions to descriptive works. Separate subheadings have been set aside under a given unit heading if there are separate keys to families (and genera) and/or dictionaries, but in practice this has been done only at regional level and above. The same has been done with atlases of distribution maps and like chorological works, save for a few, such as *Pacific plant areas*, given under **001** as they are not readily referable elsewhere.

Under unit headings, any 'local' or 'partial' work deemed important enough for inclusion has been treated as a 'secondary' work and its citation and commentary appear in *smaller type*, usually following a subheading. The same procedure has been adopted with respect to works on the woody flora (including 'tree-books'), the ferns and fern-allies, and (in a very few cases) the grasses which have been included in the *Guide* due to general interest or where these groups are not well accounted for in available standard floras. However, some outstanding works on the tropical woody flora have been given as primary contributions, especially where general works are hopelessly outdated or lacking.

Schedule of geographical entities

As already noted, the arrangement of titles is geographically systematic, following a hierarchical decimal scheme with arrays at four levels devised especially for this book. The development of such a scheme was undertaken in 1963 in the belief that other schedules of space, especially the common or universal schedules of geographical auxiliaries used in the Dewey Decimal Classification (1876 onwards) but also the more specific and purely enumerative 'traditional' schemes of the QK (Botany) section of the Library of Congress Classification (1901 onwards) and of Holden and Wycoff in *Bibliography relating to the floras* (1911–14, itself based upon Jackson's *Guide* of 1881) were obsolete, lacking in method, or not particularly suited to the material at hand – particularly in a truly international sense. Alphabetical systems of geographical entities, such as those of Blake and Atwood in the *Geographical Guide* (1942–61) or the scheme developed

by Travis *et al.*[29] for entomological literature, were similarly not considered although in both the wealth of geographical detail made them very useful in constructing a new schedule. Subsequent studies in later years, particularly from 1967 onwards, only confirmed these beliefs. Moreover, with but few exceptions existing schedules were all rooted in nineteenth-century 'Euro-centred' notions of history and geography, past and present. There was thus a clear need for a workable geographical scheme for floristic literature independent of schemes based upon rather different criteria, but one which might also be applicable to the regional literature of the rest of biology and also of the earth sciences.

The possibility that the universality of the geographical facet, or classificatory element, as accepted in standard classifications was as a canon unworkable appears first to have been raised by de Grolier in 1953 with respect to history and geography.[30] He argued that a schedule suitable for physical geography would not suit economic geography, and even less would suit history (upon which most current general schemes are based). It is here argued *a posteriori* that the regional literature of botany (and zoology) is in like fashion more closely related to that of physical geography than to economic geography or history, thus supporting de Grolier's claim. However, no new scheme tailored to the biological or earth sciences regional literature and at the same time compatible with one or more of the existing widely used classifications – particularly the Universal Decimal Classification – has been seen. The only radically new geographical schedule published in recent years appears to be the *Geocode* of S. W. Gould,[31] but due to its rigid structuring it lacks geographical contiguity as well as any sense of relationship to the existing (and likely) literary warrant; its use for the present book proved impossible.

What was really required was a more representative and uniform geographical schedule suitable in the first instance for floristic (and, by extension, faunistic) literature. An early discovery was that the structure of existing (and expected) floristic literature – known in librarianship as the 'literary warrant'[32] – was such that it could be grouped into successive hierarchical arrays, thus enabling construction of a 'decimal' system along the lines of the Dewey Decimal System or the Universal Decimal System. The actual geographical arrangement has, however, almost from the beginning diverged widely from that used in these standard systems.

Primary concerns in constructing a necessarily linear schedule were logic, practicality, contiguity, mnemonic value, and physical and biogeographical relationships. The schedule now presented differs greatly from that used in the 1964 *Guide*, but is yet not considered final.

Particularly notable with respect to remaining problems in the *Guide* geographical schedule has been the classification of physiographic and synusial isolates, for which many key floras worthy of inclusion as 'standard' works already exist and/or which are potentially attractive subjects for future floristic treatment. No logical, generally acceptable schedules or sets of common auxiliaries appeared to exist, including the 'symmetrical' scheme of Ranganathan,[33] and moreover none was well-suited to floristic and faunistic considerations. At a late date, finally, following an earlier empirical attempt at listing works not conveniently included in a geopolitical unit, a system of common auxiliaries based upon those used for the Universal Decimal System (B.S. 1000, 5th edn., 1961) was developed for the *Guide*, with the following structure:

–01	Vague areas (e.g., Patagonia, Tropical Africa)
–02	Islands
–03	Alpine and upper montane areas (e.g., the Andes, the Pamir)
–04	Major uplands or highlands (e.g., the Guayana Highland, the Ural)
–05	Deserts and plains (e.g., the Sahara, the Gobi)
–06	Oceans and the oceanic littoral
–07	Deep lakes and their littoral (e.g., Lake Baikal, the Great Lakes)
–08	Wetlands
–09	Rivers and riverbanks

The usage of these auxiliaries has been comparatively sparing, save for –03 and –08, where because of the existence of a more or less considerable and much-called-upon body of literature the opportunity has been taken to refer all (or most) such works as were included in the *Guide* to these auxiliaries, even where the geographical area concerned was fairly limited (as in *Kosciusko Alpine Flora* and *Rocky Mountain Flora*; wetland floras of subregional level, however, have been omitted from the *Guide*). The nine auxiliaries are in theory definable in all ten divisions of the *Guide*'s geographical system, but in practice do not appear unless there are appropriate works to be covered.

Auxiliary –06 in particular is recognized here largely against the possibility of future expansion of the *Guide* into coverage of non-vascular plant groups, although a work such as C. den Hartog's *Sea-grasses of the world* falls within our present scope and properly belongs in **006**, just as *Rheophytes of the world* by C. G. G. J. van Steenis appears here under **009**.

The highest category in the system adopted here is the *division*. These are numbered from 0 through 9; general floristic works with a division-wide coverage are designated by the numbers 100, 200, etc., up to 900. The category below is the *region*; these are numbered from 01 through 99, according to the division into which they fall (00 being used notionally for world-wide floras, world synusial works, and a few chorological works). Some regions are grouped together into superregions, with separate principal headings; these are designated by hyphenated figures, such as 14–19, 42–45, or 91–93, indicating those regions which they encompass. Very large single regions comprising more than nine units, of which examples include the northeastern USA, Brazil, and the European part of the Soviet Union, are designated by a stroke between two figures, such as 14/15, 35/36, or 68/69. Individual regional floras, enumerations, etc., are always given a three-digit number ending in a single zero, viz: 160, 220, 560, 830, or 990, except that floras of *superregions*, such as *Flora orientalis* or · *Index florae sinensis*, are designated by 'inclusive' unit numbers such as 770–90, 910–30, etc.

The lowest category, or the 'species' of the system, is the *unit*; these are designated by figures running from 001 through 999 excluding those ending in a zero. Units as recognized here generally correspond to geographical areas such as states, countries of small or medium size, large provinces, or significant islands or island groups; it is for these that the bulk of the so-called standard floras have been written. By contrast, regions comprise large countries (or natural groups of smaller countries or states) or comparable areas of large size; while divisions consist of continents, parts of continents, giant island empires, or combinations of these. No category has been devised for the relatively small number of local or partial floras included in this work.

Examples of *divisions* are North America, Europe, or Greater Malesia and Oceania. The polar zones beyond the 'tree-lines' of north and south, together with some isolated oceanic islands, have been allocated to Division 0. Representative *superregions* include the West Indies, south Asia, Greater Malesia, and Australia (with Tasmania). Areas such as the southeastern United States, Argentina, south central Africa, Madagascar, Western Australia, central Europe, the British Isles, the Soviet Far East, southeast Asia, Papuasia, and the Hawaiian Islands have been designated as *regions*. To the *unit* level are assigned such areas as Macquarie Island, St Helena, Alberta, New York State, Puerto Rico, Mato Grosso, Buenos Aires Province, South Australia, Mauritius, Natal, Nigeria, France, Finland, the Ukraine, Yakutia, Iraq, Uttar Pradesh, Nepal, Korea, Szechwan Province, Java, the Solomon Islands, and the Marquesas.

Certain physiographic areas for which there are important standard floras have also been included in the present scheme. These have been designated by three-digit numbers with a *middle* zero, e.g., **201**, **703**; examples of such areas are the Sonoran Desert, the Andes, the Afroalpine zone, and the Altai and Sayan mountains. In general, this class comprises areas which are too awkward to fit into geopolitical regions, or which otherwise deserve special emphasis. For convenience, the floras of high mountain areas, including the now fairly numerous *Alpenfloren*, have in this Guide been by and large treated as representing physiographic entities, and thus appear near the beginning of their respective divisions.

The ten primary divisions are all listed in the Table of Contents, but for ready reference are repeated below:

Division 0 World floras, isolated oceanic islands and polar regions
Division 1 North America (north of Mexico)
Division 2 Middle America
Division 3 South America
Division 4 Australasia and Malagassia
Division 5 Africa
Division 6 Europe
Division 7 Northern, central, and southwestern Asia
Division 8 Southern, eastern, and southeastern Asia
Division 9 Greater Malesia and Oceania

The full classification scheme for each division appears under the divisional heading as a conspectus. The spread and limits of the primary divisions are depicted in Map I.

Bibliographies and indices

A special feature of this *Guide* is the systematic inclusion of references to more detailed local, regional, and general botanical and floristic bibliographies, so that any person seeking more detailed information on any given area will know where to turn to. Reference to major local, national, or regional bibliographies and indices are included under their appropriate headings, although for practical reasons this coverage may not be uniform. For general bibliographies (such as those of Blake and Atwood, Hultén, or Jackson) and indices (such as *Excerpta Botanica* or the *Kew Record*), abbreviated references or mnemonic devices appear throughout the text at divisional and regional levels; full citations of these works are given in the General Bibliographies and General indices lists located at the end of the introductory section of this work.

Under the appropriate headings are also included references to significant reviews of the state of floristic knowledge for given major geographical entities, but no attempt is here made at exhaustive coverage of such literature.

Limitations

In order to make this *Guide* as compact and practical as possible, various limitations have had to be imposed. These are set out below:

1. The *Guide* is limited to works covering **vascular plants**, either exclusively or as part of their total coverage. Extension of coverage to the nonvascular plants would have entailed much additional bibliographic research for which time and facilities have not been available; furthermore, except to some extent among the bryophytes (and the higher fungi, if these be considered plants), regional floras and checklists for nonvascular plant groups are comparatively few in number. Nevertheless, because of gradually increasing interest in nonvascular plants, guides similar to the present one, written according to appropriate ground rules, should in due course be undertaken.

2. Few **superseded** floras or enumerations are included, except in some instances where they may be mentioned as subsidiary titles. Comprehensive inclusion of these would greatly increase the bulk of this *Guide* and cause unnecessary overlap with the older standard bibliographies of Pritzel, Jackson, Rehder, Holden and Wycoff, and Blake and Atwood. However, a more lenient view towards the inclusion of such works has

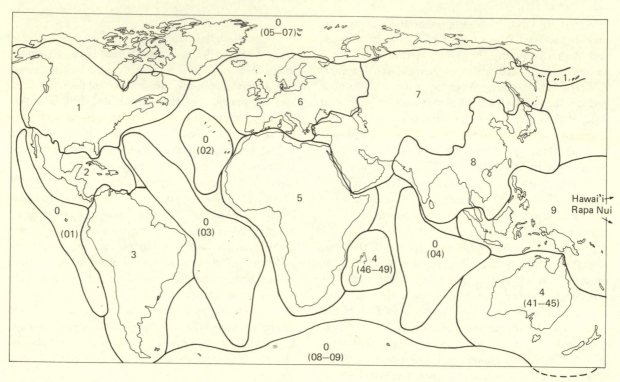

Map I. The spread of Divisions 0–9 as used in this book. For explanation see text (Antarctica has been excluded for technical reasons).

been taken with respect to those parts of the world (eastern Europe and Asia) not covered by the last-named.

3. With but few exceptions, no literature dating from **before 1840** is included. As discussed in the chapter on the style of floras, it is here held that only from about this time did the format of floras begin to assume recognizably 'modern' forms (as exemplified by W. J. Hooker's *Flora boreali-americana* (1829–40), *Flora brasiliensis* (begun in 1840), Grenier and Godron's *Flore de France* (1848–56), Torrey's *Flora of the state of New-York* (1843), and J. D. Hooker's *Flora antarctica* (1843–7)). Furthermore, it is expected that the forthcoming *Bibliographia Huntiana* will provide a detailed review of all pre-1840 botanical literature, inclusive of floristic works.

4. No purely **popular** works are included, nor does coverage extend to lexica and other works on vernacular names. To do so would greatly increase the bulk of this *Guide*. In a number of instances, though, especially with regard to popular works, limits have been hard to draw; moreover, there is some evidence that the distinction between 'scientific' and 'popular'

floras is becoming more frequently blurred, as in the recent *Kosciusko Alpine Flora* by Costin and others (1979). Consequently, some exceptions have had to be made, especially for areas for which no good recent standard floras exist. An instance is provided by the European Alps; for this important and well-studied physiographic unit no separate complete flora is available – a lacuna already noted a quarter-century ago by Gams[34] – and various more or less popular works have had to be cited as substitutes. In addition, some more or less popular works on trees (and woody plants in general) have also been included as explained under §5 below.

5. With regard to works dealing only with **trees** (or **woody plants**), coverage varies according to the importance of these life-forms in the total vascular flora. Thus, speaking generally, within the largely Holarctic Divisions 1, 6 and 7 only works which cover areas the size of regions or larger have been accounted for. This eliminates among others nearly all of the large number of states and provincial 'tree-books' which have been produced in North America; their systematic inclusion would lead to a serious imbalance in the overall depth

of coverage of this *Guide* and moreover they have recently been the subject of a separate bibliography (Little and Honkala, 1976, for the United States). Some exceptions have been made where deemed useful. Except for works at regional level or for the largest countries, a similar limitation has been applied in Europe and northern Asia. Many works dealing with the woody flora (or the trees) therein also include a substantial number of introduced park and garden trees, reflecting the long European interest in dendrology (which has also manifested itself in the many tropical 'forest floras'). For other parts of the world, particularly those lying within the humid tropics, dendrological works, woody floras, semi-popular 'tree books', and the like have been selected on the same criteria as full floras and enumerations; it is here that works on woody plants attain their greatest significance, a fine recent example being *Tree flora of Malaya* (1972–).

6. Works on **ferns and fern-allies** (i.e., the pteridophytes) have been selected in the same manner as works on the woody flora of given entities, with the qualification that, because of the sweeping changes to fern taxonomy and nomenclature which have taken place since World War II,[35] older fern floras are now largely obsolete, and those published from 1939 through the 1960s are presently in need of considerable revision. Pre-1939 works have thus largely been excluded unless no other coverage is available. In a number of instances, 'fern floras' of a given area have been cited where there is no corresponding standard work or works on the whole vascular flora.

7. Works on **applied botany**, i.e., regional treatises on economic, medicinal, or poisonous plants and on weeds have generally been omitted. It is my belief that, important though many of these works are, they should not come within the scope of a basic guide to floras; moreover, as with other classes of regional works referred to above, their inclusion would greatly increase the size of this work. There is, however, scope for a separate guide along the same lines as the present work which would cater for such literature.

8. With few exceptions, no works covering **single families** of seed plants are included. It should be noted, though, that for the Gramineae, Leguminosae and Orchidaceae (and for some other groups such as Cactaceae in the New World and Dipterocarpaceae in Malesia) a more or less extensive regionally oriented literature exists, which might warrant the preparation of separate guides for publication in appropriate media or in issues of *Regnum Vegetabile*.

Some concluding remarks

The author has often been asked 'but why not a continuation and revision of Blake and Atwood's *Geographical Guide*?' in reaction to discussion of the present work, or has heard arguments similarly favoring a more comprehensive coverage of floras. While this may be regarded as an ideal goal, two major points should be considered.

The first, generally evident, is that the mass of floristic literature, like that in all other fields of botany and including works considered as 'nominally useful' floras in the sense of Blake and Atwood, has increased manyfold since 1939. At the same time, increasing fragmentation of as well as other changes to the system of botanical information reporting, processing, and indexing have taken place. Among these have been the serious disruptions, due to World War II, in Europe, the establishment of additional regional indices, the greater numbers of people and organizations involved, and, significantly, the discontinuance in 1952 (through alleged lack of funds) of a prime source for the *Geographical Guide*, the botany subject catalogue of the present United States National Agricultural Library. Neither the largest nor the most specialized of the indexing and abstracting journals relevant to floristic botany completely covers the field, contributors to which have become far more numerous and scattered in the last 40 years. Moreover, most of the general and regional periodical bibliographies are still not, or not effectively, computerized; *Excerpta Botanica*, for instance, remains manually prepared and produced. Retrospective coverage on the scale necessary for a new comprehensive work or even an extension of the *Geographical Guide* would now require very substantial institutional support, financing, and personnel. The apparent lack of provision made by the appropriate authorities for completion of the *Geographical Guide* after Blake's death is perhaps indicative of the priority which might be assigned to such a project, and very few other institutions would have the necessary resources.

The second point, perhaps less obvious but subtly more important, revolves around the *need* or *desire* for such a work, especially when measured against the mechanics involved. With increasing specialization and changing interests and methodologies, there comes the necessity to review the *scope* and *style* of publications, including reference works, with regard to *function* and *efficiency*. For Floras this has been done by Aymonin,[36] Heywood[37] and the present author,[38] with the growing recognition that they serve two or more functions and that in relation to these there are required publications of differing scope rather than one all-purpose work. Similarly, in bibliographical compilation and writing in the field of botany it is now clearly evident that a like analysis leading to functional differentiation is necessary, or has again become necessary.[39] Thus, a single, comprehensive work covering floras of the world, as conceived by Blake and Atwood, or east Asian botany, as conceived by Merrill and Walker, while reflecting the changes in bibliographic methodology which characterized the first third of the twentieth century and, perhaps, a valid statement in terms of the 1930s (and the decade after World War II), may no longer be in current terms satisfactory as a *methodological*, let alone practical, solution. Apart from the volume of literature to be assessed and the cost of production, much of the material which would perforce be included even in simple extensions of the older works would be of relatively local or specialized interest. Moreover, the standards of coverage adopted for the *Geographical Guide*, while perhaps relatively satisfactory as an index of the status of knowledge in entities such as Europe, North America, and a scattering of others elsewhere where good floras are more or less numerous, fail on the other hand to reflect accurately the actual standard of floristic knowledge over the greater part of the earth's surface – where there may exist a considerable 'literature' but comparatively few substantial floras or checklists. Enumerative bibliography has in the last several decades been 'absorbed' into information science, and the whole approach towards fields of knowledge – and the questions asked – have become more systematic.

Given these two points, therefore, the realization of a new comprehensive guide to floras, while desirable as an ideal, would be difficult or impossible, especially in the present depressed social and economic climate, and perhaps in the final reckoning not really what was most needed, given the methodological and technological changes in bibliography since the 1930s. There was, however, clearly a need for a convenient general-interest guide to floras, and thus a theoretical solution had to be found which was at the same time practical and indicative of the 'state of the art'. This was done by recognizing that the functions of comprehensiveness and general utility could no longer be considered as necessarily congruent with regard to coverage of floristic literature – a reiteration, so to speak, of the differentiation of the period after the 1914–18 War described by Malclès[40] – and that in opting for general utility as the only practical philosophy (given available means) an attempt had to be made to discern relatively objective criteria by which to guide the fairly rigorous selection of titles necessary so as to keep the projected work within reasonable bounds – and also attractive as an 'elbow-book' for individuals (to use an expression of Joseph Ewan). It was thus fortunate that in the 'standard' flora concept was a measure positively correlated with widely recognized intuitive selections for general usefulness, which, to paraphrase some well-known concepts originally introduced by Gilmour,[41] tended to distinguish 'general-purpose' from 'special-purpose' works. This measure, which *a posteriori* can be seen to conform to the parsimony principle[42] and moreover is broadly congruent with the bibliometric Bradford 'law' (actually an axiom) of 'scatter' (recently likened to a comet),[43] and its inverse, Garfield's 'law' of 'concentration',[44] has governed the development of the present book. It remains for this measure to be tested quantitatively by recognized procedures,[45] but it is likely that these will merely confirm the perceived pattern of usage and its broad conformity with the above-mentioned bibliometric 'laws', as already demonstrated in a large number of widely varying contexts.[46]

Assembly, review and synthesis of the information presented in this *Guide* has, nevertheless, been a time-consuming task; work on the present version began over 15 years ago. The greatest problem, apart from the remote geographical location of the author for much of this period (itself a serious handicap) in compilation of the work was as already noted the present structure of the botanical (and biological) information system.[47] As a consequence, it was found more efficient to work 'downwards', beginning with available 'short lists' of floras, leaving searches through more extensive bibliographic and indexing sources until later. At the same time, there had to be an opportunity for 'shelf hunting'. These procedures required the use of a well-organized, comprehensive botanical library with a wide, readily accessible range of floras. Few such libraries meet these criteria, and it has been the good fortune of the author to have been able to make extensive use of that at the Field Museum of Natural History, Chicago, during preparation of the original version, and those of the British Museum (Natural History) and the Royal Botanic Gardens, Kew, between them the most comprehensive in the world, for the present version. For some parts of the world searches were made in libraries with a good representation of regional literature, such as that of the New York Botanical Garden (for North America) or the Komarov Botanical Institute (for the Soviet Union). Nevertheless, many titles as well as associated information sources came to light through chance discoveries. Personal contacts, regrettably sporadic due to remoteness from major centers, were also extensively relied upon for further information and 'insights'. These last-named were felt to be of particular benefit in relation to the style and philosophy of this *Guide*; they have been extensively supplemented through perusal of available 'progress' literature, itself very scattered, as well as histories of botany of various periods, regions, countries, and institutions. Over 90 per cent of the items included in the *Guide* have been seen at first hand and, as far as possible, all commentaries are original.

The final result as here presented has a number of advantages. With a more limited scope (based upon well-tempered selection criteria) than the form of comprehensive bibliography usually considered as the ideal in systematic botany and with the formal listings supplemented by historical and other commentary related to the genesis of the standard works selected, it has been possible to fashion this *Guide* as a kind of critical analytico–synthetic systematic bibliography, which to me is something more communicative than a purely empirical work, scholarly but lifeless. In this respect, it represents something of a return to the bibliographic styles of Linnaeus and von Haller. Moreover, because this *Guide* has always been conceived of in relation to the nonspecialist and student – the *tyrones* of the Linnaean *Bibliotheca botanica* – as well as the practicing professional, I believe it should be not a mere list of books, but also, as was advocated by F. A. Ebert in the first volume of his *Allgemeines bibliographisches Lexikon* (1821),[48] a *codex diplomaticus* which in this context would seek to indicate and interpret the relative state of floristic knowledge and levels of activity in different parts of the world through an account of the 'most useful and representative works' as culled from the whole corpus of floristic literature. The value of critical selectivity has

been well demonstrated also for rain forest vegetation by Webb and others[49] and for information in science by Ziman.[50] Where the means exist, quantitative procedures, including the use of computers, can (and should) be used in support of the overall study, but never so mindlessly that they dominate the final form and thrust of the work. The ultimate judgments should remain with the author upon whom rests responsibility for the work's value and effectiveness.

A communicating bibliography may also be looked upon as a form of scientific monograph or treatise, whose function is to transmute information into knowledge. Like other such works, communicating bibliographies require a great deal of research and study, much of it time-consuming and tedious and of little value for publication in a pleiad of precursory papers. Yet, as Ziman[51] has noted in his eloquent defense of the scientific monograph, the research involved is of the kind paradoxically not considered 'research' in many contemporary circles, but something 'to be done in one's own time', and as a result serious monographic bibliographies, even of comparatively restricted scope (by comparison with the artisan-bibliographies of the nineteenth century), are now less often attempted, as with monographic studies in general. Corner,[52] Jacobs,[53] Mabberley,[54] Stace[55] and others have pointed up this same lack of understanding (and associated lower 'status') in relation to monographic studies of families and genera – with Stace in particular critical of the amount of effort being invested in floras – which leads inevitably to the consequence that botanical science fails to advance, or only does so in a fragmented manner. These attitudes also reflect present patterns of financing, a comparative lack of 'big thinking', the influence of 'fads', and preoccupation with form rather than structure, not to mention a seeming trend towards academic neo-scholasticism[56] and wordly alienation, but on a deeper plane might also be related to a current widespread lack of appreciation of quality as well as to what the writer V. S. Naipaul has called a 'cult of stupidity' with concomitant intellectual debasement.[57] The present book is here offered in the hope that at some time in the future there will return cultural mastery and integrity (albeit most likely with different paradigms),[58] and as part of it a greater awareness of the value of critical monographs and treatises, including analytical bibliographies, as instruments for scientific advance and communication.

Notes

1 For commentary, see Heller, 1970; Simon, 1977, pp. 36–9.
2 Stearn, 1957.
3 Cain, 1958, 1959*a*.
4 Stearn, 1959.
5 Davis and Heywood, 1963, p. 18.
6 The Banks Library; now in the British Museum, London.
7 Simon, 1977, pp. 44–5.
8 For commentary, see Stafleu, 1973.
9 Cf. Malclès, 1961.
10 Sutton, 1970, pp. 171–2.
11 Established as a private foundation by the Lloyd family (including the mycologist C. G. Lloyd) in Cincinnati, Ohio, towards the end of the nineteenth century.
12 Simon, 1977, pp. 68 ff.
13 E. L. Little, Jr., personal communication.
14 Turrill, 1964.
15 For description, see Atwood, 1911.
16 In *Annals of the New York Academy of Sciences* 5: 237–300.
17 In *Bull. Wild Flower Preserv. Soc.* 1: 1–16.
18 European Science Foundation, 1978– .
19 Garfield, 1979.
20 Prance, 1977 (1978).
21 An instance of the inadequacy of some parts of the biological information system, cf. Wyatt, 1967.
22 As in *Flora Malesiana* and *Flora Zambesiaca*.
23 Jonsell, 1979.
24 Chapter 4 (pp. 91–111) in the UNESCO synthesis report *Tropical forest ecosystems* (1978, Paris) is an example.
25 Jäger, 1978, 1979.
26 In *Progress in Botany*, 38: 317 (1976).
27 Heywood, 1957.
28 Heywood, 1958, 1960.
29 Travis *et al.*, 1962.
30 Cf. Vickery, 1975, pp. 46–7.
31 Gould, 1968–72.
32 Kumar, 1979, pp. 266–7, 283.
33 Ranganathan, 1957.
34 Gams, 1954.
35 Pichi-Sermolli, 1973; Wagner, 1974.
36 Aymonin, 1962.
37 Heywood, 1973*a*.
38 Frodin, 1976 (1977).
39 Malclès, 1961.
40 Malclès, ibid. pp. 109–10.
41 Gilmour, 1952.
42 Ziman, 1968, p. 125; Dunbar, 1980.
43 Garfield, 1980.
44 Bradford, 1953, pp. 144–59; Garfield, 1979, pp. 21–3.
45 Cf. Leimkuhler, 1967; Bulick, 1978.
46 Garfield, 1980.
47 Bamber in Downs and Jenkins, 1967; Bottle and Wyatt, 1971; Simon, 1977.
48 Simon, 1977, p. 1.
49 Webb *et al.* 1970, 1976.
50 Ziman, 1968.
51 Ziman, 1968.
52 Corner, 1946, 1961.
53 Jacobs, 1969, 1973.
54 Mabberley, 1979.
55 Stace, 1980.
56 Drucker, 1979.
57 Interview with E. Behr in *Newsweek* 96(7): 38 (18 August 1980).
58 Borgese, 1964.

2

Floras since 1939: progress and prospects

As concerns the flowering plants we may say that we live again in an age of floras and floristic work. [It is part of] a cyclic development [with several phases].

Stafleu, *Syst. Zool.* **8**: 66 (1959).

Seit der Mitte des Jahrhunderts hält eine Epoche des Florenschreibens an.

Jäger, *Prog. Bot.* [Fortschr. Bot.], **40**: 413 (1978).

The preparation and publication of floras and related works has been a constant feature of systematic botany since the end of the sixteenth century. In that time, this activity has spread from western and central Europe all over the world, in some cases absorbing (and being influenced by) local traditions, notably in eastern Asia. At times, including those in which we now live, floristic work has dominated; Jäger[1] in his review of floristic geobotany for 1977 affirmed a trend first noted in 1959 by Stafleu[2] and later by Thorne,[3] and Gómez-Pompa and Nevling.[4] In this brief overall review will be presented only the most salient features of flora preparation and production in the four decades since Blake and Atwood signed their introduction to the first volume of their *Geographical Guide*. Reviews relating to developments in specific geographical regions are given at their appropriate places in the text, mainly under divisional and supraregional headings. Moreover, this review is limited to matters of narrative; format and style are left to the next chapter.

There can be little doubt that since 1939 the preparation and publication of floras and related works around the world have surged forward to an unprecedented level, in spite of certain major setbacks, and appear still to be on the rise even in the face of increasing constraints on research financing. In many instances, increasing public concern with the environment has boosted political support for these works and thereby aided their realization, even in the more

Map. II. Five-grade map of the approximate state of world floristic knowledge as of 1979. Based upon (1) quantity, quality, age and completeness of floras, (2) collecting density, (3) an estimation of the percentage of undescribed and/or unreported species, and (4) status of distribution mapping. [From Jäger in *Progress in Botany* **38**: 317 (1976); revised by him for this book.]

difficult climate for basic research which has generally prevailed since 1973. Evidence of the current widespread activity and interest in floristic work is also furnished by the goodly number of progress reviews which have appeared since the 1950s, many of them parts of symposia and conferences, and by the annual survey of developments, trends, and new literature which has appeared for several years in *Progress in Botany* [Fortschritte der Botanik]. Moreover, resolutions urging preparation of further floras and enumerations have been adopted at more than one recent international meeting.[5] Curiously, however, hardly a mention of floras and enumerations was made in a widely disseminated American review of trends and priorities in systematic biology,[6] in spite of the obvious need for them in relation to assessing the status of species and ecosystems,[7] or as sources for the teaching of botany in the tropics, long ago emphasized by van Steenis.[8]

The most significant general reviews of floristic knowledge to have appeared in the last decade are by Thorne and others for the north-temperate zone,[9] Prance for the tropics,[10] and groups of papers presented at Uppsala in 1977[11] and Aarhus in 1978,[12] the former covering the Americas, Africa, and Europe, the latter with a scattering of tropical lands. It is a matter for regret that a similar collection covering the lands of the Pacific Basin, presented at Vancouver in 1975, has yet (early 1983) to be published.[13] Many treatments having a narrower circumscription are also available; these are accounted for in the introductions to their respective divisions and superregions. Salient floras of the period from 1947 to 1972 have been briefly reviewed by Raven as part of a general survey of plant taxonomy,[14] and reviews relating to the lower vascular plants have been presented by Pichi-Sermolli[15] and Wagner.[16]

A graphical survey of the current state of world floristic knowledge (to 1975) was attempted for the first time in 1976 by Jäger. This is here reproduced, with revisions, as Map II. The five grades delineated are based upon a blending of four criteria: (1) quantity and quality of floristic works, (2) collecting density indices, (3) an estimation of the percentage of undescribed and/or unreported species, and (4) status of distribution mapping. Other attempts at scaling, with or without maps, include two successive versions of a three-grade classification (with map) for infratropical and southern Africa,[17] and a five-grade classification for central Pacific islands.[18] For the Holarctic zone, the actual geographical coverage of a large number of floristic works was mapped by Hultén.[19] More such maps to complement those published are needed.

The progress of flora writing and publication, upon which the first of Jäger's criteria is based, has itself been subject to different types of measurement. Symington[20] used a simple, three-step scale which although crude remains generally useful. More elaborate was the seven-step scale proposed by van Steenis[21] with special reference to the tropics. For North America, Shetler[22] proposed a five-step scale which, rather than being largely historical *per se*, was based upon approaches and methodologies. None of these proposals, however, is sufficiently strongly based in breadth or depth to be generally applicable in relation to objective historical and/or bibliometric studies, although positive relationships exist between them.

Whatever the scale and criteria used, however, it is evident that within the last 40 years the bulk of the vascular floras of the north temperate and the two polar zones have by and large become reasonably well-known as to inventory, with information upon them variously consolidated into more or less readily accessible forms. The same applies for scattered areas in the tropical and south temperate zones, usually where a substantial history of local botanical endeavor has existed. On the other hand, over a fair part of the south temperate zone and the bulk of the tropics, a combined area of the greatest importance to a proper understanding of the earth's vascular flora, floristic progress has been uneven, with efforts by individual persons, institutions, or other organizations playing an exceptional role, more often than not in the absence of general movements as well as official indifference or suspicion; large areas still remain imperfectly studied and documented. Moreover, what literature is available is often so out of date as to be all but valueless for anything save professional revisionary work. Yet even in these zones significant progress has been made since 1939, notably in Africa, various parts of Asia and Malesia, Middle America, and

the Pacific but also, more gradually, in Australia and South America. This contrasts very positively with the opinion of Blake and Atwood that outside of Europe and parts of northern Asia only Greenland, Australasia (in my view partly mistakenly), and some islands could be considered floristically relatively well known, the more so as standards of knowledge and documentation have increased substantially in the 40-year period.

Among notable events of the period may be mentioned the following:

1. The completion of *Flora SSSR* and the publication of an all-but-comprehensive range of regional floras and manuals for the Soviet Union (see Region 68/69 and Superregion 71–75). A similar development is currently taking place in the People's Republic of China, headed by the large-scale *Flora reipublicae popularis sinicae* which, after a slow beginning and suspension of most work between 1966 and 1972, is currently appearing rapidly (see Superregion 86–88 and Region 76).

2. The initiation and successful prosecution of work on *Flora Europaea*, of which the last of five volumes appeared in 1980 after 25 years of work. The history of the project, the most important of its kind in Europe for a century or more, has been well described in a number of papers (see Division 6). A number of continuing projects are under way (*Atlas florae europeae*, *Med-Check List*) or are being organized (European Floristic Information System).

3. Commencement of a new flora of Australia after 20 years of agitation. Formally approved in 1979, this 'state-of-knowledge' flora is presently scheduled for completion in 15 years with 48 volumes projected; by 1981–2 the first volumes had been published (see Superregion 41–45).[23]

4. The publication of numerous modern, relatively concise, and more or less critical floras and manuals for many different parts of North America and Europe, with a more scattered representation of such works in other parts of the world. It has furthermore been claimed by van Steenis[24] that the general standard of such works, although viewed by him as largely 'routine' in character, has improved in the present era.

5. The initiation of a significant number of large-scale flora-serials for various parts of the tropical zone as well as some extra-tropical areas (apart from Australia) of which the most notable are, perhaps, *Flora of Turkey* (1964–), *Flora of Tropical East Africa* (1952–), *Flora Zambesiaca* (1960–), *Flore de*

Madagascar et des Comores (1936–), successfully continued despite unfavorable prognostications by Blake and Atwood, *Flora of Guatemala* (1946–77), *Flora of Panama* (1943–80), *Flora Iranica* (1963–), *Flora Malesiana* (1948–), and *Flora Neotropica* (1966–). A great many of these works are for developing countries, representing a renewal and continuation of the colonial-flora traditions which arose in many of the large taxonomic centers in Europe during the nineteenth century and later were adopted under other guises in North America. The continuance of such efforts for botanically lesser-known areas has been advocated by several authors and symposia, but more attention to improvement of 'southern connections' as has been done by *Flora Neotropica* is in order. By and large, these works represent primary contributions to the generation or synthesis of botanical knowledge, albeit in too many cases at the expense of monographic studies, and have been termed 'creative' floras by van Steenis[25] as antithetic to the largely 'routine' floras of Europe (and parts of North America). Nevertheless, there remain large areas, particularly in the inner latitudes, for which knowledge is imperfect or poorly consolidated and searching for information can be time-consuming and difficult.

6. The increasing development of inter-institutional and international links between botanists, largely replacing the old nationally oriented schools (and one-time colonial rivalries). Among these links, apart from the umbrella International Association for Plant Taxonomy (established in 1951), are AETFAT (1951, for Africa south of the Tropic of Cancer), the Flora Europaea Organization (1955–80, partially succeeded by the European Science Research Councils (ESRC) *Ad Hoc* Group on Biological Recording, Systematics and Taxonomy), the Flora Malesiana Foundation (1950), the Organization for Flora Neotropica (1966), the Flora North America Program (1966–73), and OPTIMA (1975, for the Mediterranean basin). Some publish periodic communications of various kinds, including literature indices, and/or conduct, or have conducted, periodic symposia. Their bases of organization vary widely from small committees (and even individuals) to regular societies, and their ethea from centrist to all-inclusive. Similar groups may exist within larger international or regional associations, among them the Committee on Pacific Botany of the Pacific Science Association. Official and unofficial agreements aimed at coordination of activities and reducing or eliminating duplication also exist.

Offsetting these progressive developments have been a number of setbacks. Some projects have appeared just to 'run out of steam' on account of the death (or retirement) of their moving spirits, loss of institutional interest, lack of finance, or for other reasons.[26] Others have been variously affected by political or other external circumstances, either temporarily or permanently. The initiation of some projects has been more or less long drawn-out due to factionalism, government attitudes, and other factors.[27]

The most spectacular setback in the modern era is, however, without doubt the sudden termination in early 1973 of the Flora North America Program. Attempts to revive it have at this writing (early 1980) ended in failure. This event was all the more dramatic because the project, aimed at development of a computerized information system as well as production of a conventional multivolume flora, had successfully completed its 'feasibility' phase and from late 1972 had advanced into full 'operation', with a central secretariat, a team of editors around the continent, an advisory committee, and scores of actual or potential contributors of treatments (including the present writer) more or less fully organized.[28] By the project's leaders it was considered to represent an entirely new approach to flora-writing, as bold for its time as Torrey and Gray's *Flora of North America* had been for the mid-nineteenth century – a point acknowledged even by its several critics. Central to the concept of the Program was the use of large-scale systems technology, including a central computer and a specially designed operating system. But then, in a flash, all was gone: six months' grace was allotted to pick up the pieces.[29] Later attempts at revival were rather less ambitious, in keeping with changes in the American mood, with nothing envisaged beyond a synthesis-flora akin to *Flora Europaea* or even its great predecessor, the *Synoptical Flora* begun by Asa Gray; and with the reported collapse of even that plan (in late 1979)[30] only one checklist (now joined – 1980 – by another produced under private auspices) remains from a decade and a half of ambitious effort – along with much loss of face.

Interest continues, though, in the use of modern systems methodology in the writing of floras and in related activities, although it has been less than might be expected. Successful applications to date have largely been on a smaller scale, i.e., at state or smaller country or even lower level. Much depends upon a better understanding of the methodology and equipment and its rapidly changing nature, as well as improvements in availability, technology and cost. The largest scheme currently envisaged is for a European floristic, taxonomic, and biosystematic information system as an extension of the work of the Flora Europaea Organization;[31] an investigation of existing documentation systems is planned in searching for an optimal system for the scheme which may have wider implications[32]. Such methodology might well be considered for those concerned with the various large infratropical zones where flora–projects are currently in progress, perhaps in specified areas such as literature control and specimen documentation.[33]

33 For an introduction to recent literature in the rapidly developing field of information technology, see Wright, W. T. and Hawkins, D. T., 1981. Information technology: a bibliography. *Special Libraries*, **72**: 163–74. [86 references, mainly 1976–80.] A more recent survey, with emphasis on the use of minicomputers and microcomputers, is Griffiths, José-Marie (comp.), 1981. *Application of minicomputers and microcomputers to information handling.* Paris: UNESCO (General Information Programme and UNISIST, document no. PGI-81/WS/28).

Notes

1 Jäger, 1978.
2 Stafleu, 1959.
3 Thorne, 1971.
4 Gómez-Pompa and Nevling, 1973.
5 Hedberg, 1979; Larsen and Holm-Nielsen, 1979.
6 Anonymous, 1974.
7 Fosberg, 1972.
8 van Steenis, 1962.
9 Thorne, 1971.
10 Prance, 1977 (1978).
11 Hedberg, 1979.
12 Larsen and Holm-Nielsen, 1979.
13 Submitted to *Allertonia* but delayed; *see* Gentry in *Brittonia* **30**: 134 (1978).
14 Raven, 1974.
15 Pichi-Sermolli, 1973.
16 Wagner, 1974.
17 Léonard, 1965; for revision *see* Hepper, 1979.
18 Fosberg, in Prance, 1977 (1978).
19 Hultén, 1958.
20 Symington, 1943.
21 van Steenis, 1949.
22 Shetler, 1979.
23 *Australian Systematic Botany Society Newsletter*, **23**: 5–6 (1980).
24 van Steenis, 1954.
25 van Steenis, *ibid*.
26 De Wolf, 1963, 1964.
27 Department of Science, Australia, 1979.
28 Shetler *et al.*, 1973.
29 Shetler and Read, 1973.
30 C. Humphries, personal communication.
31 European Science Foundation, 1978– .
32 Viewdata, offered by Prestel International (UK), has been chosen (V. H. Heywood, letter to the writer, 27 April 1982). If successful, It would be the first botanical bibliographic and information utility to be offered in videotex form. [For reviews of the 'state-of-the-art' in videotex, the several articles in *Microcomputing*, 5(10) (October 1981) and *Which Computer?* (May 1982) on the subject may be consulted.] The consequences and benefits to bibliography of the advent of on-line information retrieval are further considered in Appendix A.

3

On the style of floras: some general considerations*

The whole question of the design of Floras requires considerable attention. Little advance has been made in practice during the last century.

> Heywood, European floristics: past, present and future; in *Essays in plant taxonomy* (ed. Street, 1978).

It happens that nearly every tropical flora is fundamentally unsuited to its subject...they not merely discourage the aspirant by so aggravating his difficulties but they expose their authors to unlearned ridicule.

> Corner, *New Phytol.* **45**: 187 (1946).

The writing of an analytical bibliography of the kind here presented has seemed to call not only for a review of current developments and future progress in the actual preparation and writing of floras, manuals and enumerations insofar as they add to our still-imperfect knowledge of the plant cover of the earth, but also a discussion of the purpose, design, and content of such works. We hear much these days from the conferential circuit of the need to write floras, more floras, and still more floras, especially for the fast-disappearing flora of tropical regions and oceanic islands, but rather less – with some notable exceptions – of what floras are for, how they should be written, the way in which their message can best be spread, and, perhaps most significantly from where I write, their relationship to the political processes of the different countries concerned.

Whereas in the preceding chapter emphasis was placed upon substantive developments in flora writing, especially for the period since 1939, here an attempt is made to consider the diversity of aims, styles, and content which lie behind the term 'flora' – a diversity with many stages[1] – and to relate this to methodological, philosophical and historical movements in botany and beyond. Both orthodox and alternative styles will be examined, with reference both to professional respect-

---

* Revised and expanded from the author's essay of the same title in *Gardens' Bulletin, Singapore,* **29**: 239–50 (1976 [1977]).

ability and user experience, and some suggestions put forward. At the end are given some personal views regarding what I believe to be a prime aim of floristic writing: effective communication. To the author, this function is of equal importance to those more usually expressed – identification and documentation.

The present chapter also represents a synthesis of a series of papers on different aspects of flora writing which have appeared since the evolution of the *Flora Europaea* project in 1955,[2] as well as contributions in the 1940s by Symington and Corner[3] and in 1954 by van Steenis.[4] Together, these writings represent the first serious attempts at reconsideration of the principles and the style of floras and related works since the close of the 'patristic' era around 1880, although in the two decades before World War II pertinent remarks had been made by Hitchcock, and Blake and Atwood.[5] Of particular interest in the post-war group of papers, all written by men with long experience of the tropics, was a concern for the effective communication of botanical knowledge; for two of the authors this meant radical alternatives to accepted flora-writing practices while for the third a return to 'first principles', as laid down by the great nineteenth-century systematic botanists, was necessary and desirable. Later papers have also been influenced by a consideration of the 'information explosion' in systematics, the introduction of the new methodologies of taximetrics (numerical taxonomy) and automatic data processing (ADP), the work of the late Flora North America Program, and, notably, the increasing demands on the systematics profession made by other biologists (notably ecologists), environmental scientists, and professional conservationists as plants, even those in the rain forest, begin to 'gather political power'.

Historical background

Most of the standard floristic works in current use around the world adhere to any one of a number of stereotyped formats (or combinations thereof): (1) the more or less detailed semi-monographic or 'research' flora, (2) the handbook- or manual-flora (corresponding to the 'concise' or 'field' floras of Davis and Heywood,[6] (3) the manual-key or 'excursion flora' or (in the Soviet Union) *opredelitel'*, (4) the 'illustrated' flora, often of coffee-table dimensions and then known as an atlas, (5) the more or less extensively annotated enumeration (occasionally provided with keys), and (6) the (usually unannotated or but barely so) 'handlist'

with one- or two-line entries for species. All of these had by 1914 assumed their present forms; only in recent years have there been serious attempts to devise new formats or (more commonly) significant variations on the older ones in response to new types of information (e.g., karyotypes, population biology, phytosociological classes) and/or functional and methodological changes (including in many cases a greater use of illustrations, once notably frowned upon).

While there have been arguments in favor of other solutions, the professional 'standard of excellence' has long been by general consent the critical descriptive or 'phytographic' flora with the species as the main working unit: essentially the form so eloquently advocated by van Steenis.[7] With variations, this classical form, by and large, adheres to empirical principles gradually evolved after 1815 by the leading botanists of that century (here termed the 'patristic' writers), notably A. P. de Candolle and W. J. Hooker and succinctly summarized by Bentham[8] and A. de Candolle.[9] Bentham's principles, contained in the first five of the 247 aphorisms of his *Outlines of Botany* (1861), which appeared in nearly all of the colonial floras in the original series issued from the Royal Botanic Gardens, Kew, as well as some other contemporary English-language works, have enjoyed widespread and lasting influence. The first three aphorisms are particularly apropos to the present discussion and I repeat them here:

1. The principal object of a Flora of a country, is to afford the means of determining (i.e. ascertaining the name of) any plant growing in it, whether for the purpose of ulterior study or of intellectual exercise.

2. With this view, a Flora consists of descriptions of all the wild or native plants contained in the country in question, so drawn up and arranged that the student may identify with the corresponding description any individual specimen which he may so gather.

3. These descriptions should be clear, concise, accurate, and characteristic, so as that each one should be readily adapted to the plant it relates to, and to no other one; they should be as nearly as possible arranged under natural divisions, so as to facilitate the comparison of each plant with those nearest allied to it; and they should be accompanied by an artifical key or index, by means of which the student may be guided step

by step in the observation of such peculiarities or characters in his plant, as may lead him, with the least delay, to the individual description belonging to it.

The second part of the fifth aphorism is also of some interest and is likewise quoted:

The botanist's endeavours should always be, on the one hand, to make as near an approach to precision as circumstances will allow, and, on the other hand, to avoid that prolixity of detail and overloading with technical terms which tends rather to confusion than clearness. In this he will be more or less successful. The aptness of a botanical description, like the beauty of a work of imagination, will always vary with the style and genius of the author.

It is in the first of these aphorisms that Bentham's concept of a flora is most clearly stated; the rest are concerned largely with matters of intrinsic style (including the use of analytical keys and the need for technical terms for the sake of precision). Similar precepts, although less explicitly stated, had been employed in the floras of the British Isles written by him (*Handbook of the British Flora*, 1858) and W. J. Hooker (*British Flora*, 1830; 5th edn., 1842, on the natural system; 6th–8th edns., 1850–60, with G. A. Walker-Arnott), and were perpetuated by J. D. Hooker (*Student's Flora of the British Islands*, 1870; 3rd edn., 1884); the main differences between these works lie in the use of analytical keys by Bentham and the concern with geographical distribution and habitat shown by the younger Hooker. No effective successor to these works was produced until after World War II and even today they are still appreciated for their method and conciseness. Bentham's basic concept of a flora has continued to be the ideal among the majority of English-language writers, in contrast to others who believe, with more or less reason, that floras should be encyclopedic in scope.

The popular success of the above-mentioned British floras and an unquestioned, typically Victorian confidence in their governing precepts caused them to be adopted as standards for the growing British Empire by W. J. Hooker upon the commencement of his colonial floras (or 'Kew floras') scheme,[10] conceived in 1857 and launched three years later (and which, with modifications, continues to the present time as part of the work of the Kew Herbarium). Indeed, J. D. Hooker was to write in the preface to his *Flora of British India*

in 1872 that in style and phraseology he was specifically following 'my *Flora of the British Islands*', originally written, like his father's *British flora*, with a view to the requirements of the Scottish universities but which, with characteristic singlemindedness, he believed was capable of serving as a model for a much longer work for a very different part of the world. Bentham's experience with his own *Handbook* similarly doubtless influenced his two major contributions to the imperial botanical survey, *Flora hongkongensis* (1861) and *Flora australiensis* (1863–78). So influential were these colonial floras and such was the spirit of the era within which most of them (as originally planned) were published that it was not until the key Imperial Botanical Conference of 1924 that the approach and style represented by these works began to be questioned.[11] Even stronger criticisms appeared from Symington and Corner in Malaya in the 1930s[12] and from Symington in west Africa in the 1940s.[13]

Although Bentham and the younger Hooker were between them largely responsible for perfecting the phytographical utilitarianism considered characteristic of the 'English school'[14] and expressed in the many home, colonial, and Indian handbook-floras, much of the credit for its development should go to W. J. Hooker, until 1841 professor of botany at Glasgow University, as well as Robert Brown and G. A. Walker-Arnott, the latter a later incumbent of the Glasgow chair. It was in Scotland that the real evolution of utilitarianism in relation to flora writing appears to have taken place, along with the development of a professional interest in geographical botany.[15] The senior Hooker first employed a utilitarian style in an overseas flora in the late 1820s (*Flora boreali-americana*, 1829–40), a work contemporaneous with the early editions of his *British Flora*. Following the fashion set by *Nova genera et species plantarum* (1816–25) and other contemporary continental works, this and the three parts of the *Botany of the Antarctic Voyage* written by his son (1843–59) were published as 'prestige' works in quarto format, but by the 1850s the senior Hooker was to argue that the new series of colonial floras being projected should be in octavo, 'botany [not being] what it once was, a science confined to the learned, and of little or no benefit to the people at large'.[16] Precedents had already been set in 1834 by Wight and Walker-Arnott's *Prodromus florae peninsulae Indiae orientalis* and in 1855 by J. D. Hooker and T. Thomson's *Flora indica*, the latter aborted due to lack of interest on the part of the then-senescent East India Company as well as being in a style too detailed to enable expeditious completion.

W. J. Hooker's *Flora boreali-americana* also left a lasting imprint on botany in the United States. It certainly influenced Torrey and Gray in their preparation of *Flora of North America* (1838–43), the first continental work following the natural system, Torrey's *Flora of the state of New-York* (1843), the first 'modern' state flora, and, with concessions to conciseness, Gray's *Manual of botany of the northern United States* (1848). Utilitarianism, however, had already come to characterize a number of American botanical works, among them A. Eaton's *A manual of botany for the northern states* (1817; 8th edn., 1840) and Torrey's *Compendium of the flora of the northern and middle states* (1826); this may be connected in part with the limited means available but also with the early recognition that a popular demand existed for botanical works. In this way the standard manual–flora format of most state and regional works in North America was established. Not until the Britton era, beginning around 1890, were significant variations introduced, and these have been in part transitory.

In parallel to the developing precepts of the Anglo–North American 'school' of phytography there were evolved those of A. P. de Candolle, beginning from his work on the third edition of Lamarck's *Flore française* (1805; supplement 1815) through his experimental *Regni vegetabilis systema naturale* (1818–21) to the *Prodromus* (1824–73; from 1841 continued by A. de Candolle). The concise style of the *Prodromus*, closely resembling that of the British 'school', came to be very influential on the Continent; it is so acknowledged by Grenier and Godron in their *Flore de France* (1848–56), Willkomm and Lange in their *Prodromus florae hispanicae* (1861–80; supplement, 1893), Ledebour in his *Flora rossica* (1842–53), Bossier in his *Flora orientalis* (1867–88), and, for an overseas entity, Miquel in his *Flora indiae batavae* (1855–9). However, analytical keys are absent or but partially employed in these works; as in the floras of the Hookers, Torrey, and Gray, separation in larger groups was achieved through synoptic devices, necessitating close reading of descriptions to achieve identification (and even then, without authentically named specimens one could not always be certain, especially if the flora was at all imperfectly known – which was very often the case).

Both the Anglo-North American and Franco-Swiss 'schools' of flora-writing were the outcome of developments in French botany in the latter part of the eighteenth century, during which Linnean philosophy was largely rejected and modern systematics founded.[17] Among the many new precepts and methods which came into being was the use of more less concise but nevertheless detailed descriptions with supporting information as plant 'portraits', a phytographic style first developed by Lamarck around 1782 for the botanical volumes of Panckoucke's epochal *Encyclopédie méthodique* (1783–1808; supplement 1810–17), which for plants was the definitive work of the Revolutionary and Napoleonic eras. Lamarck evidently devised it as an antidote not only to the telegraphic Linnean diagnostic style of phytography but also to the rambling 'herbalist' mode of documentation and writing of many pre-Linnean authors which then was here and there undergoing a revival.[18] This 'Lamarckian' style, which was maintained by Poiret after 1791 as he completed Lamarck's project, soon became very influential, steeped as it was in the ideals of the Age of Reason, and was first used in an extra-European flora by Kunth (then also working in Paris) when writing *Nova genera et species plantarum*, the major part of the botanical results of Humboldt and Bonpland's expedition to Spanish America; as already noted, the excellence of this work served greatly to diffuse the new style yet more widely. It was also adopted by A. P. de Candolle for the greater part of his published output, notably *Regni vegetabilis systema naturale* (1817–21) and, later, the *Prodromus*, both works being 'species plantarum' on the 'natural' system, and expounded by him in his *Théorie élémentaire de la botanique* (1813; 2nd edn., 1819). The demands of time, resources, and democratization were in many instances to force a condensation of this 'French style' both on the Continent (even for de Candolle) and in the British Isles and other English-speaking lands, while in other cases generous financial and other support, project institutionalization, prestige, and a continuing sense of 'elitism' have led to a sometimes substantial degree of expansion (or 'inflation'). Nevertheless, even with the addition of many new classes of information and the appearance of all kinds of stylistic variations, the Lamarckian phytographic formula, a product of reasoned essentialism framed in a nominalist superstructure, continues to the present time as the accepted standard for descriptive floras of all kinds from the concise to the elaborate. Few significant additions have in nearly two centuries been incorporated into this Aristotelean legacy: among them have been increasingly extensive taxonomic and biological commentary, greater ecological and geographical detail, and, most significantly, the now commonplace analytical keys for identification – another of Lamarck's revolutionary developments.

For it was through his development and systematic application of the analytical dichotomous key in something approaching its present form, a step which also marks the entry of empiricism in its own right into systematics, that Lamarck set in train the evolution of another class of floristic work, the 'manual-key'. The great French biologist, believing that identification was functionally a different problem in systematics to that of cataloging and ordering plant diversity, first introduced this 'analytical method' in his original *Flore françoise* (1778; 2nd edn., 1795), stating his principles in its *Discours préliminaire*. It was designed almost entirely as a handy means of plant identification and, although species were included under their respective genera (as in the works of Linnaeus), no families or other suprageneric categories were used – deliberately.[19] Although evidently not originating with Lamarck, there being antecedents in Morison's *Plantarum umbelliferarum distributio nova* (1672) and Ray's *Historia piscium* (1686), the successful introduction of analytical devices for easy identification must be credited to the author of the *Flore françoise*, who at the same time projected the flora into one of its modern and lasting roles.[20]

The cleavage in the function of systematic works, especially floras, so effectively espoused by Lamarck gradually found practical expression on the continent of Europe (but, it appears, not or but slightly so in the English-speaking world) as well as other parts of the globe where continental European ideas prevailed (such as the Russian Empire and in many European colonial territories). As it later evolved (and as Lamarck would have wished), the 'manual-key' came largely to supplement or complement larger descriptive works which continued to be viewed essentially as belonging to the herbarium or library and ideally exhibiting a more or less encylopedic (and sometimes 'prestigious') character. Social and recreational changes during the nineteenth century also brought about a great increase in the number of 'manual-keys', especially for smaller areas too limited for the scope of this *Guide*. That this is a continuing tradition is shown not only by many recent Soviet works – where Larmarck's practical aim

is embodied in the term *opredelitel'* – but also by the field manual-key *Bestimmungsschlüssel zur Flora der Schweiz* by Hess, Landolt, and Hirzel which complements their comprehensive three-volume *Flora der Schweiz*. A comparable example from outside Eurasia is demonstrated by the two Sénégal floras of Berhaut: *Flore du Sénégal*, a manual-key, and the still-uncompleted *Flore illustrée du Sénégal*, an atlas-flora with descriptive text.

On the European continent, great impetus was given to the growth of the manual-key as a genre of floristic writing due to the rise in interest in field botany brought about by the transportation revolution and the growth of a large middle class with sufficient leisure. The requirements of secondary and tertiary education also generated a strong demand for this type of work, sometimes in a very condensed form and with very rare species not accounted for. As the Victorian epoch proceeded, numerous manual-keys were written not only for national units but also for a great many provinces, regions, physiographic areas of special interest, and the like. By and large, however, the writing of such school- and excursion-floras was in the hands of amateur or semi-professional botanists, among them numerous clerics and schoolmasters; for the latter, who in central Europe were normally university educated if they held posts in gymnasia or middle schools of comparable standard, local floristic work was a means of steering clear of mental stagnation and professional oblivion,[21] and in some cases could lead in the end to significant recognition and full professional employment. Nevertheless, wherever they have been produced, manual-keys are to a large extent based upon more comprehensive 'research' or 'creative' floras; because of their largely derivative nature and (in some parts of the world) frequent issue to meet public demand, they (along with local descriptive manuals) have been termed 'routine' floras.[22]

Away from the European heartland, however, the manual-key has been a comparatively uncommon form of floristic publication, save in the Soviet Union where due to strong central European influence and the aforementioned concept of 'division of labor' among floristic works, this formula penetrated long before the 1917 Revolution, notably into the Baltic provinces, Ingria (the Leningrad region) and Muscovy (middle Russia). Manual-keys eventually became an ubiquitous feature in the comprehensive network of regional floras which developed under Soviet rule from the mid-1920s, in which the 'division of labor' principle was maintained. (The Russian term for such works is *opredelitel'*, sometimes translated as 'the keys' or 'determinator' but better rendered in English, I believe, as 'manual-key', being more expressive and idiomatic.) Elsewhere, lack of a market and/or the continuing influence of the omnibus descriptive flora or manual (or at the least its basic format) is the likely reason for the rarity of manual-keys. Middle school curriculum policies and structures have also played their role; in many parts of the world in the present century the acquisition of a sound knowledge of natural history by the scholar has been less valued – a development to my mind also associated with increasing urbanization. Cultural constraints appear to represent a further significant factor.[23] Good recent examples by non-Continental authors include *Flora of the Sydney Region* by Beadle, Evans and Carolin (2nd edn., 1972), *Flora of the Pacific Northwest: an illustrated manual* by Hitchcock and Cronquist (1973, based upon a multivolume 'parent' work), *Flora of Illinois* by Jones (1963; also associated with a larger work), *Flore du Sénégal* by Berhaut (2nd edn., 1967), and *Rocky Mountain Flora* by Weber (1953; 5th edn., 1976).

An interesting link between the two formulae of Lamarck – a link which later was to become standard practice – appeared for the first time in the handbook-floras written by Bentham (and those influenced by him). Although his works were, like most of those written by the de Candolles, the Hookers, Torrey, and Gray, basically concisely descriptive in accordance with French models, analytical keys were consistently used in place of (or in addition to) the synoptical devices which characterized the works of the other authors and their contemporaries and successors. In taking this practical step, Bentham *amalgamated* hitherto separate formulae, creating the genre which has now become commonly accepted. The usefulness of such dichotomous keys (or 'indexes', to use Bentham's term) had been demonstrated during his formative years as a botanist which were largely spent in France (1817–26), a country with a flora of more than twice the number of species in the whole of the British Isles, and thus influenced by de Candolle's edition of Lamarck's *Flore françoise*[24] wherein, as already noted, analytical keys were a central feature. By contrast, J. D. Hooker evidently believed that such keys made things too easy in that students would pay little attention to diagnoses and descriptions. Such a view may well have been

shared by A. de Candolle who as late as 1880 failed to mention them in his *Phytographie*.[25]

In spite of the pragmatic influence of the Anglo–North American and Franco–Swiss 'schools' on the writing of floras, there yet remained during much of the nineteenth century the belief that a descriptive flora, particularly of a new 'exotic' area, should continue to act as a detailed compendium and repository of informaton about the plants covered – in short, a specialized encyclopedia. The sub-monographic accounts of families and genera presented had to contain detailed descriptions, synonymy, specimen citations, extensive notes, and (often) illustrations in large plates, and, for reasons of 'prestige', called for publication in a sumptuous format in the manner of the many sets of 'scientific results' of contemporary voyages and expeditions (which as a genre enjoyed their zenith in the period 1770–1850 and again in the late-nineteenth and early-twentieth centuries). This concept of a flora seems to have taken hold most strongly in the central European intellectual sphere, and cannot fail to have been influenced by the Germanic predilection for detail rather than conciseness. This seemingly ingrained trait can in spirit be seen to have descended directly from the herbals and other botanical compilations of preceding centuries, the very works against whose frequent verbosity and mindlessness Lamarck so strongly reacted. The Linnean system had simply represented for many central European writers a new and improved framework for the preparation of such general compendia of the plant kingdom for which there was a continuing demand,[26] and it comes as no surprise that (with some exceptions) the Linnaean sexual system persisted as dogma longer in the German Confederation than elsewhere, due largely to the strength of scholastic traditions (and the *ex cathedra* professorial system). The prevailing state of mindlessness in the philosophy of classification was also strongly attacked by Sachs,[27] whose writings had great influence on the later development of botany. The stark differences which existed between central European and French botany in the early-nineteenth century are exemplified in the philosophy and styles of the respective treatments of Humbolt and Bonpland plants by Kunth for *Nova genera et species plantarum* and by Roemer and Schultes for their version of Linnaeus' *Systema vegetabilium*.[28] The first truly original systematic work in central Europe professing a 'natural' system was *Genera plantarum* by Endlicher

(1836–40). Shortly afterwards, Endlicher became chief assistant to Martius on the latter's *Flora brasiliensis*.

While a number of large-scale semi-monographic descriptive floras written on the French model and following 'natural' systems had been commenced prior to 1840 – among them Moris's *Flora sardoa* (1837–59, not completed), Webb and Berthelot's *Phytographia canariensis* (1836–50), Blume and Fischer's *Flora javae* (1828–51), and, notably, the Reichenbachs' *Icones florae germanicae* (1834–1914) – the first significantly 'modern' work in this genre is the king-sized (and regally sponsored) *Flora brasiliensis* (1840–1906) which at the same time is the first comprehensive, wide-area 'tropical' flora, benefiting of course from the sheer size of Brazil. Contemporary reviews greeted the first fascicles of this work as a major step forward in floristic phytography, and it soon became widely influential as the best central European systematic work of the period.

This greatest of the comprehensive nineteenth-century floras was to drag its detailed pages slowly on for 66 years, a time span exceeding that of most contemporary British overseas floras, and together with the Candolles' *Prodromus* and *Monographiae phanerogamarum* was for long a dominant factor in European phytography, eliciting contributions from nearly all the leading European botanists of the day and fostering the spirit of international collaboration which ever since has characterized systematic botany. Through these and other works the endemic central European mania for large compendia was shifted into new and fruitful channels, with lasting results in later decades under the strong influence of Engler. Both the *Prodromus* (from 1841 under A. de Candolle) and *Flora brasiliensis* were administered on a collaborative basis; the latter, as with *Flora Europaea* a century later, possessed an organization consisting of editors, professional co-workers or flora writers (*Privatassistenten*) and an ultimate total of 65 specialist contributors. Closely involved with the project at different times were Eichler (second editor), Engler (a *Privatassistent*) and Urban (third and final editor), all later to make the Berlin Botanical Garden and Museum enormously influential.

Flora brasiliensis thus established the tradition, still with us today, of large-scale, multi-volume, descriptive regional floras, buoyed up by their suitability as vehicles for submonographic studies although now usually published in octavo format. Most of them continue to be more or less encyclopedic, and moreover retain an aura of prestige. For many botanists, they represent the ideal in floristic phytography, particularly for areas not intensively well known, and in addition make suitable institutional projects. Their merits and inadequacies are further discussed later in this chapter.

The economy of presentation of systematic information, as established by Linnaeus' *Systema naturae* and its successors, the *Systema vegetabilium* of Murray and later authors (a branch of the Linnaean tree), yet continued to find favor among many botanists, even including Kunth who wrote two major synopses of this kind. These are now usually known as **enumerations**. In such works, only essential information of greater or lesser brevity is provided, and descriptions and (for the most part) keys are lacking. Early adaptations of this formula to the floristic context include Kunth's summary of Humboldt and Bonpland's American collections, the four-volume *Synopsis plantarum* (1822–5) and Blume's *Enumeratio plantarum Javae* (1827–8, not completed). Such works were prepared with an eye towards rapid and convenient publication of results, but at first they were regarded as summaries of, or precursors to, larger descriptive undertakings. Gradually, however, the synopsis or enumeration (and its even more telegraphic relative, the checklist) developed into an independent genre of floristic writing.

Although the major part of the self-contained floristic enumerations which appeared in greatly increased numbers from the mid-nineteenth century onwards were for local or insular areas of relatively limited extent, a number were written for whole countries, groups of countries, or even subcontinents. An important step forward towards respectability for the floristic enumeration was Thwaites' *Enumeratio plantarum Zeylaniae* (1864); and the last third of the century saw the publication of two giant annotated enumerations: *Biologia centrali-americana: Botany* (1879–88) and *Index florae sinensis* (1886–1905), both wholly or largely prepared by W. B. Hemsley and to this day fundamental to floristic work in their respective areas. Others, many familiar to a wide range of botanists, have continued to appear up to the present time.

While enumerations (and checklists) have been sharply criticized by some writers, such works, particularly if an attempt has been made at critical evaluation of taxa, should be regarded as better than no consolidated work at all. In many instances they represent the only serious work for botanically poorly known areas, especially in the tropics, and more than once have fared, or may well fare, better than semi-monographic floras for corresponding areas. Many of their authors and/or editors lacked the means and/or the time to prepare full descriptive works but believed some kind of consolidated publication, even if imperfect, to be necessary. This reasoning has also been a motivation for several checklists, in which information on individual species is tersely presented in one or two lines for the most part. However, these highly condensed works are chiefly a phenomenon of the late-nineteenth and twentieth centuries, and often represented, as for instance in Natal, efforts to cope with large vascular floras within limited means.

Floras in the 'imperial' era, 1870–1930

By the time of the Franco–Prussian War in 1870–1 and the coeval foundation of the Second German Empire, the main genres of floristic writing had nearly all precipitated themselves into essentially their modern forms. The following six or seven decades were ones of increasing orthodoxy as well as refinement: floristic works became more sophisticated through critical research and inclusion of new classes of information but at the same time intrinsically more 'remote' from their audiences (save for herbarium botanists), despite the definitive substitution of modern languages for Latin in the majority of works. Little further consideration was given to the style and philosophy of floras, despite their continuing great importance as a means of phytographic communication;[29] the last consequential writings on the subject were contained in Bentham's presidential address of 1873 to the British Association[30] and by de Candolle in 1880.[31] Van Steenis[32] has noted that Diels in his *Methoden der Phytographie* (1921)[33] somewhat oddly did not give special attention to this question but concluded that perhaps the older writers, especially Bentham, 'had at the time exhausted the subject in such an admirable way that nobody found occasion to discuss it any further'. It may be also noted that Diels, while contributing many monographic and revisionary treatments for the *Pflanzenreich* and other collective works, wrote few if any floras.[34] While Diels was exceptional as a systematist in the broad sense, he was, however, active in an era characterized by an evident lessening of interest in the production of major descriptive floras (save in certain circles in North America, western Europe, and the Russian Empire) after 1880 as many key projects of the mid-nineteenth century were completed or became far advanced.[35] On the other hand, this was offset by a rise in the publication of state and regional floras in increasingly better known areas, among them the United States, European Russia (and adjacent areas), south Asia, the Japanese Empire, and Australia where strong official and/or public demand prevailed; but the works of this new cycle were to a considerable extent 'derivative' in character and bore some of the hallmarks of 'routine' floras. During this period, smaller units (up to regional level) could still be adequately catered for by one or two botanists working alone, something no longer possible with respect to subcontinents or other areas of corresponding size. Serious advances in systematics and in floristic botany now required specialization and coordination, along with some form of organized information processing system.[36]

Fortunately for posterity, the period 1870–1930 was one of outstanding progress in synthetic systematics. In large measure this was due to the work and influence of the integrated 'Englerian school' of taxonomy, phytogeography, and comparative morphology at Berlin whose heyday almost spans this period.[37] This leadership from Germany was no isolated phenomenon: in the decades before World War I the 'teuton' had come to dominate most branches of science as well as assume leadership in scientific bibliography.[38]

The 'Englerian school' was actually founded by A. W. Eichler, author of the famed *Blüthendiagramme* (1875–8), in the 1870s upon his appointment at Berlin[39] but because of his long tenure as Eichler's successor (1889–1921) and organizational ability this Berlin 'school' will always bear Engler's name. Together with Urban, who like Eichler and Engler had been, as already noted, seriously involved with the running of *Flora brasiliensis* and in Berlin had connections in high places, and Diels, Engler's later associate and successor and 'einer der letzten grossen, in der ganzen Welt geachteten deutschen Pflanzensystematiker', Engler was largely responsible for his 'school' becoming imbued with the *Weltanschauung* and scholarship which were to make it so influential. However, rather than having a direct interest in flora production, the *Berliner Kreis* specialized in large-scale monographic works, among them that supreme monument *Die natürlichen Pflanzenfamilien* (1887–1915; 2nd edn., 1926– , not completed), detailed series of regional revisions and other studies, notably for Africa and the western Pacific (*Monographien afrikanischer Pflanzen-Familien und -Gattungen, Beiträge zur Flora von Afrika, Beiträge zur Flora Papuasiens*, etc.), plant–geographical and vegetatiological studies (especially the series *Die Vegetation der Erde*), and other critical encylopedic treatises such as *Pflanzenreich* and *Synopsis der mitteleuropäischen Flora*. In their work, the Berlin group was strongly supported by the botanical circle at Breslau (Wrocław), from 1884 until the end of the period led successively by Engler and (after 1889) F. Pax. Other centers in central Europe active in systematics included Hamburg, Vienna, Prague, Zurich, and Munich, with the latter under Hegi being responsible for another leading encyclopedic work, *Illustrierte Flora von Mitteleuropa* (1906–31; 2nd and 3rd edn., 1935– , not completed).

However, even with the stimulus provided by the formation of the German overseas empire, the widespread emphasis on synthetic work in the Englerian era resulted in less attention being paid to floras as such. Few, if any, concise practical works ever appeared; there was nothing comparable to the 'Kew series' or the range of regional manuals in North America, south Asia or Australia. By contrast, Germans in overseas possessions were enouraged to send their material to Berlin for elaboration. One of the few German colonial floras is Schumann and Lauterbach's privately sponsored *Flora der deutschen Schutzgebiete in der Südsee* (1900–5) for German New Guinea, Micronesia, and Samoa. This is essentially an enumeration, containing a useful repository of geographical and other data but lacking in methodical organization and largely innocent of keys. It is all but useless for identification and cannot be compared with, for example, Merrill's *Flora of Manila*. Indeed, the central European predilection for detail was perhaps of such a nature as to have precluded (or retarded) the development of a practical philosophy towards floras, at least outside the context of 'routine', non-progressive domestic 'excursion-floras'. Writing in 1874 of German botany, Bentham[40] remarked 'The country abounds in those plodding minds which revel in the working out of minutiae of detail, and, to find their way, are satisfied with a sexual, alphabetical, or any other artificial index'. A similar lack of method, elsewhere commented on by Bentham, also marred much of von Mueller's writing on the Australian and New Guinean floras, and the same could be said of some Dutch works

on the East Indies (the latter reflecting a period when taxonomy, in its own right, was hardly taught in the Netherlands).[41] Disorganization and confusion of objectives likewise characterized an otherwise extensive effort by Russians in the writing of floras, although high standards were attained by some works such as those by Krylov and Schmal'hausen; only at the end of the 1920s were significant reforms imposed under the guidance of Komarov, thereby creating the framework for *Flora SSSR* and the complementary network of Soviet regional floras and handbooks (the latter itself of varying quality).

Among circles of floristic writing less directly influenced by central Europe, notice may be taken of those in France, Britain, and the United States. In France, after the era of Lamarck, the elder de Candolle, and the de Jussieus, systematic botany declined through the middle of the century along with much else in national life. Grenier and Godron's *Flore de France* was not succeeded by any major work until the 1890s and for several decades comparatively few overseas floras were to follow Weddell's *Chloris andina* (1855–61). The suppression of any chair at the Sorbonne or the Paris Museum specifically responsible for systematics and classification between 1853–73 and the associated loss of the Delessert Herbarium to Geneva in 1869 were serious setbacks.[42] Associated with this development was the continued formation and maintenance of several large private herbaria by wealthy amateurs, among them de Francqueville, Drake del Castillo, and Cosson as well as the Delesserts, and it was largely upon them that French systematic botany, with its strong tradition of monographic writing, largely rested in the mid-nineteenth century. At the same time, much of the abundant local and regional floristic work and publication within France was, as in other countries, the work of clerics and schoolmasters – some of whom later produced national floras of the greatest importance.[43] Professional revival, led by Baillon, Bureau and others, was gradual, with a strong stimulus from achievements in other countries and an awareness of the extensive botanical resources of the developing Second French Empire, but until early in the present century the writing of overseas floras remained in the hands of individuals, sometimes with grants-in-aid from metropolitan or colonial authorities. The concept of organization and teamwork in flora-writing was evidently first effectively implemented by H. Lecomte at Paris from 1906 when, influenced by

Flora of British India and other 'Kew floras'[44] and, perhaps, the completion in that year of *Flora brasiliensis*, he initiated, under the sponsorship of the colonial administration, *Flore générale de l'Indochine*. With this work there began the 'Paris school' of semi-monographic floristic writing which continues to the present and has doubtless influenced the conduct of similar works in Belgium and Portugal. While the usual format recalls the 'Kew series' tradition, in scope these works appear to represent a fusion of the French predilection for monographic writing with the precepts of *Flora brasiliensis*. The 'Paris floras' were born in the atmosphere of national optimism – and deliberate overseas expansion and development – which distinguished *La Belle Epoque*,[45] and given the cultural and political tendencies of the day came to exude a certain chauvinism, rather marked in the *Flore générale de l'Indochine* but also breaking the surface in the later works, with unfortunate consequences of the kind strongly criticized by van Steenis: they considered the given areas of coverage in isolation, without serious relation to adjacent territories and broad floristic regions. Also pervaded by the general spirit of optimism, floristic botany in France itself in the late-nineteenth and early-twentieth centuries was marked by the production of three major national floras, two by non-professionals and the third by a noted Sorbonne professor/educator; that by Coste, based in part on the style established by Britton and Brown in their *Illustrated flora of the northern United States, Canada and the British possessions* (1896–8), was perhaps the most influential. Detailed national floras were also published in Belgium, Holland, and Italy at this period, likewise largely through the efforts of amateur and para-professional botanists.

In Great Britain, few significant new developments in flora writing took place after 1870; until the 1920s Bentham's address of 1873 was, as for Diels, viewed as virtually the last word on the subject. Production of 'Kew floras' continued throughout the period, along with the two great enumerations for mainland Middle America and eastern Asia previously mentioned. In south Asia, following the completion of *Flora of British India* in 1897 at the zenith of the Empire, several regional floras, all in the combined Hooker and Bentham traditions, were written during the remaining years of the Raj. Forest botany began to develop as a distinct field, likewise with roots in south Asia and Burma where major early contributions were

made by the Germans Brandis and Kurz, with some aid from Thistleton-Dyer at Kew, in the last third of the nineteenth century (which in turn influenced Koorders in his later development of forest botany in present-day Indonesia), and the descriptive forest flora emerged as a new form; early ones are *Forest flora of British Burma* by Kurz (1877) and the illustrated *Indian trees* by Brandis (1906). However, the new forest floras, which gradually spread across the Asia–Pacific region and tropical Africa, remained for a long time largely traditional in format. Only after World War I was there some fresh consideration of the purpose and style of colonial botanical works and their methods of preparation, evidently first expressed at the 1924 Imperial Botanical Conference;[46] this paralleled a general reappraisal of colonial responsibilities.[47] Yet, in spite of the production of the very concise *Flora of West Tropical Africa* by Hutchinson and Dalziel (1927–36) and, through the new Imperial Forestry Institute at Oxford, of a number of forest floras and checklists (especially for African territories), the changes were cosmetic rather than real, taking little account of workers in the field.[48]

Across the Atlantic, where the botanical frontier until 1900 was largely domestic, the elegant style of floristic writing embodied in the works of Torrey and Gray (and their associates and students) had caused the manual-flora to become firmly established (at least until recent years) as a genre for most state and regional (and, later, some overseas) works. The modern floras of California, Texas, the Carolinas, south Florida, and some other states as well as the more recent regional floras differ but slightly from their mid-nineteenth-century predecessors in style. However, the efforts of the Torrey–Gray 'school' of botanists to establish a lasting tradition of sound scholarship (as well as to complete Gray's *Synoptical flora of North America* (1878–97), a continuation of the original *Flora of North America* project begun 40 years before (but by 1890 becoming a task beyond the efforts of one or two men)) failed, firstly in the face of regionalism (notably on the West Coast) and then, as the era of the 'robber barons' progressed, through the rise to a dominant position of the Britton 'school' at New York.[49] From a new base (the New York Botanical Garden in Bronx Park, founded in 1891 with assistance from some of the new tycoons) Britton and his group, with a strong view towards the geographically and numerically rapidly increasing commercial market, authored numerous works both for different parts of the mainland United States (and southern Canada) and for Bermuda and the Caribbean; most, however, were written in a dull mechanical style with more concern for quantity than quality. Moreover, they also employed the pretentious and jingoistic 'American Code', introduced in 1892 and widely used in the United States until 1930. On the other hand, some of the methodological contributions of the Britton 'school' were lasting, although in themselves the end products have been sharply criticized. Among these were: (1) the value of organization, teamwork, and public relations (Britton being, like Engler at Berlin or Komarov at Leningrad and Moscow, something of a botanical entrepreneur), best exemplified by *North American Flora* (begun in 1905) and the series of regional American floras, and (2) the systematically illustrated flora, introduced in 1896–8 with the multivolume *Illustrated Flora* (co-authored by Addison Brown; 2nd edn., 1913) which, while taxonomically slovenly in the eyes of Shinners and other critics,[50] achieved wide acceptance for the style, both at home (inspiring the four-volume *Illustrated flora of the Pacific States* (1923–60) by Abrams and Ferris), and abroad (through Coste's *Flore descriptive et illustrée de la France* of 1901–6, and Makino's *Nippon shokubutsu sôran* [Illustrated flora of Japan] of 1925). To the many floras prepared by Britton and his circle must also be added those written by federal botanists (and others) and published through federal outlets; for at this period, which also was the heyday of the United States Biological Survey, much botanical (and zoological) work in the federal lands-rich western (and, to a lesser extent, southern) states was sponsored from Washington. However, most of these publications also were encumbered with the 'American Code', then official in government circles as was perhaps only natural, given the prevailing mood of national chauvinism (and concomitant isolationism).

For it was in part this chauvinism, acting through the influence of the Britton 'school', which was to have deleterious and long-lasting effects upon the science of systematic botany in the United States. Although the Britton 'school' never achieved total dominance – several centers, among them Harvard and the Bureau of Science (Manila), remained aloof and at the same time loyal to the internationalism represented in the Torrey and Gray tradition – and by the 1920s had become discredited, the field lost scientific credibility and many professional cadres became 'locked' into a kind of insularity tinged with conservatism which continues in part to persist.[51] Moreover, American participation in external floristic activity and the writing of foreign floras has largely remained limited to the New World, save for some notable work in the Asia–Pacific region in the first half of this century. The prevailing attitudes also had broader international implications which have been touched upon by Verdoorn;[52] moreover, it is my belief that subconsciously they may have contributed to the non-institutionalization of Blake and Atwood's *Geographical Guide*, causing it to lapse after the death of its authors.

Altogether, the 'imperial' or 'post-Darwinian' phase of systematics, which is scarcely considered in most general histories of science or even biology (since within commonly accepted but oversimplified notions of progress in science[53] it is evidently of no significance) and has, to date, seldom been the subject even of contemporary or retrospective special studies, was one of continuing great publication efforts, reflecting the 'silver age' of exploration and natural history which accompanied imperial expansion, attempts at global influence through major efforts in consolidation (as exemplified by the Englerian school), international competition, American (and other national) chauvinism, and attempts to merge evolutionary theory with taxonomy. At the same time, floras (and other works) became increasingly stylized and academic; their role in communication gradually was lost sight of through misapplication of old formats to new different situations along with philosophical drift – seen here as forming a botanical parallel to the architectural developments of the period.

The 'modern' era (mid-twentieth century)

The 'modern' era of flora-writing is for convenience here dated from 1930. Apart from being the year of the key Fifth International Botanical Congress at Cambridge (England), at which a unified nomenclatural code was effectively promulgated, it was also the year of Engler's death (with that of Urban following early in 1931) and of the organization of the *Flora SSSR* project. In the preceding year, Britton had retired from the directorship of the New York Botanical Garden. From this time onward, stimulated in part by the appearance (from 1934) of the successive volumes of *Floras SSSR*, there was a revival of interest in flora-writing which, on account of the loss during World War II of most of the Berlin Herbarium and subsequently – and perhaps more significantly – the greatly increased interaction of science with the public and with governmental financing, was largely to supplant serious monographic work, except where the new cycle of floras provided suitable outlets and/or opportunities – a development which continues to be regretted in some quarters.[54] By 1958 systematic botany was well and truly into a new 'age of floras', with activity extending over most of the earth's land surface;[55] this 'boom' has continued to the present time.

From the early 1930s there also came major philosophical and other changes to systematic biology, including the doctrine of neo-Darwinism as well as the methodologies and techniques collectively grouped under 'biosystematics'. These developments have been best reviewed by Huxley,[56] Davis and Heywood,[57] and Raven[58] with a further consideration of their application to systematics provided in the textbook of Radford *et al.*[59] Under the 'New Systematics', phytography moved from being a static typological exercise based largely upon comparative morphology, with yet some creationist, even essentialist aspects – the 'Darwinian revolution' hitherto having had little effective influence on much of taxonomic practice[60] – to something more dynamic, accounting for parallelism, variability, genomes, chromosomes, biochemistry, ecology, and biology. In an increasingly technological age, this movement improved the credibility of systematics in the eyes of the majority of mainly laboratory-based scientists, notably in the United States.[61]

Nevertheless, the influence of these changes on the actual writing of floras was in practice selective and only gradual. Analytical keys had become usual (except in most enumerations and checklists), there were improvements in nomenclature so as to conform with the International Code, and more critical commentary as well as ecological and biological data manifested themselves. The preferred minimum standards for information content in a flora were summarized by Hitchcock[62] and by Blake and Atwood.[63] Subsequent later refinements have included information on life-forms, karyotypes, and phenology; a reduction in the use of formal infraspecific categories; and increased consideration of habitats, variability, hybridism, introgression, clines, and phyletic relationships among other topics (although the formal treatment of hybrids remains an epistemological problem in flora-writing). Information from pollen analysis, comparative phyto-

chemistry, anatomy, and reproductive biology have also been introduced but these are as yet mainly at generic level, especially in the larger 'research' floras. Published or circulated 'guidelines' for flora projects, such as the *Flora Europaea* 'Green Books',[64] are becoming more of a requirement in all parts of the world as the writing of any but local floras becomes increasingly a collective effort.[65] The ideal contents of floras have been discussed with increasing thoroughness by a number of authors,[66] and in the author's view the best modern 'cookbook' for traditional manual-flora writing (as practiced in North America) is provided by the textbook by Radford *et al.* referred to above.

Yet what that textbook prescribes would still be clearly recognizable to the patristic writers of the nineteenth century – a point also made by Meikle[67] in relation to current British floristic work. Many modern floras are not so very different from those written by Torrey, Gray, Bentham, J. D. Hooker, and their contemporaries; the main differences are the almost universal use of analytical keys and the inclusion of many more classes of data. One who compares some current American state floras with Torrey's *Flora of the state of New-York* might be forgiven for thinking that the modern works had been written by descendants of Rip van Winkle; but only a generation ago the classical style of flora-writing was passionately defended by van Steenis who argued that 'besides...ecology and distribution, and nomenclature, nothing needs to be added to Bentham's classic exposition of a purely descriptive Flora'.[68]

It is thus evident that, consciously or not, the classical descriptive formula still finds wide acceptance, even after more than 140 years; the variations introduced by the Britton 'school', especially in the mechanistic *North American Flora*, never gained popularity on that continent and after 1930 the older precepts imparted in the name of Torrey and Gray enjoyed renewed support. The most significant changes were in the handling of distributional data, notably in state floras where the use of small dot maps for all species – first introduced by Deam in his *Flora of Indiana* (1940) – has gradually gained popularity. More or less amplified versions of the classical formula, following the precepts of *Flora brasiliensis* but now usually presented, like contemporary French overseas floras, in the more convenient octavo format, have become standard for 'research' floras; these include the greater part of those written for the tropics, both by

American authors (*Flora of Guatemala*, *Flora of the Lesser Antilles*, *Flora of Panama*) and by Europeans, as well as *Flora SSSR* (and many other Soviet and eastern European floras), *Flora reipublicae popularis sinicae* (in style and philosophy modeled after *Flora SSSR*) and Chinese provincial floras, a number of other modern Asian floras, *Flora of Southern Africa*, and the regional floras of the Argentine. Among the most concise of modern 'research' floras – and at the same time very true to the Benthamian tradition – are *Flora of Turkey* by Davis and collaborators (1964–) and *Flora of Cyprus* by Meikle (1978–) and, for the tropics, *Flora Zambesiaca* (1960–).

This conservatism evidently also extended to the incorporation of 'biosystematic' information. This curious lag, to which Raven[69] has drawn attention, was especially evident in the United States in spite of early efforts by W. H. Camp and others in the 1930s and 1940s to bring about a break from traditional taxonomic and floristic approaches. One of the first manual-floras demonstrating consistent use of the new classes of information becoming available was *Flora of the British Isles* by Clapham, Tutin, and Warburg (1952; 2nd edn. 1962). A corresponding American pioneer was *A California flora* by Munz and Keck (1959; supplement, 1968). Neither work, however, has wholly displaced their more traditional predecessors; in California Jepson's *Manual* (1923–5) was in 1973 still selling, due in large part to its many illustrations. More widespread acceptance of 'biosystematic' data came only towards the end of the 1960s, with representative works being *Manual of the vascular flora of the Carolinas* by Radford *et al.* (1968), *Manual of the vascular plants of Texas* by Correll and Johnston (1970), and *A flora of tropical Florida* by Long and Lakela (1971).

It was in central Europe, however, that the first 'critical' floras appeared, some as complements to field manuals. About the earliest was *Flora der Schweiz* by Schinz and Keller (1900), whose *Kritische Flora* had by 1914 appeared as a separate volume; this remains something of a landmark.[70] An important work of more recent date, based upon extensive cooperative research from the 1930s to the 1950s, is the series *Exkursionsflora von Deutschland* by Rothmaler; its part IV, *Kritischer Band*, first appeared in 1963. The first North American work in this class, though differing sharply in style, is *Flora of the Queen Charlotte Islands* by Calder and R. Taylor (1968), covering an island group on the coast of British Columbia. In New Zealand, substantial

critical commentary, covering hybridism, introgression, and other patterns of variability, was incorporated into a more conventional format in the first volume of *Flora of New Zealand* by Allan (1961). Somewhat more specialized is the series of cytotaxonomic conspectuses which have been prepared and published for various groups and areas by Löve and collaborators. On the whole, however, the influence of what has been termed 'camp-ing' in floristic writing has come through gradual diffusion into the traditional stylistic framework, and radical departures have been uncommon.

Even less common, to date, have been floras written with the assistance of computerized information banks. Among the few currently appearing is *Flora de Veracruz* (1978–); in this work, however, the available information is reprocessed into a traditional format for the published fascicles. The projected descriptive flora of British Columbia will also make use of a data bank. However, the most elaborate project of this kind, the Flora North America Program of the late 1960s and early 1970s, was suspended well before anything had been published, and when revival was attempted in the latter 1970s a return to traditional conceptions not very different from Gray's *Synoptical Flora* was envisioned.

Other long-established and stereotyped formulae of flora writing have also remained in vogue, despite criticism from certain quarters, especially of the enumeration.[71] Practical considerations, however, dictate that for many areas enumerations will continue to be produced. Among notable recent works in this genre are *Prodromus einer Flora von Südwestafrika*, edited by H. Merxmüller (1966–72), *Enumeration of the flowering plants of Nepal* by Hara *et al.* (1978–), and the computer-produced *Vascular plants of British Columbia: a descriptive resource inventory* by R. Taylor and B. MacBryde (1977). The former contains keys, a feature unusual in an enumeration but much to be commended, while the latter is the first substantial work in the genre to be created from a computerized data base. However, less critical works continue to appear, sometimes in attractive guises like the works of Angely for southern Brazil. Similarly, the manual-key formula remains in wide use for works of identification, including the greater part of the 'routine' floras of Europe, the *opredel'itely* of the Soviet Union, and occasional works in other continents such as *Flora of the Sydney Region* by Beadle *et al.* (1972), *Flora of the Pacific Northwest: an illustrated manual* by Hitchcock and

Cronquist (1973), and, unfortunately, the three-volume *Flora of Java* by Backer and Bakhuizen van den Brink, Jr. (1963–8). The illustrated flora, in various guises including the costly atlas format, is also becoming more common – to the benefit (or, to some, detriment) of the botanical education process.

The most important development of the 'modern' era, and to which it largely owes its sobriquet as *eine Epoche des Florenschreibens*, an 'age of Floras',[72] has been the gradual revival of interest in the large-scale descriptive flora. This trend, although apparent from early in the century with *North American Flora* and *Flore générale de l'Indochine*, has become especially prominent only in the last four decades, with numerous projects begun in the quarter-century after 1947. Like their nineteenth-century counterparts they have, by and large, been, and continue to be, prepared for the many countries, regions, large islands, and island groups – more often than not within, or partly within, the tropics – where knowledge of the plant life is as yet imperfect in one way or another. On the other hand, though, these modern 'research' floras (or critical enumerations) have usually been issued in serial parts, frequently without regard to systematic sequence although in some cases publication is in volumes or parts thereof. The growth of this genre is strongly related to increasing professional specialization as well as concomitant increases in the amounts of herbarium material, literature, and other observations to be accounted for, but it may also be explained by their evident attraction as 'acceptable' long-term institutional projects, as botanical contributions to overseas 'development', or as projections of national or supranational prestige. At the same time, however, some at least have become partial substitutes for serious monographs, for which there seem today to be too few satisfactory publication outlets and which in some quarters, notably many university circles, seem to possess a low academic 'status'.[73] This attitude, while partly autochthonous, may perhaps also be due to their lesser ability, at least in the modern 'Western' world, to attract research funds, particularly where funding agencies are 'mission-oriented'. A serial flora has the advantage of being a kind of botanical work still comprehensible to the modern *apparatchik*, who in some cultural milieux is most probably less well informed about (or interested in) plant life than his Victorian counterpart.

At the present time, large-scale floras are *une vague courante*, the two-fold rise and fall of the Flora North America Program notwithstanding, and most larger botanical institutions in North America, Europe, and (to a lesser extent) elsewhere have one or more of these projects under way. In many cases they are legacies of former colonial ties, whose value has very often been recognized by successor governments. While some of these serial works cover areas comparable to those of the great floras of the nineteenth century and additionally may possess more or less natural botanical limits, many others have been set up for smaller, mostly politically delimited units (in which language may also be a criterion), in spite of the strong advice of van Steenis against such fragmentation.[74] The variety of styles exemplified in the modern serial floras is paralleled only by their modes of preparation and administration; however, in most cases successful completion has depended upon the efforts of one to a few men and/or continuing institutional commitment, despite ostensible collectivism.[75] The energy with which these projects have been, or are being, pursued likewise varies widely, and the same is to some extent true of their critical standards.

Some of these large-scale works are meaningful, and as they progress represent real contributions to botanical knowledge, although in some cases unfolding rather too slowly; examples include *Flora SSSR* (1934–64), *Flora Malesiana* (1948–   ), *Flore d'Afrique Centrale* (1948–   ), *Flora Zambesiaca* (1960–   ), *Flora Reipublicae Popularis Sinicae* (1959–   ), *Flora Iranica* (1963–   ), *Flore de Madagascar et des Comores* (1936–   ), *Flora Neotropica* (1966–   ), and *Flora Europaea* (1964–80). Others are too detailed or otherwise long-winded, too grandiose, cover unnecessarily small areas, or have an insecure basis. Of special concern is the length of time required for execution; for many works this has been, or is likely to be, very prolonged, inevitably raising questions about financing as well as institutional and individual motivation.[76] The 66 years taken by *Flora brasiliensis* has already been noted, but against this must be set its coverage of over 22 000 species at a time when available manpower was very much less. Even the energetically pursued *Flora SSSR*, with something over 17 000 species, required over 30 years for completion. Other works requiring long periods of time have been *Flore générale de l'Indochine* (1907–51), *Flore de Madagascar et des Comores*, *North American Flora* (1905–   , but now run as a serial), *Flora of Suriname* (1932–   ), *Flora of Peru* (1936–   ), *Flora Polska* (1919–80), *Flora brasilica* (1940–   ), *Flora capensis* (1859–1933, with a 31-year break from 1865 to 1896), and Hegi's *Illustrierte Flora von Mitteleuropa* (1906–31; 2nd edn., 1935– , 3rd edn. 1971–). More recent undertakings having more than half of their species still to be published include *Flore d'Afrique Centrale* and *Flora of Tropical East Africa* (between 40 and 50 per cent complete), *Flora Zambesiaca* (about 35 per cent), *Flora Malesiana*, *Flora Reipublicae Popularis Sinicae*, and *Flora of Southern Africa* (1963–) (between 10 and 20 per cent), and *Flora Neotropica* (less than 10 per cent).[77] It is only those works with limited objectives, in which completeness has been sacrificed to a fairly marked degree, which, wherever the flora is at all imperfectly known or synthesized, have been fully realized within a generation or less; among these are *Prodromus einer Flora von Südwestafrika* (1966–72), *Flora of West Tropical Africa* (1927–36; 2nd edn., 1954–72), and *Flora Europaea*.

A corollary to the increasing length of time taken per species in 'research' floras is an evident decline in average taxonomic productivity. Whereas in the nineteenth century perhaps 250 species per year could be written up for a handbook-flora by one author (Bentham achieving a still higher rate), by the 1930s the optimum for critical semimonographic floristic work was estimated at 80 species per man-year.[78] This had by 1963 further declined to 50 species,[79] and the latest estimate is a mere 15–20.[80] The only large-scale floras of modern times achieving outputs comparable to those in the mid-nineteenth century are those with limited objectives, as already noted, or the major Soviet and Chinese floras; if the great new Chinese flora is completed by 1985[81] or even by 1990 (as is more likely; B. Bartholomew, personal communication), a record rate exceeding even that of *Flora australiensis* will have been established – but with more than 100 times the manpower.

It is not to be wondered, then, that De Wolf[82] and Jacobs[83] have questioned the wisdom of many large-scale projects, suggesting that more attention be paid to the preparation of 'concise' works. There is some indication that attention is being paid to these suggestions, partly through force of circumstances. Moreover, some over-elaborate works such as *Genera et species plantarum argentinarum* (1943–56), physically larger even than *Flora brasiliensis*, have (fortunately!) been discontinued. However, for many botanically little-known or lesser-known areas (below stage 3 on Jäger's map) where existing documentation is poor and/or scattered, any general flora would in Symington's sense[84] represent 'first' or 'second' coverage and should therefore possess adequate documentation and references; moreover, such works should always have priority. The objection would also be raised that for 'first' floras, a substantial amount of basic monographic and revisionary work is required and this must be expressed in some way in the published work, because there may be no alternative. Thus, with any advocacy of conciseness in floristic publication, careful consideration must be given to the satisfactory disposal of what is not included; too often this is relegated to a plethora of scattered outlets. Within the tropics, for example, a good compromise has been struck by *Flora Zambesiaca*; the inadequate specimen and distributional data commonly encountered in *Flora Malesiana* necessitates access to at least a large file of associated precursory papers and other records for effective use. At a more local level, *Flora of Java*, a 'second' flora like *Flowering plants of Jamaica* and *Flora of West Tropical Africa*, is regrettably not well documented, lacking even the consistent citation of standard revisions, monographs and other precursory papers as has been done in many other works. On the other hand, conciseness is more easily achieved in works on the north-temperate flora and in parts of the temperate southern hemisphere, but even here many manuals are deficient in documentation; one of the better ones is Willis' two-volume *A handbook to plants in Victoria* (1960–72). The question of information content and handling, while not seriously considered at the beginning of the 'modern' era, can thus be seen to have grown to acute proportions in the last half century; with respect to floras, no generally acceptable guidelines have yet emerged.

The most radical criticisms of traditional formats of floristic writing since 1930 arose, however, not from academia or the 'hayloft' but from the rainforests of Africa and Malaya. In strong critiques of existing floras including even those of fairly recent date (*Flora of West Tropical Africa* and Ridley's *Flora of the Malay Peninsula*), Symington[85] and Corner[86] argued that the conventional handbook-flora, by then long established as orthodox in the British Empire, was in fact comparatively uninformative and useless for the man in the field, on the tropical roadside, or the student. Even the 'new wave' arising from the Imperial Botanical Conference had by the 1940s effected little real advance in the practical coordination and dissemination of botanical information in the tropics, a difficulty left unexplained by Symington but very likely resulting from a combination of economic factors, obsolete professional structures, academic inbreeding, and especially cultural and intellectual sclerification. While the amount of botanical (and ecological) work published was superficially of a size beguiling to undereducated administrators, it by and large, at least in Symington's view, lacked definition. This mindlessness may also have contributed to what Corner called the 'enormous humbug of tropical botany'. Behind the absence of definition was the persistence of traditional philosophical and epistemological notions considered to be part and parcel of a science which had grown up in the northern temperate zone and which had hitherto resisted change, despite strong efforts by such pioneer tropical botanists as Adanson and Griffith.[87] Much of this framework, already as early as 1922 considered to be riddled with the dry rot of esotericism,[88] yet continues to persist – and be exported to the tropics.

These notions strongly influenced the writing of floristic works in the tropics, even at local level, with the result that, consciously or unconsciously, most conformed *a priori* to formats and styles evolved for other conditions. A doubtless culturally-induced view also existed that, to quote Symington, 'the European herbarium worker could solve the field biologist's problem without the latter's full cooperation'. Such misconceptions and distortions, in Corner's view, led to works that were far from realistic. Particular points included: (1) ignorance of vegetative characters, (2) ambiguous descriptions, (3) faulty nomenclature, and (4) errors resulting from repeated copying and/or lack of critical investigation. For the compilation of *Wayside Trees of Malaya* (1940; 2nd edn., 1952), so imperfect and heterogeneous was the available information that a substantial amount of locally-based, original research was required, a situation also faced by other writers of tropical field manuals including the present author (a task made somewhat easier, however, by the latterly gradually increasing number of smaller 'local' herbaria). Only in recent years has the value of such field manuals slowly come to be recognized; the need to develop unorthodox styles as was done by Corner and later by Allen in his *Rain Forests of Golfo Dulce* (1956) has for long remained almost incomprehensible, with the result that most conventional works published have remained effectively closed books to all but initiates – and thus lacking substantive political impact although they may be academically 'correct'. Even I was

somewhat baffled by *Wayside Trees of Malaya* in the early stages of compiling this *Guide*; experience and students' preferences have changed that. The impact of such works would be even greater if it were not for the excessive orthodoxy and esotericism, often imported, of biological curricula in many parts of the world – which, in spite of technical and other improvements, remains almost as it was 60 years ago.[89]

In summary, then, the 'modern' era, following the nomenclatural and theoretical reforms of the 1930s, has been in floristic botany characterized largely by an enormous expansion in the production of descriptive floras and related works, largely conforming to established formulae and, in addition, for lesser-known regions frequently serving as partial substitutes for wide-ranging monographic and revisionary studies now considered somewhat unfashionable. Many new and not-so-new sciences as well as modes of thought and techniques of analysis have yielded results drawn upon for floras along with those traditionally used, but 'critical' floras complementing conventional manuals have been rare outside central Europe. Some consideration has been given to the identification process, but dichotomous analytical (or Lamarckian) keys are still usual. Elaborate use of infraspecific categories has sharply declined as the importance of variability has come to be appreciated but the proper treatment of this as well as hybridism, 'microspecies', and speciation remains a problem, seriously tackled in but few works either within or without the traditional framework. Some botanists perhaps consider these topics not really the province of floras with their aura of formality – but then overlook the information problem thus created, although this is partially alleviated for the period 1945–64 by *Biosystematic Literature* by Solbrig and Gadella (1970). Moreover, only in the last decade has it been realized that 'biosystematic' problems also exist in the lowland as well as montane tropics. On the whole, the products of the second 'Age of Floras' remain fairly tradition-bound, even with respect to the tropics (and some other lesser-known areas) despite strong criticism as well as attempts, latterly increasing, at alternative solutions to communication as well as documentation (the latter usually having been considered to be the proper role of a flora of a lesser-known area). There has, moreover, existed a certain mindlessness, with the aims and purpose of many a project not clearly worked out (or done so in a one-sided manner), and a goodly number of works appear to lack 'personality' and 'direction' and instead give an impression of monumental dullness, even in their more egalitarian octavo coats. This has in recent years, especially with the renewed interest in phenetic (and cladistic) systematics, the introduction of automatic data-processing and a growing concern for information problems, occasioned a small flurry of contributions which are considered in the paragraphs which follow.

Floras at the present time

At the beginning of this chapter, I noted that in recent years there has been an information 'explosion' in systematics, from which floras have not been spared; most recent efforts at coping with this development, where they have been made at all, have been inadequate or have resulted in little more than placebos. The realization of this serious problem, together with the introduction of ADP methods and their rapidly growing sophistication, has led to considerable fresh discussion of the content and style of floras and the philosophy and methodology of flora-writing – the first substantial debate for some 100 years in this field, with, as has been noted, only few key contributions in between. As this has been a recent issue, still very important (in spite of a lull in fresh contributions) and deserving of being termed as of 'crisis' proportions,[90] it seems useful to consider in broad terms the question of flora-writing at the present time and to make some suggestions about the future, with particular reference to infra-tropical regions.

The continued acceptance, in part mindlessly, of long-standing and stereotyped formats and sets of questions relating to floras and related works by generations of botanists is not only evidence of their general utility but also a reflection of the conservatism inherent in much of the taxonomic profession. In other words tradition, as much as any intrinsic worth in the parameters governing format and contents, has contributed to the longevity of present floristic styles. Taylor[91] has stated that these parameters are some 200 years old but, as described earlier in this chapter, it seems evident that the design principles and content of most modern types of works were largely crystallized betwen 1820 and 1860, with some classes such as manual-keys and ligneous floras evolving later to meet particular needs or, in the case of the large-scale enumeration, to cope with difficult situations like the inventorying and classification of floras of humid tropical regions.

An examination of the relevant literature, largely cited at the beginning of this chapter, as well as personal observations together suggest that there have been essentially two views, both of long standing and to some extent at odds, concerning the central purpose of descriptive floras (and other kinds of floristic compilations). This in some respects runs parallel to the view of van Steenis[92] that most floras are 'dualistic' in nature. That is, they attempt to serve two different ends, the one communication (through keys, descriptions, illustrations, and commentary), the other archival or encyclopedic. To him this problem could be resolved in north temperate regions but, due to the sheer size of the flora in most areas, not in the tropics. A similar theme has been central to the discussion of the 1970s.

The first philosophy – one which sees floras as tools for communication (through identification) – harks right back to the first aphorism of Bentham quoted early in this chapter. The relative value of this philosophy has once more been emphasized by Heywood[93] as well as indirectly by Watson.[94] It was argued by the former that floras were not necessarily intended to serve as sources of strictly comparative data, and that their main function was to address themselves to certain questions about the plants of an area: (1) what there is, (2) how they may be recognized, and (3) where they may be found. To this end they should include keys, descriptions, necessary auxiliary information, and essential nomenclature, synonymy, and citations. This same philosophy has in general been adhered to by Brenan[95] and Jacobs,[96] whose papers are particularly oriented towards the 'Third World', as well as by Shinners.[97]

The second philosophy – in which floras are seen as essentially archival or encylopedic – has its origins in the *Flora brasiliensis* tradition and, beyond that, to the herbalist tradition of central Europe within which also lay the philosophical origins of such works as *Hortus malabaricus* by Rheede and *Herbarium ambionense* by Rumphius, both written in the seventeenth century. It characterizes a goodly number of recent and current large-scale flora projects, with the justification (if any) that floras should be 'a physical repository of descriptive data about plants which are organized and formatted, usually in book form, so as to answer a time-tested set of prescribed questions'.[98]

The differences between these two philosophies as related to developments in the nineteenth and twentieth centuries have already been discussed, with many examples given. It was also noted that for floristically lesser-known parts of the world the encyclopedic flora has been much favored, but that the realization of such works, largely still by traditional means, involved a considerable investment of time and manpower, with gradually falling productivity. In the words of one commentator,[99] such works were veritable botanical 'booby-traps', with progress measurable in decades, not years. By contrast, projects with more limited objectives stood a greater chance of successful completion within comprehensible time-spans, and more often than not were in spirit 'tighter'; as Webb[100] has noted in his *Flora Europaea* valediction, close editorial control was at all times essential.

Sometimes, though, the two philosophies have been confused. In the 'introductory notes' to one flora project currently under way in the region where these words are being written, it is stated that, in order to make available 'information' on the flora (which is presently very scattered beyond that available in a more inclusive standard regional work), the sponsoring institutions have 'embarked on a project to produce, in a handbook format, a *concise* Flora' [my italics]. By contrast, the first volume suggests that the work, even with some information having been relegated to 'technical supporting papers', will be somewhat repand in nature, similar in many ways to its more inclusive sister; in one family, four pages of text are required to deal with three relatively easily recognized species. This is hardly 'concise' in the Benthamian sense, still apparent in works such as the *Flora of Turkey*, but rather brings to mind larger-scale works in the Martian tradition such as *Flora of Panama* or *Flore d'Afrique Centrale* – and in some respects is less informative than either of these works. Other examples demonstrating confusion of objectives or the more underlying effects of academic bondage could be given. Yet without a strong taxonomic and documentary foundation the achievement of conciseness in floras of tropical areas has been difficult and but rarely overcome; works such as *Flora of West Tropical Africa*, *Flowering plants of Jamaica*, *Flora of Java*, and *Tree Flora of Malaya* (the last not yet completed), all of comparatively recent date, have normally been dependent upon major parent works or a long tradition of overseas and/or local botanical work or both.[101] Even these, however, are still fairly traditional or have traditional elements, although they have had, or will have, a greater effective impact that many other works (the awkward format of *Flora of Java* is nevertheless a severe drawback).

One of the central questions confronting floristic works of whatever stripe is the handling of data. In the past, guidelines for floristic works were so fixed by tradition that little consideration was given to this issue; that considerable subjectivity existed in the selection process appears simply not to have been perceived. With the growth of the phenetic (and cladistic) approaches in taxonomy, both claimed to be more 'objective', the issue was forced; at the same time attention was drawn seriously to the sheer proliferation of data becoming available to modern systematists due to the many new approaches and techniques latterly introduced.[102]

From this grew the idea that different kinds of floristic works or other forms of information storage should be brought in to serve different purposes: universality was no longer practical. Watson,[103] in calling for a return to the Benthamian tradition of 'concise' floras, made the significant point that the kind of information which now tends to go into elaborate 'archival' floras was in reality more appropriate to other kinds of taxonomic publication or for storage and retrieval through data banks or other non-print media; as things then stood, selection of information for floras and its processing was somewhat mindless and, what was worse, a good deal of what was left out, though potentially useful in some contexts (especially to local workers) was simply not accounted for or ended up in the rubbish bin. Believing that the two main philosophies of flora writing – the 'archival' and the 'practical' – should be separated and that a given work should follow one or the other, he considered that confusion of objectives in current floras was frequent and that many represented unhappy compromises, failing in both areas: they were neither useful sources of comparative data nor practical tools for identification. There was no recognition of the need under modern conditions to all but separate these functions (and that, in many cases, their value did not justify their high cost). His final words were that 'we have all these advantages [computerization, philosophical analysis, masses of data, etc.], yet have more difficulty in getting to grips with real problems than Bentham did'.

In addition to creating an awareness of the exponential proliferation of new information, Fisher[104] also drew attention to the possible inadequacy of language alone for plant description, and stressed the importance of illustrations; for the humid tropics this has special import in view of the presence of so many different kinds of plants (relative to the cool-temperate zones where most botanical thought has been shaped) and where the perception of most people is much more visually than literarily oriented. This point has been clearly recognized in such Asian works as *Cay-co mien-nam Viêt-Nam* by Pham Hoang Hô (1960; 2nd edn., 1970–2), *Iconographia cormophytorum sinicorum* by the Botanical Institute, Peking (1972–6), and *Choson singmul myongchip* by Chŏng (1956–7); all are atlas-floras comprising small figures with parallel text, with analytical keys playing a supporting role (although the indirect influence of Britton and Brown's *Illustrated Flora* should be accounted for through its impact on Makino and his *Illustrated flora of Japan*, first published in the 1920s). While in form they may be influenced by 'Western' models, I believe that something of the Asian (and particularly Chinese) botanical tradition has permeated these works. At the same time, some are as floras concise, thus owing something to the traditions of Bentham and his contemporaries. It is thus strange (or not so strange?) that Backer, the leading Dutch student of the Javanese flora, was so opposed to illustrations in a formal flora; the *Flora of Java* by him and the younger Bakhuizen van den Brink is entirely innocent of them as to Backer they had no proper place in such a work. In a number of other tropical floras authored by persons of European extraction illustrations yet seem mere appendages, whether in lesser or (in consequence of recent major changes in printing technology) greater quantity, rather than as an integral part of the conception of the work concerned. There are signs of change, however: *Flora del Avila* by Steyermark and Huber (1978) and *Flora of the Rio Palenque Science Center* by Dodson and Gentry (1978). Good illustrations are in my view very important to field botanists and others without ready access to a large herbarium, despite nineteenth-century views to the contrary.

In recent years, there has been considerable interest shown in the use of computer-based information processing systems for floristic work, and this is one of the principal future directions suggested by Watson and other writers. A strong movement to apply such methods for information processing in taxonomy arose in the latter half of the 1960s with the increasing spread and sophistication of computers and was publicly developed over a number of international meetings from 1969 to 1975. The first practical application on a large scale was embodied in the Flora North America Program,[105] wherein a relatively sophisticated information system was to be developed as a basis not only for a concise conventional flora in some five or six volumes (similar to *Flora Europaea*) but also for a 'dynamic' continuing service. Ancillary programs were also developed for a type specimen register, bibliographic information storage and retrieval, and identification assistance. A close reading of the various descriptions of the Program suggests, though, that the production of a hard-copy flora was viewed as much as a recognition of the strength of convention and tradition in flora-writing as it was nominally a primary goal of the Program. Unfortunately, the project was killed in 1973 as a result of internal and external politics and administrative pressures of the day on science; later it became evident that the new methodology had, as some skeptics had suggested, threatened to become the master rather than the servant of the operation. As of 1975, when the forerunner of this chapter was written, it seemed a moot point whether automated information processing would become an effective tool in flora-writing in the way hoped for by its advocates; in 1980, judging from the near-absence of substantial new contributions on the subject following publication of the proceedings of a 1973 NATO conference at Kew,[106] this would still seem so. Tradition runs strong in much of the profession (as Shetler himself admitted) and, moreover, few appear to have an effective understanding of computers, software and information science (the latter often academically grouped with library science) and of the rapid developments taking place, especially where it is most needed. The generally high cost of the new technology has been another deterrent, but for some types of applications use might well be made of such machines as the Tandy TRS-80 Model 12* micro-computer which is in the upper part of the lower price range, particularly as less expensive mass storage in the form of Winchester-technology hard disk drives and, in the future, videodisk units, becomes more widely available. Interest continues here and there, however, in the application of automated data processing,[107] and a register of projects has for some years been maintained at Notre Dame University in Indiana.[108] Smaller-scale projects involving floristic work, some inspired by the FNA Program, have been developed since the late 1960s; among the best-known are the Flora of Vera Cruz (Mexico) Program, the Flora of British Columbia Program, and the British Antarctic Survey floristic survey (the latter of course mainly

---

* This does not constitute specific endorsement of this model or of the manufacturer concerned.

dealing with non-vascular plants). Some form of automated information processing system is planned for the Med-Check List project and for the planned generic flora of Columbia. Variously automated procedures have also been used for different floristic mapping projects, of which the most famous is the *Atlas of the British Flora* by Perring and others (1962, 1968). The most ambitious project currently under development, representing an extension of the work of the Flora Europaea Organization, is a 'European Floristic, Taxonomic and Biosystematic Information System'.[109] A somewhat similar network is reportedly under development in Brazil.[110] The creation of such data banks was among the prime recommendations of a 1974 Missouri Botanical Garden workshop.[111] Related schemes have involved the complete encoding of essential data from specimens in the Queensland Herbarium (Brisbane), the Herbario Nacional Colombiano (Bogotá), and the South African National Herbarium (Pretoria); this will enable production *inter alia* of preliminary floristic lists or provision of information towards preparation of descriptive floras, as is being done in all of the centers named. Nevertheless, an 'acceptance barrier' remains to be overcome before the new technology gains a widespread foothold and becomes standard. Among other factors such as availability, cost, and user capability, many systems (and packages therefrom) are not suitably structured – and many are those who remain 'unrepentant supporters of print'.

By contrast, despite the above-mentioned calls for a 'division of labor' among floristic works, relatively little notice seems to have been taken of one previously advocated solution: 'generic' floras. Originating in the *genera plantarum* of the eighteenth (Linnaeus, de Jussieu, Lamarck and Poiret) and nineteenth (de Candolle, Endlicher, Bentham and Hooker, and later Engler and Prantl) centuries, the 'generic' flora was evolved as a medium for some form of communication about the floras of continents where a full account (to species level) was not practicable but where a desire for a descriptive account existed. Important early examples were the pre-Candollean *Genera of North American Plants* by Nuttall (1818), *Genera florae americae boreali-orientalis illustrata* by Gray and Sprague (1848–9, not completed), and *Genera of South African plants* by Harvey (1838; 2nd edn., 1868, by J. D. Hooker). In later years, however, following one major Asian work, *Handleiding tot de kennis der flora van*

Nederlandsch-Indië by Boerlage (1890–1900, not completed), which was very closely modeled on Bentham and Hooker's *Genera plantarum*, the genre became peculiarly African. The leading modern representative is *The genera of South African flowering plants* by Phillips (1926; 2nd edn., 1951; 3rd edn., 1975–6, by R. A. Dyer). Mention should also be made of the revisionary series on the genera of the southeastern United States produced through the Arnold Arboretum of Harvard University under the direction of C. Wood (1958–   ). Nevertheless, although Just[112] made a strong plea for more 'generic' floras in place of (or as well as) big semi-monographic 'research' floras, such works are evidently still viewed with skepticism.

In summarizing the present 'state of the art', then, it may be noted that while attention has been drawn to an 'information crisis' and to associated inadequacies in current, mostly traditional formats and styles of floras, not a great deal of change has yet taken place. Generic floras have been viewed with skepticism as being 'imperfect' and automated information systems technology has so far had only comparatively limited acceptance. Illustrations are becoming more widespread, partly due to the use of offset lithography and other modern printing methods, and new classes of information have been introduced. The distinction between the two prime purposes of floras remains, however, often unacknowledged, as does the fact that these roles are no longer compatible in one all-purpose work – and even demand different, and perhaps unorthodox, methods of treatment. 'Shakers' and a few movers have been active, but mindless conservatism is still widespread, augmented by austerity and feelings of resignation.

Future developments

The above comments on the introduction of modern systems methodology – claimed by some to be the most important change affecting the philosophies of flora-writing for a century – as well as other current developments lead naturally to the final question: what of floras in the future? The impact of new methodologies could eventually bring about the revolution hoped for by their advocates, but firstly some key philosophical (and practical) questions must be resolved.

Floras today, as Watson has noted and the present author has come to believe, often seem confused in their philosophy and as a result are deficient in many ways. Most of them, unless they are really elaborate, large-scale works with a consistent format and standard of information content, are of little use for comparative data because of the pull of traditional essentialist conventions in the writing process; most authors still appear to see identification as a principal aim (supplemented by limited information of relatively general interest such as habitat, distribution, life-form, phenology, karyotypes, and sociology) but in many cases are obliged, or feel obliged, to include more comprehensive information of the kind formerly considered proper to monographs or revisions, resulting in a confusion of objectives. Keys are often highly selective, too, and in floras where the manual-key format prevails (such as *Flora of Java*) it becomes very difficult to extract useful comparative data. Moreover, most keys remain steeped in the shadows of the herbarium: to use them very often involves fiddling with minute and not always available parts of the reproductive structures – even when good vegetative characters exist in abundance (especially after a little experience). As already noted, the importance of illustrations, so valuable to the student, remains insufficiently appreciated; often the rare is emphasized at the expense of the more common. In tropical regions, the use of even moderately comprehensive works often requires access to a host of associated literature, including many older works. Unlike the Code of Nomenclature, no internationally recognized set of conventions for flora-writing effectively exists, as will soon become apparent from a perusal of the annotations for individual works. This *Guide*, among other things, will enable those unfamiliar with the literature as to what to *expect*.

What, then, might be the best way to resolve the apparent impasse? Firstly, there should be much more effort given in planning new projects to the *philosophy* and *objectives* of the proposed work as well as to the *means*, *manpower*, and *motivation* available (especially for larger works which may take, even in a concise form, many years to complete). Secondly, more concern should be given to the standardization of data accumulation and organization and the avoidance of the losses which usually occur when work is published – a problem frequently encountered in my floristic studies in Port Moresby. In this connection, much depends upon continuing improvements in and greater professional acceptance of automated systems methodology and its applications (which might include approaches to and cooperation with sympathetic computer software (and hardware) firms in developing more suitable systems and packages) along with appreciation of the advantages and limitations of this methodology (and the kinds of equipment concerned).

Such advances, however, depend upon improved *organic* education of future systematists (as well as an improvement of their public image!), which, at least until recently, was largely left to chance,[113] and moreover has been the victim of serious erosion (particularly at undergraduate level). Although recognized as necessary early in the 1960s, it is only recently that significant studies have been made on the education and training of systematists.[114] It has even been suggested that certain advances in the philosophy of systematics have been distorted if not retarded by widespread mindlessness,[115] this being at least partly a product of conventional scientific cosmology as well as of the position of the discipline within commonly held perceptions of the progress of science[116] and its systematization.[117] Related to this mindlessness may be much of the current 'serendipitious' pattern of systematic work, remarked upon by Just and by Heywood but also having deeper roots in patterns of behaviour.[118] Since large floras have been beyond the reach of one or two men for the past 100 years, the presence or absence of a specialist for a given group (or groups) has been a matter for chance and a significant problem in large flora-projects[119] where totally imposed central planning is limited or absent.

Personally, I believe that the best role for a flora remains the utilitarian practical one: inventory, identification, and essential related data. To the 'essential data' of Bentham's time there should now be added that from ecology and karyology[120] as well as plenty of illustrations. Moreover, there should be a clear indication of where taxonomic or 'biosystematic' problems occur, including the question of hybrids. If lesser-known areas are involved (as is the case with most of the humid tropics), it may be desirable to expand supporting data and commentary (including specimen citations) somewhat, as has been done with *Flora zambesiaca* and with the *Flora of Turkey* (which in my opinion are among the best of the current larger floras dealing with lesser-known areas and very kindred in spirit to the famous 'Kew Series' of the Hookerian era. In addition, concise floras should always have a clear indication of references to standard monographs, revisions, floras and other contributions under each family and genus heading, a method very well handled in *Tree flora of Malaya* but in *Flora Malesiana* regrettably obfuscated within an often bewildering jumble of nomenclatural citations.

By contrast, large-scale floras should be viewed as having an entirely separate function; they should not be undertaken except for large natural regions such as Malesia (and for some groups even this is unsuitable, with awkward results) or for very large political entities such as the USSR. Some should perhaps be run as open-ended serials rather than as 'finite' works, like Series II of *North American Flora* and *Flora Neotropica*. They should also be made more 'open', with each family involving input by a large number of people rather than being limited to one or two 'experts'. Furthermore, much of the 'archival' function of such works, with their often elaborate synonymy, could be handled outside the traditional (and increasingly expensive) print medium, either as processed output or on microcard or microfiche or in information systems (with suitable controls). Detailed information in this form could then be used for the preparation and publication in print media of conventional 'concise' floras (as well as for the production of major systematic treatments). At least the often extensive materials accumulated in the course of systematic research could be preserved rather than discarded, as too often happens.

For some little-known areas where time or local conditions may not permit the preparation of more extensive works, I believe it desirable to continue to produce annotated enumerations or checklists. These should preferably be in the manner of Merrill's *Enumeration of Philippine flowering plants* or the more recent *Enumeration of the flowering plants of Nepal*, though if keys can be added, so much the better. Such a format would have perhaps been the best means for a complete account of the Papuasian flora to species level, especially as an extensive file had already been built up before 1968 by one specialist. An excellent example of what can be done in a relatively short time for a comparatively little-known area with limited manpower is *Prodromus einer Flora von Südwestafrika*, referred to previously. The checklist format was also utilized for the only tangible floristic record of the ill-fated FNA Program, *A provisional checklist of species for Flora North America* (revised) by Shetler and Skog (1978). The enumeration, though imperfect, is perhaps

the best means for coping with the need for urgency in documenting what is known about plant species in many parts of the world where forests have disappeared or are under threat, both for scientific reasons as well as for political action. Only lately have there been signs that this is being more widely realized.[121]

There will certainly be instances where it is necessary or desirable to make encylopedic information on a given group (or groups) of plants readily available. In these cases, this is better done outside the realm of formal floras, i.e., as separate publications or in serials. The best systematic encyclopedia ever produced was *Die natürlichen Pflanzenfamilien*, and it would be highly desirable if the means and manpower could be found to complete the second edition of this work or undertake a new version in English. It is therefore a pleasure to record here that moves are being made to mount such a new work.[122] Any such work, however, should avoid becoming too bogged down in detail, a fault shown here and there by the second edition of the *Pflanzenfamilien*. Two versions of the proposed work could perhaps be made: a more concise one in printed form and one more elaborately documented in some other form such as microcard or microfiche. Heywood has argued that much of the information contained in such works as the *Pflanzenfamilien* could be handled more flexibly in non-taxonomic publications such as *Biology and Chemistry of the Umbelliferae*.[123] This latter work has indeed established a new trend – the 'family symposium' – which is at this writing still continuing, with several publications already having resulted (covering Compositae, Cruciferae, Solanaceae, etc.). Their nature as symposium contributions, however, inevitably leads to some inconsistency in treatment.

In closing, I should like to comment on four matters of particular interest concerning floras. Firstly, floristic works, or at least some of them, should be written in such a way as will have a wide public appeal and impact, a very important consideration in most tropical regions (and developing countries generally) for both cultural and educational reasons. Atlas-floras, such as those already mentioned, may well enjoy a much wider audience than more 'conventional' works. Keys should be simple and practical, with quick and reliable identification the overriding aim; descriptions should be concise, clear, and provide the essentials (a task made easier if illustrations are used consistently). In these areas, it will be but a small number of persons who would prefer a detailed treatment, and this could be provided from other sources. Where the total flora is very large (and comprehensive works often correspondingly costly, especially in local terms), there is also scope for a number of works of more limited scope. Thus continuing attention should be given to forest floras and tree books (which often have considerable public appeal) as well as works on grasses, orchids, lianas, pteridophytes, the 'decorative' families, weeds, etc. One country in the humid tropics, Malaysia (together with Singapore), is particularly well supplied with such partial works. For teaching purposes, these may adequately be supplemented by a compact, illustrated school manual covering a range of more easily accessible species, such as *Flora untuk sekolah di Indonesia* (and its Dutch-language predecessor) by van Steenis.[124] One of the finest of tropical manuals ever published remains E. J. H. Corner's *Wayside trees of Malaya*; this is considered a favorite by my students in Papua New Guinea because of its interesting text, many illustrations, and clear keys. This should be (and has been) revised and updated, and more of its kind should be written (there being lamentably few in the tropics, although lately this is starting to change). Sometimes this has had to be done in the face of continuing official interest in more traditional types of works. In any case, it is, as I see it, an important function, if not even a responsibility, of university botanists to prepare works aimed at local people. Regrettably this is something not always easy to accomplish, as apart from official indifference and technical problems, imported academic attitudes and preconceptions, often non-organismal in orientation, have to be faced. To realize this ideal, we at the Herbarium in the University at Port Moresby commenced some time ago to create a critical data base to serve as a foundation for a planned series of illustrated teaching booklets on the local flora, each dealing with a given habitat, life-form, or small area. The first of these, published in 1975 and covering the woody mangrove flora, due to demand has recently been revised and reissued. Similar handbooks for both plants and animals have been prepared and issued through the departmental museum as well as through other organizations in the country. Very much tedious background research, however, has been required to achieve a measure of criticality, and this may prove a serious if not insurmountable obstacle to those in less favorable circumstances.

Secondly, more attempts should be made at 'collegialism' in floras as was the case in *Flora Europaea*. As a contributor myself to that work, it meant repressing a certain amount of individual pride and expression, but the tight editorial and advisory control imposed made in the end for a more consistent product, with corresponding stature (and prestige).[125] A similar collective organization was planned for the Flora North America project. In the currently appearing general flora of China, many if not most families are being treated by cadres of senior and junior workers, some geographically spread out, under the direction of a central editorial board (but without, it appears, a specific network of 'regional advisors').[126] On the other hand, most current large-scale floras are still run like *Flora brasiliensis*, with a central editor or editorial committee and cooperating individual specialists, sometimes largely 'in-house'. This, although time-honored, is in large measure old-fashioned and to the increasing number of local workers possibly demeaning. No significant internal control is possible (save where the flora is largely 'in-house' as was the *Flora of Guatemala*) and the only critique can come through post-publication peer review.[127] Treatments may be (and often are) very variable in standard and philosophy. If consistency of the kind achieved by one (or two) men in the great regional floras of the last century is really desired, then 'collegiality' rather than a form of scientific colonialism is in order. It is conceded that some distinctive approches are needed where tropical floras are concerned, and moreover that methodological problems and needs differ from family to family, but is there any good reason to run such projects wholly 'from the top'?

Thirdly, monographic treatments, which must remain the basis for sound critical work but have, as already noted, long been 'unfashionable', must be given a new lease of life. Apart from families, this must include the study of large genera, covering from many hundreds up to thousands of species, which are demonstrably convex[128] and as natural units have had evolutionary lives of their own. Serious study of such groups can often result in greatly altered ideas of their classification which naturally has consequences for writers of floras, not to mention other areas of biology; such was my experience with *Schefflera* (Araliaceae), a convex though paraphyletic group. But, as Mabberley has most recently commented, 'which "evolutionary biologists" find the time for monographs of large genera? It is impossible to grapple with a group of this size in a Ph.D. thesis or short-term research periods, or in the writing of regional floras, as important as this may be. Who has time to use his intellect to unravel the stories of *Cassia*, *Eugenia*, *Euphorbia*, *Piper*, *Solanum*, or *Vernonia*?'.[129] This short-term mentality has, alas, largely been forced upon researchers by equally short-sighted funding policies and politics (academic and otherwise);[130] few projects – one of them being *Flora Europaea* – in our time have enjoyed consistent long-term support unless they be purely institutional. Perhaps there is, apart from public and official attitudes, also within the profession a shunning of the implications involved in such serious research: a form of intellectual castration seems to be preferred, favoring the familiar small steps, because bold thoughts and actions may be too much to handle in a group which is (or sees itself) on the defensive. Yet the methods of *Flora Europaea* might also be applied to studies of great natural groups. Collaboration was first advanced as a solution by Jacobs[131] a decade or so ago under the sobriquet 'Large families – not alone!' and has been again strongly emphasized by Heywood,[132] who noted that for revisions and monographs covering a wide territorial range 'it is seldom materially possible to review (or even to trace) personally all the relevant literature and materials'. The collaboration concept has been given effect for some groups but is as yet not widely practiced.

Lastly, let there be floras which are not permeated with learned and prosy dullness. Such sanitization may be saving for the professional collegian, but, as Corner[133] has so forcefully argued, nothing of the living plants or their essential features is thereby conveyed. In this there has been relatively little change since the rise of botanical rationalism and the Victorian empiricism that followed it, notwithstanding Wheeler's witty and pointed critique of 60 years ago[134] and, later, the fresh air of biosystematic 'camp-ing'. Flora-writing should be more sentient, as in the three-centuries-old *Herbarium amboinense* and, in our day, that 'parascientific' book *Wayside trees of Malaya*. The spark of enlightenment should pervade the presentation of factual information: plants should somehow be described with feeling, as when Conrad has his narrator Marlow introduce the coast of Patusan.

The coast...is straight and sombre, and faces a misty ocean. Red trails are seen like cataracts of rust streaming under the dark-green foliage of bushes and creepers clothing the low cliffs.

Swampy plains open out at the mouth of rivers, with a view of jagged blue peaks beyond the vast forests. In the offing a chain of islands, dark, crumbling shapes, stand out in the everlasting sunlit haze like the remnants of a wall breached by the sea.[135]

Floras require a new ethic,[136] and the foregoing suggestions, arising from examining and meditating upon a veritable constellation of works, large and small, as well as the author's own experiences as botanist and educator, are offered as contributions to that end. The standard of content and presentation, though, must ultimately be guided by 'correctness and clearness of method and language [as] the first qualities requisite',[137] with, in our day, an eye to 'care and coordination, and a clear sense of the priorities both from the producer's and user's point of view'.[138]

Notes

1 Brenan, 1979.
2 Heywood, 1957; Davis and Heywood, 1963; Fisher, 1968; Shetler, 1971; Taylor, 1971; Watson, 1971; Heywood, 1973*a*; Jacobs, 1973; Frodin, 1976 (1977).
3 Symington, 1943; Corner, 1946.
4 van Steenis, 1954.
5 Hitchcock, 1925, chapter 5; Blake and Atwood 1942. Hitchcock's work was written as a successor to de Candolle 1880.
6 Davis and Heywood, 1963, p. 296.
7 van Steenis, 1954.
8 Bentham, 1861, 1874.
9 de Candolle, 1880.
10 For historical account, see Thistleton-Dyer, 1906; it was written in 1905, the last year of his directorship of Kew during which period the Gardens reached its apogee as the apex of an imperial botanical system (Brockway, 1979, pp. 100–2).
11 Burtt-Davy *et al.*, 1925.
12 Corner, 1946.
13 Symington, 1943.
14 de Wit, 1949, p. cxviii.
15 Fletcher and Brown 1970, pp. 168–71.
16 W. J. Hooker to Colonial Office, 14 May 1857; quoted in Thistleton-Dyer, 1906, p. 11.
17 Stafleu, 1971*a,b*.
18 de Wit, 1949, p. xcvi.
19 Voss, 1952; Stafleu, 1971*b*.
20 Pankhurst, 1978; Stace, 1980, pp. 223–30; Stafleu, 1971*a*, pp. 399, 404–8. Lamarck made one other 'practical' concession: his *Tableau encyclopédique des trois règnes de la nature Botanique* (1791–9; 1819–23), another of Panckoucke's projects and also completed by Poiret, was at the publisher's request organized and written in accordance with Linnean principles (Stafleu, p. 409). This clearly brings out the transitional character of the late eighteenth century in botanical writing, even in France and even in the works of a single author..
21 Stafleu and Cowan, 1979.

22 van Steenis, 1954.
23 Prance, 1977 (1978).
24 Bentham, 1874.
25 de Candolle, 1880.
26 Bentham, 1874.
27 Sachs, 1875.
28 McVaugh, 1955.
29 van Steenis, 1954; Heywood, 1973*a*.
30 Bentham, 1874.
31 de Candolle, 1880.
32 van Steenis, 1954.
33 Diels, 1921.
34 For list of Diels' contributions, see Mildbraed, 1948.
35 de Wit, 1949, p. cxviii.
36 Simon, 1977.
37 Davis and Haywood, 1963, p. 33.
38 Simon, 1977, p. 153; Malclès 1961, p. 84.
39 Eckardt, 1966, p. 168.
40 Bentham, 1874.
41 van Steenis, 1979.
42 Léandri, 1967.
43 Jovet, 1954.
44 Léandri, 1962.
45 Kahn and Pepper, 1980; by them termed *La Première Belle Époque*.
46 Hill *et al.*, 1925.
47 Grattan, 1963, p. 420.
48 Symington, 1943.
49 Shinners, 1962, p. 14, 19.
50 Shinners, *ibid.* p. 7.
51 Cf. Shinners, 1962.
52 Verdoorn, 1945, Preface.
53 Wheeler, 1923; Ravetz, 1975.
54 Corner, 1961; Jacobs, 1973; Mabberley, 1979, p. 274.
55 Stafleu, 1959; see especially the map, p. 67.
56 Huxley, 1940; for plants see chapter by Turrill, pp. 47–71.
57 Davis and Heywood, 1963.
58 Raven, 1974.
59 Radford *et al.*, 1974.
60 Heywood, 1974.
61 Shinners, 1962, p. 22.
62 Hitchcock, 1925.
63 Blake and Atwood, 1942.
64 Heywood, 1958, 1960.
65 Radford *et al.*, 1974; Yü, 1979 (1980).
66 van Steenis, 1954; Brenan, 1963; Turrill, 1964.
67 Meikle, 1971.
68 van Steenis, 1954. For another view of the art of description, see Jacobs, 1980.
69 Raven, 1974.
70 Endress, 1977; Hj. Eichler, personal communication.
71 van Steenis, 1954; Davis and Heywood, 1963, p. 299.
72 Jäger, 1978, p. 413.
73 Jacobs, 1969.
74 van Steenis, 1949.
75 De Wolf, 1963; Shetler, 1967.
76 De Wolf, 1963, 1964.
77 African figures from Jäger, 1979; others from various sources, especially Prance 1977 (1978). Writing of *Flora Zambesiaca* is rather further advanced than figures suggest (P. Osborne, personal communication). See also Brenan, 1979.
78 van Steenis, 1938; recalled in van Steenis, 1979, p. 73.
79 De Wolf, 1963.
80 van Steenis, 1979, p. 73.
81 Yü, 1979 (1980).
82 De Wolf, 1963.
83 Jacobs, 1973.
84 Symington, 1943.
85 Symington, *ibid.*
86 Corner, 1946.
87 For Adanson *see* Stafleu, 1971, pp. 310–20; for Griffith *see* Burkill, 1965.
88 Wheeler, 1923.
89 Wheeler, *ibid.*
90 Heywood, 1973*b*.
91 Taylor, 1971.
92 van Steenis, 1962.
93 Heywood, 1973*a*.
94 Watson, 1971.
95 Brenan, 1963.
96 Jacobs, 1973.
97 Shinners, 1962.
98 Shetler, 1971.
99 Jacobs, 1973.
100 D. A. Webb, 1978.
101 Symington, 1943.
102 Fisher, 1968.
103 Watson, 1971.
104 Fisher, 1968.
105 Thorne, 1971; Shetler, 1974.
106 Brenan *et al.*, 1975.
107 Keller and Crovello, 1973.
108 T. J. Crovello, 1977, personal communication.
109 European Science Foundation, 1980, pp. 43–7.
110 Prance, 1977 (1978).
111 Anonymous, 1974.
112 Just, 1953.
113 Just, *ibid.*; Cain 1959*b*, p. 243.
114 Royal Society of London, 1963; European Science Research Councils, 1977; Advisory Board for the Research Councils, 1979.
115 Corner, 1946; Heywood, 1976; Kubitzki, 1977, p. 192.
116 Wheeler, 1923; Ziman, 1968; Ravetz, 1975.
117 Dolby, 1979.
118 Skinner, 1975.
119 Valentine, 1971.
120 van Steenis, 1954; Raven, 1974.
121 Heywood, 1975.
122 Bates *et al.*, 1980.
123 Heywood, 1973*a*, 1976.
124 van Steenis, 1962.
125 Heywood, 1964.
126 Bartholomew *et al.*, 1979; Yü, 1980.
127 Heywood, 1976.
128 Estabrook, 1978 (1979); see also Wiley, 1981*a,b*.
129 Mabberley, 1979, p. 274.
130 Just, 1953; Janzen, 1977 (1978), p. 731; Department of Science, Australia, 1979, p. 9; Kim, 1980.
131 Jacobs, 1969.
132 Heywood, 1976.
133 Corner, 1946. The question of changes over historical time in approaching plant description is further considered by Jacobs, 1980.
134 Wheeler, 1923.
135 Joseph Conrad, *Lord Jim* (1900; Penguin edn. 1957).
136 Ravetz, 1975.
137 Bentham, 1874.
138 Brenan, 1979, p. 57.

References

ADVISORY BOARD FOR THE RESEARCH COUNCILS, UNITED KINGDOM, 1979. *Taxonomy in Britain*. London: HMSO.

ANONYMOUS, 1974. Trends, priorities, and needs in systematic and evolutionary biology. *Syst. Zool.* **23**: 416–39 (also in *Brittonia*, **26**: 421–44).

ATWOOD, A. C., 1911. *Description of the comprehensive catalogue of botanical literature in the libraries of Washington* (United States Department of Agriculture, Bureau of Plant Industry Circular, 87). Washington.

AYMONIN, G., 1962. Où en sont les flores européennes? *Adansonia*, II, **2**: 159–71.

BARTHOLOMEW, B. *et al.*, 1979. Phytotaxonomy in the People's Republic of China. *Brittonia*, **31**: 1–25. [See also Thorhaug, A. (ed.), 1978 (1979). *Botany in China* (United States–China Relations Report, 6). Stanford, Calif.: US–China Relations Program, Stanford University.]

BATES, D. *et al.*, 1980. A prospectus for a proposed new work: The families and genera of vascular plants. *Taxon*, **29**: 318–20.

BENTHAM, G., 1861. Outlines of botany. In *idem, Flora hong-kongiensis*, pp. i–xxxvi. London: Reeve.

BENTHAM, G., 1874. On the recent progress and present state of knowledge of systematic botany. *Reports Brit. Assoc. Adv. Sci.* (1874): 27–54.

BLAKE, S. F. and ATWOOD, A. C., 1942. *Geographical guide to floras of the world*, 1. Washington, D.C.: Government Printing Office. (Repr. 1963, New York: Hafner.)

BORGESE, E. M., 1964. The universalization of Western civilization. In *Humanistic education and Western civilization: essays for Robert M. Hutchins* (ed. A. A. Cohen), pp. 75–86. New York: Holt, Reinhart & Winston.

BOTTLE, R. T. and WYATT, H. V., 1971. *The use of biological literature*. 2nd edn. London: Butterworths. (Botanical taxonomy, pp. 141–56.)

BRADFORD, S. C., 1953. *Documentation*. 2nd edn., with introduction by J. H. Shera and M. E. Egan. London: Crosby Lockwood.

BRENAN, J. P. M., 1963. The value of floras to underdeveloped countries. *Impact* (Paris), **13**: 121–45.

BRENAN, J. P. M., 1979. The flora and vegetation of tropical Africa today and tomorrow. In *Tropical Botany*, (ed. K. Larsen and L. B. Holm-Nielsen), pp. 49–58. London: Academic Press.

BRENAN, J. P. M. *et al.* (eds.), 1975. *Computers in botanical collections* (Kew, 1973). London: Plenum.

BROCKWAY, L. H., 1979. *Science and colonial expansion: the role of the British Royal Botanic Gardens*. London: Academic Press.

BROOKS, F. T. (ed.), 1925. *Imperial Botanical Conference* (London 1924): *report of proceedings*. Cambridge: Cambridge University Press.

BULICK, S., 1978. Book use as a Bradford-Zipf phenomenon. *College Res. Libraries*, **39**: 215–19.

BURKILL, I. H., 1965. *Chapters on the history of botany in India*. Calcutta: Manager of Publications, Government of India.

BURTT-DAVY, J. *et al.*, 1925. Correlation of taxonomic work in the dominions and colonies with work at home. In Brooks, F. T., *op. cit.*, pp. 214–39.

CAIN, A. J., 1958. Logic and memory in Linnaeus's system of taxonomy. *Proc. Linn. Soc., London*, **169**: 144–63.

CAIN, A. J., 1959*a*. Deductive and inductive methods in post-Linnean taxonomy. *Ibid.*, **170**: 185–217.

CAIN, A. J., 1959*b*. The post-Linnean development of taxonomy. *Ibid.*, **170**: 234–44.

CANDOLLE, A. DE, 1880. *La phytographie*. Paris: Masson.

CORNER, E. J. H., 1946. Suggestions for botanical progress. *New Phytol.* **45**: 185–92.

CORNER, E. J. H., 1961. Evolution. In *Contemporary botanical thought* (ed. A. M. MacLeod and L. S. Cobley), pp. 95–114. Edinburgh: Oliver & Boyd.

DAVIS, P. H. and HEYWOOD, V. H., 1963. *Principles of angiosperm taxonomy*. Edinburgh: Oliver & Boyd. (Reprinted with corrections, 1973, Huntington, N.Y.: Krieger.)

DEPARTMENT OF SCIENCE, AUSTRALIA, 1979. *Australian Biological Resources Study 1973–78* (Commonwealth Parliamentary Paper 354/1978). Canberra: AGPS.

DE WOLF, G. P., JR., 1963. On the *Flora Neotropica*. *Taxon*, **12**: 251–3.

DE WOLF, G. P., JR., 1964. On the sizes of floras. *Ibid.*, **13**: 149–53.

DIELS, L., 1921. Die Methoden der Phytographie und der Systematik der Pflanzen. In *Handbuch der biologischen Arbeitsmethoden* (ed. E. Abderhalden), 2nd edn., vol. 11, part 1, pp. 67–190. Berlin: Urban & Schwarzenberg.

DOLBY, R. G. A., 1979. Classification of the sciences: the nineteenth-century tradition. In *Classifications in their social context* (ed. R. F. Ellen and D. Reason), pp. 167–93. London: Academic Press.

DOWNS, R. B. and JENKINS, F. B. (eds.), 1967. *Bibliography: current state and future trends*. Urbana, Ill.: University of Illinois Press. [Agriculture, pp. 542–57; biology, pp. 491–7.]

DRUCKER, P. F., 1979. Science and industry, challenges of antagonistic interdependence. *Science*, **204**: 806–10.

DUNBAR, M. J., 1980. The blunting of Occam's Razor, or to hell with parsimony. *Can. J. Zool.* **58**: 123–8.

ECKARDT, T., 1966. 150 Jahre Botanisches Museum Berlin (1815–1965). *Willdenowia*, **4**: 151–82.

ENDRESS, P. K., 1977. Das Institut für Systematische Botanik der Universität Zürich. In Die Botanischen Institute der Universität Zürich (ed. H. Wanner and P. K. Endress). *Vierteljahrsschr. Naturf. Ges. Zürich*, **122**: 143–50.

ESTABROOK, G. F., 1978 (1979). Some concepts for the estimation of evolutionary relationships in systematic botany. *Syst. Bot.* **3**: 146–58.

EUROPEAN SCIENCE FOUNDATION, 1978– . *ESF Reports 1977–* . Strasbourg. (Annual; reports for 1977–80 consulted for this *Guide*.)

EUROPEAN SCIENCE FOUNDATION, 1977. *Taxonomy in Europe* (European Science Research Councils, Rev. no. 13). Strasbourg. (Revised, definitive edition, 1982.)

FISHER, F. J. F., 1968. The role of geographical and ecological studies in taxonomy. In *Modern methods in plant taxonomy* (ed. V. H. Heywood), pp. 241–59. London: Academic Press.

FLETCHER, H. R. and BROWN, W. H., 1970. *The Royal Botanic Garden Edinburgh 1670–1970*. Edinburgh: HMSO.

FOSBERG, F. R., 1972. The value of systematics in the environmental crisis. *Taxon*, **21**, 631–4.

FRODIN, D. G., 1976 (1977). On the style of floras: some general considerations. *Gardens' Bull. Sing.* **29**: 239–50.

GAMS, H., 1954. Flores européennes modernes: resultats acquis, objectifs à atteindre. In *Huitième Congrès International de Botanique*, Paris (1954), Rapports et Communications parvenus avant le Congrès aux sections 2, 4, 5 et 6, pp. 101. Paris: SEDES.

GARFIELD, E., 1979. *Citation indexing – its theory and application in science, technology, and humanities*. New York: Wiley.

GARFIELD, E., 1980. Bradford's Law and related statistical patterns. *Current Contents/Life Sci.* **23** (19): 5–12.

GILMOUR, J. S. L., 1952. The development of taxonomy since 1851. *Adv. Sci. (UK)*, **9**: 70–4.

GÓMEZ-POMPA, A. and BUTANDA, C. A., 1973. *El uso de computadoras en la 'Flora de Vera Cruz'*. Mexico City.

GÓMEZ-POMPA, A. and NEVLING, L. I., 1973. The use of electronic data-processing methods in the *Flora of Veracruz* program. *Contr. Gray Herb.*, N.S. **203**: 49–64.

GOULD, S. W., 1968–72. *Geo-code*. 2 vols. Maps. New Haven, Conn.: Gould Fund. [Vol. 1, Western hemisphere; vol. 2, Eastern hemisphere.]

GRATTAN, C. H., 1963. *The southwest Pacific since 1900*. Ann Arbor: University of Michigan Press.

HEDBERG, I. (ed.), 1979. *Systematic botany, plant utilization and biosphere conservation*, Uppsala, 1977. Stockholm: Almqvist & Wiksell.

HELLER, J. L., 1970. Linnaeus's *Bibliotheca botanica*. *Taxon*, **19**: 363–411.

HEPPER, F. N., 1979. Second edition of the map showing the extent of floristic exploration in Africa south of the Sahara published by AETFAT. In *Taxonomic aspects of African economic botany*, Las Palmas, 1978 (ed. G. Kunkel), pp. 157–62, map. Las Palmas, Gran Canaria.

HEYWOOD, V. H., 1957. A proposed flora of Europe. *Taxon*, **6**: 33–42.

HEYWOOD, V. H., 1958. *The presentation of taxonomic information: a short guide to contributors to 'Flora Europaea'*. Leicester: Leicester University Press.

HEYWOOD, V. H., 1960. *The presentation of taxonomic information: supplement*. Alcobaça, Portugal: Flora Europaea Organization.

HEYWOOD, V. H., 1964. *Flora Europaea* and the problems of organization of floras. *Taxon*, **13**: 48–51.

HEYWOOD, V. H., 1973a. Ecological data in practical taxonomy. In *Taxonomy and ecology* (ed. V. H. Heywood), pp. 329–47. London: Academic Press.

HEYWOOD, V. H., 1973b. Taxonomy in crisis? or Taxonomy is the digestive system of biology. *Acta Bot. Acad. Sci. Hung.* **19**: 139–46.

HEYWOOD, V. H., 1974. Systematics – the stone of Sisyphus. *Biol. J. Linn. Soc.* **6**: 169–78.

HEYWOOD, V. H., 1975. Contemporary philosophies in biological classification. *Informatics*, **2**: 57–60.

HEYWOOD, V. H., 1976. Contemporary objectives in systematics. In *Proceedings, Eighth International Conference on Numerical Taxonomy*, Oeiras, Portugal, 1974 (ed. G. F. Estabrook), pp. 258–83. San Francisco: Freeman.

HILL, A. W. *et al.*, 1925. The best means of promoting a complete botanical survey of the different parts of the Empire. In Brooks, F. T., *op. cit.*, pp. 196–213.

HITCHCOCK, A. S., 1925. *Methods of descriptive systematic botany*. New York: Wiley.

HULTÉN, E., 1958. *The amphi-Atlantic plants and their phytogeographical connection* (Kongl. Svenska Vetenskaps-akad. Handl., IV, 7(1)). Stockholm: Almqvist & Wiksell.

HUXLEY, J. S. (ed.), 1940. *The new systematics*. Oxford University Press.

JACOBS, M., 1969. Large families – not alone! *Taxon*, **18**: 253–62.

JACOBS, M., 1973. Flora-projecten als intellectuele uitdaging. *Vakblad Biol.* **53**(15): 252–5.

JACOBS, M., 1980. Revolutions in plant description. In *Liber gratulatorius in honorem H. C. D. de Wit* (ed. J. C. Arends *et al.*), pp. 155–81. (Misc. Paper Landbouwhogeschool Wageningen, vol. 19.) Wageningen, the Netherlands.

JÄGER, E. J., 1978. Areal- und Florenkunde (floristische Geobotanik). *Prog. Bot.* [*Fortschr. Bot.*] **40**: 413–28.

JÄGER, E. J., 1979. Areal- und Florenkunde (floristische Geobotanik). *Ibid.*, **41**: 310–23.

JÄGER, E. J., 1980. Areal- und Florenkunde (floristische Geobotanik). *Ibid.*, **42**: 331–45.

JANZEN, D., 1977 (1978). Promising directions of study in tropical animal–plant interactions. *Ann. Missouri Bot. Gard.* **64**: 706–36.

JONSELL, B., 1979. Europe. In Hedberg, I., *op. cit.*, pp. 34–40.

JOVET, P., 1954. Flore et phytogéographie de la France. In *Histoire de la botanique en France* (coord. A. Davy de Virville), pp. 243–68. Paris: SEDES.

JUST, T., 1953. Generic synopses and modern taxonomy. *Chron. Bot.* **14**(3): 103–14.

KAHN, H. and PEPPER, T., 1980. *Will she be right?* St Lucia, Queensland: University of Queensland Press.

KELLER, C. and CROVELLO, T. J., 1973. Procedures and problems in the incorporation of data from floras into a computerized data bank. *Proc. Indiana Acad. Sci.* **82**: 116–22.

KIM, KE CHUNG, 1980. Role and management of systematics collections within universities. *Assoc. Syst. Coll. Newsletter*, **8**(3): 39–41.

KUBITZKI, K., 1977. Systematics and evolution of seed plants. *Prog. Bot.* [Fortschr. Bot.] **39**: 192–237.

KUMAR, K., 1979. *Theory of classification*. New Delhi: Vikas.

LARSEN, K. and HOLM-NIELSEN, L. B., 1979. *Tropical botany*. London: Academic Press.

LÉANDRI, J., 1962. Deux grands artisans de la floristique tropicale: Henri Lecomte (1856–1934) et Achille Finet (1863–1913). *Adansonia*, II, **2**: 147–58.

LÉANDRI, J., 1967. La fin de la dynastie des Jussieu et l'éclipse d'une chaire au Muséum (1853 à 1873). *Ibid.*, **7**: 443–50.

LEIMKUHLER, F. F., 1967. The Bradford distribution. *J. Documentation* **23**: 197–207.

LÉONARD, J., 1965. Statistiques des progrès accomplis en 10 ans dans la connaissance de la flore phanérogamique africaine et malgache (1953–1962). *Webbia*, **19**: 869–75 (map reproduced in Hedberg, I., *op. cit.*, p. 44).

MABBERLEY, D. J., 1979. Pachycaul plants and islands. In *Plants and islands* (ed. D. Bramwell), pp. 259–77. London: Academic Press.

MALCLÈS, L.-N., 1961. Bibliography (trans. T. C. Hines). New York: Scarecrow (reprinted 1973). (Originally pub. 1956, Paris, as *La bibliographie* in the 'Que sais-je' series.)

McVAUGH, R., 1955. The American collections of Humboldt and Bonpland as described in the 'Systema Vegetabilium' of Roemer and Schultes. *Taxon*, **4**: 78–86. (Reprinted in *Humboldt, Bonpland, Kunth and tropical American botany*, ed. W. T. Stearn), pp. 32–43. Lehre, W. Germany: Cramer.)

MEIKLE, R. D., 1971. Co-ordination of floristic work and how we might improve the situation. In *Plant life of South-West Asia* (ed. P. H. Davis *et al.*), pp. 313–31. Edinburgh: Botanical Society of Edinburgh.

MILDBRAED, J. 1948. Ludwig Diels. *Bot. Jahrb. Syst.* **74**: 173–98.

PANKHURST, R. J., 1978. Biological identification. London: Arnold.

PICHI-SERMOLLI, R. E. G., 1973. Historical review of the higher classification of the Filicopsida. In *The phylogeny and classification of the ferns* (ed. A. C. Jermy *et al.*), pp. 11–40. London: Academic Press.

PRANCE, G. T., 1977 (1978). Floristic inventory of the tropics: where do we stand? *Ann. Missouri Bot. Gard.* **64**: 659–84.

RADFORD, A. E., *et al.*, 1974. *Vascular plant systematics*. New York: Harper & Row.

RANGANATHAN, S. R., 1957. *Prolegomena to library classification*. 2nd edn., London: Library Association.

RAVEN, P. H., 1974. Plant systematics 1947–1972. *Ann. Missouri Bot. Gard.* **61**: 166–78.

RAVETZ, J. R., 1975. '…et augebitur scientia.' In *Problems of scientific revolution: progress and obstacles to progress in the sciences* (ed. R. Harré), pp. 42–57, 1 pl. Oxford: Oxford University Press.

ROYAL SOCIETY OF LONDON, 1963. *Taxonomy*. London.

SACHS, J., 1875. *Geschichte der Botanik von 16. Jahrhundert bis 1860*. Munich. (Translated into English, 1890, as *History of botany (1530–1860)*). Oxford: Oxford University Press; reprinted 1906, 1967.)

SHETLER, S. G., 1967. *The Komarov Botanical Institute: 250 years of Russian research*. Washington, DC: Smithsonian Institution Press.

SHETLER, S. G., 1971. Flora North America as an information system. *BioScience*, **21**: 524, 519–32.

SHETLER, S. G., 1974. Demythologizing biological data banking. *Taxon*, **23**: 71–100.

SHETLER, S. G., 1979. North America. In Hedberg, I., *op. cit.*, pp. 47–54.

SHETLER, S. G. and READ, R. W., 1973. *Flora North America/International index of current research projects in plant systematics*, 7 (Flora North America Report, 71). Washington, DC.

SHETLER, S. G. *et al.*, 1973. *A guide for contributors to 'Flora North America' (FNA)*. Provisional edition (Flora North America Report, 65). Washington, DC.

SHINNERS, L. H., 1962. Evolution of the Gray's and Small's manual ranges. *Sida*, **1**: 1–31.

SIMON, H.-R., 1977. *Die Bibliographie der Biologie*. Stuttgart: Hiersemann.

SKINNER, B. F., 1975. The steep and thorny way to a science of behaviour. In *Problems of scientific revolution: progress and obstacles to progress in the sciences* (ed. R. Harré), pp. 58–71. Oxford: Oxford University Press.

STACE, C. A., 1980. *Plant taxonomy and biosystematics*. London: Arnold.

STAFLEU, F. A., 1959. The present status of plant taxonomy. *Syst. Zool.* **8**: 59–68, map.

STAFLEU, F. A., 1971a. *Linnaeus and the Linnaeans* (Regnum Veg., 79). Utrecht: Oosthoek.

STAFLEU, F. A., 1971b. Lamarck: the birth of biology. *Taxon*, **20**: 397–442.

STAFLEU, F. A., 1973. Pritzel and his *Thesaurus*. *Taxon*, **22**: 119–30.

STAFLEU, F. A.and COWAN, R. S., 1979. The making of a book: an interim report on *TL-2. Taxon*, **28**: 77–86.

STEARN, W. T., 1957. *An introduction to the 'Species plantarum' and cognate botanical works of Carl Linnaeus*. London. (Reprinted in Linnaeus, C., 1957. '*Species Plantarum': a facsimile of the first edition 1753*, vol. 1, pp. v–xiv, 1–176. London: Ray Society.)

STEARN, W. T., 1959. The background of Linnaeus' contribution to the nomenclature and methods of systematic biology. *Syst. Zool.* **8**: 4–22.

STEENIS, C. G. G. J. VAN, 1938. Recent progress and prospects in the study of the Malaysian flora. *Chron. Bot.* **4**: 392–7.

STEENIS, C. G. G. J. VAN, 1949. De Flora Malesiana en haar betekenis voor de Nederlandse botanici. *Vakblad Biol.* **29** (2): 24–33.

STEENIS, C. G. G. J. VAN, 1954. General principles in the design of Floras. In *Huitième Congrès International de Botanique*, Paris (1954), Rapports et Communications parvenus avant le Congrès aux sections 2, 4, 5 et 6, pp. 59–66. Paris: SEDES.

STEENIS, C. G. G. J. VAN, 1962. The school-flora as a medium for instruction in the tropics. *Proceedings, Ninth Pacific Science Congress*, 4 (Botany): 139–40.

STEENIS, C. G. G. J. VAN, 1979. The Rijksherbarium and its contribution to the knowledge of the tropical Asiatic flora. *Blumea*, **25**: 57–77.

SUTTON, S. B., 1970. *Charles Sprague Sargent and the Arnold Arboretum*. Cambridge, Mass.: Harvard University Press.

SYMINGTON, C. F., 1943. The future of colonial forest botany. *Empire For. J.* **22**: 11–23.

TAYLOR, R. A., 1971. The *Flora North America* project. *BioScience*, **21**: 521–3.

THISTLETON-DYER, W. T., 1906. Botanical survey of the Empire. *Bull. Misc. Inform. (Kew), 1905:* 9–43.

THORNE, R. F., 1971. Introduction: North temperate floristics and the *Flora North America* project. *BioScience*, **21**: 511–12.

TRAVIS, B. V. *et al.*, 1962. *Classification and coding system for compilations from the world literature on insects and other arthropods that affect the health and comfort of man* (US Army Quartermaster Research and Engineering Center, Technical Report, ES-4). Natick, Mass.

TURRILL, W. B., 1964. Floras. In *Vistas in botany* (ed. W. B. Turrill), vol. 4, pp. 225–38. Oxford: Pergamon.

VALENTINE, D. H., 1971. Floristics in Europe. *BioScience*, **21**: 512–14.

VERDOORN, F., 1945. *Plants and plant science in Latin America*. Waltham, Mass.: Chronica Botanica.

VICKERY, B. C., 1975. *Classification and indexing in science*. 3rd edn. London: Butterworths.

VOSS, E. G., 1952. The history of keys and phylogenetic trees in systematic biology. *J. Sci. Labs. Denison Univ.* **43**: 1–25. Quoted from Pankhurst, R. J., 1978. *Biological Identification*. pp. 82–7. London: Arnold.

WAGNER, W. H., 1974. Pteridology 1947–1972. *Ann. Missouri Bot. Gard.* **61**: 86–111.

WATSON, L., 1971. Basic taxonomic data: the need for organisation over presentation and accumulation. *Taxon*, **20**: 131–6.

WEBB, D. A., 1978. *Flora Europaea* – a retrospect. *Taxon*, **27**: 3–14.

WEBB, L. J. *et al.*, 1970. Studies in the numerical analysis of complex rainforest communities, V. A comparison of the properties of floristic and physiognomic-structural data. *J. Ecol.* **58**: 203–32.

WEBB, L. J. *et al.*, 1976. The value of structural features in tropical forest typology. *Australian J. Ecol.* **1**: 3–28.

WHEELER, W. M., 1923. The dry-rot of our academic biology. *Science*, **52**(1464): 61–71 (reprinted in 1970 in *BioScience*, **20**: 1008–13).

WILEY, E. O., 1981*a*. *Phylogenetics: the theory and practice of phylogenetic systematics*. New York: Wiley.

WILEY, E. O., 1981*b*. Convex groups and consistent classifications. *Syst. Bot.* **6**: 346–58.

WIT, H. C. D. DE, 1949. Short history of the phytography of Malaysian vascular plants. In *Flora Malesiana* (ed. C. G. G. J. van Steenis), ser. 1, vol. 4, pp. lxxi–clxi. Batavia (Jakarta): Noordhoff-Kolff.

WYATT, H. V., 1967. Research newsletters in the biological sciences – a neglected literature service. *J. Documentation*, **23**: 321–5.

YÜ, TE-TSUN, 1979 (1980). Special report: status of the *Flora of China. Syst. Bot.* **4**: 257–60.

ZIMAN, J., 1968. *Public knowledge: the social dimension of science*. Cambridge: Cambridge University Press.

PART II

Systematic bibliography

Conventions and abbreviations

Bibliographic conventions

Books. Author, date, title, edition, collation, place, and publisher; previous edition (if any); reprint (if any); bibliographic notes (if required). The author may be defined as an institution, or other collective body, but no more than two authors by name will be given.

Irregular serials. Author, date, title, edition, collation, serial reference (in parentheses), place, and publisher (if obscure). Authorship convention as above; Roman numeration/letters for series and Arabic numeration for volumes. [Defined separately as many book-length floras and enumerations are published in various kinds of irregular serials, treatment of which varies in libraries and other book collections.]

Periodicals. Author, date, title, serial reference, and collation. Authorship convention as above; numeration the same as for irregular serials. [Used for more or less regularly periodic serials, which are always treated as such in libraries.]

Symposia, congresses, etc. Author of article, date, title of article, 'in', title of whole work, year and place of meeting (if known and applicable), editor(s), collation of article, serial reference (if applicable), place, and publisher. Authorship and numeration conventions as above.

Publication

As far as possible, only the primary point of publication (assumed to be the first or only city or town given) is indicated, and superfluous portions of the titles of printeries or publishing houses (where relevant) are omitted, including *izdatel'stvo* (Russian for 'publishing house') in Soviet books consequent to the general adoption of acronyms or cognomina from 1964 (or where such designations had previously been used). Where necessary for clarity, the term 'press' has been used in the sense of a publishing house, as in Cambridge University Press or Academia Sinica Press. In some cases, precisely who has been the publisher has been hard to ascertain; it is hoped that this task will be eased somewhat for future editions of this *Guide* through the completion of the monumental second edition of *Taxonomic Literature* by F. A. Stafleu and R. A. Cowan (1976– , Utrecht) [*TL-2*], of which at this writing (1981) three volumes, covering authors from A through O, have been published.

The designation of a work as *mimeographed* indicates that, as far as can be ascertained, it has been reproduced from stencils with the use of a duplicating machine, but in recent years the introduction of more sophisticated forms of reproduction has made the application of this designation somewhat arbitrary. Whether such works are properly published or not has long been a matter for argument among bibliologists; to me, however, effective circulation, usage, and citation irrespective of reproductive process constitutes a form of publication, although not necessarily *definitive* publication.

Transcription and transliteration

All titles in non-roman alphabets or in ideographic characters or in both have been transcribed or transliterated into their nearest equivalents in the modern roman alphabet (with diacritical marks used as necessary), where no equivalent title in the roman alphabet is given in the work. Transcription of Cyrillic characters follows ISO recommendation R9 (1954; see also *Unesco Bulletin for Libraries* 10: 136–7 (1956)), except that Х is rendered as *kh* and Ц as *ts* as being more familiar to Anglophone users, and that personal names are as far as possible rendered according to the author's own preference or to conventional usage. For Mandarin (Chinese), the system used in *Bibliography of Eastern Asiatic Botany* (1938) and its *Supplement* (1960) has been followed as far as possible; this applies also to Japanese and Korean. For Chinese works, no set of transliterations on the *pinyin* system has been available; however, most works published since 1972

have had an alternative title, usually in Latin. Few problems have been presented by other languages; where works have been written in the Jawi script (Arabic or Farsi; none has been published in Malay), alternative titles in a European language have nearly always been provided.

Periodical and serial abbreviations

The system of periodical and serial abbreviations here followed is based upon that in Lawrence, G. H. M. *et al.* (eds.), 1968. *Botanico-Periodicum Huntianum.* Pittsburgh, Pa.: Hunt Botanical Library, with the abbreviations of titles not included in that work as far as possible following its guidelines. The use of *Botanico-Periodicum Huntianum* (or *B-P-H*) brings the present work into line with the second edition of *Taxonomic Literature* previously referred to. Full titles are given in Appendix B.

General bibliographies and indices: explanation of abbreviations

Only short titles are given here. For fuller details, see Appendix A.

General bibliographies

Bay, 1909.   Bay, *Bibliographies of botany*.
Blake and Atwood, 1942.   Blake and Atwood, *Geographical guide to floras of the world*, 1.
Blake, 1961.   Blake, *Geographical guide to floras of the world*, 2.
Frodin, 1964.   Frodin, *Guide to the standard floras of the world*.
Goodale, 1879.   Goodale, *The floras of different countries*.
Holden and Wycoff, 1911–14.   Holden and Wycoff, *Bibliography relating to the floras*.
Hultén, 1958.   Hultén, *The amphi-Atlantic plants*, [bibliography].
Jackson, 1881.   Jackson, *Guide to the literature of botany*.
Pritzel, 1871–7.   Pritzel, *Thesaurus literaturae botanicae*.
Rehder, 1911.   Rehder, *The Bradley bibliography*, 1: Dendrology, part 1.

Sachet and Fosberg, 1955, 1971.   Sachet and Fosberg,
    Island bibliographies and *Island bibliographies:*
    supplement.
USDA, 1958. US Department of Agriculture,
    Botany subject index.

General indices
The dates give years of coverage, not publication.

BA, 1926– . *Biological abstracts* (and *Biological*
    abstracts/RRM).
BotA, 1918–26. *Botanical abstracts.*
BC, 1879–1944. *Botanisches Centralblatt* (from 1938
    Botanisches Zentralblatt).
BS, 1940– . *Bulletin signalétique* (through 1955
    Bulletin analytique).
CSP, 1800–1900. *Catalogue of scientific papers.*
EB, 1959– . *Excerpta Botanica*, section A: *Taxono-*
    mia et chorologia.
FB, 1931– . *Progress in Botany/Fortschritte der*
    Botanik.
IBBT, 1963–9. *Index bibliographique de botanique*
    tropicale.
ICSL, 1901–14. *International catalogue of scientific*
    literature, section M: *Botany.*
JBJ, 1873–1939. *Just's botanischer Jahresbericht.*
KR, 1971– . *Kew record of taxonomic literature.*
NN, 1879–1943. *Naturae novitates.*
RZ 1954– . *Referativnyj žurnal:* (04) *Biologija.*

Conspectus of divisions and superregions

Division 0   World floras; isolated oceanic islands
and polar regions
    Superregion 01–04 Isolated oceanic islands
    Superregion 05–07 North Polar regions
    Superregion 08–09 South Polar regions

Division 1   North America (north of Mexico)
    Superregion 11–13 Boreal North America
    Superregion 14–19 Conterminous United States

Division 2   Middle America
    Superregion 21–23 Mexico and Central America
    Superregion 24–29 The West Indies

Division 3   South America

Division 4   Australasia and islands of the
southwestern Indian Ocean (Malagassia)
    Region (Superregion) 41 New Zealand and
        surrounding islands
    Superregion 42–45 Australia (with Tasmania)
    Superregion 46–49 Islands of the south-
        western Indian Ocean (Malagassia)

Division 5   Africa

Division 6   Europe

Division 7   Northern, central, and southwestern
(extra-monsoonal) Asia
    Superregion 71–75 USSR-in-Asia
    Region (Superregion) 76 Central Asia
    Superregion 77–79 Southwestern Asia

Divison 8   Southern, eastern and southeastern
(monsoonal) Asia
    Superregion 81–84 South Asia (Indian sub-
        continent)
    Region (Superregion) 85 Japan, Korea, and
        associated islands
    Superregion 86–88 China (except Chinese
        central Asia)
    Region (Superregion) 89 Southeastern Asia

Division 9   Malesia and Oceania (tropical Pacific
islands)
    Superregion 91–93 Malesia
    Superregion 94–99 Oceania

Division

O

World floras, isolated oceanic islands and polar regions

It may appear paradoxical, at first sight, to associate the plants of Kerguelen's Land with those of Fuegia, separated by 140 degrees of longitude, rather than with those of Lord Auckland's Group, which is nearer by about 50 degrees. But the features of the Flora of Kerguelen's Land are similar to, and many of the species identical with, those of the American continent, constraining me to follow the laws of botanical affinity in preference to that of geographical position.

J. D. Hooker, *Flora antarctica* (1846).

[Our main objective is] to assist the tyro in the verification of genera and species...natural habit is often a safer guide than minute microscopic characters. [Such structural details are unimportant.]

W. J. Hooker, *Species filicum*, vol. 3 (1859).

Apart from general world floras (**000**) and world-wide works relating to the vascular plants of particular physiographically defined habitats (**001–009**), this division comprises three geographical superregions: 01–04, isolated oceanic islands; 05–07, North Polar regions; and 08–09, South Polar regions. The geographical limits and the biohistory of each of these are considered beneath their respective headings.

Under unit **000** are described the more modern *species plantarum* such as the Candollean *Prodromus* and *Das Pflanzenreich*. For completeness, Hooker's *Species filicum*, Kunth's *Enumeratio* (for monocotyledons), and the 'successor' to the *Prodromus*, the *Monographiae phanerogamarum* edited by A. and C. de Candolle, are also accounted for, along with the principal *genera plantarum* relating to vascular plants and the *Pflanzenareale* series of distribution maps edited by Hannig and Winkler. The standard general keys, indices, and dictionaries relating to vascular plants are, however, merely listed; these, and the various *syllabi* of families, are discussed fully in textbooks of taxonomy.[1]

Bibliographies. See under supraregional headings. Those relating to **000** are listed above and described in Appendix A.

Indices. See under supraregional headings. Those relating to **000** are listed above and described in Appendix A.

Conspectus

000   World: general works
001   World: transoceanic elements
006   World: marine plants
008   World: aquatic and wetland plants
009   World: river plants

Superregion 01–04 Isolated oceanic islands
Region 01 Islands of the eastern Pacific Ocean
011   Guadeloupe
012   Rocas Alijos
013   Revillagigedo Islands
014   Clipperton Island
015   Cocos Island
016   Malpelo Island
017   Galápagos Islands
018   Desventuradas Islands
019   Juan Fernández Islands
Region 02 Macaronesia
020   Region in general
021   The Azores
022   Madeira Islands
023   Salvage Islands
024   Canary Islands
025   Cape Verde Islands
Region 03 Islands of the Atlantic Ocean (except Macaronesia)
030   Region in general
031   The Bermudas
032   St Paul Rocks
033   Fernando Noronha (with Rocas)
034   Ascension Island
035   St Helena
036   Trindade (South Trinidad) and Martin Vaz Islands
037   Tristan da Cunha Islands
038   Gough Island
Region 04 Islands of the central and eastern Indian Ocean
041   Laccadive Islands (Lakshadweep)
042   Maldive Islands
043   Chagos Archipelago
044   New Amsterdam and St Paul Islands
045   Cocos (Keeling) Islands
046   Christmas Island

Superregion 05–07 North Polar regions
050–70 Superregion in general
Region 05 Islands of the Arctic Ocean
051   Jan Mayen Island
052   Bear Island (Bjornøya)
053   Spitsbergen (Svalbard)
054   Franz Josef Land (Zemlja Frantsa Iosifa)

055   Novaja Zemlja
056   Severnaja Zemlja
057   New Siberian Islands (Novosibirskije Ostrova)
058   Wrangel Island (Ostrov Vrangelja)
Region 06 Palearctic mainland region
060   Region in general
061   Lapland
062   Kola Peninsula (Arctic zone)
063   Northern and Northeastern Russia (Arctic zone)
064   The Ural (Arctalpine zone)
065   Western Siberia (Arctic zone)
066   Central Siberia (Arctic zone)
067   East Siberia (Arctic zone or Arctic Yakutia)
067   Anadyr and Chukotia
Region 07 Nearctic region
071   Bering Sea Islands
072   Alaska and Yukon (Arctic zone)
073   Northwest Territories of Canada (Arctic mainland zone)
074   Canadian Arctic Archipelago
075   Ungava (far northern Québec)
076   Greenland

Superregion 08–09 South Polar regions
080–90 Superregion in general
Region 08 Circum-Antarctic islands
081   Macquarie Island
082   Macdonald (Heard) Islands
083   Kerguelen Archipelago
084   Crozet (Possession) Islands
085   Marion (Prince Edward) Islands
086   Bouvetøya (Bouvet Island)
087   South Georgia
088   South Sandwich Islands
Region 09 Antarctica
090   Region in general

000

World: general works

The descriptive works on species and genera deemed most appropriate to a guide to floras are here preceded by a selected listing of dictionaries, indices, and identification keys relating to vascular plants. It is hoped that this selection, along with the citations of the standard descriptive treatises, will be of particular value to the non-specialist. It may be noted that certain of the latter have enjoyed a very strong influence upon the systematic arrangement of floras.

General keys to vascular plants

CRONQUIST, A., 1979. *How to know the seed plants*. vii, 153 pp., 7 lvs., 337 text-figs. Dubuque, Ia.: Brown. [Analytical students' key covering the majority of seed plant families, notably those wild or widely cultivated in the USA. Terminal leads in the key are illustrated by small figures depicting representatives of the families concerned.]

DAVIS, P. H. and CULLEN, J., 1978. *The identification of flowering plant families: including a key to those native and cultivated in north temperate regions*. 113 pp., 8 figs. Cambridge: Cambridge University Press. (1st edn, 1965, Edinburgh.) [Pocket-sized; covers northern extratropical angiosperm families. Careful reading of the introductory part is advised.]

HUTCHINSON, J., 1967. *Key to the families of flowering plants of the world*. viii, 117 pp., 8 text-figs. Oxford: Oxford University Press. (Reproduced with new index, 1979, Königstein/Ts., W. Germany: Koeltz.) [World-wide in scope, based upon the general keys in his *Families of flowering plants*.]

THONNER, F., 1981. *Thonner's analytical key to the families of flowering plants*. (Revised R. Geesink, A. J. M. Leeuwenberg, C. E. Ridsdale, and J. F. Veldkamp.) xxvi, 231 pp., portrait (Leiden Botanical Series, 5). Wageningen: PUDOC; The Hague: Leiden University Press. [Analytical key to seed plant families on a worldwide basis, with an abundance of alternative leads to take account of the many taxa with features aberrant for a given family. A glossary, index, and bio-bibliography of Franz Thonner are also included. P. W. Leenhouts, in his *Guide to the practice of herbarium taxonomy* (1968, Utrecht) and other botanists considered the original version of this work to be superior to that of Hutchinson (see above) but having limited usage due to its rarity.] Based upon THONNER, F. 1917. *Anleitung zum Bestimmen der Familien der Blütenpflanzen (Phanerogamen)*. 2nd edn., vi, 280 pp. Berlin: Friedländer.

General dictionaries and indices of vascular plants

CHRISTENSEN, C., 1905–6. *Index filicum*. lix, 744 pp. Copenhagen: Hagerup. Continued as *idem*, 1913–34. *Index filicum*: supplements 1–3. Copenhagen; and PICHI-SERMOLLI, R. E. G. *et al.*, 1965. *Index filicum*: supplement 4 (Regnum Veg., 37). Utrecht. [Homosporous and heterosporous Filicinae through 1960. Since 1971, new names have been published in *Kew Record of Taxonomic Literature*, beginning in 1974, but no cumulation for 1961–70 is available.]

FARR, E. R., LEUSSINK, J. A. and STAFLEU, F. A. (eds.), 1979. *Index nominum genericorum (plantarum)*. 3 vols. (Regnum Veg., 100–2). Utrecht. [Authoritatively documented index to all generic names of plants (in traditional sense) published from 1754 onwards, with indication of

place of publication. Originally issued in card form, beginning in 1955 and continuing until 1972.]

GRAY HERBARIUM OF HARVARD UNIVERSITY, 1894– . *Gray Herbarium card index*. Cards, issued quarterly. Cambridge, Mass. (Reprinted in book form as *idem*, 1968. *Gray Herbarium index*, with preface by R. C. Rollins, 10 vols. Boston: Hall. [Index to all new plant names published in the Western Hemisphere from 1886, including those in infraspecific categories. In recent years, retrospective coverage of infraspecific names to 1753 has been undertaken. The book version published by Hall provides coverage through 1967, with a one-volume supplement for the years 1967–77 expected to appear in 1981. (Covering of infraspecific names for the world as a whole has also been assumed by the compilers of *Index Kewensis* as from 1971.)]

HERTER, W. G., 1949. *Index lycopodiorum*. 120 pp. Basel: The author. [Index to *Lycopodium sens. lat.* from 1753.]

HOOKER, J. D. and JACKSON, B. DAYDON, 1893–5. *Index Kewensis plantarum phanerogamarum*. 2 vols. Oxford. Oxford University Press. Continued as DURAND, T. and JACKSON, B. DAYDON, 1902–6. *Index Kewensis*: supplementum I. Brussels: Castaigne (for T. Durand; later issued by Oxford University Press), and THISTLETON-DYER, W. T. *et al.*, 1904– . *Index Kewensis*: supplementum II– . Oxford: Oxford University Press (reprinted 1977– , Königstein/Ts., W. Germany: Koeltz.) [The standard index to species names given to seed plants from 1753, with indication of known geographical range when first described (save for names treated as synonyms in the original work and in early supplements) as well as place of publication. A consolidated version is badly needed; as of the late 1970s fifteen separate supplements had been published. For an index to generic entries throughout the work, see ROULEAU, E., 1970. *Guide to* Index Kewensis *and its supplements I–XIV*, vi. 370 pp. Montréal: The author.]

REED, C. F., 1953. *Index Isoetales*. 72 pp. (Bol. Soc. Brot., II, 27). Alcobaça, Portugal and Baltimore, Md.: The author. [Index to names of Isoetales.]

REED, C. F., 1966. *Index selaginellarum*. 287 pp. (Mem. Soc. Brot., 18). Alcobaça, Portugal and Baltimore, Md.: The author. [Index to names of *Selaginella sens. lat.*]

REED, C. F., 1971. *Index equisetorum: index to Equisetophyta*. 2 vols. Baltimore, Md.: The author (Reed Herbarium). [Vol. 2, Extantes, indexes names in the modern genus *Equisetum*.]

WILLIS, J. C., 1973. *A dictionary of the flowering plants and ferns*. 8th edn, revised by H. K. Airy-Shaw. xxii, 1245, lxvi pp. Cambridge: Cambridge University Press. (First edn, in 2 vols., 1895–7; 6th edn, 1931.) [The standard guide to generic and family names and their synonyms, with accepted equivalents and/or brief notes on one or more lines about each, the latter including overall distribution and approximate number of species. The two most recent editions are less broad in scope than those of 1931 and before.]

General index to distribution maps

TRALAU, H. (ed.), 1969– . *Index holmensis – a world phytogeographic index*. Vols. 1– . Zurich: The Scientific Publishers. [Index to all published distribution maps of vascular plants, both recent and fossil, arranged alphabetically within major classes and sub-classes. Twelve volumes have been projected, with references to over 250000 maps; as of 1980 four had been published (covering lower vascular plants, gymnosperms, monocotyledons, and dicotyledons from A to B) and a fifth (dicotyledons, C) was reportedly in preparation. Based in great part upon data accumulated at Stockholm (Hultén) and Halle (Meusel).]

World floras (*Florae cosmopolitanae*)

Most works of this nature are now only of historical value in terms of practical usefulness, having been written before 1840 when such 'encyclopedic' undertakings were still fashionable and such a coverage of the plant kingdom (or at least the higher plants) to species level was possible for one or two men, without many other encumbrances. The one real exception is the Candollean *Prodromus*, which covers the dicotyledons and was only 'completed' in 1873 with the aid of several specialists, although beginning in the 1820s as a single-handed enterprise like its late-Linnean counterparts but distinguished by its use of the natural system of which it was planned as an exponent. It is here regarded as the last 'world flora' (along with the contemporaneous works by Kunth on monocotyledons and Hooker on the vascular cryptogams). Later 'world floras' took the form of monograph-series, a term which effectively describes *Monographie phanerogamarum* and *Das Pflanzenreich*.

CANDOLLE, A. P. DE and CANDOLLE, A., DE (eds.), 1824–73. *Prodromus systematis naturalis regni vegetabilis*. 17 vols. Paris: Treuttel & Würtz (vols. 1–7); Fortin, Masson (vols. 8–9); Masson (vols. 10–17). (Reprint of vols. 1–7, 1966, Lehre, Germany: Cramer, in connection with an offer of the remaining stock of volumes 8–17.)

BUEK, H. W., 1842–74. *Genera, species, et synonyma Candolleana alphabetico ordine disposita*. 4 parts. Berlin: Nauck (parts 1–2); Hamburg: Perthes-Besser & Mauke (part 3); Gräfe (part 4). (Reprinted 1967, Amsterdam: Asher.)

Concise descriptive formal systematic account of the families, genera, and species of Dicotyledoneae, arranged according to the Candollean system; includes notes on distribution, synonymy, some citations of

(usually representative) *exsiccatae*, and taxonomic commentary. A historical analysis is presented in volume 17, pp. 303–14 also an abridged index, pp. 323–493. Until 1841 the work was largely written by A. P. de Candolle himself in the old tradition without collaboration from specialists, but in later years, with A. de Candolle as senior editor, the work became a collaborative enterprise, like the contemporary *Flora brasiliensis*, to which most of the leading monographers of the day contributed. With the completion of the *Prodromus*, the logical step was to proceed to a series of independent monographs, which was designated *Monographiae phanerogamarum* (see below). Detailed surveys of the contents are found in Pritzel's *Thesaurus* (see Appendix A) and in Stafleu and Cowan's *Taxonomic Literature*, 2nd edn., vol. 1 (1976). For a modern commentary, see STAFLEU, F. A., 1966. *The great Prodromus*. Lehre. (Also included in volume 1 of the 1966 reprint of the main work).

CANDOLLE, A. and C. DE (eds.), 1878–96. *Monographiae phanerogamarum*. Vols. 1–9. Paris: Masson.

'Prodromi nunc continuatio, nunc rivisio.' A series of formal descriptive family monographs, not issued in any systematic order but intended as an extension and revision of the *Prodromus*. All treatments were by specialists, including the Candolles themselves. The accounts include indications of distribution, citations of *exsiccatae*, and taxonomic commentary, as well as synoptic keys. The series came to an end some time after the death of A. de Candolle in 1893. Its discontinuance may have influenced Adolf Engler, who was a significant contributor to the work (Araceae; Anacardiaceae), to initiate moves which resulted in the establishment of *Das Pflanzenreich* (see below).

ENGLER, A. *et al.* (eds.), 1900– . *Das Pflanzenreich*. Heft 1– . Illus. Leipzig: Engelmann. (Heft 1–106 published 1900–43; Heft 106 reprinted 1956, Berlin: Akademie-Verlag; Heft 107–8 published 1953, 1968, Berlin; Heft 1–105 reprinted 1957–60, Weinheim, Germany: H. R. Engelmann/Cramer.)

Comprises an illustrated series of monographs of families (or major parts thereof) of the plant kingdom, numbered according to the Engler and Prantl system but not issued in sequence (thus retaining some semblance of the 'world flora' concept). The accounts include brief descriptions, synonymy, indication of distribution, citation of *exsiccatae*, and taxonomic commentary, as well as analytical keys. The work may also be viewed as an extension of *Die natürlichen Pflanzenfamilien* as well as a successor to *Monographie Phanerogamarum*. The series is far from complete; progress on the work diminished after Engler's death in 1930 and since 1943 only two new parts have appeared. The work was formerly published with sponsorship from the Royal Prussian Academy of Sciences (current rights are held by the Akademie-Verlag in East Berlin) and its future is unclear at this writing (1981). For a detailed bibliographic summary see DAVIS, M., 1957. A guide and analysis of Engler's *Das Pflanzenreich*. *Taxon*, **6**: 161–82. (Reprinted in *Taxonomic literature* by F. A. Stafleu, 1967, and in *TL-2* by F. A. Stafleu and R. S. Cowan, vol. 1, 1976.)

HOOKER, W. J., 1844–64. *Species filicum*. 5 vols., 304 pls. London: Pamplin (vol. 5: Dulau).

Descriptive account of known fern species, with special reference to the contents of the then-Hookerian Herbarium (Kew) and provided with numerous plates. This is the last such work for the Filicinae; although it is now largely of historical value it set precedents in fern taxonomy which were not seriously challenged until the 1930s through the work of R. C. Ching, E. B. Copeland, and others (B. S. Croxall, personal communication). The impact of the work was aided through its digests; see *idem*, 1865–8. *Synopsis filicum*. London; 2nd edn., 1874, revised by J. G. Baker. xiv, 559 pp., 9 pls. London. A supplement appeared as BAKER, J. G., 1891. A summary of the new ferns which have been discovered or described since 1874. *Ann. Bot.* **5**: 181–222, 301–32, 455–500, Pl. 14 (reprinted separately, 1892, London. v, 119 pp.).

KUNTH, C. S., 1833–50. *Enumeratio plantarum omnium hucusque cognitarum*. Vols. 1–5 (in 6). Stuttgart, Tübingen: Cotta.

Systematic enumeration, with diagnoses, of a large part of the known Monocotyledoneae, but lacking among other smaller groups the Zingiberales and Orchidales; Bentham estimated that it covered 'little more than half the class (i.e., sub-class)'. The work lapsed following the author's suicide in 1850. Although now largely of historical value, it is included here as in some ways it is complementary to the contemporary *Prodromus* (see above).

Generic encyclopedias (*Genera plantarum*)

These works provide descriptions of individual families and genera of Tracheophyta and are in effect specialized encyclopedias although with their keys and indications of distribution they also act to some degree as 'world floras'. Of the *genera plantarum* presently available the most comprehensive is easily Engler and Prantl's *Die natürlichen Pflanzenfamilien* (1887–1915; 2nd edn., 1924–). The published parts of the second edition are particularly notable for their extensive organized lists of literature covering each major taxon.

Not included here is Endlicher's *Genera plantarum* (1836–41; supplements 1842–50) which, although a key work in its time, is now largely of historical interest.

In addition to the works given below, mention may be made here of a new international project, launched in 1979 and aimed at producing a concise *genera plantarum vascularium* during the 1980s (*Taxon*, **29**: 318–20, 1980). This would act as a functional successor to Bentham and Hooker's *Genera plantarum* as well as Lemée's *Dictionnaire descriptif* and, moreover, would to a large extent realize Hutchinson's goal of a revision of the former work, attempted towards the end of his life in *The genera of flowering plants* (1964–7, not completed).

BAILLON, H., 1866–95. *Histoire des plantes*. Vols. 1–13, illus. Paris: Morgand/Hachette. English edn.: *idem*, 1871–88. *The natural history of plants*. Trans. M. M. Hartog. Vols. 1–8. London: Reeve.

A systematically arranged, partly derivative work containing descriptions of families and genera with extensive commentary and numerous fine illustrations, but without keys. Extensive references are given in the footnotes. The family arrangement follows that of A. P. de Candolle, but with a very wide conception of genera. A final fourteenth volume would have included Musaceae, Zingiberaceae, and Orchidaceae, but the author's death in 1895 precluded realization. The English version covers the Polypetalae and part of the Gamopetalae.

BENTHAM, G. and HOOKER, J. D., 1862–83. *Genera plantarum*. 3 vols. London: distributed by Black, Pamplin, Reeve, and Williams & Norgate (vol. 1, part 1); Reeve and Williams & Norgate. (Reprinted 1965, Codicote near Hitchin, England: Wheldon & Wesley, and Weinheim, W. Germany: Cramer.)

This nineteenth-century botanical *tour de force* comprises an unillustrated, formal descriptive account (in Latin) of all families and genera of seed plants then known, arranged on a modified version of the Candollean system and including synonymy, notes on distribution and special features, and limited but pointed critical commentary. The generic accounts in each family are preceded by a concise synopsis, but no analytical keys are provided. As far as possible, the work was based upon original investigation of specimens and other plant materials, largely those available at Kew. It has served as the basis for the arrangement used in many floras, notably the 'Kew series' covering the British Empire (and Commonwealth), but it is now long out of date. A revision was undertaken and partly realized by Hutchinson as *The genera of flowering plants* (see below).

COPELAND, E. B., 1947. *Genera filicum: the genera of ferns.* xvi, 247 pp., 10 pls. (Annales cryptogamici et phytopathologici, 5). Waltham, Mass.: Chronica Botanica.

A systematic descriptive treatment of the families and genera of the Filicinae, with keys, synonymy, references and citations, indication of types, geographical distribution, statistics, and taxonomic commentary; references to monographs, revisions, etc., under each entry; index. The most recent comprehensive survey of the group, but now in need of a thorough revision in the light of marked advances in fern taxonomy. For corrections, see *idem*, 1951. Additions and corrections to the *Genera Filicum. Amer. Fern J.* 50: 16–21.

ENGLER, A. and PRANTL, K. (eds.), 1887–1915. *Die natürlichen Pflanzenfamilien.* Teil I–IV, Nachträge zu Teil I/2, Nachträge I–IV zu Teil II–IV. 33 parts in 23 vols. Illus. Leipzig: Engelmann.

ENGLER, A. *et al.* (eds.), 1924–   . *Die natürlichen Pflanzenfamilien.* 2. Aufl. Vols. 1b–21, *passim.* Illus. Leipzig: Engelmann (1924–43); Berlin: Duncker & Humblot (1953–   ). (Partly reprinted 1959–61, Berlin.)

One of the greatest of all plant taxonomic works, this is a copiously illustrated, scholarly, encyclopedic systematic account of the plant kingdom (as traditionally defined), covering all hierarchical levels down to infrageneric categories with mention or discussion of a great number of individual species. Analytical keys to all genera are provided, and much attention is paid to distribution, habitat, special features, properties, and uses. The general family accounts include lists of references, extensive considerations of general taxonomy, morphology, anatomy, significant attributes, distribution, flower biology and dispersal, properties, uses, etc. The supplements in the first edition cover new data up through 1912. Many floras throughout the world are arranged according to this work. An extensive

review and analysis was prepared by F. A. Stafleu (*Taxon* **21**: 501–11, 1972; see also *Taxonomic Literature*, 2nd edn. (*TL-2*), pp. 764–9, 1976).

The second edition, commenced in 1924, is even more extensive in scope than the first. As of this writing (early 1981), 27 parts covering a miscellany of higher taxa from the Schizophyta through the Angiospermae/Gentianales have appeared. Of these, however, only four have appeared since World War II with a gap of more than 20 years from 1959 to 1981. Work is reported to be actively in progress again, with preparation of several more part-volumes in hand. With publication of volume 28bI in 1980, the work has become more international in character through a change to English as the language of publication and recruitment of authors from a wider circle; stylistic changes have also been made. Of special note is Engler's general essay on the flowering plants in volume 14a (1926). For full details, see *TL-2* pp. 769–83.

HUTCHINSON, J., 1964–7. *The genera of flowering plants.* Vols. 1–2. Oxford: Oxford University Press.

A relatively concise but formal descriptive systematic account of the families and genera of flowering plants, based partly on *Genera plantarum* of Bentham and Hooker but arranged according to the author's so-called phylogenetic system as enunciated in his *Families of flowering plants* (2nd edn., 1959). Generic accounts include synonymy, type species, key references, the number of species, and overall distribution but little if any critical dicussion, extensive infrageneric synopses, or mention of species of botanical interest (although this latter subject is covered in the author's companion *Evolution and phylogeny in the flowering plants* (1969). Family accounts include sections on phylogeny, unusual features, and uses as well as lists of references. Analytical keys are given to all genera. With the author's death in 1972, work on this series has effectively lapsed, although extensive materials are preserved at the Kew Herbarium. Its effective successor may be the projected concise work announced in 1980 (see above under subheading). Nevertheless, the two versions of the Hutchinson system, as expressed in *Families of flowering plants*, should be noted as they have served as the basis for the arrangement of several modern floras, notably in Africa.

LEMÉE, A., 1929–59. *Dictionnaire descriptif et synonymique des genres des plantes phanérogames.* 10 vols.

(in 11). Brest, France: Imprimerie. Commerciale et Administrative (for the author); Paris: Lechevalier (vol. 10).

Comprises an extensive but largely compiled descriptive account of all genera of seed plants, with indication of their geographical distribution. The arrangement of genera in vols. 1–7 and 8b is alphabetical; volumes 9 and 10 are supplements. Volume 8a includes keys to the genera in each family. The work is the most recent of existing *genera plantarum* but lacks the scope of *Die natürlichen Pflanzenfamilien*.

Distribution maps

Only one extensive general series of world distribution maps of individual plant taxa exists: the *Pflanzenareale* edited by Hannig and Winkler (1926–40).

HANNIG, E. and WINKLER, H. (eds.), 1926–40. *Die Pflanzenareale*. Vols. 1–4, 5, parts 1–2. Maps. Jena: Fischer.

A series of world distribution maps of recent and fossil families, genera and species, with explanatory text and references giving sources. [Not completed and largely outdated.]

001

World: transoceanic elements

Included here are three key systematic works dealing with transoceanic distributions not readily placed elsewhere.

Atlantica

HULTÉN, E., 1958. *The amphi-Atlantic plants and their phytogeographical connection*. 340 pp., 279 maps (Kongl. Svenska Vetenskapsakad. Handl., IV, vol. 7, part 1). Stockholm. (Reprinted 1973, Königstein/Ts., W. Germany: Koeltz.)

A series of annotated distribution maps (mainly of species or groups of species) distributed on both sides of the Atlantic Ocean, organized to show the ranges of presumed vicariant species and provided with critical commentary including synonymy, taxonomic notes, etc. (partly compiled from other literature). In the general part there is given an overall review of transatlantic phytogeography and its problems, in which along with among other matters the question of long-distance dispersal is considered. At the end (pp. 298–330) there is an extensive bibliography with a remarkable map showing the areal coverage of nearly every work. Complementary to the same author's *The circumpolar plants* (see **050–70**).

Holarctica

MEUSEL, H., JÄGER, E., RAUSCHERT, S., and WEINERT, E. (eds.), 1965–78. *Vergleichende Chorologie der zentral-europäischen Flora*. Bde. 1–2 (each in 2 vols.). 15 figs., 1648 maps. Jena: Fisher (not yet completed).

Comprises a systematically arranged atlas of distribution maps of vascular plants native to central Europe depicting their *overall* range world-wide, with descriptive and tabular commentary covering details of distribution, ecology, and chorological classification or *Arealtypen*. The introductory section to the first volume of commentary includes chapters on the principles and approaches to the study of plant areas and their classification, sources of data, and the floristic regions and provinces of extratropical Eurasia. Text and maps are in separate volumes within each *Band*. Three *Bände* in all are projected; of these Bd. 1 covers (on the traditional Englerian system) the pteridophytes, gymnosperms, monocotyledons, and 'early' dicotyledons (Salicaceae through Leguminosae), Bd. 2 extends from Oxalidaceae through Plantaginaceae, and Bd. 3 (still in preparation) will complete the work (and include a general index). [In this work, unlike any other distribution atlas – save for the preceding work and that on the circumpolar flora (see **050–70**) by Hultén and, to a lesser extent, the following work by Steenis and associates – the habitat preferences, life-forms, floristic background, and phytogeography of each species is given detailed discussion.]

Indo-Pacifica

Bibliography
STEENIS-KRUSEMAN, M. J. VAN, 1963–75. Bibliography of Pacific and Malesian plant maps of phanerogams. In *Pacific plant areas* (ed. C. G. G. J. VAN STEENIS, and M. M. J. VAN BALGOOY), 1–3. Manila, Leiden. [Concisely annotated bibliography in vol. 1, with supplements in each of the succeeding volumes.]

STEENIS, C. G. G. J. VAN and BALGOOY, M. M. J. VAN (eds.), 1963– . *Pacific plant areas*. Vols. 1–3.

Maps 1–293. Manila: National Institute of Science and Technology, Philippines (vol. 1); Leiden: Rijksherbarium (vols. 2–3). (Vol. 1 also published 1963 as *Monogr. Philipp. Inst. Sci. Tech.* 8(1); vol. 2, 1966, as *Blumea*, suppl. 5; vol. 3 separately published. Volume 3 edited by van Balgooy alone.)

A series of annotated distribution maps of genera and some species partly or wholly occurring within the Indo-Pacific region, with special reference to the Pacific Basin. Each map, which is not gridded, depicts distribution by lines, dots, or solid areas or combinations of these. Associated text includes systematic position of the plant(s) concerned, synonymy, taxonomic background, notes on habit, habitat, ecology, diaspore type, and modes of dispersal, and indication of sources of information, signed by the contributor. No fixed sequence is followed.

006

World: marine plants

See also **008** (Cook). **006** is proposed to encompass more specifically works on *marine* plants (sea-grasses, mangroves, and, by extension, marine algae). Only the first-named enjoy a recent full monograph, described below.

DEN HARTOG, C., 1970. *The sea-grasses of the world.* 275 pp., 63 text-figs. (including 10 maps), 31 halftone pls. (Verh. Kon. Ned. Akad. Wetensch., II. reeks, Afd. Natuurk., vol. 59, no. 1). Amsterdam: North-Holland.

Conservatively styled monographic treatment of all marine Hydrocharitaceae and Potamogetonaceae ss. *11.*, with keys, full synonymy including references and literature citations, generalized indication of overall range, citations of *exsiccatae* with localities, critical taxonomic commentary, and ecological notes including biology, reproduction, spread, habitat, and variability; addenda and index to all scientific names at end. All species are illustrated. The introductory section includes maps, a key for sterial specimens, and a bibliography (pp. 36–8). [Now in need of thorough revision.]

008

World: aquatic and wetland plants

COOK, C. D. K. *et al.*, 1974. *Water plants of the world.* viii, 561 pp., 261 text-figs. The Hague: Junk.

Comprises a manual for identification of the genera (and families) of fresh-water and benthic macrophytes (helophytes and hydrophytes) encompassing Charophyta, Bryophyta, and Tracheophyta; includes analytical keys, short descriptions, indication of distribution, approximate number of species, general ecology, and representative illustrations. Principal references are cited under family and generic headings. The introductory section is devoted mainly to two general keys, respectively based mainly upon reproductive and upon vegetative attributes, while at the end are a glossary and an index. A concise list of principal works upon aquatic macrophytes, with inclusion of several specialized floras, is also given.

009

World: river plants

The monograph *River plants* by S. M. Haslam (1978, Cambridge: Cambridge University Press) should also be consulted for general information on biology and ecology of river plants, although it is largely limited to those of the Holarctic zone and moreover includes no systematic enumeration or individual species accounts.

STEENIS, C. G. G. J. VAN., 1981. *Rheophytes of the world.* xv, 407 pp., 47 text-figs., 23 photographs. Alphen aan den Rijn (Netherlands): Sijthoff & Noordhoff.

This work includes a systematic census, without keys, of known vascular rheophytes (riverbed and riverbank plants growing within reach of riverine flash floods); for each species are given a concise description, synonymy with references, literature citations, generalized indication of distribution and altitudinal range (including for some species, chiefly Malesian, citations of *exsiccatae*), habitat and biological notes (based on available information), and taxonomic commentary. In the general chapters preceding the census appear discussions of habitats, morphology and autecology, regional floristics and environmental factors, phytogeography, cultivation, the phenomenology of the habitat and the willow-leaf (the latter characteristic of the land-rheophytes), and their place in the author's theory of 'autonomous evolution', which latter is considered at some length in chapter 8. A systematic conspectus appears on pp. 70–2 preceding the phytogeographical treatment, while a list of references and a glossary are given at pp. 143–8. All plant names as well as concepts are indexed at the end.

Superregion

01–04

Isolated oceanic islands

Within this supperregion are included all those islands of the Atlantic, Indian, and Pacific Oceans which can neither be grouped into a large quasi-continental series (such as the West Indies) nor be associated very readily with a nearby continent. Moreover, botanical work on these islands (as well as those of the far southern hemisphere) has often been carried out independently from similar work relating to continents or parts thereof, often because many, if not most, of these islands are of considerable biological and biogeographic interest.

The history of botanical exploration and documentation of the islands included here is to a large extent associated with voyages of discovery, exploration, and oceanic documentation and research and/or specialized field work. Some of the islands have been made known through specifically organized expeditions. In most cases, a great deal of 'chance' has been involved. To some, like the Desventuradas west of Chile, access is difficult; and what is found also depends upon the season of the visit.

Nevertheless, by the late-nineteenth century, especially through the efforts of the 'Challenger' expedition, most of the islands had botanically become tolerably well known. Of the remainder, Christmas Island was studied by a British Museum expedition at the end of the century, and many of the eastern Pacific groups were visited by teams from the California Academy of Sciences in the first half of the century

following. Willis and Gardiner, with others, pioneered extensive atoll research on the Indian Ocean islands, including those from the Laccadives to the Chagos Archipelago; and Tristan da Cunha, Gough, and New Amsterdam and St Paul were visited by some of the Antarctic expeditions of the day.

This phase of primary floristic exploration and writing of checklists or floras has been followed by more intensive study on single islands or island groups. Early students included Britton at Bermuda and especially Skottsberg at Juan Fernández (and other islands, e.g., Hawaii and the Falklands). Since World War II, intensive floristic exploration has been carried out in the Galápagos and in Macaronesia, especially the Canaries. The presence of local field stations or botanical gardens in these latter groups, both of exceptional importance, has greatly facilitated such work. Similar work has also been done on some of the smaller islands.

Revised floras and/or checklists are now available for many of the island units in the four regions accounted for here. The most outstanding of these contributions is at present *Flora of the Galápagos Islands* by Wiggins and Porter (1970). Serious lacunae, however, remain; for the majority of units available accounts are more or less antiquated, some nearly 100 years old. Good modern treatments are, for instance, needed for the Cocos (Keeling) group, Christmas Island, New Amsterdam/St Paul, Trindade/Martin Vaz and Bermuda; above all, the realization of the *Flora Macaronesica* project is awaited. A useful review of the present state of island floristics, phytogeography, vegetation, and ecology is provided by the several papers in BRAMWELL, D. (ed.), 1979. *Plants and islands.* x, 459 pp., illus. London: Academic Press.

Bibliographies: Bay, 1910; Blake and Atwood, 1942; Frodin, 1964; Rehder, 1911; Sachet and Fosberg, 1955, 1971; USDA, 1958.

Indices. BA, 1926– ; BotA, 1918–26; BC, 1879–1944; BS, 1940– ; CSP, 1800–1900; EB, 1959– ; FB, 1931– ; IBBT, 1963–9; ICSL 1901–14; JBJ, 1873–1939; KR, 1971– ; NN, 1879–1943; RŽ, 1954– .

Region

01

Islands of the eastern Pacific Ocean

This region incorporates the following islands and island groups: Guadelupe, Rocas Alijos, the Revillagigedos, Clipperton, Cocos, Malpelo, the Galápagos, the Desventuradas, and Juan Fernández.

The eastern Pacific islands are mostly small and scattered, with floras more or less closely related to those of the nearest continental areas. Some, such as Clipperton, are very impoverished and consist almost entirely of oceanic 'wides', while others, such as Juan Fernández, have a high percentage of endemism, many 'relicts', and more generalized affinities. The best-known and most thoroughly studied group is the Galápagos, associated with Darwin (who also made the first notable plant collection in the group).

Post-1939 treatments are available for Clipperton, Cocos (checklist only), and the Galápagos. Skottsberg's floras of the Desventuradas (1937) and the Juan Fernández group (1922) were supplemented in the 1950s. No treatments are available for Rocas Alijos, which in any case lacks vascular plants, and Malpelo.

Bibliographies. General bibliographies as for Superregion 01–04.

Indices. General indices as for Superregion 01–04.

011

Guadelupe Island

Bibliography

See the Mexican bibliographies of JONES, and LANGMAN under Region 21/22.

EASTWOOD, A., 1929. List of the plants recorded from Guadelupe Island, Mexico. *Proc. Calif. Acad. Sci.*, IV, **18**: 394–420, pls. 33–4.

Systematic enumeration of vascular plants, with indication of local range, citations of *exsiccatae*, and notes on habitat, special features, occurrence, etc. The plant list is preceded by accounts of physical features,

vegetation, plant geographical relationships, and past botanical exploration as well as a special list of species originally described from the island.

012

Rocas Alijos

No vascular plants have been recorded (so far as known) from this very low-lying group of barren rocks, which lie about halfway between Guadelupe and the Revillagigedo group (see below).

013

Revillagigedo Islands

Bibliography

See the Mexican bibliographies of JONES, and LANGMAN under Region 21/22.

JOHNSTON, I. M., 1931. The flora of the Revillagigedo Islands. *Proc. Calif. Acad. Sci.*, IV, **20**: 9–104

Systematic enumeration of vascular plants, with descriptions of new taxa; synonymy, with references and pertinent citations; indication of localities with *exsiccatae* and general summary of extralimital range (if applicable); critical commentary and extensive notes on life-form, habitat, occurrence, etc. The plant list is preceded by accounts of the geography and physical features of the several islands (with tabular summaries of their floras), the origin and general features of the flora as a whole, plant–geographical relationships, and past botanical exploration.

014

Clipperton Island

SACHET, M.-H., 1962. Flora and vegetation of Clipperton Island. *Proc. Calif. Acad. Sci.*, IV, **31**: 249–307, 12 halftones, map.

Includes a systematic enumeration of native and introduced vascular and non-vascular plants, with localities, citations of *exsiccatae*, and general indication of extralimital range; possible methods of inward migration; and notes on habitat, status, biology, etc. The remainder of the work deals with prior botanical investigation, the history and present condition of the vegetation, and dispersal mechanisms, drift seeds collected, and plant geographical relationships.

015

Cocos Island

See also **236** (STANDLEY).

FOSBERG, F. R. and KLAWE, W. L., 1966. Preliminary list of plants from Cocos Island. In *The Galápagos: Proceedings of the Galápagos International Scientific Project* (ed. R. I. BOWMAN), pp. 187–9. Berkeley: University of California Press.

Compiled list of all known vascular and non-vascular plants (except algae), with essential synonymy. See also FOURNIER, L. A. Botany of Cocos Island, Costa Rica. *Ibid.*, pp. 183–6. [General description of main features of the island and its vegetation; account of the origin of the flora; notes on visiting scientific expeditions; list of references.]

STEWART, A., 1912. Notes on the botany of Cocos Island. *Proc. Calif. Acad. Sci.*, IV, 1: 375–404, Pls. 31–4.

Systematic enumeration of vascular plants and mosses, based mainly on collections made by the author; includes synonymy (with references), localities and citations of *exsiccatae*, general summary of extralimital range, critical commentary, and notes on habitat, special features, etc. The plant list is preceded by accounts of the physical features, vegetation, and plant geographical relationships of the island and its flora. For additions, see SVENSON, H. K., 1935. Plants of the Astor expedition, 1930 (Galápagos and Cocos Islands). *Am. J. Bot.* **22**: 208–77, 9 illus.

016

Malpelo Island

No vascular plants seem as yet to have been recorded, although 'scrub' has been reported. For a general description of this nearly barren small island, see NAVAL INTELLIGENCE DIVISION, 1943. *Pacific Islands*, 2: *Eastern Pacific*, p. 21. [London].

017

Galápagos Islands

The genesis and preparation of the Galápagos flora is recounted in PORTER, D. M., 1968. The flora of the Galápagos Islands. *Ann. Missouri Bot. Gard.* 55: 173–5.

Bibliography
SCHOFIELD, E. K., 1973. Annotated bibliograpny of Galápagos botany, 1836–1971. *Ann. Missouri Bot. Gard.* 60: 461–77. [Includes 286 references, arranged by author within major divisions of plant kingdom.]

WIGGINS, I. L. and PORTER, D. M., 1970. *Flora of the Galápagos Islands.* xx, 998 pp., 268 text-figs., 16 color pls., maps. Stanford: Stanford University Press.

Illustrated descriptive flora of native, naturalized, and adventive vascular plants, with keys to all taxa, full synonymy (with references), detailed indication of local range, citation of some *exsiccatae* and inclusion of many maps, general summary of extralimital distribution (if applicable), critical taxonomic commentary, and notes on variability, habitat, life-form, ecology, etc.; glossary, bibliography, and index to all botanical names. The introductory section includes accounts of the geography, physical features, geology, climate, soils, vegetation, and animal life of the archipelago, together with historical reviews of botanical exploration and human influences. Encompasses 702 species and subspecies.

018

Desventuradas Islands (San Ambrosio and San Felix)

See also **390** (MUÑOZ PIZARRO). The islands are noted for their difficulty of access, and the flora most likely has not been completely accounted for, although it is doubtless very small.

SKOTTSBERG, C. J. F., 1937. *Die Flora der Desventuradas-Inseln (San Felix und San Ambrosio) nach den Sammlungen F. Johows.* 87 pp., 46 text-figs. (incl. map) (Göteborgs Kungl. Vetensk. Vitterh. Samhälles Handl., ser. V/B, vol. 5, part 6). Spanish edn: *idem*, 1949. *Flora de las islas San Felix y San Ambrosio* (trans. A. Horst). 64 pp., 39 text-figs. (incl.

map) (Bol. Mus. Nac. Hist. Nat., Santiago, vol. 24). Santiago.

Critical illustrated descriptive flora, without keys; includes synonymy, references and citations, detailed indication of local occurrences with citations of *exsiccatae* or other pertinent sources, general indication of overall distribution (if applicable), extensive taxonomic commentary, and occasional notes on habitat, biology, etc.; illustrations of all species; general summary of the flora and its relationships. An introductory section gives a general description of the islands and their topography and climate together with an account of botanical exploration. See also *idem*, 1950. Weitere Beiträge zur Flora der Insel San Ambrosio (Islas Desventuradas, Chile), pp. 453–69, illus. (Ark. Bot., N.S., 1). Stockholm. [Enumeration with descriptions of new taxa; revised summary and added references.]

019

Juan Fernández Islands

See also **390** (MUÑOZ PIZARRO). The home of an endemic plant family, these islands have now been, in part, named after Robinson Crusoe.

SKOTTSBERG, C. J. F., 1922 (1921). The phanerogams of the Juan Fernandez Islands. In *Natural history of Juan Fernandez and Easter Islands* (ed. C. J. F. SKOTTSBERG), vol. 2 (Botany), pp. 95–240, 39 text-figs., Pls. 10–20 (1 colored). Uppsala: Almqvist & Wiksell.

CHRISTENSEN, C. and SKOTTSBERG, C. J. F., 1920. The Pteridophyta of the Juan Fernandez Islands. *Ibid.*, 1–46, 7 text-figs., Pls. 1–5.

SKOTTSBERG, C. J. F., 1951. A supplement to the pteridophytes and phanerogams of Juan Fernandez and Easter Island. *Ibid.*, 763–92, Pls. 55–7.

The original contributions respectively constitute systematic enumerations of native and introduced phanerogams and pteridophytes, with a limited number of keys and including full synonymy, references and citations, fairly detailed indication of local distribution (with citations of *exsiccatae* and literature sources), general summary of extralimital range (where applicable), critical taxonomic commentary, and notes on habitat, biology, occurrence, etc.; general features of the flora and its indigenous and introduced components; bibliography and index to botanical names.

The supplement is in the same style. All three contributions form part of a comprehensive monograph of the group, together with Easter Island (**988**).

Region

02

Macaronesia

Included herein are all the islands of Macaronesia, i.e., the Azores, Madeira, the Salvages, the Canaries, and the Cape Verdes, which lie in the eastern part of the North Atlantic. The rest of the Atlantic Ocean islands are grouped into Region 03, save those relatively closely associated with nearby continents or island systems.

Macaronesia is at present documented by a large variety of individual floras and checklists, but not until 1974 was there any comprehensive account. In that year, *Flora of Makaronesia: checklist of vascular plants*, by Eriksson *et al.*, was published as a contribution to the *Flora Macaronesica* project, with a revised edition in 1979. The work is a consolidation of lists for the individual groups by its various authors. Island bibliographies were also produced. These works serve to bring together a great deal of scattered information, particularly important as the existing floras (except for the Azores) all date from before 1939 (and the majority from 1914 or before) and do not reflect the considerable amount of new floristic work accomplished, particularly in the Canaries where there is now a local botanical unit. They also represent steps towards the realization of a descriptive *Flora Macaronesica*, for which plans were announced in 1972.

Bibliographies. General bibliographies as for Superregion 01–04.

Indices. General indices as for Superregion 01–04.

020

Region in general

Apart from the recent checklist described below, no comprehensive floras are available for these islands as a single entity. In 1972, however, plans were announced in the United Kingdom for a new general

Flora Macaronesica and an editorial committee organized (Bramwell, D., 1972. Flora of Macaronesia Project. *Taxon* **21**: 730–1). Useful current reviews of the flora may be found in the papers by P. Sunding (pp. 13–40) and C. J. Humphries (pp. 171–199) in Bramwell, D. (ed.), 1979. *Plants and islands*. London: Academic Press.

– Vascular flora: about 3200 native species in 223 genera (Humphries in Bramwell).

ERIKSSON, O., HANSEN, A., and SUNDING, P., 1979. *Flora of Macaronesia: checklist of vascular plants* (2nd edn., by A. Hansen and P. Sunding). 2 parts. vi, 93, 55 pp. Oslo: Oslo University Botanic Garden and Museum. (First edn., 1974, Umeå (Sweden).)

Annotated tabular checklist of vascular plants, arranged alphabetically within the major classes; against the accepted name of each species is a notation of its distribution by archipelago and island (or island group) and an indication of endemicity in the region (if applicable). Nomenclatural changes from the first edition appear on pp. 92–3. All synonymy has been relegated to part 2, which takes the form of an index. The list has been prepared from punch cards using automated data-processing techniques, and represents a consolidation of earlier lists for each of the Macaronesian archipelagoes by the various authors.

and to vernacular names. The introductory section includes a list of references. In support of this work, illustrations of selected Azorean species, with detailed descriptive text, full synonymy, references, and annotations, have commenced to appear as: FERNANDES, A. and FERNANDES, R. B. (eds.), 1980. *Iconographia selecta florae Azoricae*. Fasc. 1. Pls. 1–25. Coimbra, Portugal: Secretary for Culture, Azorean Autonomous Region.

TRELEASE, W., 1897. Botanical observations on the Azores. *Annual Rep. Missouri Bot. Gard.* **8**: 77–220, Pls. 12–66.

Annotated systematic checklist of non-vascular and vascular plants and fungi, with citations, synonymy, localities, indication of *exsiccatae*, some critical commentary, and occasional notes on habitat; abbreviated designation of status (common, rare, weedy, etc.); list of references and index to generic names. Adventive and cultivated species are given in smaller type. The plates are representative and include at least some endemics. A brief introduction covers physical features, climate, agriculture, gardening, land use and the status of the flora, and sources for and methodology of the work, along with remarks on origin and dispersal of the plants treated. [Based on two field trips by the author in 1894 and 1896.]

021

Azores

See also **600** (TUTIN *et al.*); **611** (FRANCO).

Bibliography

HANSEN, A., 1970. *A botanical bibliography of the Azores*. 9 pp. Copenhagen: [Botanical Museum]; continued by HANSEN, A., 1975. *A botanical bibliography of the Azores: additions 1975*. 6 pp. Copenhagen.

PALHINHA, R. T., 1966. *Catalogo das plantas vasculares dos Açores* (rev. and ed. A. R. Pinto da Silva). xi, 186 pp., portrait, maps (end-papers). Lisbon: Sociedade de Estudos Açorianos 'Alfonso Chaves'.

Systematic enumeration of native, naturalized, and commonly cultivated vascular plants, with synonymy, references and citations, vernacular names, general indication of local distribution (with some localities given), summary note of extralimital range, and brief notes on habitat, occurrence, etc.; indices to botanical

022

Madeira Islands

Bibliography

HANSEN, A., 1975. *A botanical bibliography of the Madeira Archipelago*. 31 pp. Copenhagen: [Botanical Museum]. [523 titles.]

CHRISTENSEN, T. B. *et al.*, 1970. *Oversigt over Madeiras flora*. 167 pp., 1 text-fig., cover illus. Copenhagen: University of Copenhagen. [Mimeographed.]

Concise diagnostic manual (in Danish) of native, naturalized, and commonly cultivated vascular plants, with keys to all taxa; symbolic indication of status, local range, habitat, and phenology; list of references; index to generic and family names.

HANSEN, A., 1969. Checklist of the vascular plants of the archipelago of Madeira. *Bol. Mus. Munic. Funchal*, **24**: 1–62, 2 maps.

Systematic list of native, naturalized, and commonly cultivated vascular plants, with synonymy;

symbolic indication of local range and distribution elsewhere within Macaronesia; list of references; index to families.

MENEZES, C. A. DE, 1914. *Flora de archipelago da Madeira* (phanerogamicas e cryptogamicas vasculares). 282 pp. Funchal: Bazar del Povo.

Brief preface; annotated enumeration of native, naturalized, and commonly cultivated vascular plants, with synonymy and citations of major references; vernacular names, phenology, and brief indication of local range; descriptive notes on less common and endemic species; indices to vernacular and to generic and family names. An appendix includes descriptive accounts of local climate, vegetation zones, and botanical contributors, as well as a list of references. For additions and corrections, see MENEZES, C. A. DE, 1922. Subsidios para o estudo da flora. *Brotéria*, Bot. 20: 113–19; and MENEZES, C. A. DE, 1926. Novos subsidios para o estudo da flora. *Ibid.*, 22: 20–7.

023

Salvage Islands

PICKERING, C. H. C. and HANSEN, A., 1969. List of higher plants and cryptogams known from the Salvage Islands. *Bol. Mus. Munic. Funchal*, 24: 63–71, map.

Systematic list of cellular and vascular plants, with limited synonymy, indication of earlier records, and brief statement of local range; short list of references.

024

Canary Islands

Apart from the works described or noted below, mention should be made of the classical *Phytographia canariensis* of P. B. Webb and S. Berthelot (1836–50), which appeared as tome III, part 2 of their *Histoire naturelle des îles Canaries;* this splendid atlas-flora is still of considerable usefulness (D. Bramwell, personal communication). A more modern, popular but well-written and -illustrated introduction to the vascular flora, with special attention to the many endemic species, is BRAMWELL, D. and BRAMWELL, Z.,

1974. *Wild flowers of the Canary Islands*. 261 pp., 118 text-figs., 205 color illus., 16 pls. London: Stanley Thornes.

Bibliography

SUNDING, P., 1973. *A botanical bibliography of the Canary Islands*. 2nd edn., 46 pp. Oslo: University of Oslo, Oslo Botanic Garden.

ERIKSSON, O., 1971. *Check-list of vascular plants of the Canary Islands*. [i], 35 pp. Umeå, Sweden: Section of Ecological Botany, University of Umeå.

Systematic list of vascular plants, with essential synonymy and tabular summary of local range. Special indication is made of additions to the flora since Lems' *Floristic botany of the Canary Islands* (see below).

LEMS, K., 1960. *Floristic botany of the Canary Islands*. 94 pp. (Sarracenia, 5). Montreal: University of Montreal (mimeographed).

Annotated enumeration of vascular plants (with families and genera alphabetically arranged within the major classes), incorporating limited synonymy, citations of the author's *exsiccatae*, some taxonomic commentary, notes on habitat and biology, and symbolic indication of distribution, life-form, etc.; list of references and addenda. The introductory section gives *inter alia* a list of abbreviations, an analysis of the flora and its origin, and statistical tables. The work is provisional and incomplete, and now is of most use for its physiognomic, ecological, and biological notes.

PITARD, J. and PROUST, L., 1908. *Les Canaries: flore de l'archipel*. 502 pp., 19 pls. Paris: P. Klincksieck. (Rpt. 1972, Königstein/Ts., Koeltz).

Systematic enumeration of native and naturalized vascular plants and Bryophyta, with descriptions of new taxa; limited synonymy, with references and citations of major source works; fairly detailed indication of local range (no *exsiccatae* listed), and general summary of extralimital distribution (where applicable); notes on habitat, phenology, etc.; index to all botanical names. The introductory section contains accounts of prior botanical exploration, physical geography, climate, soils, vegetation zones and formations, characteristic features of the flora and its probable history and geographic relationships (with statistical tables), and endemic taxa. For substantial additions to the flora, see LINDINGER, L., 1926. *Beiträge zur Kenntnis von Vegetation und Flora der Kanarischen Inseln*. pp. 135–350 (Abh. Auslandsk., Hamburg Univ., 21; Reihe C, 8). [See particularly part 2: Flora der

Kanarischen Inseln. Berichtigungen und Nachträge zu Pitard und Proust...]

Partial work: Gran Canaria

KUNKEL, M. A. and KUNKEL, G., 1974–9. *Flora de Gran Canaria*. Vols. 1–4. Pls. 1–200. Las Palmas: Ediciones del Excmo. Cabildo Insular de Gran Canaria.

Large-scale color-plate atlas of vascular plants, with detailed descriptive text (including references, vernacular names, distribution, ecology, phenology, nature of propagules, variability, and source of the material together with plant description) facing each plate; introductory matter and indices in each volume. The species chosen are being allocated to volumes according to synusiae (woody plants, lianas, succulents, ferns, etc.) rather than systematically, with fifty species in each volume; in all, 11 volumes are projected. [Not a flora as such, but valuable for its illustrations.]

025

Cape Verde Islands

CHEVALIER, A., 1935. *Les îles du Cap Vert. Géographie, biogéographie, agriculture, flore de l'archipel.* 358 pp., 11 text-figs. (incl. map), 16 pls. Paris: Muséum National d'Histoire Naturelle, Laboratoire d'Agronomie Coloniale. [Also published as *Rev. Int. Bot. Appl. Agric. Trop.* **15**: 733–1090, Figs. 22–32, Pls. 1–16.]

Part 3 of this work comprises a systematic enumeration of vascular and non-vascular plants, with descriptions of new or noteworthy taxa; limited synonymy, with citations of significant literature; vernacular names; detailed indication of local range, with notation of *exsiccatae*; general summary of extralimital distribution; taxonomic commentary and notes on special features, habitat, occurrence, etc.; indices to vernacular, generic, and family names at end. The introductory sections (parts 1 and 2) cover geography, physical features, climate, geology, soils, other biota, effects of human disturbance, agriculture, vegetation and floristics, and botanical exploration; bibliography, pp. 138–40 [870–2].

SUNDING, P., 1973. *Check-list of the vascular plants of the Cape Verde Islands.* 36 pp. Oslo: University of Oslo, Oslo Botanic Garden.

Systematic list of vascular plants, with essential synonymy and tabular summary of local range; bibliography. Additions to the flora since the enumeration of Chevalier (see above) are specially indicated. For supplement, see SUNDING, P., 1974. Additions to the vascular flora of the Cape Verde Islands. *Garcia de Orta*, Bot. **2**(1): 5–29.

Region

03

Islands of the Atlantic Ocean (except Macaronesia)

This region is construed to encompass all isolated Atlantic islands not normally considered part of Macaronesia, i.e., the Bermudas, St Paul Rocks, Fernando Noronha (and Rocas), Ascension, St Helena, Trindade (South Trinidad) and Martin Vaz, the Tristan da Cunha group, and Gough.

Nearly all islands in this region were visited on the voyage of the *Challenger* in the 1870s, and consolidated botanical reports prepared by Hemsley and published in the voyage reports. Subsequent work has been patchy, with concern usually for individual islands. Bermuda was studied by Britton early in in the present century, and a manual-type *Flora of Bermuda* appeared from him in 1918. This, however, uses the obsolete 'American Code', and is now in need of revision. Fernando Noronha was studied separately by Ridley in the 1880s (*Notes on the botany of Fernando Noronha*, 1890), and Melliss' *St Helena* (1875) covers among other aspects of the island the native and naturalized plants, with illustrations. The only detailed work in modern times has been done on the Tristan group and on Gough by Wace and Dickson (*The terrestrial botany of the Tristan da Cunha Islands*, 1965). Although this has an ecological orientation it accounts completely for the flora and its local and extralimital distribution, with references to earlier work. Investigation of the status of the native flora of St Helena has also been carried out in recent years, but only scattered reports have been published. No modern reports on Trinidade and Ascension have been seen.

Bibliographies. General bibliographies as for Superregion 01–04.

Indices. General indices as for Superregion 01–04.

030

Region in general

HEMSLEY, W. B., 1885 (1884). Report on the botany of the Bermudas and various other islands of the Atlantic and Southern Oceans, part 2. In *Reports on the scientific results of the voyage of 'HMS Challenger' during the years 1873–6* (ed. J. Murray), *Botany*, vol. 1, part 2, pp. 1–281, Pls. 14–53. London: HMSO.

This stately work includes individual annotated enumerations of the vascular and non-vascular plants of the following islands or island groups: St Paul Rocks (pp. 1–7); Fernando Noronha (pp. 9–30, Pls. 14–15, 47); Ascension (pp. 31–48, Pls. 16–17); St Helena (pp. 49–122, Pls. 18–22, 48–51); Trinidade (South Trinidad) (pp. 123–32, Pls. 23–4), and the Tristan da Cunha group (pp. 133–85, Pls. 25–39, 46). Each area account includes synonymy, references and citations, descriptions of new taxa, generalized indication of local and extralimital range (with citations of *exsiccatae*), extensive critical commentaries, and often ample notes on habitats, special features, biology, uses, etc. (in part compiled from other sources). The large-format plates depict new or noteworthy plants. Each treatment is prefaced by remarks on physical features, botanical exploration, features of the flora, and vegetation formations along with tabular phytogeographic analyses. [The remainder of part 2 deals mostly with sub-Antarctic islands; these are covered in Region 08. Part 1 is devoted to the Bermudas (see **031** below), but for practical purposes this has been superseded by Britton's *Flora of Bermuda*.]

031

The Bermudas

This group was treated by Hemsley in part 1 of the botanical report of the *Challenger* voyage (see **030** above) prior to the still-current manual by Britton.

BRITTON, N. L., 1918. *Flora of Bermuda*. 585 pp., numerous text-figs., color frontispiece. New York: Scribners. (Reprinted 1965, New York: Hafner.)

Descriptive manual-flora of non-vascular and vascular plants, with keys to all taxa, limited synonymy, generalized indication of local and extralimital range,

vernacular names, occasional critical remarks, and notes on habitat, phenology, etc.; glossary and short cyclopedia of collectors; indices to all botanical and vernacular names. Figures for most species of vascular plants and bryophytes are provided to aid identification. [Employs the obsolete 'American Code' of Nomenclature.]

032

St Paul Rocks

See **030** (HEMSLEY, pp. 1–7). No vascular plants have so far been recorded from this group, which consists largely of barren rocks.

033

Fernando Noronha (with Rocas)

See also **030** (HEMSLEY, pp. 9–30).

RIDLEY, H. N., 1890. Notes on the botany of Fernando Noronha. *J. Linn. Soc., Bot.* **27**: 1–95, Pls. 1–4.

Systematic enumeration of vascular and non-vascular plants (based mainly on a field trip made by the author), with descriptions of new or noteworthy taxa; synonymy, with references and citations; generalized indication of local and extralimital ranges, with mention of some localities; descriptive notes on habitat, occurrence, special features, biology, properties and uses; account of local geology (by T. Davies). The introductory section includes accounts of physical features, geography, history of human contact and settlement, origin of the flora, pollination and dispersal, fresh-water life, and past botanical work on the island (together with the author's itinerary).

034

Ascension Island

See **030** (HEMSLEY, pp. 31–48). Additional information may be found in DUFFEY, E., 1964. The terrestrial ecology of Ascension Island. *J. Appl. Ecol.* **1**: 219–51.

035

St Helena

See **030** (HEMSLEY, pp. 49–122). Much information on the peculiar native flora is also included in MELLISS, J. C., 1875. *St Helena*. xiv, 426 pp., plates, maps. London: Reeve. This is the classic illustrated monograph on the island (botany, pp. 221–383). Unfortunately, no recent consolidated account of the flora with a review of the conservation status of the native species is available.

036

Trindade (South Trinidad) and Martin Vaz Islands

See **030** (HEMSLEY, pp. 123–32). No more recent consolidated account of the flora is available, and it is not known if any botanical surveys have been made since the nineteenth century. The treatment by Hemsley is rather sketchy.

037

Tristan da Cunha Islands

See also **030** (HEMSLEY, pp. 133–86). In addition to the work described below, there are some scattered floristic papers by Christophersen based upon Norwegian exploratory work in the 1930s.

WACE, N. M. and DICKSON, J. H., 1965. The terrestrial botany of the Tristan da Cunha Islands. *Phil. Trans.*, B, **249**: 273–360, Figs. 14–16, Pls. 35–42.

Appendix A (pp. 336–54) of this mainly biogeographical work comprises a systematic list of known native and introduced vascular and non-vascular plants of the Tristan da Cunha group and Gough, with synonymy, citations of pertinent literature or other sources, and an indication of the presence or absence of each species on each of the four islands involved. The main part of the paper includes sections on botanical exploration, general features of the flora and its component groups (with statistical tables), phyto-geography and the origins of the flora, dispersal mechanisms, vegetation formations, the history and external geographical relationships of the vegetation, special features of the plant life, and human influence. A bibliography is also included.

038

Gough Island

See also **037** (WACE and DICKSON). No coverage of this island was provided in Hemsley's regional work (**030**); the first significant plant collections evidently date from 1904 with the visit of the *Scotia*. For a separate but unannotated list of vascular plants, see WACE, N. M., 1961. The vegetation of Gough Island. *Ecol. Monogr.* **31**: 337–67, 22 figs.

Region

04

Islands of the central and eastern Indian Ocean

Islands incorporated under this heading include the Laccadives, the Maldives, the Chagos Archipelago (with Diego Garcia), New Amsterdam and St Paul, the Cocos (Keeling) group, and Christmas Island.

These scattered and mostly very small islands have had very few new treatments since 1939, with most of the standard accounts dating from before 1914. The 1885 account of Forbes for the Cocos (Keeling) group was based partly on work by Darwin. Some recent treatments prepared under the auspices of the Atoll Research Programme have given new coverage of the Maldives and the Chagos group, but are not full floras for the units concerned. It is particularly important that new treatments for New Amsterdam and St Paul as well as Christmas Island be published, as they are rather poorly documented.

Bibliographies. General bibliographies as for Superregion 01–04.

Indices. General indices as for Superregion 01–04.

041

Laccadive Islands (Lakshadweep)

See also **042** (WILLIS and GARDINER).
– Vascular flora: 121 species (PRAIN) but actually somewhat higher (Renvoize, 1979, p. 115).[2]

PRAIN, D., 1893–4. Botany of the Laccadives. *J. Bombay Nat. Hist. Soc.* **7**: 268–95, 460–86; **8**: 57–86. (Reprinted 1894, Calcutta, as part of the author's *Memoirs and memoranda, chiefly botanical.*)

Includes a systematic enumeration of the known native, naturalized, and commonly cultivated non-vascular and vascular plants, with synonymy, references and citations, English vernacular names, localities (with collectors), overall range outside the archipelago, and sometimes copious notes on habitat, special features, varibility, uses, etc.; statistics and characteristic features of the flora, remarks on phytogeography. The first installment comprises extensive remarks on local geography, physiography, vegetation, and methods of plant introduction as well as an account of botanical contributions and a list of references. See also WILLIS, J. C., 1901. *Notes on the flora of Minikoi.* pp. 39–43 (Ann. Roy. Bot. Gard. Peradeniya, 1). Peradeniya, Ceylon. [Supplementary list for this most southerly of the Laccadives; 134 species.]

042

Maldive Islands

– Vascular flora: 323 species (FOSBERG).

WILLIS, J. C. and GARDINER, J. S., 1901. *The botany of the Maldive Islands.* pp. 45–164, Pl. 2 (Ann. Roy. Bot. Gard., Peradeniya, 1). Peradeniya, Ceylon.

Systematic enumeration of native, naturalized, and commonly cultivated vascular plants, with vernacular names; detailed indication of local range, with citations; generalized summary of overall distribution, including the Laccadives, the Chagos Archipelago, and beyond; notes on habitat, ecology, uses, etc. The introductory section includes a brief general description of the geography, physical features, and climate of the archipelago, while a lengthy concluding section includes notes on vegetation of various Maldivian atolls, the origin of the flora, order of plant succession, economic products, and a discussion of floras of oceanic islands in general. In addition to the Maldives, this enumeration incorporates all known records from the Laccadives and the Chagos Archipelago. For a revised Maldives list, partly based upon a two-day visit in 1956 and accounting for 323 species, see FOSBERG, F. R., 1957. *The Maldive Islands*, *Indian Ocean.* 37 pp. (Atoll Res. Bull. 58). Washington. [Briefly annotated systematic checklist, with vernacular names, revised nomenclature, citations of the author's *exsiccatae*, and references to Willis and Gardiner; references, p. 37. An introductory general part is also included.[3]]

043

Chagos Archipelago

See also **042** (WILLIS and GARDINER); **490** (HEMSLEY). The group includes the atoll of Diego Garcia with its strategic air base.
– Vascular flora: about 150 species (FOSBERG and BULLOCK).

WILLIS, J. C. and GARDINER, J. S., 1931. Flora of the Chagos Archipelago. pp. 301–6 (Trans. Linn. Soc. London, II, Zool., 19). London.

Systematic list of native, naturalized, adventive and widely cultivated plants, with brief indication of local range and notes on habitat, ecology, taxonomic problems, etc. The introductory section includes notes on previous botanical work in the group and on the origin and general features of the flora. Supplemented, with special reference to Diego Garcia, by FOSBERG, F. R. and BULLOCK, A. A., 1971. List of Diego Garcia vascular plants. In *Geography and ecology of Diego Garcia Atoll, Chagos Archipelago* (eds. D. R. STODDART and J. D. TAYLOR), pp. 143–60 (Atoll Res. Bull. 149). Washington. [Systematic list, with localities and brief notes.]

044

New Amsterdam and St Paul Islands

The phytogeographical view most acceptable here is that these islands are cool-temperate rather than sub-antarctic. St Paul, the more southerly, lacks woody vegetation and also has been actively volcanic.

HEMSLEY, W. B., 1885 (1884). Amsterdam and St Paul Islands. In his Report on the botany of the Bermudas and various other islands of the Atlantic and Southern Oceans, part 2. In *Reports on the scientific results of the voyage of 'HMS Challenger' during the years 1873–76* (ed. J. Murray), *Botany*, vol. 1, part 2, pp. 259–81, Pls. 39, 41–5, 52. London: HMSO.

Systematic enumeration of vascular and non-vascular plants, with descriptions of new taxa; synonymy, with references and citations; details of local distribution for each island with indication of *exsiccatae*, etc.; general summary of extralimital range; critical taxonomic commentary and notes on habitat, biology, etc.; figures of new or noteworthy species. The introductory section includes accounts of physical features, botanical exploration, vegetation formations, and introduced plants together with a tabular analysis of the flora.

SCHENCK, H., 1905. Über Flora und Vegetation von St Paul und Neu-Amsterdam. In *Wissenschaftliche Ergebnisse der deutschen Tiefsee-Expedition auf dem Dampfer 'Valdivia', 1898–1899* (ed. C. CHUN), vol. 2, part 1, pp. 179–218, 14 text-figs., Pls. 11–15. Jena: Fischer.

Comprehensive illustrated account of the floristic phytogeography and the vegetation of this island group, including tabular lists of vascular and non-vascular plants with essential synonymy and distributional details. The remainder of the work deals with botanical exploration (with references), physiography, climate, vegetation formations, ecological/morphological adaptations (reflecting the work of A. F. W. Schimper, a member of this expedition), and comparisons with the flora and vegetation of the Tristan da Cunha group and Gough (037, 038). Based largely on specimens and other data obtained during the visit of the *Valdivia*. Supplemented by SCHENCK, H., 1906. Die Gefässpflanzen der deutschen Südpolar-Expedition 1901–1903 gesammelt auf der Possession-Insel (Crozet-Gruppe), Kerguelen, Heard-Insel, St Paul und Neu-Amsterdam. In *Deutsche Südpolar-Expedition 1901–1903* (ed. E. VON DRYGALSKI), vol. 8 (Botanik), pp. 97–123, 10 text-figs. Berlin: Reimer. [Includes on pp. 120–3 an annotated list of vascular plants collected on St Paul and New Amsterdam during the visit of the *Gauss*.]

045

Cocos (Keeling) Islands

For a background survey of this island group, in which the existence of groves of *Pisonia grandis* originally described by Darwin is referred to, see GIBSON-HILL, C. A., 1950. A note on the Cocos-Keeling Islands. *Bull. Raffles Mus. Singapore* **22**: 11–28.

– Vascular flora: 57 species of which 43 are native (RENVOIZE, 1979, p. 116).[4]

FORBES, H. O., 1885. List of the Keeling Atoll plants. In *idem. A naturalist's wanderings in the Eastern Archipelago*, appendix to part 1, pp. 42–3. London: Sampson Low, Marston, Searle & Rivington; New York: Harper.

Tabular checklist of vascular and nonvascular plants observed respectively by Charles Darwin (in 1836) and the author, with indication of status. For additions, see GUPPY, H. B., 1890. The dispersal of plants as illustrated by the flora of the Keeling or Cocos Islands. *J. Trans. Victoria Inst., London,* **24**: 267–301. [List of new records, pp. 272–3.]

046

Christmas Island

See also **910–30** (VAN STEENIS). Although some fresh botanical surveys have been made in recent years, no modern consolidated account is yet available.

– Vascular flora: 145 native species (BAKER *et al.*, in ANDREWS).

RIDLEY, H. N., 1906. An expedition to Christmas Island. *J. Straits Branch Roy. Asiat. Soc.* **45**: 137–271.

Systematic enumeration of vascular plants, with descriptions of new taxa and citations of earlier literature, vernacular names, general indication of local and extralimital range (with mention of some localities) as well as means of dispersal, notes on taxonomy, habitat, biology, etc., and general summary on dispersal of plants to the island; no separate index. The enumeration is preceded by accounts of botanical exploration, the author's field investigations and lists of endemic, adventive, naturalized, and cultivated plants. For additions, see RIDLEY, H. N., 1907. Christmas Island flora: additional notes. *Ibid.*, **48**: 107–8. [For a complementary but less detailed list, based on the results of an earlier expedition, see BAKER, E. G. *et al.*, 1900. Botany. In *A monograph of Christmas Island* (ed. C. W. Andrews), pp. 171–95, Pls. 17–18. London: British Museum (Natural History).]

Superregion

05–07

North Polar Regions

Generally speaking, the southern limits of the Arctic region as delimited for the purposes of this book may be considered to be equivalent to the 'tree line', except that the Aleutian and Commander Islands as well as Iceland are excluded. These limits, with the exception of Iceland, closely conform with those proposed by YOUNG (see **071**). They also conform broadly to those proposed by LÖVE and LÖVE (1975), but are wider than those of POLUNIN (1959), for both references see **050–70**). For practical reasons, Iceland has been incorporated with the rest of Scandinavia (region 67).

The history of Arctic botany is fairly lengthy, with early contributors including Linnaeus in Lapland and Bering and Steller around the Bering Strait, but until the present century exploration was fairly sporadic and localized. One of the earliest areas to become reasonably well known was Lapland, for which the first modern general flora was written in the early-nineteenth century by Wahlenburg; curiously, it has not as such been superseded. Later, much exploration was carried out in Greenland, with its first general flora appearing in 1880–4. (The latest manual, *Grønlands Flora* by Böcher *et al.*, appeared in 1966.)

A first effort at a consolidated treatment for the whole of the Arctic was *Flora arctica* by C. H. Ostenfeld, of which the first volume appeared in 1902, covering vascular plants except the dicotyledons. Nothing further, however, was published.

In the early part of this century the Arctic islands from Jan Mayen to Novaya Zemlya were explored botanically, and treatments of differing dates published. After World War I, however, serious survey of the bulk of the superregion – in Alaska, Canada, and the Soviet Union – was begun, with prominent contributors including Hultén, Polunin, Porsild, and Tolmatchev. Local accounts first began to appear in the 1930s, with the first large standard work, *Botany of the Canadian Eastern Arctic*, appearing in 1940.

Exploration has been continuing since World War II, with major contributions also coming from Wiggins and Yurtsev among others. Several standard works have been published, covering Alaska (and the Alaskan Arctic), the Canadian Arctic, and the whole of the Soviet Arctic. Some of this exploration was doubtless influenced and aided because of the increased strategic importance of the Arctic zone after the war. In both hemispheres much logistical support was provided by the development of defence bases and early-warning stations. Within the Soviet Arctic, further contributing factors were the importance of the northern shipping route and the fossil mammoth excavations.

The period from 1946 to 1975 also saw the publication of three more or less comprehensive works, widely varying in scope. These include *Circumpolar arctic flora* (Polunin, 1959), *The circumpolar plants* (Hultén, 1964–71) and *Cytotaxonomical atlas of the Arctic flora* (Löve and Löve, 1975). None of these represents a critical descriptive flora, still a desideratum. Polunin's descriptive work includes keys and good illustrations but employs a narrower definition of the Arctic and lacks any real documentation. Its species concept is, furthermore, rather wider than the admittedly narrow one of the Löves. The Löves' work is a checklist with karyotypes and indication of distribution. Hultén's work is a critical distribution atlas with extensive commentary.

The Arctic is at present comparatively well-documented internally, although *Arktičeskaja flora*, begun in 1960, has yet to be completed. The flora of the mainland Northwest Territories in Canada by Porsild has now been published. However, the depth of coverage, including local floras, varies widely, being greatest by far in Greenland and followed by Alaska, Lapland, and the Kola Peninsula.

One significant problem over which some

argument remains is the delimitation of the Arctic zone. In this Guide, the more comprehensive limits of the Löves and Young, based approximately on the 'tree-line', have been more or less followed in place of the narrower limits of Polunin, who omits most of the partly shrub-dominated 'low-Arctic' sub-zone.

Progress

The only accounts dealing with large stretches of the North Polar zone, neither less than twenty years old, are MAGUIRE, B., 1958. Highlights of botanical exploration in the New World. In *Fifty years of botany* (ed. W. C. Steere), pp. 209–46. New York (particularly 'Boreal America', pp. 210–16), and TOLMATCHEV, A. I., 1956. K izučeniju arktičeskoj flory SSR. *Bot. Zurn. SSSR*, **41**: 783–96. [In the latter article the preparation of more 'regional' floras for different parts of the Arctic is advocated.] A more recent review is represented by the papers, read to the Twelfth International Botanical Congress in Leningrad (1975), in YURTSEV, B. A. (ed.), 1978. *The Arctic floristic region*. 166 pp., illus., maps. Leningrad: 'Nauka'.

Bibliographies. Bay, 1910; Blake and Atwood, 1942; Frodin, 1964; Goodale, 1879; Holden and Wycoff, 1911–14; Hultén, 1958; Jackson, 1881; Pritzel, 1871–7; Rehder, 1911; USDA, 1958. The bibliography in LÖVE and LÖVE, *Cytotaxonomical atlas of the Arctic flora* (050–70) should also be consulted.

Regional bibliographies

ARCTIC INSTITUTE OF NORTH AMERICA, 1949–69. Arctic bibliography. Vols. 1–16. lxiv, 10, 408 pp.

HULTÉN, E., 1971. [Bibliography.] In *idem. The circumpolar plants*, II, pp. 405–46 (Kungl. Svensk. Vetenskapsakad. Handl. IV, vol. 13, part 1). Stockholm. [Geographically arranged, unannotated list of floristic literature for the entire circum-Arctic floristic and vegetational zone, with some extension southwards (emphasizing upper-montane and alpine areas). Compiled in Stockholm from sources also used for *Index Holmensis* and to be viewed as an extension of the bibliography in the same author's *The amphi-Atlantic plants* (1958; see Appendix A under General Bibliographies); however, unlike that work there is here no map showing areal coverage of the works cited.]

Indices. BA, 1926– ; BotA, 1918–26; BC, 1879–1944; BS, 1940– ; CSP, 1800–1900; EB, 1959– ; FB, 1931– ; ICSL, 1901–14; JBJ, 1874–1939; KR, 1971– ; NN, 1879–1943; RZ, 1954– .

050–70

Superregion in general

In addition to the two major circum-Arctic works accounted for below, users should be aware of HULTÉN's two-volume *The circumpolar plants* (050–70). There is also a semipopular illustrated guide of relatively restricted scope in Italian, viz. ZUCCOLI, T., 1973. *Flora artica*. 209 pp. 96 text-figs., 6 color pls. Bologna: Edagricole.

LÖVE, Á. and LÖVE, D., 1975. *Cytotaxonomical atlas of the Arctic flora*. xxiii, 598 pp., map. Vaduz, Liechtenstein: Cramer.

Detailed systematic list of vascular plants, with abbreviated indication of distribution together with chromosome numbers. Synonyms and taxa without known karyotypes are also included, so in effect this work constitutes a complete enumeration of the known Arctic flora (1629 species in 404 genera). The introductory section includes considerations of geographical limits, distribution of taxa, taxonomic concepts, botanical progress, and sources of information (including references); comprehensive bibliography (pp. 507–94) and index given at end.

POLUNIN, N., 1959. *Circumpolar arctic flora*. xxvii, 514 pp., 900 text-figs., map. Oxford: Oxford University Press.

Illustrated descriptive manual–flora of vascular plants, with keys to all families and genera (and to species in the larger genera), essential synonymy (without authorities, although the latter appear in the picture captions), generalized indication of circum-Arctic distribution, as well as regional and local occurrences, vernacular names, and some taxonomic commentary but only limited information on habitats, ecology, biology, etc.; glossary and general index to all botanical and vernacular names. The introductory section includes a definition of the limits of the Arctic zone for the purposes of this flora and a delimitation of botanical regions (with map). [The coverage of species is only some 55 per cent of that in Löve and Löve, partly explained by narrower geographical limits and a broader species concept deliberately adopted by the author.]

Distribution maps

HULTÉN, E., 1964–71. *The circumpolar plants*, I–II. 228 pp., 301 maps (Kongl. Svenska Vetenskapsakad. Handl., IV, vol. 8, part 5, vol. 13, part 1. Stockholm.

Critical two-part atlas of distribution maps of Arctic vascular plants in large format, with appropriate literature references and accompanying notes on local ranges and taxonomic commentary; essential synonymy is also given. The work is concluded in part 2 with a comprehensive, geographically arranged bibliography of floristic and related works on the Arctic zone and regions to the south. This atlas forms a valuable complement to the two other major circum-Arctic works described here, in which distribution is more sketchily indicated. Its coverage is markedly extended by the complementary work, *The amphi-Atlantic plants and their phytogeographical connection* (see **001**).

Region

05

Islands of the Arctic Ocean*

This region incorporates all the major islands and island groups of the Arctic Ocean from Jan Mayen eastward to Wrangel, thus also including Bear Island, Spitsbergen, Franz Josef Land, Novaja Zemlja, Severnaja Zemlja, and the New Siberian Islands.

Floristic coverage of these scattered island groups is relatively complete for the European islands, but is less so for those north of Asia. No separate accounts have been seen for Severnaja Zemlja. The remaining units each have separate floras, although that for Novaja Zemlja (Lynge, 1923) is now relatively incomplete. The level of botanical exploration parallels that of floristic coverage.

Bibliographies. General bibliographies as for Superregion 05–07.

Indices. General indices as for Superregion 05–07.

* For a lexicon of Russian words used in titles see Table 1 p. 348.

051

Jan Mayen Island

LID, J. 1964. *The flora of Jan Mayen*. 108 pp., 26 text-figs., 1 color pl., 64 maps (Skr. Norsk Polarinst., 130). Oslo.

Briefly descriptive illustrated flora of vascular plants, with keys to all taxa; limited synonymy; detailed accounts of local range, with citations and dot maps; general indication of overall range; notes on habitat, special features, etc., as well as some critical remarks; summary of major features of the flora and its phytogeographic relationships; lists of species present in selected sample plots; list of references and index to species. The introduction includes an account of botanical exploration together with a synopsis of the flora.

052

Bear Island (Bjornøya)

All vascular plant species known from this island are included in *Svalbards flora* (see **053** below); however, more detailed information on local range, ecology, variability, etc., may be found in RØNNING, O., 1958. *Vascular flora of Bear Island*. 62 pp., maps (Acta Borealia A, Sci, 15). Tromsø.

053

Spitsbergen (Svalbard)

RØNNING, O., 1964. *Svalbards flora*. 123 pp., 60 illus., maps (endpapers) (Polarhåndbok 1). Oslo: Norsk Polarinstitutt.

Briefly descriptive, illustrated manual–flora of native and introduced vascular plants of Spitsbergen and Bear Island, with keys to all taxa, limited synonymy, vernacular names, generalized indication of local range, and notes on habitat, ecology, special features, etc.; indices to vernacular and generic names. The introductory section includes an illustrated descriptive account of the vegetation.

054

Franz Josef Land (Zemlja Frantsa Iosifa)

See also **060** (TOLMATCHEV); **680–90** (KOMAROV *et al.*; STANKOV and TALIEV); **688** (PERFIL'EV).

HANSSEN, O. and LID, J., 1932. *Flowering plants of Franz Josef Land.* 42 pp., 5 halftones, map (Skr. Svalbard Ishavet, 39). Oslo: Norsk Polarinstitutt.

Enumeration of all known seed plants from this desolate archipelago, with limited synonymy, detailed indication of local range (including citations of *exsiccatae*), some critical remarks, and notes on habitat, ecology, special features, etc.; list of references. An introductory section gives an account of botanical exploration, while an appendix provides a summary and analysis of general features of the flora.

055

Novaja Zemlja

See also **060** (TOLMATCHEV); **680–90** (KOMAROV *et al.*; STANKOV and TALIEV); **688** (PERFIL'EV),

LYNGE, B., 1923(–24). Vascular plants from Novaya Zemlya. In *Report on the scientific results of the Norwegian expedition to Novaya Zemlya in 1921* (ed. O. HOLTEDAHL), 13: Botany. 151 pp., 47 pls. (including maps). Kristiania [Oslo]: Videnskapsselskapet i Kristiania.

Descriptive account of vascular plants, based principally on the expedition's collections but also incorporating earlier records; localities with citations of *exsiccatae*, etc., and dot maps of local range; discursive notes on habitat, life-forms, ecology, etc., as well as taxonomic commentary; photographs of vegetation and interesting plants; tabular summary of the flora and list of references; index to botanical names. The introduction includes an account of previous botanical exploration along with the itinerary of the 1921 expedition.

056

Severnaja Zemlja

See **060** (TOLMATCHEV); **710–50** (KOMAROV *et al.*). No separate general floras or lists appear to have been published.

057

New Siberian Islands (Novosibirskie Ostrova)

See also **060** (TOLMATCHEV); **710–50** (KOMAROV *et al.*); **727** (all works). The following paper appears to be the first significant separate contribution for any part of these islands, which lie in the Arctic Ocean north of Yakutia.

SAFRANOVA, I. N., 1980. K flore ostrova Kotelinyj (Novosibirskie ostrova). *Bot. Žurn. SSSR*, **65**: 544–51.

This checklist includes for each of the 89 species treated notes on habitat, chorological type, and ecology (including associates); these are preceded by a brief introduction to the area (Kotel'nyj is the largest of three main islands, and the most northerly) along with notes on previous botanical work and followed by a floristic analysis and summary (with four references).

058

Wrangel Island (Ostrov Vrangelja)

See also **060** (TOLMATCHEV); **710–50** (KOMAROV *et al.*). The island is fairly high, reaching an elevation of 1096 m.

PETROVSKIJ, V. V., 1973. Spisok sosudistykh rastenij ostrova Vrangelja. *Bot. Žurn. SSSR*, **58**: 113–26.

Systematic checklist of 312 vascular plant species confirmed as occurring on Wrangel, with brief descriptive notes on each recording habitat, occurrence, etc.; this is followed by a supplementary list of species reported in the literature as being on the island but not supported by herbarium records. A list of references is also given. [The work documents the progress of exploration: 42 species known in 1933, 185 in 1964, and about 330 in the present work. Not seen by this reviewer; recorded by Jäger in *Prog. Bot.* 35: 308, 319 (1973) and abstracted in *Biological Abstracts*, 57: 1904 (1974).] Succeeds the elaborate enumeration of NAZAROV, M. I., 1933. K flore ostrova Vrangelja. *Trudy Gosud. Okeanogr. Inst.* (Moscow), 3(4): 3–21.

Region

06

Palearctic mainland region

Within this region are incorporated the Arctic and sub-Arctic zones of Eurasia north of the 'tree line', comprising lands stretching from Lapland eastwards to the Chukotsk Peninsula. Almost the whole of the area lies within the Soviet Union, and therefore the *Arktičeskaja flora SSSR*, edited by A. I. Tolmatchev, has been listed under **060** as if it were a 'regional' work.

Apart from that large-scale work, which is yet to be completed but must be regarded as a thorough and critical contribution ranking with Hultén's works, few other floras covering significant sectors of the Eurasian Arctic are available. Those within the Soviet Union are, however, covered by the floras for the Kola peninsula, northern Russia, western Siberia, Krasnoyarsk Krai, and Yakutia, and the Anadyr-Chukotsk subregion is largely covered by Hultén's *Flora of Alaska and neighboring territories*. For Lapland, reference should similarly be made to standard Scandinavian floras; the latest separate work is *Flora lapponica* by Wahlenberg (1812; supplement, 1826)! The European sectors, notably west of the White Sea, are botanically the best-explored; local coverage in Asia is rather patchy, the central Taimyr peninsula and the Anadyr-Chukotsk subregion evidently having had the most attention. For an introduction to the flora and vegetation, see TOLMATCHEV, A. I. (ed.), 1966. *Rastenija Severa Sibiri i Dal'nego Vostoka.* 223 pp. Moscow/Leningrad: 'Nauka'. English edn.: *idem*, 1969. *Vascular plants of the Siberian north and the northern Far East*, trans. L. Phillips. 340 pp., map. Jerusalem.

Bibliographies. General bibliographies as for Superregion 05–07. Soviet bibliographies (**680**) should also be consulted.

Indices. General indices as for Superregion 05–07.

060

Region in general

See also **680–90** (KOMAROV *et al.*).

TOLMATCHEV, A. I., 1960– . *Arktičeskaja flora SSSR* [Flora arctica URSS]. Vyp. 1– . Illus., maps. Moscow/Leningrad: AN SSSR Press (later Leningrad: 'Nauka').

Detailed, critical enumeration of vascular plants of the Soviet Arctic, with descriptive keys to all taxa, full synonymy, references and citations (including illustrations), very full accounting of internal range and of distribution inside and outside the Arctic zone as a whole, individual dot maps for each species, critical taxonomic commentary, and extensive notes on habitat, ecology, biology, karyotypes, etc.; tables of distribution, lists of references, and indices to botanical names at end of each fascicle. A brief introduction to the whole work is given in volume 1. By 1981 eight fascicles had appeared, running from the pteridophytes through the Scrophulariaceae (Englerian system).

061

Lapland

See also **675, 676, 678** (all works). Lapland is here considered to encompass the treeless parts of northern Norway, Sweden, and Finland, but includes for convenience the arctalpine 'enclaves' of the Scandinavian Mountains. No modern general flora has evidently been written for Lapland; the most recent available is evidently WAHLENBERG, G., 1812. *Flora lapponica*. lxvi, 550 pp., 30 pls., 2 tables, map. Berlin: Reimer; supplemented by SOMMERFELT, S. C., 1826. *Supplementum 'Florae lapponicae'.* xii, 331 pp., 3 color pls. Christiania (Oslo): Borg. [The main work comprises an annotated descriptive account of non-vascular and vascular plants, without keys; localities and many critical remarks included. The 1826 supplement is in essentially the same format.]

062

Kola Peninsula (Arctic zone)

See **687** (Gorodkov and Pojarkova). The only markedly elevated areas center around Khibiny Gora (1191 m), for which a floristic and geobotanical account appeared some time ago: MISCHKIN, B. A., 1953. *Flora Khibinskikh Gor, es analiz i istorija.* 114 pp. Moscow/Leningrad: AN SSSR Press.

063

Northern and Northeastern Russia (Arctic zone)

See also **688** (PERFIL'EV; TOLMATCHEV). The offshore islands of Kolgu'ev and Vajgač (Vaigach) are included within this area. No overall flora of this sector is available as a separate work, but the following provides a useful partial account.

LESKOV, A. I., 1937. *Flora malozemel'skoj tundry.* 105 pp. (*Trudy Severnoj bazy AN SSSR*, 2). Moscow/Leningrad.

Comprises a systematic enumeration of 392 vascular plant species, with indication of localities and *exsiccatae* (including earlier records) and notes on habitat, taxonomy, etc. An introductory section includes sources for the work, a survey of collectors, and a list of references. [The area covered lies on the mainland south of Kolgu'ev Island and west of the northern Urals.]

064

The Ural (Arctalpine zone)

See also **604** (GOVORUCHIN). The following work provides a concise floristic checklist of the arctalpine and transitional zones.

IGOSHINA, K. N., 1969. Flora of the mountain and plain tundras and open forests of the Urals. In *Vascular plants of the Siberian north and the northern Far East* (ed. A. I. TOLMATCHEV, trans. L. Phillips), pp. 182–340 (Vegetation of the far north of the USSR and its utilization (ed. B. A. Tikhomirov), fasc. 6). Jerusalem: Israel Program for Scientific Translations. (Russian version publ. 1966, Moscow: 'Nauka', in *Rastenija Severa Sibiri i Dal'nego Vostoka* (being vyp.

6 of *Rastitel'nost' Krainego Severa SSSR i ee osvoenie*, ed. B. A. Tikhomirov.)

Analytical floristic checklist of vascular plants (769 species in 245 genera) of the semi- to non-forested arctalpine areas of the Ural, including the tundra (*goltsy*) which becomes gradually discontinuous to the south; details in brief of ecology and occurrence are given. The main list is preceded by a historical account. [The author here calls for a new comprehensive flora of the Ural; Govoruchin's manual (see **604**), good in its day, is now viewed as obsolete.]

065

Western Siberia (Arctic zone)

See **710** (KRYLOV). This area, traversed by the Ob' estuary and its branches, is topographically almost featureless.

066

Central Siberia (Arctic zone)

See also **721** (both works). This area is here understood to comprise the Arctic zone of Krasnojarsk Krai, including the Taimyr Peninsula (with elevations to 1146 m), as well as the arctalpine 'enclave' of the Putorany Plateau which rises to a maximum of 1701 m from the northern Krasnojarsk *taiga* (**723**). Apart from general works on the Arctic or on Krasnojarsk Krai, only scattered florulas or lists are available. In addition to the Taimyr flora described below, generally regarded as an Arctic botanical classic, the following work on the Putorany Plateau may also be useful: MALYSCHEV, L. I., (ed.), 1976. *Flora Putorana. Materialy k poznaniju osobennostej sostava i genezisa gornykh subarktičeskikh flor Sibiri.* 248 pp., 6 figs., 19 tables. Novosibirsk: 'Nauka'.

TOLMATCHEV, A. I., 1932–35. *Flora tsentral'noj časti Vostočnogo Tajmyra.* 3 parts (Trudy Poljarn. Komiss. AN SSSR 8, 13, 25). Leningrad.

Sections 5 and 6 of this work include a systematic checklist (194 species) of the flora of the central part of the eastern Taimyr, with synonymy, localities, critical commentary, and notes on habitat, biology, etc. Based mainly on collections made on a 1928 expedition. [Prepared with the assistance of M. E. Kirpicznikov, Leningrad.]

067

Eastern Siberia (Arctic zone or Arctic Yakutia)

See **728** for general works on Yakutia. Only scattered florulas and checklists otherwise exist covering the plants of this area, which is here taken as extending to include arctalpine 'enclaves' on the mountains of northeastern Siberia (particularly the Verkhoyansk Ranges, the Suntar-Khayat Range, and the Chersk Range, with elevations to 2389 m, 2959 m and 3147 m respectively). The collection of papers on the far north of the USSR published in 1966 under Tolmatchev's editorship (see IGOSHINA under **064**) includes two contributions from along the coast of Yakutia as well as one on the Verkhoyansk Range, but perhaps the key recent work is YURTSEV, B. A., 1968. *Flora Suntar-Khajata*. Leningrad: 'Nauka'. [Not seen by this reviewer; cited by E. E. Jäger in *Prog. Bot.* **35**: 308, 318 (1973).]

068

Anadyr and Chukotia

See also **730** (KOMAROV and KLOBUKOVA-ALISOVA; VOROŠILOV). HULTÉN'S *Flora of Alaska and neighboring territories* (**110**) is also of value, as records from this area are included in its species distribution maps. The unit as delimited here corresponds approximately to the limits of the Chukotsk National District of the Magadan Oblast'; topographically it is rather broken, with a number of discrete ranges reaching to 1843 m. No separate flora fully covering it is yet available, but many scattered lists have been published over several decades and a useful local florula by Koževnikov, described below, has lately appeared. A general survey of the vegetation over much of this area is given in VASIL'EV, V. N., 1956. *Rastitel'nost' Anadyrskogo kraja*. 218 pp., maps, tables. Moscow: AN SSSR Press.

KOŽEVNIKOV, JU. P., 1979. *Flora osnovanija Čukotskogo poluostrova*. 240 pp. Vladivostok/Magadan: 'Depot'.

[This 'basic' flora of the Chukotsk Peninsula, the most easterly part of the Soviet Union, has not been seen. It is referred to by E. E. Jäger in *Prog. Bot.* **42**: 333, 343 (1980), who considers it as contributing greatly to an improved understanding of the flora of the Soviet Eastern Arctic zone.]

Region

07

Nearctic region

This region includes the mainland Arctic and sub-Arctic zones of Alaska, Canada, and Greenland, together with the Canadian Arctic Archipelago and the Bering Sea Islands (except for the Aleutian chain).

Until the late-nineteenth century, the region, except for Greenland, was comparatively poorly explored botanically. Most of the detailed exploration has taken place since World War I, and especially in the last thirty-five years. Maguire has given a review of the work of the various collectors and the areas visited. Present botanical coverage is at a high level.

The first consolidation of information, at that time rather fragmentary, was accomplished by W. J. Hooker in his *Flora boreali-americana* (1829–40). A more meaningful basis was, however, available to Macoun who provided the next consolidation through his *Catalogue of Canadian Plants* (1883–1902), which covers both vascular and non-vascular plants and includes Alaska and Greenland. The most recent consolidation is found in *Flora of Canada* by Scoggan (1978–9).

Unlike Eurasia, however, subregional treatments are fairly plentiful; current standard works, with one exception, have all been published since World War II. More or less detailed documentary works are available for the Alaskan North Slope and the Canadian Arctic Archipelago and adjacent areas of the mainland within the Arctic, and illustrated field manuals have been written for the Archipelago and Greenland, both in two or more editions. Ungava is covered by ROUSSEAU's *Géographie floristique du Québec-Labrador* (**133**) and Alaska in general by HULTÉN's and ANDERSON's floras (**110**). A checklist for the mainland Northwest Territories by Porsild and Cody, which has been followed by a descriptive flora, is also available. However, considerable variation in local documentation exists as in Eurasia.

Bibliographies. General bibliographies as for Superregion 05–07.

Indices. General indices as for Superregion 05–07.

071

Bering Sea Islands

See also **110** (HULTÉN, 1968; WELSH). These include the islands of St Lawrence and St Matthew as well as the Pribilofs. The only recent floristic work is that by S. B. Young on St Lawrence (see below).

YOUNG, S. B., 1971. *The vascular flora of St Lawrence Island with special reference to floristic zonation in the Arctic regions.* pp. 11–115, 24 figs., incl. map (Contr. Gray Herb., 201). Cambridge, Mass.

Systematic enumeration, with synonymy, citation of *exsiccatae*, extralimital range, and notes on habitat and ecological/geographical limits of species together with critical commentary; no separate index. The introductory section provides accounts of physical features, climate, vegetation, history of contact, botanical exploration, and the Bering 'land bridge', while concluding parts deal with general floristic and biogeographic questions and considerations, including a new attempt at a delimitation of floristic limits on the southern periphery of the Arctic zone.

072

Alaska and Yukon (Arctic zone)

See also **110** (HULTÉN, 1968; WELSH).

WIGGINS, I. and THOMAS, J. H., 1962. *A flora of the Alaskan Arctic slope.* 425 pp., frontispiece, maps. Toronto: University of Toronto Press.

Briefly descriptive flora of vascular plants of that part of Alaska north of the Brooks Range, with keys to all taxa, limited synonymy, detailed accounts of local range with citations of *exsiccatae* and distribution maps for all species, critical taxonomic commentary, and notes on habitat, special features, etc.; gazetteer of localities, glossary, and index to all botanical names. The introductory section includes accounts of physical features, climate, soils, permafrost, biological factors, and botanical exploration and research in the area.

073

Northwest Territories of Canada (Arctic mainland zone)

See **074** (POLUNIN; PORSILD, 1955) and **123** (PORSILD and CODY 1968, 1980). The Arctic zone encompasses the Melville and Boothia Peninsulas (in Franklin District), the greater part of Keewatin District and the northern part of Mackenzie District (Great Bear Lake northwards), i.e., the northeastern half of the area covered by the two works of Porsild and Cody (see especially the map given in their 1980 flora).

074

Canadian Arctic Archipelago

The best field flora is the illustrated manual by Porsild (1957; 2nd edn., 1964). The other two works by Porsild and Polunin, respectively dealing with the western and eastern parts of the Archipelago with their adjacent mainland areas, are basically works of documentation, without keys or many illustrations.

PORSILD, A. E., 1957. *Illustrated flora of the Canadian Arctic Archipelago.* 209 pp., 70 text-figs., 332 maps (Bull. Natl. Mus. Canada, 146). Ottawa. (Reprinted with supplement, 1964, 1973.)

Illustrated manual of vascular plants, with keys to all taxa, essential synonymy, English vernacular names, generalized indication of formation-zones and local range with individual maps for all species (in a separate section), and notes on habitat, occurrence, frequency, ecology, phenology, etc.; glossary and index to all botanical and vernacular names. The introductory section includes accounts of physical features, climate, soils, habitats, formation-zones, vegetation, and botanical exploration. Does not include any part of the mainland. [A supplement, published in 1964, contains revisions of maps 1–332 and additional maps 333–44 as well as amended pp. 205–11 (revised index) and additional pp. 213–18 with figures 71–2. Notes on new collections are also given.]

– The work was designated as a 'Basic Flora' for the former *Flora North America* Program.

Partial works

POLUNIN, N., 1940. *Botany of the Canadian Eastern Arctic, I: Flora*. 408 pp., 8 text-figs, 2 maps (Bull. Natl. Mus. Canada, 92). Ottawa.

Systematic enumeration, without keys, of vascular plants of the eastern Canadian Arctic Archipelago as well as Ungava (extreme northern Québec) and the Melville Peninsula, with descriptions of new taxa; synonymy, with references but no citations; detailed accounts of local range, with indication of *exsiccatae*; general summaries of eastern Arctic and overall distribution; notes on habitat, ecology, biology, etc., as well as critical remarks; summary (with statistical tables) of the flora and its distribution; bibliography and index to all botanical names. The introductory section includes notes on geography and botanical exploration, together with lists of collecting localities.

PORSILD, A. E., 1955. *Vascular plants of the western Canadian Arctic Archipelago*. 266 pp., illus., maps (Bull. Natl. Mus. Canada, 135). Ottawa.

Systematic enumeration of vascular plants of the western Canadian Arctic Archipelago and adjacent mainland area, complementing Polunin's work (see above); limited synonymy, with references and citations; detailed indication of local range, with citation of *exsiccatae*, and general summary of overall distribution; taxonomic commentary and notes on habitat, ecology, biology, and phytogeographic relationships; bibliography and index to botanical names. The introductory section includes accounts of physical features, geology, climate, glaciation and the history of the flora and vegetation, the composition of the flora, phytogeographic divisions, and botanical exploration.

075

Ungava (far northern Québec)

See **074** (POLUNIN). The area is not effectively covered by any of the standard floras relating to Québec (**133**) except for that by ROUSSEAU (1974) and then only partially.

076

Greenland

BÖCHER, T. W., HOLMEN, K. and JAKOBSEN, K., 1966. *Grønlands flora*. 2nd edn. 313 pp., illus., map.

Copenhagen: Haase (1st edn., 1957). English edn.: *idem*, 1968. *The flora of Greenland*, trans. T. T. Elkington and M. C. Lewis. 312 pp., 66 text-figs, map. Copenhagen: Haase.

Illustrated manual of vascular plants, with keys to all taxa, limited synonymy, abbreviated indication of local and extralimital range, karyotypes, and notes on habitat, special features, etc.; illustrated glossary, bibliography, and index to generic and family names at end. The introductory section includes chapters on earlier published flora accounts, general floristics, and the major plant communities. For a systematic checklist, condensed from this manual, see HOLMEN, K., 1968. *Checklist of the vascular plants of Greenland*. 2nd edn. 40 pp. Copenhagen: Universitets Arktiske Station/Grønlands Botaniske Undersøgelser.

Superregion

08–09

South Polar Regions

The limits of this subdivision are as defined by GREENE and WALTON in their *Checklist* (1975; under **080–90**). It conforms to the southern 'boundary' of woody plant growth, thus incorporating all land below 45 °S (with the exclusion of the Falkland Islands, southern South America, and southern New Zealand as well as the islands in the New Zealand region from Campbell Island northwards). The Tristan da Cunha group, Gough Island, and the New Amsterdam and St Paul group, formerly also thought to be sub-Antarctic, are now properly included in the southern cool-temperate zone.

Antarctic botany, which in the guise of 'southern zone' botany has had a scope somewhat wider than as limited here, appears to have had its serious beginnings with two voyages by Bougainville, respectively with the naturalists Pernetty and Commerson; the second voyage was also a global circumnavigation which brought Commerson to the southwestern Indian Ocean (Superregion 46–49). However, in the years until 1843 British contributions became the most substantial, culminating in the collections and observations by Hooker and Lyall on the voyage of the *Erebus* and *Terror* in 1839–43 which the young J. D. Hooker used together with other available evidence for the preparation of his classic *Flora antarctica* (1844–7), still the only comprehensive descriptive account. This, however, omits some islands not yet discovered or on which a landing was impossible (the latter including the Marion group and the Crozets).

The remaining islands were explored chiefly by the *Challenger* expedition in 1873 and the German *Valdivia* and *Gauss* expeditions at the turn of the century, thus completing 'primary' documentation. Detailed exploration by temporarily resident students dates from the early part of the present century, but did not become widespread until after World War II and a revival of interest in Antarctic exploration and scientific research. In recent years, within the superregion under review, monographic accounts have been published for Macquarie (Taylor, 1955), Kerguelen (Chastain, 1958), the Marion group (Huntley, 1971),

South Georgia (Greene, 1964) and, just to the outside, the Falklands (Moore, 1968) and the Tristan da Cunha group (Wace and Dickson, 1965). A short account of the miniscule vascular flora of Antarctica itself (with the South Shetlands and South Orkneys) was given by Skottsberg (1954) and extended by Greene and Holtom (1971).

A new, but brief, consolidated account of the superregion was produced in the form of a checklist by Greene and Greene (1963; revised version, 1975).

As a whole, the vascular flora is now relatively thoroughly known; but for Heard and the Crozets there are as yet no detailed accounts and there is also a need for a critical comprehensive flora as a successor to Hooker's work. Also of interest will be the monitoring of new aliens and of other changes to the vegetation.

Progress
Exploration to 1843 is considered in detail in GODLEY, E. J., 1965. Botany of the southern zone, exploration to 1843. *Tuatara* **13**: 140–81, and in the introduction to *Flora antarctica* (Hooker, 1845). No consolidated account of subsequent work has been seen.
Bibliographies. Bay, 1910; Blake and Atwood, 1942; Frodin, 1964; Goodale, 1879; Holden and Wycoff, 1911–14; Jackson, 1881; Pritzel, 1871–7; USDA, 1958.
Indices. BA, 1926– ; BotA, 1918–26; BC, 1879–1944; BS, 1940– ; CSP, 1800–1900; EB, 1959– ; FB, 1931– ; ICSL, 1901–14; JBJ, 1873–1939; KR, 1971– ; NN, 1879–1943; RŽ, 1954– .

080–90

Superregion in general

The only modern work accounting for the entire superregion is the checklist by Greene and Walton (1975). The user should also consult HOOKER, J. D., 1844–7. *Flora antarctica*. 2 vols. xii, 544 pp., 198 pls., map (The botany of the Antarctic voyage of HM Discovery Ships *Erebus* and *Terror*, I). London: Reeve. This classic 'primary' work has been supplemented for most islands by the accounts of Hemsley (1885) and Schenck (1905, 1906).

GREENE, S. W. and WALTON, D. W. H., 1975. An annotated checklist of the sub-Antarctic and Antarctic vascular flora. *Polar Rec.* **17** (110): 473–84.

Tabular list of native vascular plants of the six

islands or island groups of the sub-Antarctic floristic zone – Macquarie, Macdonald (Heard), Kerguelen, Crozet, Prince Edward (Marion), and South Georgia – with presence or absence indicated for each unit; essential synonymy indicated in a separate table. The work also includes descriptive floristic remarks on individual islands and island groups, with notes on aliens and sources of information; a summary of records of vascular plants in the Antarctic floristic zone proper (the South Sandwich, South Orkney, and South Shetland groups together with the Antarctic Peninsula and adjacent islands); and a list of references. Succeeds GREENE, S. W. and GREENE, D. M., 1963. Checklist of the sub-Antarctic and Antarctic vascular flora. *Ibid.*, 11 (73): 411–18.

HEMSLEY, W. B., 1885 (1884). Report on the botany of the Bermudas and various other islands of the Atlantic and Southern Oceans, part 2. In *Reports on the scientific results of the voyage of HMS Challenger during the years 1873–76* (ed. J. Murray), *Botany*, vol. 1, part 2, pp. 1–281, Pls. 14–53. London: HMSO.

In this stately work are incorporated annotated enumerations of non-vascular and vascular plants of the Marion (Prince Edward) Islands (pp. 187–206, Pl. 53), the Crozet Islands (pp. 207–10), the Kerguelen Archipelago (pp. 211–43, Pl. 40), and the Macdonald (Heard) Islands (pp. 245–58, Pl. 53), with descriptions of new taxa as well as synonymy, references, and citations; also included are general indications of local and extralimital range (with citation of *exsiccatae*), extensive taxonomic commentary, and notes on habitats, special features, biology, etc. (partly compiled from earlier sources), with figures of new and noteworthy plants appended. Each formal treatment is preceded by general notes on physical geography, features of the flora, vegetation formations, and botanical exploration together with tabular analyses of phytogeographical relationships in relation to the island group concerned. The remainder of this second part deals with the South Atlantic Islands (Region 03) along with New Amsterdam and St Paul (**044**).

SCHENCK, H., 1905. Vergleichende Darstellung der Pflanzengeographie der subantarktischen Inseln insbesondere über Flora und Vegetation von Kerguelen, I–VI. In *Wissenschaftliche Ergebnisse der deutschen Tiefsee-Expedition auf dem Dampfer 'Valdivia', 1898–1899* (ed. C. CHUN), vol. 2, part 1, pp. 1–178, 33 text-figs., Pls. 1–10, map. Jena: Fischer.

Comprises separate treatments, with a strong ecological, vegetatiological and phytogeographic bias,

of the following areas: I, 'Kerguelenbezirk', encompassing the Kerguelen, Marion (Prince Edward), Crozet, and Macdonald (Heard) island groups (pp. 1–82); II, 'Sudgeorgien', including also Bouvetøya and the South Sandwich chain (pp. 82–96); III, 'Falkland-Inseln' (pp. 96–106); IV, 'Feuerland', or Tierra del Fuego (pp. 106–30); V, 'Inseln südlich von Neuseeland', including the Snares, Antipodes, and Auckland groups and Campbell Island as well as Macquarie Island (pp. 130–61); VI, 'Antarktisches Polargebiet', comprising various localities on and near the Antarctic (or Palmer) Peninsula (pp. 161–78). Each section consists of accounts of botanical exploration, physical geography, climate, general features of the flora and its 'Charakterpflanzen', and ecological adaptations as given by morphology and anatomy (reflecting the interests of A. F. W. Schimper who had been a member of this expedition), plant formations, floristics, and phytogeography; included also (in parts I, II, V, and VI) are tabular lists of vascular and non-vascular plants occurring on individual islands or island groups, with essential synonymy and distributional details (based upon all available information). The only islands within the circum-Antarctic zone actually visited by the *Valdivia* were Bouvetøya and the Kerguelen Archipelago, so that to a goodly extent this work has been compiled from other sources; it was in fact Schenck's expressed intention to provide a comprehensive survey of the floristics, phytogeography, vegetation, and ecology of the whole South Polar zone. Supplemented by SCHENCK, H., 1906. Die Gefässpflanzen der deutschen Südpolar-Expedition 1901–1903 gesammelt auf der Possession-Insel (Crozet-Gruppe), Kerguelen, Heard-Insel, St Paul und Neu-Amsterdam. In *Deutsche Südpolar-Expedition 1901–1903*, (ed. E. VON DRYGALSKI), vol. 8 (Botanik), pp. 97–123, 10 text-figs. Berlin: Reimer. [Includes annotated lists of vascular plants collected during the stay of the *Gauss* in each island group (Possession Island, Crozets, pp. 99–102; Kerguelen, pp. 102–19; Heard Island, pp. 119–20).]

Region

08

Circum-Antarctic Islands

Included under this heading are the following islands or island groups: Macquarie, Macdonald (Heard), Kerguelen Archipelago, Crozet (Possession), Marion (Prince Edward), Bouvetøya, South Georgia, and South Sandwich. These islands are distinguished from others in the 'southern oceans' by their remoteness from continents and by their absence of woody vegetation, a definition narrower than that adopted by J. D. Hooker but more in keeping with current usage. The Falkland Islands (Islas Malvinas) are relatively close to South America and may be found at **389**, while woody vegetation is present in Tristan da Cunha, Gough, and the New Amsterdam and St Paul group, classified by Good as 'sub-Antarctic'.

The various islands, although few in number and collectively limited in area, are of great biographical interest. Their main botanical features and phytogeographical affinities had become known by the 1840s, and available information was first consolidated by J. D. Hooker in his *Flora antarctica*. The Heard group was not, however, discovered until 1853 and it remained for the *Challenger* expedition and others to complete primary investigation there and on the other islands from Macquarie to the Marion group in the decade up to 1880. In the present century, and particularly since World War II, detailed biological surveys have been conducted on the majority of the island units; for some, illustrated reports have been published. Much attention was also paid to life-forms, structure and biology of the flora by Schimper and Schenck on the German *Valdivia* and *Gauss* expeditions at the turn of the century.

Bibliographies. General bibliographies as for Superregion **08–09**.

Indices. General indices as for Superregion **08–09**.

081

Macquarie Island

See also **080–90** (GREENE and WALTON; SCHENCK, 1905, pp. 130–61); **410** (ALLAN; CHEESEMAN); **415** (CHEESEMAN).

TAYLOR, B. W., 1955. *The flora, vegetation and soils of Macquarie Island*. 192 pp., 11 text-figs., 42 pls. (Rept. Austral. Natl. Antarctic Res. Exped., B, 2). Melbourne.

Includes an ecologically oriented enumeration of known vascular plants (pp. 105–56), with general accounts of local distribution and extensive notes on habitat, life-forms, phenology, floral biology, dispersal mechanisms, and associations; several species illustrated. The list is preceded by extensive chapters on physical features, climate, soils, biotic factors, vegetation formations and plant communities. No keys included.

082

Macdonald (Heard) Islands

See **080–90** (GREENE and WALTON; HEMSLEY, pp. 245–58; SCHENCK, 1905, pp. 1–82 (with Kerguelen); SCHENCK, 1906, pp. 119–20). No separate lists appear to be available. The group comprises two main islands (Heard and Macdonald) and some small islets.

083

Kerguelen Archipelago

See also **080–90** (GREENE and WALTON; HEMSLEY, pp. 211–44; SCHENCK, 1905, pp. 1–82; SCHENCK, 1906, pp. 102–19).

CHASTAIN, A., 1958. *La flore et la végétation des Îles Kerguelen*. 136 pp., 6 text-figs, 36 pls., 25 tables, 2 maps (Mém. Mus. Natl. Hist. Nat. II/B (Bot.), vol. 11, part 1). Paris.

Chapter 3 of this work incorporates an ecologically oriented enumeration of the known species, with general indication of local range and detailed notes on habitat, variability, etc.; each species illustrated. The remainder is devoted to physical features, geology, climate and wind, fauna, history of exploration, general features of the flora, phytogeography, vegetation formations and plant communities. A summary list of the flora (pp. 31–3) and a bibliography are also provided, but there are no keys.

084

Crozet (Possession) Islands

See **080–90** (GREENE and WALTON; HEMSLEY, pp. 207–10; SCHENCK, 1905, pp. 1–82 (with Kerguelen); SCHENCK, 1906, pp. 99–102). No separate work generally covering the group appears to be available.

085

Marion (Prince Edward) Islands

See also **080–90** (GREENE and WALTON; HEMSLEY, pp. 187–206; SCHENCK, 1905, pp. 1–82 (with Kerguelen)).

HUNTLEY, B. J., 1971. Vegetation. In *Marion and Prince Edward Islands* (ed. E. M. VAN ZINDEREN BAKKER, Sr., J. M. WINTERBOTTOM and R. A. DYER), pp. 98–160. Cape Town: Balkema.

Table 3 comprises a tabular checklist of the vascular flora, based on earlier records and one year's fieldwork by the author; includes indication of presence on other groups (Crozet, Kerguelen and South Georgia in particular). Twenty-two native and 13 introduced species recorded for Marion Island; 21 native and one introduced species on Prince Edward Island. Background information on the flora appears on pp. 105–7 and includes historical data; the rest of the chapter deals with vegetation and phytosociology. [Part of a monographic treatment of this island group; other chapters deal with non-vascular plants, animal life, etc.]

086

Bouvetøya (Bouvet Island)

See **080–90** (GREENE and WALTON; SCHENCK, 1905. pp. 82–96 (with South Georgia)). No vascular plants have yet been recorded from this almost wholly glaciated island, despite recent exploration (which, however, has yielded four moss species).

087

South Georgia

See also **080–90** (GREENE and WALTON; SCHENCK, 1905, pp. 82–96).

GREENE, S. W., 1964. *The vascular flora of South Georgia*. 58 pp., 8 text-figs. (incl. map), 6 pls., 31 text-maps, 1 folding map (*Scient. Rep. Brit. Antarctic Surv.*, 45). London.

Detailed descriptive large-format flora of native, naturalized, and adventive vascular plants, with keys to all taxa; synonymy, with citations only; detailed distribution given for each species with accompanying gridded maps, with (if appropriate) general indication of extralimital range; taxonomic commentary and notes on habitat, altitudinal distribution, ecology, variability, etc.; glossary and list of references at end. The introductory chapters deal with the history of the island since contact, physical features, geology, climate, botanical exploration and location of collections, vegetation formations and plant communities, status of individual species and records of those introduced, and general comments on the maps. A summary list of species appears on pp. 29–30.

088

South Sandwich Islands

See also **080–90** (GREENE and WALTON; SCHENCK, 1905, pp. 82–96 (with South Georgia)).

LONGTON, R. E. and HOLDGATE, M. W., 1979. *The South Sandwich Islands*, IV: Botany. 53 pp., 13 text-figs., 1 pl., tables (*Scient. Rep. Brit. Antarctic Surv.* 94). Cambridge.

Monographic treatment of the flora and vegetation of this volcanic island group, covering physical features, climate, geological history, exploration, plant habitats, and treatments of individual islands, concluded with a discussion of different vegetation types and the origin of the flora; references (pp. 50–1). A table of the flora and its internal (and external) distribution appears on pp. 8–11; *Deschampsia antarctica* Desv. (*D. elegantula* (Steud.) Parodi) is the only vascular species, recorded from only one of the islands (Candlemas). Some islands are barren or nearly so, and glaciation is extensive. A sub-Antarctic affinity is demonstrated on the basis of the bryophyte flora.

Region

09

Antarctica

In addition to the Antarctic mainland and the Antarctic Peninsula, the following nearby islands or island groups are included herein: South Orkney, South Shetland, Palmer Archipelago, Peter I, Balleny, and Scott. Only two vascular plants are certainly native, according to Skottsberg (1954; see below), but there is a considerable non-vascular flora in which mosses and lichens are prominent.

The first collection of a vascular plant in Antarctica was made by Eights at New South Shetland Island, near the Palmer Peninsula, in 1829.
Bibliographies. General bibliographies as for Superregion 08–09.
Indices. General indices as for Superregion 08–09.

090

Region in general

See also **080–90** (GREENE and WALTON). Since 1954, additional vascular plants have become naturalized in the region.

SKOTTSBERG, C. J. F., 1954. Antarctic flowering plants. *Svensk. Bot. Tidskr.* **51**: 330–8, illus., map.

Comprises a descriptive account, with map, of the three vascular plant species (two certainly native) then recorded from the Antarctic mainland along with the Palmer Archipelago, South Shetland Islands, and South Orkney Islands. For further information on the native species, see GREENE, D. M. and HOLTOM, A., 1971. Studies in *Colobanthus quitensis* (Kunth) Bartl. and *Deschampsia antarctica* Desv., III: Distribution, habitats and performance in the Antarctic botanical zone. *Bull. Brit. Antarctic Surv.* **26**: 1–29.

Notes

1 Examples include Lawrence, G. H. M., *Taxonomy of vascular plants* (1951); Leenhouts, P. W., *A guide to the practice of herbarium taxonomy* (1968); and Radford, A. E., et al., *Vascular plant systematics* (1974).
2 Renvoize, S. A. 1979. The origins of Indian Ocean island floras. In *Plants and islands*, ed. D. Bramwell, 107–29. London: Academic Press.
3 Fosberg has noted that the word 'atoll' is derived from Maldivian.
4 Renvoize, S. A., *op. cit.* The figures cited are from Wood-Jones, F., 1912. *Coral and atolls*. London: Reeve.

Compactness being essential, only the leading synonymy and most important references are given, and these briefly.

> A. Gray, preface to *Synoptical flora*, vol. 2, part 1 (1878)

Watson's death will make a big gap in American botany...There is no one now to go on with the flora [*Synoptical Flora*] and the possibility of our having a North American continental flora seems very remote, a not very creditable state of things for American botanists to contemplate.

> C. S. Sargent to W. T. Thistleton-Dyer, March 15, 1892; quoted from S. B. Sutton, *Charles Sprague Sargent and the Arnold Arboretum*, pp. 130–1 (1970).

A synoptical Flora of North America [on the lines of *Flora Europaea*] is both feasible and desirable at this time.

> S. G. Shetler, *Taxon*, **15**: 257 (1966).

[FNA is] a new concept of linking modern information systems technology with time-honored means of scientific research and publication to produce a flora – a species-based repository of information on plants – *as an electronic data bank and information system.*
...[In the] 6-year first phase, an intense effort will be mounted to produce the [synoptical] flora.

> Advertisement for FNA, *BioScience*, **21**: 527–8 (1971).

The Flora North America project was recently revitalized...to produce a conventional flora of the vascular plants of North America north of Mexico *using traditional methods* [emphasis mine]... It is hoped that the flora project will be completed by 1990.

> Announcement of the 'new' FNA, *Brittonia* **31**: 124 (1979).

Division

1

North America (north of Mexico)

This division encompasses the entire North American continent north of Mexico, except for the tundra zone and Greenland (works referring entirely to these latter areas are covered under Region 07). Certain floras accounted for hereinunder are referred to as 'basic Floras' according to a short list[1] adopted by the former Flora North America Program (1966–73) in connection with its aborted project for a new continental flora.

It is difficult in a few paragraphs to present an adequate summary of the background to the present network of standard regional, state, and provincial floras now blanketing most of the continent. This network assumes particular importance in the only major part of the northern temperate or Holarctic floristic zone still lacking a modern comprehensive flora, a glaring gap only moderately alleviated by the recent completion of a general flora of Canada. Indeed, no continental flora has ever been completed within the 'modern' (post-1840) era of floristic botany.

However, this absence of a comprehensive flora should perhaps be set against a long-standing historical trend, evident since the latter part of the nineteenth century: the gradual increase in specialization at local, state, or (less often) regional level, along with increasing emphasis on the solution of systematics problems of smaller scale through the use of increasingly diverse and sophisticated methodologies and subdisciplines. Added to this of course is the division of the continent into two large, geographically diverse federal states each comprising a number of unofficial 'regions' (as well as political entities) which among much else have had an effect on professional life. Indeed, in the developmental phase of the Flora North America Program, the importance of these 'regions', with their concentrations of specialists and key herbaria, was acknowledged in a provision for seven regional editorial offices in addition to the central unit at Washington.[2]

The overall evolution of floristic botany in the continent and the development of regional activities may be related not only to geography but also to personalities and has been shown to mesh well with general social and cultural developments. From the emergence of botany as a separate profession until the late 1860s – the 'Torrey and Gray epoch' of Ewan – was largely dominated by a few 'big men', comprising professionals (Torrey and especially Gray), popularizers (Eaton and Wood) and their respective associates, the latter often acting in a 'field office' capacity. In this period, all of the 'big men' were based in New England or New York; the earlier importance of Philadelphia and Charleston as botanical centers had waned by 1840. It was during this period that, partly through the work of government surveys and expeditions, there occurred the great expansion of floristic knowledge of the continent, and, with an assist from Hooker's *Flora boreali-americana* (1829–40) on British North America, that Torrey and Gray in their *Flora of North America* (1838–43), Torrey in his *Flora of the state of New-York* (1843), and Gray in his *Manual of botany of the northern United States* (1848) established the basic North American manual-flora style which since has been more or less customary.

With increasing settlement in other parts of the continent and the formation of more tertiary institutions, particularly after the Morrill Act of 1862, the botanical profession, at first largely floristically oriented but later greatly diversifying, grew considerably, and it was (not surprisingly) the botanists in the San Francisco area, notably E. L. Greene, who first seriously 'challenged' the northeastern oligarchy, having developed at least a partially independent capability for floristic work and the preparation of local floras. Similar nuclei emerged in the Midwest, initially in St Louis, and in eastern Canada at Ottawa through the work of the Macouns. After 1873 and his retirement, Gray in the 'decades of transition' returned to the ideal of a continental flora, and before his death completed two parts (corresponding to one volume) of *Synoptical flora of North America* and

began work on another, which was partly completed in two fascicles by Watson (until 1892) and Robinson by 1897 but afterwards lapsed through changing institutional interests, although the indexing work begun under Watson has carried on to the present day.

The growing decentralization of floristic botany, as well as the development of taxonomy in North America, was strongly influenced by the dominance from 1890 to 1930 by N. L. Britton and his 'school', based at New York. Indeed, it has been said that the New York Botanical Garden, founded through his efforts in 1891, was established mainly for flora production. Britton's group was responsible for several regional floras, including some in direct competition with Gray's *Manual*, which between them blanketed the United States and adjacent parts of Canada except for the Great Basin, the Southwest, and parts of Texas and Oklahoma, and a goodly number of state and local floras together with bibliographic and other tools. All of them became vehicles for the propagation of the so-called 'American Code' of nomenclature, a set of rules largely conceived by Britton who with a precise but rigid mind wanted no part of the 'sensible...Kew Rule' and other such nomenclatural subtleties. The 'American Code' was adopted by the United States government after 1900 as it was, perhaps fortuitously, *American*, but doubtless also as some members of the Britton 'school' were involved at the time in the expanding Federal scientific establishment, which with the impact of the United States Biological Survey under Merriam enabled the United States National Herbarium (from the 1890s attached to the Smithsonian Institution) to become a significant center in its own right. It was thus that the 'American Code' was used in a large number of the many floras and other contributions on western (and to a lesser extent, southern) states, as well as publications on dendrology and other topics, prepared by Federal or other botanists and published from the 1890s to the 1940s in the *Contributions from the United States National Herbarium* and other government outlets.

The use of the 'American Code' was of course also extended to Britton's ambitious but never-completed comprehensive project, *North American Flora*, launched in 1905 partly as an answer to Gray's *Synoptical Flora* but with a different methodology (involving among other things collective effort, like *Flora brasiliensis*) and with more extensive limits (it included the whole of Middle America in a gesture smacking more of botanical imperialism than common sense). In many respects, the Britton era was the great age of American descriptive floristics, as well as the one exhibiting the greatest number of herbarium starts. During this period a number of other independent 'floristic centers' came into being, notably on and near the Pacific coast, in Wyoming and Colorado, and in the Midwest and, in later years, elsewhere.

Although the Britton 'school' had become discredited by the 1920s and the various floras were provocative of much criticism, the wide collective coverage of the latter helped greatly to promote local exploration and collecting in the period of the second transportation revolution, thus laying the groundwork for the later revival of floristic work (linked in part with ecology and 'biosystematics') and the marked proliferation of state floras and related contributions from the 1930s onwards, a development continuing to the present partly under its own momentum. This promotion was also accomplished by the floras from those cadres, among them Hall and Jepson in California, remaining loyal to the International Code and broader species and generic limits characteristic of Torrey and Gray, which gained renewed influence after unification of the nomenclatural codes in the 1930s and the findings of early 'biosystematic' work. This period between the world wars also saw the further development of independent capabilities for floristic work and the writing of floras, firstly in eastern Canada and in Texas and later (mainly after World War II) in the Plains, the mountain states, and most notably in the Southeast. These have contributed greatly to a much improved knowledge of the flora, important in local terms, and to the waning of botanical 'colonialism' within North America north of Mexico.[3] The genesis of the popular modern 'documentary' state flora appears to have been in the topographically relatively level Middle West, several of which from the time of Deam's *Flora of Indiana* (1940) included county dot maps, still the most commonly used system of floristic mapping in the United States.

By the 1950s, if not before, floristic botany in the continent had become effectively decentralized. The various fields of 'biosystematics', spreading mainly from California and to many possessing a greater 'scientific respectability' with their seemingly more exact approaches to and hopes of 'solving' the 'species problem', attracted great attention in the rapidly expanding graduate schools of the day and, by and large, surpassed classical floristic work. Within the relatively diminishing floristic field, 'extralimital' involvement became comparatively limited, relating mainly to Greenland, Alaska, the Canadian North, and parts of the American West, Southwest, and South, some of which at the same time remain the least adequately known floristically and where interesting discoveries have been made, and can still be expected. Indeed, much exploration has continued over most of the continent, now strongly influenced by the needs of monographic and revisionary work, 'biosystematic' studies, and ecological and environmental research (including investigation of the status of threatened and endangered species), which comprise the three latter of Shetler's five 'states' of plant taxonomy. However, with some exceptions, the involvement in North American floristics and flora-writing by central national institutions had become comparatively limited.

Yet changes in the style and range of data presented in floras came slowly, with the evidence and conclusions from 'biosystematics' – including karyotypes, population structure and gene flow, hybridization, apomixis, and pollination and dispersal biology – having little influence until the late 1950s, and then at first mainly in the western United States. Even now, the style of many North American floras would still be clearly recognizable to Torrey and Gray. There has, however, been a greatly increased emphasis on mapping and on distribution, ecological and environmental data in some works, with a corresponding increase in bulk. Significant departures from transition have been made in British Columbia, however, where both a 'biosystematic' flora (of the Queen Charlotte Islands) and a state checklist based on the methodology of the Flora North America Program have been published. Other examples of non-traditional approaches have appeared patchily elsewhere, but in general little serious thought appears to have been given to the problem of presentation of data.[4]

It was not until the mid-1960s, that, under the stimulus of overseas developments in flora-writing such as (notably) the appearance of the first volume of *Flora Europaea* but probably also influenced by a contemporaneous reappraisal of 'biosystematics', that a mood for synthesis and 'big thinking' again appeared. This led to the initiation in 1966 of the most important and best-documented, but still chimeric, recent development in North American floristic botany: the Flora North America Project (later Program). The genesis, development, and subsequent vicissitudes of FNA have been extensively recorded elsewhere. In a period of nearly 15 years, the FNA concept underwent marked ideological and methodological oscillations, as indicated by the quotations under the divisional heading, and reflecting, it would seem, the changing socio-economic and political atmosphere of the day. The project has been involved in controversy from time to time, particularly with respect to the projected data-bank system. A revival on a more modest scale begun in 1976 has at this writing (January 1980)[5] again flickered out: the dream of Torrey and Gray nearly 150 years ago remains yet a dream – while modern comprehensive floristic documentation for most of the rest of the north temperate zone is now available or is in course of publication.

This development must be regarded with the gravest concern, and will once again, as in 1973, cause acute embarrassment. To this writer, even more so now than in the 1970s, it is symptomatic of the more fundamental problems facing North America and especially the United States. The slow continuance of *North American Flora*, reorganized in the 1950s as a serial, is no substitute; moreover, its pedantic style has been the target of some recent criticism. The existence of many state, provincial, and regional works throughout the continent likewise is no substitute for a synoptic continental flora, although their aggregate coverage is considerably improved in contrast to the gloomy picture painted by Blake and Atwood forty years ago.

Even with a higher level of coverage, it has nonetheless been asserted by Heiser that 'many areas of the United States [and Canada] have no manuals, or at least no up-to-date manuals, for the identification of their floras...'[6] In the intervening ten years, additional new and revised works have been published, but the statement remains largely true for much of the Southeast and parts of the West, the 'Old North', and Canada. Improvements have been piecemeal and for a number of states the only available general works are antiquated. The effort put into writing floristic works for Illinois, a state here considered oversupplied, would have been better directed towards one of the areas more in need, such as Kentucky. However, flora projects are in progress for a number of these areas, at regional or state level, as well as for areas presently better documented.

At regional level, current projects include the *Intermountain Flora*, two volumes of which have now been published, and *Flora of the Prairie Provinces*, of which four of the five parts planned have appeared as installments in *Phytologia*. Work is also under way on regional floras for the Southeast, Southwest, and Northern Plains within the United States and for the Northwest Territories in Canada. At an earlier date were published *Vascular plants of the Pacific Northwest*, coordinated by Hitchcock and Cronquist, as well as the final volumes of *Illustrated flora of the Pacific States* by Abrams and Ferris. All of these projects have been, or are, based at large, well-founded herbaria.

With respect to states, provinces, and territories, new general works have appeared for Arkansas, Alaska, British Columbia, Texas, Utah, Wyoming, Kansas, South Dakota, the Carolinas, and Quebec among others and tangible progress is being made on floras for New York, Virginia, Tennessee, Florida, Louisiana, and Mississippi, to name those where documentation is lacking or markedly outdated. Projects have also been initiated in New Jersey and Pennsylvania. A considerable number of state checklists has also been published, along with some substantial partial floras (e.g., south Florida, northeastern Minnesota, northern Arizona, and southern California). Notable gaps yet remain, however; even where the resources are available, professional or official interest or both may be lacking. Some areas may be considered adequately covered for most needs by the regional floras.

As for the woody flora, there is a wide range of more or less modern works now available, with trees being more fully treated than shrubs. A bibliography was published by Little and Honkala in 1976 as a successor to Dayton's list of 1952, and more recently a revised edition of the Forest Service *Check List* was produced by Little to succeed the edition of 1953. All tree species in the United States have now been mapped in some detail in the five-volume *Atlas of United States trees* (1971–8), a work also useful for adjacent parts of Canada and Mexico. However, no successor has yet appeared for the standard descriptive technical manual by Sargent, last revised in 1922.

Grasses have been covered nationally by Hitchcock and Chase (1950) and for the pteridophytes a number of regional and many state guides are available, recently listed in detail by Miasek (1977). The latter group is now covered in a new continental guide by Mickel (1979), published in the well-known 'How to Know' field guide series. Manuals on aquatic plants (see **108**) have been listed by Stuckey (1975) for the conterminous United States.

No bibliography of North American floristic works as detailed as the listing in Blake and Atwood has appeared since completion of that list in 1939, although new works have been recorded in *Index to American Botanical Literature*, the *Taxonomic Index* (1939–67), and elsewhere. State and regional floras have been selectively covered by Gunn (1956), United States only, the present author (1964), Shetler (1966), and most recently by Laywer *et al.* (in press in *Memoirs of the Torrey Botanical Club* at the time of writing). Popular and semi-popular floras were listed by Blake (1954) as well as in a nine-page leaflet, *Selected guides to the wildflowers of North America* by E. R. Shetler (1967, Washington: Department of Botany, Smithsonian Institution).

A final noteworthy recent development was the publication of the sumptuous six-volume set (with separate general index), *Wild flowers of the United States* (1966–73), produced, like many other North American floras this century, at the New York Botanical Garden. This work, edited by H. W. Rickett, is now the only 'modern' professionally prepared floristic work collectively covering the conterminous United States, but it is not, and was never intended as, a comprehensive scientific manual-flora.

Progress

For early and recent history, see particularly EWAN, J., 1969. *A short history of botany in the United States.* New York: Hafner; MAGUIRE, B., 1958. Highlights of botanical exploration in the New World. In *Fifty years of botany* (ed. W. C. Steere), pp. 209–46. New York: McGraw-Hill; SHINNERS, L. H., 1962. Evolution of the Gray's and Small's manual ranges, *Sida* 1: 1–31; and UNDERWOOD, L. M., 1907. The progress of our knowledge of the flora of North America. *Popular Sci. Monthly*, **70**: 497–517. More recent developments in general have been reviewed in RAVEN, P. H., 1974. Plant systematics, 1947–1972. *Ann. Missouri Bot. Gard.* **61**: 166–78, and in SHETLER, S. G., 1979. North America. In *Systematic botany, plant utilization and biosphere conservation* (ed. I. HEDBERG), pp. 47–54. Stockholm: Almqvist & Wiksell.

Flora North America. The following is a representative selection from the considerable literature: SHETLER, S. G., 1966. Meeting of Flora of North America Committee. *Taxon*, **15**: 255–7,* idem, 1968.

Flora North America project. *Ann. Missouri Bot. Gard.* **55**: 176–8;* *idem*, 1971. Flora North America as an information system. *BioScience* **21**: 524, 529–32;* and TAYLOR, R. L., 1971. The Flora North America project. *Ibid.*, **21**: 521–3.*

On the first suspension, see IRWIN, H. 1973. Flora North America: austerity casuality? *Ibid.*, **23**: 215;* MACBRYDE, B., 1974. Flora of North America Program suspended. *Biol. Conservation*, **6**(1): 71; and WALSH, J., 1973. Flora North America: project nipped in the bud. *Science*, **179**: 778.

For the projected revival, see REVEAL, J. L., 1979. Announcement: Vascular plants of North America north of Mexico. *Brittonia*, **31**: 124 (and other journals). [A complete bibliography and classified index to the series of *Flora North America Reports* (nos. 1–83) may be found in ROHR, B. R. *et al.*, 1977. The *Flora North America Reports – bibliography and index. Brittonia*, **29**: 419–32. Papers listed above forming part of this series are asterisked *.]

General bibliographies. Bay, 1910; Blake and Atwood, 1942: Frodin, 1964; Goodale, 1879; Holden and Wycoff, 1911–14; Hultén, 1958; Jackson, 1881; Pritzel, 1871–7; Rehder, 1911; USDA, 1958.

Divisional bibliographies

BLAKE, S. F., 1954. *Guide to popular floras of the US and Alaska.* 56 pp. (Bibliogr. Bull. USDA, 23). Washington: Government Printing Office. [Listing, by states and regions, of popular and semi-popular floristic works, including 'wildflower books'.]

GUNN, C. R., 1956. An annotated list of the state floras. *Trans. Kentucky Acad. Sci.* **17**: 114–20. [Comprises a select list of 'standard' works, arranged by states, with terse commentary; does not include regional floras.] See also *idem*, 1956. A guide to some recent state floras. *Castanea*, **21**: 33–8.

LAWYER, J. I. *et al.*, in press. *A guide to selected current literature on vascular plant floristics for the contiguous United States, Alaska, Canada, Greenland, and the US Caribbean and Pacific Islands* (*Mem. Torrey Bot. Club*, N.S.) New York. [Selected list of floras, manuals, state checklists, and works on rare, threatened, or endangered taxa, with brief annotations and cross-references; includes 'regional' works.]

MIASEK, M. A. 1977. *Regional, state, and local fern floras, manuals, checklists, and new fern and fern-ally records for the United States and adjacent Canada since*

1950. 21 pp. New York: Library, New York Botanical Garden (mimeographed).

LITTLE, E. L., Jr. and HONKALA, B. H., 1976. *Trees and shrubs of the United States: a bibliography for identification*. ii, 56 pp. (USDA Misc. Publ., 1336). Washington: Government Printing Office. [Index to selected references on the woody flora of the United States and territories, mainly emphasizing more recent works; the 470 titles are classified into several categories. Supersedes Dayton, W. A., 1952. *United States tree books: a bibliography of tree identification*. 32 pp. (Bibliogr. Bull. USDA, 20). Washington.]

SHETLER, E. R., 1966. *Floras of the United States, Canada, and Greenland: a selected bibliography with annotations*. 12 pp. Washington: Department of Botany, Smithsonian Institution (mimeographed). [Concisely annotated list of major national, regional, state, and some local works, with indication of availability and prices. Provided the nearest equivalent in the former Flora North America Program to a working list of 'standard' floras similar to that used for preparation of *Flora Europaea*.]

STUCKEY, R. L., 1975. A bibliography of manuals and checklists of aquatic vascular plants for regions and states in the conterminous United States. *Sida*, **6**: 24–9.

General indices. BA, 1926– ; BotA, 1918–26; BC, 1879–1944; BS, 1940– ; CSP, 1800–1900; EB, 1959– ; FB, 1931– ; ICSL, 1901–14; JBJ, 1873–1939; KR, 1971– ; NN, 1879–1943; RZ, 1954– .

Divisional indices

TORREY BOTANICAL CLUB, NEW YORK, 1969. *Index to American botanical literature, 1886–1966*. 4 vols. Boston: Hall. [Originally published serially in *Bulletin of the Torrey Botanical Club* as well as on index cards. In this reprint, taxonomic and other non-author entries appear in vol. 4. The *Index* has been continuing in the *Bulletin*, and the first 10-year cumulative supplement (1967–76) has recently been published (1977, Boston: 740 pp.).]

AMERICAN SOCIETY OF PLANT TAXONOMISTS, 1939–67. *The taxonomic index*. Vols. 1–30. New York (later Cambridge, Mass.: from vol. 20 (1957) published serially in *Brittonia*). (Vols. 1–9 mimeographed.) [Begun on the initiative of W. H. Camp, this index had an existence of 28 years but from the mid-1940s it was reproduced from the appropriate parts of the *Index to American Botanical Literature*. In 1957 it was consolidated with *Brittonia* although retaining separate volumation, and in 1967 discontinued.]

Conspectus

100

Division in general

Unless otherwise noted, all works described below encompass the entire North American continent north of the Mexican border. General works comprise two floras, neither complete, and one recent checklist. A large selection of works on various special groups, however, is available and is accounted for here. For works on boreal North America (Canada, Alaska, and Greenland) as a whole (Superregion 11–13), see **120–30**. For *Wild flowers of the United States*, see **140–90**.

– Vascular flora: 16274 species in 2350 genera (SHETLER and SKOG, 1978); figures subject to change through more 'definitive' checklists currently in preparation.

Comprehensive works

BRITTON, N. L. *et al.*, (eds.), 1905–57. *North American flora*. Vols. 1–34. New York: The New York Botanical Garden. [Most vols. incomplete or not published.]

ROGERSON, C. T. (ed.), 1954– . *North American flora*, ser. II. Parts 1– . New York: The New York Botanical Garden.

Briefly descriptive, rather strictly formatted flora of the vascular and non-vascular plants of North and Middle America and the West Indies, with keys to genera and species, type localities, extensive synonymy (with references and citations) and concise indication of local and regional distribution; no taxonomic commentary or ancillary remarks included. Each completed volume separately indexed. The original work, with families and higher groups arranged according to the Englerian system, was terminated by 1957 with some 99 parts in 24 volumes published; it has been replaced by a new series II, which has a similar format but is now published serially without reference to a taxonomic sequence. In addition, some revisions now omit Middle America and the West Indies (B. Maguire, personal communication). By 1979 ten parts of series II had appeared.

GRAY, A. *et al.*, 1878–97. *Synoptical flora of North America*. Vol. 1, part 1, fascicles 1–2; vol. 1, part 2; vol. 2, part 1. New York: Ivison, Blakeman, Taylor & Co. (for vol. 1, part 2 and vol. 2, part 1); American Book Co. (for vol. 1, part 1). (Vol. 1, part 2 and vol.

2, part 1 reprinted 1886, Washington: Smithsonian Institution; vol. 1, part 1 not completed.)

Concise descriptive flora of seed plants of North America north of Mexico, with synoptical keys to families, genera, and species or groups of species; includes full synonymy (with references), generalized indication of internal and extralimital range, extensive taxonomic commentary, and notes on phenology, special features, etc.; indices at end of each part. A special supplement to vol. 1, part 2, containing addenda, corrigenda, and an index, appeared in 1886. Volume 1, part 1, of which only two fascicles were published, extends from Ranunculaceae through the end of the Disciflorae (Polygalaceae) on the Bentham and Hooker system (slightly modified), thus omitting the Calyciflorae; vol. 1, part 2 from Caprifoliaceae through Compositae; and vol. 2, part 1 from Goodeniaceae through Plantaginaceae (thus reaching the end of the Gamopetalae). [As originally conceived, the work was to have had two volumes, each of some 1200 pages; volume 1 'was to go over the old ground', i.e., the contents of the old *Flora of North America* as far as completed by Torrey and Gray, and volume 2 was to deal with families not covered in that work. Only part of this plan was realized; when work lapsed towards 1900, revision of the Calyciflorae, though advertised, had not materialized and the Apetalae, Gymnospermae, Monocotyledoneae and vascular cryptogams had been ignored.]

SHETLER, S. G. and SKOG, L. E., 1978. *A provisional checklist of species for Flora North America* (revised). xix, 199 pp. (Missouri Bot. Gard. Monogr. Syst. Bot. 1/FNA Rep., 84). St Louis (Original edn., 1972, Washington, as FNA Rep., 64).

Briefly annotated, computer-generated comprehensive checklist of vascular plants of the North American continent (north of Mexico) and Greenland (16274 spp. in 2350 genera); each one-line species entry includes its accepted name, authority, coded indications of life-form and status as well as regional distribution, and a source of information. An extensive introductory section gives a good review of the FNA Program (suspended in 1973 by the 'President's Men', with subsequent attempts at revival unsuccessful as of 1981) and the genesis of the checklist together with a discussion of the data system used and other technical matters and an explanation of the various classes of information and their codes; four appendices respectively index references/sources and reviewers, present statistics of the flora by families, and provide a sample of data forms. For index to families, see pp. 197–9. Succeeds HELLER, A. A., 1900. *Catalogue of North American plants north of Mexico, exclusive of the lower cryptogams.* 2nd edn. 252 pp. Lancaster, Pa. (1st edn., 1898, Lancaster, Pa., and Minneapolis: the author). [Accounts for 16673 species and infraspecific taxa. A 3rd edn. of 1912–14 (1909–14) dealt with on pp. 13–276 *Ophioglossum* (Filicinae) to *Uva-ursi* (Ericaceae) on the Englerian system, with 15319 numbered taxa (species and varieties).] Reference should also be made to KARTESZ, J. T. and KARTESZ, R., 1980. *A synonymized checklist of the vascular flora of the United States, Canada, and Greenland,* xlviii, 498 pp. (Biota of North America, 2). Chapel Hill: University of North Carolina Press. [Useful for synonymy and for infraspecific taxa, but without geographical or other notations. The work is in some respects a latter-day successor to the unfinished *Bibliographical index to North American botany* by S. Watson (Vol. 1, *Polypetalae*; 1890, Washington, DC).]

Distribution maps

Bibliography: distribution maps

PHILLIPS, W. L. and STUCKEY, R. L., 1976. *Index to plant distribution maps in North American periodicals through 1972.* 752 pp. Boston: Hall. [29000 cards.]

LITTLE, E. L., Jr., 1971–8. *Atlas of United States trees.* Vols. 1–5. 943 maps (Misc. Publ. USDA, 1146, 1293, 1314, 1342, 1361). Washington: Government Printing Office.

This monumental series provides distribution details of all known tree species in the continental United States in some 943 maps, with the exception of the critical genus *Crataegus*; the maps also include Canadian and northern Mexican distribution where possible. For Alaska, shrubs are also included. Details of the volumes are as follows:

1. *Conifers and important hardwoods* (1971)
2. *Alaska trees and shrubs* (1975)
3. *Minor western hardwoods* (1976)
4. *Minor eastern hardwoods* (1977)
5. *Florida* (1978)

Each volume is prefaced by technical notes together with commentary on distribution patterns, conservation status, sources of information, etc., and is separately indexed. The series represents the culmination of perhaps 20 years' work and fulfills an old dream of

Charles Sprague Sargent, who himself contributed much to our knowledge of tree distribution.

Special group – woody plants (including trees)

The North American literature on woody plants (and trees), especially that of a popular and semi-popular nature, is so considerable that a fairly rigorous selection has had to be made for this *Guide*. For a more comprehensive list of references (USA only), including the many state tree books, the recent bibliography by Little and Honkala (1976; see Divisional Bibliographies) should be consulted.

LITTLE, E. L., Jr., 1979. *Checklist of United States trees (native and naturalized)*. iv, 375 pp. (Agric. Handb. USDA, 541). Washington, DC: Government Printing Office.

Detailed, annotated, alphabetically arranged list of tree species, with full synonymy (including references), citations of standard botanical treatments, earlier checklists, and the author's *Atlas of United States trees*, vernacular names, extensive indication of internal and extralimital range, critical commentary, and needs for future study; index to all vernacular names. An extensive introductory section gives details of previous checklists, remarks on nomenclature, species ranges, vernacular names, and naturalized and rare and local species as well as a statistical summary and a list of major references. Eight appendices, including a concise systematic list, are also given. The work includes all Canadian species, but not those of Hawaii or the external US territories (Puerto Rico, Guam, Samoa, etc.). Supersedes *idem*, 1953. *Check list of native and naturalized trees of the United States (including Alaska)*. 472 pp. (Agric. Handb. USDA, 41). Washington, DC.

SARGENT, C. S., 1922. *Manual of the trees of North America* (exclusive of Mexico). 2nd edn. xxvi, 910 pp., 783 text-figs., map. Boston: Houghton. (Reprinted, with corrections, 1926; this latter reprinted in 2 vols., 1961, New York, Dover. First edn., 1905.)

Copiously illustrated, amply descriptive technical 'tree book', with keys to all taxa, limited synonymy, vernacular names, fairly detailed indication of internal range, summary of extralimital distribution (where appropriate), and notes on diagnostic features, habitat, wood, bark, ornamental and other uses, etc.; remarks on hybrids; glossary and complete index. Almost all species recognized have been provided with figures by C. E. Faxon. The work is essentially a condensed and revised version of *idem*, 1891–1902. *Silva of North America*. 14 vols., 740 pls. Boston: Houghton.

VAN DERSALL, W. R., 1938. *Native woody plants of the United States: their erosion-control and wildlife values*. ii, 362 pp., 44 pls., 3 maps (Misc. Publ. USDA, 303). Washington: Government Printing Office.

Comprises a concise, alphabetically arranged enumeration of woody plants (including trees and shrubs) of the conterminous United States, with limited synonymy, vernacular and 'standardized' plant names, indication of growth-zone within the USA, and extensive notes on habitat, ecology, phenology, biology, fruit types, special features, uses in plantings, etc.; copious bibliography; lexicon of vernacular names, with botanical equivalents. The introductory section covers plant growth-zones, relationship of vegetation to soil conservation, and evaluation of plants with respect to wild life, planting out, etc.

Special groups – ferns and grasses

As with woody plants, especially trees, there is considerable popular interest in ferns and fern-allies which has led to a host of fern floras covering individual states/provinces and/or larger areas. To preserve balance, only those covering the continent or designated regions will be separately accounted for in this *Guide*. State fern floras will be mentioned in passing if there is no state flora; most recent ones are accounted for in the new bibliography by Lawyer *et al.* (1979) referenced under Divisional Bibliographies. The grass flora of Hitchcock and Chase, long a standby, has been included but no attempt has been made at a more detailed listing in that field.

BROUN, M. (ed.), 1938. *Index to North American ferns*. 217 pp. Orleans, Mass.: The author.

Annotated checklist of native, naturalized, and adventive ferns and fern allies, with full synonymy (including references), vernacular names, fairly detailed indication of distribution in the continent, and notes on habitat; statistical tables; list of authors and index to all botanical names. The nomenclature and some taxonomic concepts are by now somewhat out of date.

HITCHCOCK, A. S., 1950. *Manual of the grasses of the United States*. 2nd edn., revised A. Chase. 1051 pp. (Misc. Publ. USDA, 200). Washington: Government Printing Office. (Reprinted 1972 in 2 vols., New York: Dover. First edn., 1935.)

Illustrated descriptive manual of grasses occurring in the conterminous United States, with keys to all genera and species, limited synonymy, vernacular names, indication of internal distribution, and notes on habitat, special features, etc.; many distribution maps; complete indices at end.

– Designated as a 'Basic Flora' for the *Flora North America* Program.

MICKEL, J. T., 1979. *How to know the ferns and fern-allies.* 229 pp., illus., maps. Dubuque, Iowa: Brown.

Illustrated field guide to all species of pteridophytes in North America, with complete keys, illustrations depicting diagnostic features, and distribution maps in a manual-key format for field use; general systematic conspectus, bibliography of 54 state and regional treatments, glossary, and index at end. Vernacular names are included. An introduction discusses relationships of pteridophytes, organography, life-cycles, cytology, hybridization, spore culture, fern gardening, and preparation of herbarium specimens.

103

Alpine and upper montane zones

In contrast to certain other parts of the world, notably Europe and northern Asia, North America has as yet very few serious floras of high-mountain areas, at least in recent decades. Almost the only notable *Alpenflora* is Weber's *Rocky Mountain flora*, an outstanding work first produced in 1953 and now in its fifth edition.

Rocky Mountains

WEBER, W. A., 1976. *Rocky Mountain flora.* 5th edn. xii, 479 pp., illus. (some in color). Boulder, Colo.: Colorado Associated Universities Press. (First edn., 1953, under title *Handbook of plants of the Colorado Front Range.*)

Field manual-key to vascular plants, with synonymy, vernacular names, and notes on habitat, occurrence, biology, variation, uses, etc.; includes numerous diagnostic figures as well as a glossary, list of references, and index at the end. An introductory section includes a description of the region and its phytogeograpic features as well as an illustrated organography. [Limited to the Colorado Rockies, whose highest peak is Mt Elbert (4399 m) and most famous feature is Pike's Peak. The author remarks particularly on the affinity

of the area with the Altai Mountains in Siberia (703) and notes that works of this type are much more numerous for Eurasian mountain regions.]

105

Deserts

To date, comparatively few significant works have been written which relate specifically to the desert areas of North America, which are extensive and typologically diverse. For most, recourse must be made to state and regional floras. Only the Sonoran Desert has had special treatment; but as the greater part of this physiographic unit lies within Mexico, the important *Flora of the Sonoran Desert* by WIGGINS appears at **205**. Listed below, however, is a work on southwestern desert trees and shrubs by Benson and Darrow (1954).

Southwestern deserts – woody plants

BENSON, L. and DARROW, R. A., 1954. *The trees and shrubs of the southwestern deserts.* Revised edn. 437 pp., illus., plates, maps. Tucson: University of Arizona Press; Alberquerque: University of New Mexico Press. (First edn., 1944, as *A manual of southwestern desert trees and shrubs.*)

Briefly descriptive, illustrated treatment of woody plants (except Cactaceae), with keys to all taxa, essential synonymy, vernacular names, generalized indication of local and extralimital range (with some localities as well as distribution maps for most species), and notes on habitat, phenology, special features, properties, uses, etc.; index. The introductory section includes remarks on climate, vegetation, floristic regions, and medicinal properties, along with an illustrated organography.

108

Wetlands

Of all comparable areas in the world, North America currently enjoys the most complete coverage of aquatic and wetland plants, although no work covering the whole continent is available. The manuals of Muenscher and Prescott encompass the widest area, but are essentially limited to more or less open-water aquatic species (hydrophytes). A greater depth of coverage as well as variously more extensive documentation is employed in the regional and state manuals selected for inclusion here. No regional coverage is yet in existence for the northern and central plains and mountain United States or for Alaska and Canada, save for parts of the southern fringe of the last-named; for these areas, standard general floras should be consulted.

MUENSCHER, W. C., 1944. *Aquatic plants of the United States.* x, 374 pp., 154 figs., 400 maps, 3 tables. Ithaca, NY: Comstock (at Cornell University).

Descriptive illustrated manual of vascular plants, with keys to all taxa, vernacular names, and brief notes on habitat, frequency, distribution, etc.; descriptions of genera; dot maps depicting distribution of species by states; references to revisionary treatments; glossary and general index to all names. A brief introductory section defines the scope of coverage (hydrophytes) and discusses distribution, biology, dispersal, and seed characteristics (with list of references). The Englerian system is followed except that pteridophytes are placed last.

PRESCOTT, G. W., 1969. *How to know the aquatic plants.* viii, 171 pp., 229 illus. Dubuque, Iowa: Brown.

Illustrated field-key to aquatic macrophytes of North America, with small figures accompanied by often extensive morphological, biological, and ecological notes. [Limited to hydrophytes and 'obligate' helophytes; encompasses 165 of the 306 North American genera reported as having at least one or more marginally aquatic or marsh species. Intended primarily for student use.]

I. Northeastern and North Central United States

FASSETT, N. C., 1960. *A manual of aquatic plants.* 2nd edn., with revision appendix by E. C. Ogden. ix, 405 pp., illus. Madison: University of Wisconsin Press. (First edn., 1940, New York: McGraw-Hill.)

Manual-key to marsh and aquatic vascular and non-vascular macrophytes, with concise indication of habitat and distribution as well as essential synonymy; includes citations of significant monographs and revisions under genera and families and concludes with a complete index as well as an appendix on uses by wildlife (the latter with a special list of references). The revision appendix in the second edition (pp. 363–84) is cross-referenced to the main text. In the introductory section are directions for use of the work, key source works, and an illustrated artificial general key; for the

scope of this work, see the preface. Vernacular names are given only at generic and family levels. Covers most of the *Gray's Manual* range, but omitting Newfoundland and not accounting fully for Virginia.

II. Southeastern United States

GODFREY, R. K. and WOOTEN, J. W., 1979. *Aquatic and wetland plants of southeastern United States,* [1]: *Monocotyledons.* xii, 712 pp., 397 text-figs., frontispiece. Athens, Ga: University of Georgia Press.

Copiously illustrated manual-flora, with keys to all taxa, vernacular names, very limited synonymy, brief indication of regional distribution, some taxonomic commentary, and notes on habitat; glossary and indices to vernacular and to scientific names at end. The introductory section includes a consideration of the format adopted – essentially the North American 'manual-flora' style, and very much like that of the work by Correll and Correll (see next entry) – as well as remarks on scope of coverage (here hydrophytes and helophytes), distribution, habitats, environments and land use, and future perspectives and an artificial key to families. Covers states from North Carolina and Tennessee south to Florida and west to Arkansas and Louisiana, the same as for Small's *Manual* of 1933.

III. Southwestern United States

CORRELL, D. S. and CORRELL, H. B., 1975. *Aquatic and wetland plants of southwestern United States.* 2 vols. xv, 1777 pp., 785 text-figs., map. Stanford, Calif.: Stanford University Press. (Also published in 1 vol., 1972, Washington: Government Printing Office, for Environmental Protection Agency.)

Illustrated descriptive manual-flora of vascular plants, with keys to all taxa, occasional synonymy, vernacular names, generalized indication of distribution and habitat, and notes on variability and infraspecific forms; illustrated glossary, list of abbreviations, and index to all vernacular and scientific names at end. The introductory section includes remarks on the background and development of the work, a consideration of what aquatic plants are – here, a broad definition is taken, as in the work by Godfrey and Wooten (see above), to which this treatise acted as a 'godfather' – and discussion of habitats, special features, distribution, economics and control, significance for wildlife, and pollution problems, followed by a general key to families. Covers Arizona, New Mexico, Texas, and Oklahoma. The Washington printing is in a slightly reduced format.

IV. Pacific Coast United States

MASON, H. L. 1957. *A flora of the marshes of California*. [ix], 878 pp., 367 figs. (incl. maps). Berkeley/Los Angeles: University of California Press.

Descriptive flora of marsh and aquatic vascular plants, with keys to all taxa, synonymy and references (no literature citations), vernacular names, critical remarks, and generalized indication of distribution, status and habitat; illustrated glossary and general index to all names at end. The introductory section includes a definition of scope of coverage (hydrophytes and helophytes), and considerations of the nature of habitats (including a classification thereof), marshland geography, species distributions, floristics, and vegetation, the phenomenon of rapid change characteristic of marsh and aquatic ecosystems, and stylistic matters, followed by an illustrated general key to families; no separate bibliography included. [A rather methodical and conservative work in a somewhat generous format, but highly regarded and very influential in relation to the current generation of wetland floras.]

STEWARD, A. N., DENNIS, LaR. J. and GILKEY, H. M., 1963. *Aquatic plants of the Pacific Northwest*. 2nd edn. ix, 261 pp., 27 pls. (Studies in botany 11). Corvallis, Ore.: Oregon State University Press. (First edn., 1960.)

Illustrated manual-flora to aquatic plants of Oregon, Washington, British Columbia and Alaska, with keys to all taxa, limited synonymy, vernacular names, generalized indication of distribution within and without the area covered, and brief indication of habitat (variously within or without the keys); glossary, list of references, and complete general index to names. The introductory section is merely prefatory, and precedes the general key to families. The figures are somewhat crudely executed. [The critical and technical standards set for aquatic macrophyte floras generally by Mason's work, see above, are here not attained – C. D. K. Cook, personal communication.]

Superregion

11–13

Boreal North America

In this superregion is included that part of the North American continent and associated islands between the 'tree-line' to the north and the boundary of the conterminous United States to the south. For the polar zone (here considered also to include Greenland), see Region 07. It thus accounts for the greater part of both Alaska (Region 11) and Canada (Region 12/13).

The principal all-Canada floristic works actually account for the whole of this area as well as the contiguous polar zone, but on account of their nominal titles are listed under 120–30. These are in addition to the works covering the entire continent, described under **100**.

The history of floristic work and flora-writing is in part the same as that for the continent as a whole, but has certain distinctive features. In the general essay of Maguire (1958; see 'Progress' under the divisional heading), a separate section deals with 'Boreal North America'. At the present time, considerable areas remain only sketchily explored and documented, despite significant progress in recent decades.

Bibliographies. Any coverage is accounted for under the divisional heading.

Indices. Any coverage is accounted for under the division heading.

Region

11

Alaska

Comprises the state of Alaska, including the mainland and offshore islands, the Pribilof Islands, and the Aleutian chain. Floristic works dealing only with the Bering Strait Islands or the Arctic Slope are listed respectively under **071** and **072**.

The interesting history of earlier botanical exploration in this region, which was until 1867 Russian territory and thus was treated in Ledebour's

Flora rossica and other Russian works, has been reviewed in detail in HULTÉN, E. 1940. History of botanical exploration in Alaska and Yukon Territories from the time of their discovery to 1940. *Bot. Not.* [93]: 289–346. The modern critical basis for Alaskan taxonomy was subsequently laid down by Hultén in his *Flora of Alaska and Yukon* (1941–50). From World War II, activity has expanded greatly within the state and the large collections which have resulted formed an important additional basis for the definitive modern manual-floras of Hultén (1968) and Anderson/Welsh (1974), both of which also cover the Yukon. Partial floras are available for the Aleutians (Hultén, 1960) as well as some parts of the Polar zone and the trees and shrubs have been covered in an illustrated semi-popular work (Viereck and Little, 1972). Distribution maps have been published in Hultén's manual of 1968, as well as by LITTLE in the 1970s for the trees as part of his *Atlas of United States trees* (**100**). Knowledge of the Alaskan flora has thus become well-integrated at a high level, representing a great improvement over 1939.

Bibliographies. General and divisional bibliographies as for Division 1.

Indices. General and divisional indices as for Division 1.

110

Region in general

See also **121**/I (TAYLOR).

HULTÉN, E., 1968. *Flora of Alaska and neighbouring territories: a manual of the vascular plants.* xxii, 1008 pp., text-figs., maps. Stanford: Stanford University Press.

Illustrated, briefly descriptive, generously formatted manual-flora of vascular plants, with keys to all taxa, essential synonymy, vernacular names, distribution maps showing local and overall range for most species, taxonomic commentary, and remarks on habitat, special features, etc.; large maps in end papers; glossary, list of authors, lengthy bibliography, and indices to botanical and to vernacular names. An introductory section includes chapters on climate, geology, ecological zones, infraspecific taxa, and botanical exploration. The area covered by this work encompasses Alaska, Yukon, the Chukotsk district

(USSR), the Commander and Aleutian Islands, northwestern British Columbia, and the western part of the Mackenzie District.

– Designated as a 'Basic Flora' for the *Flora North America* Program.

The detailed technical base for this manual-flora may be found in *idem, Flora of Alaska and Yukon.* 10 parts (Acta Univ. Lund., afd. II, N.S. 37–46, [Kongl. Fysiogr. Sällsk, Handl., 52–61]). Lund, 1941–50. [Systematic enumeration, with detailed synonymy, references and citations of literature, and account of local distribution (but no maps!); taxonomic commentary. Much of this detail is omitted from the 1968 work.]

WELSH, S. L., 1974. *Anderson's Flora of Alaska and adjacent parts of Canada.* xvi, 724 pp., illus., col. frontispiece, endpaper maps. Provo, Utah: Brigham Young University Press.

Descriptive manual of vascular plants, with keys to all taxa, an abundance of good illustrations (all new), limited vernacular names, generalized indication of local and extralimital range, some taxonomic commentary, and indication of status, habitat, etc.; list of references, glossary and general index at end. No karyotypes or extensive ecological or biological notes are included. An introductory section includes a brief history of Alaskan botanical work and notes on the genesis of the book. Based on J. P. Anderson's *Flora of Alaska and adjacent parts of Canada* (1959), but so completely revised as to constitute virtually a new work. [Intended as a field manual, but the format renders it too bulky to be practical.]

Special groups – woody plants

VIERECK, L. A. and LITTLE, E. L., Jr., 1972. *Alaska trees and shrubs.* vii, 265 pp., 128 text-figs., map (Agric. Handb., USDA, 410). Washington: Government Printing Office.

Copiously descriptive and illustrated treatment, with keys to all species; vernacular names, indication of local and extralimital range, with distribution maps; some critical remarks; notes on habitat, biology, special features, properties, uses, etc.; list of references and index to all botanical and vernacular names. The introductory section includes *inter alia* accounts of vegetation formations, general features of the flora, and phytogeography. For distribution maps of species, see vol. 2 in the junior author's *Atlas of United States trees* series, listed under **100**.

111

Aleutian Islands

HULTÉN, E., 1960. *Flora of the Aleutian Islands and westernmost Alaska Peninsula, with notes on the flora of the Commander Islands.* 2nd end. 420 pp., 32 pls., maps (50 pp.). Weinheim: Cramer. (First edn. 1937, Stockholm.)

Critical systematic enumeration of vascular plants, with detailed synonymy and references, indication of local range (with distribution maps for all species), and some taxonomic commentary; bibliography and index to generic names. A comprehensive introduction, containing sections on physical features, climate, floristics, vegetation, etc., is also provided.

Region

12/13

Canada

In addition to the Dominion of Canada, this enlarged region also includes the French territory of St Pierre and Miquelon (near Newfoundland). However, the Northwest Territories as a whole (except for the southwestern half of Mackenzie District) and northernmost Québec (Ungava), as well as nearly the whole of Hudson's Bay (with James Bay), are considered to be part of the North Polar zone and accordingly appear as units 073–075. Floristic works dealing wholly or for the most part with these areas are described under those headings.

Certain regional works relating to the western and central provinces appear under **121**. In the populated areas of eastern Canada, the regional manuals of the northern and northeastern United States (region 14/15), with substantial Canadian coverage added, are widely used and few separate works in English are available.

The first major flora dealing with what is now largely within Canada is Hooker's *Flora boreali-americana* (1829–40). The subsequent 150 years of botanical exploration have been somewhat piecemeal, with early sketchy exploration succeeded by more intensive work from a few isolated centers. Federal involvement did not of course begin until after 1867, and in subsequent years the Macouns built up a reference collection in Ottawa, upon which was based their *Catalogue of Canadian plants* (1883–1902), the first major 'domestic' effort in floristic botany. During the present century, more intensive work began in a number of the provinces, notably in Québec through the efforts of Marie-Victorin, Louis-Marie and others, as well as in the Canadian north through Porsild and others, often in collaboration with visitors from south of the border and overseas. However, much of Canada remained poorly known and documented until after World War II; for identification, standard American manuals were (and still are) widely used. The first modern provincial manual is Marie-Victorin's *Flore laurentienne* (1935), partly revised by E. Rouleau and reissued in a more convenient format in 1964, but few, if any, others had been published up to 1939.

Since World War II, however, a number of provincial and territorial floras and checklists have been published. Modern provincial floras are available for the Yukon (covered in the Alaskan floras of Anderson/Welsh and Hultén), Alberta (Moss, 1959), Manitoba (Scoggan, 1957), the Prairie Provinces (Looman and Best, 1979; Boivin, 1967–79), Québec (Rousseau, 1974, not covering the whole flora and without keys), and Nova Scotia (Roland and Smith). The Gaspé Peninsula and Anticosti Island in Québec as well as the Queen Charlotte Islands in British Columbia are treated in important partial floras produced since 1950. Most other provinces and territories are covered by more or less detailed checklists of recent vintage, except Ontario, Québec and New Brunswick, where checklists are lacking, antiquated, or only partial in coverage. In thinly populated northern Ontario and northern Québec (within the tree line), no separate works of any kind are available; and in the more populous areas to the south, the English-speaking community has long relied on Gray's *Manual* in its various editions or the works from the New York Botanical Garden. One of the most interesting recent developments has been the application of a modified Flora North America Program in British Columbia, where R. L. Taylor and his associates have recently produced *Vascular plants of British Columbia: a descriptive resource inventory* (1977) as a basis for a descriptive provincial flora.

At the national level, significant recent publications include *Énumeration des plantes du Canada*

(Boivin, 1966–9) and *Flora of Canada* (Scoggan, 1978–9) which effectively supersede Macoun's *Catalogue*, but neither of these is a descriptive flora. The trees are additionally covered in a long-used illustrated work, *Native trees of Canada*, most recently revised by R. C. Hosie (1969, 1979).

Thus, it may be said that, in spite of the considerable increase in new floras and other publications since 1939, including two nation-wide accounts, knowledge of the Canadian vascular flora remains unevenly documented; this reflects the influence of the regional floras of the northern USA as well as population distribution. For a general review of the flora and its characteristics, see TAYLOR, R. L. and LUDWIG, R. A. (eds.), 1966. *The evolution of Canada's flora*. viii, 137 pp., illus., tables. Toronto: University of Toronto Press.

Bibliographies. General and divisional bibliographies as for Division 1.

Indices. General and divisional indices as for Division 1.

120–30

Region in general

In addition to the recent enumeration in French by Boivin, there is also a new, keyed, non-descriptive flora in English by Scoggan, the first of its kind for Canada as a whole. Both supplant the long-outdated catalogue of Macoun.

BOIVIN, B., 1966–9. *Énumeration des plantes du Canada*. Parts 1–7, index (Provancheria, 6). Québec: L'Université Laval [Reprinted from *Naturaliste Canad.* **93** (1966): 253–74, 371–437, 583–646, 989–1063; **94** (1967): 131–57, 471–528, 625–55. Index published separately, 54 pp., Québec, 1969.]

Bibliographic enumeration of vascular plants of Canada (together with Alaska, Greenland, and St Pierre and Miquelon), with limited synonymy, numerous citations of appropriate floristic and systematic papers, and extensive notes on regional distribution; statistical summary and lists of species limited respectively to Alaska, Greenland, and St Pierre and Miquelon; separate index to all botanical names.

SCOGGAN, H. J., 1978–79. *Flora of Canada*. 4 parts, xiii, 1711 pp. (Natl. Mus. Nat. Sci. Canada, Publ. Bot., 7). Ottawa.

Concise, non-descriptive flora of native and naturalized vascular plants (4153 species in 934 genera), with keys to all taxa (that for families in part 2), essential synonymy, critical commentary, indication of internal and overall distribution, biogeographic affinities and ecological preferences, lifeform symbols, references to distribution maps, and Dalla Torre and Harms numbers (for genera). The general index appears in part 4. Part 1 comprises a general survey of the vascular flora, with sections on floral regions, factors affecting plant distribution and distribution patterns (accompanied by biogeographic tables), life-form spectra in different zones in relation to 'bioclimates' (including comparisons with other parts of the northern hemisphere), and technical notes; glossary and bibliography at end. Supplants Macoun's *Catalogue of Canadian plants* (1883–90), now long out of date. [The lack of differentiation of type faces in this book is regrettable and wearing for the user.]

Special groups – trees

HOSIE, R. C., 1979. *Native trees of Canada*. 8th edn. 380 pp., illus., maps (some in color). Hull, Québec: Canadian Government Publishing Centre. (First edn., 1917, by B. C. Morton.)

Photographic atlas-manual with descriptive text, the latter giving details of tree form, size, botanical features, properties, uses, habitat, etc.; vernacular names and maps provided for every species. Appendices include a pictorial key to all species, etymology of tree names, historical sketches and a list of references; a complete index concludes the work. The introductory section gives an account of forest regions in Canada. See also MINISTÈRE DES FORÊTS, CANADA, 1966. *Arbres indigènes du Canada*. 289 pp. Ottawa: Imprimerie de la Reine.

121

Western subregions of Canada

Included here are certain floras and other works relating to the Pacific and Rocky Mountain zone (mainly British Columbia and the Yukon) and the 'prairie belt' (covering large portions of Alberta, Saskatchewan and Manitoba). Provincial headings should also be consulted, as well as **175** (RYDBERG), **181** (RYDBERG), and **191** (HITCHCOCK *et al.*) in the conterminous United States.

I. Pacific and Cordilleran zone

TAYLOR, T. M. C., 1970. *Pacific Northwest ferns and their allies.* ix, 247 pp., text-figs., maps. Toronto: University of Toronto Press. (First edn., 1956, Victoria, as *The ferns and fern allies of British Columbia.*)

Illustrated descriptive treatment, with keys to all taxa, synonymy (with references and many literature citations), distribution maps, taxonomic commentary, and notes on variation, habitat, special features, etc.; list of excluded species; list of karyotypes; phytogeographic summary; bibliography, glossary, and index to botanical names. The area covered extends from Oregon north to Alaska (but centers on British Columbia).

II. 'Prairie Belt'

BOIVIN, B., 1967–79. *Flora of the prairie provinces.* Parts 1–4 (Provancheria, 2–5). Québec: L'Université Laval. (Reprinted from *Phytologia* 15(1967): 121–59, 329–446; **16**(1967–8): 1–47, 219–339; **17**(1968): 58–112; **18**(1969): 281–93; **22**(1972): 315–98; **23**(1972): 1–140; **42**(1979): 1–24, 385–414; **43**(1979): 1–106, 223–51.)

Descriptive manual of vascular plants, with keys to all taxa, essential synonymy, English and French vernacular names, generalized indication of local and extralimital range, extensive taxonomic commentary, and notes on habitat, phenology, etc.; index to genera at end of each part. As of mid-1979 the work was complete except for the Gramineae, projected for part 5.

LOOMAN, J. and BEST, K. F., 1979. *Budd's Flora of the Canadian praire provinces.* 863 pp., 230 figs. (incl. halftones) (Publ. Canada Dept. Agric. Res. Br., 1662). Hull, Québec: Canadian Government Publishing Centre. (Originally pub. 1957, Ottawa, as *Wild plants of the Canadian prairies* by A. C. Budd; 2nd edn., 1964, by A. C. Budd and K. F. Best.)

Briefly descriptive manual-flora with keys to all taxa, limited synonymy, vernacular names (to family and generic level only), indication of distribution, and brief notes on frequency, status, habitat, and behavior; glossary and indices to scientific and vernacular names at end (along with a note on spelling of vernacular names). The introductory section includes a *curriculum vitae* of the original author, a note on the expansion of the work (now covering the whole of the three Prairie Provinces, like the formally more detailed enumeraion of Boivin cited above), a description of vegetation formations, pedagogical notes, and the general key to families.

122

Yukon Territory

See also **110** (all works). The following enumeration, although mainly concerned with the southeastern part of the territory, includes records of all vascular plants recorded from the Yukon and is therefore described below.

PORSILD, A. E., 1951. *Botany of southeastern Yukon adjacent to the Canol Road.* 400 pp., 39 pls., 3 text-figs. (including maps) (Bull. Natl. Mus. Canada, 121 (Biol. Ser. 41)). Ottawa.

Annotated systematic enumeration of vascular plants (894 species, subspecies, and major varieties) recorded from the Yukon Territory, with emphasis on the southeastern part as defined; includes descriptions of new or little-known taxa, essential synonymy and citations, locality records, and extensive notes on overall range, frequency, phenology, altitudinal zonation, taxonomy, etc.; bibliography and index to all botanical names. The introductory section includes accounts of geography, exploration, climate, soils and topography as well as botanical collecting, plant communities, floristics, and the origin of the flora. See also *idem*, 1966. *Contributions to the flora of southwestern Yukon Territory.* 86 pp., illus., map (*Ibid.*, 216 (Bot. Ser., 4)). Ottawa; and *idem*, 1975. *Materials for a flora of the central Yukon Territory.* xiii, 77 pp. (Natl. Mus. Nat. Sci. Canada, Publ. Bot., 4). Ottawa.

123

Northwest Territories (forested zone)

Included here are general works relating to the whole of the Northwest Territories of Canada, although more specifically this unit stands for the forested parts of Mackenzie and Keewatin Districts, southwest of the 'tree-line'. The more northerly portions of the Territories are designated as **073** (mainland) and **074** (the Arctic Archipelago); under the latter will be found the listing of Porsild's *Illustrated flora of the Canadian Arctic Archipelago*.

PORSILD, A. E. and CODY, W. J., 1980. *Vascular plants of continental Northwest Territories, Canada*. viii, 667 pp., 978 text-figs, 1155 distribution maps. Ottawa: National Museums of Canada (for the National Museum of Natural Sciences).

Copiously illustrated manual-flora of vascular plants (1113 species), with keys, limited synonymy, vernacular names (not below generic level), indication of internal and extralimital distribution (with dot maps for each species), and notes on habitat and other points (e.g., infraspecific forms); glossary and full general index at end [on green paper]. Preceding the formal treatment are *curricula vitarum*, an abstract, and an introductory section covering physical features, climate, biomes, major source works, the history of botanical exploration, and general keys to families. A general bibliography is also included, but publication or revisionary references are not given in the text. Covers mainland Northwest Territories save for the mainland portion of Franklin District. Succeeds *idem*, 1968. *Checklist of the vascular plants of continental Northwest Territories, Canada*. 102 pp., map (at end). Ottawa: Plant Research Institute, Department of Agriculture, Canada. [Systematic list, with essential synonymy, literature citations, and abbreviated indication of internal range and relative abundance; no index; list of references, pp. 6–10.]

Partial work

RAUP, H. M., 1947. *Botany of southwestern Mackenzie*. 275 pp., 16 text-figs., 37 pls. (Sargentia, 6). Jamaica Plain, Mass.

Enumeration of vascular plants, without keys; includes synonymy (with references and citations), indication of *exsiccatae* (with localities), distribution maps for most species, critical remarks, and notes on habitat, frequency, associates, special features, etc.; list of references and index to all botanical names. The introductory section gives accounts of physical features, climate, geology, soils, plant communities, agriculture, phytogeography, the origins of the flora, and botanical exploration. The area covered extends noth to 64° 30′ N, while the eastern limit follows the Marian and Great Slave Rivers (passing through Great Slave Lake).

124

British Columbia

See also **121**/I (TAYLOR), **191** (HITCHCOCK *et al.*). Lists of standard and other references for the province are given by Taylor and MacBryde (see below).
– Vascular flora: 3137 species and infraspecific entities (Taylor and MacBryde).

TAYLOR, R. L. and MACBRYDE, B. 1977. *Vascular plants of British Columbia: a descriptive resource inventory*. xxiv, 752 pp., col. map (Univ. BC Botanical Garden, Tech. Bull., 4). Vancouver: University of British Columbia Press.

Computer-generated enumeration in tabular format of vascular plants (3137 taxa), with abbreviated indications of karyotypes, habitat, distribution, reproductive biology, life-span, economics, conservation status, etc. – all shown uniformly (as far as possible) for each entry. Taxa are alphabetically arranged within the major classes, with an index to scientific and vernacular names at end. Six appendices account for literature references and standard floristic works on British Columbia, taxon-reference links, a lexicon of authorities, and a sample data form, while the colored map depicts biogeoclimatic zones in the province. An introductory section gives technical notes and a description of the computer program (the latter based on the Flora North America Program, with which both authors were associated before its suspension in 1973). The computer program is projected to be a continuing one, allowing for updating of the inventory, and is to serve as a basis for a descriptive flora (Shetler, 1977. *Syst. Bot.* **2**: 226.

Partial and special-group works

Included here are a work on the woody plants (Garman) and an outstanding island flora (Calder and Taylor).

CALDER, J. A. and TAYLOR, R. L., 1968. *Flora of the Queen Charlotte Islands*. 2 vols., illus., maps (Canada Dept. Agric. Res. Br. Monogr., 4). Ottawa: Queen's Printer.

Volume 1 comprises a detailed enumeration of vascular plants, with keys, appropriate synonymy, citations of *exsiccatae* with localities, critical remarks, and extensive notes on habitat, ecology, variation, biology, etc., accompanied by numerous figures and distribution maps; index to all botanical names at end. A fairly comprehensive introductory section is also provided. Volume 2 is a 'cytotaxonomic supplement', containing a synopsis of species with their chromosome numbers where known. [An uncommon example of a 'biosystematic' flora.]

GARMAN, E. H., 1963. *Pocket guide to the trees and shrubs of British Columbia*, 3rd edn. 137 pp. illus. (Brit. Columbia For. Serv. Spec. Publ., B-28). Victoria. (First edn. 1937.)

Pocket manual of trees and shrubs (covering the whole province), with keys to all taxa, essential synonymy, generalized indication of local range (with some details), and notes on habitat, ecology, etc.; map of forest regions; glossary and index to all vernacular and botanical names.

125

Alberta

See also 121 (BOIVIN; LOOMAN and BEST).

MOSS, E. H. 1959. *Flora of Alberta.* 546 pp. Toronto: University of Toronto Press (reprinted 1964).

Descriptive manual of vascular plants, with keys to all taxa, essential synonymy, vernacular names, generalized indication of local range, and brief notes on habitat, frequency, etc.; some taxonomic commentary; glossary and index to all vernacular and botanical names at end. Two significant lots of additions have appeared since 1959: PACKER, J. G. and DUMAIS, M., 1972. Additions to the flora of Alberta. *Canad. Field-Nat.* **86**: 259–74; and SMOLIAK, S. and JOHNSTON, A., 1978. Additions to the flora of Alberta and new records. *Ibid.* **92**: 85–9.

126

Saskatchewan

See also 121/II (BOIVIN; LOOMAN and BEST). No separate keyed descriptive flora is available.

BREITUNG, A. J., 1957. Annotated catalogue of the vascular flora of Saskatchewan. *Amer. Midl. Naturalist*, **58**: 1–72; also *idem*, 1959. Supplement. *Ibid.*, **61**: 510–12.

Concisely annotated systematic checklist, with synonymy, vernacular names, and indication of localities (with citation of some *exsiccatae* and other authorities), habitat, frequency, etc.; occasional critical remarks; discussion of doubtful and excluded species at end. An introductory section gives limits of the flora, main features of the vegetation, history of collecting,

etc., together with abbreviations. Largely replaces FRASER, W. P. and RUSSELL, R. C., 1953. *An annotated list of the plants of Saskatchewan*, revised R. C. Russell *et al.* 47 pp., illus., Saskatoon. (First edn., 1937.)

127

Manitoba

See also 121/II (BOIVIN; LOOMAN and BEST).

LOWE, C. W., 1943. *List of the flowering plants, ferns, club-mosses and liverworts of Manitoba.* 110 pp., 1 halftone, map. Winnipeg: Natural History Society of Manitoba.

Systematic census of vascular plants, with notes on local range, habitat, frequency, and phenology; limited synonymy; vernacular names; list of Bryophyta; index to family names. A folding map of vegetation zones in the province is also provided.

SCOGGAN, H. J., 1957. *Flora of Manitoba.* 619 pp. (Bull. Natl. Mus. Canada, 140). Ottawa.

Critical enumeration of vascular plants, with keys to all taxa; essential synonymy; vernacular names; detailed indication of local range, with many localities (and authorities); summary of extralimital range; critical remarks and notes on habitat, etc.; discussions of doubtful and excluded taxa; list of references and index to all botanical names. The introductory section includes accounts of physical features, climate, vegetation, and affinities of the flora as well as lists of new additions and excluded taxa and a tabular summary of the flora.

128

Northern Ontario

This area comprises that part of Ontario north of the international boundary and Lake Superior to the West and the 48th parallel to the east. No separate coverage is available, and recourse must be had to manuals or lists for surrounding areas, including those to the south. Encompassing somewhat more than half the province, it is a sparsely populated, largely wilderness region tending to act as a psychological barrier between eastern and western Canada. For other works on Ontario, see **132**.

131

Eastern subregions of Canada

No overall works are available, mainly as a result of the language division and the long dominance of the standard manuals for the northern United States and 'adjacent parts of Canada' (see **140–50**). These latter, with their usual northern limit of the 49th parallel (some also extend to Anticosti and Newfoundland), cover most populated areas. French-Canadian naturalists have in addition a number of works in French on Quebec or parts thereof.

132

Southern Ontario (and Ontario in general)

Included here along with general works on southern Ontario are references relating to the province as a whole. For northern Ontario alone, see **128**. Full coverage of southern Ontario is also provided by the standard manuals for the northeastern United States (**140–50**).

Bibliography

HODGINS, J. L., 1977. *A guide to the literature on the herbaceous vascular flora of Ontario.* 25 pp. Toronto: The author.

SOPER, J. H. 1949. *The vascular plants of southern Ontario.* vi, 95 pp. Toronto: Department of Botany, University of Toronto/Federation of Ontario Naturalists. (Mimeographed.)

Provisional checklist of vascular plants, without notes; includes limited synonymy (where differing from that in standard regional manuals), vernacular names, and an index to family and generic names. The introductory section (with a map) includes among other matters a definition of geographical limits and a list of major references relating to the Ontario flora.

133

Quebec (except Ungava)

Only one general work, not representing a complete flora, purports to cover the whole province. However, four useful partial works are available, treating respectively the southern part below 49° and west of 68° (Louis-Marie; Marie-Victorin), the eastern Anticosti-Manganie area (Marie-Victorin and Rolland-Germain), and the Bic-Gaspé Peninsula area (Scoggan). All works except Scoggan's are in French. Reference should also be made to the standard manuals in English under **140–50**. For Ungava, in the polar zone, see **075**.

ROUSSEAU, C., 1974. *Géographie floristique du Québec-Labrador. Distribution des principales espèces vasculaires.* xiii, 799 pp., 1016 area maps. Québec: Les Presses de l'Université Laval.

Atlas-flora of the more significant vascular plants of Quebec (covering just under 60 species of the total vascular flora), with up to three separate maps for each species covered along with descriptive text. Individual species accounts include notes on ecology, biology, geographical range and limits, floristic classificaton, and history, but no keys or descriptions are provided. The maps do not follow the sequence of the species accounts, but for technical reasons are grouped into three sections depending on the map limits (upper St Lawrence; southern Quebec and the Gaspe Peninsula; Quebec as a whole). The introductory section gives accounts of general geology and physiography, and bioclimatic zones as well as the philosophy and methodology of the work, and there is also an extensive bibliography (pp. 559–615) and general index (at end of work). A general chapter on floristic elements and their distribution as well as floristic history is also provided. Although it is not a complete flora, this work is the only modern treatment covering the whole province.

Partial works: southwestern Quebec

LOUIS-MARIE, Père, [1931]. *Flore-manuel de la province de Québec.* 321 pp., text-figs., 90 pls. (some in color). Montréal: Centre de Psychologie et Pédagogie.

Well-illustrated students' manual of vascular (and some cellular) plants, with keys to all taxa and short descriptions; includes limited synonymy, English and French vernacular names, brief notes on local and extralimital range, uses, pests and diseases, etc., together with indices to family, generic, and vernacular names. An introductory section includes chapters on descriptive terminology, collecting methods, nomenclature, etc., while appendices include a glossary and accounts of medicinal plants and the essential principles of ecology. Despite its title, this much-appreciated didactic work essentially covers the southwestern part of the province.[7]

MARIE-VICTORIN, Frère, 1964. *Flore laurentienne.* 2nd edn., revised by E. Rouleau. 925 pp., 324 text-figs., maps. Montreal: Les Presses de l'Université de Montréal. (First edn., 1935.)

Briefly descriptive flora of vascular plants, with keys to all taxa; essential synonymy; French and English vernacular names; generalized indication of local range, with some localities; taxonomic commentary and notes on habitat, frequency, occurrence, uses, etc.; figures of diagnostic features; glossary, list of authors, and index to all botanical and vernacular names. The rather lengthy introductory section includes an account of the history of botany in Québec together with a general survey of the flora and its phytogeography and Quaternary history. The work encompasses southern Québec up to 49° N and east to the western limits of the Gaspé Peninsula.

Partial works: other parts of Quebec

MARIE-VICTORIN, Frère and ROLLAND-GERMAIN, Frère, 1969. *Flore de l'Anticosti-Minganie.* 527 pp., 26 text-figs., portrait, map. Montréal: Les Presses de l'Université de Montréal.

Systematic enumeration of vascular plants, without keys, of Anticosti Island and the adjacent mainland to the north; includes synonymy, citation of *exsiccatae* with localities, and taxonomic commentary, with bibliography and index to botanical names at end. A lengthy introductory section includes accounts of geography, geology, botanical exploration in the area, etc., while pp. 399–498 are devoted to floristic analyses and phytogeographic considerations.

SCOGGAN, H. J., 1950. *The flora of Bic and the Gaspé Peninsula, Québec.* 399 pp. (Bull. Natl. Mus. Canada, 115). Ottawa.

Systematic enumeration of vascular plants, with keys to all taxa; essential synonymy, with some references and citations; vernacular names; localities given in some detail, together with generalized summaries of local and extralimital range; brief notes on habitat and life-form; bibliography and index to botanical names. The introductory section includes accounts of physical features, soils, life-form spectra, Quaternary and post-Quaternary history, and phytogeography, together with a tabular summary of the flora and a list of Quebec plants absent from the Peninsula.

134

Labrador

See 133 (ROUSSEAU), 135 (ROULEAU). No general floristic accounts of recent date specifically relating to the territory are available.

135

Newfoundland

See also 140–50 (BRITTON and BROWN; FERNALD).

ROULEAU, E., 1956. *A checklist of the vascular plants of the province of Newfoundland.* Pp. 41–106 (Contr. Inst. Bot. Univ. Montréal, 69). Montreal.

Systematic list of vascular plants (based to a large extent on Fernald's edition of *Gray's Manual of Botany*), with references to subspecific taxa; terse, mainly nomenclatural notes; index to families. Crosses opposite species notes respectively indicate presence in Newfoundland, Labrador, and/or St Pierre and Miquelon.

136

St Pierre and Miquelon

See also 135 (ROULEAU). These islands are French metropolitan territory, and not part of Canada.

LE GALLO, C., Père, 1954. Les plantes vasculaires des îles Saint-Pierre et Miquelon. *Naturaliste Canad.* 81: 105–32, 149–64, 181–96, 203–42.

Systematic enumeration, with indication of frequency and habitat as well as citation of *exsiccatae* and other authorities (627 species and infraspecific entities recognized). For less common or more unusual plants, some notes on uses, distribution, and special features as well as critical remarks are provided. The list proper is preceded (in the first part) by an introduction covering generalities, previous work on the flora, and floristic statistics along with technical notes.

137

Nova Scotia

See also **140–50** (all works). The flora by Roland and Smith is the only work with keys founded in the Maritime Provinces.

ROLAND, A. E. and SMITH, E. C., 1966–9. *The flora of Nova Scotia*. Revised edn. 2 parts (Proc. Nova Scotian Inst. Sci., 26, part 2, pp. 3–244; 26, part 4, pp. 277–743). Halifax. (First edn., 1949 (1944–5).)

Enumeration of vascular plants, with diagnostic keys to all taxa; references to key literature on individual plant groups; vernacular names; generalized indication of local and extralimital range, with some citations; numerous figures of characteristic features together with distribution maps; some taxonomic commentary and notes on habitat, frequency, biology, etc.; index to all botanical and vernacular names in each part. The work also accounts for all species in Prince Edward Island.

138

Prince Edward Island

See also **137** (ROLAND and SMITH); **140–50** (all works).

ERSKINE, D. S., 1960. *The plants of Prince Edward Island*. 270 pp., illus., map (Canada Dept. Agric. Res. Br. Publ., 1088). Ottawa: Queen's Printer.

Systematic enumeration of native and naturalized vascular plants, with limited synonymy and some vernacular names; detailed indication of local range, with citation of *exsiccatae* and other records; distribution maps given for all species; notes on habitat, frequency, special features, uses, etc.; index to families. The introductory section includes accounts of botanical exploration, physical features, floristics, vegetation associations, and phytogeography, together with a list of references.

139

New Brunswick

See also **140–50** (all works).

FOWLER, J., 1885. Preliminary list of the plants of New Brunswick. *Bull. Nat. Hist. Soc. New Brunswick*, 1(4): 8–84. (Reprinted separately, 82 pp., St John, N.B.: Ellis, Robertson & Co.)

Systematic enumeration of vascular plants, with limited synonymy, vernacular names, detailed indication of local range (including *exsiccatae* and other records), and notes on habitat, frequency, etc.; occasional taxonomic comments; list of adventive species on railway and other ballast; no index. An account of previous botanical work in the province appears in the introductory section.

Superregion

14–19

Conterminous United States

This superregion comprises all of the forty-eight conterminous United States of America, 'from sea to shining sea'. Alaska has been designated as region 11, and Hawaii, being in the Pacific Ocean, is here treated as region 99.

The various 'national' floras, works on special groups (woody plants, pteridophytes, grasses, etc.), checklists, and bibliographies more often than not overtly or, in effect, cover Canada, St Pierre and Miquelon, and (in some cases) Greenland as well as the conterminous United States and Alaska, and therefore, for convenience, have all been listed under **100**.

The long, expansive, many-faceted and personality-rich history of botanical exploration and floristic study in the United States is in large measure synonymous with that of the continent north of Mexico. It has therefore been treated, together with current developments and some indications of the present state of knowledge of the flora, under the divisional heading. It may suffice to note here that the largest amounts of new data are coming from the Southeast (region 16) and the mountain states (region 18).

The only general work treated under this heading is *Wild flowers of the United States* (1966–73), directed by H. W. Rickett.

Bibliographies. Any supraregional works, as well as divisional and general bibliographies, are accounted for under the divisional heading.

Indices. Any supraregional works, as well as divisional and general indices, are accounted for under the divisional heading.

140–90

Superregion in general

No purely 'national' floras covering the conterminous United States have been published since 1840, the nominal base line for this *Guide*. The great floras begun by Gray and by Britton (alas, never completed)

and the recent checklist by Shetler and Skog all cover at least the whole continent, and the same will be true for *Flora North America* should it ever see fruition. The nearest approach to a modern 'national' flora, although not scientific in the strict sense, is *Wild flowers of the United States* (see below).

RICKETT, H. W. and collaborators, 1966–73. *Wild flowers of the United States*. 6 vols. (in 14 parts), illus. (some in color). New York: McGraw-Hill.

Large-scale, lavishly illustrated, popularly oriented guide to wildflowers in the conterminous United States, divided into six volumes (each with two or three parts) representing different regions, viz: The Northeastern States; The Southeastern States; Texas; The Southwestern States; The Northwestern States; The Central Mountains and Plains. Each 'volume' includes descriptions, synoptic keys, explanations of terms, glossaries, and complete indices, but no systematic arrangement by families is followed. A complete general index is projected. [Written under professional auspices – with an editorial office at the New York Botanical Garden – and published with gusto and fanfare under prominent private patronage, this work is a fine monument to a great age of American affluence and to the botanical sensibilities of its public.]

Region

14/15

Northeastern and North Central United States

This large region comprises the twenty-two states (and one federal district) stretching from Maine to Virginia and west to Missouri, Iowa and Minnesota. Two 'subregions' may be recognized: the Northeast forest region (which floristically continues further south along the Appalachians), and the Midwest, comprising basically the lake forest region and the woodland/prairie belt. Both subregions overlap into adjacent Canada. Much of the southern fringe, however, has greater affinities with the Southern forest belt (region 16); the traditional mutual boundary between the ranges of Northern and Southern manuals is patently arbitrary, in part a result of nineteenth-century commercial pressures.

The rivalry between the two leading centers of Northeastern floristic studies became muted towards mid-century, but their respective manuals were both thoroughly revised after 1939: Gray's *Manual* by Fernald (1950) and the Britton and Brown *Illustrated Flora* by Gleason (1952), with the text of the latter revised further by Cronquist and published as *Manual of vascular plants of northeastern United States and adjacent Canada* (1963), succeeding the corresponding 1907 work by Britton. Both works are of high quality and the choice is mainly a matter of taxonomic philosophy and personal taste. A continuing demand has, however, existed for the 1913 *Illustrated Flora* and it was reissued by Dover in 1967. Many guides to the woody flora as well as 'tree books' are also available, the most elegant perhaps being Hough's *Handbook* (1907; reissued 1947) but the field guides of Harlow (1942) and Petrides (1958) among the most widely used; the former was reissued with revisions in 1957 by Dover.

At state level, progress has been very mixed. For the Midwest, the publication of the widely acclaimed *Flora of Indiana* by Deam (1940), accounted for by Blake and Atwood just before press time in a footnote, inaugurated a tradition of more or less comprehensive state floras, now available for several states: Ohio (Braun, 1960, 1967); Illinois (several works, notably *The illustrated flora of Illinois*, Mohlenbrock, 1967–); Michigan (Voss, 1972–); northeastern Minnesota (Lakela, 1965); and the well-regarded *Flora of Missouri* (Steyermark, 1963). A number of less comprehensive floras or enumerations are also available for these and other states. However, the critical revision of the Wisconsin flora, begun in 1929, has yet to be consolidated, and in Minnesota coverage is very heterogeneous, with no modern state flora but a number of partial works. In the old Northeast, few modern state floras are available; the documentary example of the Midwest has yet to be widely emulated. Notable works include those for West Viriginia (Strausbaugh and Core, 1952–64; 2nd edn., 1970–7); New York (state flora contributions by Mitchell *et al.*, 1978–), and New England (Seymour, 1969). The latter author has also written the fourth successive state flora of Vermont, and other floras, partial floras, or checklists are now available for other states. Since 1961, a number of works have appeared on Virginia, long deficient in this respect; but for New York, Maryland, Pennsylvania, New Jersey, and some New England states the available floras are mostly antiquated, in particular the 'Empire State' whose only general flora is Torrey's classic of 1843. The use of detailed distribution maps on the Midwest model is not yet widespread, except in recently initiated series in New York and Virginia, and there has been little evidence of any trend towards the production of concise 'critical' manuals or field manual-keys of the kinds familiar in Europe. Moreover, in nomenclatural terms a tendency exists for little notice to be taken of ongoing developments east of the Atlantic.

Bibliography. General and divisional bibliographies as for Division 1.

Indices. General and divisional indices as for Division 1.

140–50

Region in general

Except as indicated, all works described below cover the whole of the region as outlined. Most of these also cover mainland Canada south of the 49th parallel and the St Lawrence River, with the older Britton and Brown *Illustrated Flora* and Fernald's edition of *Gray's Manual of Botany* extending as far as Newfoundland and Anticosti Island. In addition to the regional manuals, there is a goodly number of semipopular treatments of woody plants (or simply the trees); a selection of these is described under a separate subheading following the general works.

– Vascular flora: 5523 species (Fernald).

Keys to families and genera

BATSON, W. T., 1977. *A guide to the genera of the eastern plants.* 203 pp. New York: Wiley. [Illustrated keys to families and genera, with general statements of distribution; covers an area from Key West to the Arctic and from the Atlantic to the Plains and southern Texas.]

FERNALD, M. L., 1950. *Gray's manual of botany.* 8th edn., lxiv, 1632 pp., illus. New York: American Book Co. (First edn. by A. Gray, 1848, New York, entitled *A manual of the botany of the northern United States.*)

Briefly descriptive, critical manual-flora of

vascular plants, with keys to all taxa, essential synonymy, English and French vernacular names, generalized indication of internal and extralimital range, sometimes spicy taxonomic commentary, and notes on habitat, frequency, special features, phenology, etc.; glossary, list of authorities, and indices to all botanical and vernacular names at end. An introductory section includes a detailed statistical table (5523 species, with some 80 per cent native) as well as technical and general notes. Certain 'critical' taxa are here treated in more detail than in the works by Gleason (and Cronquist), and there is a concomitant tendency for species to be more narrowly delimited.

– Designated as a 'Basic Flora' for the *Flora North America* Program.

GLEASON, H. A., (ed.) 1952. *The new Britton and Brown illustrated flora of the northeastern United States and adjacent Canada*. 3 vols. Illus. New York: The New York Botanical Garden. (Reprinted 1958 with slight revisions.)

Briefly descriptive atlas-flora of vascular plants, with keys to all taxa; limited synonymy; vernacular names; generalized indication of local and extralimital range; taxonomic commentary and notes on habitat, phenology, variation, special features, etc.; figures of individual species grouped into plates; glossary and index to all botanical and vernacular names in vol. 3. Partially supersedes the following: BRITTON, N. L. and BROWN, A., 1913. *An illustrated flora of the northern United States, Canada, and the British possessions*. 2nd edn. 3 vols. 4666 text-figs. New York: Scribners. (Reprinted 1967, Dover; 1st edn., 1898.) [Similar to Gleason's version in content but with a more satisfying arrangement of text and illustrations in which each species is provided with a discrete figure and parallel text. In spite of its obsolete 'American Code' nomenclature, this version is still considered useful in a number of quarters. Furthermore, its geographical coverage is rather wider than that of the 1952 edition, extending northeast to Newfoundland and west to the 102nd degree of longitude (i.e., the Kansas/Colorado border). The style of the work reflects Britton's precise mind and his early education in engineering, and its first edition, through the French publisher Klincksieck, directly influenced the preparation of *Flore descriptive et illustrée de la France*, (1901–6), that European classic by H. Coste,[8] and the two works together served as models for a number of similar illustrated floras in North America, Europe and eastern Asia.]

GLEASON, H. A. and CRONQUIST, A., 1963. *Manual of vascular plants of northeastern United States and adajacent Canada*. li, 810 pp. New York: Van Nostrand.

Briefly descriptive manual-flora of native, naturalized, and adventive vascular plants, with keys to all taxa; limited synonymy; vernacular names; concise indication of internal range and notes on habitat, etc.; complete general index. The introductory section includes a glossary. Based for the most part on Gleason's *New Britton and Brown illustrated flora*, of which it is in effect a condensation; however, it omits coverage of the Gaspé Peninsula in Québec as well as southern Missouri.

– Designated as a 'Basic Flora' for the now-suspended *Flora North America* Program.

Guides to woody plants, including trees

Only a selection from the abundant popular and semipopular literature is given here. Fuller coverage may be had in the various bibliographies listed at the divisional heading, especially that by LITTLE and HONKALA (1976).

BLACKBURN, B., 1952. *Trees and shrubs in eastern North America*. xv, 358 pp., illus. New York: Oxford University Press.

Illustrated manual-key to trees and shrubs, including many cultivated forms; includes notes on regional distribution, habitat, special features, etc., as well as a glossary and index.

BROWN, H. P., 1938. *Trees of northeastern United States, native and naturalized*. Revised edn. 490 pp., illus. Boston: Christopher.

Essentially a dendrological treatment, with large plates accompanied by descriptive text relating to botanical features, habitat, distribution, uses, timber properties, etc.; includes analytical keys to all species together with a systematic conspectus arranged by genera and families; index.

GRAVES, A. H., 1956. *Illustrated guide to trees and shrubs; a handbook of the woody plants of the northeastern United States and adjacent regions*. Revised edn. 271 pp., illus. New York: Harper.

Illustrated descriptive treatment, with essential synonymy, some taxonomic commentary, and notes on habitat, distribution, special features, uses, etc.; includes keys to all species as well as a glossary and index. Covers only the northeasthern states from Delaware and Pennsylvania to Maine.

HARLOW, W. M., 1942. *Trees of the eastern (and central) United States and Canada*. xiii, 288 pp., 152 text-figs., plates (some in color). New York: McGraw-Hill. (Reprinted with slight revisions, 1957, Dover Publications.)

Illustrated pocket guide to native and naturalized trees,

with much emphasis on field features, uses, folklore, etc.; includes keys to all species and indication of regional distribution; index.

HOUGH, R. B., 1907. *Handbook of the trees of the northern states and Canada east of the Rocky Mountains*. x, 470 pp., 498 text-figs. (incl. maps). Lowville, NY: The author. (Reprinted 1947, New York, Macmillan.)

Copiously illustrated descriptive treatment, with numerous fine photographs based on freshly collected material and field characters; includes keys to all species, distribution maps, and notes on habitat, biology, winter recognition features, uses, etc., as well as a glossary and index.

LI, HUI-LIN, 1972. *Trees of Pennsylvania, the Atlantic States and the Lake States*. x, 276 pp., 724 illus. Philadelphia: University of Pennsylvania Press.

Illustrated treatment of 118 species, with vernacular names, distribution, and notes on habitat, cultivation, etc.; keys for summer and winter identification; glossary and index. Each species is illustrated, with accompanying text.

PETRIDES, G. A., 1972. *A field guide to trees and shrubs*. 2nd edn. xxxii, 428 pp., illus. (some in color) (Peterson Field Guide Series, no. 11). Boston: Houghton Mifflin. (First edn., 1958.)

Pocket manual, with picture-keys to species and notes on diagnostic features, phenology, distribution, biology, etc.; includes also a winter key as well as a glossary and index.

141

New England

New England comprises the states of Maine, New Hampshire, Vermont, Massachusetts, Connecticut, and Rhode Island. Works relating to individual states are listed under separate headings following the main entry.

Seymour's manual is a welcome addition; however, a better standard of documentation would be desirable in a region with a tradition of floristic work beginning in the seventeenth century.

SEYMOUR, F. C., 1969. *The flora of New England*. xvi, 596 pp., illus., map (end papers). Rutland, Vt.: Tuttle.

Manual-key to vascular plants of the subregion, with very limited synonymy, vernacular names, fairly detailed indication of local range, and notes on phenology, special features, etc.; includes figures depicting critical and diagnostic features, a county map (endpapers), a glossary, and an index to all vernacular and botanical names. The introductory section

provides brief accounts of physical features, climate, etc., as well as addenda and corrigenda. [Reduced and printed offset from typescript, and hence somewhat wearing for the user.] The nomenclature of ferns has been revised in TRYON, A. F., 1978. New England ferns (Filicales). *Rhodora* **80**: 558–69.

Maine

BEAN, R. C., RICHARD, C. D. and HYLAND, F., 1966. *Revised check-list of the vascular plants of Maine*. [ii], 71 pp. (Bull. Josselyn Bot. Soc. Maine, 8). Orono, Me.

Tabular systematic checklist of vascular plants (2137 species and 714 additional infraspecific entities), with indication of local range (by counties) in 16 columns; index to genera and families at end. Nomenclature follows the 8th edn. (1950) of *Gray's Manual of Botany*. [Supersedes the 1st edn. (1946) by E. C. Ogden, F. H. Steinmetz, and F. Hyland.]

New Hampshire

No modern work covering the state as a whole is available. However, the biologically important northern part is thoroughly covered by the following: PEASE, A. S., 1964. *A flora of northern New Hampshire*. v, 278 pp., 2 maps. Cambridge, Mass.: New England Botanical Club. [Systematic enumeration, with vernacular names, indication of localities with *exsiccatae*, and notes on habitat, frequency, etc.; maps and index provided. There is also a substantial introductory section.]

Vermont

SEYMOUR, F. C., 1969. *The flora of Vermont*. 4th edn. ix, 393 pp., illus. (Vermont Agric. Exp. Sta. Bull., 660). Burlington. (First edn., 1900.)

Briefly descriptive manual-flora of vascular plants, with keys to all taxa, very limited synonymy, indication of county distribution and/or localities with *exsiccatae*, notes on habitat, frequency, phenology, etc.; addenda and index to family and generic names. An introductory section provides brief accounts of physical features, climate, and floristics and phytogeography.

Massachusetts

No recent general floras or enumerations for the senior state of New England are available.

Rhode Island

PALMATIER, E. A., 1952. *Flora of Rhode Island.* 75 pp. Kingston: Department of Botany, University of Rhode Island.

Systematic list of vascular plants, with cross-references to the 8th edn. of *Gray's Manual of Botany*, vernacular names, and limited synonymy; includes a list of species properly excluded from the flora of the state.

Connecticut

GRAVES, C. B. *et al.*, 1910. *Catalogue of the flowering plants and ferns of Connecticut growing without cultivation.* 569 pp. (Connecticut State Geol. Surv. Bull., 14). [Hartford.]

Annotated enumeration of vascular plants, with essential synonymy and principal literature citations, vernacular names, localities with indication of *exsiccatae*, and notes on habitat, frequency, phenology, status, biology, medicinal values, uses, etc.; lists of authorities for botanical names as well as of species excluded from the state flora; index to all botanical and vernacular names. An introductory section includes accounts of physical features, botanical exploration, and floristics. For additions and corrections, see HARGER, E. B. *et al.*, 1931. *First supplement.... Additions to the flora of Connecticut.* 94 pp. (Connecticut State Geol. Surv. Bull., 48). [Hartford]; also EAMES, E. H., 1931. Further additions to the Connecticut flora. *Rhodora*, 33: 167–70.

142

New York State

For the New York City region and Long Island, see also **143**. In addition to the annotated checklist of House (with now-obsolete nomenclature), Torrey's classical state flora of 1843, and the new series of state-wide revisions by Mitchell and others produced since 1978 from the State Museum, one useful local work, *Flora of the Cayuga Lake Basin* by Wiegand and Eames, has been included. The Mitchell series is intended to form the basis for a new state flora.

Bibliography

HOUSE, H. D., 1941–2 (1942). *Bibliography of botany of New York State, 1751–1940.* 2 parts, 233 pp. (New York State Mus. Bull., 328, 329). Albany. [Detailed bibliography, with about 3000 entries (Besterman in *Biological sciences: a bibliography of bibliographies*, 1971, Totowa, N.J.).]

HOUSE, H. D., 1924. *Annotated list of the ferns and flowering plants of New York State.* 759 pp. (New York State Mus. Bull., 254). Albany.

Systematic enumeration of known vascular plants, with full synonymy and references, vernacular names, fairly detailed indication of local ranges (often with citations), and notes on habitat, frequency, special features, etc., as well as occasional descriptive remarks; index to generic and family names at end. An introductory section gives details of previous state floras or lists. [Nomenclature follows the obsolete 'American Code'.]

MITCHELL, R. S. (ed.), 1978– . *Contributions to a flora of New York State*, I– . Illus., maps. (New York State Mus. Bull., 431, passim). Albany.

Comprises a series of illustrated descriptive revisionary accounts of individual families with county distribution maps in an atlas format, with one page for each species; text includes key synonymy, indication of phytogeographic element, habitat, habit, phenology, and overall range, notes on special features, biology and behaviour, uses, properties, etc., and critical remarks on variability, forms, hybridization, and the like, together with vernacular names and keys to all taxa. Appendices on associated fungi and insects, a list of references, and an index appear in each fascicle. The work, projected to appear serially without regard to family sequence as treatments are completed, is guided by a six-man 'Flora Committee'. At this writing (1980), two treatments, Polygonaceae and Magnoliaceae-Ceratophyllaceae have been published. [The work represents the first fruits of many years of preparation and field work by the State Botanist's office, notably by the late S. J. Smith, the writer's earliest mentor in systematic botany.]

TORREY, J., 1843. *A flora of the state of New-York.* 2 vols. 157 pls. (Natural history of New-York, 2: Botany). Albany: State of New-York.

Detailed descriptive flora of vascular plants in a large format, with synoptic keys to genera and species, vernacular names, full synonymy, with references and citations, generalized indication of local range, taxonomic commentary and notes on habitat, etymology, uses, etc. (with references to related European plants); figures of representative species; list of probable additions to the flora and statistics; complete general index in vol. 2. An introductory section in vol. 1 gives accounts of botanical exploration, floristic regions, and sources for the work, with a list of authorities. [No modern New York flora of comparable extent has taken the place of this classic, the earliest 'modern' American state flora. The young Asa Gray was among the significant contributors of information to the work. Copies occur with either plain or hand-colored plates.]

Partial work

While there are a number of florulas for various parts of the state, by far the most significant is that by Wiegand and Eames for the south-central Finger Lakes district and adjacent areas.

WIEGAND, K. and EAMES, A. J., 1926. *Flora of the Cayuga Lake basin, New York. Vascular plants.* 491 pp., map (Cornell Univ. Agric. Expt. Sta. Mem., 92). Ithaca.

Concise manual-key to vascular plants in south-central New York and adjacent northern Pennsylvania, with essential synonymy, vernacular names, local and extralimital range, critical remarks, and notes on habitat, frequency, phenology, etc.; map and index. Extensive appendices account for physical features, geological history, soils, climate, floristics and vegetation, botanical exploration, etc. [Originally designed as a students' manual.]

143

New Jersey, New York City area, and Long Island

See also **142** (HOUSE, MITCHELL, TORREY) as well as the works by Gleason and Taylor listed below for the northern part of this area, including those portions politically part of New York State. The remainder of New Jersey has for general works only the antiquated catalog by Britton and the fairly substantial partial flora by Stone.

New Jersey

Bibliography

FAIRBROTHERS, D. E., 1964. An annotated bibliography of the floristic publications of New Jersey from 1753 to 1961. *Bull. Torrey Bot. Club* **91**: 47–66; [Supplement to the annotated bibliography], 1962–65. *Ibid.*, **93**: 352–6.

BRITTON, N. L., 1889 (1890). *Catalogue of plants found in New Jersey.* pp. i, 25–642 (New Jersey, Geological Survey: Final report of the state geologist, vol. 2, part 1). Trenton: Murphy Publishing.

Systematic enumeration of vascular (on Bentham and Hooker system) and non-vascular plants, with limited synonymy, vernacular names, distributions with county ranges, and notes on habitat, frequency, etc.; some critical remarks; tabular summary of the flora (1919 species and varieties) and index to genera. An introductory section incorporates remarks on arrangement and nomenclature (in this work an early version of the later 'American Code' is utilized).

Partial work: southern New Jersey

STONE, W., 1911 (1912). The plants of southern New Jersey, with especial reference to the flora of the pine barrens and the geographic distribution of the species. *Annual Rep. New Jersey State Mus.* 1910: 23–828, figs. 1–5, pls. i–cxxix (also issued separately; reprinted 1973, Boston, Quarterman Publications).

Includes a systematic enumeration of vascular plants (pp. 213–779), with keys to all larger genera, synonymy, citation of *exsiccatae* with localities, and extensive notes on local occurrence, habitat, ecology, phenology, etc., together with gazetteer, bibliography, and index. An introductory section includes remarks on local herbaria, phytogeography, floristic districts, vegetation, etc. Covers the area from Burlington and Monmouth Counties southward.

New York City area and Long Island

GLEASON, H. A., 1962. *Plants of the vicinity of New York.* Third edn. 307 pp., 32 text-figs. New York: Hafner. (First edn. 1935.)

Semipopular manual-key to vascular plants in small format, with vernacular names and notes on occurrence, habitat, and frequency; index to family, generic, and vernacular names. An introductory account of basic organography and a glossary are also provided. Covers the environs of New York City to a radius of some 100 miles together with Long Island. Partially supersedes the following: TAYLOR, N., 1915. *Flora of the vicinity of New York.* vi, 683 pp., 9 maps (Mem. New York Bot. Gard., 5). New York.

144

Pennsylvania

In addition to the rather elderly state flora by Porter (to which there is a substantial lot of additions by Moldenke), one significant partial work, covering the western third of the state, has been included. Part of the northern fringe is additionally covered by *Flora of the Cayuga Lake Basin* (see **142**).

– Vascular flora: just over 3100 species (Wherry *et al.*).

PORTER, T. C., 1903. *Flora of Pennsylvania.* Edited with the addition of analytical keys by J. K. Small. xv, 362 pp., map. Boston: Ginn.

Systematic enumeration of seed plants, with keys to all taxa but no synonymy; vernacular names; detailed indication of local range (by counties) and general summary of extralimital distribution; brief notes on habitat and miscellaneous features; summary of the flora and indices to vernacular, generic, and family names. For substantial additions, see MOLDENKE, H. N., 1946. A contribution to the wild and cultivated flora of Pennsylvania. *Amer. Midl. Naturalist,* **35**: 289–399. [Annotated catalogue of non-vascular and vascular plants, based on over 4300 collections.]

WHERRY, E. T., FOGG, J. M., Jr. and WAHL, H. A., 1979. *Atlas of the flora of Pennsylvania.* xxx, 390 pp. (maps), frontispiece. Philadelphia: Morris Arboretum, University of Pennsylvania.

Comprises an atlas of distribution maps (eight on a page) showing exact localities for species records (and not, as in many other North American works, by counties). The base map of the state used for each taxon indicates broad geological zones. In the introductory section are given considerations of geology in relation to plant distribution, plants of unusual habitats, general biogeography, endemic plants, abbreviations, synonymy (tables, pp. xvii–xxviii), and list of families. The Englerian system is followed. [Prepared in part from sample pages sent by the New York Botanical Garden Library.]

Partial work

JENNINGS, O. E., 1953. *Wild flowers of Western Pennsylvania and the upper Ohio basin.* With watercolors by A. Avinoff. 2 vols. lxxv, 574 pp., 200 col. pls. Pittsburgh: University of Pittsburgh Press.

[Volume 1 of this large quarto work comprises a comprehensive descriptive flora of seed plants, with keys to all taxa, concise synonymy, vernacular names, rather detailed indication of local range (with many maps), summary of extralimital distribution, and notes on habitat, ecology, phenology, special features, etc.; some critical remarks; index to all botanical and vernacular names. The extensive introductory section includes a bibliography, gazetteer, and glossary together with accounts of botanical exploration, physical features, soils, geology, floristics, and phytogeography. Volume 2 constitutes an atlas of 200 watercolors.]

145

Delaware (and the Eastern Shore)

See also **146** and **147** for additional coverage of those parts of Maryland and Virginia east of Chesapeake Bay, which, together with the state of Delaware, make up the so-called 'Delmarva' peninsula (corresponding to the limits of this unit).

TATNALL, R. R., 1946. *Flora of Delaware and the Eastern Shore; an annotated list.* xxvi, 313 pp., 9 pls., map. Wilmington: Society of Natural History of Delaware.

Briefly annotated, concise enumeration of vascular plants, with essential synonymy, vernacular names, more or less detailed indication of local range, some critical remarks, and notes on habitat, frequency, phenology, special features, etc.; tabular summary of the flora, bibliography, and index to all botanical and vernacular names. The introductory section includes *inter alia* accounts of botanical exploration, floristics, and phytogeographic regions.

146

Maryland (and District of Columbia)

For the 'Eastern Shore' of Maryland, see also **145** (TATNALL); for the Baltimore area, see also the District of Columbia subheading below.

SHREVE, F. *et al.*, 1910. *The plant life of Maryland.* 533 pp., 15 text-figs., 39 pls. (Maryland Weather Serv. Spec. Publ., 3.) Baltimore: Johns Hopkins University Press. (Reprinted 1969, Lehre, Cramer.)

Part VII of this work (pp. 381–497) comprises a systematic enumeration of vascular plants, with very limited synonymy, vernacular names, more or less detailed indication of local range, and notes on habitat, frequency, etc.; no separate index (see general index to work, pp. 507–33). The remaining parts (I–VI) deal with physical features, floristics, ecology, agriculture, forestry, etc. For an updated checklist, see NORTON, J. B. S. and BROWN, R. G., 1946. A catalog of the vascular plants of Maryland. *Castanea* 11: 1–50. [Includes citations of representative *exsiccatae* and literature records as well as essential synonymy; list of references.]

Partial works: District of Columbia

HITCHCOCK, A. S. and STANDLEY, P. C., 1919. *Flora of the District of Columbia and vicinity*. 329 pp., 42 pls. (Contr. US Natl. Herb., 21). Washington: Smithsonian Institution.

Manual-key to vascular plants, with essential synonymy, vernacular names, generalized indication of local and extralimital range, and notes on habitat, frequency, phenology, special features, etc.; statistics of flora, glossary, and index to generic, family and vernacular names. An introductory section provides background information on the region, its botanical features, and previous investigations. Its coverage extends for a radius of 15 miles from the Capitol (an area now to a large extent built-up). Partly supplanted by HERMANN, F. J., 1946. *A checklist of plants in the Washington-Baltimore area*. 2nd edn. 134 pp. Washington: Conference on District Flora. (First edn., 1941.)

147

Virginia

For the 'Eastern Shore', see also **145** (TATNALL). Arlington County and adjacent areas are also covered by works on the District of Columbia (**146**). The recently appearing state floras and atlas are the first since the eighteenth century.

HARVILL, A. M., Jr. 1970. *Spring flora of Virginia*. xxx, 240 pp., illus., map (inside front cover). Parsons, W.Va.: McClain Printing Co.

Illustrated, briefly descriptive students' manual of seed plants flowering before 1 June, with keys to all taxa, vernacular names, distribution (by county or physiographic province), and commentary on habitat, occurrence, phenology, etc.; figures of representative species; map; index to generic and vernacular names.

An introductory section covers botanical history, plant classification, use of keys, etc., and includes a glossary (botanical terms also illustrated on inside back cover).

HARVILL, A. M., Jr. *et al.*, 1977. *Atlas of the Virginia flora*, 1: *Pteridophytes through monocotyledons*. iv, 59 pp., maps. Farmville, Va.: Virginia Botanical Associates.

Comprises county dot maps for each species, without annotations (but with inclusion for each family of census information); no index (the Englerian sequence has, however, been followed). In the introductory section (pp. 1–17) are accounts of the history of botanical exploration in the Old Dominion as well as its physical features, climate, geology, soils, vegetation formations and plant communities, and phytogeography and vegetation history (with a separate list of references). [The remainder of this work is still in preparation, D. M. Porter, personal communication.]

MASSEY, A. B., 1961. *Virginia flora*. 258 pp. (Virginia Agric. Exp. Sta. Tech. Bull., 155). Blacksburg.

Concise systematic enumeration of vascular plants, with essential synonymy, abbreviated indication of local range (by counties and physiographic regions), and notation of habitat and frequency; bibliography and index to family and generic names. The introductory section includes a descriptive account of Virginian physiography (with its consequences for the flora) and a list of counties (with index). To a large extent, this work has been based on the 8th edition of *Gray's Manual of Botany*. It is only in recent years that serious resident work has been carried out (or, better to say, resumed) on the flora of the 'Old Dominion'. For additions, see JOHNSON, M. F., 1970. Additions to the flora of Virginia. *Castanea*, 35: 144–9; and UTTAL, L. J. and MITCHELL, R. S., 1970. Amendments to the flora of Virginia, I. *Ibid.*, 35: 293–301. [Both include county localities for specimens studied.]

148

West Virginia

In contrast to Virginia, West Virginia has had, at least since the 1860s, a succession of general state floras, of which that by Strausbaugh and Core is now well-established as a standard over much of the east-central United States.

– Vascular flora: 2200 species (Strausbaugh and Core).

Bibliography

CORE, E. L. GILLESPIE, W. H. and GILLESPIE, B. J., 1962. *Bibliography of West Virginia plant life.* 46 pp. New York: Scholar's Library. [Encompasses 900 entries (Besterman, in *Biological sciences: a bibliography of bibliographies*, 1971, Totowa, N.J.)]

STRAUSBAUGH, P. D. and CORE, E. L., 1978. *Flora of West Virginia.* 2nd edn., revised E. L. Core. xl, 1079 pp., illus. Grantsville, W.Va.: Seneca Books. (Originally published in 4 parts., 1970–7, Morgantown, W.Va., as *West Virginia Univ. Bull.*, ser. 70, no. 7–2; ser. 71, no 12–3; ser. 74, no. 2–1, and ser. 77, no. 12–3.; 1st edn., 1952–64.)

Copiously illustrated, briefly descriptive atlas-manual of vascular plants, with keys to all taxa, essential synonymy and citations of relevant papers in headings, vernacular names, indication of local distribution (by counties), critical remarks, and notes on habitat, phenology, variation, special features, etc.; addenda, corrigenda and index to all botanical and vernacular names at end. Text is given on one side, with facing illustrations. An introductory section includes accounts of West Virginian climate, physical features, and vegetation. Covers 2200 species. [This second edition, though somewhat expanded and in a slightly larger format, differs little in style from its predecessor.]

149

Kentucky

No state flora is available, and the only reliable general checklist is that by Braun (1943). This has been usefully supplemented by a pair of semi-popular works by M. E. Wharton and R. W. Barbour. Also included here is a flora of the Louisville district by C. R. Gunn.

Bibliography

MEIJER, W., 1970. The flora and vegetation of Kentucky as a field for research and teaching. *Castanea*, 35: 161–76. [Includes a bibliography of Kentucky floristic literature, with classified index.]

BRAUN, E. L., 1943. *An annotated catalog of spermatophytes of Kentucky.* 161 pp., 6 maps. Cincinnati: The author.

Concise systematic enumeration of seed plants, with very limited synonymy, vernacular names, indication of local range (by counties), some critical remarks, and notes on habitat, associates, etc.; summary of the flora; description of geographical and physiographic regions of Kentucky (with map); index to family names. Almost contemporaneously with this work there appeared an unannotated list, based largely on herbarium labels without verification, by McFARLAND, F. T., 1942. A catalogue of the vascular plants of Kentucky. *Castanea*, 7: 77–108. [Includes the ferns and fern-allies, absent from Braun's list.]

WHARTON, M. E. and BARBOUR, R. W., 1971. *The wildflowers and ferns of Kentucky.* viii, 344 pp., col. illus., maps (Kentucky nature studies, 1). Lexington, Ky.: The University Press of Kentucky.

WHARTON, M. E. and BARBOUR, R. W., 1973. *Trees and shrubs of Kentucky.* x, 582 pp., illus. (some in color), maps (Kentucky nature studies, 4). Lexington.

These two similarly organized, essentially popular illustrated works each comprise three sections: an introduction, an atlas with explanatory text, and systematically arranged 'natural history accounts' (the latter with many attractively written details of habitat, occurrences, uses and other features, especially for the trees and shrubs). The atlas sections have species grouped artificially by gross features, without recourse to keys, and give both vernacular and scientific names. Technical details and background information appear in the introduction to each work, while at the end are given in each case an illustrated glossary and general index. [Apart from the checklist of Braun cited above, these works, albeit incomplete, provide the best available coverage of the Kentucky flora.]

Partial work

GUNN, C. R., 1968. *The floras of Jefferson and seven adjacent counties, Kentucky.* 322 pp. (Ann. Kentucky Soc. Nat. Hist., 2). Louisville, Ky.

Annotated flora of the Louisville district in northern Kentucky, with keys and indication of local range. [Included here in the absence of a state flora. Not seen by this writer; cited from the divisional bibliography of Lawyer *et al.* (in press).]

151

Midwest subregion

This category is used for works falling within the 'American Middle West' (the area bounded by Ohio, Michigan, Minnesota and Missouri) but not specifically related to a given state.

HARTLEY, T. G., 1966. *The flora of the 'Driftless Area'*. 174 pp., map (Univ. Iowa Stud. Nat. Hist., 21 (part 1)). Iowa City, Ia.

Annotated systematic checklist of vascular plants, with vernacular names and brief notes on distribution and usual habitat, occurrence (with indication of counties) if less common, and occasional citations of *exsiccatae* for rarities; statistical summary (1639 species, of which 1344 are native), list of excluded and doubtful records as well as those based purely on literature, bibliography, and index. An introductory section deals with the plan of the work, earlier studies, geography, topography, geology, climate, and glaciation along with the central topic, the relationship of the present flora with Quaternary history. The area, almost centering on La Crosse, Wis., straddles southwestern Wisconsin, eastern Iowa, southeastern Minnesota, and a corner of Illinois, and is of considerable biological interest as not having been glaciated during the Pleistocene.

ROSENDAHL, C. O., 1955. *Trees and shrubs of the upper Midwest*. 411 pp., illus. Minneapolis: University of Minnesota Press.

Copiously illustrated, briefly descriptive treatment with keys to all taxa, synonymy (with references for accepted species), vernacular names, generalized indication of local and extralimital range, and notes on habitat, phenology, special features, etc.; glossary and index to all botanical and vernacular names. Vegetation regions are described in the introductory section. The work, which covers Iowa, upper Michigan, Minnesota, and Wisconsin, originally appeared in 1928 as *Trees and shrubs of Minnesota*, by C. O. Rosendahl and F. K. Butters.

152

Ohio

As the descriptive flora by Braun is incomplete and Weishaupt's *Vascular plants of Ohio* is functionally a students' manual, the 1930s checklist (with supplements) by Schaffner has been included. The upper Ohio basin in the east is also covered by JENNINGS' Western Pennsylvania flora (see **144**).

BRAUN, E. L., 1961. *The woody plants of Ohio: trees, shrubs and climbers, native, naturalized and escaped*. viii, 362 pp., text-figs., maps. Columbus: Ohio State University Press. (Reprinted 1969, New York, Hafner.)

Illustrated, briefly descriptive atlas-flora of woody plants, with keys to all taxa; essential synonymy and citations; vernacular names; indication of local and extralimital range, with dot maps for all species; taxonomic commentary and notes on occurrence, habitat, uses, properties, special features, etc. (including special references); glossary, bibliography, list of authors, and indices to subjects and to all botanical and vernacular names. The introductory section includes an illustrated organography as well as an account of Ohio vegetation. The work may be considered complementary to the same author's *Vascular flora of Ohio* (see next entry).

BRAUN, E. L., 1967. *The Monocotyledoneae: cat-tails to orchids*. viii, 464 pp., illus., maps (The vascular flora of Ohio, 1). Columbus: Ohio State University Press.

Briefly descriptive, illustrated atlas-flora, with keys to all taxa; essential synonymy; vernacular names; generalized indication of local and extra-limital range, with dot maps for all species; notes on habitat, biology, associates, special features, uses, etc., as well as critical commentary (including also references to standard monographic and revisionary treatments); bibliography and indices to subjects as well as to all botanical and vernacular names. The introductory section gives accounts of physiographic features (emphasizing their impact on vegetation) and of general aspects of the flora and its paleohistory. [The author died in 1971 and no continuation of this work appears to be in sight.]

SCHAFFNER, J. H., 1932. Revised catalog of Ohio vascular plants. *Bull. Ohio Biol. Surv.* 25 (= vol. 5, no. 2): 87–215, 3 maps. [Columbus.] (Also pub. as *Ohio State Univ. Bull.* 36(9).) Columbus, Ohio.

Annotated checklist of vascular plants (some 2300 species plus varieties), with synonymy and indication of county distribution; based mainly on records in the 'state herbarium' at Ohio State University. A bibliography and discussion of floristic regions and phytogeography are also included. For supplements, see the author's Additions.... I–VII. *Ohio J. Sci.* **33**: 288–94 (1933): **34**: 165–74 (1934); **35**: 297–303 (1935); **36**: 195–203 (1936); **37**: 260–65 (1937); **38**: 211–16 (1938); **39**: 232–4 (1939; with C. H. Jones). [Not seen; references and description based on BLAKE and ATWOOD, 1942.]

WEISHAUPT, C. G., 1971. *Vascular plants of Ohio.* 3rd edn. iii, 292 pp. [5] pls. Dubuque, Iowa: Kendall/Hunt. (First edn., 1960, Columbus, Ohio.)

Concise students' manual-key to vascular plants in large format, with short descriptions of genera and families but not to species; vernacular names; indication of status and miscellaneous notes; glossary (with supporting illustrations) and partial index. An artificial key on vegetative features to woody plants is also included. Synonymy, local distribution, and ecological data are, however, lacking (for these, recourse must be had to Schaffner and other works).

153

Indiana

The classic comprehensive treatment of Deam (1940), the first 'modern' state flora in the Midwest, has long set a standard for more recent floristic work and manual writing in the region, although by now it is in need of revision.

DEAM, C. C., 1940. *Flora of Indiana.* 1236 pp., 1 pl., 2200 + maps. Indianapolis: Indiana Department of Conservation. (Reprinted 1970, Lehre, Germany: Cramer.)

Detailed but nondescriptive atlas-flora of vascular plants, with keys to all taxa, limited synonymy, citation of standard monographic/revisionary treatments under family and generic headings where appropriate, vernacular names, extensive discussion of local distribution (with references) and summary of extra-limital range; and notes on habitat, occurrence, ecology, phenology, uses, etc., as well as critical remarks; dot distribution maps provided for all species; lists of excluded species, collectors, and authorities, a tabular summary of the flora, and a bibliography, gazetteer, and

index to all botanical and vernacular names at end. The introductory section includes general accounts of the flora and vegetation of the state. [This state flora, by common consent regarded as a classic, was written by a man without formal botanical education.]

154

Illinois

In recent decades, the 'Great State' has been well-served through the indefatigable efforts of two rival flora-writers (and their respective associates) with the result that it is one of the best-documented polities in North America. For 'Chicagoland' the excellent local flora of Swink has also been included.

– Vascular flora: 3107 species (Mohlenbrock, 1975).

JONES, G. N., 1963. *Flora of Illinois.* 3rd edn. 401 pp., maps (Monogr. Amer. Midl. Naturalist, 7). South Bend, Ind.: University of Notre Dame Press. (First edn. 1945.)

Manual-key to vascular plants, with essential synonymy, vernacular names generalized indication of local range (by counties or state regions), and notes on habitat, phenology, and special features; conspectus of work, county map, glossary, and index to all botanical and vernacular names. The introductory section includes a general review of the flora and vegetation of the state (with map).

JONES, G. N. and FULLER, G. D., 1955. *Vascular plants of Illinois.* xii, 593 pp., 1375 maps. Urbana: University of Illinois Press.

Comprehensive, nondescriptive flora of vascular plants, with detailed synonymy (including references and citations), vernacular names, indication of local and extralimital range (with dot maps for all species), and notes on habitat, phenology, special features, etc.; list of principal collectors, extensive bibliography, and index to all botanical and vernacular names. The introductory section includes a general description of Illinois flora and vegetation. For supplement, see WINTERRINGER, G. S., and EVERS, R. A., 1960. *New records for Illinois vascular plants.* ix, 135 pp. (Illinois State Mus. Sci. Pap. Ser., 11). [Springfield.]

MOHLENBROCK, R. H., 1975. *Guide to the vascular flora of Illinois,* xii, 494 pp., 2 maps. Carbondale: Southern Illinois University Press.

Very concise field manual of vascular plants

(3107 spp.) in small format, with somewhat telegraphic keys to all taxa, vernacular names, and brief indication of habitat, phenology, distribution, etc.; notes on hybrids; glossary, statistics, addenda and indices at end. Arrangement follows the Englerian system, and the brief species accounts are separate from the keys. An introductory section accounts for habitat factors and the natural regions of the state (14 recognized), and includes an artificial key to families. [In format and style, this manual closely resembles its counterparts in Europe, but is unusual in the American context where many state manuals are for practical purposes desk- or herbarium-bound. However, Jones's *Flora of Illinois* most likely served as a model.]

MOHLENBROCK, R. H., 1967– . *The illustrated flora of Illinois.* [Vols. 1– .] Illus. Carbondale: Southern Illinois University Press.

Detailed descriptive illustrated flora of vascular plants, with keys to all taxa; limited synonymy, with some citations; vernacular names; generalized indication of local and extralimital range, with dot maps for all species; notes on habitat, occurrence, phenology, ecology and biology, karyotypes, special features, etc., as well as some critical remarks; bibliography, glossary, and index to all botanical and vernacular names in each volume. At this writing (1980) eight volumes have appeared encompassing the pteridophytes (one volume), the larger part of the monocotyledons including the grasses (five volumes), and the 'willows to mustards' (Salicaceae–Cruciferae) and 'hollies to loasas' (Aquifoliaceae–Loasaceae) (two volumes). The plan of the work follows the Thorne system, but individual volumes have not been appearing in systematic order. [A flora with such scope and detail would have been more appropriate if designed to cover the whole Middle West, not just relatively well-documented Illinois. In addition, the pseudo-popular presentation of a serial work in no fixed order but arranged on a still relatively unfamiliar (and in floras so far little-used) system is jarring to say the least. The work, planned eventually to extend to the lower cryptogams and higher fungi, is nominally guided by an advisory committee of five.]

MOHLENBROCK, R. H. and LADD, D. M., 1978. *Distribution of Illinois vascular plants.* vii, 282 pp., maps. Carbondale: Southern Illinois University Press.

Comprises an atlas of dot distribution maps of the vascular plants of Illinois, with 251 pp. (each with 12 maps) accounting for 3001 taxa (species and varieties).

Arrangement is alphabetical, and only accepted names as used in the senior author's *Guide* (see above) are used. A table of synonymy (relating to other major floras) and a systematic name list appear as appendices.[9]

MYERS, R. M., 1972. *Annotated catalog and index for the Illinois flora.* 64 pp. (Biol. Sci. Ser. Western Illinois Univ. 10). Macomb.

Numbered systematic list of native, naturalized, and adventive vascular plants, with indication of status and bibliographic references; statistical table and addenda; no index. Notes on the flora of McDonough County are also appended. The introductory section includes remarks on alien, weedy, and extinct species.

Partial work: Chicago region

SWINK, F., 1969. *Plants of the Chicago region.* xxii, 445 pp., maps. Lisle, Ill.: The Morton Aboretum.

Alphabetically arranged (by genera), dictionary-style enumeration of vascular plants, with limited synonymy, vernacular names, references to other floras, detailed account of local distribution with county dot maps for most species, and extensive commentary on habitat, frequency, associates, etc.; some taxonomic commentary; no index. The introductory section accounts for the scope, basis and plan of the work and the area covered (22 counties in three states). Over 2000 species are accounted for.

155

Michigan

See also **151** (ROSENDAHL). The early checklist of Beal is included as it has not effectively been superseded; to date (mid-1979), only one volume of the new *Michigan flora* has been published. Gleason's work is a students' manual.

Bibliography

VOSS, E. G., 1962–3. Michigan plants in print; guide to literature on the Michigan flora. *Michigan Botanist*, **1**: 43–5, 91–2; **2**: 55–9.

BEAL, W. J., 1904. *Michigan flora:* [*a list of the fern and seed plants growing without cultivation.*] 147 pp. Lansing, Mich.: Michigan State Board of Agriculture. (Originally published as Annual Rep. Mich. Acad. Sci., 5, pp. 1–147.)

Annotated checklist of vascular plants (2365 species, including those added in 1908), with synonymy and county records; also includes remarks on climate, phytogeography, etc., in an introductory section. For supplement, see the author's Additions.... Annual Rep. Mich. Acad. Sci., 10, pp. 85–9 (1908). [Not seen; references and description based on Blake and Atwood, 1942.]

GLEASON, H. A., 1939. *The plants of Michigan.* 204 pp. Ann Arbor: George Wahr.

Concise unannotated students' manual-key to seed plants; includes brief descriptions of families, vernacular names, and a glossary and index to generic, family, and vernacular names. Technical notes on the use of the work are given in an introductory section. No synonymy or details of distribution in the state are accounted for.

VOSS, E. G., 1972– . *Michigan flora.* Vols. 1– . Illus. (some in color), maps. Bloomfield Hills, Mich.: Cranbrook Institute of Science; Ann Arbor: University of Michigan Herbarium.

Concise annotated manual of seed plants, with diagnostic keys to all species; limited synonymy; vernacular names; general indication of local range, with dot maps for all species; critical remarks and notes on habitat, phenology, occurrence, biology, variation, special features, etc.; glossary and index to all botanical and vernacular names. The introductory section includes a historical survey of botanical work together with descriptive remarks on vegetation, phytogeography, and paleohistory of the flora. By mid-1979 one volume, covering the Gymnospermae and Monocotyledoneae, had appeared. For a companion work on pteridophytes, see BILLINGTON, C., 1952. *Ferns of Michigan.* 240 pp., map. Bloomfield Hills, Mich.: Cranbrook Institute of Science.

156

Wisconsin

See also **151** (HARTLEY, ROSENDAHL). No recent state flora is available; the nearest approach to it is represented by the long series of *Preliminary reports* begun by N. C. Fassett in 1929 with which substantial monographs on the grasses, legumes and ferns can be associated. For student use, there is Fassett's *Spring flora of Wisconsin* (4th edn., 1975).

Bibliography

GREEN, H. C. and CURTIS, J. T., 1955. *A bibliography of Wisconsin vegetation.* 84 pp. (Milwaukee Public Mus., Publ. Bot., 1). Milwaukee. [Includes references to local floristic works; classified by methodology. Also features a short history of Wisconsin botany.]

FASSETT, N. C. *et al.*, 1929– . Preliminary reports on the flora of Wisconsin, I– . Illus. *Trans. Wisconsin Acad. Sci.* **24**, etc., *passim*. [Madison].

Comprises a long series of descriptive (sometimes illustrated) revisions of families of Wisconsin vascular plants, each with keys to genera and species, concise synonymy, vernacular names, generalized indication of local range (with commentary and dot distribution maps), extensive critical discussion, and notes on habitat, frequency, karyotypes, biology, special features, etc.; prefatory remarks and lists of references are also included. Most treatments have been contributed by specialists. At this writing (mid-1979), the series has reached pt. LXVI (1974) but no further. Related contributions, all published in book form by University of Wisconsin Press, include: FASSETT, N. C., 1939. *The leguminous plants of Wisconsin.* xii, 157 pp. Madison; *idem*, 1951. *Grasses of Wisconsin.* 173 pp. Madison; and TRYON, R. M., Jr. *et al.*, 1953. *The ferns and fern allies of Wisconsin.* 2nd edn. 158 pp. Madison.

FASSETT, N. C. and THOMSON, O. S., 1975. *Spring flora of Wisconsin.* 4th edn. 416 pp., 564 figs., 4 maps. Madison: University of Wisconsin Press. (First edn., 1931.)

This well-known, widely used work, the doyen of American 'spring floras', comprises a copiously illustrated, concise student's field manual in small format of seed plants flowering before 15 June, with complete keys, vernacular names, local distribution, and notes on habitat, special features, etc.; glossary, selected list of references, and index at end. An introductory section includes four maps depicting counties and major geoclimatic and vegetation zones. Includes Cyperaceae (omitted from earlier editions), but Gramineae are merely listed (for fuller treatment by Fassett (1951), see preceding entry) and *Salix* and Juncaceae are omitted.

157

Minnesota

See also **151** (HARTLEY, ROSENDAHL); **175** (RYDBERG). The only state-wide general work is the annotated but antiquated checklist by Upham (1884–7). There is also a more recent but slim name list by Moore and Tryon, and a students' spring flora by Morley. Two substantial partial works respectively covering the northeastern triangle and southern third of the state, the main population centers, are also included here.

MOORE, J. W. and TRYON, R. M., Jr., 1946. *A preliminary checklist of the flowering plants, ferns, and fern allies of Minnesota*. 99 pp. Minneapolis: University of Minnesota Press.

Unannotated systematic list of vascular plants, with vernacular names; index.

MORLEY, T., 1969. *Spring flora of Minnesota*. 2nd edn. 283 pp., 2 maps. Minneapolis: University of Minnesota Press. (First edn., 1966.)

Descriptive students' manual of seed plants flowering before 7 June, with complete keys, essential synonymy, vernacular names, local range, and notes on habitat; map of vegetation formations, gazetteer, county map, glossary, and index at end. Patterned after Fassett's *Spring flora of Wisconsin* (*see* **156**).

UPHAM, W., 1884. *Catalogue of the flora of Minnesota*. 193 pp. (Annual Rep. Geol. & Nat. Hist. Surv. Minnesota, 12 (1883), part 6). Minneapolis.

Annotated checklist of native, naturalized and adventive vascular plants, with indication of state and county distributions and descriptions for some taxa; index. Includes also accounts of botanical exploration, geography and topography along with a list of references. For additions, see *idem*, 1887. *Supplement to the flora of Minnesota*. pp. 46–54 (Bull. Geol. & Nat. Hist. Surv. Minnesota, 3). St Paul, Minnesota. [Not seen by this author; references and description based on Blake and Atwood, 1942.]

Partial works

LAKELA, O., 1965. *A flora of northeastern Minnesota*. 541 pp., 110 text-figs., maps. Minneapolis: University of Minnesota Press.

Descriptive manual of vascular plants, with keys to all taxa; vernacular names; generalized indication of local range, with maps for all species (as well as localities with *exsiccatae*), and summary of extralimital range; notes on habitat and phenology; figures of representative species; glossary and complete index. The introductory section includes accounts of physical features, geology, soils, and vegetation. The area covered is the triangle between Duluth, Lake Superior, and the Canadian border, and includes Isle Royale and the Mesabi Range.

MACMILLAN, C., 1892. *The Metaspermae of the Minnesota Valley*. xiii, 826 pp., 2 maps (Rep. Geol. Surv. Minnesota, Bot. Ser., 1). Minneapolis.

Detailed enumeration of vascular plants, including full synonymy (with references and citations), indication of localities and *exsiccatae*, summary of extralimital range, and notes on habitat, special features, etc.; bibliography and index to all botanical names. An introductory section gives a delimitation of the area, while several appendices account for physical features, climate, floristics, phytogeography, and statistics of the flora. The area covered, centering on the Minnesota River, covers approximately the southern third of the state as well as adjacent parts of Iowa and South Dakota.

158

Iowa

See also **151** (HARTLEY, ROSENDAHL); **175** (RYDBERG). For a useful review of progress to 1954, see THORNE, R. F., 1954. Present status of our knowledge of the vascular plant flora of Iowa. *Proc. Iowa Acad. Sci.* **61**: 177–83. No general compendium has yet appeared to replace the aging checklist of Cratty; the works of Conard and Pohl are students' keys.

– Vascular flora: 1785 species (THORNE).

CONARD, H. S., 1952. *Plants of Iowa*. 94 pp. Ames: Iowa State College Book Store.

Concise, unannotated manual-key to vascular plants, with vernacular names; illustrated glossary and list of rare and unusual plants; no index. Complements Cratty's *Iowa flora* and is designed to be used together with that work.

CRATTY, R. I., 1933. The Iowa flora. *Iowa State Coll. J. Sci.* **7**: 177–252.

Concise, annotated systematic enumeration of native and naturalized vascular plants, with essential synonymy, vernacular names, and notes on habitat and occurrence; statistical tables; index to genera. No keys are provided (for these, see Conard's *Plants of Iowa*). The work was largely based upon herbarium holdings at Iowa State College (now University), Ames.

POHL, R. W., 1975. *Keys to Iowa vascular plants*. 198 pp. Dubuque, Iowa: Kendall/Hunt.

Students' manual to vascular plants; includes all but a few very rare species known from the state. [Not seen; cited from Lawyer *et al.* 1979.]

159

Missouri

See also **175** (RYDBERG). The state is the beneficiary (like Indiana) of one of the finest of modern comprehensive state manual-floras, Steyermark's *Flora of Missouri* (1963), which grew out of his earlier but still-useful *Spring flora of Missouri* (1940).

STEYERMARK, J. A., 1963. *Flora of Missouri.* 1725 pp., 2300 illus., 2400 maps. Ames, Ia.: Iowa State University Press.

Comprehensive illustrated atlas-flora and manual of vascular plants, with essential synonymy (including some literature citations), vernacular names, general indication of local and extralimital range (with county dot maps for all species), taxonomic commentary, and extensive notes on habitat, frequency, occurrence, phenology, biology, special features, properties, uses, etc.; tabular summary of the flora; glossary and index to all botanical and vernacular names. The introductory section includes a historical sketch as well as accounts of the flora and vegetation in general and lists of species with distributions bordering on Missouri; a general county map is also provided. See also *idem*, 1940. *Spring flora of Missouri.* viii, 582 pp., 66 figs., 163 pls., map. St Louis: Missouri Botanical Garden; Chicago: Field Museum of Natural History. (Reprinted 1954, Columbia, Mo.: Lucas Brothers.)

Region

16⁵

Southeastern states

This region, the heart of Dixie, encompasses the nine states from North Carolina to Florida and west to Arkansas and Louisiana, although larger or smaller parts of most neighboring or nearby states are floristically included. However, southern Florida effectively belongs to the Caribbean or West Indian region (Superregion 24–29).

This floristically richest part of the conterminous United States is also that which has visibly shown the least progress overall in the production of new floristic works since 1939. The region has long needed a new general flora to replace Small's nearly obsolete manuals, and but few state floras are available. Despite a long history of collecting, botanical knowledge lags behind most of the rest of the country; more recent documentation is presently very scattered although much more extensive than forty years ago. For long there were but few resident botanists, and the region was to a greater or lesser extent looked upon as a botanical *Kolonialgebiet* in rather the same way as did northern Europe with respect to the Mediterranean basin; the Harvard *Generic flora of the Southeastern United States* has among more inherent deficiencies been criticized for 'carpetbagging' although with more than 70 contributions now published it now constitutes a solid basis for any new regional manual.

It is perhaps, therefore, indicative of the considerable development of the 'Old South' since World War II that such a project for a new regional manual should have been initiated, and from a Southern base – in North Carolina. This project, *Vascular Plants of the Southeastern United States*,[10] is under the direction of A. E. Radford, and represents an extension of the work put into the successful *Manual of the vascular flora of the Carolinas* (1968), the first 'modern' Southern flora (which also followed the Midwestern tradition of county distribution maps for all species). This work has proved a great stimulus to floristics in the Southeast.

Other key works in the region have appeared for south Florida (Long and Lakela, 1971) and Arkansas (Smith, 1978). Projects in varying forms are underway in Louisiana, Mississippi, Tennessee, and Florida although some, unfortunately, are only slowly being published. Interim checklists are available for Tennessee and Florida, and a state bibliography for Louisiana. The woody flora of Alabama was recently revised (Clark, 1971) to supplement the antiquated account by Mohr (1901), and a similar work for Arkansas is reportedly in press. However, the overall position is still not satisfactory; in some respects, it resembles the Mediterranean, West Indies or parts of Australia with respect to level of coverage. The vascular floras are comparatively large, and the number of active workers (as well as financial support) often limited.

For the tree flora, there are two regional guides;

that by Harrar and Harrar (1946; reissued with revisions, 1962), a companion to that by Harlow on the northern states, is comparatively widely used. The woody vines have been more recently treated by Duncan (1973), and pteridophytes by Wherry (1964). No general treatment of the woody flora, however, is available.

Bibliography. General and divisional bibliographies as for Division 1.

Indices. General and divisional indices as for Division 1.

160

Region in general

Small's 1913 *Flora* covers the entire region as here defined, together with eastern Texas and Oklahoma, whereas his 1933 *Manual* encompasses only the states or parts thereof east of the Mississippi River. Of the fairly considerable semi-popular literature on woody plants and pteridophytes, only a selection, given here under a separate subheading, can be described in this *Guide*.

Much new floristic information on the region has been appearing in *Castanea*, the journal of the Southern Appalachian Botanical Club, published since 1936. Another significant outlet is *Sida*, published from Southern Methodist University in Texas.

Keys to families and genera

BATSON, W. T., 1972. *Genera of the southeastern plants.* iv, 151 pp., 1089 text-figs. Columbia, SC: The author. [Comprises keys to all taxa of vascular plants down to the genus, with representative figures and notes on distribution, habitat, etc., as well as synonymy and vernacular names; glossary, twig key, list of poisonous plants, and index are also included. A revised, expanded edition appeared in 1977 under the title *A guide to the genera of the eastern plants*. New York: Wiley; see **140–50**.]

RADFORD, A. E. *et al.* (eds.), 1980. *Vascular flora of the southeastern United States.* Vol. 1, Asteraceae, by A. Cronquist. xv, 261 pp. Chapel Hill, NC: University of North Carolina Press.

This first of five projected volumes comprises a descriptive manual-flora of vascular plants, with keys to all taxa, limited synonymy with citations, vernacular names (where known), more or less generalized indication of distribution, and notes on karyotypes,

phenology, habitat, special features, etc.; illustrations and maps absent; index. An introductory section gives what is in effect a style-manual for the whole flora, with notes on format, geographical and ecological notation, and synonymy as well as abbreviations. Limited by the Atlantic Ocean, the Gulf of Mexico, the Mason-Dixon line, the Ohio and Mississippi Rivers, the northern boundary of Arkansas, and the western boundaries of that state and Louisiana, but presence in all adjacent states is mentioned. Key monographs and revisions are cited in generic and other headings. The work will gradually supersede the two floras of Small (see below). [Reviewed by T. M. Barkley in *Brittonia*, **32**: 103–4 (1980) with the comment that it resembles Correll and Johnston's *Manual of the vascular plants of Texas*; to this writer the treatment of Asteraceae also somewhat recalls *North American Flora*. Description prepared from above review as well as sample pages sent from New York Botanical Garden Library.]

SMALL, J. K., 1913. *Flora of the southeastern United States.* 2nd edn. xii, 1394 pp. New York: The author. (First edn. 1903.)

Briefly descriptive manual-flora, with somewhat awkward keys covering all taxa; limited synonymy; vernacular names; generalized indication of internal and extralimital range; miscellaneous notes on habitat, special features, etc.; indices. Now superseded by the same author's *Manual of the southeastern flora*, except for Louisiana and Arkansas.

SMALL, J. K., 1933. *Manual of the southeastern flora.* xxii, 1554 pp., illus. New York/Chapel Hill: The author. (Reprinted 1953, Chapel Hill, University of North Carolina Press.)

Illustrated, briefly descriptive manual-flora of seed plants, with keys to all taxa; limited synonymy; vernacular names; generalized indication of internal and extralimital range; miscellaneous notes; indices to all botanical and vernacular names. Does not cover the areas west of the Mississippi River; a projected companion volume to cover Arkansas, Louisiana, and eastern Texas never materialized. For Pteridophyta east of the Mississippi, see *idem*, 1938. *Ferns of the southeastern states.* 517 pp., illus. (incl. map). New York: The author. (Reprinted 1965, New York: Hafner.)

WOOD, C. E. Jr. *et al.*, 1958– . A generic flora of the southeastern United States, [I–]. *J. Arnold Arbor.* **39**, etc.

As of this writing (late 1980) some 85 parts of this

series have been published, each with one or more family treatments. Each treatment is written along the lines of *Die natürlichen Pflanzenfamilien*, with family description, discussion of morphological, anatomical, palynological, karyological and other attributes, classification, and keys to and treatments of the genera in the region, with figures of representative species. References are also included. [The series was originally begun as part of a project to produce a new regional manual-flora; it was thought that a critical basis at supraspecific level would be desirable in this relatively poorly-known area. It has, however, drifted on in an open-ended fashion.]

Special groups – woody plants and pteridophytes

The works described below represent a selection from a larger literature, and will generally be useful over the region. However, Florida with its peculiar tropical tree flora deserves special consideration, and under the state heading will be found further references.

COKER, W. C. and TOTTEN, H. R., 1945. *Trees of the southeastern states*. 3rd edn. 419 pp., illus. Chapel Hill, NC: University of North Carolina Press. (First edn., 1934.)

Illustrated semi-popular descriptive treatment of native and naturalized trees, with keys to all species, vernacular names, taxonomic commentary, and extensive notes on distribution, habitat, phenology, special features, etc.; remarks on special trees, list of larger shrubs; glossary, bibliography, and index to all names at end. Includes Virginia but does not effectively cover southern Florida with its tropical tree flora.

DUNCAN, W. H., 1973. *Woody vines of the southeastern United States*. 87 pp., illus., maps. Athens, Georgia: University of Georgia Press. (Originally published 1967 as *Sida* 3(1): 1–79.)

Semi-popular treatment of the rich (for North America) woody vine flora, with keys, descriptive text, dot distribution maps, and illustrations; index.

HARRAR, E. S. and HARRAR, J. G., 1946. *Guide to southern trees*. 709 pp., illus. New York: McGraw-Hill. (Repeated with slight revisions and altered nomenclature, 1962, New York, Dover Publications.)

Copiously illustrated semipopular descriptive treatment of trees, with simple keys to all taxa, limited synonymy, vernacular names, and extensive remarks on occurrence, habitat, diagnostic features, variation, properties, uses, wood, etc.; glossary, list of references, and complete index. Covers the entire region, including southern Florida.

WHERRY, E. T., 1964. *The southern fern guide*. 349 pp., illus., map. New York: Doubleday. (Reissued with corrections and nomenclatural changes, 1972, New York: New York Chapter, American Fern Society.)

Illustrated field guide to ferns and fern-allies, with keys, descriptions, vernacular names, indication of distribution, and notes on habitat, occurrence, frequency, etc.; limited synonymy and some critical commentary. Includes an etymological lexicon and indices as well as an introduction with glossary, chapters on morphology, life-cycles, cytology, classification, biogeography, and cultivation, references (list of fern floras, pp. 35–6) and general keys (pp. 37–56). [Included here in view of the comparative antiquity of Small's treatment. Not seen by this writer; description prepared from the companion guide to northeastern ferns as well as sample pages sent from the New York Botanical Garden Library.]

161

Tennessee

The last formal state flora is that by Gattinger (1901), although separate works on woody plants and on ferns have appeared since then. Tragically, Gattinger's collection was destroyed by fire at Knoxville in 1934. In the 1950s some work was done under the direction of A. J. Sharp towards a new flora, and two preliminary checklists resulted. In recent years, however, activities have been orientated towards production of a distribution atlas, of which the first installments appeared in 1979 and 1980.

GATTINGER, A., 1901. *The flora of Tennessee and a philosophy of botany*. 296 pp., illus., portrait. Nashville: Tennessee Dept. of Agriculture.

Includes on pp. 1–184 an enumeration of vascular plants, with concise synonymy, very brief indication of local range (including mention of some localities), critical remarks, and notes on habitat, phenology, special features, cultivation (if applicable), uses, etc.; summary tables; index to generic names. The introductory section gives remarks on general features of the flora, floristic zones, and lists of plants in particular habitat-types. The abbreviation 'OS' (not defined in the text) stands for 'over the whole state'.

SHARP, A. J. *et al.*, 1956. *A preliminary checklist of monocots in Tennessee*. 33 pp. Knoxville, Tenn.: Department of Botany, University of Tennessee (mimeographed).

SHARP, A. J. *et al.*, 1960. *A preliminary checklist of dicots in Tennessee*. 114 pp. Knoxville: Department of Botany, University of Tennessee (mimeographed).

Briefly annotated systematic checklists, with

indication of distribution by regions or counties; precursory to a projected state flora.

WOFFORD, B. E. and EVANS, A. M., 1979–80. Atlas of the vascular plants of Tennessee, 1–3. *J. Tennessee Acad. Sci.* **54**: 32–8, 75–80; **55**: 110–14 (part 3 by B. E. Wofford alone).

Comprises county dot maps for each species, without annotations; arrangement of angiosperms follows the Cronquist system (to date, maps cover Aristolochiales through Ranunculales, the Caryophyllales, and the Polygonales). The introductory section (in part 1) provides background information, including the present stand of floristic coverage of the state, a definition of the 'major works' on the state (essentially those given under **161**), comments to the effect that the atlas was superior to a manual-flora for handling a unit as floristically diverse as Tennessee, and emphasizes the ground work of Sharp and others; separate list of references.

Special groups – woody plants and pteridophytes

SHANKS, R. E., 1952. Checklist of the woody plants of Tennessee. *J. Tennessee Acad. Sci.* **27**: 27–50.

Systematic list, with limited synonymy, vernacular names, and indication of local range; index to genera.

SHAVER, J. M., 1954. *Ferns of Tennessee, with the fern allies excluded.* xviii, 502 pp., 243 text-figs. Nashville, Tenn.: George Peabody College. (Reprinted 1970, New York: Dover, as *Ferns of the Eastern Central States.*)

Detailed descriptive treatment, with keys, synonymy, vernacular names, and extensive documentation of distribution (including *exsiccatae*); all species illustrated; index. [One of the most thoroughly documented of all fern floras in North America.]

162

The Carolinas

Under this heading are included both North and South Carolina. The superb bi-state flora by Radford and collaborators, the most modern in the Southeast, is a worthy successor to a botanical tradition beginning with Catesby in the early eighteenth century.

– Vascular flora: 3360 species (Radford *et al.*, 1968).

RADFORD, A. E., AHLES, H. E. and BELL, C. R., 1968. *Manual of the vascular flora of the Carolinas.* lxi,

1183 pp., illus., maps. Chapel Hill, NC: Univ. of North Carolina Press.

Descriptive manual of vascular plants, with keys to all species; essential synonymy; vernacular names; concise indication of local and extralimital range, with dot maps for all species; notes on habitat, frequency, phenology, and karyotypes; figures of diagnostic features and representative species; list of authors; index to all vernacular and botanical names. The introductory section includes historical remarks, a summary of the flora, and a county map.

– Designated as a 'Basic Flora' for the *Flora North America* Program. [The success of this work may be indicated by its having had five printings by 1978.] Important precursors include *idem*, 1964. *Guide to the vascular flora of the Carolinas.* 392 pp., map. Chapel Hill: The Book Exchange; and *idem*, 1965. *Atlas of the vascular flora of the Carolinas.* 208 pp., maps (Bot. Dept. Univ. North Carolina Tech. Bull., 165). Chapel Hill.

163

Florida (in general)

For works dealing only with the (sub)tropical southern part of the state, see **164**. No general state flora is available, and the checklist initiated by Ward (who has now begun another series, *Keys to the flora of Florida*) has yet to be completed. Some works on special groups have also been noted.

WARD, D. B., 1968. *Checklist of the vascular flora of Florida.* Part 1. 72 pp. (Univ. Florida Agric. Exp. Sta. Tech. Bull., 726). Gainesville, Fla.

Systematic list of vascular plants, with essential synonymy and vernacular names; list of recent literature on the flora of the state; index to generic, family, and vernacular names. Covers the Pteridophyta, Gymnospermae, and Monocotyledoneae.

WARD, D. B. (ed.), 1977– . *Keys to the flora of Florida*, 1– (Phytologia, 35, part 6, etc.)

A series, not issued in systematic order, comprising revisions of families or genera in manual-key format, with diagnostic leads also incorporating notes on habit, phenology, ecology, biology and county distribution; some synonyms and much critical commentary included. In part 1 is the plan of the work, a state map, and definition of vegetation associations (19 recognized). Preceded by WARD, D. B. *et al.*, 1962–75. Contributions to the flora of Florida, 2–9. *Castanea*, **28** (1962)–**40**(1975), *passim*. [Part 1 evidently not published.]

Special groups – woody plants and pteridophytes

For tree distribution, see also vol. 5 (Florida) of LITTLE's *Atlas of United States trees* (under **100**).

LONG, R. W. and LAKELA, O., 1976. *Ferns of Florida*. xiii, 178 pp., 117 text-figs. (including photographs). Miami: Banyan Books.

Illustrated descriptive semi-popular treatment, with keys, indication of distribution, local names, synonymy, commentary, etc.; glossary and index. Covers 135 spp.

SMALL, J. K., 1913. *Shrubs of Florida*. x, 140 pp. New York: The author.

Field-manual to native and naturalized species, with keys, limited synonymy, local names (where known), generalized indication of distribution with some details, and notes on special features, etc.; index. [Nomenclature follows the old 'American Code'. A companion work on trees by the same author has been superseded by the following.]

WEST, E. and ARNOLD, L., 1956. *The native trees of Florida*. 2nd edn. xx, 218 pp., illus. Gainesville, Fla.: University of Florida Press. (First ed. 1946.)

Illustrated descriptive atlas-type treatment, with keys (mostly to genus only), local names, general and specific indication of distribution, and commentary, including notes on uses; county maps in end-papers; glossary, references and index. Less common or 'borderline' trees are merely noted. [The coverage for Florida in HARRAR and HARRAR (**160**) or SARGENT (**100**) is nearly as inclusive.]

164

Southern Florida

This unit relates particularly to the southern one-third of the state, characterized by an Antillean tropical and subtropical flora unique within the conterminous United States.

LONG, R. W. and LAKELA, O., 1971. *A flora of tropical Florida*. xvii, 962 pp, text-figs., map, portrait. Coral Gables: University of Miami Press. (Reprinted 1976, Miami: Banyan Books.)

Descriptive manual of native, naturalized, and commonly cultivated vascular plants of southern peninsular Florida, with limited synonymy, vernacular names, general indication of local and extralimital

range, and notes on habitat, phenology, special features, etc.; illustrations of representative species; list of authors, glossary, and index to all botanical and vernacular names. The introductory section gives an account of botanical exploration (by J. Ewan) together with chapters on geology, plant communities, and the origin of the flora; included also are statistical tables and a select list of (recent) references.

– Designated as a 'Basic Flora' for the now-suspended *Flora North America* program.

165

Georgia

Until recently, no state-wide floras or checklists had been issued since a brief list published in 1849 in *Statistics of Georgia* by G. White. The pteridophytes, however, have been treated by McVaugh and Pyron (1951) and for the southwestern counties a good enumeration by Thorne (1954) is available. Nevertheless, the relative standard of coverage for the state has been low and for identification work reference should be made to the flora of the Carolinas (**162**). [Latterly there have been two works issued from the University of Georgia at Athens: a list of Georgia plants in the University of Georgia Herbarium, compiled by S. B. Jones, Jr. and N. M. Cole and accounting for 4673 species, and a general checklist by W. H. Duncan and J. T. Kartesz, described below.]

DUNCAN, W. H. and KARTESZ, J. T., 1981. *Vascular flora of Georgia; an annotated checklist*. xi, 147 pp., map. Athens, Ga.: University of Georgia Press.

Comprises a brief annotated modern checklist of vascular plants (3686 accepted specific and non-duplicating infraspecific taxa), with some synonymy and indications of geographical occurrence (by botanical provinces). [Not seen; annotation adapted from notice in *Plant Sci. Bull.* (Botanical Society of America) **27**(3): 22 (1981).]

Partial work

THORNE, R. F., 1954. The vascular plants of southwestern Georgia. *Amer. Midl. Naturalist*, **52**: 257–327, maps.

Regional 17-county checklist of vascular plants, based on floristic surveys by the author in the 1940s and encompassing among other areas the peanut country around Plains; includes essential synonymy and general and more specific indication of distribution; statistics (1750 species). The area, of some 5000 square miles, spreads across the two major physiographic zones. [Account derived partly from *Biological Abstracts*.].

Special group – pteridophytes

McVaugh, R. and Pyron, J. H., 1951. *Ferns of Georgia*. xviii, 195 pp., maps. Athens: University of Georgia Press.

Illustrated semi-popular descriptive treatment of pteridophytes, with keys, vernacular names, synonymy, indication of local and extralimital distribution with range maps, and commentary; index and glossary. An introductory section is also provided.

166

Alabama

In addition to the old but still definitive state flora by Mohr, the best in the South until the 1960s, relatively recent treatments of woody plants and of pteridophytes are accounted for here.

Mohr, C. T., 1901. *Plant life of Alabama*. 921 pp., 13 pls., incl. map (Contr. US Natl. Herb., 6). Washington: Smithsonian Institution. (Reprinted 1901, Montgomery, Ala.: Alabama Geological Survey; 1969, Lehre, Germany: Cramer.)

Detailed, non-descriptive flora of non-vascular and vascular plants, without keys; synonymy, with references and some citations; vernacular names; fairly extensive indication of local range; summary of extralimital distribution; critical remarks together with notes on habitat, life-form, frequency, phenology, special features, etc.; catalog of cultivated plants; statistics of flora and index to all vernacular and botanical names and subjects. The extensive introductory section features accounts of physiography, climate, botanical exploration, floristics, phytogeography, vegetation, and agriculture. [A fine work in its day, still considered definitive by Clark (see below) but now in need of extensive revision. It may be noted here that, according to Clark, Mohr travelled largely on horseback for transport, and, while in the countryside, to conform with social etiquette.]

Special groups – woody plants and pteridophytes

Clark, R. C., 1971. The woody plants of Alabama. *Ann. Missouri Bot. Gard.* 58: 99–242, maps.

Systematic enumeration, with keys, relevant synonymy, general and specific indication of distribution and physiographic provinces, and commentary; county dot maps for each species; list of references. An introductory section accounts for the history of Alabama botanical work, sources of information, descriptions of physical features, climate, soils, and natural regions, and discussion of plant distribution and floristics. Partly succeeds Harper, R. M., 1928. *Economic botany of Alabama*, 2: *catalogue of the trees, shrubs and vines of Alabama*. 357 pp., 66 figs., 23 maps. (Alabama Geol. Surv. Monogr., 9). University, Ala.

Dean, B. E. 1969. *Ferns of Alabama*. Revised edn. xxiv, 214 pp., illus. Birmingham: Southern Universities Press. (First edn., 1964.)

Illustrated semi-popular treatment, with keys, synonymy, local names, and indication of distribution (by counties for less common species); index. A general introduction is also provided.

167

Mississippi

Much scattered information has become available in the last fifteen years, particularly through the work of S. B. Jones, Jr., who is reportedly preparing a new state guide as a successor to the now very incomplete checklist of Lowe (1921), and T. M. Pullen.

Jones, S. B., Jr. 1974–6. Mississippi flora, I–VI. *Gulf Res. Rep.* 4(3): 357–79, 1974 (I); 5: 7–22, 1975 (IV); *Castanea* 39: 370–9, 1974 (II); 40: 238–53, 1975 (III); 41: 41–58, 189–212, 1976 (V–VI).

Comprises miscellaneous contributions, with keys and county distribution maps, towards a new state flora, *Guide to the flora of Mississippi*, indicated as being in preparation. For a related contribution on pteridophytes, with keys, indication of distribution by Lowe's physiographic zones, and county maps, see Evans, A. M., 1978. Mississippi flora: a guide to the ferns and fern allies. *Sida* 7: 282–97.

Lowe, E. N., 1921. *Plants of Mississippi. A list of flowering plants and ferns*. 292 pp., map (Mississippi State Geol. Surv. Bull., 17). Jackson.

Enumeration of vascular plants, with synonymy, vernacular names, general indication of local range (with some citations and other details), and brief notes on habitat, phenology, etc.; county map and gazetteer; no index. The introductory section includes accounts of life-zones (Merriam system), ecological types, habitat factors, and floristic and topographic regions in the state. The work was to a large extent based on Mohr's *Plant life of Alabama*. For substantial additions, see PULLEN, T. M. *et al.*, 1968. Additions to the vascular flora of Mississippi. *Castanea*, 33: 326–34. [Based on over 29000 collections made from 1963 through 1967; adds 150 taxa not in Lowe, with county distributions.] A related treatment on Compositae is TEMPLE, L. C. and PULLEN, T. M., 1968. A preliminary checklist of the Compositae of Mississippi. *Ibid.*, 33: 106–15, map.

168

Arkansas

The recently published annotated atlas/checklist of Smith (1978) effectively supersedes the enumerations by Branner and Coville (1891, with supplement by Buchholz and Palmer, 1926) and Demaree (1943). A guide to the woody flora, with keys, by G. E. Tucker was in press for *Annals of the Missouri Botanical Garden* at the time of writing (August 1979).

– Vascular flora: 2338 species and infraspecific taxa (Smith).

SMITH, E. B., 1978. *An atlas and annotated list of the vascular plants of Arkansas*. iv, 592 pp., maps. Fayetteville, Ark.: distr. by University of Arkansas Bookstore. [Reproduced from typescript.]

Comprises an atlas/checklist of county dot maps, one for each species (nine maps per page), with facing annotated text; the latter includes synonymy, vernacular names, and notes on distribution, occurrence, karyotypes (with vouchers), origin of aliens, etc.; some critical remarks; references to revisions/monographs and floristic papers scattered throughout text. A total of 2338 entities has been recognized; and collecting statistics are given along with a bibliography (pp. 551–62). Supersedes DEMAREE, D., 1943. A catalogue of vascular plants of Arkansas. 88 pp. *Taxodium*, 1(1) (all published).

169

Louisiana

Although there is a long botanical tradition in the state, which includes the early *Flora ludoviciana* of C. S. Rafinesque (1817), the large size of the flora as now known (between 3000 and 4000 species; Brown, 1972) has been an impediment to realization of a modern state flora here as elsewhere in the Deep South where botanical resources until recently have been scanty. The first part of a new state checklist by Thieret appeared in 1972. Some popular works on special groups are also available.

Bibliography

EWAN, J., 1967. A bibliography of Louisiana botany. *Southwestern Louisiana J.* 7: 2–81. (In *Flora of Louisiana*, 1, ed. J. W. Thieret, and collaborators.) [Annotated chronological list, with subject and name indices.]

THIERET, J. W., 1972. *Checklist of the vascular flora of Louisiana*, 1: Ferns and fern-allies, gymnosperms and monocotyledons. 48 pp. (Lafayette Nat. Hist. Mus. Tech. Bull., 2). Lafayette, La.

Unannotated checklist, with synonymy related to that given in major floristic works impinging upon the state. Nomenclature follows the Texas manual of CORRELL and JOHNSTON (**171**). A brief introductory section gives the plan of and basis for the work (only *vouchered* records included), and a table of family statistics (799 non-dicotyledonous species) and index appear at the end. [A second part is to cover the dicotyledons.]

Special groups

BROWN, C. A., 1945. *Louisiana trees and shrubs*. 262 pp., 147 halftones, 1 color pl., map (Louisiana Forestry Commiss. Bull., 1). Baton Rouge, La.

Semi-popular illustrated descriptive treatment of trees and shrubs, without keys; includes limited synonymy, vernacular names, generalized indication of local range, critical remarks, and notes on habitat, biology, uses, etc.; glossary, list of references, and complete index at end. An introductory section includes *inter alia* details of tree regions (with map). See also *idem.*, 1957. *Check list of woody plants of Louisiana, native, naturalized and cultivated*. 16 pp. (Louisiana Forestry Commiss. Bull., 8). Baton Rouge, La. [A 2nd edn. of this list, not seen by the author, appeared in 1964.]

Brown, C. A., 1972. *Wildflowers of Louisiana and adjoining states.* xl, 247 pp. *c.* 450 col. illus. Baton Rouge: Louisiana State University Press.

Popular illustrated guide, copiously illustrated by photographs but without keys; serves as a companion to the author's book on trees and shrubs (see above). The introductory section contains a size estimate of the Louisiana flora (3000 to 4000 spp.).

Region

17

Central Plains states and Texas

In this region are the states of Texas, Oklahoma, Kansas, Nebraska, South Dakota, and North Dakota. The four northern states comprise the Northern Central Plains subregion, the core territory for two major works which collectively extend west to the Rockies, east to the woodland 'ecotone', and south into Oklahoma, i.e., the total extent of the northern Great Plains. All states are also covered by state floras and/or checklists, largely of relatively recent date.

This region in fact contains the cores of two major floristic areas: the northern plains and the *llanos* of Texas respectively, with Oklahoma containing a transitional belt. The work of botanical circles has largely conformed to this pattern, with the result that there as never been a flora for the plains as a whole. With some exceptions, the eastern regional manuals do not extend into this territory. For general works on the northern plains subregion, see **175**.

No successor to Rydberg's *Flora of the prairies and plains of central North America* (1932) has yet appeared. However, a project for a new manual-flora for the northern plains subregion, based at Lawrence, Kansas as the Great Plains Flora Association, is presently under way and two key works have appeared: *Atlas of the flora of the Great Plains* (Barkley, 1977) and *Woody plants of the North Central Plains* (Stephens, 1973). The outstanding post-war event in the region, however, is the publication of *Manual of the vascular plants of Texas* (1970), which was in part distilled from the larger, incomplete *Flora of Texas* and superseded a heterogeneous collection of partly obsolete partial floras. Floras or checklists, mostly of modest dimensions,

have been published for each of the remaining five states, although for Nebraska the most recent work dates from the 1930s and none has achieved critical stature in a regional context. However, the new Texas manual is widely regarded as outstanding, and it and the Great Plains *Atlas* will set new standards for the region.

Bibliographies. General and divisional bibliographies as for Division 1.

Indices. General and divisional indices as for Division 1.

171

Texas

The large-scale *Flora of Texas* apparently ceased publication in 1969 with volume 3; nevertheless, the state is, considering its size and large, diverse flora, well-served by the modern manual (Correll *et al.*) from the same stable and by Gould's checklist. For eastern Texas trees, see also **160** (Harrar and Harrar); for the Panhandle, see also **175** (Great Plains Flora Association); for Trans-Pecos Texas, see also **180**, **181/II** (all works).

– Vascular flora: 4862 species (Correll *et al.*, 1970–2).

Correll, D. S., Johnston, M. C. and collaborators, 1970. *Manual of the vascular plants of Texas.* xv, 1881 pp., col. frontispiece, 3 maps. Renner, Tex.: Texas Research Foundation (distributed by Stechert-Hafner Service Agency, New York).

Descriptive manual of vascular plants, with keys to all taxa, essential synonymy, vernacular names, generalized indication of local and extralimital range or area of origin, concise critical remarks, and notes on habitat, phenology, special features, etc.; glossary, list of authorities, selected references, and complete index. An introductory section includes descriptions of floristic zones in the state (with maps) and a summary account of the flora.

– Designated as a 'Basic Flora' for the Flora North America Program. For amendments, see Correll, D. S., 1972. Manual of the vascular plants of Texas: 1. Additions and corrections. *Amer. Midl. Naturalist*, **88**: 490–6.

Gould, F. W., 1975. *Texas plants: a checklist and ecological summary.* [3rd] revised edn. 124 pp., map

(Texas Agric. Expt. Sta. Publ., MP-585). College Station, Texas.

Systematic list of vascular plants, with limited synonymy, vernacular names, and tabular indication of local range (ten floristic districts) as well as life-form, phenology, and status; bibliography, supplement, and index to family, generic and vernacular names. An introductory section includes accounts of physical features, climate, plant communities, and vegetation.

LUNDELL, C. L. and collaborators, 1942–69. *Flora of Texas*. Vols. 1–3. Illus. Dallas: Southern Methodist University (vol. 3); Renner: Texas Research Foundation (vols. 1, 2). (Distributed by Stechert-Hafner Service Agency, New York.)

This incomplete semi-monographic 'research' flora comprises a series of family revisions for the state, with keys to all lower taxa, full synonymy, with references, vernacular names, citation of *exsiccatae* and authorities for localities, generalized summary of internal and extralimital range, extensive taxonomic commentary, and notes on habitat, special features, uses, etc.; complete indices at end of each volume. Originally planned as a 'closed' work in ten volumes but from 1955 conducted in serial form; after publication of volume 3 suspended with the dissolution of the Texas Research Foundation. Families, as in *Flora Malesiana*, not published in systemtic sequence although presented in bound volumes.

172

Oklahoma

See also **175** (GREAT PLAINS FLORA ASSOCIATION). The aging, incomplete general flora of this botanically 'in between' state is complemented by the checklist and students' manual of Waterfall.

STEMEN, T. R. and MYERS, W. S., 1937. *Oklahoma flora*. xxix, 706 pp., 494 text-figs. Oklahoma City: Harlow Publishing Corporation.

Manual of vascular plants, with keys to genera and species; essential synonymy; vernacular names; generalized indication of local range; notes on habitat, frequency, life-form, etc.; representative illustrations; lists of endangered flowers, hay-fever plants, drug plants, useful water plants, and edible plants; glossary and indices to botanical and vernacular names. Does not include Gramineae, Cyperaceae, or Juncaceae; for

the former, see FEATHERLY, H. I., 1946. *Manual of the grasses of Oklahoma*. 137 pp. (Bull. Oklahoma Agric. Mech. Coll., 43, part 2). Stillwater, Okla.

WATERFALL, U. T., 1952. *Catalogue of the flora of Oklahoma*. 91 pp. Stillwater: Oklahoma A & M University Research Foundation.

WATERFALL, U. T., 1972. *Keys to the flora of Oklahoma*. 5th edn. 246 pp. Stillwater: the author. (First edn. 1960.)

The first of these complementary works comprises a systematic list of vascular plants, with limited synonymy and some literature citations; index to family and generic names at end. The second work, a students' manual, includes analytical keys to all taxa, a glossary, and generic index.

175

Northern Central Plains subregion

See also **140–50** (BRITTON and BROWN). This area, the heart of the 'Great Plains', comprises Kansas, Nebraska, and the two Dakotas, but the two major subregional works (Rydberg's flora and the *Atlas*) both cover substantial parts of adjacent states, including (for Rydberg) southern Manitoba and a corner of Saskatchewan. A substantial work on the woody flora is also cited below.

Bibliography

BROOKS, R. E., 1976. *A bibliography of taxonomic literature of the Great Plains flora*. 74 pp. (Rep. State Biol. Surv. Kansas, 2). Topeka, Kan.

THE GREAT PLAINS FLORA ASSOCIATION, 1977. *Atlas of the flora of the Great Plains*, coordinator R. L. McGregor; ed. T. M. Barkley. xii, 600 pp., 2217 maps. Ames, Ia.: Iowa State University Press.

Distribution atlas of the Great Plains vascular flora, with coverage from northern Oklahoma and the Texas panhandle to the Canadian border and from the Rocky Mountains to the woodland 'ecotone' in Minnesota, Iowa and Missouri; includes some 3000 spp. on 2217 dot distribution maps, with four maps per page, each captioned with accepted scientific and vernacular names. A list of infrequent taxa (pp. 557–78) and complete index appear at the end. A small general part contains introductory remarks, a list of families, and a county map showing limits of the area covered. Intended as a precursor to a projected new regional manual-flora of the Great Plains, superseding Rydberg's work.

RYDBERG, P. A., 1932. *Flora of the prairies and plains of central North America.* vii, 969 pp., 600 text-figs. New York: The New York Botanical Garden. (Reprinted 1965, New York: Hafner; 1972, Dover Publications.)

Illustrated descriptive manual of vascular plants, with keys to all taxa; essential synonymy; vernacular names; sketchy general indication of internal range and floristic zones; notes on habitat, phenology, etc.; glossary, list of authors, summary of the flora, and index to all botanical and vernacular names. Includes the states from Kansas north to North Dakota as well as Iowa, Minnesota, and adjacent parts of Canada. [Now considered to be largely of historical interest, according to McGregor *et al.* (see **176**), but useful for its illustrations.]

Special groups – pteridophytes, woody plants

PETRIK-OTT, A. J., 1979. The *pteridophytes of Kansas, Nebraska, South Dakota, and North Dakota.* 332 pp., 122 pls. (Beih. Nova Hedwigia, 61). Vaduz, Liechtenstein: Cramer.

Lavish illustrated treament of ferns and fern-allies (65 species), with keys, descriptions, indication of distribution (with maps), synonymy, vernacular names, karyotypes, critical remarks, and indication of habitat, occurrence, frequency, etc.; glossary and complete index.

STEPHENS, H. A., 1973. *Woody plants of the North Central Plains.* xxx, 530 pp., 2472 figs. on 255 pls., 2+255 maps. Lawrence: University Press of Kansas.

Atlas-flora of woody plants of Kansas, Nebraska, and the Dakotas, with a general key to species and separate keys in some 'critical' genera; each species text includes limited synonymy, vernacular name(s), an extensive description, dot maps, summary of extralimital range, and remarks on taxonomy, biology, special features, etc. A glossary, list of references, and complete index are given as well as an appendix on adventive species and doubtful and erroneous records. In the introductory section are remarks on distribution and floristic zones (with maps).

176

Kansas

See also **175** (all works).
BARKLEY, T. M., 1978. *A manual of the flowering plants of Kansas.* 2nd edn. vi, 402 pp., map. Manhattan,

Kan.: Kansas State University Endowment Association. (First edn. 1968.)

Manual-key to seed plants, with limited synonymy, vernacular names, generalized indication of local range, and notes on habitat, frequency, etc.; county map; glossary and index to family and generic names. A list of principal references appears in a brief introductory section. [Basically a corrected reprint of the first edition.]

MCGREGOR, R. L., BROOKS, R. E. and HAUSER, L. A., 1976. *Checklist of Kansas vascular plants.* 168 pp. (State Biol. Surv. Kansas, Techn. Publ., 2). [Lawrence], Kansas.

Systematic name list, with synonymy; indication of naturalization if applicable. The introductory portion covers previous lists along with technical matters; from p. 112 are lists of non-naturalized aliens and excluded species, a list of references (2 pp.), an appendix, and index. The order of families evidently follows the Cronquist system, but generic and specific arrangement is alphabetical. [Not seen; prepared from sample pages sent from the New York Botanical Garden Library.] See also GATES, F. C., 1940. *Annotated list of the plants of Kansas: ferns and flowering plants.* 266 pp., 1853 maps. Topeka: Kansas State Printing Plant. [Includes range maps for all species, not represented in either of the more modern works.]

177

Nebraska

See also **175** (all works). Nothing has as yet taken the place of the now-obsolescent flora of Winter (1936).

WINTER, J. M., 1936. *An analysis of the flowering plants of Nebraska, with keys.* 203 pp. (Bull. Conservation Dept. Univ. Nebraska, 13). Lincoln.

Concise, non-descriptive flora of flowering plants, with diagnostic keys to all taxa, limited synonymy (including literature citations), vernacular names, and notes on local range, habitat, life-form, phenology, critical features, biology, etc.; index to family, generic, and vernacular names.

178

South Dakota

See also 175 (all works).

VAN BRUGGEN, T., 1976. *The vascular plants of South Dakota.* xxvi, 538 pp., 3 pls., frontispiece (maps). Ames, Ia.: Iowa State University Press.

Concise students' manual to vascular plants, with descriptive keys which include essential synonymy, vernacular names, and brief indication of local distribution, habitat, phenology, etc.; selected taxonomic references in family headings; glossary, list of general references, and complete index at end. An introductory section gives details of physical features, vegetation, and floristic elements together with technical notes and statistics. Some adventive and cultivated plants are also included. [The concise style of this work resembles that of Jones' Illinois manual (see 154), but the presentation (offset from typescript) is not very satisfactory.]

WINTER, J. M., WINTER, C. K. and VAN BRUGGEN, T., 1959. A checklist of the vascular plants of South Dakota. 176 pp. Vermillion, SD: Department of Botany, State University of South Dakota.

This mimeographed 'provisorium' includes more technical taxonomic details than the students' manual, and gives an indication of previous work on the flora. [Description based upon notes furnished by P. F. Stevens, Cambridge, Mass.]

179

North Dakota

See also 175 (all works).

STEVENS, O. A., 1963. *Handbook of North Dakota plants.* Revised edn. (3rd printing). 324 pp., appendix. Fargo, ND: North Dakota Institute for Regional Studies. (First published, 1950.)

Descriptive manual of vascular plants, with illustrated keys to genera and families, essential synonymy, vernacular names, generalized indication of local range (with some county details here and there), and notes on habitat, phenology, frequency, special features, behaviour of alien species, etc.; glossary, addenda, and index to all botanical and vernacular

names. The introductory section includes chapters on physiography, soils, vegetation, floristics, botanical exploration, and plant names; an illustrated organography is also provided. The 'third printing' has an appendix with new data to 1962. See also HOAG, D. G., 1965. *Trees and shrubs of the northern plains.* xi, 376 pp., illus. (some in color), map (endpapers). Fargo.

Region

18

Western states

Included herein are the states of Montana, Idaho, Wyoming, Colorado, New Mexico, Arizona, Utah, and Nevada. The region thus encompasses such subregions as the Rocky Mountains proper, the Southwest desert areas, and the Intermountain Plateau or Great Basin. The subregions also can be taken to include western (Trans-Pecos) Texas, southeastern California, southeastern Oregon, and the Black Hills in South Dakota. Apart from the subregional floras, all states are now covered by state manuals and/or checklists, although a few are relatively old.

The Western United States is physiographically and floristically very heterogeneous and no single manual covers the entire area, except for the trees, covered by Preston's guide (1970). The field manual of Coulter and Nelson (1909) and the manual-flora of Rydberg (1922) are primarily limited to the Rocky Mountains subregion and the plains to the east (181/I). Both works are obsolescent, with Rydberg's manual having the added disadvantage of obsolete nomenclature. A new work to replace them is badly needed. Elsewhere, a new, important *Intermountain Flora* (Cronquist *et al.*, 1972–) has begun to appear and a similar work is projected for the Southwestern subregion under the editorship of Reveal. The latter area is at present well-supplied with works on its peculiar woody plants as well as 'cactus books' (not included here).

At state level, a goodly number of new floras and checklists have been published since 1939, covering most if not all states. Descriptive floras have appeared for Montana (Booth, 1950–66); Idaho (Davis, 1952), Wyoming (Porter, 1962–72; Dorn, 1977); Colorado (Harrington, 1964), Arizona (Kearney and Peebles,

1942, 1960; Tideström and Kittel, 1941, with inclusion of New Mexico); and Utah (Welsh & Moore, 1973). For Nevada there is a long series of contributions published by the United States National Arboretum (1940–65), one of the last 'state floras' published under Federal auspices. A number of partial floras, notably McDougall's *Seed plants of Northern Arizona* (1973), and some state checklists have also been published. However, none of these works has been widely influential, in part because of the region's heterogeneity, although the most critical appear to be those by Harrington, and by Kearney and Peebles. Some older standard works, notably the floras of New Mexico by Wooton and Standley (1915) and of Utah and Nevada by Tideström (1925), both recently reprinted, have yet to be fully superseded. Parts of the region are also covered by works centered outside, such as *Vascular plants of the Pacific Northwest* (see **191**) and *Flora of the Sonoran Desert* (see **201**).

Coverage may thus be considered moderately good, but it is of rather disparate standard and, as yet, is poorly consolidated in modern terms, except in one of the three subregions.
Bibliographies. General and divisional bibliographies as for Division 1.
Indices. General and divisional indices as for Division 1.

180

Region in general

The region is, as a whole, too large and botanically diverse to be covered by a single general flora, save the comprehensive North American works. Larger works on the western flora have centred on three subregions, here grouped under **181**. However, there is one general 'tree book', described below.

Guide to trees

PRESTON, R. J., Jr., 1970. *Rocky Mountain trees.* 3rd edn. pp. i–lix, 1–285, lxi–lxxxi, illus., maps. New York: Dover Publications. (First edn. 1940, Ames, Ia.: Iowa State College Press.)

Illustrated, semi-popular descriptive treatment, with keys to all genera and species; essential synonymy; vernacular names; indication of range by life-zones, with distribution maps for all species; extensive notes on diagnostic features, habitat, ecology, biology, wood, silvical characteristics, and

related shrubby species; glossary, list of references, and index to all subjects together with botanical and vernacular names. The introductory section gives an account of the region's life-zones (with map), notes on descriptive morphology and use of the work, and checklists of trees for each individual state. The Dover reprint incorporates corrections and updated nomenclature.

181

Subregions in general

Three are recognized here: I. *The Rocky Mountains and adjacent plains*, centering on the states of Montana, Wyoming and Colorado and extending into northern and eastern Idaho, eastern Utah, northeastern Arizona, northern New Mexico (and the Black Hills of South Dakota). II. *The Southwest*, centering on Arizona and New Mexico and extending into southeastern California and western (Trans-Pecos) Texas (as well as northern Mexico); and III. *The Intermountain Plateau*, containing the 'Great Basin' and centering on the states of Utah and Nevada, extending into southeastern Oregon, southern Idaho, southwestern Wyoming and small parts of eastern California and northern Arizona. All subregional works are here entered under their respective headings.

I. The Rocky Mountains and adjacent plains

See also **182** through **185**, and **188** (all works); **191** (HITCHCOCK *et al.*; HITCHCOCK and CRONQUIST). The work by Coulter and Nelson (1909) is included for its coverage of the full area defined here and its use of the International Code as then adopted; but both it and Rydberg's flora are for practical purposes largely obsolete.

COULTER, J. M., 1909. *New manual of botany of the central Rocky Mountains (vascular plants).* Second edn. revised A. Nelson. 646 pp. New York: American Book Co. (First edn. 1885, under title *Manual of botany...of the Rocky Mountain regions.*)

Field manual of vascular plants, with concise keys to all taxa; essential synonymy, with references; vernacular names; generalized indication of regional distribution; taxonomic commentary and notes on habitat, special features, etc.; statistical summary; list of new names and combinations; list of authors' names, glossary, and indices to all vernacular and botanical names. Nomenclature in this work follows the International Code of 1905, and the species concept adopted is broader than in Rydberg's *Flora of the Rocky Mountains and adjacent plains* (see below). The area covered includes Colorado, Wyoming, northern New Mexico, northeastern Arizona, eastern Utah, southeastern Idaho, and most of Montana, along with the Black Hills in South Dakota.

RYDBERG, P. A. 1922. *Flora of the Rocky Mountains and adjacent plains*. 2nd edn. xii, 1144 pp. New York: The New York Botanical Garden. (Reprinted 1954, 1961, New York, Hafner; 1st edn. 1917.)

Briefly descriptive manual-flora of vascular plants, with keys to all taxa; limited synonymy; vernacular names; concise indication of regional range, with indication of life-zones (Merriam system); notes on habitat, phenology, etc.; brief sketch of life-zones, glossary, and index to all botanical and vernacular names. Encompasses Montana, Idaho, Wyoming, Colorado, and Utah as well as adjacent parts of Canada. The second edition differs from the first only by the incorporation of an addendum (pp. 1111–44). For a very concise field version of this work, see *idem*, 1919. *Key to the Rocky Mountain flora*. 305 pp. New York: The author.

II. The Southwest

See also **105, 171, 186, 187, 195** and **196** (all works); **205** (SHREVE and WIGGINS). Works relating specifically to desert areas appear at **105** and **205**. General treatments for the southwestern United States are given here, although the subregion more particularly is defined as comprising Arizona and New Mexico. Subregional works include the general flora by Tideström and Kittel (1941), now largely superseded, and works by Little (1950) on the trees and Vines (1960) on woody plants.

TIDESTRÖM, I. and KITTEL, T., Sister, 1941. *Flora of Arizona and New Mexico*. xxvi, 897 pp., map. Washington, DC: Catholic University of America Press.

Concise descriptive manual-flora of vascular plants, with keys to all taxa, limited synonymy, vernacular names, generalized indication of local and extralimital range along with modified Merriam life-zones as appropriate, and notes on habitat, etc.;

index to generic and vernacular names at end. The introductory section contains, among other material, an account of early explorers in the area, together with a map. [An early example of a book reproduced by offset from typescript. Now largely superseded by more recent state floras in both Arizona and New Mexico.]

Guides to woody plants (and trees)

LITTLE, E. L., Jr., 1950. *Southwestern trees; a guide to the native species of New Mexico and Arizona*. ii, 109 pp., illus. (Agric. Handbook USDA, 9). Washington: Government Printing Office.

Descriptive, illustrated popular account, without keys; includes synonymy, vernacular names, local range, some critical remarks, and notes on habitat, ecology, uses, etc.; list of references and index.

VINES, R. A., 1960. *Trees, shrubs and woody vines of the Southwest*. 1104 pp., illus. Austin: University of Texas Press.

Copiously illustrated, descriptive systematic treatment of native and naturalized woody plants, with vernacular names, indication of local and extralimital range, and notes on habitat, variation, special features, uses, derivation of names, etc.; glossary, bibliography, and complete indices. All species are figured, but no keys are provided.

III. The Intermountain Plateau (Great Basin)

See also **182, 188, 189**, and **193** (all works). Previously without a general work of any kind and still unevenly known botanically, this subregion is now being treated by the fine modern *Intermountain Flora*, produced at the New York Botanical Garden under the direction of A. Cronquist.

CRONQUIST, A. *et al.*, 1972– . *Intermountain flora: vascular plants of the Intermountain West, USA*. Vols. 1– . Illus., maps. New York: Hafner (vol. 1); Columbia University Press (vol. 6).

Large-scale, illustrated descriptive 'research' flora, with keys to all taxa, synonymy (with references), vernacular names, generalized indication of local and extralimital range, taxonomic commentary, and notes on habitat, special features, karyotypes, variation, etc.; glossary and comprehensive indices in each volume. The introductory section (vol. 1) includes accounts of physical features, history of the flora, phytogeography, ecology, and botanical exploration. At this writing (July 1979), volumes 1 and 6 (of 6 projected) have been published, respectively covering the pteridophytes, gymnosperms and 'early' monocotyledons and the

Compositae – arranged according to the Englerian [!] sequence.

– Designated as a 'Basic Flora' for the Flora North America Program.

HOLMGREN, A. H. and REVEAL, J. L., 1966. *Checklist of the vascular plants of the Intermountain Region.* iv, 160 pp., map (U.S. Forest Serv. Res. Pap., INT-32). Ogden, Utah.

Systematic list of vascular plants, with rather extensive synonymy as well as vernacular names; map; list of references and complete index. The introductory section gives remarks on the limits of the checklist, physiography of the region, and botanical exploration. Serves as a precursor for Cronquist's *Intermountain flora* (see above).

182

Montana

See also **181**/I (all works); **175** (GPFA, 1977); **191** (HITCHCOCK and CRONQUIST, 1955–69, 1973).

Bibliography

HABECK, J. R. and HARTLEY, E., 1965. The vegetation of Montana: a bibliography. *Northw. Sci.* **39**(2): 60–72. [Accounts for local floristic works.]

BOOTH, W. E., 1950. *Flora of Montana*, I: *Conifers and monocots.* 232 pp. Bozeman, Mont.: Montana State College Research Foundation.

BOOTH, W. E. and WRIGHT, J. C., 1966. *Flora of Montana*, II: *Dicotyledons.* 305 pp., illus., maps. Bozeman: Montana State University.

Descriptive manual of seed plants, with keys to all taxa, essential synonymy, vernacular names, general indication of local range (with inclusion of distribution maps in the second volume), and notes on habitat, etc.; figures and photographs of diagnostic features; indices to all botanical and vernacular names in each volume. The introductory section in the first volume includes an illustrated organography. See also the partial successor to the first volume, which, however, does not cover grasses: HAHN, B. E., 1973. *Flora of Montana: conifers and monocots.* 143 pp. Bozeman: Montana State University.

183

Idaho

See also **181**/I, **181**/III (all works), **191** (all works except Gilkey). To the relatively long-standing flora by Davis has recently been added a checklist by Fulton.

Bibliography

DAUBENMIRE, R. F., 1962. Vegetation of the state of Idaho: a bibliography. *Northw. Sci.* **36**: 120–2. [Includes references for local floras.]

DAVIS, R. J. and collaborators, 1952. *Flora of Idaho.* iv, 828 pp. Dubuque, Ia.: Brown.

Descriptive manual-flora of vascular plants, with keys to all taxa but lacking illustrations; includes essential synonymy, vernacular names, rather sketchy indication of local range as well as summary of extralimital distribution, and notes on habitat, phenology, etc.; glossary and index to all botanical and vernacular names. The introductory section includes accounts of physiography, vegetation zones, phytogeography, and the origin of the flora (in part by R. F. Daubenmire). For revised nomenclature, see FULTON, D., 1976. *List of scientific and common plant names for Idaho.* 82 pp. Boise, Id.: Soil Conservation Service, US Department of Agriculture.

184

Wyoming

See also **181**/I, **181**/III (all works); **175** (GPFA, 1977). Local botanical work has a relatively long history dating back to Aven Nelson but only recently has a complete state flora been published, replacing Nelson's catalogue of 1890. The serial flora of Porter has evidently ceased publication.

– Vascular flora: 2144 species (Dorn).

DORN, R. G., 1977. *Manual of the vascular plants of Wyoming.* 2 vols. viii + viii, 1498 pp., 27 illus. New York: Garland.

Partly compiled, briefly descriptive manual-flora, with keys to all taxa, essential synonymy, vernacular names, notes on habitat and local distribution, and references to appropriate taxonomic papers at family and generic headings; a few figures of details; some special keys; glossary and complete index at end. A brief introductory section includes a key to families, which in the work are alphabetically arranged. Covers 2144 species in 605 genera. [Prepared between 1971 and 1974 and hence not claimed as 'critical', this work is unduly bulky; it is reproduced by offset lithography from typescript on thick paper with over-generous margins, a feature justly criticized by Weber (*Syst. Bot.* **2**: 230–1 (1977)) who claimed that, without loss, the manual could have been compressed to 631 pp. in one volume, a view supported by this writer.]

PORTER, C. L., 1962–72. *A flora of Wyoming.* Parts 1–8 (Wyoming Agric. Exp. Sta. Bull., 402, 404, 418, 434 (parts 1–4); Wyoming Agric. Exp. Sta. Res. J., 14, 20, 64, 65 (parts 5–8).) Laramie, Wyo.

Briefly descriptive (species only enumerated) flora of vascular plants, with keys to all taxa; essential synonymy; references to more detailed papers where relevant; vernacular names; generalized indication of local range (with some details); taxonomic commentary and notes on habitat, frequency, variation, biology, properties, uses, etc.; indices in each part (from part 3 onward). The introductory section includes accounts of physical features, climate, major floristic elements, and vegetation zones as well as a list of 108 families. As of 1979, eight parts had appeared, containing families up through the Fumariaceae (no. 45) on the classical Englerian system, but nothing more has appeared since 1972. See also the precursory series by *idem*, 1942–61. *Contributions toward a flora of Wyoming.* 34 parts. Laramie: Rocky Mountain Herbarium (mimeographed). [Includes many range maps as well as keys.]

185

Colorado

See also 181/I (all works); 175 (GPFA, 1977). In addition to Harrington's manual of 1964, reference is also made here to one of the first state checklists produced from a computerized data storage/retrieval system (compare with the British Columbia inventory, 124). For Weber's outstanding *Rocky Mountain flora*, see 103.

HARRINGTON, H. D., 1964. *Manual of the plants of Colorado.* 2nd edn. 666 pp. Chicago: Swallow Press. (First edn., 1954, Denver.)

Descriptive manual of vascular plants, with keys

to all taxa; limited synonymy; vernacular names; generalized indication of local and extralimital range, with some details; taxonomic commentary and notes on habitat, etc.; state map, with counties; glossary, summary of the flora, and index to vernacular, generic, and family names. The introductory section includes a summary account of the vegetation and its zonation (by D. Costello). The second edition differs only slightly from the original version.

– Designated as a 'Basic Flora' for the Flora North America program.

WEBER, W. A. and JOHNSTON, B. C., 1976. *Natural history inventory of Colorado*, I. Vascular plants, lichens and bryophytes. ii, 205 pp. Boulder, Colo.

Tabular checklist, printed from computer-generated 'printout' (vascular plants, pp. 1–148). Each taxon entry contains 12 descriptors – taxon number, genus, species, family, status, provenance, type of record, literature reference, Colorado references, remarks, major group (vascular plants, lichens, bryophytes), and rarity – on two lines. Arrangement of genera and species is alphabetical. Based on specimens at COLO (Boulder) along with literature and other reports and processed by the 'EXIR' program (a derivative of TAXIR).

186

New Mexico

See also 181/II (all works). The pioneer state flora of Wooton and Standley (1915) has as of this writing (late 1980) been supplanted by a new and compendious work by Martin and Hutchins, the result of a long organized program of state exploration which was summarized in interim fashion by Martin and Castetter's checklist of 1970. Also largely superseded by the new work is the combined Arizona and New Mexico manual-flora by Tideström and Kittel (1941; see 181/II) which is, however, rather less bulky.

– Vascular flora: 3728 species and infraspecific taxa (Martin and Hutchins).

MARTIN, W. C. and HUTCHINS, C. R., 1980–1. *A flora of New Mexico.* 2 vols. Illus., maps. Vaduz, Liechtenstein: Cramer (*apud* Gantner). (Vol. 1: pp. i–xii, 1–1276, figs. 1–755; vol. 2 not available at time of writing – January 1981.)

Briefly descriptive, well-illustrated flora of vascular plants (3728 entities recognized, many at infraspecific level), with keys to all taxa, limited synonymy, 'standardized vernacular names', generalized indication of internal and extralimital distribution (supplemented by dot distribution maps, each depicting three entities), altitudinal zonation, and notes on habitat, phenology, etc.; citations of pertinent literature for each genus after the last of its species; glossary, list of authorities and abbreviations, bibliography, and general index (these last are given in vol. 2). The introductory section (in vol. 1) encompasses physical features, climate, geology, and vegetation of the state, along with an explanation of the unit distribution maps and family key. [The contents of this over-generously produced work could have been spatially considerably compressed without loss of content, and, with the use of less substantial paper, contained in less bulky volumes.] Preceded by an interim work: MARTIN, W. C. and CASTETTER, E. F., 1970. *A checklist of gymnosperms and angiosperms for New Mexico*. 245 pp. Alberquerque, NM: The authors (through Department of Biology, University of New Mexico).

WOOTON, E. O. and STANDLEY, P. C., 1915. *Flora of New Mexico*. 794 pp (Contr. US Natl. Herb., 19). Washington: Smithsonian Institution.

Relatively detailed flora of vascular plants, with keys to all taxa and descriptions of genera and families (but not species), synonymy (with references), vernacular names, indication of local range (including use of Merriam life-zones) and summary of extralimital distribution; taxonomic commentary, and notes on habitat, special features, uses, etc.; gazetteer and complete index at end.

187

Arizona

See also 181/II, 181/III, 201 (all works). In addition to the complementary state floras by Kearney and Peebles and a recent checklist (Lehr, 1978), a substantial manual on the northern part of the state is also included here. This latter area, which takes in the Grand Canyon and the San Francisco Peaks, is the 'type locality' for C. H. Merriam's 'life-zone' biotic classification, widely used in natural history writing including a number of American floras.

Bibliography

SCHMUTZ, E. M., 1978. Classified bibliography on native plants of Arizona. xii, 160 pp. Tucson: University of Arizona Press. [Comprehensive though unannotated treatment with several topical headings; floras, pp. 34–53, and trees, pp. 130–2.]

KEARNEY, T. H., PEEBLES, R. H. and collaborators, 1960. *Arizona flora*. 2nd edn., with supplement by J. T. Howell and E. McClintock. 1085 pp., illus. Berkeley: University of California Press. (First edn. 1951.)

Concise manual of vascular plants, with abbreviated keys to all taxa, very limited synonymy, vernacular names, fairly detailed indication of local range and summary of extralimital distribution, extensive taxonomic commentary, and notes on habitat, frequency, biology, uses, etc.; glossary, list of references, and complete indices at end. An introductory section, with a state map, includes accounts of physical features, geology, soils, climate, vegetation, life-forms, floristic elements, and botanical exploration. The above 'descriptive' work is complemented in a more 'technical' form by *idem*, 1942. *Flowering plants and ferns of Arizona*. 1069 pp. (Misc. Publ. USDA, 423). Washington: Government Printing Office. [Similar to the above, but with more detailed keys, synonymy, and accounting of local distribution and lacking vernacular names and extensive commentary.]

LEHR, J. H., 1978. *A catalogue of the flora of Arizona*. vi, 203 pp. Phoenix: Desert Botanical Garden.

Systematic checklist of vascular plants, with occasional notations, relating to taxonomic problems, etc., as references; vernacular names (for genera, families, and species) where available. [Not seen; description based on materials sent from New York Botanical Garden Library.]

Partial work

McDOUGALL, W. B., 1973. *Seed plants of Northern Arizona*. 594 pp., frontispiece. Flagstaff, Ariz.: Museum of Northern Arizona.

Concise manual-flora for the five northern counties, with abbreviated keys to all taxa, brief descriptions, scant synonymy, vernacular names, indication of status and of Northern Arizona and overall distribution, and notes on habitat, altitudinal zonation, phenology, etc.; a few critical remarks; glossary and index to vernacular, generic and family names. Based partly on Kearney and Peebles (see above) but written to provide more detail on this botanically important area (centering on the Grand Canyon).

188

Utah

See also 181/I, 181/III (all works). The recent students' manual by Welsh and Moore (1973) does not cover the total flora, and Tideström's manual is by now rather outdated.

Bibliography

CHRISTENSON, E. M., 1967. *Bibliography of Utah botany and wildlife conservation.* 136 pp. (Sci. Bull. Brigham Young Univ., Biol. Ser., 9). Provo, Utah. [Includes local floristic works.]

TIDESTRÖM, I., 1925. *Flora of Utah and Nevada.* 665 pp., illus. (Contr. US Natl. Herb., 25). Washington: Smithsonian Institution. (Reprinted 1969, Lehre, West Germany: Cramer.)

Manual-key to vascular plants, with synonymy and references; vernacular names; generalized indication of local and extralimital range, including life-zones (Merriam system); notes on habitat, etc.; index to all botanical and vernacular names. An introductory section features descriptions of life-zones and plant communities, as well as statistics of the flora (in part by H. L. Shantz and A. W. Sampson).

WELSH, S. L. and MOORE, G., 1973. *Utah plants: Tracheophyta.* iv, 474 pp., illus. Provo: Brigham Young University Press. (Second edn. 1965, as *Common Utah plants.*)

Concise manual-key to some 2500 native, naturalized, and commonly cultivated vascular plants; includes descriptions of genera and families, brief notes on habitat, occurrence, and local range for each species, illustrations of representative taxa, and vernacular names (where known). An illustrated glossary, bibliography (pp. 435–6) and complete index conclude the work. Does not represent a complete account of the state's flora.

189

Nevada

See also 181/II, 181/III (all works especially Cronquist *et al.*); 188 (Tideström). Tideström's Utah manual covers the state only sketchily; and the nearest approach to a modern state flora is the 50-part series of contributions issued by the United States National Arboretum. Partial works are available for the northeastern and central-southern areas of the state.

Bibliography

TUELLER, P. T., ROBERTSON, J. H. and ZAMORA, B., 1978. *The vegetation of Nevada – a bibliography.* 28 pp. Reno, Nev.: Agricultural Experimental Station, College of Agriculture, University of Nevada. [Includes local floristic references.]

UNITED STATES NATIONAL ARBORETUM, 1940–65. *Contributions toward a flora of Nevada.* 50 parts. Maps. Washington: Crops Research Division, Agricultural Research Service, USDA (for the National Arboretum). (Mimeographed.)

Comprises a series of fairly detailed descriptive revisions of seed plant families and other groups, with keys, full synonymy and references, localities with citations of *exsiccatae*, general summary of local and extralimital range (including some distribution maps), taxonomic commentary, and notes on habitat, phenology, special features, etc.; each part separately indexed. Many treatments have been contributed by W. A. Archer, with others by specialists; for full list, see the state bibliography above.

Partial works

BEATLEY, J. C., 1976. *Vascular plants of the Nevada test site and central-southern Nevada: ecologic and geographic distributions.* viii, 308 pp., 28 figs. (incl. 4 maps, 24 halftones). Washington: Energy Research and Development Administration (distributed by US National Technical Information Service, Springfield, Va.).

Includes a systematic enumeration (pp. 110 ff.) of vascular plants, with details of local distribution, altitudes, life-form, phenology, and associates; no vernacular names. The lengthy general part includes details of physical features, climate, vegetation, floristics (with floristic regions), disturbed vegetation, and threatened and endangered species, with a list of references. [Represents in effect an evaluation of the flora and vegetation in an area subjected to atomic radiation in varying degress, having long been used for bomb tests.]

HOLMGREN, A. H., 1942. *A handbook of the vascular plants of northeastern Nevada.* vi, 214 pp. [Logan], Utah: Utah State Agricultural College and Experiment Station/ Grazing Service, US Department of the Interior.

A keyed, annotated systematic enumeration of vascular plants, with 'official' (SPN – Standardized Plant Names) vernacular names and notes on habit, special features, frequency, habitat, etc.; indication of overall distribution.

The introductory section covers geography, physiography, climate, history of botanical exploration, statistics of the flora, and technical matters; list of references (p. vi). An index and glossary conclude the work. Compiled under the direction of the Intermountain Herbarium (Laramie, Wyo.), this work covers the Elko grazing district (Elko County and northern portions of Eureka and Lander Counties).

Region

19

Pacific Coast states

This region nominally comprises the states of Washington, Oregon and California, and is here considered to include the whole of the Cascades and the Sierra Nevada. Geographically and floristically, however, it is rather diverse, and in practice three subregions can be recognized: the Pacific Northwest, here formally designated as **191**; Northern and Central California, including the Central Valley; and Southern California with the Channel Islands. The first-named extends into British Columbia, while the last includes part of the Southwestern desert flora (**181/II**) and extends into Mexico. Parts of eastern California and Oregon fall within the Intermountain Plateau.

Since 1939, several major floras have been published, bringing consolidated knowledge of the vascular plants to a level equalled only by the Northeast and Midwest. The *Illustrated Flora of the Pacific States*, a West Coast answer to 'Britton and Brown' begun by L. Abrams in 1923, was completed in 1960 by R. S. Ferris and covers the entire region. A number of books on trees covering the states in the region also are available, none of them recent although *Illustrated Manual of Pacific Coast Trees* (McMinn and Maino, 1935; 2nd edn., 1946) is perhaps the most generally useful.

The region's physiographic and biological heterogeneity has led to a number of works of more restricted scope being produced. The most influential recent contributions are *Vascular plants of the Pacific Northwest* (1955–63), directed by Hitchcock and Cronquist (condensed in 1973 into a one-volume manual) and *A California Flora* (1959; supplement 1968) by Munz and Keck, the latter one of the first manuals to introduce 'biosystematic' information. However, for practical use the latter has not fully

supplanted Jepson's *Manual* (1923–5), still in demand due to its abundance of illustrations. Oregon also features a state flora (Peck, 1941; 2nd edn., 1961). Many partial and local floras are also available for different parts of the region, the most substantial being *A flora of Southern California* (Munz, 1974), which succeeded a more modest manual of 1935 by the same author. Unfortunately, no separate descriptive treatment for the botanically distinctive California Channel Islands is available, apart from Eastwood's rather sketchy checklist of 1941 and a number of florulas and lists for individual islands.

A feature of most floras in this region, as well as in the Western States, is the relatively limited use of county distribution maps. This may be related not only to insufficient information but also to the rugged topography which would require that other criteria be used in species mapping.

Bibliographies. General and divisional bibliographies as for Division 1.

Indices. General and divisional indices as for Division 1.

190

Region in general

The whole area is covered by one general work of the same class as Britton and Brown's *Illustrated flora of the northern United States*. For more detail, the mostly relatively recent subregional or state works should be consulted.

ABRAMS, L. and FERRIS, R. S. 1923–60. *Illustrated flora of the Pacific States*. 4 vols. Illus. Stanford: Stanford University Press. (Reprint of vol. 1, 1940.)

Briefly descriptive atlas-flora of vascular plants, with keys to all taxa; rather full synonymy, with references; vernacular names; generalized indication of local, regional, and extralimital distribution, including life-zones; notes on habitat, frequency, phenology, special features, etc.; indices in each volume (complete general index in vol. 4). The introductory section includes *inter alia* descriptions of physical features, Merriam life-zones, and floristic elements. In the reprint of vol. 1, nomenclature was altered to conform to the International Code in place of the 'American Code'. [As noted above, inspiration for this work came from the successful atlas-flora of Britton and Brown, see **140–50**.]

Guides to trees

ELIOT, W. A., 1938. *Forest trees of the Pacific Coast.* 565 pp., illus. New York: Putnam's (reissued 1948 as a revised edn.).

Concise, popularly oriented, illustrated descriptive treatment of trees, with some keys to species, vernacular names, very detailed indication of internal range, and notes on habitat, frequency, special features, etc.; full synoptic list of trees, with details of distribution; complete index.

MCMINN, H. F. and MAINO, E., 1946. *An illustrated manual of Pacific Coast trees.* 2nd edn. xii, 409 pp., 415 text-figs., 1 col. pl. Berkeley: University of California Press. (First edn. 1935.)

Fully keyed, briefly descriptive semi-popular treatment, including native, naturalized, and many cultivated species; includes synonymy, vernacular names, generalized indication of regional distribution or country of origin, and notes on habitat, phenology, cultivation, special features, etc.; lists of trees recommended for particular uses; glossary, list of references, and index to all botanical and vernacular names.

SUDWORTH, G. B., 1908. *Forest trees of the Pacific slope.* xv, 455 pp., 207 illus., 2 maps. Washington: Govt. Printing Office. (Reprinted with revised nomenclature, 1967, New York, Dover Publications.)

Illustrated, semi-popular descriptive systematic treatment, without keys; includes synonymy, vernacular names, and extensive notes on distribution, habitat, ecology, special features, wood, bark, properties, uses, etc.; complete index. All species are figured. (The Dover reprint features a table of necessary nomenclatural changes provided by E. S. Harrar.)

191

Washington (and Pacific Northwest in general)

See also **121/I** (Taylor). The last flora specifically treating Washington State is C. V. Piper's *Flora of the state of Washington* (1906), now effectively superseded. The later general works by Hitchcock (with Cronquist and others) have extended coverage to the geographical entity known as the Pacific Northwest, an area which, along with Washington, includes the greater part of Oregon, northern Idaho, far western Montana, and southern British Columbia. Two useful students' manuals, respectively covering western Washington and Oregon and southeastern Washington and adjacent parts of Idaho, are also accounted for here.

Bibliography (Washington State only)

DAUBENMIRE, R. F., 1962. Vegetation of the state of Washington: a bibliography. *Northw. Sci.* **36**: 50–4. [Includes local floristic works.]

HITCHCOCK, C. L. *et al.*, 1955–65. *Vascular plants of the Pacific Northwest.* 5 vols. Illus. Seattle: University of Washington Press.

Comprehensive, illustrated descriptive 'research' flora of vascular plants with keys to all taxa; extensive synonymy and references; vernacular names; relatively detailed indication of local range; general summary of extralimital distribution; extensive taxonomic commentary and notes on habitat, phenology, etc.; well-executed figures of most species; indices to all botanical and vernacular names in each volume. The introductory section includes an index to families as well as a glossary. The area covered by this work encompasses Washington, northern Oregon, northern Idaho, western Montana, and southern British Columbia.

– Designated as a 'Basic Flora' for the *Flora North America* program.

HITCHCOCK, C. L. and CRONQUIST, A., 1973. *Flora of the Pacific Northwest: an illustrated manual.* xix, 730 pp., illus. Seattle: University of Washington Press.

Manual-key to vascular plants, with limited synonymy, vernacular names, concise indication of local range, and notes on habitat, phenology, life-form, etc.; diagnostic figures; complete index. Essentially a condensation, with revisions, of the preceding work; covers the same area.

Partial works

GILKEY, H. M. and DENNIS, LaR. J., 1967. *Handbook of Northwestern plants.* [vi], 505 pp., illus. Corvallis, Ore.: Oregon State University Bookstores.

Briefly descriptive students' manual of vascular plants, with keys to all taxa, vernacular names, distinguishing features, and notes on local distribution, frequency, habitat, etc.; glossary and complete index. Covers western Oregon and Washington from the Cascade Range to the Pacific coast, and includes the Olympic Peninsula with its 'rain forest'.

ST JOHN, H., 1963. *Flora of southeastern Washington and adjacent Idaho.* 3rd edn. xxix, 583 pp., illus. Escondido, Calif.: Outdoor Pictures. (First edn. 1937.)

Students' manual of seed plants, with keys to all taxa, brief descriptions, partial synonymy, vernacular names, local distribution (with indication of Merriam life-zones), and notes on habitat, frequency, biology, etc., with some critical commentary; glossary, lexicon and index at end. Covers an area in two states centering on Pullman, Wash. (Washington State University), an important agricultural region.

193

Oregon

For the northern part, see also **191** (all works except ST JOHN). The state as a whole is covered by Peck (1961).

Bibliography

FRANKLIN, J. F. and WEST, N. E., 1965. Plant communities of Oregon: a bibliography. *Northw. Sci.* **39**: 73–83. [Includes floristic papers.]

PECK, M. E., 1961. *A manual of the higher plants of Oregon.* 936 pp. [Portland]: Binfords & Mort (in cooperation with Oregon State University Press, Corvallis).

Descriptive manual of vascular plants, with keys to all taxa; essential synonymy; vernacular names; concise indication of local and extralimital range; miscellaneous notes; illustrated glossary and index to all botanical and vernacular names. The introductory section gives a description of floristic regions in Oregon, with map.

195

California (in general)

Under this heading are listed works pertaining to the state as a whole. For southern California, see also **196**; for the Californian Channel Islands, see **197**. With a large population and having a long-autonomous botanical tradition, California is well-supplied with general and partial works.

– Vascular flora: about 6000 native, naturalized and adventive species.[11]

Bibliography

HOWARD, A. Q., 1974. *An annotated reference list to the native plants, weeds, and some of the ornamental plants of California.* 34 pp. Berkeley, Calif.: Cooperative Extension, University of California. [Includes coverage of local floras.]

JEPSON, W. L., 1909– . *A flora of California.* Vols. 1–4 (in parts). Illus. Berkeley: the author (vols. 1–3; distributed through Associated Students' Store at the University of California and other outlets); Jepson Herbarium and Library, University of California (vol. 4, part 2). (From vol. 4, part 2 edited by L. R. Heckard and R. Ornduff.)

Comprehensive descriptive 'research' flora of seed plants, with keys to genera and species, full synonymy with references and citations, indication of localities with mention of *exsiccatae* and other records along with generalized summary of local and extralimital range, extensive taxonomic commentary, and brief notes on habitat, special features, etc.; large figures and photographs of representative species and/or diagnostic features thereof (480 to date); no indices (unpublished save that to family and generic names in volume 2, although in the recent volume 4, part 2, there is an index to families treated as of 1979). A historical sketch of Californian botany appears in volume 2. Arrangement follows the classical Englerian system; volume 1 (1909–22, completed save for an introduction and indices) covers families from Pinaceae to Fumariaceae (parts 1–7; parts 8(1), pp. 1–32, and 8(2), pp. 579 ff., i.e. the index, lacking); volume 2 (1936, complete) covers Capparidaceae to Cornaceae (in 3 parts); volume 3 (1939–43, not completed) covers Lennoaceae to Solanaceae (parts 1–2; the introductory pages 1–16 and a final part as well as the index are lacking); and volume 4 (begun 1979) covers at this writing only the Rubiaceae (in part 2). [The recent resumption of publication of this work, one of the most important state floras in North America, is to be welcomed. Approximately 20–25 per cent of the seed plants have still to be covered, however, and revision of the earlier parts is now necessary. For announcement of volume 4, part 2, not seen by this writer, see *Taxon*, **30**: 547 (1981).]

JEPSON, W. L., 1923–5. *A manual of the flowering plants of California.* pp. 1–24, 24a–24h, 25–1238; 1023 text-figs. Berkeley: Sather Gate Bookshop and Associated Students' Store. (Reprinted 1938; 1951, 1960, Berkeley/Los Angeles: University of California Press.)

Illustrated descriptive manual of vascular plants, with keys to all taxa; includes essential synonymy, vernacular names, generalized indication of local and extralimital range, with occasional details, and notes on habitat, phenology, etc.; glossary and index to all botanical and vernacular names. The introductory section includes chapters on phytogeography, Merriam life-zones, endemism, variation, and taxonomic philosophy. The many figures form an especially useful feature of this work, which is still widely used and appreciated despite the existence of the newer *A California flora* by Munz (see below).

MUNZ, P. A. (in collab. with D. D. Keck), 1959. *A California flora.* 1681 pp., 134 text-figs., 1 color pl.,

7 maps. Berkeley/Los Angeles: University of California Press. Continued by MUNZ, P. A., 1968. *Supplement to 'A California Flora'*. 224 pp. Berkeley/Los Angeles. (Both works subsequently reprinted and bound together as one volume.)

Descriptive manual of vascular plants, with keys to all taxa; limited synonymy; vernacular names; generalized indication of local range (organized according to community-units) as well as extralimital distribution; taxonomic commentary and notes on habitat, karyotypes, special features, variation, etc.; figures representative for families and subfamilies; glossary, list of vernacular names with botanical equivalents, and index to all botanical names. The introductory section includes chapters on geology and vegetation, with descriptions of 29 community-units.

– Designated as a 'Basic Flora' for the *Flora North America* program.

196

Southern California

See also 181/II, 195 (all works). This area, botanically distinctive, features a substantial recent manual-flora here separately accounted for.

MUNZ, P. A., 1974. *A flora of southern California*. 1086 pp., 103 pls., map. Berkeley/Los Angeles: University of California Press.

Illustrated manual-flora of vascular plants, with keys to all taxa down to species, concise descriptions, limited synonymy (without references), vernacular names, and brief notes on habitat, occurrence, frequency, local and extralimital range, altitude, karyotypes, and vegetation formations; glossary and indices to all vernacular and botanical names. The short introductory chapter includes sections on geography, geology, vegetation formations, the Californian Channel Islands, phytogeography, and the plan of the work. Replaces *idem*, 1935. *A manual of southern Californian botany*. Claremont, Calif.

197

Channel Islands (California)

See also 195, 196 (all works, especially MUNZ on Southern California). These islands, whose flora exhibits several unusual features, are located in the Pacific Ocean just off the southern Californian coast. Somewhat related to them floristically is the more distant island of Guadelupe (011). The only work covering the group as a whole appears to be Eastwood's checklist of 1941; however, certain recent enumerations for individual islands have also been noted here.

EASTWOOD, A., 1941. *The islands of Southern California and a list of the recorded plants*, I–II. *Leafl. W. Bot.*, **3**. 27–36, 54–78.

Part II of this work comprises a systematic list of vascular plants of the Channel Islands, with tabular indication of distribution by islands (and including some citations of authorities); list of references but no index. The introductory section (part I) includes brief descriptions of each island with encapsulated histories of botanical exploration. Some enumerations of more recent date for individual islands have appeared, viz.: RAVEN, P. H., 1963. A flora of San Clemente Island, California, *Aliso* **5**: 289–347; and THORNE, R. F., 1967. A flora of Santa Catalina Island, California. *Ibid.* **6**(3): 1–77, Fig. 1–28 (incl. map).

Notes

1 See Shetler, S. G. *et al.*, 1973. A guide for contributors to Flora North America (FNA). Provisional edition. ix, 28 pp., appendices A–E (unnumbered) (Flora North America Rep., 65). Washington. [Short list, app. D.]

2 In the FNA organization, the central unit was located in the Smithsonian Institution. The projected revival called for an editorial base at another location.

3 One of the last 'colonial' works was the series of treatments of families and genera of Southeastern plants begun at Cambridge, Mass., in 1954 and published serially in *Journal of the Arnold Arboretum* (see 160). Initially well-conceived, this partly privately-supported venture has been overtaken by events and is now somewhat of a *perpetuum mobile*, still far from evident completion at this writing (1981).

4 The average American manual-flora is bulkier and less well adapted for field use than its European counterpart. This is perhaps as much due to tradition as to factors such as larger areas and floras, reinforced by widespread use of the automobile for botanizing. [Standley's hopes to start a Honduras flora after World War II largely depended on when gasoline and tires would be available.]

5 The news of the second suspension was communicated to the writer by C. J. Humphries, December 1979.

6 In Ewan, J., 1969. *A short history of botany in the United States*, p. 113. New York.

7 For obituary of Père Louis-Marie and further information on his *Flore-manuel*, see Boivin, B., 1979. Père Louis-Marie (1896–1978). *Taxon*, **28**: 432.

8 See Aymonin, G. G., 1974. La naissance de la 'Flore descriptive et illustrée de la France' de l'Abbé Hippolyte Coste. *Taxon*, **23**: 607–11.

9 Although in this work specific reference is made to the atlas supplement of Winterringer and Evans (1960), no mention is made of the original work by Jones and Fuller (1955). As with Java (**918**), Illinois floristics has been marked by 'controversy'.

10 Described in Radford, A. E. *et al.*, 1967. Contributor's guide for the Vascular Plants of the Southeastern United States. Chapel Hill, NC; also *idem*, 1974. *Vascular plant systematics*, pp. 501–21. New York.

11 See Ornduff, R., 1974. *An introduction to California plant life*. Berkeley, Los Angeles. The total of native species is just over 5000, with thirty per cent endemism.

Division

2

Middle America

What I would venture to suggest is a work in 8vo, without plates, scientific yet intelligible to any man of ordinary education; and, the country that I particularly have in view is the British West Indian Islands, so rich in useful vegetable products. I have reason to know that a very able botanist, Dr Griesbach [*sic*], is only deterred from publishing this Flora, by the fact that such works are not remunerative to the author...A sum of £300 would be required.

> W. J. Hooker to the Colonial Office, May 14, 1857; quoted from Thistleton-Dyer, Botanical survey of the Empire, in *Bull. Misc. Inform.* (Kew), *1905*: 12 (1906). [This was the first of the 'Kew floras.']

In tropical America...the flora has been studied from isolated centers with little regard for the species accepted at other centers, but with the assumption that each area is floristically distinct. Correlation through monographic work, covering a genus throughout its range, will reduce the species that have been multiplied unnecessarily.

> Standley in *Flora of the Panama Canal Zone* (1928); quoted from Prance in *Ann. Missouri Bot. Gard.* **64** (1978).

This division may be defined as including Mexico, the Central American countries from Guatemala to Panama, and the West Indies from Cuba and The Bahamas east and south around the Caribbean arc to Trinidad and the southern Netherlands Antilles (Curaçao, Aruba, and Bonaire). Apart from the seven geopolitical regions here delimited, one heading has been established for the Sonoran Desert, a physiographic entity for which a significant flora is available.

Introductory remarks on the history and progress of floristic botany in Middle America are here given under two superregional headings: 21–23, Mexico and Central America, and 24–29, The West Indies. This separation arises from their largely mutually distinctive patterns of exploration, botanical 'development' and floristic writing, which moreover have involved to a considerable extent different sets of personalities – a situation which has also existed, and continues to exist, within each superregion, rendering the creation of new overall floras difficult if not impossible. However, the greater part of both superregions is currently being encompassed by the supranational *Flora Neotropica* undertaking, which began work in 1964; for descriptions of this project, see MAGUIRE, B., 1973. The organization for 'Flora Neotropica'. *Nature and Resources*, **9**(3): 18–20, and GENTRY, A., 1979. Flora Neotropica news. *Taxon*, **28**: 647–9. [The latter represents the first of a current series covering this long-term project; the latest report seen is in *Taxon*, **30**: 81–7 (1981).]

General bibliographies. Bay, 1910; Blake and Atwood, 1942; Frodin, 1964; Goodale, 1879; Holden and Wycoff, 1911–14; Jackson, 1881; Pritzel, 1871–7; Rehder, 1911; Sachet and Fosberg, 1955, 1971; USDA, 1958. See also under the superregions.

Divisional bibliography

AGOSTINI, G., 1974. Taxonomic bibliography for the neotropical flora. *Acta Bot. Venez.* 9: 253–85. [Covers the six families from Acanthaceae through Anacardiaceae in alphabetical order. No continuation yet seen.]

General indices. BA, 1926– ; BotA, 1918–26; BC, 1879–1944; BS, 1940– ; CSP, 1800–1900; EB, 1959– ; FB, 1931– ; IBBT, 1963–9; ICSL, 1901–14; JBJ, 1873–1939; KR, 1971– ; NN, 1879–1943; RŽ, 1954– . See also under the superregions.

Divisional indices

TORREY BOTANICAL CLUB, NEW YORK, 1969. *Index to American botanical literature, 1886–1966.* 4 vols. Boston: Hall. [Originally published serially in *Bulletin of the Torrey Botanical Club* as well as on index cards. In this reprint, taxonomic and other non-author entries appear in volume 4. The Index has been continuing in the *Bulletin*, and the first 10-year cumulative supplement (1967–76) has now been published (1977, Boston; 740 pp.).]

AMERICAN SOCIETY OF PLANT TAXONOMISTS, 1939–67. *The taxonomic index.* Vols. 1–30. New York (later Cambridge, Mass.; from vol. 20 (1957) published serially in *Brittonia*). (Vols. 1–9 mimeographed.) [Begun on the initiative of W. H. Camp, this index had an existence of 28 years but from the mid-1940s it was reproduced from the appropriate parts of the *Index to American Botanical Literature*. In 1957 it was consolidated with *Brittonia* although retaining separate voluminination, and in 1967 discontinued.]

FORERO, E. (coord.), 1978– . *Boletín botánico latinoamericano.* Nos. 1– . Bogotá. (Offset-printed.) [Primarily a newsletter, but includes news of floras and flora projects as well as a limited selection of other literature references. Eight numbers published at time of writing (late 1981).]

Conspectus

200

Division in general

See 100 (BRITTON).

201

Tropical Middle America

See 301 (ORGANIZATION FOR FLORA NEOTROPICA; PITTIER).

205

Deserts

Of the extensive areas in Middle America which can be considered as deserts, only the Sonoran Desert, which extends across southeastern California, southwestern Arizona, the central (and largest) part of the Baja California peninsula, and most of the northern and central parts of the Mexican state of Sonora, has well-consolidated floras. The portions within the United States are additionally covered by the floras and related works on the Southwest in general (105, 181), Arizona (187), and California (195), but no similar coverage, apart from what is recorded in general works (210), exists for the Mexican portions except in the work described below and in *Flora of Baja California* (211), both based upon substantial field work by their author along with many other collectors.

WIGGINS, I. L., 1964. Flora of the Sonoran desert. In *Vegetation and flora of the Sonoran desert* (ed. F. SHREVE and I. L. WIGGINS). 2 vols. 1740 pp., 37 pls., 27 maps. Stanford: Stanford University Press.

The greater part of this work comprises a descriptive flora of vascular plants of those parts of Baja California and Sonora, together with corresponding parts of the states of California and Arizona, which comprise the Sonoran Desert; included are keys to all taxa, concise synonymy, references and some citations of significant revisions and monographs, fairly detailed indication of local and extralimital range, taxonomic commentary, and notes on habitat, special features, etc., and an index to all botanical names. The introductory section of the *Flora* incorporates a history of botanical exploration in the area as well as itineraries of the two authors. The *Flora* proper is preceded by an extensive illustrated descriptive account of the vegetation by F. SHREVE (originally published in 1951 as Publ. Carnegie Inst. Wash. **591**). Complementing this flora is the following: HASTINGS, J. R., TURNER, R. M. and WARREN, D. K., 1972. *An atlas of some plant distributions in the Sonoran Desert*. 255 pp., maps (Techn. Rep. Meterol. Climatol. Arid Regions, Inst. Atmos. Phys., Univ. Arizona, 21). Tucson. [Detailed range maps of 238 species.]

Superregion

21–23

Mexico and Central America

This superregion comprises the continental portion of Middle America, together with closely associated islands. It thus extends from the northern boundaries of Mexico to the southern boundary of Panama. For the nearby islands of the Eastern Pacific Ocean, including Guadelupe, Rocas Alijos, the Revillagigedo group, Clipperton, and Cocos, see Region 01.

The southern portion of the superregion, from the Mexican state of Chiapas southwards, is estimated as having 18 000–20 000 species of vascular plants (L. O. Williams in GENTRY, 1978). It is thus likely that the whole area, with its extremes of relief and climate, has at least 30 000 species. In spite of a long history of botanical exploration, particularly in Mexico, knowledge of the different parts of the superregion is very uneven, and no more than one third is covered by floras published since 1930. An additional proportion is covered by incomplete or older works.

In Mexico, the gathering of botanical information dates back to the Aztec Herbal and the work of Hernández, through which some pre-Columbian traditions were preserved; serious field work, however, only began at the end of the eighteenth century when, as part of the Spanish king, Charles III's, botanical programme, a seventeen-year 'Real Expedición Botánica á Nueva España', which ultimately ranged from Guatemala through Mexico to California and Vancouver Island, was conducted by M. Sessé with a number of associates, chief among them Mociño, from 1787 to 1804. Much of the value of this work was lost, however, through lack of publication until late in the nineteenth century when under Porfiro Díaz a cultural and scientific revival awakened interest, and two preliminary manuscripts left in Mexico were published. Evaluation of the names in these works with regard to those now used began in 1936 but until recently the results have only been published haphazardly. From the 1800s until the present day, field work in Mexico has been more or less continuous but in the absence, until recently, of central planning rather unevenly distributed; in particular, the lowland tropics have remained poorly known,[1] with few logistical bases to hand. During the nineteenth century, collecting was dominated by outside field workers but since then an increasing amount has been undertaken by Mexicans. Significant foci of activity have included Baja California, the Northwest and Northeast, Nueva Galicia, the Mexican highlands, Vera Cruz, the Yucatán Peninsula and (partly in connection with work in Central America, floristically closely related) the southern states of Oaxaca and Chiapas. Major summaries have included the enumeration by Hemsley (1879–88) and the floras by Standley (1920–6) and Conzatti (1939–47, incomplete), but at present the flora is considered too insufficiently known for a new general work and current activity is somewhat decentralized, with a number of 'state' projects published or in progress.

Floristic work in Central America, except in central Panama and at scattered points along the coasts, began seriously only in the latter third of the nineteenth century, but it remained relatively localized until after 1920, with the greatest amount of work being done in Costa Rica and Panama. From 1921 until the 1950s virtually the entire region was dominated by the collecting and publications, including several floras and checklists, of P. C. Standley,[2] who from 1928 was attached to the Field Museum of Natural History at Chicago and began its continuing tradition as a center for Central American botany. He was also partly responsible for the establishment in Honduras of the first major Central American herbarium, now one of sixteen which have in recent years acted as a marked spur to locally based field work, vitally necessary in the face of widespread vegetation degradation and destruction.[3] With continuing activity by a large number of local and overseas botanists, prominent among the latter those from the Field Museum and the Missouri Botanical Garden, most countries, with the exception of Nicaragua and patches elsewhere, are considered more or less moderately well-collected, with an overall average somewhat exceeding Malesia.[4] These successors of Standley have also been largely responsible for three modern national floras, one new local work, and a host of other contributions.

The most recent development in Central American floristic botany, and one of a significance comparable to *Biologia centrali-americana* in the nineteenth century, is the launching of the *Flora Mesoamericana* project. Under plans published in 1981, the projected work will be a descriptive flora (along the lines of *Flora Europaea*, but with some variations such as selected specimen citations) written in Spanish and encompassing seven volumes. Preparation of the new flora would be supported by programs of field work focused on inadequately known areas (such as Nicaragua, where related collecting has already been in progress). It is considered that among other advantages the new work would cut through a host of synonyms, the result of a long past preoccupation with single polities, as noted already by a number of writers – including Standley himself in 1928 as quoted in one of the divisional epigraphs. Nevertheless, some 16000 vascular plant species would be included within the projected compass of this work, which extends from the southern boundary of Panama north to the Veracruz–Oaxaca and Tabasco–Chiapas state boundaries within Mexico (thus including the whole of the state of Chiapas and the Yucatán Peninsula). Project sponsors include the Missouri Botanical Garden, the British Museum (Natural History), and the Universidad Nacional Autónoma de México. (I thank C. J. Humphries of the British Museum (Natural History), one of the four project coordinators, for advance information on this project.)

Progress

For early and recent history, see: VERDOORN, F., 1945. *Plants and plant science in Latin America*. Waltham, Mass., especially the 'Historical sketch' by F. W. Pennell on pp. 35–48; MAGUIRE, B., 1958. Highlights of botanical exploration in the New World. In *Fifty years of botany* (ed. W. C. STEERE), pp. 209–46. New York; and SHETLER, S. G., 1979. North America. In *Systematic botany, plant utilization and biosphere conservation* (ed. I. HEDBERG), pp. 47–54. Stockholm. Further remarks appear in the general review of tropical floristics by Prance (1978). The current situation for Central America is well summarized in GENTRY, A. H., 1978. Floristic knowledge and needs in Pacific tropical America. *Brittonia*, **30**: 134–53, but no similar separate review has yet been seen for Mexico. On the *Flora Mesoamericana* project, see SOUSA, M., 1981. Flora Mesoamericana. *Bol. Bot. Latinoamer.* **8**: 11; also ANON., 1981. Flora Mesoamericana. *Syst. Bot.* **5**: 447.

Bibliographies. Bay, 1910; Blake and Atwood, 1942; Frodin, 1964; Goodale, 1879; Holden and Wycoff, 1911–14; Jackson, 1881; Pritzel, 1871–7; Rehder, 1911; Sachet and Fosberg, 1955, 1971; USDA, 1958. See also the divisional bibliography by AGOSTINI, 1974 (full reference under Division 3).

Indices. BA, 1926– ; BotA, 1918–26; BC, 1879–1944; BS, 1940– ; CSP, 1800–1900; EB 1959– ; FB, 1931– ; IBBT, 1963–9; ICSL, 1901–14; JBJ, 1873–1939; KR, 1971– ; NN, 1879–1943; RŽ 1954– . See also the divisional indices (full references under Division 1).

210–30

Superregion in general

No work has yet appeared to supersede Hemsley's comprehensive enumeration of vascular plants for the monumental *Biologia centrali-americana*, now, in part, over a century old but still a standard reference for all of Mexico and Central America, just as Bentham's *Flora australiensis* has remained for that continent. A gauge as to the increase in botanical knowledge is given by the fact that Hemsley's work, which admittedly was to a large extent compiled, covers perhaps only one-third of the presently known and estimated number of vascular plants. Its projected successor, *Flora Mesoamericana*, will cover only a portion of the area covered by the earlier work, where perhaps the need is greatest: the countries and states from the Yucatán Peninsula and the Isthmus of Tehuantepec eastwards and southwards, with their patchwork of state and national floras (or lack of them) of which only three are more or less current (with none of them in Spanish, the planned language of the new work).

HEMSLEY, W. B., 1879–88. *Biologia centrali-americana: or, contributions to the knowledge of the fauna and flora of Mexico and Central America* (ed. F. D. Godman and O. Salvin), Botany. 5 vols. (incl. atlas), 111 pls. (incl. map; some in color). London: R. H. Porter; Dulau. [Text, vols. 1–4; atlas, vol. 5.]

Systematic enumeration of vascular plants, with descriptions of new taxa; includes synonymy, references, citations of significant revisions and monographic works under family and genus headings, detailed indication of internal distribution, with citations of *exsiccatae*, and general summary of extralimital range, critical commentary, and complete indices to botanical names in each volume (general index in vol. 4). An introductory section includes chapters on phytogeography and the composition of the flora, while in vol. 4 an Appendix (pp. 116–332) by J. D. Hooker includes accounts of geography, vegetation formations, botanical exploration, the high-mountain flora, distribution of the more prominent families, and florulas of certain offshore islands together with a summary and statistical analysis of the flora (with a bibliography). A total of 12 233 species is covered in this work, which was based largely on collections in the Kew Herbarium. Baja California, however, is *not* included.

Region

21/22

Mexico

This comprises the entire Federal Republic of Mexico, including Baja California and the Yucatán Peninsula. For Guadelupe, Rocas Alijos, and the Revillagigedo Islands, see respectively **011**, **012**, and **013**.

The history of floristic work in Mexico to the end of the nineteenth century has been summarized by León, N., 1895. *Biblioteca botánico-mexicana*. 372 pp. Mexico City (especially pp. 297–368). Some of the subsequent activity has been covered in the reviews of VERDOORN (1945), MAGUIRE (1958), and SHETLER (1979) (for all references, see Superregion 21–23), but no relatively comprehensive account of recent history and current work has been seen. In recent decades, floristic work has tended to become somewhat fragmented, with a few active centers covering 'work areas' of logistically manageable size, and there are a number of 'state' herbaria now in existence alongside the large central unit in Mexico City. Shetler notes that 'as yet no serious effort to write a Flora of Mexico is underway' although the Flora of Veracruz program has been visualized as a precursor to such a project. Although scattered fieldwork is continuing, the country remains unevenly explored[5] and is but patchily documented. Partial floras are few, with the only ones of significance being *Flora of Baja California* (Wiggins, 1979), *Flora of the Sonoran Desert* (Wiggins, 1964, covering parts of Baja California and Sonora), *Flora novo-galiciana* (McVaugh, 1974–), the bibliographically complicated *Flora del Estado de México* (Martínez *et al.*, 1953–), *Flora de Veracruz* (INIREB, 1978–), and *Flora of Yucatán* (Standley, 1930; actually an enumeration). One students' manual is available: *La flore del Valle de México* (Sánchez Sánchez, 1969); to

this must now be added the Rzedowskis' *Flora fanerogámica del Valle de México* (1979–   ). Miscellaneous local florulas and other works also exist.

No nominally complete accounting has been made since 1888. Of the two 'national' works, the most useful by far remains the outdated and now difficult-to-obtain *Trees and shrubs of Mexico* by P. C. Standley (1920–6).

Bibliographies. General and divisional bibliographies as for Divison 2.

Regional bibliographies

JONES, G. N., 1966. *An annotated bibliography of Mexican ferns*. xxxiii, 297 pp. Urbana: University of Illinois Press. [Includes geographical, botanical, subject, and author cross-references as well as indices.]

LANGMAN, I. K. [1964]. *A selected guide to the literature on the flowering plants of Mexico*. 1015 pp. Philadelphia: University of Pennsylvania Press.

Indices. General and divisional indices as for Division 2.

210

Region in general

See **210–30** (HEMSLEY). Only two general works for Mexico as a whole are available: Conzatti's *Flora taxonomica mexicana* (never completed) and Standley's well-known *Trees and shrubs of Mexico*, still the most widely used work (it was three times reprinted in the 1960s) in spite of its age, obsolete nomenclature, and relatively limited basis. A field key for tropical lowland trees is also available.

Dictionary

CONZATTI, C., 1903. *Los géneros vegetales mexicanos*. 449 pp. Mexico City.

CONZATTI, C., 1939–47. *Flora taxonomica mexicana*. Vols. 1–2. Mexico City: Sociedad Mexicana de Historia Natural (vol. 1 reprinted 1946, Mexico City).

Briefly descriptive flora, with keys to all taxa; synonymy, with some references and citations; vernacular names; fairly detailed indication of local range, with citation of some collections; indices to all botanical and to vernacular names in each part. The introductory section includes a brief summary of botanical exploration in Mexico, a glossary, and a selected bibliography. Not completed; includes Pteridophyta (vol. 1) and Monocotyledoneae (vol. 2).

STANDLEY, P. C., 1920–26. *Trees and shrubs of Mexico.* 5 parts, 1721 pp. (Contr. US Natl. Herb., 23). Washington. (Parts 1–3, 5 reprinted 1961 as Smithsonian Inst. Publ., 4461; complete work reprinted in 2 vols., 1967, 1969, Smithsonian Institution Press).

Briefly descriptive flora of native, naturalized, and widely cultivated woody vascular plants, with keys to all taxa, synonymy, references, citations (of relevant monographs and revisions under each genus), vernacular names, concise indication of local and extralimital range, more or less extensive notes on properties, uses, cultivation, local significance, etc., some critical remarks, and complete indices (for each part, with general index to families and genera at end of work). An introductory section includes a brief general account of early botanical exploration and research in Mexico, with special attention to Hernandez, Sessé and Mociño, etc. This still-useful work, to a large extent compiled (principally from literature sources and the large Mexican collections of the United States National Herbarium) and regrettably also with nomenclature following the then official (but now obsolete) American Code for Nomenclature, was written mostly by Standley alone, but several family treatments were contributed by specialists. Its status as a modern botanical classic has been underscored by several reprintings, beginning in 1961. For the lowland tropical areas in the country, the *Trees and shrubs* has been supplemented by PENNINGTON, T. D. and SARUKHÁN, J., 1968. *Manual para la identificación de campo de los principales árboles tropicales de México.* vii, 413 pp., illus., maps. Mexico City: Instituto Nacional de Investigaciones Forestales. [Illustrated field manual of the more important lowland trees, with keys, range maps, vernacular names, and notes on ecology, properties, uses, etc.; bibliography; illustrated glossary.]

211

Baja California

See also **205** (WIGGINS). The area encompasses the political units of Baja California Norte and Baja California Sur. Wiggins' new flora represents the first complete work on the peninsula, the fruit of some fifty years of study.

WIGGINS, I. L., 1980. *Flora of Baja California.* [xiii], 1025 pp., 970 text-figs., 3 tables, 4 maps. Stanford: Stanford University Press.

Copiously illustrated (nearly 1000 figures) descriptive manual-key of native, naturalized and adventive vascular plants, with keys to all taxa, synonymy with references, citations of localities, indication of local and extralimital range, critical taxonomic commentary, and notes on phenology, habitat, special features, etc.; index. An introductory section encompasses remarks on physical features, geology, soils, climate, major features of the flora, plant communities, endemism and the history of botanical exploration. Covers 2705 species (2958 entities). [Description based partly on cover advertisement in *Taxon* 28(5/6) (1979) and other sources. Already incomplete following field use, S. Bladt, personal communication.]

212

Sonora

See **205** (WIGGINS) for coverage of those parts of the state falling within the Sonoran Desert (amounting to about three-fifths of the area). No general works specifically relating to the state are available. In the far southern (extra-Sonoran Desert) zone around the Río Mayo, the following will be found useful: GENTRY, H. S., 1942. *Río Mayo plants.* 328 pp., 29 pls. (Publ. Carnegie Inst. Wash., 527). Washington.

213

Chihuahua

The sole general work on this state deals only with the pteridophytes (see below). However, in the northern part general floras of Texas (**171**) will be useful.

Special group – pteridophytes
KNOBLOCH, I. W. and CORRELL, D. S., 1962. *Ferns and fern allies of Chihuahua, Mexico.* 198 pp., 57 pls. Renner, Texas: Texas Research Foundation (distributed by Stechert-Hafner Service Agency, New York).

Detailed illustrated descriptive treatment, with keys to all taxa, complete synonymy, citations of *exsiccatae*, notes on habitat and special features, and taxonomic commentary; gazetteer, glossary, and index at end.

214

Coahuila

No separate general works for this state are available. However, for the northern part, general works on Texas (171) may be found useful.

215

Nuevo Leon and Tamaulipas

No separate general floras or enumerations for either of these states are available. However, as with Coahuila, the northern parts are at least partially provided for through general works on Texas (171).

216

San Luis Potosí

No separate general works for this state are available.

217

Zacatecas

No separate general flora or enumeration is available. However, the southern parts within the Sierra Madre Occidental are covered by the recently-commenced *Flora novo-galiciana* (221).

218

Durango

No separate general flora or enumeration is available. However, the southernmost part is encompassed by the recently-commenced *Flora novo-galiciana* (221).

219

Sinaloa and Nayarit

No separate general floras or enumerations for either of these states exist. However, Nayarit is partly covered by *Flora novo-galiciana* (221), and two partial works (one for the Tres Marías Islands), exist.

Partial works
RILEY, L. A. M., 1923–4. Contributions to the flora of Sinaloa. *Bull. Misc. Inform.* (Kew), *1923*: 103–15, 163–75, 333–46, 388–401; *1924*: 206–22.

Enumeration, with citation of *exsiccatae*, vernacular names, and a few critical remarks. Covers only Polypetalae (Bentham and Hooker system).

EASTWOOD, A., 1929. A list of plants recorded from the Tres Marias Islands, Mexico. *Proc. California Acad. Sci.*, IV, 18: 442–68.

Enumeration of vascular plants, with citation of *exsiccatae*, localities, and critical and other notes; preceded by an account of botanical exploration and list of species first described from the group. Covers 324 species. [The islands lie in the Gulf of California due west of Nayarit.]

221

'Nueva Galicia'

This former Spanish colonial province, centering on Guadalajara, includes the modern states of Jalisco, Colima and Aguascalientes as well as parts of adjoining states. In 1974, the first fascicle of the long-awaited *Flora novo-galiciana*, directed by R. McVaugh, was published; this followed precursory accounts on botanical history and the vegetation published 1966–72 in the same serial as the flora itself.

McVAUGH, R., 1974– . *Flora novo-galiciana*. Published in fascicles. Illus. (Contr. Univ. Michigan Herb., 12, part 1, no. 3, etc.). Ann Arbor.

Large-scale serial flora, with rather long descriptions, keys, full synonymy, references and citations, detailed indication of localities (with citations of *exsiccatae*), extralimital distribution, vernacular names, critical remarks, and notes on habitat, altitude, phenology, etc.; numerous illustrations; index. Apart from Jalisco, Colima and Aguascalientes, this work covers parts of Nayarit, Durango, Zacatecas, Guanajuato, and Michoacán. At this writing (late 1979), only one fascicle (of 15 planned), covering Fagaceae, has been published. For related contributions on botanical exploration and on vegetation, see *ibid.*, vol. 9: pp. 1–123 (1966) and pp. 205–357 (1972).

222

Guanajuato

No separate general works for this state are available. However, it borders on 'Nueva Galicia' (221) with its recently-commenced *Flora novo-galiciana*, and parts of the state are included within the limits of that work.

223

Michoacán

No separate general works for this state are available. Part of its area, however, is encompassed by *Flora novo-galiciana* (221).

224

Guerrero

No separate works for this state are available.

225

'Central Highlands' (with the Valle de México)

This subregion, within which is located the Valle de México and Mexico City, is here considered to encompass the states of Mexico, Morelos, Puebla, Tlaxcala, Hidalgo, and Queretaro, together with the Distrito Federal. The most useful partial works are described below. For a useful introduction to the flora and vegetation, with special reference to the plain, see RZEDOWSKI, J., 1975. Flora y vegetación en la cuenca del Valle de México. In *Memoria de las obras del sistema del drenaje profundo del Distrito Federal*, ed. Anon, Vol. 1, pp. 79–134. Mexico City.

– Vascular flora (Valle de México): 2150 species (Guzmán in RZEDOWSKI, p. 85).

Partial works

MARTÍNEZ, M., MATUDA, E. *et al.*, 1953–74. [*Familias de la flora del Estado de México*: *Pinaceas*, etc.] Published in fascicles. Illus. Toluca: Dirección de Agricultura y Ganaderia (later Dirección de Recursos Naturales), Edo. de México [here referred to as Series A].

MARTÍNEZ, M., 1956–8. *La flora del Estado de México*. 5 fasc. Illus. Toluca [here referred to as Series B].

Semi-popular illustrated flora of vascular plants; each family treatment includes keys to genera, limited synonymy, and local distribution: fascicles separately indexed. Not completed. 'Series A' as designated here comprises single family (or generic) fascicles, mainly dealing with larger families; of these, 23 have been published as of this writing (late 1979), the most recent in 1974. 'Series B', the so-called *Flora*, is actually complementary, and in its five fascicles deals with a large number of families with but few representatives in the state. Progress on 'Series A' was relatively rapid until the early 1960s but since 1962 only five fascicles have been published. The work was one of the results of a state botanical survey; related works by Martínez include a lexicon of vernacular and scientific names (1956) and a treatise on medicinal plants (1958). [Rzedowski and Rzedowski in the list of sources for their flora (next entry) consider 1972 to have been the date of termination for this work.]

RZEDOWSKI, J. and RZEDOWSKI, G. C. DE (eds.), 1979. *Flora fanerogámica del Valle de México*. Vol. 1. 403 pp., 56 text-figs. (incl. maps). Mexico City: Continental.

Descriptive manual-flora of seed plants, with keys to all taxa, indication of occurrence, habitat, altitudinal range, and special features; some taxonomic commentary; references to revisionary treatments of families and genera given in text; index (at end). The introduction includes sections on botanical exploration, geography, geology, hydrology, climate, floristics and vegetation, uses, and conservation conditions; literature (pp. 57–60). The physiographically defined area of 7500 square kilometers centers on Mexico City and includes parts of the states of Mexico, Hidalgo and Tlaxcala as well as the Federal District. The work, which is to encompass three volumes, will effectively succeed REICHE, K. F. 1926. *Flora excursoria en el Valle Central de México*. 303 pp. Mexico City.

SÁNCHEZ SÁNCHEZ, O., 1969. *La flora del Valle de México*. viii, 519 pp., 368 text-figs. Mexico City: Herrera.

Illustrated manual of seed plants, with keys to all taxa, vernacular names, local range, and notes on habitat, phenology, etc.; glossary, list of references, and full index. Designed as a students' manual.

226

Veracruz

This state has a long history of botanical exploration and much scattered literature is available. However, a detailed survey by A. Gómez-Pompa and his associates under the aegis of the 'Flora of Veracruz Program', in which collaborators from Harvard University and the Field Museum of Natural History have also been involved,[6] has been in progress for some years (at present from a base at Jalapa), and publication of a state flora has now commenced (see below). An important precursor, which includes a partial treatment of the families and genera of woody plants in Veracruz, is GÓMEZ POMPA, A., 1966. *Estudios botánicos en la region de Misantla, Veracruz.* xvi, 173 pp. Mexico City.

INSTITUTO DE INVESTIGACIONES SOBRE RECURSOS BIOTICOS [INIREB]. 1978– . *Flora de Veracruz* (ed. A. GÓMEZ POMPA and V. SOSA). Publ. in fascicles. Illus. Xalapa: INIREB.

Descriptive flora with illustrations, published serially; each family fascicle, as in the *Flora of West Pakistan* (**793**), includes keys to all taxa, synonymy, references and citations (with stress on significant monographs, revisions, and floras), localities with citations of *exsiccatae* and altitudes, generalized distribution, some critical notes, and many remarks on habitat, vegetation type, phenology, special features, etc. Vernacular names and typification are also included, along with an index, and each species is individually mapped. The emphasis is very much towards distribution and ecology.[7] At this writing (1981) nine fascicles have been issued, covering a miscellany of generally 'small' families; further fascicles will be issued without regard to systematic sequence. [A computer-generated state checklist has also been circulated but not formally published.]

227

Oaxaca

No separate general works for this state are available.

228

Chiapas

No separate general floras or enumerations for this state, which lies east of the Isthmus of Tehuantepec, are available. The Chiapas flora, however, is closely related to that of Guatemala so that the now-completed *Flora of Guatemala* (**231**) will be found useful. Moreover, in recent years field work and research for a separate state flora have been in progress under the direction of D. Breedlove of the California Academy of Sciences (Shetler, 1979; see Superregion 21–23, introduction, p. 153). For an introduction to the vegetation and plant communities, see MIRANDA, F., 1952–3. *La vegetación de Chiapas.* 2 vols. Illus. Mexico City: Tuxtla Gutierrez.

229

Yucatán Peninsula and Tabasco

Included here are the three polities of the Yucatán Peninsula (Campeche, Yucatán, and Quintana Roo), a center of the Mayan civilization, as well as the state of Tabasco. The greater part of the area is covered by the now-outdated *Flora of Yucatán* by Standley (see below). However, field work and research towards a new general flora, involving INIREB in Veracruz (**226**) and the Field Museum in Chicago, has lately been initiated. Part of the area is moreover contiguous with the northern province of Petén in Guatemala, so that *Flora of Guatemala* (**231**) will also be found useful.[8]

STANDLEY, P. C., 1930. *Flora of Yucatán.* Pp. 157–492 (Publ. Field Columbian Mus., Bot. Ser., 3, part 3). Chicago.

Briefly descriptive, systematic enumeration of fungi, Bryophyta, and vascular plants of the three states and territories of the Yucatán Peninsula, without keys; limited synonymy, with references; vernacular names; general indication of local range; taxonomic commentary and notes on habitat, frequency, special features, properties, and uses; index to families, genera, and vernacular names. The introductory section includes accounts of physical features, climate, vegetation, botanical exploration, and vernacular nomenclature together with a floristic analysis and list of references.

Region

23

Central America

This region is here considered to include the countries from Guatemala and Belize to Panama inclusive, along with the Swan Islands and the San Andrés and Providencia archipelago in the nearby parts of the Caribbean. For Cocos Island, a dependency of Costa Rica, see 015.

The history and progress of botanical work in the region has been very well reviewed by VERDOORN (1945) and MAGUIRE (1958) and the current status of knowledge and present activities by GENTRY (1978) (for all references, see Superregion 21–23). For additional discussion, see D'ARCY, W. G., 1977. Endangered landscapes in Panama and Central America: the threat to plant species. In *Extinction is forever* (ed. G. T. PRANCE and T. S. ELIAS), pp. 89–104. New York. Current belief is that present and future efforts should be directed towards adequate collecting in threatened areas, lands poorly explored in the past (particularly Nicaragua), and for more specialized studies as well as to the preparation of florulas for areas of special interest, enlargement of the growing literature in Spanish, and the achievement of a new regional synthesis (the latter is now under way).

The present standard of floristic documentation in terms of major works is quite variable from country to country. At opposite ends of the region, Guatemala and Panama are each covered by nominally complete 'research' floras produced over many years. However, that for Panama was begun on a less satisfactory collection basis and early parts are often rather outdated. Panama also has *Flora of the Panama Canal Zone* (Standley, 1928) and the recent *Flora of Barro Colorado Island* (Croat, 1979), the latter written for a small area of special scientific interest. For Costa Rica, Standley's partly compiled *Flora of Costa Rica* is gradually being supplanted by *Flora costaricensis* (Burger, 1971–), a critical work comparable with the *Flora of Guatemala* and *Flora of Panama*. Unfortunately, no such works are available for the remaining countries. Nicaragua is effectively without a separate flora, while Honduras and El Salvador are each covered by a mere checklist over and above a miscellany of florulas and other works of more limited scope, such as Standley's *Flora of the Lancetilla Valley, Honduras* (1931). Belize is wholly covered by the *Flora of Guatemala* as well as provided with a separate enumeration (Standley & Record, 1936) now supplemented for the monocotyledons (Spellman *et al.*, 1975), while the first work can be used effectively in El Salvador. Two contributions are available for San Andrés and Providencia in the Atlantic (Toro, 1929–30).

On the whole, Central America is now a comparatively well-collected part of the tropics, far ahead of tropical South America and, according to PRANCE (1978), nearly half again as thoroughly covered as Malesia (save peninsular Malaysia and Java). However, apart from differences between units, variations also exist between different plant life-forms: the tree flora is evidently better-known in Malesia than Central America, whereas the non-woody flora in certain parts of Malesia has received comparatively little attention. It is only comparatively recently that there have begun to appear publications on the tree flora in Mexico and Central America more or less comparable to those long available in Malesia; one of the first was Allen's *Rain forests of Golfo Dulce* (1956), and others, all in Spanish, have followed for humid lowland Mexico, Costa Rica (in general), and Panama.

The vascular flora estimates given for each polity are based on those provided by GENTRY (1978; see Superregion 21–23, p. 153).

Bibliographies. General and divisional bibliographies as for Division 2 or Superregion 21–23.

Indices. General and divisional indices as for Division 2 or Superregion 21–23.

230

Region in general

HEMSLEY, W. B. *Biologia centrali-americana*; Botany. See **210–30**.

231

Guatemala

The comprehensive *Flora of Guatemala* begun by Standley has now been completed after 31 years of publication. It is considered to have had a relatively good collecting basis; experience has shown it to cover most of the presently known flora. Early parts are, after three decades, still relatively complete. Related works cover orchids and pteridophytes, otherwise omitted.

– Vascular flora: about 8000 species (Gentry).

STANDLEY, P. C. *et al.*, 1946–77. *Flora of Guatemala.* 13 parts. Illus. (Fieldiana, Bot., 24). Chicago.

STOLZE, R. G., 1976–81. *Ferns and fern allies of Guatemala*, I–II. 80 text-figs. (*ibid.*, 39, N.S. 6).

The main work comprises a copiously descriptive, illustrated flora of native, naturalized, and commonly cultivated seed plants of Guatemala and Belize, with keys to genera and species, appropriate synonymy, references and citations (with indication of appropriate revisions and monographs under family and generic headings), vernacular names, typification, detailed indication of local range (with citation of *exsiccatae*) and general summary of extralimital distribution, taxonomic commentary, and often extensive notes on habitat, occurrence, frequency, special features, ethnobotany (properties, uses, cultivation), etc.; index to genera in each part. Part 13 comprises a comprehensive index. Representative illustrations appear in all except parts 3–6 (the earliest published). The classical Englerian sequence of families is largely followed. Substantial contributions have been made by J. A. Steyermark and L. O. Williams, but few 'outside' collaborators have been involved. Excludes Orchidaceae; for this family see AMES, O. and CORRELL, D. S., 1953–4. *Orchids of Guatemala.* 2 parts. 739 pp., 200 text-figs. (*ibid.*, 26). Chicago; also CORRELL, D. S., 1965. *Supplement to Orchids of Guatemala and British Honduras.* Pp. 177–221, illus. (*ibid.*, 31, part 7). The associated treatment of the pteridophytes is similar in format to the main work and includes keys to all taxa as well as representative illustrations.

232

Belize (British Honduras)

This area is at present considered relatively well-collected. A new checklist (Spellman *et al.*, 1975) has been initiated as a partial successor to that of Standley and Record (1936), but the area is also entirely covered in *Flora of Guatemala* (231).[9]

– Vascular flora: 2500–3000 species (Gentry).

SPELLMAN, D. L., DWYER, J. D. and DAVIDSE, G., 1975. A list of the Monocotyledoneae of Belize including a historical introduction to plant collecting in Belize. *Rhodora* 77: 105–40, illus.

Nomenclatural checklist, with citations of representative *exsiccatae* and occasional notes, but with no literature citations, notes on distribution, or indication of habitat; families, genera, and species alphabetically arranged. The introductory section treats physical features, climate, vegetation, and the history of botanical exploration. The work, slanted towards reporting the work of 16 recent collectors, covers 721 species of monocotyledons.

STANDLEY, P. C. and RECORD, S. J., 1936. *The forests and flora of British Honduras.* 432 pp., 16 pls. (Publ. Field Mus. Nat. Hist., Bot. Ser., 12). Chicago.

Partially annotated enumeration of vascular plants, with keys and descriptions relating largely to the woody species; limited synonymy, with some references; general indication of local and extralimital range, with some citations of *exsiccatae*; vernacular names; notes on habitat, frequency, special features, properties, uses, etc.; index to genera and to vernacular names. The introductory section includes general remarks on geography, geology, soils, climate, agriculture, forest formations, principal timber species, other forest products, vernacular nomenclature, and relationships of the flora, together with a bibliography.

233

El Salvador

As the one separate checklist by Standley and Calderón is now considered valueless, the only really useful general work, which contains many Salvadorean records, is the *Flora of Guatemala* (231).[10]

– Vascular flora: about 2500 species (Gentry).

STANDLEY, P. C. and CALDERÓN, S., 1944 (1941). *Flora salvadoreña. Lista preliminar de las plantas de El Salvador*. 2nd edn. 450 pp., 2 portraits. San Salvador: Imprensa Nacional. (First edn., 1926 (1925).)

Pages 1–378 of this work comprise an annotated enumeration of known fungi, bryophytes, and vascular plants, with brief general indication of local range or country of origin, vernacular names, and notes on habitat, uses, etc.; index to all botanical and vernacular names. An introductory section includes an account of local botanical exploration.[11]

234

Honduras

An unannotated general checklist has recently been published by Molina (1975), but apart from the northwestern lowlands, other coverage is rather scanty. Some florulas and other works became available for the Tela area on the north coast consequent to the establishment there of 'El Pulpo'.[12] The country is considered still to be rather unevenly collected, with a basis insufficient for a general flora. In western parts, the *Flora of Guatemala* (**231**) will be found useful.

– Vascular flora: about 5000 species (Gentry).

MOLINA R., A., 1975. *Enumeración de las plantas de Honduras*. 118 pp. (Ceiba, 19, part 1). Tegucigalpa.

Unannotated systematic list of vascular plants, with indication if only cultivated. A brief introduction outlines the present status of botanical exploration and knowledge. Based on the 'Standley' herbarium of the Esculea Agricola Panamericana (EAP), Tegucigalpa.

Partial works

STANDLEY, P. C., 1931. *Flora of the Lancetilla Valley, Honduras*. 418 pp., 68 pls. (Publ. Field Mus. Nat. Hist., Bot. Ser., 10). Chicago.

Annotated enumeration, without keys, of vascular plants, with descriptive notes and vernacular names; list of non-vascular plants; index. (The valley lies near Tela on the Atlantic coast.)

YUNCKER, T. G., 1938. *A contribution to the flora of Honduras*. pp. 283–407, pls. 1–18 (Publ. Field Mus. Nat. Hist., Bot. Ser., 17, part 4). Chicago.

Enumeration of collections made by the author in 1934 and 1936 from various points along a route from Tegucigalpa to San Pedro Sula and the Lancetilla basin around Tela. Includes additions to Standley's florula (see above). See also

idem, 1940. Flora of the Aguan Valley and the coastal regions near La Ceiba, Honduras. *Ibid.*, 9(4): 245–346, pls. 1–4. [Similar to preceding; the area covered is about 75 km east of Tela.]

Special group – trees

RECORD, S. J., 1927. Trees of Honduras. *Trop. Woods*, 10: 10–47.

STANDLEY, P. C., 1930. A second list of the trees of Honduras. *Ibid.*, 21: 9–41.

Briefly annotated, systematic lists of known tree species, with indication of vernacular names, uses, etc.; indices (vernacular names only). For supplement, see STANDLEY, P. C., 1934. Additions...*Ibid.*, 37: 27–39.

235

Nicaragua

The flora of Nicaragua has until recent years been very poorly collected and documented. Aside from its other defects, the only nominally general work, Goyena's *Flora nicaragüense*, in fact deals mainly with Honduran plants! The realization of an adequate general flora or enumeration is presently one of the greatest *desiderata* in Central American floristic botany, particularly as the area lies on a floristic transition zone, and it is good news that the Flora of Nicaragua project initiated in 1977 will be continued in spite of the 1979 popular revolution; formal writing has begun (W. D. Stevens, personal communication). In the east, works on Costa Rica (**236**) will be found useful.

– Vascular flora: about 5000 species (Gentry).

RAMÍREZ GOYENA, M., 1909–11. *Flora nicaragüense*. 2 vols. 1064 pp. Managua.

Pages 123–1064 of this work comprise a briefly descriptive flora of vascular plants (and charophytes), with synoptic keys to families, genera, and groups of species as well as vernacular names, also includes an appendix on medicinal plants and indices to generic, family, and vernacular names. The remainder of the work is pedagogic, incorporating among other material an extensive glossary. Of limited or no value as a flora (Blake and Atwood 1942: 153) and actually treating Honduran plants (L. Williams in GENTRY, 1978; see Superregion 21–23, p. 153.)

SEYMOUR, F. C. (comp.), 1980. *A check list of the vascular plants of Nicaragua*. x, 314 pp. (Phytologia, Mem. 1). Plainfield, New Jersey.

Comprises a systematic checklist of vascular plants, with citations of key floristic works relating to Central America and indication of *exsiccatae* together with departments and places of deposit. [Based largely upon collections made by the author and others from 1968 to 1976 and a number of precursory papers, mainly in *Phytologia* from 1973 onwards.] See also SALAS E., J. B., 1966. *Lista de la Flora nicaraguensis con especimenes en la Herbario de la Esculea Nacional de Agricultura y Ganaderia, Managua.* 60 pp. Managua (mimeographed).

236

Costa Rica

The flora of Costa Rica is the best-collected and among the best-documented in Central America. The partly compiled descriptive enumeration of Standley (1937–9) is gradually being supplanted by a descriptive flora directed by Burger (1971–). For the tree flora, two works are available: Holdridge and Poveda A.'s *Arboles de Costa Rica* (1975–) and Allen's *The rain forests of Golfo Dulce* (1956; reissued 1977). The pteridophytes are being studied by D. B. Lellinger at the Smithsonian Institution.

– Vascular flora: about 8000 species (GENTRY, p. 153).

BURGER, W. (ed.), 1971– . *Flora costaricensis.* In parts. Illus. (Fieldiana, Bot., 35, etc.). Chicago.

Designed as a large-scale, detailed descriptive flora of seed plants, with keys to genera and species, full synonymy, references and citations, general indication of local and extralimital range (with some citations of *exsiccatae*), critical commentary, and notes on habitat, phenology, variability, biology, etc.; index to botanical names and subjects and list of references in each part. At this writing (1981) three parts, respectively covering Graminae (family 15, by R. W. Pohl), Casuarinaceae through Piperaceae (families 40–41) and Chloranthaceae through Urticaceae (families 42–53) on a modified Englerian system, have been published.

Families will continue to appear in related groups under single covers, but unlike the *Flora of Guatemala* they will not form parts of a single volume of Fieldiana, Botany.

STANDLEY, P. C., 1937–9. *Flora of Costa Rica.* 4 parts, index. 1616 pp., map (Publ. Field Mus. Nat. Hist., Bot. Ser., 18). Chicago.

Largely compiled descriptive enumeration of seed plants, also covering Cocos Island (015), with a few keys in larger genera and families; includes concise synonymy (with some references), vernacular names, generalized indication of local and extralimital distribution (with citations of some *exsiccatae*), critical commentary, and notes on special features, properties, uses, etc.; index to all botanical and vernacular names at end. An introductory section includes accounts of physical features, climate, vegetation formations, relationships of the flora, botanical exploration, and regional geography, together with a bibliography. Now becoming outdated, this work is gradually being supplanted by the more critical *Flora costaricensis* (see above). A slightly amended Spanish edition (never completed) has also appeared, viz.: *idem*, 1937–40. *Flora de Costa Rica*, vol. 1, parts 1–4 (Mus. Nac. Costa Rica, Ser. Bot., 1). [Covers Cycadaceae through Araceae on Englerian system.]

Special group – trees

ALLEN, P. H., 1956. *The rain forests of Golfo Dulce.* xi, 417 pp., 22 text-figs., 34 pls. Gainesville: University of Florida Press. (Reprinted with new preface, 1977, Stanford: Stanford University Press.)

Illustrated manual of forest trees, with artificial keys, vernacular names, synonymy, local range, and notes on taxonomy, habitat, biology, timber, uses, etc.; glossary. All family, generic, specific, and vernacular names entries are alphabetically arranged in the main text, with appropriate cross-referencing. Designed for the non-specialist, this work, containing much original field information, ranks with ROCK's *Indigenous trees of Hawaii* (990), CORNER's *Wayside trees of Malaya* (911), and WORTHINGTON's *Ceylon trees* (829) as among the best introductory works to tropical trees. The 1977 reprint contains a new preface by P. H. Raven. [Golfo Dulce is in the southeastern part on the Pacific coast.]

HOLDRIDGE, L. R. and POVEDA A., L. J., 1975. *Arboles de Costa Rica.* Vol. 1. 646 pp., 527 text-figs. (largely half-tones). San José: Centro Cientifico Tropical.

Artificially arranged treatment of trees in atlas form, with illustrations accompanied by descriptive text (cf. Worthington's *Ceylon trees*); the latter includes details of habit, leaves, flowers, and fruit (with essential features in bold face) together with notes on habitat, distribution, uses, properties, special features, and taxonomy. Artificial keys for identification (using vegetative features as far as possible) and indices are also provided. The scope of the work and its style are discussed in an introductory section, where the main key to the artificial groups is given. The first volume treats palms, other monocotyledonous trees, and dicotyledonous trees with lobed or compound leaves; another volume will treat trees with simple leaves.

237

Panama

The vascular flora of Panama, in the last decade or more shown to be far larger than was commonly believed, is likely to be increased still further as detailed exploration continues in previously unbotanized areas. Earlier plant hunting, though considerable, was by and large limited to a comparatively small number of areas, notably around the Panama Canal; the species-richness of the poorly-known wet forest zone facing the Caribbean, including much of the Darien district, was not suspected. Although a substantial number of collections had accumulated by the 1930s when the *Flora of Panama* project was definitively mounted by the Missouri Botanical Garden, it is now a matter of general consent, following a lead given by Standley as early as 1945, that the undertaking was premature. Commenced in 1943 and 'completed' in 1981, the *Flora* almost lives up to Standley's comment (in a letter to C. V. Morton) that he doubted 'they have half the total number of species in the published flora'.[13]

Apart from many newly discovered local endemic species, much of the increase has come from range extensions from northern South America (and notably northwestern Colombia). Panama has also been the beneficiary of some good local floras as well as a tree manual, and these have been included here as 'support' for the larger flora. The reader should peruse the lately published introduction to that work, which contains a brief guide to key literature, as well as the following general survey of plant life (which appeared as part of a monograph, *The Panamic Biota*): DRESSLER, R. L., 1972. Terrestrial plants of Panama. *Bull. Biol. Soc. Washington*, **2**: 179–86.

– Vascular flora: 9000–10 000 species (D'Arcy in *Flora of Panama*, Introduction, p. vii (1981)).

WOODSON, R. E., Jr., SCHERY, R. W. (and collaborators). 1943–81. *Flora of Panama*. Introduction (pp. i–xxxiii) and parts 2–9 (in 41 issues). Illus. (Ann. Missouri Bot. Gard., vols. 30–67, *passim*). St Louis. (At this writing distributed by Allen Press, Lawrence, Kansas; available in separate issues or as a whole.)

Illustrated descriptive flora of seed plants, with keys to genera and species, extensive synonymy, references and citations (including indication of appropriate revisions and monographs under family and generic headings), typification, vernacular names, general indication of local range (with citations of *exsiccatae*) and extralimital distribution, critical commentary, and notes on habitat, special features, uses, etc.; index to taxa at end of each part (i.e., each issue; after 1965 family treatments within issues are separately indexed). Numbering of families (and division of the work into parts) essentially follows the traditional Englerian sequence. An index to families has been published as part of the Introduction (1981), along with a list of authors. In the first two decades of the project, publication of families was systematic; but in later years families appeared when ready, with elements from any one or more of the originally designated formal parts present in a given issue (the practice of Garden's *Annals* being to give over one or more of the four annual issues entirely to the *Flora*). Early issues of the *Flora* carried dual pagination, but this practice was discontinued in 1965. For some family treatments issued early in the history of the project, supplements have been published, but much of the work produced in the 1940s and 1950s is now rather outdated. As of 'completion' in 1980, some 6200 seed plant species had been covered, amounting to perhaps two-thirds of the total considered likely to occur in Panama.

The long and somewhat tortuous history of the *Flora of Panama* project has been well documented by W. G. D'Arcy (Introduction, pp. v–viii) as well as in LEWIS, W. H., 1968. The *Flora of Panama*. *Ann. Missouri Bot. Gard*. **55**: 171–3. A critical bibliographic analysis of the first phase of the project along with its precursors may be found in ROBYNS, A., 1965. Index to the 'Contributions toward a Flora of Panama' and to the 'Flora of Panama' through March 1965. *Ibid.*, **52**: 234–47. Notable in this work is its evolution from a largely 'in-house' project utilizing little more than the Garden's own resources to a semi-monographic collaborative enterprise based upon a multiplicity of sources. The importance of long-term institutional commitments, which in turn depend upon relative societal stability, is also demonstrated, as are also the likely practical limits of such large-scale floristic projects if they are not to drag on forever (unless non-conventional circumstances exist). Within this work also is an admission that a flora can never truly be *completed*: this contrasts with attitudes common before 1900, as exemplified by the *Flora of British India* and other Victorian-era works.

Partial works

In addition to the two key works on the Canal Zone, a florula is also available for San José Island in the Gulf of Panama.

CROAT, T. B., 1979. *Flora of Barro Colorado Island.* 960 pp., illus. Stanford: Stanford University Press.

Descriptive flora of this well-known 14.8 km² 'research' island in Gatun Lake (actually formed when the Lake was created in the 1910s), covering 1369 species; includes keys to all taxa (with separate key to woody plants in sterile condition), references and citations, synonymy, indication of local and extralimital range, critical remarks, and often extensive notes on habitat, phenology, and biology (including pollination and dispersal mechanisms), with index at end. An introductory section deals with physical features, climate, geology, soils, life-forms, floristics, and phytogeography as well as general and botanical history of the area. Supersedes STANDLEY, P. C., 1933. *The flora of Barro Colorado Island.* 178 pp., map, 21 pls. (Contr. Arnold Arbor. 5). Jamaica Plain, Mass.

JOHNSTON, I. M., 1949. *The botany of San José Island* (Gulf of Panama). ii, 306 pp., 17 pls. (Sargentia 8). Jamaica Plain, Mass.

This 'biological florula' (compare with that by Croat for Barro Colorado Island, cited above) includes a systematic enumeration of vascular plants, with very detailed descriptions and notes written in semi-popular style; supporting the plant descriptions are English vernacular names, citation of *exsiccatae* as well as some literature (mainly Standley, 1928; see below), localities, critical commentary, and remarks on overall distribution, habitat, and biology; index to vernacular and scientific names. The introductory section contains background historical, physical and biotic information, along with descriptions of vegetation types and synusiae and special biological features, beach drift, and leaf fall and renewal; keys to tree and shrub species and to orchids are also provided. [Based upon field work in 1943–6; treats 627 indigenous and alien species.]

STANDLEY, P. C., 1928. *Flora of the Panama Canal Zone.* x, 416 pp., 7 text-figs., 67 pls. (Contr. US Natl. Herb., 27). Washington. (Reprinted 1968, Lehre, Germany: Cramer.)

Manual-key to seed plants, with essential synonymy, local range, vernacular names, and notes on special features, uses, etc.; extensive introductory section treating physiography, floristics, land uses, and botanical exploration; index at end. Long a standard reference in the area, this work, although covering in a compact fashion some 2000 species and possessing one of the best available keys to neotropical plant families, unfortunately is written in the now-obsolete American Code of Nomenclature and with the passage of time has become somewhat incomplete.

Special group – forest trees

HOLDRIDGE, L. R., 1970. *Manual dendrológico para 1000 especies arbóreas en la República de Panamá.* xi, 325 pp., 9 text-figs. (Panamá: Inventariación y demonstraciones forestales.) Panamá: FAO, United Nations.

Artifically arranged tree-manual for foresters and ecologists, similar in style to his *Arboles de Costa Rica* (236) but without illustrations; species entries include vernacular names as well as habit, botanical, dendrological, and other data. The introductory section gives the plan and scope of the work, a selected bibliography, an illustrated glossary, and the general analytical keys; indices to scientific and vernacular names appear at the end (pp. 293 ff.).

238

San Andrés and Providencia Islands

These two low islands lie east of Nicaragua in the Caribbean Sea, and are a Colombian possession. In addition, 238 also includes the scattered islets and cays in the south and southeast near San Andrés (Courtown Cays and Cayos de Albuquerque) and to the northeast towards Jamaica (North Cay and Southwest Cay on Serrana Bank; Middle Cay and East Cay on Serranilla Bank; Bajo Nuevo; Roncador Cay).

TORO, R. A., 1929. Una contribución a nuestro conocimiento de la flora silvestre y cultivada de San Andrés. *Revista Soc. Colomb. Ci. Nat.* **18**: 201–7.

TORO, R. A., 1930. Una contribución a nuestro conocimiento de la flora de San Andrés y Providencia. *Ibid.*, **19**: 56–8.

Unannotated systematic lists of flowering plants (respectively 96 and 40 species), with inclusion of some English vernacular names. Most of the records appear to relate to San Andrés.

239

Swan Islands

These two small islands lie in the Caribbean at 84 °W to the northeast of Honduras, and are a United States possession. No floristic accounts are available so far as is known.

Superregion

24–29

The West Indies

As delimited here, this includes all the islands from the Bahamas and Cuba through the Greater and Lesser Antilles to Curaçao, Aruba, and Bonaire. Barbados, Tobago, Trinidad, Margarita, and Coche and their associated islets are also included. For Bermuda, see Region 02.

Due to their economic importance in the seventeenth and eighteenth centuries as well as their relative accessibility, the West Indies have among tropical lands one of the longest histories of botanical exploration, rivalled only by southern India, Sri Lanka, and the East Indies. The earliest serious work was done in the 1680s by Sloane in Jamaica and elsewhere and then over several years from 1689 by Plumier in Haiti and Martinique. Others, including Catesby in the Bahamas, followed in the eighteenth century, so that by 1753 Linnaeus was able to include (and thus typify) hundreds of West Indian species. This first consolidation was followed by the field work and publications of Browne, Jacquin, and Swartz.

However, important as this early legacy is in view of the changes which have taken place in the vegetation, these efforts were somewhat uncoordinated and incidental and have left considerable difficulties with respect to typification, nomenclature and geography, not, in view of a prevailing emphasis on local floristics, fully resolved by later scholarship.[14]

This lack of coordination persisted through most of the nineteenth century, although a gaggle of local efforts in the British-controlled islands (many in economic difficulties from the 1830s) was consolidated from afar by Grisebach in his landmark *Flora of the British West Indian islands* (1859–64), prepared as part of the 'Kew Series'. In the latter part of the century, many collectors, both resident and visiting and including both amateurs and professionals, were active in the superregion, to a large extent in the Greater Antilles although Bélanger and Duss were active in the French Lesser Antilles. The first important gatherings in Cuba (Wright), Hispaniola (Eggers), Puerto Rico (Bello y Espinosa, Krug, and Sintenis) and the Virgin Islands (Eggers and others) were made at this time.

Towards the end of the century, this scattered activity began to resolve itself around a limited number of foci overseas, with field work in the hands of a comparatively small number of professionals. The overseas foci included Berlin, with Krug and Urban; New York and Chicago, with Britton and Millspaugh respectively; and London, with Fawcett and Rendle at the British Museum (Natural History). It is the men at these centers, and the collectors and local workers whom they sponsored (including substantial contributions by Britton, Fawcett and Millspaugh themselves), who have been responsible for the greater part of botanical work in the West Indies in the first part of the present century, and to whom are owed a large number of the standard floras of the superregion. Urban also attempted to bring some synthesis of knowledge through his privately financed, nine-volume *Symbolae antillanae* (1898–1928), which included many family revisions as well as some island floras (e.g., for Hispaniola and Puerto Rico), along with a bibliography, historical account and a cyclopedia of collectors.

In more recent years, activities have again tended to become more locally based, notably in Cuba, Hispaniola, Puerto Rico, Jamaica, Guadeloupe, Barbados, and Trinidad, and several local herbaria, some of many decades' standing, are at present in existence. Additional 'standard' floras have been published or are now in progress. The flora, estimated as having from 12 000 to 15 000 vascular species, is now generally considered comparatively well-known, and, although novelties continue to be found here and there, it 'cannot be regarded as . . . needing immediate study or a massive collecting program' (Howard, 1977). Yet, as with Central America, there is a need for further efforts at consolidation, including monographic studies (within the framework of *Flora Neotropica* or otherwise), to overcome the legacy of fragmentation with its excess of 'endemics'. As the superregion is now perhaps about as well-collected and documented as much of the Mediterranean Basin, it would also not be unreasonable to suggest the compilation of a work similar to the projected 'Med-Check List' (see remarks under **601**) which would among other benefits draw attention to problems requiring solution.

Progress

For early and recent history, see URBAN, I., 1902. Notae biographicae peregrinatorum Indiae occidentalis botanicorum. In *idem*, *Symbolae antillanae*, vol. 3, pp.

14–158. Berlin: VERDOORN, F., 1945. *Plants and plant science in Latin America*. Waltham, Mass., especially pp. 35–48 ('Historical sketch', by F. W. Pennell); MAGUIRE, B., 1958. Highlights of botanical exploration in the New World. In *Fifty years of botany* (ed. W. C. STEERE), pp. 209–46. New York; PROCTOR, G. R., 1961. Our knowledge of the flora of the West Indies. In *Ninth International Botanical Congress. Recent advances in botany*, vol. 1, pp. 929–32. Toronto; and STEARN, W. T., 1980. Swartz's contributions to West Indian botany. *Taxon*, **29**: 1–13.

Other papers, in part dealing with the current situation, are HOWARD, R. A., 1977. Conservation and the endangered species of plants in the Caribbean Islands. In *Extinction is forever* (ed. G. T. PRANCE and T. S. ELIAS), pp. 105–14. New York; HOWARD, R. A., 1979. Flora of the West Indies. In *Tropical botany* (ed. K. LARSEN and L. B. HOLM-NIELSEN), pp. 239–50. London; and in the more general reviews of SHETLER (1979; see Superregion 21–23, introduction, p. 153) and PRANCE (1978).

Bibliographies. Bay, 1910; Blake and Atwood, 1942; Frodin, 1964; Goodale, 1879; Holden and Wycoff, 1911–14; Jackson, 1881; Pritzel, 1871–77; Rehder, 1911; Sachet and Fosberg, 1955, 1971; USDA, 1958. See also the divisional bibliography by Agostini.

Supraregional bibliography

URBAN, I., 1898. Bibliographia indiae occidentalis botanica. In *Symbolae antillanae* (ed. I. URBAN), vol. 1, pp. 3–195. Berlin, Leipzig. Supplemented in *idem*, 1900–04. Continuatio I–III. *Ibid.*, vol. 2, pp. 1–7; vol. 3, pp. 1–13; vol. 5, pp. 1–16. (Reprinted collectively in one volume, 1964, Amsterdam: Asher.)

Indices. BA, 1926– ; BotA, 1918–26; BC, 1879–1944; BS, 1940– ; CSP, 1800–1900; EB, 1959– ; FB, 1931– ; IBBT, 1963–9; ICSL, 1901–14; JBJ, 1873–1939; KR, 1971– ; NN, 1879–1939; RŽ, 1954– . See also the divisional indices.

240–90

Superregion in general

The only comprehensive works are Grisebach's *Flora of the British West Indian islands* (1859–64) and the more detailed *Symbolae antillanae* edited by Urban (1898–1928). Although by now much out of date, these works serve as a foundation for all other current floras in the superregion. For many Lesser Antillean islands, they have yet to be superseded. A comprehensive later revision of the Gramineae is found in HITCHCOCK, A. S., 1936. *Manual of the grasses of the West Indies*. 439 pp. (Misc. Publ. USDA, 243). Washington.

GRISEBACH, A. H. R., 1864 (1859–64). *Flora of the British West Indian islands*. xvi, 789 pp. London: Reeve. (Reprinted 1963, Weinheim, Germany: Cramer.)

Concise, briefly descriptive flora of vascular plants of those islands then under British hegemony, with partly synoptic keys to genera and groups of species, limited synonymy (with some citations), indication of distribution by island or island group, brief summary of extralimital range, vernacular names, critical commentary, and notes on phenology, uses, etc.; indices to all botanical and vernacular names at end. An introductory section includes a note on geographical limits of the flora as well as chapters on collectors, sources of information, etc. The work forms part of the 'Kew Series' of British colonial floras initiated by W. J. Hooker in the 1850s. Although by now largely of historical interest, with old-fashioned non-analytical keys, this fundamental work has yet to be superseded. For a retrospective review of this work, see STEARN, W. T., 1965. Grisebach's *Flora of the British West Indian islands*: a biographical and bibliographical introduction. *J. Arnold Arbor.* **46**: 243–85.[15]

URBAN, I. (ed.), 1898–1928. *Symbolae antillanae*. 9 vols., portrait. Berlin: Borntraeger. (Reprinted 1964, Amsterdam: Asher.)

This monumental serial work, in addition to complete or partial floras of a number of West Indian islands, is largely devoted to critical revisions of many families (in part by specialists) on a regional basis together with reports of various collecting trips, especially (in later years) those of E. Ekman. The revisions include keys, descriptions, extensive synonymy (with references and citations), detailed indication of West Indian range (with citation of *exsiccatae*), and critical taxonomic commentary together with miscellaneous notes. For complete index, see CARROLL, E. and SUTTON, S. (compil.), 1965. *A cumulative index to the nine volumes of the 'Symbolae Antillanae' edited by I. Urban*. 272 pp. Jamaica Plain: Arnold Arboretum of Harvard University. [Pp. 1–5 comprise a retrospective review by R. A. Howard of the author's career and of the basis for the *Symbolae antillanae*.]

Region

24

Bahama Archipelago

This comprises the entire chain of islands from Grand Bahama, Great Abaco and Bimini southeast to the Caicos and Turks, thus encompassing the two polities of the Bahamas and the Turks and Caicos Islands. The islands, about half extratropical, are all relatively low, with no elevation higher than 63 m.

Exploration began early in the eighteenth century and by the present time they are considered reasonably well-known. One of the most extensive series of surveys was that made over the first two decades of the present century by Britton and his associates; the resulting collections and observations formed the main basis for the still-current standard flora by Britton and Millspaugh (1920). Further extensive collecting has been carried out in the last decade or two, especially under the aegis of the Fairchild Tropical Garden outside Miami, with the goal of a new descriptive flora and related works; to this end some additions and corrections to the *Bahama Flora* had been published from 1973 onwards. The final result was expected to appear in 1982 from Stanford University Press as *Flora of the Bahamian Archipelago* by D. S. Correll. For a recent review of floristics and major communities, see CORRELL, D. S., 1979. The Bahama archipelago and its plant communities. *Taxon*, 28: 35–40.

Bibliographies. General, divisional and supraregional bibliographies as for Division 2 and Superregion 24–29.

Regional bibliography

GILLIS, W. T., BYRNE, R., and HARRISON, W., 1975. Bibliography of the natural history of the Bahama Islands. [ii], 123 pp., map (Atoll Res. Bull., 191). Washington. [Botany, pp. 17–29. Section headings in the text do not stand out well due to the use of typescript reproduction.]

Indices. General and divisional indices as for Division 2 or Superregion 24–29.

240

Region in general

BRITTON, N. L. and MILLSPAUGH, C. F., 1920. *The Bahama flora*. viii, 695 pp. New York: The authors. (Reprinted 1963, Hafner).

Concise descriptive flora of native and naturalized cellular and vascular plants, with keys to all taxa of the latter; extensive synonymy, with references; vernacular names; local range indicated by island; general summary of extralimital range; taxonomic commentary and notes on habitat, etc.; summary accounts of botanical exploration and general features of the region; annotated bibliography and index to all vernacular and botanical names. Treatments of nonvascular plants by specialists. For supplements, see: GILLIS, W. T., HOWARD, R. A., and PROCTOR, G. R., 1973. Additions and corrections to the Bahama flora since Britton and Millspaugh, I. *Rhodora*, 75: 411–25; also GILLIS, W. T. and PROCTOR, G. R., 1975. *Idem*, II. *Sida*, 6: 52–62, and GILLIS, W. T. 1977. *Idem*, III, *Phytologia*, 35: 79–100. [New records, reductions, critical notes, etc.] Due to the use of the now-obsolete American Code of Nomenclature in the main work, reference should also be made to the following: GILLIS, W. T., 1974. Name changes for the seed plants in the Bahama flora. *Rhodora*, 76: 67–138.

Special group – trees

PATTERSON, J. and STEVENSON, G. 1977. *Native trees of the Bahamas*. 128 pp., illus. Hope Town, Abaco (Bahamas): Patterson.

Popular account, with artificially arranged species accounts including vernacular and botanical names, local distribution and extralimital range, and features enabling distinction from related species; indices to all names at end. An introductory section gives background information as well as a preliminary artificial key (p. 7). Appendices cover very rare species, figures of very common shrubs and herbs, tree geography, and a list of families. The illustrations are partly in color, and naturalized and commonly cultivated trees are also included.

Region

25

The Greater Antilles (except the Bahamas)

As delimited here, this region comprises the islands from Cuba and the Cayman Islands eastward to Puerto Rico and the Virgin Islands. The last-named group, although often considered as belonging to the Lesser Antilles, is included here for convenience and biohistorical reasons.

Politically and culturally diverse, the region has had a long history of botanical activity and publication and by now is relatively well-known. However, while much was already known from seventeenth- and eighteenth-century collections and publications, such as Sloane's *Natural history of Jamaica* and the works of Browne, Jacquin, and Swartz, serious work in the mountainous interiors of the islands remained for Wright, Eggers, Sintenis, Duss, and others in the nineteenth century, followed in the early twentieth century by Harris, Fawcett, Britton, Wilson, and Hitchcock and culminating with Ekman's extensive botanizing up to the 1930s. Much of this was synthesized by Grisebach (especially for Cuba) in the 1850s and 1860s and from 1898 to 1925 on a grand scale by Urban in his *Symbolae antillanae*, which included separate floras for Puerto Rico (1903–11) and Hispaniola (1920–1), the first for these islands, as well as many family revisions covering the whole of the West Indies. Contemporaneously with Urban's publications there was initiated a detailed *Flora of Jamaica* by Fawcett and Rendle (1910–36), unfortunately still incomplete although work on it had some years ago been resumed at the British Museum (Natural History). With the help of this botanical foundation, locally based activity (and collections) gradually developed during the twentieth century, first in Jamaica and Cuba and later in the other large islands, so that standard works published since 1940 are, or contain substantial contributions, by resident botanists, and in some cases locally published.

All islands are now covered by more or less adequate floras or enumerations, the most recent being Adams' *Flowering plants of Jamaica* (1972), which by general consent is considered to be one of the finest of

its kind among tropical floras. Bulkier, more or less documentary floras or enumerations include *Flora de Cuba* (León and Alain 1946–62, supplemented in 1969); *Flora of Jamaica* (noted above); *Catalogus florae domingensis* (Moscoso, 1943, with supplement by Jiménez, 1967) and *Flora domingensis* (Urban, 1920–1); and *Botany of Porto Rico and the Virgin Islands* (Britton and Wilson, 1923–30), the last-named with the obsolete 'American' nomenclature but also covering the Virgin Islands. Coverage of the pteridophytes, here usually treated apart from the seed plants, is also available for all units.

Bibliographies. General, divisional and supraregional bibliographies as for Division 2 and Superregion 24–29.

Indices. General and divisional indices as for Division 2 or Superregion 24–29.

251

Cuba

The present standard work by Brothers Léon and Alain (1946–62; supplement 1969) covers only seed plants (5785 species in 1296 genera). More recently, pteridophytes have been accounted for in an enumeration by Duek (1971). Since the advent of the Castro government, a central herbarium has been assembled from a number of smaller collections and it is understood that work is in progress on a new national flora (J. McNeill, personal communication); to this end, a number of papers have appeared, mostly in Cuban and eastern European journals. The earlier history of botanical exploration was covered by Brother Léon in *Las exploraciones botánicas de Cuba* (1918, Havana) and again in his *Flora de Cuba* (see below). Recent progress, particularly after 1959, has been reviewed in HOWARD, R. A., 1977. Current work on the flora of Cuba – a commentary. *Taxon* **26**: 417–23.

– Vascular flora: about 6000 species (Brother Alain in PRANCE, 1978).

Bibliography

SAMKOVA, H. and SAMEK, V., 1967. *Bibliografía botánica cubana (téorica y aplicada) con enfasis en la silvicultura*. 36 pp. (*Acad. Ci. Cuba, Ser. Biol.*, 1). Havana. [Literature from 1900 through 1967.]

LEÓN, Hermano and ALAIN, Hermano, 1946–62. *Flora de Cuba*. 5 vols. 794 text-figs., portrait, maps.

Havana: Cultural (vol. 1), Fernandez (vols. 2–4); Rio Pedras, Puerto Rico: Universidad de Puerto Rico (vol. 5). (Vols. 1–4 also pub. as *Contribuciones occasionales del Museo de Historia Natural del Colegio de 'La Salle'* 8, 10, 13, and 16; reprinted in 2 vols., 1974, Königstein/Taunus, Germany: Koeltz, together with reissue of vol. 5 as a third volume.) (Vol. 1 by León alone; vol. 5 by Alain alone.)

Copiously illustrated, briefly descriptive concise flora of seed plants, with keys to all taxa, limited synonymy, abbreviated indication of local and extralimital range, vernacular names, critical commentary, and brief notes on habitat, properties, uses, etc.; bibliography, summaries, and indices to botanical and vernacular names at end of each volume. An introductory section (volume 1) includes a glossary along with accounts of phytogeography and collectors and the history of floristic botany in Cuba. A bibliography is also included in the work. For additions, see ALAIN, Hermano (LIOGIER, A. H.), 1969. *Flora de Cuba: suplemento*. 150 pp. Caracas: Sucre (distributed by the author). Pteridophytes have been covered in DUEK, J. J., 1971. Lista de las especies cubanas de *Lycopodiophyta, Psilotophyta, Equisetophyta, y Polypodiophyta (Pteridophyta). Adansonia*, sér. 2, **11**: 559–78, 717–31. [Systematic list, with synonymy, references, vernacular names, typification, indication of distribution in the Greater Antilles and in Florida, and karyotyptes; bibliography.]

252

Cayman Islands

This latter-day Caribbean 'tax shelter' comprises a group including the islands of Grand Cayman, Little Cayman, and Cayman Brac together with several islets, all of low elevation. For additional coverage, see 253 (FAWCETT and RENDLE). The now-antiquated separate flora by Hitchcock is said to be in process of being supplanted by a new flora by Proctor (HOWARD, 1979; see Superregion 24–29, introduction, p. 166) but at this writing (1981) nothing has been seen.

HITCHCOCK, A. S., 1893. List of plants collected in the Bahamas, Jamaica and Grand Cayman. *Annual Rep. Missouri Bot. Gard.* **4**: 47–179, 4 pls.

Systematic enumeration of vascular plants, with descriptions of new taxa, limited synonymy (with references), citations of *exsiccatae* with localities, and index to generic names. The introductory section includes the author's itinerary, while at the end is a statistical summary of the floras concerned together with a discussion of the affinities of the Bahaman flora. This paper, based mainly upon the author's own collections, is now superseded save for its coverage of Grand Cayman.

253

Jamaica

Work resumed at the British Museum (Natural History) several years ago on the continuation of the long-lapsed *Flora of Jamaica*, but although W. T. Stearn and others have published some precursory papers and area revisions no new definitive volumes have so far (1981) been published. The gap is ably bridged for general use by Adams' *Flowering plants of Jamaica*, while for pteridophytes there is Proctor's list.

– Vascular flora: about 3100 species (based on a total of 2888 native or fully naturalized flowering plants reported by Adams).

ADAMS, C. D., 1972. *Flowering plants of Jamaica*. 848 pp. Mona, Jamaica: University of the West Indies (distributed by Maclehose, Glasgow).

Concise manual-flora of native, naturalized, and commonly cultivated flowering plants, with keys to genera and species; essential synonymy; vernacular names (where known); brief indication of local and extralimital range, with citations of representative *exsiccatae*; notes on habitat, frequency, phenology, altitude, etc., as well as some taxonomic commentary; indices to all botanical and vernacular names. The introductory section includes a summary of the composition of the flora as well as the plan of the work and its sources, while at the end are found a general key to monocotyledons, references to keys for dicotyledons, and a list of relevant references. Pteridophytes are covered in PROCTOR, G. R., 1953. *A preliminary checklist of Jamaican pteridophytes*. 89 pp., 3 pls., 2 maps. (Bull. Inst. Jamaica, Sci. Ser., 5). Kingston. [Annotated checklist, with local range, *exsiccatae*, synonymy, references and critical notes. Pteridophyta similarly are not covered by the larger *Flora of Jamaica*.]

FAWCETT, W. and RENDLE, A. B., 1910–36. *Flora*

of Jamaica. Vols. 1, 3–5, 7. Illus. London: British Museum (Natural History). (Vol. 1 reprinted in enlarged format, 1963, Kingston, as *Orchids of Jamaica*.)

Detailed descriptive flora of native and naturalized seed plants, with keys to families, genera, and species; full synonymy, with references and citations; vernacular names; fairly detailed indication of local range, with citation of *exsiccatae*; general summary of extralimital range; taxonomic commentary and notes on habitat, properties, uses, etc.; index to vernacular and botanical names in each volume. The work is complete apart from the monocotyledons (except Orchidaceae) and the 'early' metachlamydeous dicotyledons, respectively intended for vols. 2 and 6; the work as a whole follows Rendle's system. For supplement to the published volumes, see PROCTOR, G. R., 1967. *Additions to the Flora of Jamaica*. 84 pp., 35 halftones. Kingston: Institute of Jamaica.

254

Hispaniola

The island of 'Little Spain' comprises two states: Haiti and the Dominican Republic (Santo Domingo). However, published floras have generally covered the island as a whole. The most recent work is the enumeration of seed plants by Moscoso (evidently published in exile), to which a significant supplement was contributed by Jiménez. The latter is now associated with others in work towards a new flora in preparation under Brother Alain. Recent progress was reviewed by JIMÉNEZ, J. DE Js., 1961. A new catalogue of the Dominican flora. In *Ninth International Botanical Congress. Recent advances in botany*, vol. 1, pp. 932–6. Toronto.

– Vascular flora: about 5000 species (Brother Alain in PRANCE, 1978); Jiménez (1961) credits the island with 5747 non-cultivated species of seed plants, which seems high.

MOSCOSO, R. M., 1943. *Catalogus florae domingensis*, I: *Spermatophyta*. xlviii, 732 pp., 3 pls., 2 maps. New York: University of Santo Domingo.

Systematic enumeration of seed plants of Hispaniola, with synonymy and references, vernacular names, rather detailed indication of local range (by botanical districts) and general summary of extralimital distribution, brief descriptive notes for some taxa and

indices to botanical and to Haitian and Dominican vernacular names. The introductory section includes a general description of Hispaniola, including remarks on its botanical districts (with maps), a summary account of botanical exploration, and a bibliography. For additions and corrections, see JIMENEZ, J. DE Js., 1967. *Catalogus florae domingensis*. Suplemento 1. 278 pp., 3 pls. Santiago de los Caballeros, DR: Association para el Desarrollo. [Originally published in *Arch. Bot. Biogeogr. Ital.* **39**(1963): 81–132; **40**(1964): 54–149; **41**(1965): 47–87; **42**(1966): 46–97, 107–128; **43**(1967): 1–18, pls. 1–3.]

URBAN, I., 1920–1. Flora domingensis. In *Symbolae antillanae* (ed. I. URBAN), vol. 8, pp. 1–860. Leipzig: Borntraeger. (Reprinted, 1964, Amsterdam: Asher.) Accompanied by *idem*, 1925. Pteridophyta domingensis. *Ibid.*, vol. 9. pp. 273–397 (text), 544–68 (index).

Systematic enumerations respectively of seed plants and pteridophytes of Hispaniola, with full synonymy, references and citations, vernacular names, detailed indication of local range (with citation of *exsiccatae*), general summary of extralimital distribution, and brief notes on phenology, special features, etc.; critical remarks; indices to all vernacular and botanical names. For accounts of botanical exploration, vegetation formations, and plant geography, see *Symbolae antillanae*, vol. 9. pp. 1–54 (1923).

Haiti

BARKER, H. D. and DARDEAU, W. S., 1930. *Flore d'Haiti*. viii, 456 pp. Port-au-Prince: Départment de l'Agriculture.

Briefly descriptive generic flora of seed plants, with keys to families and genera together with lists of species; vernacular names; indication of some localities; occasional notes on uses, etc.; glossary and index to vernacular and botanical names. The introductory section includes a synopsis of families as well as a bibliography.

Dominican Republic

All works cover Hispaniola as a whole (see above).

256

Puerto Rico

Puerto Rico is here considered to encompass the main island together with Mona, Vieques and Culebra. The manual-flora of Britton and Wilson (1923–30), a good work in its day, is now outdated and moreover is encumbered with the obsolete American Code of Nomenclature; with this in view, two supplements were produced by Alain in 1965 and 1967. For the trees, this work has also been supplanted by the detailed pair of books by Little and co-workers (1964, 1974). Both general works cover also the Virgin Islands (257).

– Vascular flora: 2582 species (Brother ALAIN, 1965).

BRITTON, N. L. and WILSON, P., 1923–30. Botany of Porto Rico and the Virgin Islands. In *Scientific survey of Porto Rico* (NEW YORK ACADEMY OF SCIENCES), vols. 5–6. 626, 663 pp. New York.

Briefly descriptive, concise but rather detailed flora of vascular plants, with keys to all taxa; limited synonymy, with references; general indication of local and extralimital range; vernacular names; taxonomic commentary and notes on habitat, etc.; annotated bibliography and general index to all vernacular and botanical names at end of work. The introductory section includes general descriptive remarks on the islands. Mona Island, between Puerto Rico and Hispaniola, is also encompassed by this work. See also ALAIN, Brother (LIOGIER, A. H.), 1965. Nomenclatural changes and additions to Britton and Wilson's Flora of Porto Rico and the Virgin Islands. *Rhodora*, 67: 315–61; and *idem*, 1967. Further changes and additions...*Ibid.*, 69: 372–6.

LITTLE, E. L., Jr. and WADSWORTH, F. H., 1964. *Common trees of Puerto Rico and the Virgin Islands*. x, 548 pp., 250 text-figs., 4 maps (Agric. Handb. USDA, 249). Washington: Government Printing Office.

LITTLE, E. L., Jr., WOODBURY, R. O. and WADSWORTH, F. H., 1974. *Trees of Puerto Rico and the Virgin Islands: second volume*. xiv. 1024 pp., 460 pls., 4 maps (*Ibid.*, 449). Washington.

These complementary works comprise a copiously illustrated descriptive account in atlas form of all trees in Puerto Rico and the Virgin Islands, with keys to all taxa (including artificial direct keys), outline systematic lists of species, limited synonymy, generalized indication of local range (supplemented by individual distribution maps in the first volume), vernacular names, a few critical remarks, and extensive notes on natural history, phenology, timber properties, uses, etc.; indices to vernacular and botanical names (in the second volume these are complete for the whole work). In the introductory section are remarks on forests in the area as well as on earlier studies of the forest flora. The second volume covers the less common trees. For a Spanish edition of the first volume, see LITTLE, E. L., Jr., WADSWORTH, F. H. and MARRERO, J., 1967. *Arboles comunes de Puerto Rico y las Islas Virgenas*. xxxix, 827 pp., 247 pls. (some in color), 4 maps. Rio Piedras, PR: University of Puerto Rico. [Like English edition, but more lavishly produced, with somewhat expanded text and many colored illustrations.]

257

Virgin Islands

These islands are divided between the United States (St Thomas, St Croix and St John) and Great Britain (Tortola, Virgin Gorda, and Anegada). For general coverage of the group see 260 (STEHLÉ; VÉLEZ) and particularly 256 (all works). The most recent separate flora dates from the nineteenth century and for the American Virgin Islands the latest work is that of Britton (1918). In recent years, however, much new collecting has been done and published results include additions for St Croix as well as florulas for each of the three British Virgin Islands. It is to be hoped that a consolidated flora will be produced.

– Vascular flora: about 1500 species (estimated).

American Virgin Islands

BRITTON, N. L., 1918. The flora of the American Virgin Islands. *Mem. Brooklyn Bot. Gard.* 1: 19–118.

Systematic enumeration of cellular and vascular plants, with synonymy, brief indication of local range, and notes on habitat, special features, etc.; encompasses native, naturalized, and commonly cultivated species. The introductory section incorporates a general description of the islands, an annotated list of significant botanical literature, and an account of botanical exploration; at the end a table of statistics of the flora and a list of endemic species are given. For changes relating to St Croix, see FOSBERG, F. R., 1976. Revisions in the flora of St Croix, US Virgin Islands. *Rhodora*, 78: 79–119.

British Virgin Islands

D'Arcy, W. G., 1967. Annotated checklist of the dicotyledons of Tortola, Virgin Islands. *Rhodora*, **69**: 385–450, illus. (map).

Systematic checklist, with localities, *exsiccatae*, and literature references together with notes on habitat, biology, etc. Background information as well as floral statistics and floristic affinities are treated in an introductory section. Tortola is the largest of the islands and the most poorly known.

D'Arcy, W. G., 1971. *The island of Anegada and its flora.* 21 pp., illus. (Atoll Res. Bull., 139). Washington.

Systematic checklist, with localities, *exsiccatae*, and literature references as well as notes on habitat and biology; includes critical taxonomic remarks and coverage of cultivated species. The island is low and relatively flat.

Little, E. L., Jr., Woodbury, R. O., and Wadsworth, F. H. 1976. *Flora of Virgin Gorda* (British Virgin Islands). 36 pp., illus. (incl. map) (US Forest Serv. Res. Pap., ITF-21). Rio Piedras, PR.

Annotated systematic checklist, with limited synonymy and one-line notes on distribution, habitat, etc.; a few literature references. An introductory section provides background information together with remarks on rare and endemic species and available collections; a list of references appears at the end. A statistical table is also given.

Region

26/27

The Lesser Antilles (Leeward and Windward Islands)

Within this region are the Windward and Leeward Islands, from Sombrero and Anguilla in the north to Grenada in the south. The 'vanishing' island of Aves, somewhat west of the main island arc at 15 °40 ′N, 63 °30 ′W, is also included. For the Virgin Islands, see **257**. The horstian islands along the coast of South America, sometimes also considered as part of the Lesser Antilles, make up Region 28.

The general flora by Grisebach (1857), now 125 years old, is gradually being supplanted by a new, fairly elaborate *Flora of the Lesser Antilles* (Howard *et al.*, 1974–) of which three volumes (of six planned) have so far appeared. Other floristic reviews exist for the

herbaceous plants (Vélez, 1957) and the trees (Beard, 1944; Stehlé and Stehlé, 1947). Botanically the area, like the Greater Antilles, is relatively well-known and the new regional flora will help place species distribution in a proper perspective. Of the individual islands, only the 2½ Dutch and 4½ French islands have had extensive separate coverage, a state of affairs perhaps as much due to politico–cultural and linguistic reasons as to the sketchy coverage and old-fashioned keys of Grisebach's floras. Modern coverage has been provided for the French islands by *Flora illustrée des phanérogames de Guadeloupe et de Martinique* (Fournet, 1978) and has been in train for the Dutch islands through *Flora of the Netherlands Antilles* (Stoffers, 1962–), the latter but partly completed. Work is also continuing on a treatment of dicotyledons of Dominica to complete the *Flora of Dominica* begun by Hodge (1954). Separate floras for St Bartholomew (Questel, 1941) and the Grenadines (Howard, 1952) as well as checklists for St Vincent (Kew Gardens, 1893) and Anguilla (Boldingh, 1909) are also available.

Bibliographies. General, divisional and supraregional bibliographies as for Division 2 and Superregion 24–29.

Indices. General and divisional indices as for Division 2 or Superregion 24–29.

260

Region in general

Grisebach, A. H. R., 1857. Systematische Untersuchungen über die Vegetation der Karaiben, insbesondere der Insel Guadeloupe nach der Sammlungen Duchassaing's. *Abh. Königl. Ges. Wiss. Göttingen*, **7**: 151–286. (Reprinted separately, 1857, Göttingen, 138 pp.)

Systematic enumeration of vascular plants of the Lesser Antilles, with descriptions of new taxa, partial synoptic keys, synonymy (with some citations), indication of *exsiccatae* and other reports together with local range, critical commentary, and some notes on habitat, special features, etc.; no index. An introductory section contains accounts of previous botanical investigations, general features of the islands, and the composition of the flora. Now mainly of historical interest.

Howard, R. A. (ed.), 1974– . *Flora of the Lesser Antilles: Leeward and Windward Islands.* Parts 1– .

Illus., maps. Jamaica Plain, Mass.: Arnold Arboretum of Harvard University.

Moderately concise illustrated systematic treatment of vascular plants, with keys, short descriptions, synonymy (with references and citations), indication of types, local and extralimital range (with inclusion of some *exsiccatae*), critical remarks, and notes on habit, phenology, special features, etc.; indices in each part. An introductory section in the first part (on Orchidaceae) deals with the background of the work, an indication of the area covered, and a chapter on phytogeography and floristics (by R. A. Howard). At this writing (late 1981), three of the six parts planned have been published: Orchidaceae, by L. A. Garay (1974); pteridophytes, by G. R. Proctor (1977); and Monocotyledoneae, by R. A. Howard (1979).

Special group – woody plants

BEARD, J. S., 1944. Provisional list of trees and shrubs of the Lesser Antilles. *Caribbean Forest*, **5**: 48–67.

STEHLÉ, H. and STEHLÉ, M., 1947. Liste complementaire des arbres et arbustres des Petites Antilles. *Ibid.* **8**: 91–123.

Tabular lists of trees and shrubs recorded from the Lesser Antilles, with presence or absence indicated for each island (the smaller Leeward Islands being grouped for this purpose). The introductory sections include summary tables and lists of references. The lists, complementary to one another, include the Virgin Islands but not Barbados.

Special group – herbaceous flowering plants

VÉLEZ, I., 1957. *Herbaceous angiosperms of the Lesser Antilles*. [iv], 121 pp., 51 illus. San Juan, PR: Inter-American University of Puerto Rico.

This work, mainly devoted to an illustrated descriptive account of herbaceous and subwoody vegetation, includes a tabular list of known herbaceous and subwoody flowering plants of the region, arranged alphabetically by families with indication of presence or absence (with authorities) for each island or island group from the Virgin Islands to Grenada. Extralimital distribution in the Greater Antilles is also indicated. A bibliography concludes the work.

261

Sombrero

See 257 (D'ARCY, 1971) for general remarks on this island, which apparently has only a limited vascular flora. No specific checklists appear to be available.

262

Anguilla (and Dog Island)

See also 265 (BOLDINGH). For this 'mouse that roared', one elderly separate checklist is available.

BOLDINGH, I., 1909. A contribution to the knowledge of the flora of Anguilla (BWI). *Recueil Trav. Bot. Néerl.* **6**: 1–36.

Systematic enumeration of seed plants, based mainly upon collections of the author but also incorporating older records; synonymy, with references; citation of *exsiccatae*, with local range, and general indication of extralimital range; short list of references.

263

St Martin

See 265 (BOLDINGH); 271 (FOURNET); 281 (BOLDINGH, 1913; STOFFERS). The island is half Dutch, half French.

264

St Bartholomew (St Barthélemy)

See also 271 (FOURNET). Almost simultaneously with the work described below there appeared the following: MONACHINO, J., 1940–1. A check list of the spermatophytes of St Bartholomew. *Caribbean Forest*. **2**: 24–66.

QUESTEL, A., 1941. *The flora of the island of St Bartholomew and its origin*. vii, 244 pp., 1 pl., 2 maps. Basse-Terre, Guadeloupe: Imprimerie Catholique. (French edn. published simultaneously under title *La flora de l'île de St-Barthélemy et son origin*.)

Systematic enumeration of seed plants, with occasional descriptive notes; limited synonymy, with some references and citations; vernacular names; general indication of local and extralimital range, with citation of the author's *exsiccatae*; notes on habitat, special features, etc.; indices to families and genera. The introductory section includes a historical review of botanical work on the island together with general comments on the flora and vegetation and on phyto-geographical relationships; the appendices cover medicinal and economic uses.

265

St Eustatius and Saba

See also **281** (BOLDINGH, 1913; STOFFERS).

BOLDINGH, L., 1909. The flora of St Eustatius, Saba and St Martin. In his *Flora of the Dutch West Indian islands*, l. xii, 321 pp., 3 maps. Leiden: Brill.

Systematic enumeration of vascular plants (including also Anguilla and St Croix), with occasional descriptive notes; limited synonymy, with references; vernacular names; rather detailed indication of local range, with citations of *exsiccatae*, and brief summary of extralimital range; separate indices to vernacular and botanical names. The remainder of this work comprises appendices on collectors, physical features, soils, vegetation, phytogeography, etc., together with a bibliography.

266

St Christopher (St Kitts) and Nevis

No separate complete floras or lists appear to be available.

267

Barbuda

No separate complete floras or lists appear to be available.

268

Antigua

No separate complete floras or lists appear to be available.

269

Montserrat and Redonda

No separate complete floras or lists appear to be available.

271

Guadeloupe and Marie-Galante

Works covering the French Antilles in general are listed under this heading.
– Vascular flora (French Antilles): 1800–1850 species (Fournet, 1978).

FOURNET, J., 1978. *Flore illustrée des phanérogames de Guadeloupe et de Martinique*. 1654 pp., 745 text-figs. Paris: Institut National de la Recherche Agrinomique (INRA).

Copiously illustrated, briefly descriptive manual-flora of seed plants, with keys to all taxa, limited synonymy (with many references to Duss's flora), local distribution, some critical remarks, notes on habitat, frequency, origin (if introduced), altitudinal range, phenology, and vernacular names; indices to family, generic, and vernacular names. Covers 1668 indigenous and 149 naturalized species; arrangement follows Emberger system (cf. Guinochet and Vilmorin, *Flore de France* (**651**)). Supersedes DUSS, A., Père, 1897. *Flore phanérogamique des Antilles françaises (Guadeloupe et Martinique)*. xxvii, 656 pp. (Ann. Inst. Bot.-Géol. Colon. Univ. Marseille 3). Marseille. (Reprinted in 2 vols. with revised index, 1972, Fort-de-France, Martinique: Société de Distribution et de Culture); and STEHLÉ, H., STEHLÉ, M., and QUENTIN, L., Père, 1937–49. Catalogue des phanérogames et fougères. In their *Flore de la Guadeloupe et dépendances (et de la Martinique)*, vol. 2, parts 1–3. Illus., maps. Basse-Terre, Guadeloupe: Imprimerie Catholique (part 1); Paris: Lechevalier (parts 2–3).

272

Dominica

The continuation of work on the flora of this island, botanically about the most diverse in the Lesser Antilles, is under the direction of D. H. Nicholson at the Smithsonian Institution, and a treatment of the dicotyledons is nearing completion (SHETLER, 1979, p. 51; see Superregion 21–23, introduction, p. 153).

HODGE, W. H., 1954. Flora of Dominica (BWI). Part 1. *Lloydia*, 17: 1–238, 112 figs.

Enumeration of vascular plants, with keys to all taxa; extensive synonymy, with references; vernacular names; citation of *exsiccatae*, with localities, and general indication of extralimital range; brief taxonomic commentary and notes on habitat, phenology, special features, uses, etc.; figures of representative species; index to botanical and vernacular names. The introductory section includes a general description of the island and its physical features together with accounts of botanical exploration, plant communities, agriculture, etc.; bibliography. Not completed; covers Pteridophyta, Gymnospermae, and Monocotyledoneae.

273

Martinique

See 271 (FOURNET). No separate works have been published.

274

St Lucia

No separate complete floras or lists appear to be available.

275

St Vincent

ROYAL BOTANIC GARDENS, KEW, 1893. Flora of St Vincent and adjacent islets. *Bull. Misc. Inform.* (Kew), *1893*: 231–96.

Systematic list of vascular plants of St Vincent and the northern Grenadines, with limited synonymy, citation of *exsiccatae*, and general indication of extralimital range; statistical summary of the flora; no index. The introductory section includes an account of the islands' physical features.

276

The Grenadines

See also 275 for an additional checklist. The Grenadines form a chain of many small islands from Bequia (south of St Vincent) through Mustique to Carriacou and Ronde (north of Grenada).

HOWARD, R. A., 1952. *The vegetation of the Grenadines, Windward Islands, BWI.* 132 pp., 12 pls. (Contr. Gray Herb., N.S., 174). Cambridge, Mass.

Includes (pp. 71–125) a list of vascular plants recorded for these islands, with vernacular names, citations of *exsiccatae* with localities, and brief notes on habitat, special features, uses, etc.; discussion of the origin and relationships of the flora. The general part includes accounts of the physical features of the islands, geology, botanical exploration, agriculture, etc., as well as a detailed descriptive treatment of the vegetation.

277

Grenada

No separate complete floras or lists appear to be available.

278

Barbados

Revision of the current standard flora of this island, by its senior author, is in progress (HOWARD, 1979; see Superregion 24–29, introduction, p. 166).

GOODING, E. G. B., LOVELESS, A. R., and PROCTOR, G. R. 1965. *Flora of Barbados*. xvi, 486 pp., 27 text-figs., frontispiece, map. London: HMSO.

Briefly descriptive flora of native, naturalized, adventive, and commonly cultivated vascular plants, with keys to all taxa; synonymy, with references and citations; vernacular names; local range, with citations of some *exsiccatae*, and general indication of extralimital range; notes on habitat, properties, uses, special features, frequency, etc.; extinct species and doubtful records; lexicon of vernacular names with botanical equivalents; index to all taxa and synonyms. The introductory section includes a select bibliography and general key to families.

279

Aves Island

ZULOAGA, G., 1955. The Isla de Aves story. *Geogr. Rev.*, New York, **45**: 172–80.

In this short general account, the author records the presence of two vascular plants: *Portulaca oleracea* and *Sesuvium portulacastrum*. The island, a sandbank, has apparently been shrinking in size ever since it was first described in the seventeenth century. At present it is recognized as Venezuelan territory.

Region

28

The Lesser Antilles (southern chain)

This comprises the horstian chain of islands running from Aruba in the west to La Blanquilla in the east, passing through Curaçao, Bonaire, Las Aves, Los Roques, and La Orchila. The first three of these constitute the major part of the Netherlands Antilles, while the remainder are part of Venezuela. For islands on the continental shelf, see Region 29.

Botanical exploration in the region, though of comparatively long duration, has been haphazard. In the three largest islands, serious efforts by Dutch botanists did not get underway until 1884 when a mixed expedition including the Leiden botanist W. F. R. Suringar visited the whole group (along with Surinam and the Dutch Leeward Islands) for several months. Later collecting and publications were accomplished largely under the aegis of Utrecht University, culminating in the floras of Boldingh produced before World War I and in Stoffers' current *Flora of the Netherlands Antilles*, for which preparatory work began in the 1950s. This latter, however, is still some way from completion. A substantial contribution has also been made by Brother Arnoldo, working independently. For the Venezuelan islands, floristic coverage is rather patchy, with evidently no work on La Blanquilla in existence.

Bibliographies. General, divisional, and supraregional bibliographies as for Division 2 and Superregion 24–29. **Indices.** General and divisional indices as for Division 2 or Superregion 24–29.

281

Curaçao, Aruba, and Bonaire (Netherlands Antilles)

The administrative unit of the Netherlands Antilles consists of these islands together with St Eustatius (Statia) and Saba (**265**) and half of St Martin (**263**). Works relating to the whole of these possessions are for convenience listed here. The modern work by Stoffers is yet to be completed and thus has not entirely displaced the treatments of Boldingh (1913, 1914).

ARNOLDO, M., Broeder, 1964. *Zakflora: wat in het wild groeit en bloeit op Curaçao, Aruba en Bonaire*. 2nd edn. 232 pp., 3 pp. of text-figs., 68 pls. (Uitgaven Natuurw. Werkgroep Ned. Antillen, 16). Curaçao. (First edn., 1954.)

Pocket-sized, semi-popular manual of seed plants (excluding Gramineae and Cyperaceae), with keys to all taxa, vernacular names, general indication of local range, figures and plates of representative species, and summary in English; indices to botanical and vernacular names. The introductory section includes a glossary.

BOLDINGH, I., 1913. *Flora voor de Nederlandsch West-Indische eilanden.* xx, 450 pp. Amsterdam: Koloniaal Institut.

Descriptive manual of vascular plants of all the Netherlands West Indian islands, with keys to all taxa, vernacular names, brief general indication of range by islands (no extralimital distribution), and indices to botanical and vernacular names. The introductory section includes a glossary.

BOLDINGH, I., 1914. The flora of Curaçao, Aruba, and Bonaire. In his *Flora of the Dutch West Indian islands*, 2. xiv, 197 pp., 9 pls., map. Leiden: Brill.

Systematic enumeration of vascular plants, with occasional descriptive notes, limited synonymy, references, vernacular names, fairly detailed indication of local range (with citations of *exsiccatae*) and general summary of extralimital distribution; indices to botanical and to vernacular names at end. Includes some notes on the plants of Margarita. The remainder of the work comprises appendices on collectors, physical features, soils, vegetation, phytogeography, etc., together with a bibliography.

STOFFERS, A. L., 1962– . *Flora of the Netherlands Antilles.* Vols. 1 parts 1–2; 2 parts 1–2, and 3 parts 1–2 (Uitgaven Natuurwet. Studiekring Suriname Ned. Antillen, 25–102, *passim.*) Utrecht.

This conservatively styled work comprises a critical descriptive flora of native, naturalized, and commonly cultivated vascular plants of all the Netherlands West Indian islands, with keys to all taxa within respective families, full synonymy (with references), vernacular names, detailed indication of local range (by island in first instance) with citations of *exsiccatae*, general summary of extralimital distribution, and brief notes on habitat, special features, etc. At this writing (1981) a substantial proportion of the three volumes projected has been published, although no volume is yet complete; the most recent part seen appeared in 1980. The sequence of the work is approximately that of the Pulle system of 1952 (as regards the angiosperms); present coverage includes the pteridophytes (treated by K. U. Kramer) and parts of the monocotyledons and dicotyledons (some families by specialists), the latter constituting orders 1, 8, 11, 13–24, 42, 45, and 47 (in part), in all amounting to somewhat more than half of the total vascular flora. [I

thank Dr A. Stoffers for advice regarding this work. It is hoped that it can be brought to an early conclusion.]

283

Territorio Colón (Las Aves, Los Roques, La Orchila)

No general floras or checklists cover these islands as a whole. However, two relatively recent partial works are accounted for below.

ARISTEGUIETA, L., 1956. Florula de la region. In *El archipelago de Los Roques y La Orchila* (ed. SOCIEDAD DE CIENCIAS NATURALES 'LA SALLE'), pp. 47–67, 4 pls. (some in color). Caracas.

Enumeration of non-vascular and vascular plants, with indication of local and extralimital range, vernacular names, and notes on habitat, frequency, etc.; general description of the vegetation.

GINÉS, Hermano and YEPEZ T., G., 1960. *Aspectos de la naturaleza de las islas Las Aves, Venezuela.* 53 pp. (Trab. Soc. Ci. Nat. 'La Salle', 20). Caracas.

Although concerned mainly with bird life, this work includes (pp. 12–20) an account of the vegetation and a short list of vascular plants observed, with vernacular names.

285

La Blanquilla (and associated islets)

No separate floras or lists appear to be available.

Region

29

West Indian continental shelf islands

Included here are those islands which lie on the South American continental shelf south of the Windward Islands but which traditionally have been considered part of the West Indies. The largest island is Trinidad, especially notable – along with Tobago – for its long association with the other anglophone West

Indian islands in regard to botanical work as in other matters cultural and political. The remaining islands, all Venezuelan, include La Tortuga, the Nueva Esparta group (Margarita, Coche, and Cubagua), and Los Testigos.

Of these islands, Trinidad and Tobago are botanically relatively well-explored after two or more centuries of activity. The flora is currently being documented in *Flora of Trinidad and Tobago* (1928–), now far advanced after a spasmodic history of progress. A companion to this work is *Orchids of Trinidad and Tobago* (Schultes, 1960). Knowledge of the Venezuelan islands is rather more uneven and deficient; certainly the largest, Margarita, is in need of more exploration (A. L. Stoffers, personal communication). Available separate documentation is moreover very old or lacking.

Bibliographies. General, divisional, and supraregional bibliographies as for Division 2 and Superregion 24–29.

Indices. General and divisional indices as for Division 2 or Superregion 24–29.

291

La Tortuga

ERNST, A., 1876. Florula chelonesiaca; or, a list of plants collected in January, 1874 in the island Tortuga, Venezuela. *J. Bot.* **14**: 176–9.

Enumeration of lichens and vascular plants, based on the author's collections of 1874, with brief indication of local range, vernacular names, and notes on habitat, frequency, special features, etc. A brief introduction includes remarks on physical features of the island and the composition of its flora.

292

Nueva Esparta (Margarita, Coche, and Cubagua)

JOHNSTON, J. R., 1909. Flora of the islands of Margarita and Coche, Venezuela. *Proc. Boston Soc. Nat. Hist.* **34**: 163–312, pls. 23–30 (incl. 2 maps). (Also issued as *Contr. Gray Herb.* N.S., 37).

Systematic enumeration of vascular plants (with those of Coche separately listed), including synonymy,

references and some citations; detailed indication of local range, with citations of *exsiccatae*, and general summary of extralimital range; taxonomic commentary. The introductory section includes accounts on the physiography and botanical exploration of the islands, while appendices deal with useful plants (with vernacular names), vegetation, and phytogeography. Now most likely very incomplete.

293

Los Testigos

No separate floras or lists appear to be available.

295

Trinidad

See also **240–90** (GRISEBACH). General works on the state of Trinidad and Tobago appear under this heading.

WILLIAMS, R. O. *et al.*, 1928– . *Flora of Trinidad and Tobago*. Vols. 1, parts 1–8, index, vol. 2, parts 1–10, vol. 3, parts 1–2. [Port-of-Spain], Trinidad: Department of Agriculture (later Ministry of Agriculture, Lands, and Fisheries).

Briefly descriptive flora of flowering plants, with keys to families, genera, and species, synonymy, citations of pertinent literature, vernacular names, indication of *exsiccatae* and authorities with localities, generalized summary of local and extralimital range, some critical remarks, and notes on habitat, properties, uses, etc.; index to all vernacular and botanical names (vol. 1 only; the two other volumes are not yet completed). Substantial contributions were made by E. E. Cheesman from the 1930s through the 1950s, and up through 1967 work continued in fits and starts, with much assistance from botanists in Jamaica and elsewhere. After a lapse of several years, work was resumed in the mid-1970s with the aim of completing the *Flora* and the latest parts, all by D. Philcox, have appeared in 1977–9. The work is arranged according to the Bentham and Hooker system; at this writing (1981) coverage encompasses all of the Dicotyledoneae, save for series 7 (in part) and 8 in the Monochlamydeae, and part of the Monocotyledoneae (including Orchidaceae). [I thank Messrs C. D. Adams and D. Philcox for advice respecting this work.]

Special group – trees

MARSHALL, R. C., 1934. *Trees of Trinidad and Tobago.* ii, 101, viii pp., 20 pls. Port-of-Spain: Government Printing Office.

Briefly descriptive account of native and commonly cultivated trees, with synoptic keys and vernacular names; general indication of local range; illustrations of representative species; notes on properties, uses, timber characteristics, etc.; index to vernacular and botanical names.

296

Tobago

See **295** (all works). The offshore islet of Little Tobago is included here.

Notes

1 Gómez-Pompa, A., 1973. The thrust of present and future research in the lowland tropics of Mexico. *Ann. Missouri Bot. Gard.* **60**: 169–73.

2 For a compendium of essays on this 'Linnaeus americanus neotropicus', see Williams, L. O. (ed.), 1963. *Homage to Standley.* 115 pp., illus. Chicago.

3 Gentry, A. H., 1978. Floristic knowledge and needs in Pacific tropical America. *Brittonia*, **30**: 134–53.

4 Prance, G. T., 1978. *Ann. Missouri Bot. Gard.* **64**: 673 and errata sheet, pp. i–ii.

5 Gómez-Pompa (1973) calls attention to the backward development of botanical work in the Mexican lowland tropics, giving pertinent reasons. The lack of local centers is viewed as a serious problem. [Hear, hear!]

6 Gómez-Pompa, A. and Nevling, L. I., 1973. The use of electronic data-processing methods in the *Flora of Veracruz* program. *Contr. Gray Herb.*, N.S. **203**: 49–64.

7 The *Flora of Veracruz* program, at present based in Xalapa, Veracruz, has led to a great increase in available collections from the state, with consequent improvements to taxonomic knowledge, local and general.

8 See also a resumé of the Flora Mesoamericana project as part of the Introduction to Superregion 21–23, p. 152.

9 Belize was intended to be covered in the series of 'Kew floras' but by 1905 it was still too much a botanical '*terra incognita*'. Serious collecting began only in 1907.

10 I thank P. Bernhardt, at the time of writing attached to the University of Melbourne, for information on Salvadorean botany. See also Bernhardt, P. and Montalvo, E. A., 1978. Selected collecting sites in El Salvador, I. Private property. *Bull. Torrey Bot. Club*, **105**: 9–13.

11 On seeing this second edition, Standley, in no way involved with its preparation, is reported to have expostulated, 'Heaven help us!' See Williams, *Homage to Standley*, p. 19.

12 The United Fruit Company (now Standard Brands).

13 The *Flora of Panama* was originally conceived 'under the flimsy pretext of the maintenance by the Missouri Botanical Garden of a small establishment in the Canal Zone' (R. E. Woodson in Williams, *Homage to Standley*, p. 34). Standley in 1945 (*ibid.*, p. 20) wrote that 'I should hate to get out a flora of a country whose vegetation is so little known.'

14 Howard, R. A., 1975. Modern problems of the years 1492–1800 in the Lesser Antilles. *Ann. Missouri Bot. Gard.* **62**: 368–79.

15 For other documents on the background to this work, see W. T. Thistleton-Dyer, 1906. *Bull. Misc. Inform.* (Kew), *1905*: 11–13.

Division

3

South America

Gross ist der Einfluss gewesen, welchen die von Humboldt und Bonpland zurückgebrachten botanischen Materialen auf die Ausbuildung der Systematik und die umfassendere Kenntniss der Gestalten im Pflanzenreiche ausgeübt haben.

> von Martius, 1860, *Akademische Denkreden* (1866); quoted in Stearn, *Humboldt, Bonpland, Kunth and tropical American botany* (1968).

At the foot of Chimborazo, with the zoning of its vegetation before his eyes, Humboldt drafted his *Essai sur la Géographie des Plantes* (1807), which established plant geography as a discipline.

> Stearn, *Ibid.*

Botanically South America is the least explored area of the world. This survey...has shown that many large areas and important habitats are still uncollected, in spite of the long history of collection and collectors in South America, and a long tradition of local botanists in some of the countries.

> Prance in *Systematic Botany, Plant Utilization and Biosphere Conservation* (ed. Hedberg, 1979).

This division comprises the entire continent from the Panamanian border southwards to Tierra del Fuego, together with the Falkland (Malvinas) Islands. Associated islands in the eastern Pacific Ocean come within Region 01, while those in the South Atlantic are included in Region 03. The chain of islands off the Caribbean coast, which with the mainland was once the 'Spanish Main' and is inclusive of Trinidad, Nueva Esparta, and other islands to Aruba, is here considered to be part of the West Indies as Region 28 and most of Region 29. Over and above the eight politico–geographic regions delimited herein, four major physiographic units, each with a history of separate floristic treatment, are individually recognized: the Guayana Highland, **304**; the Andes (Alpine zone), **303**; and Patagonia and tropical South America, **301**.

Although 'many investigators have taken part both in early times and [in] our own in the exploration...of tropical South America, the complex and enormous flora of no part of the area can [yet] be considered sufficiently known to prepare a descriptive treatise which could satisfy the standards of contemporary demand'. Thus wrote Bassett Maguire, one of the leading contemporary South American botanical explorers, in 1958. While much new work has been accomplished in the intervening quarter-century in most parts of the continent, his words in essence still apply, even considering the whole land mass. With the largest vascular flora of any continent or island aggregation – more than 90 000 species, or a third of the world total – South America at present is botanically the least explored and most imperfectly documented part of the world, and its herbaria relatively among the most inadequately stocked. Recent writers have affirmed that the level of collecting in the infratropical zone lags well behind most parts of the tropics of Asia, Malesia, and the Pacific as well as Africa, a contrast to the 1850s when, as noted by Pennell, South America was botanically among the better-known parts of the world outside of western Europe and eastern North America. Such is the size of the flora that its basic taxonomic inventory, as Prance has emphasized, is yet quite incomplete; interesting new discoveries are constantly being made while there is still time, although the continent as a whole is more or less into its second cycle of exploration in the sense of Symington. Likewise, few practical modern floras are available; the sheer size of the flora, other factors aside, over much of the continent has by and large caused the preparation of even partial floras, let alone nation-wide compilations, to be a daunting task beyond the resources of most individual institutions, even outside the continent.

To the difficulties imposed by the size and accessibility of the flora in terms of satisfactory increase and diffusion of floristic knowledge, must be added the effects of cultural traditions, superstition and official suspicion (notably in the era of Spanish and Portuguese colonial rule), chronic political instability in many countries, economic conditions, decades of scientific imperialism in the nineteenth and twentieth centuries, and, more immediately, the generally slow, limited growth of local resources, human and material, for floristic studies, including national herbaria with adequate, well-named collections and stocks of literature so necessary for serious locally based work, as stressed by Forero. In many countries, progress depended upon the interest and enthusiasm of isolated individuals or groups, with institutionalization comparatively uncommon and continuity always a stumbling block in the face of widespread official indifference and, as perceived locally, other more pressing demands upon limited national resources.

Superstition and, with it, official suspicion and secrecy were major obstacles to scientific progress in much of Latin America in the colonial period up to the Napoleonic Wars. While in some reigns (notably those of the Spanish kings Carlos III and Carlos IV) floristic exploration and research in the viceroyalties were more or less encouraged, the publication of results was but rarely so and in Spain became all but impossible with war and post-colonial impoverishment (events probably also contributing to the 35-year delay in publication of even part of Velloso's *Florae fluminensis*). The impact of much of the relatively considerable Spanish (and Portuguese) work, particularly the three extended botanical surveys (*Reales Expediciónes Botánicas*) to New Spain (México), New Granada (Colombia), and Peru and Chile, which exercised considerable influence locally and whose results *were* originally intended for publication, and the Malaspina world voyage of 1789–94 which numbered among its objectives a survey of the Philippines but also examined the viceroyalty of La Plata and parts of the Pacific coast of South America, including Ecuador, was thus lost. Most of the key early contributions to South (and Central) American botany were very largely the work of 'foreigners' with lucky opportunities: Marcgrave in Brazil (1637–44) when the Brazilian Nordéste was under Dutch rule; de la Condamine and J. de Jussieu in Amazonia, Peru, and Ecuador in the 1730s and 1740s; Loefling in Venezuela under Spanish patronage around 1760; Aublet in French Guiana in the 1760s (whose *Histoire des plantes* (1775) was an outstanding early contribution); many visitors to Rio de Janeiro (usually, however, not allowed inland travel); Commerson, Banks, Solander, the Forsters, Sparrman, and others in Fuegia in the late eighteenth century; and, most importantly, the famous expedition of Humboldt and Bonpland in northwestern South America (as well as Cuba and Mexico) in 1799–1804 which gained wide popular acceptance, exercised considerably indirect political influence, set the pattern for future scientific exploration of the continent (in much the same way as did Horsfield and Raffles in Malesia), and whose results, published with relative celerity, created a sensation in revealing the wealth of plant and animal life to be found. The most important botanical contribution of this voyage, *Nova Genera et Species Plantarum* (1816–25), largely written by Kunth on a modified Jussieuan (natural) system, is basic to all later neotropical botany as being, in the words of W. T. Stearn, 'essentially a pioneer flora of northern South America and of Central America'.[1] Possessing many of the now-standard principles and features of modern floras, it was at the time, even though written by a German author, intellectually far ahead of most of its contemporaries, a development aided by its having been written in Paris, the center of the French 'school.' In it, 3600 species were described for the first time – four-fifths of its total coverage.

The efforts of Humboldt and Bonpland were, after the Napoleonic Wars, emulated in politically stable Brazil by the expedition of Spix and von Martius in 1817–20, which among its results included the first serious biological reconnaissance of Amazonia. The botanical results, written also in the form of a *Nova genera et species plantarum* (1824–32) by von Martius, formed a primary basis for the mid-century *Flora brasiliensis* which, when begun, could draw on the results of many more expeditions – Brazil, under Pedro I and II, being a fashionable destination. As conditions stabilized in other countries, they were in turn visited by numerous, often competitive, organized expeditions. By mid-century the continent came to be considered relatively well-known in a primary sense, and general exploratory interest largely shifted to Africa, East Asia, and finally New Guinea. The prosecution of floristic work in South America became the province of more specialized field work as well as, increasingly, resident botanists, although significant areas were yet scarcely known geographically and continued to attract substantial general expeditions until the mid-twentieth century.

However, until late in the nineteenth century, the often unsettled political conditions, economic situation, and slow social and cultural development coupled with, in most cases, the absence of formal institutional ties analogous to those characterizing floristic botany as it developed in Africa and much of the Old World (with the accompanying stimulus towards development of local centers) and the corresponding decline in involvement in South American botany by many of these metropolitan centers, militated strongly against the local prosecution of major projects. Pervading all, though, was a lack of continuity. What was accomplished after 1815, at first through general expeditions, then through more specialized efforts, remained largely the work of overseas explorers until well into the present century. With the exception of the southern 'triangle', significant resident contributions have been a feature only of the last half-century or so and in two countries have as yet hardly developed. The visitors were under various types of patronage or came for commercial reasons and were aided by scientists at various European centers (with their North American counterparts increasingly involved after 1900). A large proportion of the visitors were from central Europe and

Scandinavia; some became resident for long periods or even emigrated permanently. Botanical contributions by German visitors or by residents and their descendants were to receive some official favor at German centers (*Flora brasiliensis* having set a tradition) and, along with those by Swedes from Regnell onwards (similarly favored in Stockholm), were collectively the most important up to 1945 although contributions from Americans were rapidly rising after 1920 (even earlier in Bolivia). The abundant results were embodied in floras, enumerations, and often poorly coordinated, widely scattered 'contributions' of varying quality. Along with a widespread tendency towards specimen description, this dispersion of efforts has had, as in some other parts of the tropics, the effect of creating unnecessary synonymy, a problem further aggravated by national and even continental boundaries. An outstanding exception was *Flora brasiliensis*, that splendid early example of international albeit aristocratically sponsored official and professional collaborative effort which, along with the Candollean *Prodromus*, dominated mid-nineteenth-century European systematics; and it is only in recent decades, with the establishment of a variety of bilateral and multilateral links aimed at the promotion of floristic work and flora writing, that something like a return to this tradition has taken place, albeit in many cases with, so to speak, the shoe on the other foot. For the long period of overseas scientific domination, through which vast collections including some eighty per cent of the type specimens left the continent, inevitably brought various forms of retaliation and restrictions as locally based science began serious development.

A further complicating factor, also with effects upon the level and nature of documentation, has been the distribution of resources within the continent. The bulk yet lies within areas centering on Rio de Janeiro, São Paulo, Buenos Aires, Montevideo, and Tucumán. Elsewhere they have been scattered and limited, with certain countries virtually without herbaria of any kind. Recent decades, however, have seen expansion in Venezuela, Colombia, Peru, and Amazonia, with the national herbaria in the two first-named serving as centers for significant flora projects.

Thus, much work in South American botany has been of a disordered and to some extent '*ad hoc*' nature, with major organized projects few (and often externally administered) and transnational collaboration and integration uncommon. Until recently, decentralization, largely country-centred, has been the rule, and the historical background pertaining to the various principal floristic works has therefore been recounted in the individual introductions to Divisions 31–39. Further details can be sought in the country reports collected in the references below as well as in individual papers. Nevertheless, because of the absence until recently of supranational media corresponding to the *Flora Malesiana Bulletin* or the index and congress proceedings of AETFAT, estimation of botanical progress in South (and Middle) America has been less easy to evaluate than is the case for Africa and the Old World tropics.

In recent decades, however, significant developments towards integration have taken place, involving not only South America but also Mexico, Central America, and the West Indies. The most important of these has been the establishment of *Flora Neotropica*. Begun in 1968, this series of unstructured monographs, covering the whole of the Neotropics, is at present produced under the aegis of the international Organization for Flora Neotropica (OFN), founded in 1964, and currently designated as a UNESCO Category B organization. To date (1981), 25 monographs (of families or parts thereof in all groups of plants) have been published or are in press. Although in earlier years almost exclusively concerned with monographic production, OFN is now expanding into ancillary documentation of the kind long a feature of AETFAT and the *Flora Malesiana* organization and by the latter especially regarded as overdue. Nevertheless, associated bibliographic, biohistorical and theoretical documentation is limited. *Flora Neotropica* in many ways more closely resembles *Das Pflanzenreich* than it does *Flora Malesiana* or the several African regional floras. A description of current activities is given by Gentry (1979, 1981).

Other attempts at coordination include *Boletín Botánico Latinoamericano*, at present issued from Bogotá and serving a function similar to the *Flora Malesiana Bulletin*, and *Taxonomic bibliography for the neotropical flora* by G. Agostini (1974–), of which only a small proportion has been published. In addition, Latin American botanical congresses have been instituted.

Major current flora projects include *Flora of Suriname*, begun in 1932 and now covering the three Guianas; *Flora de Venezuela*, begun in 1964; *Flora of Ecuador*, begun in 1973 and associated with the long Swedish Regnellian botanical tradition; *Flora of Peru*, begun in 1936 and now under multi-institutional sponsorship; and the several regional floras in Argentina, representing a coordinated project begun in the early 1960s after the termination of the elephantine *Genera et species plantarum argentinarum* (1943–56). A contemporary project, *Flora brasilica*, begun by Hoehne in 1940, presently appears to be dormant but a number of state floras have been published or are in progress, with the most significant being *Flora ilustrada catarinense* (1965–). Partial floras are also being produced in Venezuela, Colombia, and Chile, and some one-volume florulas of limited areas (e.g., Avila near Caracas, Rio Palenque Science Center in Ecuador, and the environs of Buenos Aires) and generic synopses (e.g., for Chile and Venezuela) have also been published. There is also a modern flora of the Falkland (Malvinas) Islands, off southeastern Patagonia, and a Fuegian flora is reportedly in preparation. Last, but not least, there is *Botany of the Guayana Highland*, covering an area botanically one of the most important in South America and embodying the results of the many expeditions to Conan Doyle's 'Lost World' conducted from the late 1920s.

Older floras of significance include, apart from *Flora brasiliensis* (1840–1906), *Symbolae ad floram argentinarum* by Grisebach (1879), *Flora chilena* by Gay (1845–54) and the incomplete *Flora de Chile* by Reiche (1896–1911), *Plantae hasslerianae* by Chodat and collaborators (1898–1907) on Paraguay, *Flora uruguaya* by Arechavaleta (1898–1911) and the later works of Herter on Uruguay, *Prodromus florae novo-granatensis* by Triana and Planchon (1862–73) on Colombia (terminated far from completion), and *Initia florae venezuelensis* by R. Knuth (1926–8). Of more local works, mention may be made of *Synopsis plantarum aequatoriensium* by Jameson (1865), Schomburgk's *Versuch* (1848) for present-day Guyana, *Flora der Umgebung der Stadt São Paulo* by Usteri (1911), and some other state and provincial works in the southern 'triangle'. Weddell's incomplete *Chloris andina* (1855–61) partially covered the Andes, and the 1899 Princeton University Expedition to Patagonia together with the contemporary field work by the Swedish botanists Dusén and Skottsberg there and in the Falkland Islands resulted in a major compilatory work by Macloskie and collaborators (1903–14) which took in the whole area south of about the fortieth parallel. Earlier, Fuegia had been treated in the second volume of *Flora antarctica* (1845–7) by J. D. Hooker. Most of these more or less represent the primary stage of modern documentation for their respective areas, comparable to *Flora of Tropical Africa* (**501**) or *Flora indiae batavae* (**910–30**), while the works of recent decades are representative of the secondary stage as outlined by Symington and may be compared with, for example, *Flora Malesiana* (**910–30**) or *Flora reipublicae popularis sinicae* (**860–80**) although, as already noted, the infratropical zone is still very sketchily covered. They remain for the most part without, or with but incomplete, successors. The most poorly documented countries are Bolivia (without any overall work save an uncritical name list), Paraguay, and (to a lesser extent) Ecuador (with no overall work prior to the initiation of *Flora of Ecuador*).

Of historical interest but benefiting from modern revision is the multivolume *Flora de la Real Expedición Botánica del Nuevo Reino de Granada*, comprising a latter-day version of the results of the eighteenth-century 'tribe of botanical adventurers' headed by Mutis and containing reproductions of the numerous fine colored plates drawn for this botanical survey. Publication began in 1954 and at present six parts (of a projected 51) have appeared.

A few words should be said about future flora writing in South America. While production of *Flora Neotropica* and other large-scale national and state/departmental works should be continued as they represent significant contributions to floristic science and documentation, consideration should also be given to the preparation of more 'generic floras', synopses and local florulas, as these will reach a wider local audience and for teaching purposes are more useful, as Gentry and Prance have stressed.[2] These may be the best tools for raising cultural and environmental awareness and hence an effective interest in conservation, so necessary in the face of extensive destruction in the name of 'development' in Amazonia and elsewhere, however 'undesirable' they may be from some idealistic botanists' point of view. That two such works, both well-illustrated, have appeared within the last few years is heartening; they serve to show clearly how rich and diverse is the tropical South American flora even over small areas. To these may be added a third, also of recent date, covering Barro Colorado Island in adjacent Panama (**237**). More of them are needed. In Colombia, a 'generic flora' has been projected which, if completed, will be valuable as an introductory work for the floristically second-richest country in the world. Over and above floristic information, efforts should also be made at increasing the collection of biological and experimental data, already a well-established activity in several other parts of the tropics and of considerable value to critical flora writing.

Progress

The most complete introduction to the state of botanical knowledge in South America, with a historical sketch (by F. W. Pennell) and many country/territory surveys, remains VERDOORN, F., 1945. *Plants and plant science in Latin America*. xxxvii, 383 pp., illus., maps (Chronica Botanica Plant Science Books, 16). Waltham, Mass. More recent surveys include MAGUIRE, B., 1958. Highlights of botanical exploration in the New World. In *Fifty years of botany* (ed. W. C. Steere), pp. 209–46. New York; and PRANCE, G. T., 1979. South America. In *Systematic botany, plant utilization and biosphere conservation* (ed. I. Hedberg), pp. 55–70. Stockholm. The infratropical zone was also reviewed in Prance's (1978) general survey of floristic progress in the tropics,[3] and the Pacific wet belt in GENTRY, A. H., 1978. Floristic needs in Pacific Tropical America. *Brittonia*, **30**: 134–53. The role of national herbaria in promotion of locally based floristic and taxonomic study has been outlined in FORERO, E., 1975. La importancia de los herbarios nacionales de America Latina para las investigaciones botánicas modernas. *Taxon*, **24**: 133–8.

Flora Neotropica. Descriptions and status reports of this project may be found in MAGUIRE, B., 1973. The organization for 'Flora Neotropica'. *Nature and Resources* 9(3): 18–20, and GENTRY, A., 1979, 1981. Flora Neotropica news. *Taxon*, **28**: 647–9, **30**, 81–7.

General bibliographies. Bay, 1910; Blake and Atwood, 1942; Frodin, 1964; Goodale, 1879; Holden and Wycoff, 1911–14; Jackson, 1881; Pritzel, 1871–7; Rehder, 1911; USDA, 1958.

Divisional bibliography

AGOSTINI, G., 1974. Taxonomic bibliography for the neotropical flora. *Acta Bot. Venez.* **9**: 253–85. [Covers the six families from Acanthaceae through Anacardiaceae in alphabetical order. No continuation yet seen.]

General indices. BA, 1926– ; BotA, 1918–26; BC, 1879–1944; BS, 1940– ; CSP, 1800–1900; EB, 1959– ;

FB, 1931– ; IBBT, 1963–9; ICSL, 1901–14; JBJ, 1873–1939; KR, 1971– ; NN, 1879–1943; RŽ, 1954– . See also under the superregions.

Divisional indices

TORREY BOTANICAL CLUB, NEW YORK, 1969. *Index to American botanical literature, 1886–1966.* 4 vols. Boston: Hall. [Originally published serially in *Bulletin of the Torrey Botanical Club* as well as on index cards. In this reprint, taxonomic and other non-author entries appear in volume 4. The Index has been continuing in the *Bulletin*, and the first 10-year cumulative supplement (1967–76) has now been published (1977, Boston; 740 pp.).]

AMERICAN SOCIETY OF PLANT TAXONOMISTS, 1939–67. *The taxonomic index.* Vols. 1–30. New York (later Cambridge, Mass.; from vol. 20 (1957) published serially in *Brittonia*). (Vols. 1–9 mimeographed.) [Begun on the initiative of W. H. Camp, this index had an existence of 28 years but from the mid-1940s it was reproduced from the appropriate parts of the Index to American Botanical Literature. In 1957 it was consolidated with *Brittonia* although retaining separate volumination, and in 1967 discontinued.]

FORERO, E. (coord.), 1978– . *Boletín botánico latinoamericano.* Nos. 1– . Bogotá. (Offset-printed.) [Primarily a newsletter, but includes news of floras and flora projects as well as a limited selection of other literature references. Eight numbers published at time of writing (late 1981).

Conspectus

300

Division in general

No floristic works cover the entire continent.

301

Major continental subdivisions

Under this heading are included works on tropical South America and on Patagonia. For the Guayana Highland, see 304; for the high Andes, see 303.

I. Tropical South America

Keys to families

PITTIER, H., 1939. *Clave analítica de las familias de plantas superiores de la América tropical.* 4th edn. vii, 94 pp. Caracas: 'del Comercio'. (First edn., 1917, as *Clave analítica de las familias de plantas fanerógamas de Venezuela y partes adyacentes de la América tropical*; 2nd edn., under current title, 1926; 3rd edn., 1937.) [Based upon Thonner's *Analytical key to the natural orders of flowering plants* (1895; see 000, p. 61); includes a glossary. This work was in turn used by Standley in his *Trees and shrubs of Mexico* (210) and *Flora of the Panama Canal Zone* (237).]

ORGANIZATION FOR FLORA NEOTROPICA, 1968– . *Flora Neotropica: a series of monographs.* Monog. 1– . Illus., maps. New York: Hafner (later New York Botanical Garden) (for the Organization for Flora Neotropica).

Comprises an open-ended series of detailed monographic or submonographic treatments of families and infrafamilial taxa of non-vascular and vascular plants (including fungi) within the tropics of the New World, including analytical keys, descriptions, synonymy with references and citations, geographically arranged citation of *exsiccatae* with localities, general summaries of internal and extralimital range, taxonomic commentary, notes on habitat, special features, etc., and illustrations and distribution maps of a large range of species; complete indices to all names and to specimens in each fascicle as well as introductory sections on morphology, biology, taxonomy, etc., of the group(s) as a whole with synoptic lists of taxa.

At this writing (late 1981), 28 monographs have been published, one (Bromeliaceae) in three parts. The Organization for Flora Neotropica is at present controlled by a 'staff committee' of four, and the *Flora* itself is under the management of an editorial committee of six.

II. Patagonia

This unit refers to geographical Patagonia (and Fuegia), i.e., all of South America from approximately 40 °S. to Cape Horn. In the Argentine this refers to the territory south of Río Negro and Río Limay, while in Chile it includes the island of Chiloë and all territory southwards. For additional coverage, see respectively 380 and 390 for general Argentine and Chilean works, and 388 for Argentinian Patagonia (CORREA, *Flora patagónica*).

– Vascular flora: 2200 species (Macloskie).

Bibliography

PÉREZ-MOREAU, R. A., 1965. *Bibliografía geobotanica patagonica.* 110 pp. (Publ. Inst. Nac. Hielo Continental Patagonica, 8). Buenos Aires. [Includes systematic, phytogeographical, and geobotanical references; no subject index.]

MACLOSKIE, G. *et al.*, 1903–6. Botany (including ''Flora patagonica'). 9 parts. In *Reports of the Princeton University expeditions to Patagonia, 1896–1899* (ed. W. B. Scott), vol. 8 (divisions 1–2). xxii, 982 pp., 106 text-figs., 31 pls. (some in color; incl. map). Princeton, New Jersey: Princeton University; Stuttgart: Schweizerbart.

MACLOSKIE, G. and DUSÉN, P. 1914 (1915). Botany: supplement (Revision of Flora patagonica). *Ibid.*, vol. 8 (division ('section') 3). 307 pp., 4 pls. Princeton/Stuttgart.

This luxuriously produced work, publication of which was supported by the J. Pierpont Morgan Publication Fund has, in the nine parts of the original two divisions, a variety of mainly floristic contributions, of which much the largest is part 5, *Flora patagonica*, by G. Macloskie (pp. 139–906). This work comprises a partly compiled, concisely descriptive flora with keys to genera and species, limited synonymy (with some references), a few vernacular names, generalized indication of internal and extralimital range, with denotion in some cases of authorities but almost no indication of *exsiccatae*, taxonomic commentary, and notes on uses, special features, etc.; brief introduction at beginning of work. Part 6 (pp. 907–20) comprises an analytical key to the 113 families of seed plants covered in *Flora patagonica*. The remaining parts 7–9 respectively constitute a historical account (with bibliography), a gazetteer, and a general review of the flora and its possible origin; these are followed by additions, corrections and errata (pp. 961–4) and an index to vernacular, family and generic names (pp. 965–82). Parts 1–3 deal with vegetation (by P. Dusén) and bryophytes (beyond our scope) while part 4 (pp. 127–38) treats pteridophytes (with determinations by L. M. Underwood). The Supplement, dated 1914 but actually published early in 1915, is devoted to revisions of the contents of parts 4 and 5 in the main work, partly based on further field work in Patagonia to 1903 (especially by P. Dusén). [In this work, the approximate northern limit of Patagonia is taken as 40 °S. Though based in the first instance upon the expedition collections of J. B. Hatcher and O. A. Peterson, with some additions from a collection of B. Brown, Macloskie used the opportunity to produce the first, and so far only, work of its kind, a synthesis flora of the whole of Patagonia, thereby pulling together the scattered literature of the eighteenth and nineteenth centuries and incorporating the results of contemporary exploration by other (mainly Swedish) workers. However, the sumptuous appearance of this flora belies its contents: the type is virtually of a size appropriate for the 'Large Print' edition of the *New York Times* whilst the text is surprisingly poorly documented for such a key regional work. By modern standards and requirements, it must be seen as something of a 'white elephant' – alas, not to be the last in southern South America.]

303

Alpine and upper montane zones: The Andes

The only comprehensive work for the extensive area covered by high-altitude vegetation on the Andean Cordillera, comprising *páramo* in the north and *puna* in the central and southern parts, remains Weddell's *Chloris andina*. Parts of the area are covered by more recent works: Fries (1905) for northwestern Argentina, and Vareschi (1970) for Venezuela.

WEDDELL, H. A., 1855–7(–1861). Chloris andina. Essai d'une flore de la région alpine des cordillères de l'Amérique du Sud. Vols. 1–2. 90 pls. In *Expédition dans les parties centrales de l'Amérique du Sud... exécutée par ordre du gouvernement français pendant les années 1843 à 1847, sous la direction du comte Francis de Castelnau*, part 6: Botanique (ed. F. L. DE LAPORTE DE CASTELNAU). Paris: Bertrand. (Reprinted 1971 in 1 vol., Lehre, Germany: Cramer.)

Detailed descriptive 'Andino-Alpine' flora of seed plants occurring along the ranges from Venezuela to Chile; includes synonymy and references, citations of *exsiccatae* with localities (not confined to Weddell's own collections), overall distribution, extensive taxonomic commentary, figures of representative species, and generic indices. Not completed; vol. 1 encompasses Compositae, vol. 2, Calyceraceae through Ranunculaceae (44 families in all). Until the publication of *Alpine flora of New Guinea* by VAN ROYEN beginning in 1979 (see **903**), this remained about the most ambitious of 'alpine floras'.

Partial works

FRIES, R. E., 1905. *Zur Kenntnis der alpinen Flora im nördlichen Argentinien*. 205 pp., 2 figs., 9 pls., map (Nova Acta Regiae Soc. Sci. Upsal., ser. IV, vol. 1, part 1.) Uppsala.

Systematic enumeration of vascular plants from the alpine regions of the provinces of Jujuy, Salta (northern part) and Los Andes (in part), with synonymy, references, localities with citation of *exsiccatae*, and summary of overall range; list of references and index at end. The introduction includes sections on climate, plant formations, and phytogeography. Much critical commentary on distinguishing features, variability, growth forms, etc., is also included. [Based upon 1901–2 field work by the author.]

VARESCHI, V., 1970. *Flora de los páramos de Venezuela*. 429 pp., 126 figs. (some in color). Mérida: Ediciones del Rectorado, Universidad de los Andes.

Illustrated field manual of the alpine flora, comprising a keyed, partly descriptive enumeration of vascular (pp. 89–408) and non-vascular plants, with brief notes on distribution, altitudinal zonation, habitat, frequency, and special features as well as vernacular names (in Spanish, German, and English); list of references (pp. 409–11) and indices to vernacular and to family and generic names at end. The introductory section deals with physical features, soils, aspect effects, geology, etc. of the region and its special character, factors influencing the flora, and life-forms (biotypes), followed by a primary key (pp. 45–55).

304

Highlands: The Guayana Highland

In addition to the comprehensive series of revisions of the flora of this remarkable floristic region by B. Maguire and collaborators, there have also appeared florulas for some individual tepuís (table mountains): the Tafelberg (Surinam), Mt Roraima (Guyana/Venezuela), Cerro Jáua (Venezuela), Auyan-tepuí (Venezuela), and Cerro Duida (Venezuela). Additional coverage is provided in STEYERMARK *et al.*, *Contributions to the flora of Venezuela* (see 315). The whole Guayana region, dominated by an ancient continental shield, contains perhaps 8000 species of higher plants, with 75 per cent endemism; of these, the Highland flora numbers some 2000, with 90–95 per cent endemism. For general reviews of the flora, vegetation, and endemicity of the Highland, both with extensive lists of references, see MAGUIRE, B., 1970. On the flora of the Guayana Highland. *Biotropica*, 2: 85–100; and STEYERMARK, J. A., 1979. Flora of the Guayana Highland: endemicity of the generic flora of the summits of the Venezuelan tepuís. *Taxon*, 28: 45–54.

MAGUIRE, B. *et al.*, 1953– . *The botany of the Guayana Highland.* Parts 1– . Illus., map (Mem. New York Bot. Gard., 8 etc.). New York.

Comprises a series of papers incorporating critical descriptive revisions of the more important families of vascular plants of this isolated, phytogeographically very significant region of sandstone mountains, with keys to species (in large genera), synonymy and references, citation of *exsiccatae*, notes on taxonomy, habitat, special features, etc., and figures of representative species. The general introductory section (part 1) gives a historical review of botanical exploration

in the Guayana Highland as well as overall remarks on the geography of the region and its flora. At this writing (late 1980) ten parts have appeared, the most recent in 1978; the back cover of part 10 gives a complete list of the series, with contents.

308

Wetlands

Only one general work on the aquatic plants of South America (with special reference to Brazil) has been published; this is described below although not strictly a flora in the formal sense.

HOEHNE, F. C., 1948. *Plantas aquaticas.* 168 pp., 81 pls. (some in color), frontispiece. São Paulo: Instituto de Botânica. (Publicação da Serie D.)

Discursive systematic account in the style of Kerner von Marilaun's *Pflanzenleben* or of ENGLER's *Die Pflanzenwelt Afrikas* (500), with notes on occurrence, special features, local references, etc.; no keys but excellent illustrations; index to scientific names at end. The introductory section covers ecology and subdivisions of aquatic plants, their special characteristics, geographical distribution, and uses. Covers non-vascular and vascular hydrophytes.

Region

31

Venezuela and the Guianas

This region consists of French Guiana, Surinam, Guyana (formerly British Guiana), and Venezuela. The islands to the north have all been treated as Regions 28 and 29.

All countries have had a long history of floristic exploration, beginning in the eighteenth century, but after two hundred years or more they remain imperfectly known due to the size of the flora, the diverse vegetation, and the relatively high percentage of endemism, a striking feature related to the presence of the Guayana Shield under much of the region. In Venezuela it is claimed that less than two per cent of the country is adequately known botanically.

In Venezuela, early contributors included Linnaeus' students Löfling, Humboldt and Bonpland, Linden, Funck and Schlim, Karsten, Fendler, Spruce, and Moritz. The modern era began with the many years of work of Pittier, who founded the first local herbarium around 1920, and since then collecting has increasingly been in the hands of resident botanists, although many outside workers were involved in the detailed exploration of the Guayana Highland area which got underway in the 1920s. Notable among recent contributors has been Steyermark, active on and off from 1943.

Apart from *Botany of the Guayana Highland* (304), key works relating to Venezuela include a national flora, *Flora de Venezuela*, begun in 1964 and now with several parts published, Steyermark's *Contributions to the flora of Venezuela* (1951–7), and two older checklists (both now very incomplete) by Knuth (1926–8) and Pittier (1945–7). Analytical keys and a treatment of the woody flora, the latter by Aristeguieta (1973), are available. Two partial florulas have also been published: the excellent *Flora del Ávila* (Steyermark and Huber, 1978), the first part of *Gehölzflora der Anden von Mérida* (Huber, 1977), and *Flora de los páramos de Venezuela* (Vareschi, 1970; see 303). However, the level of documentation, like the state of collecting, is still inadequate.

In the Guianas, each originally under a separate colonial power, botanical progress has been fragmented; few collectors have visited more than one country. This has in turn led to nomenclatural fragmentation. At the present time, French Guiana is the least well known, little work having been done in its interior; it was, however, the earliest area to be explored in some detail through the work of Aublet in the 1760s, published as *Histoire des Plantes de la Guiane Française* (1775) in which many large South American genera were described for the first time. However, after the mid-nineteenth century collecting in French Guiana tended to lapse, but continued, with emphasis on the woody plants, in the other two units, further stimulated by the work of the Guayana Highland programme directed by Maguire.

The major flora of the Guianas is unquestionably *Flora of Surinam*, begun by Pulle and Lanjouw in 1932 after a preliminary treatment of the pteridophytes by Posthumus (1928). After a lapse in the project following the death of Pulle, it was revived in the 1960s, with new contributions, some in supplementation of earlier treatments, extending to include all three of the Guianas, thus making it, with *Botany of the Guayana Highland*, the standard work on the subregion. Other key modern works are largely forestry-orientated: *Bomenboek voor Suriname* (Lindeman and Mennega, 1963), based on extensive forest-botanical work by the Forestry Service but covering only larger forest trees, *Essences forestières de Guyane* (Bena, 1960), of similar scope but less thoroughly based and effectively covering only the coastal fringe, and *Checklist of the indigenous woody plants of British Guiana* (Fanshawe, 1949).

Other works include *Flore de la Guyane française* (Lemée, 1952–6), largely compiled; *Enumeration of the vascular plants known from Surinam* (Pulle, 1906), now long out of date; and the antique but never-superseded *Versuch einer Fauna und Flora von Britisch-Guiana* (Schomburgk, 1848), to which may be added *Flora of the Kartabo Region* (Graham, 1934). Attention should also be drawn to *Contributions to the botany of Guiana* (Maguire *et al.*, 1966). Thus, while there is considerable documentation, it is rather uneven.

Progress

Venezuelan history is reviewed in the introduction to *Manual de las plantas usuales de Venezuela y su suplemento* by Pittier (under 315: Woody plants). A more recent review is STEYERMARK, J. A., 1974. Situation actual de las exploraciones botánicas en Venezuela. *Acta Bot. Venez.* 9: 241–3. No separate reports for the Guianas have been seen.

Bibliographies. General and divisional bibliographies as for Divison 3.

Indices. General and divisional indices as for Division 3.

311

The Guianas (in general)

See also 313 (more recent parts of *Flora of Surinam*; POSTHUMUS). No inclusive work is available for the Guianas, but in addition to the treatment of pteridophytes by Posthumus, recent parts of the *Flora of Surinam* include both Guyana and French Guiana as well as the former Dutch territory. However, the work described below provides additional information on the flora of the interior, to date only sketchily covered in most standard works.

Partial work

MAGUIRE, B. *et al.*, 1966. Contributions to the botany of Guiana, I–IV. *Mem. New York Bot. Gard.* 15: 50–128.

Comprises a series of revisionary treatments of families and genera, with descriptions of new taxa, synonymy, citation of *exsiccatae* with localities, critical notes, and figures of representative species. Also included are some records of plants from Amapá and other adjacent parts of Brazil.

312

French Guiana

Includes the districts of Cayenne and Inini. For additional coverage, see 311 (MAGUIRE); 313 (POST-HUMUS; PULLE, LANJOUW and STOFFERS). Lemée's general flora is to a large extent compiled.

LEMÉE, A., 1952–6. *Flore de la Guyane française*, 1–4. 3 frontispieces. Paris: Lechevalier.

Briefly descriptive flora of vascular plants, without keys; limited synonymy, with references; general indication of local range; miscellaneous notes; indices to botanical names in each volume; bibliography (vol. 2). The introductory section in vol. 1 includes a key to families. Volume 4 incorporates addenda to the main work as well as an annotated flora of useful plants (separately indexed). Only the coastal fringe of Guyana proper, on Cayenne (up to about 40 km inland), is effectively covered, although some species from adjacent Amapá and Surinam are also included in the work.

Special group – forest trees

BENA, P., 1960. *Essences forestières de Guyane.* vii, 488 pp., text-figs., 10 pls., 4 maps. Paris: Imprimerie Nationale.

Illustrated foresters' manual of the more important trees, without keys; includes synonymy, vernacular names, general indication of local range, notes on habitat, field characteristics, phenology, timber properties, uses, etc., and index to botanical and vernacular names. The introductory section includes a bibliography. As with Lemée's flora (see below), this work limits itself largely to the coastal fringe.

313

Surinam

See also 311 (MAGUIRE). More recent parts of *Flora of Surinam*, in process of completion together with revision of older contributions, cover adjacent French Guiana and Guyana as well as Surinam.

PULLE, A. A., 1906. *An enumeration of the vascular plants known from Surinam.* 8 + 555 pp., 17 pls., map. Leiden: Brill.

Systematic enumeration of known vascular plants, with descriptions of new taxa; limited synonymy, with references; vernacular names; citation of *exsiccatae*, with localities, and general indication of extralimital range; short critical notes; summary comments on the flora and its phytogeographical relationships; bibliography; indices to vernacular and to botanical names. The introductory section includes an account of botanical exploration.

PULLE, A. A., LANJOUW, J. and STOFFERS, A. L. (eds.), 1932– . *Flora of Suriname.* Vols. 1– . Illus., map. Amsterdam: Foundation 'van Eedenfonds'/Koloniaal-Institut (vols. 1–4 in part); Leiden: Brill, for Foundation 'van Eedenfonds' (continuation of vols. 1–4 as well as vols. 5– , entitled *Flora of Surinam*). Early parts of vols. 1–4 reprinted 1966–7 with additional indices, Leiden: Brill.

Detailed descriptive 'research' flora of phanerogams and non-vascular plants, with keys to genera and species, full synonymy, references and citations (including listing of standard monographs and revisions under family and generic headings), vernacular names, citation of *exsiccatae* with localities, general indication of extralimital range, critical commentary, and miscellaneous notes; index to botanical names at end of each part.

Of the original four volumes, each with two formal parts, only the first and a portion of the second part in each volume was published before termination of the work due to World War II (and subsequent administrative changes). For a time after resumption of the project in the 1960s these volumes were considered as 'terminated', but in 1976 vol. 2, part 2 was republished in a complete form, with new contributions together with additions and corrections to the older material in that volume. Present indications are that a similar procedure will be followed with vols. 1, 3, and

4; the *Flora* as a whole is far advanced but the pre-war treatments now are rather outdated and it is planned to incorporate revisions to these as well as treatments of the remaining families. The family treatments in the original volumes broadly followed Pulle's system of classification, but from vol. 5 onwards, as well as additions to vols. 1–4, no set order has been followed, and the newer contributions now include *all* the Guianas, thus increasing the value of the work. Progress with the continuation of the *Flora* has, however, been comparatively slow; as of 1979 only vols. 5, parts 1–2, vol. 6, part 1 and the remainder of vol. 2 have appeared since resumption of publication in the 1960s. For supplement, see LANJOUW, J., 1935. Additions to Pulle's *Flora of Surinam. Recueil Trav. Bot. Néerl.* **32**: 215–61.

Special group – pteridophytes

KRAMER, K. U., 1978. *The pteridophytes of Surinam.* 198 pp. (Uitgaven Natuurw. Studiekring Suriname Ned. Antillen, 93). Utrecht.

Keyed enumeration, with synonymy, references, selective citation of *exsiccatae* as well as number of available collections, general indication of local and extralimital distribution, critical commentary, and notes on habitat, frequency, etc.; special polyclave to genera; index at end. Recent monographs and revisions are also cited in the text. Designed in part for field use; partly supplants Posthumus's treatment of 1928 (see below).

POSTHUMUS, O., 1928. *The ferns of Surinam and of French and British Guiana.* [ii], 196 pp. Pasoeroean, Java: The author.

Briefly descriptive treatment of ferns and fern allies, with keys to all taxa, synonymy (with references), citation of *exsiccatae* with localities, general indication of extralimital range, notes on features of the fern flora, phytogeographic relationships, and index to botanical names. The work was intended as a companion to Pulle and Lanjouw's *Flora of Suriname*, the first part of which appeared a few years afterwards.

Special group – forest trees

LINDEMAN, J. C. and MENNEGA, A. M. W., 1963. *Bomenboek voor Suriname.* 312 pp., illus. (1 col.) (Meded. Bot. Mus. Herb. Rijks Univ. Utrecht, 200). Paramaribo: s'Lands Bosbeheer Suriname.

Copiously illustrated foresters' manual of the more important trees of Surinam, with two general keys to species based respectively on foliar, twig, and bark attributes and on wood structure; essential synonymy; vernacular names; local range; notes on special features, timber properties, and uses; bibliography; indices to vernacular and to botanical names.

The introductory section includes an illustrated glossary and explanation of botanical and wood-anatomical terms. General summaries of the work in Spanish and English are also provided.

314

Guyana

For alternative coverage of what was formerly British Guiana, see 301 (all works); 311 (MAGUIRE); 313 (PULLE, LANJOUW and STOFFERS). The checklist by Fanshawe was given limited circulation, but does not appear to have been formally published.

SCHOMBURGK, M. R., 1848. Versuch einer Fauna und Flora von Britisch Guiana. In his *Reisen in Britisch-Guiana in den Jahren 1840–44*, vol. 3, pp. 563–1260. Leipzig: Weber.

Includes (pp. 787–1212) systematic lists of cellular and vascular plants, based mainly upon the author's collections, with limited synonymy, local range, and notes on phenology and habitat; index to all taxa (pp. 1226–60). There are four main lists, based on habitat (Coastal region; Primary forest; Sandstone region; Savanna). The introductory section includes general remarks on physical features, vegetation, and phytogeographical relationships.

Partial work

Of the few local florulas in Guyana, the most significant is Graham's flora of the Kartabo region in the north-central part, west of the Essequibo and southwest of Georgetown. Its keys represent a particularly useful feature.

GRAHAM, E. H., 1934. Flora of the Kartabo region, British Guiana. *Ann. Carnegie Mus.* **22**: 17–292, 2 text-figs. (maps), pl. 3–18.

Descriptive flora of vascular plants, with complete keys, synonymy, citations of *exsiccatae* with localities, extralimital range, miscellaneous notes, and index; list of references included. An introductory section deals with general natural history of the area and with botanical exploration.

Special group – woody plants

FANSHAWE, D. B., 1949. *Check-list of the indigenous woody plants of British Guiana.* iv, 244 pp. (For. Bull., 3). Georgetown: Forest Department, British Guiana. (Typescript in printed hard covers; copy in the Library, Royal Botanic Gardens, Kew.)

Comprises a list of families and genera (alphabetically

arranged), with notes on habitat, frequency, distinguishing features, etc.; Arawak, English, and other vernacular names; lexicon of vernacular names, with botanical equivalents. An introductory section includes technical notes and a list of references.

315

Venezuela (in general)

See also **304** (all works); **303** (VARESCHI). In recent years some useful local and state manuals have begun to appear; these are listed under **316–19**. Only general Venezuelan works appear here. Knuth's *Initia florae venezuelensis* is somewhat uncritical and moreover is now rather outdated.

– Vascular flora: 20 000–35 000 species (Steyermark, 1977).[4]

Keys to families and genera

BADILLO, V. M. and SCHNEE, L., 1965. *Clave de las familias de plantas superiores de Venezuela*. 4th edn. 255 pp. (Revista Fac. Agron. (Maracay), 9). Maracay, Venezuela. (First edn., 1951.) [Analytical keys, based on Thonner's *Anleitung zum Bestimmen der Familien der Blütenpflanzen* (1917), with descriptions of families; glossary and index.]

PITTIER, H., 1939. *Genera plantarum venezuelensium. Clave analítica de los géneros de plantas hasta hoy conocidos en Venezuela*. 354 pp. Caracas: Tipografía Americana. [Analytical keys to all Venezuelan genera of vascular plants, with index.]

KNUTH, R., 1926–8. *Initia florae venezuelensis*. 768 pp., map (Repert. Spec. Nov. Regni Veg., Beih., 43). Berlin: Fedde.

Systematic enumeration of known vascular plants, with synonymy, references, vernacular names, and geographically arranged citations of *exsiccatae* with localities; cyclopedia of collectors (in German); index to genera. For additions, see SUESSENGUTH, K. and BEYERLE, R., 1939. Ergänzungen zu den 'Initia florae venezuelensis' von R. Knuth. *Bot. Arch.* **39**: 373–81.

LASSER, T. (ed.), 1964– . *Flora de Venezuela*. Vols. 1– . Illus. Caracas: Instituto Botanico, Ministerio de Agricultura y Cria (MAC), Venezuela.

Large-scale critical illustrated 'research' flora of vascular plants, with extensive descriptions, keys to genera, species, etc., synonymy, references and citations, vernacular names, geographically arranged citations of *exsiccatae* with localities, general indication

of extralimital range, critical commentary, and notes on habitat, properties, uses, etc.; indices to botanical and vernacular names in each volume. At this writing (late 1981), vol. 1, parts 1–2; vol. 3, part 1; vol. 8, parts 1–2; vol. 9, parts 1–2; vol. 10, parts 1–2; vol. 12, part 1; and vol. 15, parts 1–6 have appeared, including among other families the pteridophytes, Rubiaceae, Compositae, Bromeliaceae, and Orchidaceae. A total of 15 volumes, each with one or more parts, is planned, with a rough systematic order being followed. In the same general style, although not in this series, is LOPEZ-PALACIOS, S., 1977. *Flora de Venezuela: Verbenaceae*. 654 pp., 146 text-figs. Mérida: Consejo de Publicaciones, Universidad de los Andes.

PITTIER, H. *et al.*, 1945–7. *Catalogo de la flora venezolana*. 2 vols. 423 and 577 pp. (Third Inter-American Conference on Agriculture, Caracas. Cuadernos verdes: serie nacional, nos. 20, 52). Caracas: Vargas.

Systematic enumeration of vascular plants, with keys to genera (and some species), vernacular names, citation of *exsiccatae* with localities, taxonomic commentary, notes on habitat, etc., and index to genera in each volume; general index to vernacular names in vol. 2.

Partial works

Apart from contributions on the Guayana Highland (**304**), the most important supplementary work on Venezuela of recent decades is that by Steyermark and others described below. It usefully bridges the general enumerations of Knuth and Pittier and the modern *Flora de Venezuela*.

STEYERMARK, J. A. *et al.*, 1951–7. *Contributions to the flora of Venezuela*. 4 parts, vii, 1225 pp., 151 text-figs. (Fieldiana, Bot., 28). Chicago.

Systematically arranged critical notes and descriptions of new or little-known taxa of non-vascular and vascular plants from various parts of Venezuela, with a few keys to species; includes synonymy, references, citations of *exsiccatae* with localities, and miscellaneous notes; general index at end. An introductory section provides an account of botanical exploration, with special reference to the author's itineraries of the 1940s and 1950s.

Special group – woody plants

ARISTEGUIETA, L., 1973. *Familias y géneros de los árboles de Venezuela*. 845 pp. (Edición especial, Instituto Botánico.) Caracas: Instituto Botánico, MAC.

A 'generic flora' of woody plants, with artificial keys to families and genera, descriptions, vernacular names, and notes on number of species, representatives in Venezuela, uses, etc. Key cross-references are also given. Designed in part as a students' manual, this work represents a considerably enlarged version of the author's *Clave y descripción de las familias de los arboles de Venezuela* (Caracas, 1954). In the present work, family accounts are alphabetically arranged.

PITTIER, H., 1970. *Manual de las plantas usuales de Venezuela y su suplemento.* 2nd edn., rev. E. Mendoza. xxii, 620 pp., illus. Caracas: Fundación 'Eugenio Mendoza'. (Reprinted 1978; 1st edn., 1926; *Suplemento*, 1939.)

Included here for its list of known woody plants, with vernacular names, incorporated in the *Suplemento*.

316

Venezuelan Guayana

See **304**. Comprises Amazonas and Bolívar states.

317

Orinoco Basin and Llanos

Comprises the plain between the Andes and coastal range and the Orinoco, incorporating the states of Apure, Barinas, Portuguesa, Cojedes, Guárico, Anzoátegui, Monangas, and Delta Amacuro. No separate works have been published.

318

The Andes and Coastal Range

Comprises the mountain ranges from the Colombian border to the Paria Peninsula, and includes the states of Tachira, Mérida, Trujillo, Lara, Yaracuy, Carabobo, Aragua, Miranda, and Sucre together with the Federal District. Florulas are available for the Mérida Andes and for the Ávila range north of Caracas; the latter is especially recommended.

HUBER, H., 1977. Gehölzflora der Anden von Mérida, 1. *Mitt. Bot. Staatssamml. München*, **13**: 1–127.

Keyed enumeration of woody plants of Mérida and parts of adjacent states, with species entries including synonymy, references, citations of *exsiccatae* with localities, critical notes, and remarks on habitat, occurrence, etc.; typification; cross-references to monographs and revisions. A short introductory section is provided. Covers families alphabetically from

Acanthaceae through Asteraceae. [This work should be used with caution; errors have been noted in the treatment of Araliaceae for which a good recent revisionary study is not available.]

STEYERMARK, J. A. and HUBER, O., 1978. *Flora del Ávila*. 971 pp., 308 pls., photographs (some coloured), folding map. Caracas: Sociedad Venezolana de Ciencias Naturales/Ministerio del Ambiente y de los Recursos Naturales Renovables, Venezuela.

Copiously illustrated enumeration of vascular plants, with descriptive keys, synonymy, local range, notes on habitat, frequency, altitude, special features, properties, etc., and vernacular names; guides to determination by special features (appendices); illustrated glossary, bibliography (pp. 923–33), and complete index. An introductory section gives details of physical features (the area reaches over 2500 m), climate, vegetation and phytogeography, and introduced species, with a floristic summary; this is followed by an illustrated key to families. Families are alphabetically arranged in this work, which covers 1892 species.

319

Lake Maracaibo region

This area comprises the states of Zulia and Falcón. A flora of the latter is at this writing in preparation.

Region

32

Colombia and Ecuador

The region comprises mainland Colombia and Ecuador, together with their immediately adjacent islands. Until the middle of the nineteenth century they constituted one country, New Granada, which comprised most of the historical viceroyalty. The Pacific islands of Malpelo (Colombian) and the Galápagos (Ecuadorean) are respectively designated as **016** and **017**, while the Caribbean islands of San Andrés and Providencia (Colombian) constitute **238**.

The rich botanical history of Colombia dates back to the latter part of the eighteenth century with the outstanding explorations of Caldas and especially Mutis (1760–1808), the latter working with many assistants under sponsorship from Spain as a 'Real Expedición Botánica', and von Jacquin (1755–9). The next notable expedition was that of Humboldt and Bonpland in the 1800s. Many other well-known collectors visited Colombia throughout most of the nineteenth century, some for extended periods, and sent large collections back to European herbaria; only those of Triana remained to any extent in Colombia. In the present century, a number of collectors came from North America, with from 1932 the addition of Cuatrecasas who collected and taught on and off for 15 years. Another important contributor was Killip, beginning in 1922. A local botanical community began to develop in the inter-war period, and substantial contributions were made by Dugand, Pérez-Arbeláez, García-Barriga, Uribe-Uribe, and others. The first significant local herbarium, that of the National University, was founded in 1931.

No complete account of the Colombian flora has been written. The major accounts are the incomplete *Prodromus florae novo-granatensis* (Triana and Planchon, 1862–73) and the long-delayed *Flora de la Real Expedición Botánica* (1954–) with handsome color plates. A widely used introductory work specifically relating to Colombia is *Plantas utiles de Colombia* (Pérez-Arbeláez, 1936; 3rd edn., 1956). Recently a generic flora project has been begun by E. Forero.

The contemporary growth of the botanical profession has been accompanied by a trend towards decentralization and the development of 'regional' herbaria, of which there are now sixteen. Associated with this has been the production of a number of partial floras based on departments. The most important of these is the serial *Catalogo ilustrado de las plantas de Cundinamarca* (1966–), a project of the National University Herbarium. Other areas with such coverage, although incomplete, include Antioquia, Bolívar, and Magdalena.

While much collecting is currently in progress, the size and diversity of the flora are such that many more man-years will be required before knowledge is even sketchily adequate for more than limited areas. The Amazonian zone is rather poorly known, while in other zones are many neglected areas. Collections have tended to be concentrated around more accessible centers.

Ecuador is also still relatively little known and unevenly collected, with until recent years a very small resident botanical community. Like Colombia, it has a long history of botanical exploration, beginning with La Condamine and J. de Jussieu in the eighteenth century and followed by Humboldt and Bonpland in the 1800s. Other nineteenth-century visitors included Seemann, Spruce, and André. Resident botanists have included Jameson and Sodiro and (in the present century) Acosta-Solís. It was Jameson who contributed the first local flora, *Synopsis plantarum aequatoriensium* (1865) which deals mainly with the *altiplano*, especially in the north around Quito.

Swedish contributions began in the late-nineteenth century, but became more significant from the end of World War I with the mounting of several 'Regnellian Expeditions'. Their collections along with others have formed the basis for the modern *Flora of Ecuador* (1973–) produced under Swedish sponsorship. However, as with Panama, even the present greatly increased stock of collections is not considered at all adequate in relation to a very rich flora, with the result that early parts of the *Flora* may soon become significantly incomplete.

One important new local florula has recently been published for a site of special interest: *Flora of the Rio Palenque Science Center* (Dodson and Gentry, 1978) in western lowland Ecuador.

Progress

Colombia: BARKLEY, F. A. and GUTIERREZ, V. G., 1948. Colectores de plantas en Colombia: tal como estan representades en los herbarios del pais. *Revista Fac. Nac. Agron., Medellín*, 8: 85–107; and SCHULTES, R. E., 1951. La riqueza de la flora Colombiana. *Revista Acad. Col.* 8: 230–42. The history of the Mutis expedition is reviewed in the first volume of *Flora de la Real Expedición Botánica* (1954; see below).

Ecuador: ACOSTA-SOLÍS, M., 1968. Naturalistas y viajeros cientificos que han contribuido al conocimiento floristico fitogeográfico del Ecuador. *Contr. Inst. Ecuat. Cienc. Nat.* 65: 1–138. [Very full survey.]

Bibliographies. General and divisional bibliographies as for Division 3.

Indices. General and divisional indices as for Division 3.

321

Colombia (in general)

No complete modern floras or enumerations of the vascular plants are available for what is, for its size, perhaps botanically the richest country in the world. Some coverage, however, has been, and is continuing to be, provided by the gradual production of family treatments in *Flora Neotropica* (**301**) and in a number of miscellaneous country-wide revisions (termed by one author an unofficial 'Flora of Colombia') in *Caldasia*, *Contributions from the United States National Herbarium*, *Webbia*, and other serials (for list of these revisions, see *Guide to standard floras of the world*, 1964, p. 15). A generic flora project recently initiated at Bogotá will, hopefully, provide a further means of linking these and a vast mass of other scattered literature, old and recent.

The only nominally general works at present are the incomplete and very imperfect *Prodromus florae novo-granatensis* of Triana and Planchon (1862–73) and the family treatments in the historically important *Flora de la Real Expedición Botánica del Nuevo Reina de Granada* (1954–), the very belated realization of the results of the explorations by the Mutis group in the eighteenth century. A still-limited number of departmental floras has also been published or is under way, and, with the existence and development of a number of 'regional' herbaria, this trend may continue. As an introduction to the flora, the following has long been widely used: PEREZ-ARBELAEZ, E. 1956. *Plantas útiles de Colombia*. 3rd edn. 832 pp., 752 text-figs., 47 pls. (2 col.). Bogotá: 'Colombiana'. (First edn., 1936.)

– Vascular flora: 45000 species (Prance, 1978); an estimate of 50000 by Schultes in 1951 is considered excessive.

Until 1861 Colombia was known as the United States of New Granada, taking its name from the former viceroyalty which until 1811 included Venezuela and from 1819–30, as 'Gran Colombia', still included Ecuador (**329**).

[MUTIS, J. C.], 1954– . *Flora de la Real Expedición Botánica del Nuevo Reina de Granada*. Vols. 1– . Illus. (some in color). Madrid: 'Cultura Hispánica' (for Instituto de Cultura Hispánica, Madrid, and Instituto Colombiano de Cultura Hispánica, Bogotá).

A sumptuous folio 'Great Flower Book', this floral emerald represents the realization of the long-unpublished results of the extensive botanical surveys conducted by J. C. Mutis and his pupils, associates, and assistants in the late-eighteenth century (termed a 'tribe of botanical adventurers' by their English contemporary, J. E. Smith). Family treatments, all revised by specialists, feature extensive descriptions, keys, synonymy with references, citations, geographical range, taxonomic commentary, miscellaneous notes, and indices. Of special interest are the reproductions in color of the thousands of beautiful paintings by Mutis's artists (said to have numbered at one stage as many as 30). The work, appearing slowly and at present under the editorship of S. Rivas-Goday and E. Pérez-Arbeláez, is projected to encompass 51 volumes; at this writing (late 1979), vols. 1, 7, 8, 27, 30, and 44 have appeared. Volume 1 includes a history of the 'Real Expedición Botánica' as well as other introductory matter.

TRIANA, J. and PLANCHON, J.-E. 1862–7. *Prodromus florae novo-granatensis*. Vols. 1–2. 4 pls. Paris: Masson. (Reprinted with changed pagination from *Ann. Sci. Nat. Bot.* sér. IV, **17–20** (1862); sér. V, 1–5, 7 (1863–67).) Continued as: *idem*, 1872–3. Prodromus florae novo-granatensis. *Ann. Sci. Nat. Bot.*, sér. V, **14**(1872): 286–325; **15**(1872): 352–82; **16**(1872): 361–82; **17**(1873): 111–94.

Systematic enumeration of cellular (except algae) and vascular plants, with descriptions of new taxa, synonymy and references, vernacular names, citations of *exsiccatae* with localities, taxonomic commentary, and notes on habitat, etc.; no indices. The treatment of vascular plants generally follows the Candollean system as far as the Papayaceae (= Caricaceae), although some large families are omitted. Volume 2 is entirely devoted to cryptogams (lower vascular plants, pp. 275–396).

Special group – forest trees

DEL VALLE A., J. I., 1972. Introducción a la dendrología de Colombia. 351 pp., illus. Medellín: Centro de Publicaciones, Universidad Nacional. (Mimeographed.)

An introductory student handbook treating families, genera, and key species of trees, with keys and many figures, notes on field and wood characters, and internal and extra-limital distribution; references (pp. 300–7); illustrated lexicon to botanical features, an illustrated key to fruits, and index to vernacular and botanical names followed by figures of buttress types. [The work has a sentience not unlike *Common forest trees of Papua New Guinea* by JOHNS, R. J. (**930**).]

322

Colombia: northern departments

Included here are Guajira, Magdalena, Bolívar, Atlántico, and Córdoba. The large isolated massif of Sierra Nevada da Santa Marta, in Magdalena, is botanically still poorly known.

ROMERO CASTAÑEDA, R., 1965– . *Flora del Centro de Bolívar*. Vols. 1– . Illus. Bogota: Instituto de Ciencias Naturales, Universidad Nacional de Colombia.

Illustrated descriptive treatment of relatively common species of vascular plants, with synonymy, vernacular names, local range, and notes on uses; glossary, key to families, and general index. [As of 1979 only one volume had evidently been published.]

ROMERO CASTAÑEDA, R., 1966– . *Plantas del Magdalena*. Fasc. 1– . Illus. Bogota: Instituto de Ciencias Naturales, Universidad Nacional de Colombia.

Illustrated descriptive treatment, with synonymy, vernacular names, local range, and notes on wood and other botanical features; glossary. [As of 1979 only one family, Zygophyllaceae, had been published.]

323

Colombia: Antioquia and Caldas

Included here are Antioquia and Caldas.

URIBE, J. A. and URIBE-URIBE, L., 1940. *Flora de Antioquia*. 383 pp., 25 pls. Medellín: Imprensa Departemental.

Systematic enumeration of the more commonly encountered non-vascular and vascular plants, with some keys; includes limited synonymy, vernacular names, and notes on special features, properties, uses, and cultivation; list of references and indices to generic, family, and vernacular names.

324

Colombia: Chocó

Comprises the department of Chocó. The area, one of the wettest and apparently botanically richest in the neotropics, has until recently been very poorly known; for extensive discussion see Gentry (1978).[5]

325

Colombia: northeastern highlands

Included here are the departments of Norte de Santander, Santander, and Boyacá (in part). No separate floras are available, but works on nearby Mérida in Venezuela (318) may be of some usefulness.

326

Colombia: Central Valley

Included here are the departments of Cundinamarca, Tolima, and Huila. Within the former is the capital, Bogotá. The *Catálogo ilustrado* of Pinto-Escobar *et al.*, currently in process of publication, is the most detailed and comprehensive of present departmental floras.

PINTO-ESCOBAR, P. *et al.* (eds.), 1966– . *Catalogo ilustrado de las plantas de Cundinamarca*. Parts 1– . Illus. Bogota: Imprensa Nacional.

Illustrated enumeration of vascular plants, with descriptive keys to genera, extensive synonymy, vernacular names, local range (with maps), and notes on habitat, etc.; each fascicle separately indexed. At this writing (late 1980), seven parts have appeared, the latest in 1979; no preset family sequence has been followed, although some fascicles contain treatments of groups of related families.

327

Colombia: southwestern departments

Included here are the departments of Valle del Cauca, Cauca, and Nariño. A catalogue for El Valle is at present in preparation by J. Cuatrecasas (Gentry, 1978).[6]

328

Colombia: Oriente

This comprises all Colombia east of the Andes – more than one-half the country. Included here are the departments of Arauca, Boyacá (in part), Meta, Vichada, Vaupés, Caquetá, Putumayo, and Amazonas. A considerable part of the area lies on the Guayana crystalline shield, which terminates to the west in the Sierra de la Macarena (in Meta), botanically still poorly known although visited by one expedition.

329

Ecuador

Only the mainland is considered here; for the Galápagos, see **017**. The *Flora of Ecuador*, initiated in 1973 as a special series of the Swedish serial *Opera Botanica*, represents the first attempt at a national flora of a small but botanically very rich country. Of partial works, two are significant: the very old, never-completed *Synopsis* by Jameson (1865) for the *altiplano* and the atlas-florula of the Rio Palenque Science Center by Dodson and Gentry (1978); the latter is the only one of its kind in Pacific Northwestern South America.

For a useful introduction to the flora and vegetation (with a list of references), see DIELS, L., 1937. *Beiträge zur Kenntnis der Vegetation und Flora von Ecuador.* 190 pp., 2 text-figs., map (Biblioth. Bot., 116). Stuttgart. (Spanish edn.: *idem*, 1938. *Contribuciones al conocimiento de la vegetación....* Translator R. Espinosa. 364 pp., map. Quito.)

– Vascular flora: 10000 species (Gentry, 1978).[7]

HARLING, G. and SPARRE, B. (eds.), 1973– . *Flora of Ecuador*. In fascicles (Opera Bot., B, part 1–). Lund, Sweden.

This work is planned as a series of revisions of vascular plant families in mainland Ecuador; each treatment includes detailed descriptions, keys, synonymy, references and citations, general indication of local and extralimital range with citation of *exsiccatae*, critical commentary, notes on habitat, special features, etc.; and index. Parts are appearing as treatments are ready, although family numbers have been assigned according to the modified Englerian sequence of the 12th edition of the *Syllabus*. At this writing (late 1981) 13 parts have been published.

Partial works

DODSON, C. H. and GENTRY, A. H., 1978. *Flora of the Rio Palenque Science Center, Los Rios Province, Ecuador.* xxx, 628 pp., 22 halftones (incl. maps), 278 pls., col. frontispiece, maps (end papers) (Selbyana, 4(1–6)). Sarasota, Florida: Marie Selby Botanical Gardens.

Atlas-flora (in octavo format) of vascular plants, with keys to genera and species and including short species descriptions with notes on ecology, biology, phenology, and distribution; index to all taxa. Appendices include a list of specimens cited, acknowledgments and a list of rare species. An introductory section gives a general discription of the area, plant-geographic patterns, major features of the flora, statistics, and vegetation formations with characteristic species. Covers 1112 species in an area of some 167 ha, north of Quevedo in the central western lowlands (0 ° 35 ′S, 79 ° 25 ′W), within the 'Wet Forest' type (according to the Holdridge life-zone system).

JAMESON, W. 1865. *Synopsis plantarum aequatoriensium...viribus medicatis et usibus oeconomicis plurimarum adjectis.* Vols. 1–3. Quito: 'del Pueblo' (Joannis Paulus Sanz).

Descriptive treatment of vascular plants, without keys; includes localities (with some citations) and (at end of each family) extensive notes on properties, uses, etc. (including medicinal values) of the species concerned; indices to family and generic names in vols. 1–2. Purposely limited to the plants of the *altiplano*; botanical text in Latin, commentary in Spanish. Not completed; vols. 1–2 cover Ranunculaceae through Labiatae and vol. 3 (unfinished) Verbenaceae through *Plantago* (Candollean system).

Special group – ferns

LATORRE A., F. and PADILLA C., I., 1974. Lista preliminar de helechos del Ecuador. *Ciencia y Naturaleza* (Quito), **14**: 21–57.

Systematic enumeration of ferns (Ophioglossaceae through *Polypodium*), with synonymy, references and citations, and generation indication of local distribution (by provinces), with altitudes; preceded by a conspectus of genera. Represents the first of two parts planned (to cover a total of some 600 species).

Region

33

Peru

The region corresponds to the political limits of Peru. The country has had a long history of botanical work, but remains comparatively poorly known due to its geography, floristic diversity, and limited resident botanical community.

Early knowledge of the Peruvian flora is based largely upon the collections and observations of the illustrious and dramatic royal expedition of Ruíz and Pavón from 1778 to 1788, in which Dombey also participated; the results, never fully published, appeared as *Flora peruviana et chilensis* (1794, 1798–1802). Following the political and other changes of the early-nineteenth century, significant but sometimes not large contributions were made by Mathews, Ball, Poeppig, Weddell, Cuming, Spruce and Raimondi, with Kuntze following towards the end of the century. From 1901 until 1940 A. Weberbauer was active in Peru, finally settling in Lima. His vegetatiological monograph on the Andean zone appeared in 1911, with a revision in Spanish in 1945. Other key collectors included Macbride, Klug, Asplund, Schunke, Killip and Smith, Williams, Goodspeed, and Herrera, along with their successors after World War II.

After a period of preparation beginning in the 1920s, publication of the *Flora of Peru* by Macbride began in 1936; by 1971 it was over three-quarters completed. Dormancy befell the project for a decade from the late 1960s but it has now been revived but with emphasis on Amazonian Peru, botanically a very poorly known area where many range extensions of Brazilian plants are expected to be found. A number of other poorly known areas also exist, particularly on the eastern fall of the Andes. It is very likely that the earlier parts of the *Flora of Peru*, which were to a large extent compiled, will require extensive revision as more collections from these areas become available.

Local coverage is very limited and scattered, with a number of collection reports and minor florulas, some centered on Lima and Cuzco. The most substantial seen, which deals with trees, is *Woods of Northeastern Peru* (Williams, 1936), centered on Iquitos.

A treatment of pteridophytes (Tryon, 1964) has also been published, complementary to the *Flora of Peru*.

Progress

HERRERA, F. L., 1937. Exploraciones botánicas en el Perú. *Revista Mus. Nac. Lima*, **6**: 296–358.
Bibliographies. General and divisional bibliographies as for Division 3.
Indices. General and divisional indices as for Division 3.

330

Region in general

In addition to the comprehensive floristic coverage provided by the still-incomplete *Flora of Peru*, a good introduction to Peruvian plant life is found in WEBERBAUER, A., 1911. *Die Pflanzenwelt der peruanischen Anden in ihren Grundzügen dargestellt*. xxi, 355 pp., 63 text-figs., 40 pls., 2 maps (Die Vegetation der Erde, 12). Leipzig: Engelmann. (For a revised edition in Spanish, see *idem*, 1945. *El mundo vegetal de los Andes peruanos*. Illus. Lima.)

– Vascular flora: perhaps over 20000 species (Gentry, 1980, in Macbride, *Flora of Peru*, [N.S., fasc. 1]; when 'completed' the *Flora* will cover 13937 species in 1952 seed plant genera.)

MACBRIDE, J. F. *et al.*, 1936–71. *Flora of Peru*. 10 parts in 24 nos. Illus. (in part), map (Publ. Field Mus. Nat. Hist., Bot. Ser., vol. 13, parts 1(1–3), 2(1–3), 3(1–3), 3A(1–2), 4(1–2), 5(1–2), 5A(1–3), 5B(1–3), 5C(1), 6(1–2)). Chicago.

MACBRIDE, J. F., and collaborators. 1980– . *Flora of Peru* [N.S.]. Pub. in fascicles. Illus. (Fieldiana, Bot., N.S., vol. 5, *passim*). Chicago.

SCHWEINFURTH, C., 1958–61. *Orchids of Peru*. viii, 1026 pp., 194 text-figs., portrait (Fieldiana, Bot., vol. 30, parts. 1–4, index). Chicago; and *idem*. 1970. *First supplement to the* Orchids of Peru. 80 pp. (*Ibid.*, vol. 33).

Comprehensive, but in large part compiled, descriptive flora of seed plants (Orchidaceae separately published), with keys to genera and species, full synonymy, references and citations (including standard monographs and revisions under family and generic headings), vernacular names, citations of *exsiccatae* with localities, general indication of extralimital range, extensive and pointed taxonomic commentary (in the tradition of M. L. Fernald; cf. under **140–50**), and notes on habitat, uses, etc.; index to all botanical names in each part. The introductory section (part 1) includes a concise account of vegetation and phytogeography by A. Weberbauer. Most treatments after 1960 are by specialists. The revision of the Orchidaceae by G. Schweinfurth at the Botanical Museum of Harvard University is in most respects similar in style to the main work. Upon publication of the last part in the original series, along with completion of the Orchidaceae, the *Flora* was about 80 per cent completed, covering 11 789 of the expected 'final' coverage of 13 937 species in all save 15 families; for indices to family revisions, see the introduction by A. H. Gentry to the new series (The *Flora of Peru*: a conspectus) on pp. 1–11 in the first fascicle as well as DALY, D. C., 1980 (1981). Families of spermatophytes included in the *Flora of Peru*. *Brittonia*, **32**: 548–50. The family sequence in the main work generally followed the classical Englerian system.

Though truly a monumental effort – among the standard floras of South America only *Flora brasiliensis* is larger – the *Flora of Peru* is inevitably deficient in its coverage, particularly of the large Amazonian region (mostly in the departments of Loreto and Madre de Dios), and moreover much of the work generally is now regarded as less than critical as well as obsolescent. The project was initiated by the Field Museum in the early 1920s; Macbride himself undertook two field trips to Peru, the first in 1922, and as early as 1923 orchid specimens were loaned to Harvard for study by Schweinfurth. Until Macbride's departure from the Museum in the 1950s the *Flora* was very largely run as an 'in-house' effort; later contributions had a more diversified authorship, although Macbride himself (by then resident in California) continued to study lots of specimens (sent to him on loan) into the early 1960s and to produce contributions (up to 1962). In the later 1960s the project became dormant; however, after nearly a decade it was 'reactivated' as a cooperative venture of the Field Museum, the Missouri Botanical Garden, and two Peruvian institutions including the Universidad Nacional Mayor de San Marcos. The revived project, planned to take six years or perhaps

longer for completion, is aimed at covering the remaining seed plant families (including such important elements as the Guttiferae, Cactaceae and Compositae) as well as the remaining lower vascular plants not treated by Tryon (see next entry). No specific order is planned for the new treatments, which will be by one or more specialists; those published to date (1981) comprise two fascicles (here designated as a 'new series', and are in a style similar to that of the original series but have descriptive details in smaller type and more 'neutral' commentary. [Historical commentary based upon personal knowledge as well as upon the above-mentioned review by Gentry which introduces the 'new series'.]

Special group – pteridophytes

TRYON, R. M., 1964. *The ferns of Peru*. 253 pp., 196 text-figs., 46 maps (Contr. Gray Herb., N.S. 194). Cambridge, Mass.

Briefly descriptive flora of lower vascular plants, with keys to genera and species; includes synonymy (with references), citations of *exsiccatae* with localities, and generalized indication of extralimital range, taxonomic commentary, and notes on habitat, special features, etc.; index to all botanical names. The introductory section includes a synopsis as well as a phytogeographical analysis of the fern flora. Covers Polypodiaceae (in the wide sense) from Dennstaedtieae through Oleandreae, amounting to about one-quarter (187 species) of the total number of known Peruvian ferns. [Continuation of this revision is planned as part of the 'reactivated' *Flora of Peru* project, as noted in the preceding entry.]

Special group – forest trees

WILLIAMS, LL., 1936. *Woods of northeastern Peru*. 587 pp., 18 text-figs., 2 maps (Publ. Field Mus. Nat. Hist., Bot. Ser., 15). Chicago.

Systematic dendrological account of forest tree species and their woods, with vernacular names, citations of author's *exsiccatae* and general indication of range, and notes on botanical and wood-anatomical features, properties, uses, etc.; tables of anatomical characters, lexicon of vernacular names with botanical equivalents, list of references, and index at end. An introductory section accounts for the explorations of the author, vegetation formations, climate, physical features, characteristic species, forest products, etc. [Encompasses parts of Loreto, San Martin, and Amazonas departments, treating an area stretching from Moyobamba into the Amazon Basin towards Iquitos. No other dendrological work is available for this and adjacent parts of 'Hispanic' Amazonia, an area in need of much more botanical exploration.]

Region

34

Bolivia

The region corresponds to the political limits of Bolivia. Floristic botany is very poorly developed in the country and at the present time its collecting index is one of the lowest in the world.

The major period of outside botanical activity extended from the late-nineteenth century to the 1940s, when several significant collections were made which constitute the main basis of present knowledge of the flora. Contributors ranged from Rusby in 1885–6 through Bang, Kuntze, Williams, Buchtien, Steinbach, Herzog, and (after World War I) the Mulford Expedition (Rusby, Cardenas, and White) and C. Troll to Krukoff. Those collections upon which reports were based were all treated in isolation, a situation with parallels in other 'open-field' parts of the tropics, such as Borneo and New Guinea. The only consolidated treatments of the flora are the useful vegetatiological account of Herzog (1923) and the uncritical checklist by Foster (1958).

A few collections were made prior to the 1880s, notably by Haenke, Cuming, Mandon, and Weddell; and before and after World War II a major contributor was the late M. Cárdenas who founded at Cochabamba one of the two (from 1977, three) herbaria in the country. No sufficient basis exists for a proper national flora and the only current general coverage is thus through *Flora Neotropica*.

Progress

The period to 1923 is covered in HERZOG, T., 1923. *Die Pflanzenwelt der bolivischen Anden und ihres östlichen Vorlandes*, pp. 1–4. Leipzig. No other separately published reviews have been seen.
Bibliographies. General and divisional bibliographies as for Division 3.
Indices. General and divisional indices as for Division 3.

340

Region in general

Apart from Foster's uncritical compiled checklist, no general flora or enumeration of Bolivian vascular plants is available. The best introduction to the flora, by far, is the following: HERZOG, T., 1923. *Die Pflanzenwelt der bolivischen Anden und ihres östlichen Vorlandes*. viii, 258 pp., 25 text-figs., 3 maps (Die Vegetation der Erde, 15). Leipzig: Engelmann. For a short list of major series of 'contributions' upon which much of the present limited knowledge is based, see below under 'Partial works'.

FOSTER, R. C. (comp.), 1958. *Catalogue of the ferns and flowering plants of Bolivia*. 223 pp. (Contr. Gray Herb., N.S., 184). Cambridge, Mass.

Unannotated, uncritical systematic list of vascular plants, with references to place of publication and limited synonymy; no index.

Partial works

The following works are all lists of collections, usually systematically arranged, with descriptions of novelties and other annotations. In some cases, family treatments are by specialists, notably in Herzog's report. For further bibliographic and other details, see Blake and Atwood (1942) pp. 237–238.

BUCHTIEN, O., 1910. *Contribuciones á la flora del Bolivia*, I. 197 pp. La Paz. [Localities, without *exsiccatae*.]

HERZOG, T., 1913–22. *Die von Dr. Herzog auf seiner zweiten Reise durch Bolivien in den Jahren 1910 und 1911 gesammelten Pflanzen*, I–VI (Meded. Rijks Herb. Leiden, 19, 27, 29, 33, 40, 46). Leiden. [Many specialist treatments.]

RUSBY, H. H., 1910–12. New species from Bolivia collected by R. S. Williams. *Bull. New York Bot. Gard.* **6**: 487–517; **8**: 89–135. [Lacks pteridophytes, grasses and orchids.]

RUSBY, H. H., 1893–1907. On the collections of Mr Miguel Bang in Bolivia. *Mem. Torrey Bot. Club*, 3(3): 1–67; **4**: 203–274; **6**: 1–130 (1893–96); *Bull. New York Bot. Gard.* **4**: 309–470 (1907). [Lacks the grasses.]

RUSBY, H. H., 1927. Descriptions of new genera and species of plants collected on the Mulford Biological Expedition of the Amazon Valley, 1921–2. *Mem. New York Bot. Gard.* **7**: 205–387, figs. 1–8. [Descriptions of new taxa.]

RUSBY, H. H. and BRITTON, N. L. 1888–1902. An enumeration of the plants collected by Dr H. H. Rusby in South America, 1885–1886. [32 parts.] *Bull. Torrey Bot. Club*, **15–29**, passim. [Covers non-vascular and vascular plants, with most collections from Bolivia.]

Region

35/36

Brazil

This region corresponds to the political limits of mainland Brazil. For the Atlantic islands of Fernando Noronha (with Rocas) and Trinidade (with Martin Vaz), see respectively **033** and **036**.

Brazil, the country with the largest vascular flora on earth, has also had one of the longest traditions of botanical work in south America. The period up to the completion of *Flora brasiliensis* was thoroughly reviewed by Urban in that work in 1906. Important early contributors included Marcgrave in the northeast and Velloso around Rio de Janeiro along with several short-term visitors on exploration voyages such as Commerson, and Banks and Solander, but the modern history of Brazilian botany may be said to begin with the visit of von Martius with Spix in 1817–20. In addition to eastern Brazil, they visited the Amazon, making the first biological survey of that great river. They were soon followed by many others; Brazil was among the most popular areas for biological exploration in the first half of the nineteenth century for European visitors, and this foreign domination continued until after the end of the century, with the result that most of the key early Brazilian collections are held outside the country. Much of this collecting was carried out in the central and southern parts of the country, but Spruce (and A. R. Wallace) worked extensively in the Amazon Basin. Among visiting and resident collectors from 1855 to 1914 were Regnell, Glaziou (under contract to the Brazilian government), the ecologist Warming, E. Ule (especially in Amazonia), Dusén, the palm specialist Barbosa Rodrigues, Huber, and the young Ducke and Hoehne (the former in Amazonia, the latter with the Rondon frontier geographical commission and the Roosevelt-Rondon expedition).

The only comprehensive flora of Brazil, *Flora brasiliensis*, was begun in 1840 and finally completed in 1906. It is the largest single work of its kind, covering 22 767 species of vascular plants and Musci. However, with the great increase in known species since publication, the work is outdated and incomplete. A successor, *Flora brasilica*, was initiated by Hoehne in 1940 but only a very small proportion of the flora has been covered.

From late in the ninteenth century, there has been a steady growth of a local botanical profession, now the largest in South America and making substantial contributions to floristic knowledge in all parts of the country, sometimes in conjunction with overseas collaborators. Substantial collecting programmes have been prosecuted in Amazonia (from Huber and Ducke onwards), the Nordéste, Bahia, Mato Grosso, the Planalto, and Santa Catarina among others. To these must now be added the government's *Programa Flora*, a collecting and inventorization programme with five regional projects. This may serve to cope with and counteract a gradual tendency towards decentralization, manifested in the existence of nearly 50 herbaria as of 1977 as well as in a number of state (and local) floras which have appeared in recent decades. The most notable of the latter are the floras for Rio Grande do Sul, Santa Catarina, Paraná, and São Paulo. The first two of these are critical serial works comparable to the 'tropical' floras prepared in Europe, while the last two are largely-compiled works by Angely. In addition, there are some local florulas of points of interest within Rio de Janeiro State as well as *Flora ecologica de restingas do sudeste do Brasil*, in progress since 1965. Few, if any, comparable works are available in central and northern Brazil. For these tracts, it has been argued that florulas of selected small areas of special interest would represent the most valuable and practicable kind of contribution.

At the present time, the Brazilian flora is still considered imperfectly known botanically, with collections and documentation as yet inadequate despite the long history of floristic work. Poorly known areas, of which several are under threat, abound; these have been noted by Prance (1979). In general, they are found in the central, western, northern, and northeastern parts of the country; in the better known south they are more localized.

Progress

No overall review of the history of botany in Brazil has been seen; available information is scattered widely in Brazilian and non-Brazilian works. Local work prior to the arrival of Piso and Marcgrave in 1637, the conventionally accepted starting date for serious natural history and botanical studies in Brazil, is treated in HOEHNE, F. C., 1937. *História da botânica e agricultura do Brasil do século XVI*. São Paulo. Companhia Editôra Nacional. The principal anthologies of collectors, both exhaustive, are respectively URBAN,

I., 1906. Vitae itineraque collectorum botanicorum. In *Flora brasiliensis* (ed. G. F. P. VON MARTIUS *et al.*) vol. 1, part 1, pp. 1–152. Munich; and the *Notas bio-bibliográficas de naturalistas botânicos* in HOEHNE, F. C. (in collaboration with M. Kuhlmann and O. Handro), 1941. *O jardim botânico de São Paulo*, pp. 19–246, illus. São Paulo: Departamento de Botânica do Estado, Secretaria da Agricultura, Indústria e Comércio [do Estado de Sao Paulo]. Information on subsequent collectors and authors is scattered, and reference should be made to general contributions on South America (see Division 3, introduction, p. 184) although some area reviews exist (Amazonia generally, Paraná, Santa Catarina). Since 1957, a more or less regular botanical bibliography has been published (Anonymous, 1957–).

Bibliographies. General and divisional bibliographies as for Division 3.

Regional bibliographies

FERREIRA DA SILVA, N. M. *et al.*, 1972– . Bibliografia de botânica, I 1– . *Rodriguésia*, **27**, *Anexo, passim*. [Taxonomic bibliography, to date covering dicotyledonous families from A through L. Published as supplementary parts to successive volumes of *Rodriguésia*, with five available by 1980.]

SAMPAIO, A. J. DE, 1924–8. Bibliographia botânica, relativa e flora brasileira, com inclusão dos trabalhos indispensaveis aos estudos botânicos no Brasil. *Bol. Mus. Nac. Rio de Janeiro*, 1(1924): 111–25, 225–45; 2(3)(1926): 35–61; 2(5)(1926): 19–38; 3(1)(1927): 37–45; 4(3)(1928): 97–119. [A series of lists of works published since 1840; coverage very incomplete. Now being succeeded by the preceding.]

Indices. General and divisional indices as for Division 3.

Regional indices

ANONYMOUS. 1957– . *Bibliografia brasileira de botânica*. Vols. 1– . Rio de Janeiro: Instituto Brasileiro de Bibliografia e Documentação. [Vol. 1, years 1950–5; vol. 2, years 1956–8; vol. 3, years 1959–60; vol. 4, years 1961–9; vols. 5– , years 1970– . Complete details have not been available, but as of the present writing (1979) seven volumes in all are known to have been published, with coverage through 1973. From vol. 4 published in an enlarged (A4) format.]

350

Region in general

The only comprehensive general floras of Brazil are *Flora brasiliensis* (1840–1906) and *Flora brasilica*

(commenced in 1940, but so far covering a small fraction of the known vascular flora and without any continuation since 1968). However, much of Brazil is being covered by the monographic treatments gradually appearing in *Flora Neotropica* (1968– ; see 301/I), and a further recent development has been the launching by the Brazilian federal government in 1976 of a floristic inventorization programme, *Programa Flora*, with five regional projects (Prance, 1979; see Division 3, Introduction, p. 184). In addition, there are a number of sets of keys to families, one of which also covers genera; four of these are listed below.

– Vascular flora: over 55 000 species (Prance, 1979, p. 55).

Keys to families and genera

BARROSO, L. J., 1946. *Chaves para a determinação de gêneros indígenas e exóticas das dicotiledôneas do Brasil*. 2nd edn. 272 pp. Rio de Janeiro: Serviço de Documentação do Ministério da Agricultura.

BARROSO, L. J., 1946. Chaves para a determinação de gêneros indígenas e exóticas das monocotiledôneas do Brasil. *Rodriguésia*, **9**(20): 55–77.

GOLDBERG, A. and SMITH, L. B., 1975. *Chave para das famílias espermatofíticas do Brasil*. 204 pp., 69 pls. (In *Flora ilustrada catarinense*, I. parte (ed. R. Reitz).) Itajaí, Santa Catarina. [Illustrated key to families, with glossary and references; explanations given on plates.]

JOLY, A. B., 1977. *Botânica. Chaves de identificação das famílias de plantas vasculares que ocorrem no Brasil, baseadas em chaves de Franz Thonner*. 3rd edn. 159 pp. São Paulo: Companhia Editôra Nacional. (Provisional edn., 1968; 1st edn., 1970; 2nd edn., 1975.) [One of the Brazilian derivatives of Thonner's *Analytical key to the natural orders of flowering plants* (1895; see 000, p. 61), as discussed in the new version of that work by R. Geesink *et al.* (1981). Other derivatives have been produced by Alvim (1943, 1950) and Rawitscher and Rachid-Edwards (1956), to which reference is made by Joly but in the absence of detailed information are not accounted for here. The source of all these was a manuscript copy of Thonner's key in use before World War II at Viçosa in central Minas Gerais.]

LOEFGREN, A., 1917. *Manual das familias naturaes phanerogamas: con chaves dichotomicas das famílias e dos generos brasileiros*. xviii, 611 pp. Rio de Janeiro: Imprensa Nacional. [Keys to families and genera, with descriptions of the latter; also includes some vernacular names and notes on uses.]

HOEHNE, F. C. *et al.* 1940– . *Flora brasilica*. Fasc. 1– . Illus. (some in color). São Paulo: Instituto de Botânica.

Projected on the scale of *Flora brasiliensis*, this

work is a comprehensive, illustrated documentary descriptive flora of vascular plants, with keys to genera and species, synonymy, references and citations, vernacular names, citation of *exsiccatae* with localities, general indication of extralimital range; critical commentary, and notes on habitat, special features, biology, etc.; indices to botanical names in some parts. At this writing (late 1979) 12 fascicles have appeared (the latest in 1968), comprising vol. 2, part 2; vol. 9, part 2; vol. 12, parts 1, 2, 6, 7; vol. 15, part 2; vol. 25, parts 2, 3, 4; vol. 41, part 1 and vol. 48. [Vol. 12 is devoted to Orchidaceae.]

MARTIUS, K. F. P. VON, EICHLER, A. W., and URBAN, I. (eds.), 1840–1906. *Flora brasiliensis*. 15 vols. (in 40). 20733 pp. (i.e., columns), 3811 pls., 2 maps. Munich: Fleischer. (Reprinted 1920–4, Leipzig: Hiersemann; 1966–7 (in reduced format), Lehre, Germany: Cramer.)

Large-scale, comprehensive, generally critical documentary descriptive flora of vascular plants (and Musci) of Brazil and adjacent lands, with synoptic or analytical keys to genera and species (or groups of species), full synonymy, references and citations (including appropriate revisions and monographs), localities with citations of *exsiccatae* or authorities, and general indication of internal and extralimital range, critical commentary, and notes on habitat, phenology, properties, uses, phytogeography, etc.; figures of representative species (some executed by nature-printing, a widely-practised graphic art form in the mid-nineteenth century) on separate plates; index to all botanical names at end of each partial volume. The general introduction to the work in volume 1, part 1, partly by von Martius and partly by Urban, incorporates a cyclopedia of collectors and collaborators (the latter numbering about 60), a chronologically dated list of all 130 fascicles, floristic statistics (in all, 22767 species covered), and a synopsis of and general index to families together with the 59 famous *Tabulae physiognomicae* illustrating various aspects of the Brazilian landscape and vegetation. This, the greatest of all floras, stands as a monument to nineteenth-century botanical scholarship as *Flora SSSR* and *Flora Europaea* do in their different ways for the present era. It is one of the first of modern large-scale floras, and one of the earliest involving international specialist collaboration and the services of 'in-house' writers and editorial assistants, among them S. Endlicher and E. Fenzl. Sponsors included the Kings of Bavaria, including 'mad' Ludwig II, the last Emperor of Brazil, Dom Pedro II, and the Austro-Hungarian Emperor, Franz-Josef I.

351–356

Amazonia

Designated by Martius as his 'Naiades', this vast area comprises the states of Amazonas, Pará, and Amapá and the territories of Rio Branco and Acre. No useful floras or florulas are yet available for any part of this vast area, which contains the heart of the *selva* of South America, the largest humid tropical forest in the world and now the sadly much misunderstood last great world frontier.[8] However, the gradually appearing family treatments in *Flora Neotropica* are providing some coverage for this area in addition to a host of monographs, revisions, and 'contributions', although the present level of botanical knowledge is such that many new species are still being found, resulting in rapid 'dating' of published work, in addition to the low level of knowledge of many other species (the latter a problem also existing in most other parts of the tropics).[9] A regional collecting and inventorization program, *Projeto Flora Amazônica*, has also been initiated under government auspices. This, however, should be accompanied by florulas for limited areas, especially around Manaus and other places where training programs are active; these would represent feasible short-term projects and will provide a good introduction to the flora.[10] They may not necessarily be academically 'correct' in the eyes of some botanists but will fill a real need and stimulate interest as has the Rio Palenque florula (see **329**). It is not necessarily through orthodox solutions that progress in biological education and understanding is made. On noteworthy trees and other plants see LE COINTE, P., 1947. *Arvores e plantas uteis da Amazônia brasileira*. 506 pp. São Paulo.

Bibliography

ANONYMOUS, 1963–70. *Amazônia: bibliografia*. 2 vols. Rio de Janeiro: Instituto Brasileiro de Bibliografia e Documentação/Instituto Nacional de Pesquisas da Amazônia, Conselho Nacional de Pesquisas. (Reprinted 1975, Königstein/Ts., W. Germany: Koeltz.) [General bibliography: includes botanical section.]

357

Maranhão and Piaui

This subregion, in which extensive *carnauba* palm forests occur, is intermediate between Amazonia and the Nordéste, and also contains the northern limits of the *campos*. The two states are comparatively poorly known botanically, and no separate floras or enumerations for either are available.

358

'O Nordéste'

Containing the major part of Martius' 'Hamadryades', this area encompasses the states of Ceará, Rio Grande do Norte, Paraíba, Pernambuco, Alagoas, and Sergipe. No individual or collective complete floras or enumerations appear to be available, although documentation of various kinds has long existed, beginning with Piso and Marcgrave's *Historia naturalis Brasiliae* (1648).[11] Within the subregion, considered moderately well-known botanically but with as yet many gaps, is the dry *sertão*, an extensive area of uncertain rainfall and largely dominated by the characteristic thorny *caatinga* vegetation. For Fernando Noronha, off the northeastern tip of the continent, see 033.

359

Bahia

This large state is to a large extent still not well known, although in recent years serious work has been accomplished by a series of joint Brazilian-British (Kew) expeditions and many new taxa and records found, both in 'new' and 'old' areas. No separate floras are available, although documentation of various kinds has long been available, beginning with the seventeenth-century natural history of Piso and Marcgrave referred to above. The first results of the joint Kew-Bahian botanical working party have been embodied in the following: HARLEY, R. M. and MAYO, S. J., 1980. *Towards a checklist of the flora of Bahia*. 250 pp., 4 maps, illus. (on cover). Kew: Royal Botanic Gardens

(mimeographed). [Based on results of several collecting expeditions; covers 1596 species out of a total of 5000–6000 presently known from the state, in the form of an annotated systematic list, with appendices, list of references, and index at end. In one appendix is a suggested format for a more definitive enumeration.]

361–363

Goiás, Brasilia, and Minas Gerais

This area, which includes the heart of the *campos* of the Planalto, or Martius' 'Oreades', with its rich *flora geral* comprises the states of Goiás and Minas Gerais together with the Federal District (Brasilia). As with other parts of infratropical Brazil, no useful separate floras or enumerations are available. However, a goodly number of studies on the vegetation and its ecology, with species lists, exist, and a substantial amount of new collecting has been accomplished in recent decades in many areas, adding greatly to a substantial 'classical' foundation. With the establishment of a 'regional' herbarium in Brasilia, an improved basis now exists for a flora or inventory program of the *campos*. For an introduction to the flora of the *campo cerrado*, one of a number of characteristic vegetation formations, see ANON. 1969. *Plantas do Brasil*, 1: *Especies do cerrado*. 239 pp., 100 figs. Brasilia. [Illustrated semi-popular account in atlas format.]

364

Mato Grosso and Rondônia

In the absence of general treatments of the flora of these states, the work described below, which includes coverage of the literature, provides the best introduction available. Much collecting, however, has been done in the area since 1951, rendering Hoehne's *Indice* rather incomplete. However, some large areas like the forest belt in the north – transitional to Amazonia – and the Pantanal marshes along the Bolivian border remain poorly known.

HOEHNE, F. C. and KUHLMANN, J. G. (comp.), 1951. *Indice bibliográfico numérico das plantas colhidas pela Comissão Rondon.* 400 pp. São Paulo: Instituto de Botânica.

Pages 110–400 of this work comprise a systematic enumeration of the known vascular plants of Mato Grosso and Rondônia, based in the first instance on the collections of the Rondon Commission but incorporating other published records, with synonymy, citations, taxonomic commentary, citations of *exsiccatae* with localities, and field observations; index to families at end. See also SAMPAIO, A. J. DE., 1916. *A flora de Matto Grosso. Memória em homenagem aos trabalhos botânicos da Commissão Rondon.* 125 pp., 11 maps (Arq. Mus. Nac. Rio de Janeiro 19). Rio de Janeiro.

365

I. Rio de Janeiro (including Guanabara)

No separate works of a general character on the vascular plants of Rio de Janeiro (and formerly separate Guanabara, including the city of Rio de Janeiro), the center of the coastal forest belt or 'Dryades' of Martius, has appeared since publication of Frei Velloso's pioneer *Florae fluminensis* and its *Icones* in the 1820s and 1830s respectively as a kind of botanical declaration of Brazil's independence under Pedro I (although the text was not released in its entirety until 1881). It is included here as a standard work in the absence of any overall successors and because an interpretation of its names in modern nomenclatural terms is available.

Some partial works are available, however, for areas of special botanical interest: the peculiar 'restingas' along the coast (Segadas-Vianna, 1965–) and for the Organ Mountains (Rizzini, 1954) and the Itatiaia massif, the highest mountain range in southeastern Brazil (Brade, 1956). These are selected for inclusion here from an extensive menu of local literature. In addition, many family revisions covering the area have in recent decades appeared in *Arquivos do Jardim Botânico do Rio de Janeiro* and *Rodriguésia*.

VELLOSO, J. M. DA C., Frei, 1825. *Florae fluminensis.* Pp. 1–352. Rio de Janeiro: Imprensa National. (Reissued and completed 1881, Rio de Janeiro, as Arq. Mus. Nac. Rio de Janeiro, vol. 5, pp. 1–xii, 1–467.)

VELLOSO, J. M. DA C., Frei, 1827 (1835). *Florae fluminensis icones.* 11 vols. 1640 pls. Paris: Senefelder

(*curante* Knecht). (Indices (pp. 1–14, 1–21) in vol. 1 reprinted separately, 1929, Berlin: Junk.)

The text of this work comprises a briefly descriptive diagnostic flora of vascular (and a few non-vascular) plants, without synonymy or detailed geographical distributions but including information on habitats and phenology; cross-references to plates but no index. The arrangement of species follows the Linnean system; in the original printing of the work the text terminated at the genus *Sabbata* in the Syngenesia, corresponding to volume 8 of the atlas (the rest, including a longer preface, did not appear until 1881; in the reissue the added material extends over pp. 329–461). The folio atlas follows the text in its arrangement of plates, but in vol. 1 after the foreword are two indices: an *index alphabeticus* (14 pp.), and an *index methodicus* (21 pp.), the latter being a conspectus of the work arranged according to the Candollean (natural) system. The *Florae* was prepared rather in isolation and appeared posthumously; many genera described therein have been reduced to synonymy and some species are hard to interpret. A nomenclatural revision was produced by A. J. de Sampaio and O. Peckolt in 1943 (*Arq. Mus. Nac. Rio de Janeiro* vol. 37, 331–94) and a revised index by M. Cruz in 1946 (17 pp.; Rio de Janeiro). The history of the work, including its difficult passage to the light of day (the abdication of Pedro I in 1831 was all but catastrophic), was reviewed in BORGMEIER, T., 1937. A historia da 'Florae fluminensis' de Frei Velloso. *Rodriguésia*, 3(9): 77–96, 2 pls.[12]

Partial works

BRADE, A. C., 1956. A flora do Parque Nacional do Itatiaia. *Bol. Minist. Agric., Serv. Florest.* 5: 1–85, figs. 1–14, 8 maps.

Includes annotated lists, emphasizing endemic or otherwise noteworthy species, with references and limited synonymy. The Itatiaia range, the highest in southern Brazil, is in the southwestern part of Rio de Janeiro State.

RIZZINI, C. T., 1954. Flora organensis. *Arq. Jard. Bot. Rio de Janeiro* 13: 117–243. (Reprinted separately.)

Systematic enumeration of vascular plants and bryophytes, with limited synonymy and indication of occurrences. The Organ Mountains are located in the central part of Rio de Janeiro State, north of the former capital.

SEGADAS-VIANNA, F. (ed.), 1965– . *Flora ecológica de restingas do sudeste do Brasil.* Part II, Flora. Fasc. 2– . Illus. Rio de Janeiro: Museu Nacional/Universidade Federal.

Illustrated descriptive flora, with in each fascicle keys, synonymy, references, vernacular names, local and extralimital range, and notes on habitat, ecology, etc.; general introduction in fasc. 2. As of 1981 fascicles 2–23 in Part II had appeared, each containing one family (without regard to sequence). The 'restingas', or sandy coastal plains, are with their vegetation a special feature of Rio de Janeiro State.

II. Espirito Santo

This state, lying between Rio de Janeiro and southeastern Bahia, remains relatively poorly documented and is somewhat undercollected. No general or significant partial floras or enumerations appear to be available.

366

São Paulo

Two local florulas, one now rather old, are available in addition to the uncritical general enumeration by Angely.

ANGELY, J., 1969–71. *Flora analítica e fitogeográfica do Estado de São Paulo.* 6 vols. 1901 maps. Sao Paulo: 'Phyton'.

Largely compiled, uncritical enumeration of vascular plants, without keys; includes synonymy, references, vernacular names, superficial summaries of local and extralimital range (with many distribution maps), and miscellaneous notes on habitat, occurrence, karyotypes, etc.; index to botanical names in each volume (general index, vol. 6). An introductory section (vol. 1) includes a plan and synopsis of the work, family statistics, and a general comparison of the state's flora with that of areas elsewhere, with also a reconsideration of the publication dates of Velloso's *Florae fluminensis*. Lists 7251 species (native, naturalized, adventive, and cultivated).

Partial works

KUHLMANN, M. and KUHN, E., 1947. *A flora do distrito de Ibiti.* 221 pp., 94 illus. (some in color). Sao Paulo: Instituto de Botânica.

Annotated enumeration of non-vascular (except algae) and vascular plants, with vernacular names, citation of *exsiccatae*, localities, and ecological and phenological data; glossary, bibliography, and indices at end. The latter part of the work deals with general natural history. [The Ibiti district is about 100 km north of São Paulo City, near the Minas Gerais boundary.]

USTERI, A., 1911. *Flora der Umgebung der Stadt São Paulo.* 271 pp., 72 text-figs., 1 col. pl., map. Jena: Fischer.

Concise manual of vascular plants occurring in the São Paulo district, with keys, localities, and notes on habitat; index. An extensive introductory section includes remarks on vegetation and phytogeography.

367

Paraná

In this state is found the heart of the great Paraná pine forest zone, part of Martius' 'Napaes'. Regrettably, no alternatives to Angely's uncritical *Flora analítica* are available, except for the works on Santa Catarina (368).

Bibliography

ANGELY, J., 1964. Bibliografia vegetal do Paraná. 304 pp. Curitiba: 'Instituto Paranaense do Botânica'.

ANGELY, J., 1965. *Flora analítica do Paraná: nomenclator species plantarum paranaënsium.* xii, 728 pp., map. Sao Paulo: 'Instituto Paranaense do Botânica'.

Uncritical, largely compiled enumeration of native and naturalized vascular plants and Musci, with synonymy, references, indication of status (endemic or otherwise), and miscellaneous notes; index to genera and families at end. An introductory section includes various statistical tables as well as historical and polemical remarks. Lists 5287 native and introduced species. [Several precursory works, too numerous to list here, have also been produced by this prolific author.]

368

Santa Catarina

For a general key to families published in the *Flora ilustrada catarinense* series, see 350.[13]

REITZ, R., P.e (ed.), 1965– . *Flora ilustrada catarinense.* I. parte, Plantas. Fasc. 1– . Illus., maps. Itajai: Herbário 'Barbosa Rodrigues'.

Comprehensive, illustrated critical and descriptive flora of vascular plants, with keys to genera and species, full synonymy, references, vernacular names, citation of *exsiccatae* with localities, general indication of local and extralimital range (including distribution maps), extensive critical commentary, and notes on habitat, ecology, phenology, etc.; index to botanical names and list of references in each fascicle. At this writing (1979), close to 90 of the projected 120 fascicles have been published, with between one-half and three-quarters of the total vascular flora treated. This work, with many family revisions by specialists, will set a high standard for future floristic writing in southeastern Brazil.

Partial work

SOUZA SOBRINHO, R. J. DE and BRESOLIN, A. (eds.), 1970– . *Florula da Ilha de Santa Catarina.* [Fasc. 1– .] Illus. Florianópolis: Universidade Federal de Santa Catarina, Divisão de Botânica, etc.

Similar in style to *Flora ilustrada catarinense*; published in fascicles. At this writing (1979) at least 16 parts have appeared. [The island is a large mainland fragment parallel to the east coast, on which is situated the state capital.]

369

Rio Grande do Sul

The prevailing *campinas* vegetation is closely related to that of Uruguay (**375**). The citations and descriptions below have been partly based on information supplied by L. B. Smith, Washington, and the Library, Universidade Federal do Rio Grande do Sul, Porto Alegre.

AUGUSTO, Irmão, 1946. *Flora do Rio Grande do Sul, Brasil.* Vol. 1. 639, vii pp., 341 text-figs. Porto Alegre: Imprensa Oficial (published under gubernatorial patronage).

Copiously illustrated manual of seed plants, with keys to all taxa and descriptions of genera and families but species merely listed under the genera. Each family treatment contains two sections: an analytical key to genera and species and a systematic catalogue of all taxa known from the state with localities and collectors thereof; however, no synonymy, references, critical commentary, or habitat and biological data are provided. The work is prefaced by a short introduction and a general key to all families, and concludes with a complete index. Covers only Clethraceae through Compositae (Englerian system). Represents a consolidation and expansion of several precursory contributions of the author and Brother Edesio, viz: AUGUSTO, Irmão and EDESIO, Irmão, 1941–2. Flora de Rio Grande do

Sul: plantas catalogadas neste Estado até hoje, 1820–1940. *Revista Agron. Rio Grande do Sul* **5**: 561–70, 639–45, 731–7; **6**: 79–81, 101–3, 205–8, 309–12, 380–2, 417–20, 497–500, 635–40; *idem*, 1943. *Flora do Rio Grande do Sul: Solanaceas-Labiadas.* 43 pp. Porto Alegre: Klein; *idem*, 1943. *Escrofulariaceas.* 32 pp. Porto Alegre; *idem*, 1943. *Quadro geral das familias.* 32 pp. Porto Alegre; *idem*, 1944. *Boraginaceas*...82 pp. Porto Alegre; and *idem*, 1944. *Myrsinaceas*...120 pp. Porto Alegre.

SCHULTZ, A. R. (coordinator), 1955– . *Flora ilustrada do Rio Grande do Sul.* Fasc. 1– . Illus., maps (Boletim Instituto Central de Biociências (Pôrto Alegre), 3, etc.) Porto Alegre.

Large-scale illustrated serial flora, with lengthy descriptions, full synonymy (with references and citations), vernacular names, local range (with *exsiccatae*), dot distribution maps, extralimital range, critical commentary, and notes on habitat, phenology, etc. Representative illustrations, family summaries, and indices are also given in each fascicle. As of this writing (1979), twelve fascicles have been published, covering for the most part pteridophytes and non-sympetalous families except Fasc. 12, which treats Rubiaceae-Spermacoceae. [From Fasc. 12 (1977) the work is under the coordination of M. H. Homrich.]

Partial work

TEODORO LUIS, Irmão, 1960. *Flora analítica de Pôrto Alegre.* iv, (unpaged) pp., illus. Canoas: Instituto Geobiológico 'La Salle'.

Very concise pocket-sized field checklist-key of higher plants of the Porto Alegre metropolitan area, with keys to genera and species as well as representative figures. No synonymy, vernacular names, distributional data, or ecological notes are given, although each entry is hierarchically numbered.

Region

37

Paraguay and Uruguay

These two countries are grouped together merely for convenience. Uruguay is, however, rather better known botanically than Paraguay.

Botanical work in Paraguay dates from the

nineteenth century but has been very sporadic, with little or no resident involvement and only some thirty collectors on record. Major contributions have been made by Balansa, Hassler (with Rojas and Chodat), Swedish Regnellian explorers such as Dusén and Malme, and some Argentine botanists (in part indirectly, through revisionary studies). Only one work which could be called a flora has been published: Chodat's *Plantae hasslerianae* (1898–1907). Together with Bolivia, it is the most poorly known botanically of South American countries.

An entirely different situation exists in Uruguay, where there are substantial local collections and a small resident botanical community. In the latter part of the eighteenth and first half of the nineteenth centuries, many foreign botanists made visits, among them Commerson, St Hilaire, Sellow, and Tweedie. Locally based work began to develop with the growth of prosperity in the late-nineteenth century, and the first separate descriptive flora, *Flora uruguaya* (1898–1911, not completed) by Arechavaleta, was published. Contributions also began to be made by Argentine botanists. In the period from the mid-1920s until World War II the controversial German botanist, W. Herter, was resident for a considerable time and made many contributions, among them *Enumeratio plantarum vascularium*...(1930), *Flora ilustrada del Uruguay* (1939–57, not completed), and *Flora del Uruguay* (1949–56, not completed). None of these, however, is a descriptive work: the first is a checklist, the second an iconography, and the third an annotated enumeration with keys.

In the 1950s a new descriptive flora was begun by the Museo Nacional de Historia Natural as a successor to that of Arechavaleta, but progress to date (1979) has been very slow, only four fascicles being available. This and the present low level of collecting may be due to the present difficult economic situation, an unfortunate state of affairs as although the flora is in general relatively well-known, intensive collecting is needed in many areas and the present level of documentation is unsatisfactory. The woody flora has, however, been covered in an illustrated treatment by Lombardo (1946; 2nd edn., 1964).

Progress
The life and activities of W. G. Herter, the long-time student of the Uruguayan flora, have been reviewed by Burdet, H. M., 1978. L'oeuvre et les tribulations du botaniste Herter: une étude biographique et bibliographique germano-uruguayenne. *Candollea*, 33: 107–34. No other separate reviews have been seen.

Bibliographies. General and divisional bibliographies as for Division 3.

Indices. General and divisional indices as for Division 3.

371

Paraguay

In addition to the cumulated series of family revisions based on Hassler's collections (see below), reference should also be made to the following introductory work, which contains general surveys of various families as they occur in Paraguay: Chodat, R. and Vischer, W., 1916–26(–27). La végétation du Paraguay, I–XIV. *Bull. Soc. Bot. Genève*, II, **8–18**, *passim*. (Reprinted Geneva: 509, 49 pp.)

Chodat, R. 1898–1907. *Plantae hasslerianae, soit énumération des plantes récoltées au Paraguay par le dr. Émile Hassler d'Aarau (Suisse) de 1885 à 1902 et déterminées par le prof. dr. R. Chodat avec l'aide de plusieurs collaborateurs*, I–II. 2 vols. Geneva: Romet (for Institut de Botanique, Université de Genève). [Reprinted from *Bull. Herb. Boissier*, ser. I, vol. 6, app. 1 (1898); vol. 7, app. 1 (1899); ser. II, vols. 1–5, 7 (1901–7), *passim*. For full collation, see Blake and Atwood, 1942, p. 252, and Stafleu and Cowan, *Taxonomic literature*, 2nd edn., vol. 3, pp. 98–9.]

Polyglot enumeration (not in systematic order) of vascular plants, based mainly upon the collections of Hassler made in Paraguay from 1885 to 1902, with synonymy, references, and a few citations; notation of *exsiccatae*, with localities; brief general indication of extralimital range; taxonomic commentary and notes on distribution, special features, etc. (in part by Hassler); addenda and corrigenda; general index to families and genera at end of pt. II, together with (p. 713) full bibliographic details. Part II also includes a short account of Paraguayan floristic regions. Treatment of some families contributed by specialists. For supplement, see Hassler, E., 1917. *Addenda ad Plantas hasslerianas*. 20 pp. Geneva.

375

Uruguay

In spite of the number of mostly incomplete works available, there is no descriptive floristic treatment as yet for certain groups of families, notably in the 'Sympetalae'. As comparatively little has been published to date in the *Flora del Uruguay* series of the Museo Nacional, modern coverage is still dominated by the works of the controversial German-Uruguayan botanist W. (or G.) Herter.

– Vascular flora: 3000–3500 species (2998 in Herter, 1930–7).

ARECHAVALETA, J., 1898–1911. *Flora uruguaya.* Vols. 1–3; vol. 4, parts 1–3. Illus. (Anales del museo nacional de Montevideo, 3, 5–7.) Montevideo.

Illustrated descriptive flora of seed plants, with some synoptic keys to genera and species, full synonymy, references and citations, vernacular names, general indication of local range, critical commentary, and notes on habitat, phenology, uses, etc.; index to botanical names in each volume (except the last). The introductory sections of vols. 2–3 include sketches of botanists associated with the Uruguayan flora. Not completed; extends from Ranunculaceae through *Cuscuta* (Convolvulaceae) on the Bentham and Hooker system.

HERTER, W., 1930. *Enumeración de las plantas vasculares que crecen espontáneamente en la República oriental del Uruguay agregando las plantas adventicias, las principales plantas cultivadas, los nombres vulgares, la repartición en la República y los números de las collecciones Gibert y Herter.* 191 pp., 18 pls. (some in color), color map. Montevideo: The author [with official state support]. (Estudios botánicos en la región Uruguaya, 4; Florula uruguayensis, II.) [Full title also in Latin.]

Systematic list of native, adventive, and commonly cultivated vascular plants, with very limited synonymy, vernacular names, and concise indication of local range by provinces (with citation of some *exsiccatae*); list of collections by Gibert and Herter; indices to vernacular and botanical names at end. An introductory section includes a historical account of botanical exploration in Uruguay. For additions, see *idem.* 1935–37. Additamenta ad floram uruguayensem, I–III. *Revista Sudamer. Bot.* 2: 111–28, 2 figs., 1 col. pl.; 3:

146–78, 1 col. pl.; 4: 179–232, 1 col. pl. (Altogether 2998 species covered.)

HERTER, W., 1939–43. *Flora ilustrada del Uruguay.* Fasc. 1–5 (comprising vol. 1). pp. i–xvi, 1–13, pls. 1–256 (figs. 1–1024). Montevideo, Berlin, Krakow: The author. Continued as *idem.* 1952–57. *Flora ilustrada del Uruguay.* Fasc. 6–13. pp. xvii–xxxii, pls. 257–352, 373–600 (figs. 1025–2320). Basel (later Hamburg): The author. [Fasc. 1 also published as *Repert. Spec. Nov. Regni Veg., Beih.* **118**(1).]

Comprises an atlas of 580 plates of the vascular flora of Uruguay, with four figures to a plate; the drawing of each species is accompanied by vernacular and botanical names together with a concise indication of local range. Not completed (owing to the death of the author); extends from the lower vascular plants through the Cactaceae (Englerian system). The figures are continuously numbered; the gap in the numbering of the plates is the result of an error. Only fascicles 1–5 (vol. 1) possess a general index.

HERTER, W., 1949–56. *Flora del Uruguay.* Vol. 1. 10 fascicles. 280 pp. Montevideo (later Berne, Basel, Hamburg): The author. (Mimeographed.)

Enumeration of vascular plants, complementing the author's *Flora ilustrada del Uruguay*, with synonymy, references, vernacular names, citation of *exsiccatae*, brief general summaries of local and extralimital range, and complete index (at end of fasc. 10). An introductory section (fasc. 1) includes a short list of references. Not completed (owing to the death of the author); encompasses Pteridophyta, Gymnospermae, and Monocotyledoneae (Englerian system).

MUSEO NACIONAL DE HISTORIA NATURAL, URUGUAY. 1958– . *Flora del Uruguay.* Fasc. 1– . Illus. Montevideo.

Detailed descriptive flora of vascular plants, with each family treatment including keys, synonymy, indication of local range, miscellaneous notes, and full-page illustrations; each fascicle separately indexed. At this writing (late 1979), only four fascicles have appeared, the latest in 1972; groups covered include the pteridophytes and miscellaneous families of dicotyledons. Some treatments are by specialists.

Special group – woody plants

LOMBARDO, A., 1964. *Flora arbórea y arborescente del Uruguay, con clave para determinar las especies.* 2nd edn. 151 pp., 223 text-figs. Montevideo: Concejo Departamental de Montevideo. (First edn., 1946.)

Illustrated manual of trees and shrubs, with keys to all taxa, limited synonymy (with a few citations), vernacular names, generalized indication of local range, taxonomic commentary, and notes on habitat, uses, etc.; indices to vernacular and botanical names. A glossary appears in the introductory section.

Region

38

Argentina

This region corresponds to the political limits of the Argentine Republic, with the addition of the Falkland Islands (Islas Malvinas).

Botany in the Argentine, which like Chile includes a portion of Patagonia (and Fuegia), has had a relatively long history with early development of a resident botanical community and institutions, including a number of centers outside Buenos Aires. Eighteenth-century visitors included Mylam, Commerson, Pernetty, Banks and Solander, the Forsters, and Neé and Haenke, followed by Gaudichaud, Darwin, Bonpland, Sellow, Tweedie, J. D. Hooker, and others in the first half of the nineteenth century. In the second half of that century, characterized by greater stability and a considerable growth of the economy with attendant prosperity, many Germans collected under contract to the government, with a key contribution being made by Grisebach whose general enumeration, *Symbolae ad floram argentinarum* (1879) represented the first overall account of any kind: for here, as elsewhere in much of Latin America, colonial and early post-colonial regimes were notable for their uncertain support for basic botanical work and resident (especially native-born) botanists were few.[14] However, the *Symbolae* were largely limited to northern Argentina, the south being still very little explored save for Fuegia and the Falklands, first fully documented by Hooker in the second volume of his *Flora antarctica* (1845–7). Around the turn of the century, Swedish and American botanists were very active in Patagonia as a whole, and their collections, with earlier records, were largely accounted for in *Flora Patagonica* by Macloskie and others (1903–14; see 301/II).

The Argentine botanical profession has had its greatest development within the present century,

growing from centers at Buenos Aires and Tucumán. This development was aided by a number of foreign botanists resident for varying periods of time, among them L. Hauman, who with others began the first overall checklist, *Catalogue des phanérogames de l'Argentine* (1917–23, not completed). A similar checklist was later prepared for the woody plants by Latzina (1937).

The idea of a national flora appears to have been conceived at Tucumán in the 1930s, and under the first Peronista government this began to be realized in the massive *Genera et species plantarum Argentinarum* (1943–56), discontinued, however, after five volumes. The general flora concept then gave way to provision of floristic coverage through a network of 'regional' floras, akin to those in South Asia sponsored by the Botanical Survey of India. In Argentina, this sponsorship was taken up by the Instituto Nacional de Técnologia Agropecuaria and has continued to the present time (1981). A number of these multivolume regional floras have been completed, are in progress, or are in preparation, covering Buenos Aires Province (completed), Jujuy, Entre Rios, and Patagonia (in process of publication), and Córdoba and the Chaco (in preparation). In addition to these, there is a miscellany of older local works, but as a whole, coverage outside of Patagonia and the pampas is inadequate. The bleak Falkland (Malvinas) Islands have been, on the other hand, the beneficiary of a modern floristic treatment by D. M. Moore (1968) which added much information to that in the earlier standard works by Skottsberg and by Vallentin and Cotton. Moore is continuing with studies of the Fuegian flora.

The flora is in general considered relatively well-collected, but a number of relatively neglected areas are on record and some key localities require more intensive work.

Progress

The state of affairs in botany as of the late 1890s was reviewed in HOLMBERG, E. L., 1898. La flora de la República Argentina. In *Segundo censo de la República Argentina, mayo 10 de 1895* (ed. Comision directiva de Censo, República Argentina), vol. 1, pp. 383–474. Buenos Aires. For later reports, see TURRILL, W. B., 1920. Botanical exploration in Chile and Argentina. *Bull. Misc. Inform.* (Kew), *1920*: 57–66, 223–4; HICKEN, C. M., 1923. *Los estudios botánicos.* 167 pp. (Papeles del cincuentenario de la Sociedad científica argentina (1872–1922). Evolución de las ciencias en la República Argentina, VII). Buenos Aires: Coni; and CABRERA, A. L., 1972. Estado actual del conocimiento de la flora Argentina. In *Memórias de symposia del primero congreso latino-americano y mexicano de botánica* (ed. Sociedad Botánica de México), pp. 183–97. Mexico City.

Bibliographies. General and divisional bibliographies as for Division 3.

Regional bibliography

CASTELLANOS, A. and PÉREZ-MOREAU, R. A., 1941. Contribución á la bibliografía botánica argentina, I–II. 162, 549 pp. (Lilloa, 6, 7). Tucuman. [Part II contains lists of floristic works published up through 1937.]

Indices. General and divisional indices as for Division 3.

380

Region in general

In addition to the general works described below, the regional floras being published under the auspices of the Centro de Investigaciones de Recursos Naturales, Instituto Nacional de Técnologia Agropecuaria (INTA) are intended collectively to comprise a modern general flora of the Argentine in succession to the defunct *Genera et species plantarum argentinarum* project of the Peron era. The floras are being issued in the *Colección científica* series of INTA; those available or in progress cover Buenos Aires (vol. 4), Entre Rios (vol. 6), Patagonia (vol. 8), and Jujuy (vol. 13). Preliminary contributions have appeared for a *Flora chaqueña* and work is said to be in progress on others.[15]

DESCOLE, H. *et al.*, 1943–56. *Genera et species plantarum argentinarum.* Vols. 1–5 (in 7). Illus. (some in color), maps. Buenos Aires: Kraft.

This massive work comprises an illustrated descriptive flora of seed plants, with keys to genera and species; full synonymy, with references and citations; indication of local range, with citation of *exsiccatae*, and general summary of extralimital range (including distribution maps); indices to botanical names at end of each family revision. The families revised are as follows: vol. 1, Zygophyllaceae, Cactaceae, Euphorbia-

ceae; vol. 2, Asclepiadaceae, Valerianaceae; vol. 3, Centrolepidaceae, Mayacaceae, Xyridaceae, Eriocaulaceae, Bromeliaceae; vol. 4, Cyperaceae; vol. 5, Scrophulariaceae. Following withdrawal of official support and suspension of this work in 1956 after the fall of the first Peronista government, further family treatments appeared in *Revista del Museo de La Plata, Sección Botánica* as a series, 'Flora Argentina'.

HAUMAN, L., VANDERVEKEN, G. and IRIGOYEN, D. H., 1917–23. *Catalogue des phanérogames de l'Argentine*, [I–II]. 351, 315 pp. (Anales Mus. Nac. Hist. Nat. Buenos Aires, 29, 32). Buenos Aires.

Systematic enumeration of seed plants, with synonymy, references and citations; general indication of local range by provinces; occasional taxonomic remarks; statistical summaries; lists of references and indices to families and genera in each part. Not completed; extends through Droseraceae (Englerian system).

Partial works

GRISEBACH, A. II. R., 1879. *Symbolae ad floram argentinarum.* 346 pp. Göttingen: Dieterich. (Reprinted from *Abh. Königl. Ges. Wiss. Göttingen*, 24: 3–345, 1879.)

Systematic enumeration of vascular plants, with description of new taxa; limited synonymy, with references and a few citations; vernacular names; abbreviated summary of local range, with citations of some *exsiccatae*, and general indication of extralimital range; some short critical remarks; index to families (p. 346). The introductory section includes a phytogeographical discussion. The emphasis of this work is on the plants of northern Argentina, ranging from the provinces of Catamarca, Córdoba, Santa Fé, and Entre Rios to the north and northwest. Succeeds the author's *Plantae lorentzianae* (1874).

Special group – woody plants (including forest trees)

LATZINA, E., 1937. Index de la flora dendrológica argentina. *Lilloa*, 1: 95–211, 14 pls.

Comprises a briefly annotated systematic checklist of woody plants (839 species), with indication of localities, provinces, habit, vital statistics, and special notes (with particular emphasis on the province of Tucumán). The introductory section contains a synopsis and indices to vernacular names as well as to families and genera, and a good bibliography appears at the end (pp. 198–211). The plates are of representative tree species. For more extended accounts of the major tree species, see TORTORELLI, L. A., 1956. *Maderas y bosques argentinas.* xxvii, 910 pp., 104 figs., 111 pls., color frontispiece, map. Buenos Aires: Acme.

381

'Mesopotamia'

Comprises the northeastern provinces of Misiones, Corrientes, and Entre Rios, as well as (for convenience) the province of Santa Fé just to the west of the Rio Paraná. No recent general works are available for the area or any of the political units, except for Burkart's new flora of Entre Rios (see below).

BURKART, A. *et al.* (ed.), 1969– . *Flora ilustrada de Entre Ríos.* Parts 2, 5, 6. Illus., maps (Colecc. Ci. INTA, 6). Buenos Aires: Librart.

Illustrated descriptive flora of vascular plants, with keys to all taxa, synonymy, references and citations, generalized indication of local range, with citation of some *exsiccatae*, summary of extralimital distribution, vernacular names, taxonomic commentary, and notes on habitat, special features, uses, etc.; glossary and index to all botanical and vernacular names at end of each part. At this writing (1979) three parts (2, 5 and 6) of a projected total of six have been published.

382

'El Chaco'

Includes the provinces of Santiago del Estero and Formosa as well as the Chaco Territory. For long the only provincial treatment has been the checklist by Alvarez (see below), a rather uncritical work. However, work is in progress on a new *Flora chaqueña* in the INTA series of regional floras, and some preliminary contributions have now appeared, viz: CENTRO DE INVESTIGACIONES DE RECURSOS NATURALES, INTA, 1971– . *Notas preliminares para la flora chaqueña* (*Formosa, Chaco y Santiago del Estero*). Fasc. 1– . Castelar, Buenos Aires. [Introduction (by A. Digilio); precursory revisions of a number of families. As of 1974 six fascicles had appeared.]

ALVÁREZ, A., 1919. *Flora y fauna de la Province de Santiago del Estero.* 176 pp., illus. Santiago del Estero: publisher not given.

Includes (pp. 43–144) a poorly annotated, largely compiled systematic checklist of known cellular and vascular plants, with indication of Spanish vernacular names (where known) as well as of distribution and habitat (particularly for tree species); good photographs of representative tree species; list of references (p. 117). At the end of the plant list are classified lists of plants by uses and chapters on forestry and the principal tree species. The first section of this work (pp. 11–40) covers geography, climate, geology, and the main features of the flora and fauna, while the last section (pp. 147–70) gives lists of fauna.

383

Northwestern Argentina

Includes the provinces of Los Andes, Catamarca, Salta, Jujuy, and Tucumán, in the last-named of which is located the well-known Miguel Lillo botanical institute. No general work on the region has yet been published, but since 1977 modern coverage has begun to develop through provincial floras. Those for Jujuy (in the INTA series) and Tucumán have commenced publication, but as of early 1980 neither was far advanced. For treatment of the alpine flora, see 303/II (FRIES).

I. Jujuy

CABRERA, A. L. (coordinator), 1977– . *Flora de la provincia de Jujuy.* Parts 2, 10. Illus., maps (Colecc. Ci. INTA, 13). Buenos Aires: Librart.

Illustrated semi-monographic flora of vascular plants, with keys to all taxa, synonymy with references, literature citations, indication of localities with *exsiccatae* along with generalized summary of local and extralimital distribution and altitudinal zonation, critical commentary, and notes on habitat, special features, uses, etc.; citations of revisionary treatments under family and generic headings; index to all scientific names at end of each volume. As of 1980 two of the ten parts projected, covering pteridophytes and Compositae, had been published. Part 2, on pteridophytes, includes a general introduction to these classes of plants as well a short introduction to the work as a whole.

II. Tucumán

MEYER, T., VILLA CARENZO, M., and LEGNAME, P., 1977. *Flora ilustrada de la provincia de Tucumán.* Fasc. 1. 305 pp., 79 text-figs. Tucumán: Ministerio de

Cultura y Educación de la Nacion, Fundación 'Miguel Lillo'.

Semi-monographic descriptive flora, with full-page plates illustrating most species (one or more to a plate) and text including synonymy (with references), citations, vernacular names, indication of representative *exsiccatae*, general summary of Argentine and overall distribution, and notes on origin or occurrence, habitat, etc., along with critical remarks; references to family or generic revisions given under appropriate headings; index to all scientific and vernacular names at end. The introductory section is very brief and merely explanatory. This first of several projected fascicles covers twelve mainly 'sympetalous' families, although no specific systematic sequence is being followed.

Special group – trees

DIGILIO, A. P. L. and LEGNAME, P. R., 1966. *Los árboles indígenas de la provincia de Tucumán.* xxvii. 214, 29 pp., illus., maps (Opera lilloana, 15). Tucumán.

Descriptive account, with keys, limited synonymy (including some citations), vernacular names, local and extralimital range, illustrations, and notes on habitat, uses, etc.; list of references and indices at end.

384

West central Argentina

Includes the provinces of La Rioja, San Juan, and Mendoza. No general floras or enumerations for the subregion or any of the provinces appear to be available.

385

Central Argentina

Includes the provinces of San Luis and Córdoba. No general floras or enumerations for the subregion or either of the provinces are available, except for Seckt's aging *Flora cordobensis*. A new project, *Flora del Centro de Argentina*, is being elaborated at Córdoba as part of the INTA-series.

Bibliography

SPARN, E., 1938. *Bibliografia botánica de Córdoba.* 108 pp., 2 pls. (Revista Mus. Prov. Ci. Nat., Córdoba, 3). Cordoba. [Includes a systematic index to the literature.]

SECKT, H., 1929–30. *Flora cordobensis.* 632 pp., 22 pls. (Revista Univ. Nac. Córdoba, 16, 17). Cordoba.

Manual-key and enumeration of native, naturalized, and cultivated seed plants of Córdoba Province, with limited synonymy, vernacular names, general indication of local range, and notes on habitat, phenology, etc.; lengthy glossary; index to vernacular, family, and generic names. Species are enumerated under their respective genera but no keys to them are provided.

386

Buenos Aires

Corresponds with the limits of the province, thus including the greater part of the pampas. Cabrera's general flora was the first in the INTA regional floras program to be initiated and completed.

Bibliography

CABRERA, A. and FERRARIO, M., 1970. *Bibliografia botánica de la provincia de Buenos Aires: plantas vasculares.* 96 pp. La Plata: Comision de Investigaciones Científicas, Provincia de Buenos Aires.

CABRERA, A. (ed.), 1963–70. *Flora de la provincia de Buenos Aires.* 6 parts. Illus. (Colecc. Ci. INTA, 4). Buenos Aires: Librart.

Illustrated descriptive flora of native, adventive, and commonly cultivated vascular plants, with keys to all taxa, synonymy, references and citations, vernacular names, general indication of local range (with citations of some *exsiccatae*) and brief summary of extralimital distribution, some critical commentary, and miscellaneous notes; indices to all botanical and vernacular names in each part. A general introduction to the work appears in part 6.

Partial work: Buenos Aires metropolitan region

CABRERA, A. and ZARDINI, E. M. 1978. *Manual de la flora de los alrededores de Buenos Aires.* 755 pp., illus. Buenos Aires: Acme. (Originally publ. 1953 as *La flora des alrededores de Buenos Aires*, by Cabrera alone.)

Concise illustrated manual of vascular plants, with keys, synonymy, vernacular names, local range, and notes on habitat, etc.; index. An introduction includes general information and an introduction to floristics and descriptive terminology. [Description based on 1953 version. The new edition is considerably enlarged in length.]

387

La Pampa

No general floras or enumerations appear to be available for this south-central province. However, work is said to be in progress on a provincial flora within the INTA regional floras program.

388

Patagonian Argentina

Works relating to Patagonia in general (including the Chilean portions) are listed under **301/II**. The Argentine part is here taken to include the provinces and territories of Rio Negro, Neuquén, Chubut, Santa Cruz, and Tierra del Fuego. The fundamental modern work on the region is that by MACLOSKIE *et al.* (see **301/II**); Correa's modern *Flora patagónica*, one of the INTA regional floras, is limited to the Argentine provinces.

CORREA, M. N., 1969– . *Flora patagónica*. Parts 2, 3, 7. Illus. (Colecc. Ci. INTA, 8). Buenos Aires: Librart.

Illustrated descriptive flora of vascular plants, with keys to all taxa, synonymy, references and citations, vernacular names, citation of *exsiccatae* and general indication of local and extralimital range, critical remarks, and notes on habitat, etc.; indices to all botanical and vernacular names in each part. Encompasses Argentine Patagonia and the Islas Malvinas (Falkland Islands). At this writing (late 1979), parts 2, 3 and 7 (of a planned seven parts) have appeared, covering all the monocotyledons as well as the Compositae.

389

Falkland Islands (Islas Malvinas)

See also **388** (CORREA). These islands are administered by the United Kingdom but have long been claimed by Argentina.

MOORE, D. M., 1968. *The vascular flora of the Falkland Islands*. 202 pp., 24 text-figs. (incl. maps), 6 pls., 2 folding maps (Sci. Rep. British Antarctic Surv., 60). London.

Critical descriptive flora of vascular plants, with keys to genera and species; relevant synonymy, with references and citations; vernacular names; citation of *exsiccatae*, with localities, and general indication of local and extralimital range; taxonomic commentary and notes on habitat, special features, biology, etc.; diagnostic figures; glossary, bibliography, and indices to botanical and vernacular names. The introductory section includes descriptions of physical features, climate, and vegetation formations as well as an account of botanical exploration and analyses of floristics and phytogeography.

Partial work

Of the considerable literature on the Falklands flora, the following is exceptionally valuable for its illustrations.

VALLENTIN, E. F. and COTTON, E. M., 1921. *Illustrations of the flowering plants and ferns of the Falkland Islands*. xii, [65] pp., 64 color pls. London: Reeve.

Atlas of hand-colored plates of 72 native vascular plants, with descriptive notes and phenological information; index.

Region

39

Chile

This region corresponds to the political limits of Chile, except for its distant Pacific islands. For the Desventuradas and the Juan Fernández (Robinson Crusoe) group, see respectively **018** and **019**; for Rapa Nui (Easter Island), see **988**. The Chilean part of Patagonia, from Valdivia southwards, is also treated as **301/II**.

While important early observations, not published until 1782, were made by the Chilean Jesuit Molina before his expulsion with the order in 1768, serious botanical collecting in Chile begins with the 1782–3 visit of Dombey, who was attached, until 1784, to the 'Real Expedición Botánica' of Ruíz and Pavón to Peru and Chile. Other early collectors, all after 1815, included Gaudichaud, Bertero, Gay, and Cuming, those of Gay being perhaps the most important. Many collections were also made by expeditions 'rounding the Horn' or traversing the Straits of Magellan. The second half of the nineteenth century was dominated by the work of R. A. Philippi, the first leading resident botanist, active from 1851 to 1904 and essentially responsible for the establishment of the Santiago herbarium.

Gay's two long sojourns in Chile, which were devoted to natural history in general, resulted in a large illustrated encyclopedic work, *Historia física y política de Chile*, of which the eight-volume botanical part contained the first complete flora. This was followed by the *Catalogus* of F. Philippi (1881) and the incomplete flora of Reiche (1896–1911). The most important modern contribution is the one-volume *Sinopsis de la flora chilena* by the late C. Muñoz Pizarro (1959; 2nd edn., 1966), and more recently a flora of the Santiago district has been in process of realization by L. E. Navas-Bustamante (1973–9), with all the three volumes now (1979) published. There are, however, comparatively few other partial florulas except in Patagonia (with Fuegia); Fuegia was first covered by Hooker in his *Flora antarctica* (volume 2, 1845–7) and the whole treated by MACLOSKIE *et al.* in their *Flora Patagonica* (1903–14; see 301/II).

Chile is still in need of much detailed botanical exploration as well as improved documentation; only a beginning, albeit significant, was made by Muñoz Pizarro. It remains to be seen whether his projected five-volume 'generic' flora, in preparation as of 1976, will be realized. General collecting is extensive but uneven, and local institutions, though of long standing, have an inadequate representation of the flora. In a way, Muñoz's generic flora project was a recognition that at the present time a definitive descriptive work would be premature, as well as most likely beyond available resources.

Progress
For early history, see TURRILL, W. B., 1920. Botanical exploration in Chile and Argentina. *Bull. Misc. Inform.* (Kew), *1920*: 57–66, 223–4, and in Reiche's *Flora*, pp. 1–27.

Bibliographies. General and divisional bibliographies as for Division 3.

Regional bibliography
An extensive bibliography is given by Muñoz Pizarro in his *Sinopsis de la flora chilena* (see below).

Indices. General and divisional indices as for Division 3.

390

Region in general

The only complete descriptive flora is the one by Gay (1845–54); that by Reiche was never finished. The best modern introduction to the flora, with a valuable bibliography, is Muñoz Pizarro's *Sinopsis* (1966).

– Vascular flora: 5358 species listed by Philippi; present total around 7000 species.

GAY, C., 1845–54. *Historia física y política de Chile. Botánica (Flora chilena)*. 8 vols., atlas. 103 color pls., map. Paris: The author; Santiago de Chile: Museo de Historia Natural de Chile.

Descriptive flora of non-vascular and vascular plants, without keys; includes limited synonymy, references, vernacular names, citation of some *exsiccatae*, generalized indication of local and extralimital range, critical commentary, and notes on phenology, special features, etc.; lexicon of vernacular names, with botanical equivalents; general index to all botanical names in vol. 8. The introductory section (in vol. 1) includes an overall survey of the Chilean flora and its major features. Volumes 1–6 encompass the vascular plants and Characeae. Botanical text in Latin; commentary in Spanish.

MUÑOZ PIZARRO, C., 1966. *Sinopsis de la flora chilena (claves para la identificación de familias y generos)*. 2nd edn. 500 pp., 248 pls. (incl. 5 in color) Santiago: Ediciones de la Universidad de Chile. (1st edn., 1959.)

Illustrated introduction to the vascular flora, with keys to and descriptions of families and keys to genera; limited synonymy, with citations; indication of more important species in each family; vernacular names; full-page figures of representative species (particularly trees); glossary and extensive bibliography; indices to all vernacular and scientific names. An appendix includes a separate key to all known tree species. The work also encompasses Juan Fernández, the Desventuradas, and Easter Island.

PHILIPPI, F., 1881. *Catalogus plantarum vascularium chilensium adhuc descriptarum*. viii, 377 pp. Santiago: Ediciones de la Universidad de Chile. (Reprinted from *Anales Univ. Chile* **75** (1881).)

Systematic list of vascular plants and Characeae, with synonymy, references, and citations (the latter chiefly to Gay's *Flora*); index to genera and families. The introductory section includes a cyclopedia of collectors and others associated with the Chilean flora. Encompasses 5358 species.

REICHE, K. F. (1894–)1896–1911. *Flora de Chile*. Vols. 1–6(1). Santiago: Ediciones de la Universidad de Chile. (Reprinted in part from *Anales Univ. Chile* **88** (1894)–**116** (1905).)

Comprehensive descriptive flora of seed plants, with keys to genera and species, full synonymy (without references), citations of key source literature (Gay, Philippi, and others), generalized account of local range, and extensive taxonomic commentary; synopses of contents and indices to botanical names at end of each volume. Not completed; extends from Ranunculaceae through Chenopodiaceae, except for Cactaceae (Candollean system). For an extensive range of additions to the Chilean flora, see LOOSER, G., 1938. Catalogo de plantas vasculares nuevas de Chile. *Revista Univ.* (Santiago) **23**: 215–75. [List of about 500 species, described for the most part after 1918.]

Partial work: Santiago Plain

NAVAS-BUSTAMANTE, L. E., 1973–9. *Flora de la cuenca de Santiago de Chile*. 3 vols. Illus. (some in color), map. Santiago: Ediciones de la Universidad de Chile.

Students' manual-flora of vascular plants, with keys, brief descriptions, synonymy, references and key citations, vernacular names, general indication of local and Chilean range, and notes on habitat preferences, properties, uses, etc.; statistics (pp. 186–7), list of references (pp. 194–5), glossary, and indices in vol. 1 (other ones not seen by this reviewer). An introductory section (vol. 1) includes a map of the area covered (the environs of Santiago de Chile), floristic elements and kinds of vegetation, useful study areas, and technical and students' notes.

Prance, is also emphasized by Aubréville in a state-of-knowledge review of sub-Saharan Africa (*Adansonia*, sér. 2, **4**: 4–7, 1964).

4 Steyermark, J. A., 1977. Future outlook for threatened and endangered species in Venezuela. In *Extinction is forever* (ed. G. T. Prance and T. S. Elias), pp. 128–35. New York.

5 Gentry, A. H., 1978. Floristic knowledge and needs in Pacific tropical America. *Brittonia*, **30**: 134–53. According to Messrs. E. Forero and A. Gentry (personal communication, 1981) the mountains of Chocó are nearly always shrouded in cloud, and exceptionally heavy rainfalls, some over 10000 mm/year, have been reported.

6 *Idem*.

7 *Idem*.

8 An excellent review of environment and development in Amazonia is Goodland, R. J. A. and Irwin, H. S., 1975. *Amazon jungle: green hell to red desert?* ix, 155 pp., maps. Amsterdam: Elsevier.

9 On collectors in Amazonia see Prance, G. T., 1972. An index of plant collectors in Brazilian Amazonia. *Acta Amazonica*, **1**(1): 25–65.

10 On floristic education in Brazil, see Prance, G. T., 1975. Botanical training in Amazonia. *AIBS Education Review*, **4**: 1–4.

11 On Marggraf (or Marcgrave) see Whitehead, P. J. P., 1979. The biography of Georg Marcgraf (1610–1643/4) by his brother Christian, translated by James Petiver. *J. Soc. Bibliog. Nat. Hist.* **9**: 301–14.

12 On Velloso, see Stellfeld, C., 1952. *Os dois Vellozo: biografias de Frei José Mariano da Conceição Vellozo e padre doutor Joaquim Vellozo de Miranda*. 266 pp., portraits. Rio de Janeiro: Sousa.

13 Reitz, R. 1949. História da botânica catarinense. *Anais botânicos* [*Sellowia*] **1**: 23–110.

14 Indeed, there was no general science association until 1872, when the Sociedad Científica Argentina was formed.

15 See note by L. B. Smith in *Taxon*, **24**: 580 (1975). I also wish to acknowledge the assistance of T. Myndel Pedersen, a collaborator of the Botanical Museum and Herbarium, Copenhagen, and resident of Corrientes Province, Argentina, with choices for Region 38. Advice has also been received from E. M. Zardini of the Museo de La Plata.

Notes

1 Stearn, W. T., 1968. The plan of the 'Nova Genera'. In *Humboldt, Bonpland, Kunth and tropical American botany: a miscellany on the 'Nova Genera et Species Plantarum'*, (ed. W. T. Stearn), pp. 10–21; quotation, p. 12.

2 The educational value of skilfully illustrated florulas of limited areas is forcefully argued in two reviews, respectively by R. C. Barneby and T. B. Croat, of *Flora del Avila* (**318**) in *Brittonia*, **31**: 496–7 (1979). This contrasts strongly with the views of C. A. Backer regarding illustrations in his *Flora of Java* (**918**), although the Dutch author did make liberal use of them in some of his other works on Javan flora. Barneby notes that comparable works are in preparation for other parts of Middle and South America, thus filling long-empty shelves.

3 The lag in the knowledge of the tropical South American flora as compared with that of Africa, strongly pointed out by

Division

4

Australasia and islands of the Southwest Indian Ocean (Malagassia)

I always think that some of the other governments might have followed the example of yours. Out of the £250 I get, I have to pay £100 down to Reeve,...[and] I have much to pay in carriage of specimens from the continent, in postage and various minor expenses attending in the work, so that on the whole I scarcely clear £125 per volume, which is very poor pay for a 12-month hard work, after being nearly 40 years in the trade.

> Bentham to Mueller on *Flora australiensis*, 24 November 1864; quoted from Daley, The history of Flora australiensis, IV. *Victorian Naturalist*, **44**: 153 (1927).

It will afford me very sincere pleasure to see a beginning made [to a flora of Mauritius] during my residence here, as has been the case in regard to the two last colonies [Jamaica and Victoria] over which I have presided.

> Sir Henry Barkly, Governor of Mauritius, to the Royal Society of Mauritius, January 1864; quoted from Thistleton-Dyer, Botanical survey of the Empire; in *Bull. Misc. Inform.* (Kew), *1905*: 36 (1906). [Barkly was later to accomplish the same in his next charge, Cape Colony.]

Within this division are grouped several fragments of ancient Gondwanaland not readily placed elsewhere: New Zealand and surrounding islands (Region (Superregion) 41); the Australian continent with Tasmania (Superregion 42–45); and 'Malagassia'[1] or the islands of the southwest Indian Ocean (including Madagascar) (Superregion 46–49). New Caledonia and dependencies, although with some claim for inclusion here, have been placed in Division 9 (as Region 94) for practical and biohistorical reasons. Other Indian Ocean islands, including Christmas Island, are in Division 0.

A unifying feature of this otherwise fairly heterogeneous assemblage of islands and island-continents is the presence of peculiar floras with high endemism, odd life-forms and many relicts. The special features of Australasia are well known; but of Madagascar Commerson already in 1770 wrote that 'it seems that nature has retired to this spot, as to a special sanctuary, to work on forms other than those she has used elsewhere; at every step one comes upon the most unusual and most marvellous shapes'. However, the biohistories of New Zealand, Australia, and Malagassia have followed quite different courses and are thus here treated under the respective superregions.

General bibliographies. Bay, 1910; Blake and Atwood, 1942; Frodin, 1964; Goodale, 1879; Holden and Wycoff, 1911–14; Jackson, 1881; Pritzel, 1871–7; Rehder, 1911; Sachet and Fosberg, 1955, 1971; USDA, 1958. For other bibliographies, see under the regions and superregions.

General indices. BA, 1926– ; BotA, 1918–26; BC, 1879–1944; BS, 1940– ; CSP, 1800–1900; EB, 1959– ; FB, 1931– ; IBBT, 1963–9; ICSL, 1901–14; JBJ, 1873–1939; KR, 1971– ; NN, 1879–1943; RŽ, 1954– . For other indices, see under the regions and super-regions.

Conspectus

403

Alpine and upper montane zones

The flora of these zones, best expressed in Australia, Tasmania, and New Zealand, is of the highest botanical interest but until comparatively recently no significant separate accounts were available. These have now begun to appear in both Australia and New Zealand and are accounted for under the following two subheadings: I. New Zealand Alps; II. Australian Alps. Particularly outstanding is *Kosciusko alpine flora* (Costin *et al.*, 1979), which is at once popular and scientific. Floristic studies have also been made on the high mountains of Madagascar and Réunion, but no proper florulas have been published.

I. New Zealand Alps

Mark, A. F. and Adams, N. M., 1979. *New Zealand alpine plants*. 2nd. edn. 262, [iii] pp., 10 figs., 118 color pls., frontispiece, maps (end papers). Wellington: Reed (1st edn. 1973).

Illustrated guide in large format to vascular plants occurring above tree-line in New Zealand (varying from 900 m to 1450 m from south to north); the plates, with four to five plants each, accompanied by text containing short descriptions, indication of distribution, and notes on habitat, phenology, and locality of the voucher collection for the respective illustration. Altitudinal range is also given. An introductory section includes the definition of the area and coverage of its physical features, environment, general features of the flora and its elements, and the vegetation (with illustrations). A glossary and index appear at the end. The work is semi-scientific, with the arrangement systematic, but vernacular names are lacking. See also Salmon, J. T., 1968. *Field guide to the alpine plants of New Zealand*. 326, [i] pp., 477 color photographs, frontispiece, 2 maps. Wellington: Reed.

II. Australian Alps

Costin, A. B., Gray, M., Totterdell, C. J., and Wimbush, D. J., 1979. *Kosciusko alpine flora*. 408 pp., illus. (incl. 351 color pls.), tables, maps. Melbourne: Commonwealth Scientific and Industrial Research Organization, Australia; Sydney: Collins.

Lavishly illustrated, semi-technical ecologically oriented field guide to the alpine flora of the Kosciusko 'Primitive Area' (from 36° 20′ to 36° 35′S.) from 1830 m upwards, with systematic treatment including keys, descriptions, scientific names with references, literature citations, vernacular names, local and extralimital distribution, a few critical remarks, and notes on habitat, frequency, occurrence, biology, special features, etc.; bibliography (pp. 381–9), glossary, and complete index at end. The introductory section includes considerations of physical features, earth history, the delimitation of the alpine region, human impact and land use, and general features of the plant life: elements, life-forms, communities and vegetation formations, etc.; notes for the user (pp. 109–10). Although the actual area covered by this field guide is small (100 km²), it will be useful over much of the geographically limited Australian alpine country (located in southern New South Wales, Victoria, and Tasmania).

408

Wetlands

Although giving emphasis to Victoria (**431**), the aquatic plant flora by Aston (1973) accounts for the whole of Australia and Tasmania and is a particularly outstanding representative of its genre.

Australia (with Tasmania)

ASTON, H. I., 1973. *Aquatic plants of Australia.* xv, 368 pp., 138 text-figs. (incl. 4 maps), 81 distribution maps, 2 location maps, and 2 maps (end papers). Melbourne: Melbourne University Press.

Detailed descriptive treatment of herbaceous non-vascular and vascular macrophytes, with keys to all taxa, synonymy, vernacular names, indication of distribution (and origin, if appropriate), and notes on habitat, biology, variability, phenology, special features, etc.; some critical remarks. Distribution maps are provided only for species occurring in Victoria. The introductory section includes notes for the user and a glossary as well as a discussion of geography, climate, etc., related to aquatic plants and a consideration of the scope of coverage (mainly hydrophytes), while at the end are appendices on the menacing water-hyacinth (*Eichhornia crassipes*) and on sea-grass species. Covers 222 species (109 in Victoria). A supplementary leaflet appeared in 1978.

Region (Superregion)

41

New Zealand and surrounding islands

This region includes New Zealand proper (North, South, and Stewart Islands, and other offshore groups) and Lord Howe, Norfolk, the Kermadec and Chatham Islands, and the 'sub-Antarctic' islands to the south and southwest (except Macquarie).

Early New Zealand botanical exploration was accomplished through expeditions from Britain and France, beginning with Banks and Solander in the *Endeavour*; but until the 1820s almost all collecting was confined to the coasts. In that decade, inland field trips began with the visits of the Cunninghams. Norfolk Island was explored from the early 1800s, beginning with F. Bauer, while the 'sub-Antarctic' islands of Campbell Island and the Auckland group were first studied extensively by J. D. Hooker in 1841 while on the *Erebus* and *Terror* voyage. Other collectors visited the Chathams, the Kermadecs and Lord Howe Island, but the latter two groups, along with the Snares and the Antipodes, apparently remained poorly known until late in the nineteenth century. With the publication of *Flora antarctica* (Vol. 1, 1843–5) and *Flora novae-zelandiae* (1851–3) by Hooker, the primary stage of botanical survey in the region was largely completed.

From 1841, with the arrival of W. Colenso, there began the detailed 'secondary' phase of botanical exploration and 'alpha' taxonomy, which, aided by the appearance of Hooker's *Handbook of the New Zealand Flora* (1864–7), one of the 'Kew Series', led to Cheeseman's *Manual of the New Zealand Flora* (1906; 2nd edn., 1925), the first complete (and still much valued) 'indigenous' work. Detailed exploration was also carried out in most of the surrounding islands and island groups, including Norfolk and Lord Howe, with the main cycles of activity taking place early this century and from World War II and the 'Cape Expeditions' onwards. The first major monograph devoted to the southern islands was *The Subantarctic Islands of New Zealand* by C. Chilton (1909), with the botany by Cheeseman.

The current standard flora, which has a considerable 'biosystematic' content with great stress on hybridism and variability, is Allan's *Flora of New Zealand: Indigenous Tracheophyta* (1961–71), complete except for the Gramineae. This work extends coverage to Macquarie Island but omits Norfolk and Lord Howe Islands. Good separate coverage is also available for the woody flora. With the considerable number of local checklists and other treatments, the native vascular flora is about the best-known in the southern hemisphere. A biological flora and a series of distribution maps are in course of publication in *New Zealand Journal of Botany*.

Coverage of the outer islands, however, is uneven. Good modern treatments are available for the Kermadecs (Sykes, 1977), the Snares (Fineran, 1969) and the Aucklands (Johnson and Campbell, 1975), while more sketchy checklists exist for Norfolk (Turner *et al.*, 1968) and Campbell (Sorenson, 1951, with additions to 1975). However, no separate treatment is available for the Antipodes Islands, and the latest florulas of the Chathams and Lord Howe were published respectively in 1864 and 1917.

Progress

 COCKAYNE, L., 1967. *New Zealand plants and their story.* 4th edn., revised by E. J. Godley. Wellington, pp. 17–30; and the introductory section (pp. xv–xl) of the general flora by Cheeseman (1925). The latter is the more detailed. See also GLENN, R., 1950. *The botanical explorers of New Zealand.* 176 pp., illus. Wellington.

Bibliographies. General bibliographies as for Division 4.

Regional bibliography

 ALLAN, H. H. *et al.*, 1961–71. Annals of New Zealand botany. In *Flora of New Zealand: Indigenous Tracheophyta* (ed. H. H. Allen *et al.*), vol. 1, pp. xiii–xxxiv; vol. 2, pp. xv–xxxiv. [Chronologically arranged lists of references relating to New Zealand systematic botany, including subject index in volume 2.]

Indices. General indices as for Division 4.

Regional index

 STEENIS, C. G. G. J. VAN and JACOBS, M. (eds.), 1948– . Flora malesiana bulletin, 1– . Leiden: Flora Malesiana Foundation. (Mimeographed.) [Issued nowadays annually, this contains a substantial bibliographic section also rather thoroughly covering New Zealand floristic literature. A fuller description is given under Superregion 91–93.]

410

Region in general

 Although the handbook-flora by Allan *et al.* is the latest available for the New Zealand region, the section on grasses has yet to be published and for this and for other reasons Cheeseman's manual continues to be used. Neither work includes the now very conspicuous naturalized element in the flora, for which Allan produced a separate handbook in 1940. Certain works treating the woody flora have also been included.

ALLAN, H. H., 1961. *Flora of New Zealand*, 1: *Indigenous Tracheophyta.* liv, 1085 pp., 40 text-figs., 4 maps (end papers). Wellington: Government Printer.

 MOORE, L. B. and EDGAR, E., 1971. *Flora of New Zealand*, 2: *Indigenous Tracheophyta.* xl, 354 pp., 43 text-figs., 4 maps (end papers). Wellington.

 HEALY, A. J. and EDGAR, E., 1980. *Flora of New Zealand*, 3: *Adventive cyperaceous, petalous and spathaceous monocotyledons.* xlii, 220 pp., 31 text-figs., 4 maps (end papers). Wellington.

 Volumes 1 and 2 comprise a concise, well-documented descriptive flora of *native* vascular plants (except the Gramineae), with keys to all taxa; synonymy, with references; Maori and English vernacular names; indication of type localities; generalized indication of local (and extralimital) range, with more details for less common taxa; extensive notes on taxonomy, phenology, ecology, heteroblasty, hybridism, etc.; extensive bibliographies and, at the end, diagnoses of novelties, glossaries, addenda, and indices to all botanical and other names in each volume. In vol. 2 there is also a list of known chromosome numbers. The introductory sections in each volume include *inter alia* the bibliographies, lists of authors, a short account of floristic regions, and synopses of families. A special feature of this work is the presence of extended biosystematic discussions on individual genera and species (with suggestions for future research). The third volume comprises a treatment, in similar style and format, of part of the now-considerable *naturalized and adventive flora* (representing a change from past practice in New Zealand, as noted in the extensive introduction); additions to the bibliographic annals in volumes 1 and 2 are also included.

 For additions and name changes, see EDGAR, E., 1971. Nomina nova plantarum Novae-Zealandiae 1960–1969: Gymnospermae, Angiospermae. *New Zealand J. Bot.* 9: 322–30; and EDGAR, E. and CONNOR, H. E., 1978. Nomina nova.... 2, 1970–1976. *Ibid.*, 16: 103–18. Naturalized plants are also covered in the following: ALLAN, H. H., 1940. *A handbook of the naturalized flora of New Zealand.* 344 pp., illus. (New Zealand Div. Sci. Indust. Res. Bull., 83). Wellington.

 CHEESEMAN, T. F., 1925. *Manual of the New Zealand flora.* 2nd edn. xliv, 1163 pp. Wellington: Government Printer. (First edn., 1906.)

 Descriptive manual of native vascular plants, with keys to all taxa; synonymy and references; generalized indication of local (and extralimital) range, with some citations for less common or interesting species; vernacular names; brief notes on habitat, etc., sometimes given; list of naturalized plants; glossary and indices to all botanical and Maori names. The introductory section gives an account of botanical exploration. Associated with this work is an excellent selective atlas: CHEESEMAN, T. F. and HEMSLEY, W. B., 1914. *Illustrations of the New Zealand flora.* Vols. 1–2. 250 (251) pls. Wellington: Government Printer. [Features large plates (by Matilda Smith of Kew) of characteristic New Zealand plants, accompanied by descriptive text giving details of distribution, habitat, and natural history.]

Special group – woody plants (including trees)

COCKAYNE, L. and TURNER, E. P., 1967. *The trees of New Zealand.* 5th edn. 182 pp., 126 figs. (mostly halftones). Wellington: Government Printer. (First edn., 1928.)

Small atlas-guide to forest trees, comprising photographs of freshly collected material accompanied by description, indication of distribution, Maori vernacular names, and miscellaneous notes; genera alphabetically arranged. Appendices cover timber species, seedlings, hybridism, borderline species occasionally becoming treelike, and monocotyledon trees; a general introduction to forest vegetation is also given. Not significantly revised since 1939.

POOLE, A. L. and ADAMS, N. M., 1963. *Trees and shrubs of New Zealand.* 250 pp., illus., map. Wellington: Government Printer.

Briefly descriptive, semipopular illustrated manual of trees and shrubs, with simple keys to genera and species, Maori and English vernacular names, concise indication of distribution, habitat, and diagnostic features, and occasional other notes; glossary and complete general index. The introductory section includes chapters on vegetation, forest lands (and their utilization and management), and biotic factors. No synonymy is given, and all illustrations are based on drawings.

411

Lord Howe Island

The recent checklist by A. N. Rodd in the environmental study edited by Recher and Clark (1974) usefully supplements, but does not effectively supplant, Oliver's 1917 work.

Bibliography

PICKARD, J., 1973. An annotated botanical bibliography of Lord Howe Island. *Contr. New South Wales Natl. Herb.* 4(7): 470–91.

OLIVER, W. R. B., 1917. The vegetation and flora of Lord Howe Island. *Trans. & Proc. New Zealand Inst.* **49**: 94–161, 3 figs., pls. 10–16.

Contains an enumeration of the known vascular plants, with limited synonymy, general indication of local and extralimital range (with citation of older records and mention of some specific localities), and notes on habitat, frequency, and life-form. In the introduction are sections on physical features, phytogeography, and vegetation formations of the island, as well as on geological history and previous botanical work.

RECHER, H. F. and CLARK, S. S. (eds.), 1974. *Environmental survey of Lord Howe Island.* viii, 86 pp., illus. (some in color), maps, graphs. Sydney: New South Wales Government Printer.

Appendix B (pp. 21–5) contains a vascular plant species checklist (by A. N. Rodd), with indication of status (naturalized, indigenous, endemic). This is followed by a vegetation account by J. Pickard (Appendix C) which features a coloured map. Other pertinent sections are notes on weeds (Appendix H) and a selected bibliography (Appendix I). About 180 species considered indigenous, with a high proportion of endemics.

412

Norfolk Island

A fuller treatment of this island's flora in a more accessible and perhaps semi-popular form, with closer reference to New Zealand and New Caledonia, would be desirable.

TURNER, J. S., SMITHERS, C. N. and HOOGLAND, R. D., 1968. *The conservation of Norfolk Island.* 41 pp., 12 halftones, map (Australian Conservation Found., Spec. Publ., 1). Canberra.

Contains on pp. 29–37 an enumeration of the known vascular plants (by R. D. Hoogland), with notes on local distribution, frequency, and habitat, vernacular names, and some taxonomic commentary. The main part of the work (pp. 1–28) deals with the survey report, problems of conservation, and recommendations; a short list of references is appended. Partly supplants LAING, R. M., 1915. A revised list of the Norfolk Island flora, with some notes on the species. *Trans. & Proc. New Zealand Inst.*, 47: 1–39.

413

Kermadec Islands

SYKES, W. R., 1977. *Kermadec Islands flora: an annotated check list.* 216 pp., 48 halftones, 9 tables (New Zealand Div. Sci. Indust. Res., Bull., 219). Wellington.

Enumeration of native, naturalized, and adventive vascular plants (195 spp., of which 113 are considered native), with commonly used synonyms, vernacular names, status, local distribution in the group, and occurrence, habitat preferences and associates; location of collections; table of local distribution (pp. 187–91), list of references, and complete index at end. The general part covers physical features, climate, introduced biota, human activities, vegetation and phytogeography, history of botanical work, and floristic statistics. Musci and Lichenes also are fully treated, but Hepaticae and Anthocerotae are merely listed in an appendix. Supersedes OLIVER, W. R. B., 1910. The vegetation of the Kermadec Islands. *Trans. & Proc. New Zealand Inst.* 42: 118–75, pls. 12–23, map.

414

Chatham Islands

It is most unfortunate that no more recent flora or checklist is available for this interesting and remote group. The practically obsolete Mueller work and the Buchanan list, both over 100 years old, are largely of historical interest, and subsequent information is rather scattered.

MUELLER, F., 1864. *The vegetation of the Chatham-Islands.* 86 pp., 7 pls., Melbourne: Victorian Government Printer.

Descriptive flora of vascular plants, with synonymy, references, and citations; vernacular names; extensive notes on habitat, phenology, biology, and taxonomy; generalized indication of local (and extra-limital) range. The rather quaint introduction includes remarks on the general features of the flora as well as

a panegyric on the need for field studies. Mueller's species concept in this work was rather broad; for a somewhat updated list (205 as against 87 species), based on further collecting and narrower species limits, see BUCHANAN, J., 1875. On the flowering plants and ferns of the Chatham Islands. *Trans. & Proc. New Zealand Inst.* 7: 333–41, pls. 12–15. [Partly annotated, with descriptions of new or little-known species and vernacular names.] For further additions, see COCKAYNE, L., 1902. A short account of the plant-covering of Chatham Island. *Ibid.*, 34: 243–325. [Mainly vegetatiological.]

415

Islands to the south of New Zealand (in general)

See also **080–90** (SCHENCK). For works relating specifically to Macquarie Island, see **081**. Much recent work on the flora and vegetation of these islands, which (except in part for Campbell) are not truly sub-Antarctic, has been accomplished during and since World War II so that Cheeseman's fundamental account is in large part superseded (apparently not, however, for the Antipodes group).

CHEESEMAN, T. F., 1909. On the systematic botany of the islands to the south of New Zealand. In CHILTON, C. (ed.), *The subantarctic islands of New Zealand*, vol. 2, pp. 389–471. Wellington: Philosophical Institute of Canterbury/Government Printer.

Systematic enumeration of vascular plants (except Gramineae), with synonymy, references, and literature citations; extensive remarks on habitat, life-form, biology, and local distribution as well as taxonomic commentary; general indication of wider range (in the first instance by island or island group). The concluding section gives a tabular summary of geographical ranges, with columns for the individual island units, and a summary of the overall affinities of the flora and its presumed historical origins. In the introductory section, *inter alia*, is an account of botanical exploration in these islands. For a recent treatment of Gramineae, superseding that by Petrie which originally accompanied the above list, see ZOTOV, V. D., 1965. Grasses of the subantarctic islands of the New Zealand region. *Rec. Domin. Mus.* 5: 101–46, 4 text-figs., 4 pls.

416

Antipodes and Bounty Islands

No separate floras or lists of recent date appear to have been published, and reference must be made to the works of CHEESEMAN and ZOTOV (**415**) or standard New Zealand floras. The land area of the Antipodes is quite limited, while the Bounty Islands are mere rocks, without vascular plants.

417

Snares Islands

See also **415** (CHEESEMAN, ZOTOV).
FINERAN, B. A., 1969. The flora of the Snares Islands, New Zealand. *Trans. Roy. Soc. New Zealand, Bot.* **3**: 237–70, 5 pls.

Enumeration of known cellular and vascular plants, with synonymy, references, and citations; *exsiccatae* listed, together with previous records in literature; extensive remarks on ecology and biology. The concluding section includes a phytogeographic and floristic summary, together with a table of extralimital distribution and a list of references.

418

Auckland Islands

See also **415** (CHEESEMAN, ZOTOV).
JOHNSON, P. N. and CAMPBELL, D. J., 1975. Vascular plants of the Auckland Islands. *New Zealand J. Bot.* **13**: 665–720, 5 figs. (incl. map).

Annotated systematic enumeration of vascular plants (257 spp.), with synonymy, localities, critical commentary, and notes on status, occurrence, special features, etc.; discussion of species of limited occurrence and uncertain status; species to be excluded from the flora and adventives; analysis of past and possible future history of the flora in relation to introduced animals and their possible increase or elimination; list of references. A brief general part accounts for past and current work in the island group and the bases for knowledge. [The biogeographical analysis in this paper is exceptionally interesting.]

419

Campbell Island

See also **415** (CHEESEMAN, ZOTOV). A new flora for this near sub-Antarctic island, now with 62 vascular plant species recorded, is in preparation (Meurk, 1975; see below).

SORENSON, J. H., 1951. Botanical investigations on Campbell Island, II: an annotated list of the vascular plants. *New Zealand Div. Sci. Indust. Res., Cape Exped. Ser. Bull.* **7**: 25–38.

Briefly annotated systematic list of native and introduced vascular plants collected between 1941 and 1947; for each species are given notes on localities, occurrence, frequency, habitat, and general range within the island, but no extralimital distribution is indicated. Section A of the list covers native species; Section B, the introductions. For additions, see GODLEY, E. J., 1969. Additions and corrections to the flora of the Auckland and Campbell Islands. *New Zealand J. Bot.* **7**: 336–48, 2 figs.; and MEURK, C. D., 1975. Contributions to the flora and plant ecology of Campbell Island. *Ibid.* **13**: 721–42.[2]

Superregion

42–45

Australia (with Tasmania)

Included here are the Australian continent and Tasmania together with immediately adjacent islands, the Torres Strait Islands, and the Ashmore and Cartier Islands in the Timor Sea. For Lord Howe and Norfolk Islands, see Region 41; for the Coral Sea Islands (Territory), see Region 96; for the Cocos (Keeling) group and Christmas Island, see Region 04; and for Macquarie Island and the Macdonald (Heard) group, see Region 08. The boundary with Papuasia (Region 93) is formed by the recently negotiated seabed boundary between Australia and Papua New Guinea through the Torres Strait and north of the Great North East Channel, thus here incorporating (as 439) all islands except those immediately off the Papuan coast.

The flora of this smallest but botanically most interesting of continents has excited botanists since the landfall of Banks and Solander in the *Endeavour* at Botany Bay in 1770. Through the subsequent fieldwork and systematic studies of Robert Brown and other workers, it also contributed greatly to the development of 'natural' classifications of angiosperms in the first half of the nineteenth century and their ouster of the Linnean sexual system. After 1810, the year of publication of Brown's epochal but loss-making *Prodromus florae novae hollandiae*, residential explorations began in earnest and from then on widened more or less steadily from the main centres of settlement, partly through expeditions and partly through the activities of amateur and professional collectors. From 1847, there developed a residential research botanical capability with the settlement of Ferdinand (later Baron Sir Ferdinand von) Mueller, who for close to half a century was to dominate Australian botany in much the same way as did his near-contemporary Asa Gray in North America. With the establishment of the Kew colonial floras scheme at the end of the 1850s, his writings and collections, as well as those of nearly all others, were drawn upon by Bentham – albeit not without some friction on the part of Mueller who had entertained a similar ambition – for his *Flora australiensis* (1863–78), a work which is the longest one-man

flora in existence and by general consent one of the best. However, even the finest flora is out of date from the day of its publication and Bentham's 8125 species of vascular plants are now considered to represent less than half of the probable total (18 000 species). A supplementary volume was more than once called for, but the continuing flow of novelties rendered this into a receding dream of overloaded state botanical officers and materials said to have been accumulated by Mueller for such a work disappeared after his death in 1896 and have never been found. However, at Bentham's behest, Mueller did produce his *Systematic census of Australian plants, with chronologic, literary and geographic annotations* (1882; 2nd edn., 1889) of which only the part on vascular plants was ever completed.[3]

With the passing of Mueller came decades of taxonomic twilight and fragmentation, pierced only here and there by a few rays of energy. Federation in 1901 did not bring unity but rather the opposite: efforts went into state herbaria and in the fifty years preceding World War II most of the standard state and territory floras and enumerations, in part still current, were published. At first largely based on *Flora australiensis* with the addition of later discoveries, these works began to feature significant original research starting with Black's *Flora of South Australia* in the 1920s. This array, however, gave in some quarters a misleading impression that the continent was relatively well documented; for instance, Blake and Atwood in the introduction to their *Geographical guide to floras of the world* (1942, p. 10) noted that 'the flora as a whole can be regarded as more satisfactorily covered by published works than that of any equally extensive division of the earth's surface except Europe'. That this was far from the truth, and that in particular *Flora australiensis* was becoming obsolete, began to be perceived seriously only after World War II, although J. H. Maiden had urged the writing of a new work as early as 1907.

Proponents of state floras nevertheless continued for some time to be in the ascendant and until mid-century or even later Australian systematic botany remained as a kind of knobbed wheel without a hub. Not until 1959 was the first formal proposal for a new Australian flora made. However, years of discussion, both ideological and technical, and fruitless proposals and negotiations then followed and only in 1973, in the face of steadily mounting local demands for such a work as well as external pressure, did preliminary investigations towards a new flora along with compilation of source materials definitively begin with support from the Australian Academy of Science and a private benefactor. Sponsorship was later assumed by the Australian Biological Resources Study of the federal government, which in 1978–9 became formally established as the 'Bureau of Flora and Fauna' with among its briefs the preparation of a national flora along lines similar to those suggested in 1959.[4] The *Flora of Australia* project was thus assured, although the concept of a semi-monographic work as originally envisaged had for practical and temporal reasons to give way to one aiming at a concise format and 'based largely on current knowledge' (George, p. 10). The first of the projected 48 volumes covering vascular plants appeared at the Thirteenth International Botanical Congress in Sydney in 1981, containing general chapters; later volumes have been numbered on the basis of the latest version of the Cronquist system (for flowering plants). The obstacles, though, remain great: the size of the flora (as noted above) with its many species still be be described, the amount of basic taxonomic work to be done, current standards of documentation, and the still limited number of specialists in various isolated centers with mainly regional collections will all make full realization a daunting task.[5] Preliminary contributions have included *Dictionary of Australian plant genera* by N. T. Burbidge (1963) and the *Australian plant name index* of the Bureau of Flora and Fauna of which publication on microfiche commenced in 1980. A work which may be regarded as a test of the cooperation required for a successful continental flora is the recently published *Flora of Central Australia* (1981) edited by J. Jessop with contributions from over 50 collaborators; this covers the central third of the continent.

Botanical work has been active in all states (and territories), particularly in recent decades, and the Australian north, much of it difficult of access, is becoming botanically better explored although there is a lot still to be done. Except in the southeast and in patches elsewhere, however, knowledge of the continent is still considered inadequate and moreover poorly documented. The several new floras and checklists for states or parts thereof which have been published since the 1950s, with more expected in the coming decade, are not a substitute for a national flora; for among other problems affecting Australian floristic botany has been a change of scientific names at state boundaries for many a given species, whether realized or not.

– Vascular flora: 18 000 species (George).

Progress

For early history see BAILEY, F. M., 1891. A concise history of Australian botany. *Proc. Roy. Soc. Queensland* 8(2): xvii–xli, xlv–xlvii; and the many papers by J. H. Maiden dating from 1907 to 1921 and cited in the *Geographical Guide* (Blake and Atwood, 1942). These have been complemented by CARR, D. J. and CARR, S. G. M. (eds.), 1981. *People and plants in Australia.* xxi, 416 pp., illus. Sydney. The current state of knowledge and recent progress towards the new *Flora of Australia* are described in ANON., 1979. Surveying Australia's plants and animals. *Ecos*, 21: 28–31; BLAKE, S. T., 1960. A new flora of Australia. *Austral. J. Sci.* 23: 173–6; BURBIDGE, N. T., 1974. Progress towards a new flora of Australia. *Annual Rep. Div. Plant Industry, CSIRO, Australia 1973*: 31–4; DEPARTMENT OF SCIENCE, AUSTRALIA, 1979. Australian Biological Resources Study 1973–8 (Commonwealth Parliamentary Paper 354/1978). Canberra; GEORGE, A. S., 1981. The background to the *Flora of Australia.* In *Flora of Australia* (ed. Bureau of Flora and Fauna, Canberra), vol. 1, pp. 3–24, illus. Canberra; and RIDE, W. D. L., 1978. Towards a national biological survey. *Search*, 9: 73–82.

General bibliographies. Bay, 1910; Blake and Atwood, 1942; Frodin, 1964; Goodale, 1879; Holden and Wycoff, 1911–14; Jackson, 1881; Pritzel, 1871–7; Rehder, 1911; Sachet and Fosberg, 1955, 1971; USDA, 1958.

Supraregional bibliography

STEENIS, C. G. G. J. VAN, 1955. Annotated selected bibliography. In *Flora malesiana*, ser. I (Spermatophyta) (ed. C. G. G. J. van Steenis), vol. 5, pp. i–cxliv. Groningen: Noordhoff. (Reprinted separately.) [Provides detailed coverage of the literature for Malesia and adjacent areas; here and there briefly annotated. The arrangement is by areas as well as by families and genera.]

General indices. BA, 1926– ; BotA, 1918–26; BC, 1879–1944; BS, 1940– ; CSP, 1800–1900; EB, 1959– ; FB, 1931– ; IBBT, 1963–9; ICSL, 1901–14; JBJ, 1873–1939; KR, 1971– ; NN, 1879–1943; RŽ, 1954– .

Supraregional indices

FERGUSON, I. K., 1970–2. *Index to Australasian taxonomic literature for 1968–70.* 3 nos. (Regnum Veg., 66, 75, 83). Utrecht. [Annual index to floristic, systematic and biohistorical literature of Australasia, Oceania and Papuasia. Superseded by *Kew Record*].

STEENIS, C. G. G. J. VAN and JACOBS, M. (eds.), 1948– . Flora malesiana bulletin, 1– . Leiden: Flora Malesiana Foundation. (Mimeographed.) [Issued nowadays annually, this contains a substantial bibliographic section also very thoroughly covering the Australian literature. A fuller description is given under Superregion 91–93.]

420–50

Superregion in general

It is only as this book went to press that the first volume of the new *Flora of Australia*, edited within the recently-constituted Bureau of Flora and Fauna, Canberra, was published (1981, Canberra: Australian Government Publishing Service); its projected 48 volumes (for vascular plants) will gradually succeed the classic *Flora australiensis* of Bentham, long the standard work but now very outdated. The various state and territory floras published in the intervening century or more have only been partial substitutes, with their coverage largely centered in the southeastern 'boomerang'; for much of the Australian west and north Bentham's work has remained the only useful treatment available. The treatment of pteridophytes has been usefully supplemented by the semipopular *Australian ferns and fern allies* by Jones and Clemensha (1976; 2nd edn., 1980). Attention should also be drawn to an early precursor of the *Flora of Australia*, the *Dictionary of Australian plant genera* by Burbidge (1963). 1981 also saw the first consolidated modern account of the flora of the arid interior, the *Flora of central Australia* prepared under the auspices of the Australian Systematic Botany Society and edited by J. P. Jessop. Supporting all current floristic work in Australia is another project of the Bureau of Flora and Fauna, the *Australian Plant Name Index*, of which publication (on microfiche) commenced in 1980, and a comprehensive taxonomic bibliography is being planned as of this writing.

Dictionaries

BURBIDGE, N. T., 1963. *Dictionary of Australian plant genera* (gymnosperms and angiosperms). xviii, 345 pp., 2 maps. Sydney: Angus & Robertson. [Annotated enumeration of seed-plant genera known from Australia, Tasmania, and Lord Howe and Norfolk Islands, including synonymy, citations, and indication of distribution. Does not cover pteridophytes, and is now considered to be in need of revision.]

BENTHAM, G., 1863–78. *Flora australiensis*. 7 vols. London: Reeve. (Reprinted 1967, Amsterdam, Asher; Brook, Ashford, Kent, England: Reeve.)

Concise descriptive flora of vascular plants, with keys to genera and species; synonymy briefly indicated, with references; detailed indication of internal range (by state and territory), with citation of *exsiccatae*, and summary of extralimital distribution; concise, informative critical commentary; notes on habitat, special features, etc. based on available information; index to all botanical names in each volume (cumulative index lacking). In the introductory section (vol. 1) are chapters on botanical exploration and sources of information, while the concluding postscript (vol. 7) consists of general remarks on the flora and its composition. The work forms part of the 'Kew Series' of British colonial floras initiated by W. J. Hooker. The 'assistance' of F. von Mueller, indicated on the title-pages, included the loan of Australian holdings in the National Herbarium, Melbourne, and much correspondence; however, actual collaboration was out of the question, as, according to Bentham, 'four months were required for an answer to the smallest query' – though in fact there was rather more to it than that. The genesis and preparation of this *Flora* are recounted in DALEY, C., 1927–8. The history of *Flora australiensis*, I–VIII. *Victorian Nat.* 44(3–10): 63–74, 91–100, 127–38, 153–64, 183–7, 213–21, 248–56, 271–8, as well as by STAFLEU, F. A., 1967. The *Flora australiensis*. *Taxon*, 16: 538–42.

MUELLER, F. VON, 1889. *Second systematic census of Australian plants*, I: *Vasculares*. 244 pp. Melbourne: Government of Victoria. (First edn., 1882.)

Tabular systematic list of known vascular plant species, with references as appropriate to Bentham's *Flora australiensis* (see above), the author's own *Fragmenta phytographiae australiae* (12 vols. and suppl., 1858–82, Melbourne), and other works; indication of geographical distribution by state (then colony) and territory; index to generic names at end. The work met with a lukewarm reception in some quarters at the time of original publication and is now chiefly of historical interest.

Partial work

Although it has not been possible to include a full account in the *Guide* at this time, mention should be made of the following important work, which covers all the arid interior parts of Australia: JESSOP, J. P. (ed.), 1981. *Flora of central Australia*. Illus. Sydney: Reed.

Special group – ferns and fern-allies

JONES, D. L. and CLEMENSHA, S. C., 1976. *Australian ferns and fern-allies*. 294 pp., 253 text-figs., 60 color photographs. Sydney, Wellington: Reed.

Well-illustrated, semi-popular descriptive treatment, without keys, of pteridophytes of the Australian continent and Tasmania (312 spp. in 101 genera), with notes on distinguishing and other special features, vernacular names, indication of geographical range, and notes on cultivation, with extensive commentaries on genera and their distribution included; glossary, list of references (pp. 286–7) and complete index at end. Four appendices encompass naturalized species, an alphabetical synopsis, a systematic lexicon (Holttum scheme), and synonymy, while the general part provides an introduction to ferns and their cultivation. Arrangement of genera in the main text is alphabetical within the Filicinae. The work, despite its popular orientation, is the only scientifically based modern review of the group for Australia as a whole. A second revised edition appeared in 1981. For keys see CLIFFORD, H. T. and CONSTANTINE, J., 1980. *Ferns, fern allies and conifers of Australia*. xvii, 150 pp., 24 figs., tables. Brisbane: University of Queensland Press.

Region

42

Tasmania

This region is considered to include the main island of Tasmania, adjacent small islands, and the Bass Strait islands of King and Flinders with their islets. Macquarie Island, administratively part of the state of Tasmania, is, however, botanically sub-Antarctic and has been included in region 08.

The primary stage in the exploration of the peculiar Tasmanian flora was concluded with the publication of J. D. Hooker's splendid *Flora Tasmaniae* in 1860, which itself reviewed botanical exploration in the what was then a colony (and in Australia generally) up to that time. From the mid-nineteenth century, collecting and observation were continued largely by

resident botanists, including many amateurs. In response to local demand for a handbook more convenient than the bulky *Flora Tasmaniae*, the first somewhat cramped field manual, *A handbook of the plants of Tasmania* by W. Spicer, appeared in 1878. This was succeeded by Rodway's *The Tasmanian flora* (1903), still a standard work although now long outdated.

Recent decades have witnessed the publication of a new state flora (yet to be completed) by Curtis, *The student's flora of Tasmania* (1956– ; 2nd edn., 1975–), which in the fashion of Allan's *Flora of New Zealand* (**410**) highlights taxonomic problems, and the magnificent *Endemic flora of Tasmania* (6 vols., 1965–78) by the same author, with over 150 paintings of peculiar and colorful species by M. Stones, one of the finest modern representatives of the 'Great Flower Book' tradition. The state is at present considered moderately well-collected and documented, although less so in the southwest, much of which is remote and difficult of access. A stimulus to more detailed study was created with the reorganization of the state herbarium in the mid-1970s.

Bibliographies. General and supraregional bibliographies as for Superregion 42–45.

Indices. General and supraregional indices as for Superregion 42–45.

420

Region in general

The recent flora by Curtis – not really a students' manual in the sense used in this Guide, but a conventional descriptive flora – has yet to be completed, and it lacks the convenient one-volume format of Rodway's manual. The *Endemic Flora of Tasmania* is another key source of information and illustrations.

CURTIS, W. M., 1956– . *The student's flora of Tasmania*. Parts 1–3, 4A. xlvii, 661 pp., 138 figs. (parts 1–3); 138 pp., illus. (some in color) (part 4A). Hobart: Tasmanian Government Printer. (Part 4A published 1980.)

CURTIS, W. M. and MORRIS, D. I., 1975. *The student's flora of Tasmania*. 2nd edn. Part 1. Illus. Hobart.

Illustrated descriptive flora of native and naturalized seed plants, with keys to all taxa; includes limited synonymy, vernacular names, generalized indication of local range, brief taxonomic commentary, and a few notes on other aspects; glossary (in part 1) and indices to all names in each part. Parts published to date (January, 1980) cover families from Ranunculaceae through Salicaceae and (in part 4A) Orchidaceae (Bentham and Hooker system), the last part with many colored illustrations. The first part of the second edition is essentially a somewhat expanded version of the first, covering Ranunculaceae through Myrtaceae.

RODWAY, L., 1903. *The Tasmanian flora*. xix, 320 pp., 50 pls. Hobart: Government Printer.

Briefly descriptive manual of vascular plants, with keys to all taxa; limited synonymy; generalized indication of local and extralimital range; phenology; glossary and index to all botanical names. Partly superseded by Curtis's *Students' flora* but handier for field use.

Special group – endemic plants

STONES, M. and CURTIS, W. M., 1965–78. *The endemic flora of Tasmania*. 6 vols. 478 pp., 155 col. pls., maps (end papers). London: Ariel.

Not a flora in the strict sense, but a sumptuous, large-scale modern 'Great Flower Book', containing 155 plates by Margaret Stones depicting the more colorful and characteristic endemic plants of the state, each with accompanying descriptive text by W. M. Curtis bearing details of form, technical details, biology, ecology, and special features; general index in vol 6. The arrangement is not systematic, but a lexicon is given in vol. 6. The introductory section, in vol. 1, includes, *inter alia*, an account of Tasmanian vegetation by the associate author. Notes on cultivation have been added by the sponsor, the late Lord Talbot de Malahide.

Region

43

Eastern Australia

Within this region are found the states of Victoria, New South Wales and Queensland, the Australian Capital Territory (here treated as part of New South Wales), and the Torres Strait Islands south of the Australia/Papua New Guinea seabed boundary. Under **430** are described *Beiträge zur Flora und Pflanzengeographie Australiens* by K. Domin and

Australian rain-forest trees by W. D. Francis; despite their titles, these works are limited almost entirely to eastern Australia. Both provide valuable supplementation for the (in part) obsolete state manuals.

As is well-known, botanical exploration in eastern Australia began with the landfall of the *Endeavour* at Botany Bay, now enveloped within metropolitan Sydney. Early collecting took place mostly near the coast, with a substantial contribution coming from R. Brown in the *Investigator*, but from approximately the end of the Napoleonic Wars serious interior exploration began, first in New South Wales proper but later extending to what are now Victoria and Queensland. Most collections, however, went overseas until the establishment of the first major local herbarium by F. Mueller at Melbourne in the 1850s.

From then onwards, with the spread of settlement and inland transportation networks, floristic exploration was accomplished increasingly by resident botanists, and locally produced floras began to appear, with a major stimulus provided by the completion of *Flora australiensis*. The first of these was Mueller's *Plants indigenous to the colony of Victoria* (1860). In the last two decades of the nineteenth century, there appeared a number of 'state floras', some still not superseded: *Key to the system of Victorian plants* (Mueller, 1886–8), *Handbook of the flora of New South Wales* (Moore and Betche, 1893), *Synopsis of the Queensland flora* (Bailey, 1883, with supplements), and finally *The Queensland flora* (Bailey, 1899–1902). Since then, the only state-wide manuals have been produced in Victoria (Ewart, 1931; Willis, 1962–72) although checklists appeared for all states in the first third of the present century. The same period also saw the production of Maiden's *Forest flora of New South Wales* (1902–25), one of a number of large-scale works by this energetic author, and a number of 'tree books', among them the still-current *Trees of New South Wales* (Anderson, 1932; 4th edn., 1968) and *Australian Rain-Forest Trees* (Francis, 1929; 3rd edn., 1970), the latter now very incomplete, especially for North Queensland. A valuable contribution is also Domin's *Beiträge* (1915–29), noted above.

The first of a new generation of state floras began with the initiation of the *Flora of New South Wales* series in 1961. This, however, is an open-ended semi-monographic work, and all of the more recent current and projected handbooks have been, or are being, written for infra-state units. The three current ones are in New South Wales: *Flora of the Sydney Region* (Beadle *et al.*, 1963; 2nd edn., 1972); *Flora of the Australian Capital Territory* (Burbidge and Gray, 1970); and *Students' flora of North Eastern New South Wales* (Beadle and Beadle, 1971–). In Queensland, multivolume handbooks have been projected for each of four subregions, with the first part of that on southeastern Queensland expected shortly as of this writing (1980). *A Handbook to Plants in Victoria* by Willis has been noted above.

The present state of floristic knowledge and documentation is thus variable, ranging from moderately good in the south to poor in the north, particularly in the rainforest and monsoon-forest areas. Indeed, in much of the north *Flora australiensis* is all that can be used.

Bibliographies. General and supraregional bibliographies as for Superregion 42–45.

Indices. General and supraregional indices as for Superregion 42–45.

430

Region in general

The region is too large and diverse to have been covered by a specific flora; however, two important works covering different aspects of the flora, especially in northern New South Wales and Queensland, should be accounted for here.

DOMIN, K., (1914–) 1915–29 (–1930). *Beiträge zur Flora und Pflanzengeographie Australiens. Teil I. Systematische Bearbeitung des eigenen sowie auch fremden besonders des von Frau Amalie Dietrich in Queensland (1863–73) und von Dr Clement in Nordwest-Australien gesammelten Materiales mit teilweiser Berücksichtigung der gesamten Flora Australiens.* Abt. 1–3 (in 12 parts). 1317 pp., 207 text-figs., 38 pls. (some in color) (Biblioth. Bot. 20 (Heft 85: Lfg. 1–4) (1913–15): 1–554, figs. 1–117, pls. 1–18; 22 (Heft 89: Lfg. 1–8) (1921–9): 1–763 (555–1317), figs. 118–207, pls. 19–38.) Stuttgart: Schweizerbart.

Systematic enumeration of vascular plants (covering all families represented), based in large part on the extensive collections (mainly from Queensland) made by the author in 1909–10 but also including records of other collectors, notably A. Dietrich and E. Clement; includes descriptions of new taxa, full

synonymy (with references and citations), localities with citation of *exsiccatae*, summary of distribution, and often extensive taxonomic and phytogeographical commentary as well as miscellaneous notes; complete index to botanical names at end of work (pp. 1245–1317). The *Beiträge* collectively comprise a work of particular importance for Queensland botany, the most important following the publication of the flora and checklist of Bailey.[6]

Special group – rainforest trees

The only purportedly general guide to this imperfectly known class of trees is now sorely in need of substantive revision. Some partial works of more recent date have been, or are being published, in both Queensland and New South Wales, but much basic taxonomic work remains to be done, especially in the north where many species remain to be named and described.

FRANCIS, W. D., 1970. *Australian rain-forest trees.* 3rd edn., revised G. Chippendale. xvi, 468 pp., 270 text-figs., frontispiece, 4 maps. Canberra: Australian Government Publishing Service. (First edn. 1929, Brisbane; 2nd edn., 1951, Sydney.)

Descriptive treatment of the more important rainforest tree species in eastern Australia, especially in northeastern New South Wales and southeastern Queensland; keys to extratropical species; limited synonymy, with occasional references and citations; vernacular names; generalized indication of distribution, with some localities; notes on special features, wood, etc.; index to all botanical and vernacular names. The introductory section includes subsections on physical features, vegetation, forest structure, plant morphology, etc. The so-called third edition differs from the second only with respect to nomenclature. Because of the imperfect coverage of this work, especially in northern Queensland, the following recent contributions for more limited areas should be consulted: FLOYD, A. G. *et al.*, 1960–

. *New South Wales rain forest trees.* Parts 1– . Illus. (Forestry Comm. New South Wales Res. Note, 3, *passim*) [Sydney]. [Comprises regional revisions of tree families, with artificial keys and diagnostic figures. By 1981 ten parts had been published.] Also HYLAND, B. P. M., 1971. *A key to the common rain forest trees between Townsville and Cooktown based on leaf and bark features.* 103+ pp., illus.; 93 punch cards in separate packet. Brisbane: Department of Forestry, Queensland. [Identification guide to some 508 species, with some notes on distinguishing features. At this writing (end of 1979) in course of revision with extension of coverage.]

431

Victoria

The recent manual-key by Willis (1970–2) is not a descriptive work in the stricter sense and thus has not entirely displaced Ewart's *Flora of Victoria*. In addition to these two works of identification, there is also a checklist/distribution index by Churchill and de Corona.

CHURCHILL, D. M. and CORONA, A. DE., 1972. *The distribution of Victorian plants.* [130] pp., 4 figs. (incl. maps). Melbourne: The authors. (Distrib. by National Herbarium of Victoria.)

Tabular census of vascular plants of the state, with letters designating presence/absence of species in each of 23 squares; genera alphabetically arranged. The introductory section gives an explanation of the work, and a table of authorities (designated only by numbers in the main part) appears at end (pp. 122 ff.).

EWART, A. J., 1930 (1931). *Flora of Victoria.* 1257 pp., 349 text-figs., 1 color pl. Melbourne: Government Printer (for Melbourne University Press).

Concise manual of vascular plants, with keys to all genera and species (including a general key to genera), brief descriptions, limited synonymy (without references), vernacular names, generalized indication of local and extralimital range, taxonomic commentary, and notes on life-form, phenology, etc.; glossary and index to all botanical and vernacular names. In the introductory section is a discussion of general features of the flora and remarks on alien plants as well as technical notes.

WILLIS, J. H., 1970–72. *A handbook to plants in Victoria.* 2 vols. Melbourne: Melbourne University Press. (First edn. of vol. 1, 1962.)

Concise diagnostic manual-key to vascular plants, with synonymy and references, citations of illustrations, vernacular names, general indication of local and extralimital range, and brief notes on taxonomy, habitat, phenology, special features, etc.; addenda and indices to all botanical and vernacular names in each volume. In vol. 2 Victorian distribution is expressed on the basis of a grid system; there are also a few minor format changes from vol. 1. The second edition of vol. 1 is in fact a reprint of the first edition with added supplement. The work is partly modeled on F. von Mueller's *Key to the system of Victorian plants* (2 vols., 1885–88, Melbourne).

432

New South Wales (in general)

The elderly handbook-flora by Moore and Betche (1893) and checklist by Maiden and Betche (1916) are now chiefly of historical interest, while the serial *Flora of New South Wales* is slowly appearing as a semi-monographic 'research' work comparable to the *Flora of Texas*, with a current tendency to cover southeastern Australia as a whole. For the needs of field work and identification in the main centres of population, however, the sizeable state flora (c. 4500 spp.) has been partly digested in a range of manual-floras published within the last 25 years; for these, see 433. Mention should also be made of a new state census: JACOBS, S. W. L. and PICKARD, J., 1981. *Plants of New South Wales.* 226 pp., maps (endpapers). Sydney: NSW Government Printer.

Two works on the tree flora should be accounted for here: Maiden's stately but disorganized *Forest flora of New South Wales*, now something of a collectors' item, and Anderson's *Trees of New South Wales*, a well-known but now perhaps old-fashioned standby.

Bibliography

PICKARD, J., 1972. Annotated bibliography of floristic lists of New South Wales. *Contr. New South Wales Natl. Herb.* 4(5): 291–317. [Briefly annotated list, including references to unpublished material. Important for the local literature.]

MAIDEN, J. H. and BETCHE, E., 1916. *A census of New South Wales plants.* xx, 216 pp. Sydney: National Herbarium of New South Wales.

Systematic list of vascular plants, with synonymy and citations of *Flora australiensis* and other works; index to generic names. No geographical distribution, vernacular names, or miscellaneous notes are included. [In 1981 superseded by a new census as noted above.]

MOORE, C. and BETCHE, E., 1893. *Handbook of the flora of New South Wales.* xxxix, 582 pp. Sydney: NSW Government Printer.

Concise manual-key to vascular plants, with limited synonymy (no references), terse summary of local range, and a few other notes; glossary and complete index. The introductory section provides a survey of previous work on the flora, while two appendices are devoted respectively to a list of alien species and to the endemic plants of Lord Howe and Norfolk Islands. This is still the only manual covering the whole state, although it is now long out of print and out of date.

NATIONAL HERBARIUM OF NEW SOUTH WALES, 1961– . *Flora of New South Wales* (by various botanists). Published in fascicles. (Until 1975 published as Contr. New South Wales Natl. Herb., Flora Series.) Sydney: National Herbarium of NSW.

Serially published semi-monographic 'research' flora, with keys to genera and species, lengthy descriptions, full synonymy (with references and citations), vernacular names, citation of *exsiccatae* with generalized indication of local range and summary of extralimital distribution, often extensive taxonomic commentary, and notes on ecology, special features, karyotypes, etc.; indices in each fascicle. The introductory fascicle (1961) includes a general description of the different parts of the state, remarks on botanical history, and a map with 'botanical districts'. Although prenumbered according to the Englerian system, families are published when ready. For the Orchidaceae, originally published separately but now considered as part of this series, see RUPP, H. M. R., 1943. *The orchids of New South Wales*. 152 pp. Sydney: Government Printer. (Reprinted 1969, Sydney, with supplement by D. J. McGillivray.)

Special group – trees

ANDERSON, R. H., 1968. *The trees of New South Wales*. 4th edn. xxvi, 510 pp., illus., 4 color pls., map. Sydney: Government Printer. (First edn. 1932.)

Descriptive manual of all native, naturalized, and commonly cultivated tree species in the state, arranged into four main sections corresponding with natural regions and an additional section for introduced trees; includes limited synonymy (without references), vernacular names, general indication of range, and notes on field attributes, habitat, uses, etc.; general analytical keys to *Eucalyptus*, *Acacia*, and all other trees (the latter in two versions); glossary and complete index. The introductory section has several parts dealing with trees in general and their ecology.

MAIDEN, J. H., 1902–25. *The forest flora of New South Wales*. Vols. 1–8. 295 pls. Sydney: Government Printer.

Large-scale descriptive atlas of trees, with full-page plates and copious accompanying texts; the latter include ample descriptions, full synonymy (with references and citations), English and aboriginal vernacular names, localities, and notes on habitat, special features, wood, properties, uses, etc.; complete index in each volume (no cumulative index). No particular system of classification has been followed. Much of the text was taken from Bentham's *Flora australiensis*.

433

New South Wales: regions (including the Australian Capital Territory)

The eleven conventional botanical/geographical districts commonly recognized are here expressed as seven regions. Only three have separate works for all or part of their area; for the others, works on neighboring states or regions can to a greater or lesser extent be utilized.

I. Central and south coast, and central tablelands

BEADLE, N. C. W., EVANS, O. D. and CAROLIN, R. C. (with TINDALE, M. D.), 1972. *Flora of the Sydney region*. 724 pp., 56 text-figs., 16 pls. (some in color), map (frontispiece). Sydney, Wellington: Reed. (Originally published 1963, Armidale, NSW, as *Handbook of the vascular plants of the Sydney district and Blue Mountains*.)

Concise manual-key to the native and naturalized vascular plants of the Sydney region (comprising the central coast and adjacent tablelands, including the 'Blue Mountains', from Newcastle to Nowra), with diagnoses, limited synonymy, vernacular names, notes on variability and hybridization, and brief indication of habitat, substrate, local range, etc.; list of authorities and complete index at end. Critical botanical details are depicted in the figures, while the photographs contain a limited representation of the district's rich flora. An illustrated glossary appears in the brief introductory section along with technical notes.

II. Southern tablelands (including the ACT)

BURBIDGE, N. T. and GRAY, M., 1970. *Flora of the Australian Capital Territory*. vii, 447 pp., 409 text-figs., map. Canberra: Australian National University Press.

Illustrated manual-flora of seed plants of the Capital Territory, with keys to all taxa, references to standard monographs or revisions in family and generic headings (but no synonymy), vernacular names, general indication of local and extralimital range, and notes on habitat, special features, etc., as well as taxonomic commentary; glossary, list of references, and complete index at end. An introductory section includes, *inter alia*, an account of the vegetation of the area. Apart from the ACT, this work is of use generally in the southern tablelands.

III. Northern tablelands, north coast and northern western slopes (including the New England district)

BEADLE, N. C. W. and L. D., 1971–80. *Students' flora of North Eastern New South Wales*. Parts 1–4. pp. i–vii, 1–686, figs. 1–300, map. Armidale: The authors. (Distributed by Department of Botany, University of New England.)

Manual-key to vascular plants, with extensive diagnoses, limited synonymy, vernacular names, and brief indication of local and extralimital distribution, habitat, phenology, etc.; includes also analytical keys to genera and to families. The illustrations mostly show critical details. Lists of references and full botanical indices are provided in each part, and Part 1 includes an illustrated glossary. The work covers the region as defined in the heading. The three parts available as of this writing (1981) cover the pteridophytes, gymnosperms, and 136 dicotyledonous families from Winteraceae through Compositae (modified Bentham and Hooker system). Part of the region is also covered by GRAY, M., 1961. A list of vascular plants occurring in the New England Tablelands, NSW, with notes on distribution. *Contr. New South Wales Natl. Herb.* 3(1): 1–82, map. [Systematic list of vascular plants, with limited synonymy, citation of *exsiccatae*, generalized distribution, and ecological notes; list of references and index to families.]

IV–VII. Other regions

These respectively comprise the central and southern western slopes, the Riverina, the western plains, and the far western district. No regional works are available and for general use the state floras of Victoria (**431**) and especially South Australia (**445**) may be most useful. One substantial regional work was in press as of this writing: CUNNINGHAM, G. M. *et al.*, 1981. *Plants of western New South Wales*. Illus. (some in color). Sydney: N.S.W. Government Printer.

434

Queensland (in general)

The *Queensland Flora* (1899–1902) and *Comprehensive Catalogue of Queensland Plants* (1913) by F. M. Bailey are both now obsolete for practical purposes, although the many figures in the latter remain useful. In recent years, work has been initiated at the Queensland Herbarium (Brisbane) on a new state flora; however, because of the size (over 6000 species) and varying state of knowledge of the flora together with geographical considerations, a decision was made to develop the work as four separate regional handbooks.[7] The first part of that for **435**, as well as a treatment of the pteridophytes of the whole state (by S. B. Andrews), are expected to appear shortly (L. Pedley and R. W. Johnson, personal communication). The flora project is considered more fully in PEDLEY, L., 1978. A new flora of Queensland. *Queensland Nat.* **22**: 8–12.

The individual handbook regions are separately numbered below as **435–438**, and the Torres Strait Islands north of Cape York as **439**. Other works of importance for the Queensland flora are given at **430**.

Keys to families and genera (flowering plants)

CLIFFORD, H. T. and LUDLOW, G., 1978. *Keys to the families and genera of Queensland flowering plants*. xiv, 202 pp., 8 pls., 1 text-fig. Brisbane: University of Queensland Press. [Analytical keys to all Queensland genera and families (both native and naturalized), with family diagnoses; glossary and index. Not easy to use without experience.]

BAILEY, F. M., 1899–1902. *The Queensland flora*. 6 parts, 2015 pp., 88 pls. Brisbane: Government Printer.

BAILEY, F. M., 1905. *The Queensland flora: general index*. 66 pp. Brisbane: Government Printer.

Descriptive flora of vascular plants, with keys to genera and species; limited synonymy, with references; vernacular names; generalized indication of local and extralimital range, with localities and *exsiccatae* cited for some species; taxonomic commentary and notes on special features, wood, uses, etc.; indices to all botanical and vernacular names in each part (the *General Index* includes only botanical names). The bulk of this work is essentially a plagiarization of the relevant parts of Bentham's *Flora australiensis*, although subsequently discovered taxa, records, and other data have of course been incorporated.

BAILEY, F. M. [1913.] *Comprehensive catalogue of Queensland plants, both indigenous and naturalized.* 879 pp., 976 text-figs., 7 halftones, 16 color pls. Brisbane: Government Printer.

Illustrated systematic list of cellular and vascular plants, with short notes relating to uses, poisonous properties, etc.; limited synonymy, with occasional references; English and aboriginal vernacular names; addenda and indices to vernacular, generic, and family names. The figures, largely by C. T. White, resulted from a suggestion made by Sir William MacGregor, who at that time was State Governor.

435–438

Queensland: regions (excluding the Torres Strait Islands)

These respectively represent southeast Queensland (including the Brisbane and Maryborough coastal belt, the Darling Downs and the Bunya Mountains), east central Queensland (centered on Rockhampton and Gladstone), inland Queensland (including the 'Channel country'), and north Queensland (including the northeastern coast, the Atherton Tableland, Cape York Peninsula, and the 'Gulf country'). They have been chosen to correspond with the four series of the new handbook-flora of Queensland under preparation at the Queensland Herbarium (see above under **434**); of these, the first part of that on southeast Queensland is due shortly for publication.

439

Torres Strait Islands

Unfortunately, only one old checklist is available for this botanically important area, which straddles the traditional boundary between the Oriental and Australian floral kingdoms (as recognized by Diels and many others) but is most likely only part of a broad zone of floristic transition involving substantial parts of northern Australia and New Guinea.[8] The area as recognized here is wholly included within *Flora Malesiana* (**910–30**).

BAILEY, F. M., 1898. A few words about the flora of the islands of Torres Strait and the mainland about Somerset. *Rpt. Meetings Australasian Assoc. Adv. Sci.* 7: 423–47.

Descriptive introduction; desultory observations on vegetation and flora and potential uses; briefly annotated list of plants from Thursday Island with miscellaneous notes on special features, phenology, habitat, behaviour, and uses; English common names. Based on collections made by the author in the dry season of 1897 together with earlier records. All observations pertain to islands south of 10 ° 30 ′S latitude, i.e., south of the Strait proper.

Region

44

Middle Australia

This region encompasses the present state of South Australia and the Northern Territory, a now self-governing Federal territory. Before 1911 the whole region constituted South Australia. Within the biogeographically diverse Northern Territory, two major areas are designated: 'Central Australia', the southern dry region with less than 300 mm rainfall/ annum, centering on Alice Springs, and 'The Top End', the northern infratropical region centering on Darwin (and including Arnhem Land).

The coastal areas, both in the north and in the south, were explored botanically at the beginning of the nineteenth century by Brown and (in the south only) Leschenault de la Tour, but not until after the establishment of South Australia in 1836 and the foundation of Adelaide was interior botanical exploration commenced. From his arrival in 1847, large contributions were made by Mueller, both through his own collecting and through reports on those made by others; this continued even after his departure for Melbourne in 1852. In the latter part of the century, public demand and the organization of higher education manifested themselves in the first 'state flora', the very concise *Handbook of the flora of extra-tropical South Australia* by Tate (1890). This in turn was succeeded in the 1920s by J. M. Black's classic *Flora of South Australia* (1922–9; 2nd edn., 1943–57 and supplement, 1965; 3rd edn., 1978–), originally published as a one-volume handbook but in its new edition now more of a reference work. Black's flora was, and continues to be, published as part of a series of state handbooks on flora and fauna produced under guidance of an advisory committee, without parallel elsewhere in Australia. For the present Northern Territory, from 1863–1911 part of South Australia, two compiled checklists were produced for 'Central Australia' in 1880–2 and 1889(–1895) by Tate, but the first separate overall account was the largely compiled, partially keyed checklist, *The flora of the Northern Territory* (Ewart and Davies, 1917), produced for the Commonwealth Government. This work, now very incomplete and of little practical value, is supplemented by the Arnhem Land expedition report by Specht and Mountford (1958) and a concise checklist by Chippendale (1972). The sub-region known as 'Central Australia', centering on Alice Springs, was provided anew in 1959 with a separate checklist, also by Chippendale, which has now been subsumed in *Flora of central Australia*, edited by J. P. Jessop (1981, Sydney; see **420–50**) which covers all of arid interior Australia.

With the establishment of the State Herbarium of South Australia in 1954 and of two herbaria in the Northern Territory, professional collecting has increased and expanded to more areas, but much of the region remains poorly known.

Bibliographies. General and supraregional bibliographies as for Superregion 42–45.

Indices. General and supraregional indices as for Superregion 42–45.

441

Northern Territory (in general)

The flora by Ewart and Davies (1917) was prepared under Commonwealth auspices shortly after takeover of the Territory from South Australia, but no successor has appeared and by now it is largely obsolete and very incomplete. However, there is a recent checklist by Chippendale, and other works are, or will be, available for each of the two biogeographic zones (**442** and **443** below).

CHIPPENDALE, G. M., 1972. Check list of Northern Territory plants. *Proc. Linn. Soc. New South Wales*, **96**: 207–67, 2 tables, map.

Systematic list of vascular plants (both native and naturalized), without synonymy or references; indication of distribution according to four main divisions;

addenda. The introductory section has accounts of floristics and vegetation, statistics of the flora, and a list of references.

EWART, A. J. and DAVIES, O. B., 1917. *The flora of the Northern Territory*. viii, 387 pp., 14 tables, 27 pls., map. Melbourne: Commonwealth Government Printer.

Systematic enumeration of known vascular plants, with descriptions of new taxa and synoptic keys to genera and species; limited synonymy; vernacular names; localities with citations of *exsiccatae*; notes on ecology, phenology, properties, and uses; appendices with detailed treatments of Cyperaceae, *Acacia*, and *Eucalyptus* (with other Myrtaceae); special lists of plants arranged by properties and uses; general index to all botanical and vernacular names. Coverage of Arnhem Land and the southern part of the Territory by this work is very imperfect, and localization of collection records is often vague – defects noticed as long ago as 1919 (*J. Bot.* **57**: 69–71).

442

'The Top End'

See also **441** (both works). The paper by Specht and Mountford described below deals only with parts of Arnhem Land, then little known, and many additions for this area and elsewhere have since been made.

Partial work

SPECHT, R. L. and MOUNTFORD, C. P. (eds.), 1958. Botany and plant ecology. In *Records of the American-Australian scientific expedition to Arnhem Land* (eds. American-Australian Scientific Expedition to Arnhem Land), vol. 3. xv, 522 pp., illus., maps. Melbourne: Melbourne University Press.

Includes among other articles separate annotated enumerations of the vascular cryptogams (pp. 171–84) by M. D. Tindale and the seed plants (pp. 185–317) by R. L. Specht and C. P. Mountford; each paper includes descriptions of new taxa, synonymy (with references and citations), citation of some *exsiccatae* and generalized indication of local and extralimital range, limited taxonomic commentary, and notes on habitat, life-form, special features, etc., and representative illustrations. The first article in this volume includes an account of previous botanical exploration in Arnhem Land, while other articles cover climate, geology, floristics, phytogeography, vegetation and ecology, and ethnobotany. The general index at the end of the work includes all botanical names. [The volume forms part of a series of four on various aspects of the expedition, produced between 1956 and 1964.]

443

'Central Australia'

See also **441** (both works). The new illustrated *Flora of central Australia* covers the whole of the arid interior of the continent and has therefore been listed under **420–50**. By Chippendale the 'Central Australia' district of the Northern Territory is bounded on the north by the twentieth parallel or the twelve-inch rainfall isohyet.

CHIPPENDALE, G. M., 1959. Check list of Central Australian plants. *Trans. Roy. Soc. South Australia*, **82**: 321–58.

Unannotated systematic list of vascular plants reported from the southern part of the Northern Territory (south of the 12-inch isohyet); includes list of references. A discussion of previous botanical work in the area is also incorporated. For supplements, see *idem*, 1960–3. Contributions to the flora of Central Australia, 1–3. *Ibid.*, **83**: 199–203; **84**: 99–103; **86**: 7–9. [All records have been incorporated into the author's 1972 checklist for the whole of the Northern Territory (**441**).]

445

South Australia

Black's state flora, widely used outside the state as well as in, has long been a standard reference in south-central and southeastern Australia. It is currently undergoing revision at the State Herbarium of South Australia. Dendrological coverage is provided by Boomsma (1972).

BLACK, J. M., 1943–57. *Flora of South Australia*. 2nd edn., 4 parts, 1008 pp., 1260 text-figs., portrait, maps. Adelaide: SA Government Printer. (Reprinted after 1960; 1st edn. 1922–9.)

EICHLER, HJ., 1965. *Supplement to J. M. Black's Flora of South Australia*. 385 pp. Adelaide.

BLACK, J. M., 1978. *Flora of South Australia*, 1. [3rd edn.,] revised and edited by J. P. Jessop. 466 pp.,

467 text-figs., 16 color pls., portrait, map (endpapers). Adelaide.

Concise, well-illustrated descriptive manual-flora of native and naturalized vascular plants, with keys to all taxa, limited synonymy, vernacular names, generalized indication of local and extralimital range, taxonomic commentary, and notes on ecology, life-forms, phenology, etc.; glossary, list of references, and complete index at end. An introductory section gives the basis for the work along with technical notes. The supplement by Eichler is completely cross-referenced to the main work, but is separately paged. The latest version, by Jessop and others, incorporates numerous amendments as well as some changes in style and content and appears in a smaller format; the first part covers Lycopodiaceae–Orchidaceae in the Englerian sequence (as used in previous editions).

Special groups: trees

BOOMSMA, C. D., 1972. *Native trees of South Australia*. 224 pp., 26 figs., 100 pls., maps (*South Australian Woods and Forests Dept. Bull.*, 19). [Adelaide.]

Dendrological atlas of more important trees, with illustrations and distribution maps for each species together with descriptive text covering botanical features, distribution, habitat, overall form, timber, properties and uses, etc.; trade names included; glossary and complete index at end. In the introductory section are general remarks on botany, distribution, phenology, timber properties, pollen and honey production along with an *artificial general key* and a list of references. Succeeds the stately work of Tate (1890).

Region

45

Western Australia

This region corresponds to the state of Western Australia with the addition of the Ashmore and Cartier Islands (a Commonwealth territory) in the Timor Sea.

In the history of floristic botany, Western Australia is something of a 'late-bloomer', with two significant cycles of internal growth related to general economic movements in the state. The desolate appearance of much of the coast, as reported by Dampier and other explorers, deterred early settlement and the Swan River colony was not founded until 1829. The earliest botanical collections were made by Labilliardière, Leschenault de la Tour and (particularly) Brown; the latter explored coasts in the tropical north as well as in the south and west. The collections of Brown caused great public interest and in the remainder of the period to 1850 several collectors visited the Swan River and other parts of the southwest, and their results were written up by Preiss, Lindley, and Lehmann and Bentham.

The initial hopes raised by the Swan River colony were not realized and settlement remained very limited until the 1890s. The remainder of the century was characterized by the criss-crossings of many exploring expeditions, the plants of which were largely written up by Mueller (who participated on some of them). Botanical and related advice to the Western Australian administration was also given by him. During this period the tropical Kimberley district was first seriously explored; in 1885 gold was found therein, followed in subsequent years by discoveries elsewhere, notably at Kalgoorlie in 1892–3.

Development and settlement were thereafter rapid and in the 25 years before World War I a new round of overseas botanists was attracted and serious resident botanical activity began. The most notable floristic contribution of this period was *Fragmenta phytographiae Australiae occidentalis* by Diels and Pritzel (1904–5) based on their field work in the southwestern region in 1900–1. This account has been fundamental to all later floristic botanical work in that area.

Locally published floristic contributions almost all date from after 1900. None of these has ever amounted to a complete state flora. Gardner's *Flora of Western Australia* (1952) never covered more than the Gramineae and the only overall works are two checklists respectively by Gardner (1930) and Beard (1965), neither of them complete and the latter in part uncritical. Until 1928 no state herbarium existed and only in recent years, consequent to extensive professional and other collecting throughout the state since the 1950s, has sufficient material accumulated to make feasible the preparation of a modern state flora. As in Queensland, however, it appears that some subdivision is contemplated: present plans 'contemplate' eventual compilation of five regional handbooks, respectively

covering the greater Perth area, the southwest, the south coast, the Eremaea, and the Kimberley district.

The first three of these units together more or less comprise the temperate extra-Eremaean zone, the only area with a current regional flora, *How to know Western Australian wildflowers* (Blackall and Grieve, 1954–75). Publication of this work is now complete, but in recent years revision of the first three parts has been undertaken. The only one of its kind in Australasia, this work consists mainly of analytical picture-keys, aimed mainly at the non-specialist. Other parts of the state, including the Kimberleys, have only scattered local documentation. Much new material, precursory to the proposed state floras, is being published in *Nuytsia* and elsewhere, but with an area four times the size of Texas – one-third of a continent – much is yet to be discovered in the state's flora. The tropical north and Eremaea are in particular still considered poorly known.

Bibliographies. General and supraregional bibliographies as for Superregion 42–45.

Indices. General and supraregional indices as for Superregion 42–45.

450

Region in general

Apart from Bentham's *Flora australiensis*, still regarded as the only equivalent of a statewide flora, two checklists are available which ostensibly account for all known vascular plants in Western Australia. Gardner's detailed *Flora of Western Australia* has never gone beyond the Gramineae. With much new information coming in from all parts of the state, including some interesting discoveries, and a large amount of basic taxonomic work to be done, a modern state flora is some way off. The Kimberley district and the southwestern zone each have significant regional works and reference should be made to them (**452, 455**). As of going to press, a new state checklist has been published to supersede that by Gardner: GREEN, J. W., 1981. *Census of the Vascular plants of Western Australia*. South Perth: Western Australian Herbarium.

BEARD, J. S. [1965.] *A descriptive catalogue of West Australian plants*. xiii, 122 pp., 16 color pls., map. Perth: King's Park Board/Society for Growing Australian Plants.

Systematic list of seed plants, with very brief indication of distribution and terse notes on habitat, life-form, flower color, etc.; index to generic names. The brief introductory section incorporates remarks on vegetation regions, especially in the southwestern part of the state. Omits at least the Cyperaceae and Gramineae; for critique, see J. H. Willis in *Muelleria*, 1(3) (1967): 240–41.

GARDNER, C. A., 1930 (1931). *Enumeratio plantarum Australiae occidentalis.* iv, 150 pp. Perth: Government Printer.

Unannotated systematic list of known Western Australian native and naturalized vascular plants, with limited synonymy and references (mainly where different from *Flora australiensis* or where subsequent additions are involved); index to generic and family names. [Until 1981 the standard floristic checklist for the state; both this and Beard's list have recently been superseded by a new census as noted above.]

GARDNER, C. A., 1952. *Flora of Western Australia.* Vol. 1 part 1. xii, 400 pp., 104 pls. (1 colored), map. Perth: Government Printer.

Copiously illustrated descriptive flora, with keys to genera and species; synonymy, with references; vernacular names; generalized indication of local distribution; notes on habitat, uses, special features, etc.; index to botanical names. Only this one part, covering Gramineae, has ever been published.

451

Ashmore and Cartier Islands

No lists of plants appear to be available for this group of low islands in the Timor Sea, officially constituted as an Australian External Territory.

452

Northeastern zone (Kimberley district)

The two works described below represent the main basis for botanical knowledge of this partly highland region, a part of the tropical monsoon zone of northern Australia, but are now very incomplete. In the fjordlike coastal belt, many important early collections were made by Brown and other botanists on exploring expeditions. Since the 1960s, much new collecting has been accomplished in the area and a better basis for a new floristic account is now available (K. F. Kenneally, personal communication).

FITZGERALD, W. V., 1918. The botany of the Kimberleys, north-west Australia. *J. & Proc. Roy. Soc. Western Australia*, 3: 102–224. (Reprinted with separate pagination.)

Systematic enumeration of vascular plants, with descriptions of new taxa; includes collecting localities and notes on habitat, life-form, associates, special features, etc.; no separate index. The work represents the results of a field trip by the author.

GARDNER, C. A., 1923. *Botanical notes, Kimberley Division of Western Australia.* 105 pp., text-figs., 18 pls., map (Bull. Forests Dept. Western Australia, 32). Perth.

Systematic enumeration of vascular plants collected by the author, with descriptions of new taxa; includes collecting localities and notes on habitat, life-form, field features, phenology, associates, and actual or potential uses. An index to botanical names is appended. The introductory section contains remarks on physiography, climate, and vegetation formations as well as on the general features of the flora.

453

The Eremaea

This vast, mainly desert region is not covered by any specific floras or other substantial floristic contributions. Some assistance can be had from the works on the southwestern zone, in which coverage extends more or less into the southerly parts of the Eremaea. Much of this area is now encompassed by the recently published *Flora of central Australia* (see 420–50).

455

Southwestern zone

Home to one of the more peculiar floras of the world, with a high endemism and wealth of wildflowers, southwestern Western Australia is now covered – albeit more or less sketchily – by two substantial floristic works: the *Fragmenta* of Diels and Pritzel (1904–5), a study of fundamental importance to the area, and the illustrated key of Blackall and Grieve (1954–), of which the elegantly produced final part appeared in 1975. Revision of parts 1–3 in complementary fashion is currently under way and parts 3A and 3B appeared respectively in 1980 and 1981.

BLACKALL, W. E. and GRIEVE, B. J., 1954–65. *How to know Western Australian wildflowers*. Parts 1–3. pp. i–xc, i–lvii, i–lxxviii, 1–595, numerous text-figs., 6 maps, 3 frontispieces, color pls. 1–30. Perth: University of Western Australia Press. (Reprinted in one volume with omission of color pls., 1974, Perth; with omission of part 3, 1981).

GRIEVE, B. J. and BLACKALL, W. E., 1975. *How to know Western Australian wildflowers*. Part 4. pp. [1–149], 596–861, 114 text-figs., 16 color pls., maps. Perth.

BLACKALL, W. E. and GRIEVE, B. J., 1980–1. *How to know Western Australian wildflowers*. Second edn. restructured and revised by B. J. Grieve. Parts 3A, 3B. Illus. (some in color). Perth.

Subtitled *A key to the flora of the extratropical regions of Western Australia*, this individualistic work is more than a mere 'wildflower book'; it comprises a rather detailed, copiously illustrated analytical key to the vascular plants of temperate Western Australia south of the 26th parallel, accompanied by vernacular names and (in parts 2–4) by concise indication of distribution. Glossaries, lists of references, user aids, and complete indices are also included in each part. Only a small range of the vast variety of wildflowers is depicted in the color plates. Arrangement of the families broadly follows the Englerian sequence but with some variations. Part 4 is published in a larger, more elegant format than the earlier parts. The reprint of parts 1–3 is a compact version for field use, and incorporates only the main illustrated key, a glossary, and botanical indices. A revised version of part 3, in two sections, appeared in 1980–1 in the new larger format.

DIELS, L. and PRITZEL, E., 1904–5. Fragmenta phytographiae Australiae occidentalis. *Bot. Jahrb. Syst.* **35**: 55–662, 70 text-figs., map.

Systematic enumeration of vascular plants, with descriptions of new taxa and some keys to species; synonymy, with references and citations; relatively detailed indication of local range, with *exsiccatae* listed; extensive taxonomic commentary; separate index to all botanical names. Botanical text in Latin; most commentary in German. The work was based principally on the large collections made by the two authors in the southwestern part of the state in 1900–1. It is in many ways a western counterpart of Domin's *Beiträge zur Flora und Pflanzengeographie Australiens* (see **430**); in the absence of a modern state flora, it forms a valuable and useful supplement to Bentham's *Flora australiensis* (R. D. Royce, personal communication).

Superregion

46–49

Islands of the southwestern Indian Ocean (Malagassia)

Within this superregion are incorporated Madagascar (Malagasy), together with the Comoros and the low islands west of Madagascar; the Mascarene high islands; the Seychelles group, *sensu stricto*; and Aldabra with all other low coralline and sandy islands and atolls to the east and southeast towards the Mascarenes and north to the Seychelles. Floristically the strongest links are with Africa but there are many groups absent, or not well developed, on that continent, some with connections with Greater Malesia/Oceania and northern (and eastern) South America, and several others peculiar to the superregion.

Although the Dutch were early entrants into this botanically curious superregion, progress in floristic botany has been largely dominated by the French, particularly in the eighteenth and twentieth centuries. The first important 'modern' collector, who worked extensively in Mauritius but also visited Madagascar, was P. Commerson. In the nineteenth century, much collecting was done both by visiting workers but also by residents, notably W. Bojer (in the Mascarenes) and the missionary R. Baron (35 years in Madagascar, and responsible also for the first general enumeration of its flora, *Compendium des plantes malgaches*, published 1901–7 in Tananarive and accounting for some 4100 species). The Mascarenes and the Seychelles had already been explored in some detail by 1877, the year of publication of the standard 'Kew-series' colonial flora by J. G. Baker, but serious professional field work in Madagascar only developed following assumption of full French control in 1895. Over the next 40 years, the many French missions, particularly those of Perrier de la Bâthie and Humbert, together with the work of Baron and the botanical contributions in the Grandidiers' lavish *Histoire physique, naturelle et politique de Madagascar* laid the foundation for the large-scale serial flora begun in 1936. Considerable additional field work has been done throughout the superregion since World War II by both resident and visiting workers, and two large flora projects are currently in progress. Notice

should also be made of the intensive field work carried out at Aldabra under the auspices of the Royal Society of London and on the low islands in general during various oceanological expeditions, so that the floristics of these islands are now known in considerable depth. Overall, the flora is considered 'moderately' well known, but, like Hawaii, much of it is very fragile and the severe pressures to which it has, and is being, subjected are causes for grave concern.

Progress

Most reports, as well as historical accounts, are incorporated with those for the African mainland (Division 5). Earlier history, mainly with reference to Madagascar, has been reviewed in HUMBERT, H., 1962. Histoire de l'exploration botanique à Madagascar. In *Comptes rendus de la IV. réunion plénière de l'AETFAT* (Lisbonne et Coïmbre, 1960) (ed. A. Fernandes), pp. 127–44. Lisbon; this, however, also contains a select bibliography covering the whole superregion. For the Mascarenes and Seychelles, the introductions to Baker's *Flora* and to more recent accounts should be consulted. Recent progress has from time to time been reviewed in reports published in the proceedings of the AETFAT Congresses, and in some overall African surveys, notably LÉONARD, J., 1965. Statistiques des progrès accomplis en 10 ans dans la connaissance de la flore phanérogamique africaine et malgache (1953–1962). *Webbia*, **19**: 869–75, map. The progress map in that paper has been revised by F. N. Hepper and presented at the ninth AETFAT Congress (Las Palmas, 1978), with publication following in its proceedings: *Taxonomic aspects of African economic botany* (ed. G. Kunkel; 1979, Las Palmas).[9] For another recent summary of current knowledge, see RENVOIZE, S. A., 1979. The origins of Indian Ocean island floras. In *Plants and islands* (ed. D. Bramwell), pp. 107–29. London.

General bibliographies. Bay, 1910; Blake and Atwood, 1942; Frodin, 1964; Goodale, 1879; Holden and Wycoff, 1911–14; Jackson, 1881; Pritzel, 1871–7; Rehder, 1911; Sachet and Fosberg, 1955, 1971; USDA, 1958. For bibliographies on particular regions, see appropriate headings.

Supraregional bibliography

LÉONARD, J., 1965. Liste des floras africaines et malgaches récentes. *Webbia*, **19**: 865–7. [List of principal floristic works, arranged by countries. Nearly all are described or referenced in this *Guide*.]

General indices. BA, 1926– ; BotA, 1918–26; BC, 1879–1944; BS, 1940– ; CSP, 1800–1900; EB, 1959– ; FB, 1931– ; IBBT, 1963–9; ICSL, 1901–14; JBJ, 1873–1939; KR, 1971– ; NN, 1879–1943; RŽ, 1954–

Supraregional indices

ASSOCIATION POUR L'ETUDE TAXONOMIQUE DE LA FLORE D'AFRIQUE TROPICALE, 1952– . *AETFAT Index*, 1– . Brussels. [Extensive coverage of African and SW Indian Ocean Islands regional and national floristic literature; includes lists of papers and books, with country index. Published annually. In addition, the periodic symposium proceedings of AETFAT, which include country progress reports with lists of references, should be consulted.]

Region

46

Madagascar and associated islands

In addition to the large island of Madagascar (Malagasy), this region also includes the Comoros and the small low islands of Juan de Nova, Bassas da India, Europa, and the Gloriosos.

For more than four decades, progress in floristic botany has been dominated by the ponderous progress of the serial *Flora de Madagascar et des Comores*, now far advanced with coverage of more than three-quarters of the known species. The first three decades of work were reviewed by Humbert, the originator of the project, in 1966. In addition to this great work, many local accounts, based on field trips of various parts of Madagascar, and precursory revisions and other contributions have been published in *Adansonia* and other French (and Malagassian) outlets. An introduction to the forest flora (Capuron, 1957) and a school-work (Cabanis *et al.*, 1969–70) are also available, the latter also designed for use in the Mascarenes. Separate florulas are also available for the Gloriosos, Europa, Juan de Nova, and the Comoros, although that for the last-named, *Flora und Fauna der Comoren* (Voeltzkow, 1917) is long outdated.

The region as a whole is considered moderately well-explored and documented, although unevenly so.

However, the peculiar indigenous flora is evidently to a great extent under threat of extinction or worse, and better documentation from a conservation point of view is highly desirable.

Bibliographies. General and supraregional bibliographies as for superregion 46–49.

Regional bibliography

GRANDIDIER, G. *et al.*, 1905–57, *Bibliographie de Madagascar*. 3 vols. (in 4). Paris (vols. 1–2); Tananarive (vol. 3). (Includes 23 003 references. A further cumulation for 1956–63 and subsequent annual continuations have been published in Tananarive.]

Indices. General and supraregional indices as for superregion 46–49.

460

Region in general

All general works on Madagascar, by far the major land mass in the region, are placed under this heading. Separate works on the Comoros and the low islands appear under **465** to **469** inclusive.

As a 'tome préliminaire' for the *Flore de Madagascar et des Comores* is still lacking, a useful general introduction to the rich, partly relict, and 80 per cent endemic Malagasy flora is KOECHLIN, J. *et al.*, 1974. *Flore et végétation de Madagascar*. xvi, 688 pp., illus., maps. Lehre, Germany: Cramer (Flora et vegetatio mundi, 5). Attention is also called to the following school-work: CABANIS, Y., CABANIS, L., and CHABOUIS, F., 1969–70. *Végétaux et groupements végétaux de Madagascar et des Mascareignes*. 4 vols. 260 pls. Tananarive.

Key to genera and families (woody plants)

CAPURON, R., 1957. *Introduction à l'étude de la flore forestière de Madagascar*. iv, 125 pp., illus. Tananarive: Inspection Générale des Eaux et Forêts. (Mimeographed.) [Analytical keys to families and genera of arborescent plants with family descriptions, accompanied by special commentaries (pp. 97–104), glossary and index.]

HUMBERT, H. *et al.* (eds.), 1936– . *Flore de Madagascar et des Comores*. In fascicles. Illus. Tananarive: Gouvernement Général de Madagascar; Paris: Laboratoire de Phanérogamie, Muséum National d'Histoire Naturelle.

Comprehensive illustrated 'research' flora of

vascular plants, with keys to genera and species, lengthy descriptions, full synonymy (with references), detailed indication of local range (with citation of *exsiccatae*) and general summary of extralimital range (if applicable!), taxonomic commentary and notes on habitat, special features, etc.; index to botanical names in each fascicle. No introductory section has been published. Families have been pre-numbered following the Englerian system but fascicles have been appearing only as family accounts were ready; however, some fascicles contain more than one family. At this writing (mid-1979) this work is about 80 per cent completed, with the latest fascicle seen having been published in 1978; on the other hand, many earlier parts are by now markedly outdated due in large measure to relatively intensive exploration by resident botanists after World War II, together with concomitant taxonomic studies and revisions. Progress to 1966 is reviewed in HUMBERT, H., 1966. La *Flore de Madagascar et des Comores*. Résultats et perspectives. *Adansonia*, II, 6: 315–17. Important precursory works relating to the *Flore* are ACADÉMIE MALGACHE, 1931–5. *Catalogue des plantes de Madagascar*. 23 fascicles. Tananarive; and CHRISTENSEN, C., 1932. *The Pteridophyta of Madagascar*. xv, 253 pp., 80 pls. (Dansk Bot. Ark., 7). Copenhagen. [Both works contain enumerations of families or of larger taxa, with synonymy, references, indication of *exsiccatae* with localities, general summaries of internal and extralimital range, and miscellaneous notes; that by Christensen also contains partial keys and illustrations, as well as a floristic and phytogeographical analysis. H. Perrier de la Bâthie was a notable contributor to the *Catalogue*.]

465

Comoro Islands

These are treated in *Flore de Madagascar et des Comores* (460) but the only separate survey remains that by Voeltzkow.

– Vascular flora: 416 native species (Voeltzkow, 1917).

VOELTZKOW, A., 1917. Flora und Fauna der Comoren. In *Reise in Ostafrika in den Jahren 1903–1905 mit Mitteln der Hermann und Elise geb. Heckmann-Wentzel-Stiftung ausgeführt. Wissenschaftliche Ergebnisse* (ed. *idem*), vol. 3, pp. 429–80. Stuttgart: Schweizerbart.

Pages 430–54 of this work, prepared with the assistance of G. Schellenberg, contain a checklist of non-vascular and vascular plants, with localities, citations of *exsiccatae*, and indication of extralimital distribution; this is preceded by an account of prior exploration of the islands and some general remarks on the flora as a whole. [Now considered to be relatively incomplete.]

466

Glorioso Islands

See also 490 (HEMSLEY, p. 145).
– Vascular flora: 48 species (Battistini and Cremers, 1972).

BATTISTINI, R. and CREMERS, G., 1972. *Geomorphology and vegetation of Îles Glorieuses*. 10 pp., illus., maps (Atoll Res. Bull., 159). Washington.

Includes in this general survey lists of plants collected from Grand Glorioso, Lily Island (Île du Lys) and other islets in 1971, with new records and changes since the appearance of Hemsley's 1919 list. [Fifteen records given by Hemsley were not reconfirmed in 1971.]

467

Juan de Nova Island

See 469 (BOSSER) for list.

468

Bassas da India Island

No separate floristic lists appear to be available, but the island is relatively close to Europa Island (see 469 below).

469

Europa Island

BOSSER, J., 1952. Notes sur la végétation des îles Europa et Juan de Nova. *Naturaliste Malg.* **4**: 41–2, illus.

Includes short lists of plants recorded from Europa and Juan de Nova as well as brief remarks on their vegetation. [Both islands, as well as Bassas da India, are low islands with mostly widespread species.] See also CAPURON, R., 1966. Rapport succinct sur la végétation et la flore de l'île Europa. *Mém. Mus. Natl. Hist. Nat.*, sér. II/A (Zool.), **41**: 19–21.

Region

47

The Mascarenes

The Mascarenes are here considered to include the islands of Réunion, Mauritius and Rodriguez, with their associated islets. For the several small low islands to the north, certain of which are dependencies of Réunion and Mauritius, see Region 49.

The standard floras of the group have long been *Flora of the Mauritius and the Seychelles* (Baker, 1877) and *Flore de l'île de Réunion* (Jacob de Cordemoy, 1895), both recently reprinted but much outdated. To replace them, a new cooperative work, *Flore des Mascareignes* (Bosser *et al.*, 1976–), has recently been commenced. Account has to be taken of additional introduced plants, a task accomplished for Mauritius by Vaughan (1937), as well as of the current status of the fragile native flora. **Bibliographies.** General and supraregional bibliographies as for Superregion 46–49. See also region 48 (Peters and Lionnet, 1973).

Indices. General and supraregional indices as for Superregion 46–49.

470

Region in general

For Baker's *Flora of Mauritius and the Seychelles*, see **472**. Successors to this and other current but largely outdated works on the region include a checklist of the pteridophytes by Tardieu-Blot (1960) as well as the modern *Flore des Mascareignes*, the latter published under joint British, French and Mauritian auspices.

BOSSER, J. *et al.* (eds.), 1976– . *Flore des Mascareignes* (La Réunion, Maurice, Rodrigues). In fascicles. Illus. Mauritius: Sugar Industry Research Institute; Kew: Royal Botanic Gardens; Paris: ORSTOM.

Descriptive 'research' flora of vascular plants, with keys, extensive synonymy (with citations and references), indication of local and extralimital distribution, notes on habitat, occurrence, status, etc., and taxonomic commentary; illustrations of representative species; index in each fascicle. Naturalized and commonly cultivated plants are also included. Families have been pre-numbered following the Kew modification of the Bentham and Hooker system but fascicles are appearing only as family accounts are ready; more than one, however, may appear at a time as in the *Flore de Madagascar et des Comores* (**460**). At this writing (January, 1981), four fascicles, with 18 families, have been published.

Special group – pteridophytes

TARDIEU-BLOT, M. L., 1960. Les fougères des Mascareignes et des Seychelles. *Notul. Syst.* (Paris), **16**: 151–201.

Systematic enumeration of known ferns and fern-allies in the whole Mascarenes/Seychelles region, with synonymy, citations of *exsiccatae* with localities, and index. Supplants the account of these plants in Baker's *Flora* (**472**). [More recently, each of the three Mascarene islands has had new surveys of its pteridophytes; preliminary accounts have now been published and are cited here under **471–473** as appropriate.]

471

Réunion

– Vascular flora: 1156 species (Jacob de Cordemoy, 1895).

JACOB DE CORDEMOY, E., 1895. *Flore de l'île de la Réunion*. xxvii, 574 pp. Paris: Klincksieck. (The consolidated issue; originally published at least in part, 1891–   , St Denis, Réunion. Reprinted 1972, Lehre, W. Germany: Cramer.)

Briefly descriptive flora of non-vascular and vascular plants (except fungi), with more detail accorded to new taxa; includes keys to genera and species (or groups of species), limited synonymy (with references), vernacular names, generalized indication of local range, and notes on habitat, phenology, properties, uses, etc.; index to vernacular, generic and family names. In the introduction are chapters on geography, physical features, climate, and the general aspects of the flora, together with statistical tables and an account of botanical exploration. See also BADRÉ, F. and CADET, T., 1978. The pteridophytes of Réunion Island. *Fern Gaz.* 11: 349–65.

472

Mauritius

– Vascular flora: No current figure seen for the native flora, but certainly more than the 869 species recorded by Baker (1877).

BAKER, J. G., 1877. *Flora of Mauritius and the Seychelles: a description of the flowering plants and ferns of those islands*. 19, 1, 557 pp. London: Reeve. (Reprinted 1970, Lehre, W. Germany: Cramer.)

Briefly descriptive flora of vascular plants, with keys to genera and species; limited synonymy, with references; vernacular names; citations of *exsiccatae*, with localities, and general indication of extralimital range (where applicable); notes on habitat, etc., and taxonomic commentary; indices to all vernacular and botanical names. The introductory section incorporates general descriptive remarks on the islands covered and notes on the main features of their floras, an account of botanical exploration, a synopsis of orders and families, and Bentham's *Outlines of Botany*. This work

is one of the 'Kew series' of British colonial floras as initiated by W. J. Hooker in the 1850s; it covers Mauritius, Rodriguez, the Seychelles proper, Aldabra, and all the other smaller islets and atolls in the region then under British administration. For additions, see JOHNSTON, H. H., 1895. Additions to the flora of Mauritius as recorded in Baker's 'Flora of Mauritius and the Seychelles'. *Trans. & Proc. Bot. Soc. Edinburgh*, 20: 391–407; also VAUGHAN, R. E., 1937. Contributions to the flora of Mauritius, I. An account of the naturalized flowering plants recorded from Mauritius since the publication of Baker's *Flora of Mauritius and the Seychelles*. *J. Linn. Soc., Bot.* 51: 285–308. For a recent review of pteridophytes, including a tabular checklist, see LORENCE, D. H., 1978. The pteridophytes of Mauritius Island (Indian Ocean): ecology and distribution. *Bot. J. Linn. Soc.* 76: 207–47, map. [A projected second contribution from Vaughan, covering newly found native species, evidently did not materialize.]

473

Rodriguez (Rodrigues)

See also **472** (BAKER).

– Vascular flora: 322 species, of which 108 seed plants viewed as introduced (Balfour 1879); Cadet (1971) records 375 + species, including over 100 introduced since 1874.

BALFOUR, I. B., 1879. Flowering plants and ferns. In *An account of the petrological, botanical and zoological collections made in Kerguelen's Land and Rodriguez during the Transit of Venus expedition carried out by order of Her Majesty's Government in the years 1874–75*, [II]: *Collections from Rodriguez* (ed. J. D. Hooker and A. Günther), pp. 302–87, Pls. 19–36. (Philos. Trans., vol. 168, extra vol.) London. (Reprinted as p. 1–86 in BALFOUR, I. B. *et al. Botany* [of Rodriguez]. 118 pp. London [for private circulation].)

Systematic enumeration of vascular plants, with descriptions of new taxa discovered; includes limited synonymy (with references), generalized indication of local and extralimital range (with mention of some localities), taxonomic commentary, and notes on habitat, uses, etc. A general part includes remarks on overall features of the flora and vegetation (already in 1874 very disturbed) and its history, together with a floristic analysis. All records have been accounted for by BAKER (472). For additions (introduced species), see CADET, T., 1971. Flore d'Île Rodrigues: espèces spontanées introduites depuis Balfour (1874). *Bull. Mauritius Inst.* 7: 1–12; for a revision of the pteridophytes, see LORENCE, D. H., 1976. The pteridophytes of Rodrigues Island. *Bot. J. Linn. Soc.* 72: 269–83, map.

Region
48
The Seychelles

The Seychelles group, home of the 'Coco de Mer', is here defined as comprising the five granitic 'high' islands together with associated islets on the Seychelles Bank. All other low islands, cays and atolls in the general region, including Aldabra – most now administratively part of the Seychelles – constitute Region 49.

Apart from treatment in Baker's *Flora of Mauritius and the Seychelles* (1877; see 472), the flowering plants are enumerated in a separate checklist by Summerhayes (1931). Pteridophytes are covered by Tardieu-Blot (1960; see 470) as well as by Christensen (1912). Although some more recent floristic surveys have been made, no new checklist or descriptive work has been published for the small but very peculiar and fragile flora.

Bibliographies. General and supraregional bibliographies as for superregion 46–49.

Regional bibliography
PETERS, A. J. and LIONNET, J. G., 1973. *Central western Indian Ocean bibliography*. 322 pp., 3 maps (Atoll Res. Bull., 165). Washington. [Centering on the Seychelles, this bibliography covers the area from 2° to 11°S and 45° to 75°E, thus covering island groups from Aldabra and the Seychelles through Agalega and Coëtivy to the Chagos Archipelago (including Diego Garcia). It is not limited to botany, and includes subject indices.]

Indices. General and supraregional indices as for superregion 46–49.

480
Region in general

See also **470** (TARDIEU-BLOT); **472** (BAKER).
– Vascular flora: 233 species (Summerhayes, 1931).

SUMMERHAYES, V. S., 1931. An enumeration of the angiosperms of the Seychelles Archipelago. *Trans. Linn. Soc. London*, II, Zool., 19(2): 261–99, table.

Systematic enumeration of native and naturalized flowering plants, with synonymy, references, and citations; detailed indication of local range, with *exsiccatae*, and general summary of overall distribution (if applicable) or country of origin. The introductory section includes an account of botanical exploration together with floristic statistics and remarks on phytogeography and the origin of the flora. Pteridophytes are covered in a related paper, viz: CHRISTENSEN, C., 1912. On the ferns of the Seychelles and the Aldabra group. *Trans. Linn. Soc. London*, II, Bot. 7: 409–25, pl. 45.

Region
49
Aldabra and other low islands north and east of Madagascar

The islands or island groups in this region include the Cargados Carajos group, Tromelin, Agalega, Platte, Coëtivy, the Amirante group, Alphonse, Providence (with St Pierre and Cerf), Farquhar, and Aldabra (with the Assumptions, the Cosmoledos, and Astove). All are low coralline or sandy islands or atolls, of which Aldabra is much the largest. In the wake of the IIOE cruises and the Aldabra research programme of the 1960s and early 1970s, most of the nine island units here delineated have modern florulas or floristic surveys, in succession to the more or less overall but in part very sketchy survey of Hemsley (1919). Of special importance in recent times has been the publication, in 1980, of a new flora of Aldabra and its neighbors by Fosberg and Renvoize which accounts for the numerous new collections and field studies of the preceding 20 years.

Bibliographies. General and supraregional bibliographies as for Superregion 46–49. See also Region 48 (Peters and Lionnet, 1973).

Indices. General and supraregional indices as for Superregion 46–49.

490

Region in general

See also **470** (TARDIEU-BLOT); **472** (BAKER); **480** (SUMMERHAYES). Reference should be made also to the *Island bibliographies* of Sachet and Fosberg (see General Bibliographies) as well as to the Peters and Lionnet bibliography (Region 48) for details of the considerable but scattered literature on these islands. The sketchy omnibus work of Hemsley (1919), in which separate accounts of the various islands and island groups were mainly given for comparative purposes in a treatise mainly devoted to Aldabra, has now to a large extent been supplanted by more recent and detailed island accounts, though some units (or parts thereof) are not so covered.

– Overall vascular flora: 228 species or more (Renvoize, 1975, pp. 134–5).

GWYNNE, M. D. and WOOD, D. 1969. *Plants collected on islands in the western Indian Ocean during a cruise of the MFRV 'Manihine', September–October 1967.* 15 pp. (Atoll Res. Bull. 134). Washington.

Incorporates a tabular list of plants, with species given on one coordinate and islands or island groups on the other; list of references included. Accounts for the vascular plants of Assumption, Astove, Coëtivy, Cosmoledo, and Farquhar as well as Daros, Desroches, and Remire in the Amirante group are thus presented, with citations of *exsiccatae* given in the table as confirmation where appropriate.

HEMSLEY, W. B., 1919. Flora of Aldabra: with notes on the flora of the neighbouring islands. *Bull. Misc. Inform.* (Kew), *1919*: 108–53, map; [correction], pp. 451–52.

Annotated list of vascular plants of Aldabra, with descriptions of some new species; includes limited synonymy, principal citations, indication of *exsiccatae* with localities, general summary of extralimital range, and occasional taxonomic commentary. The general part includes sections on principal features of the island, climate, vegetation, endemic species, phytogeo-graphy, and botanical exploration. In addition to the main Aldabra list (now superseded by the illustrated flora of FOSBERG and RENVOIZE; see **499**), separate accounts, partly discursive and all derived from literature or unpublished sources, deal with the known plants of Assumption, Astove, Cosmoledo (the first three being in the Aldabra group), Farquhar, Providence, St Pierre, Gloriosa, Alphonse and the Amirante group, Coëtivy, Agalega, and the Cargados Carajos, as well as the Chagos Archipelago (**043**). A few novelties apropos to this section of the paper are described on pp. 140–2. [For most islands or island groups now superseded, but basic to all floristic studies in the region.]

RENVOIZE, S. A., 1975. A floristic analysis of the western Indian Ocean coral islands. *Kew Bull.* **30**: 133–52.

Includes (as Table 3, pp. 146–52) a tabular systematic checklist, without keys or synonymy, of the vascular plants of 14 island units scattered over the whole region as here defined, with indication of presence or absence. This list is included as an adjunct to a semi-quantitative floristic and phytogeographic analysis of the islands treated, which were chosen on the basis of the availability of more or less adequate modern collections. [Since publication of this survey, a greater or lesser number of additions has been made for a number of the given island units.][10]

STODDART, D. R. (ed.), 1970. *Coral islands of the western Indian Ocean.* x, 224 pp. (Atoll Res. Bull. 136). Washington.

Comprises a collection of separate papers by various authors on geographical, physical, botanical, zoological and other aspects of a large number of coral atolls and sand cays in the southwest Indian Ocean. Several of these deal in whole or in part with floristics and vegetation, and include lists of the non-vascular and vascular plants of individual islands or island groups, with citations of *exsiccatae* and some general notes on the flora. As far as possible, these papers are herein accounted for under their respective geographical headings. Partially supplants Hemsley's *Flora of Aldabra* (1917; see above).

491

Cargados Carajos group

See also **490** (HEMSLEY, p. 147).

– Vascular flora: 41 species (Staub and Guého, 1968).

STAUB, F. and GUÉHO, J., 1968. The Cargados Carajos shoals or St Brandon: resources, avifauna and vegetation. *Proc. Roy. Soc. Arts Sci. Mauritius* 3(1): 7–46, Pls. 1–5 (incl. maps).

Includes (pp. 25–40) a descriptive account of vegetation and flora, with a census of the known plants.

492

Tromelin

– Vascular flora: six species (Staub, 1970).

STAUB, F., 1970. Geography and ecology of Tromelin Island. In STODDART (**490**), pp. 197–210.

Includes general remarks on the vegetation, with notes on the vascular plants observed.

493

Agalega

See **490** (HEMSLEY, p. 146; RENVOIZE) for checklist.

– Vascular flora: six species (HEMSLEY); a more realistic figure, however, is 60 (Proctor in RENVOIZE, 1979, p. 117; see Superregion 46–49, introduction, p. 240).

494

Coëtivy

See **490** (GWYNNE and WOOD; HEMSLEY, p. 146) for checklists.

– Vascular flora: 65 species of which 49 are native (GWYNNE and WOOD, 1969).

495

Platte

No separate checklists appear to be available for this islet which is close to, but not on, the Seychelles Bank.

496

Amirante group (including Desroches and Alphonse)

See also **490** (GWYNNE and WOOD, for Desroches, Daros, and Remire; HEMSLEY, p. 145, for entire group). The main group comprises several small islets on a bank, e.g., the African Banks, Daros, Poivre, Remire, and Desnoeufs; Desroches Atoll is just to the east and the Alphonse atolls lie to the south. Desroches, Remire, and African Banks are covered in the work described below.

– Vascular flora: 97 species of which 72 are native (GWYNNE and WOOD, 1969).

FOSBERG, F. R. and RENVOIZE, S. A., 1970. Plants of Desroches Atoll. In STODDART (**490**) pp. 167–70; *idem*. Plants of Remire (Eagle) Island, Amirantes. *Ibid*., pp. 183–6; *idem*. Plants of African Banks (Iles Africaines). *Ibid*., pp. 193–4.

Annotated lists of non-vascular and vascular plants recorded, with indication of *exsiccatae* and notes on distribution, ecology, dispersal, etc.

497

Farquhar group

See **490** (GWYNNE and WOOD; HEMSLEY, pp. 143–5) for other checklists. The unit as here circumscribed includes Providence Atoll (including Providence and Cerf) and St Pierre as well as the slightly removed Farquhar Island together with associated reefs and cays.

– Vascular flora: Providence, 14 species, St Pierre, 24 species; Farquhar, 44 species (RENVOIZE, 1979, p. 117; see Superregion 46–49, Introduction, p. 240).

FOSBERG, F. R. and RENVOIZE, S. A., 1970. Plants

of Farquhar Island. In STODDART, D. R. (**490**), pp. 27–33.

Enumeration of non-vascular and vascular plants, with citations of *exsiccatae* and notes on distribution, ecology, dispersal, etc.; references.

498

Rajaswaree Reef

No reports have been seen concerning plants from this isolated reef, which lies some 170 km southeast from Farquhar Island.

499

Aldabra group

As here delimited, this unit includes the large raised-coral island of Aldabra and the smaller, nearby raised atolls of Assumption, Cosmoledo and Astove. The whole group is now covered by the recent descriptive flora of Fosberg and Renvoize (1980), supplanting Hemsley's lists (see **490**) and scattered other sources. For convenience, however, modern separate accounts for Cosmoledo and Astove are also accounted for here.

– Vascular flora: 274 species and varieties, of which 185 are indigenous and 43 endemic (Fosberg and Renvoize, 1980, p. 6).

FOSBERG, F. R. and RENVOIZE, S. A., 1980. *The flora of Aldabra and neighbouring islands.* [vi], 358 pp., 55 figs., 2 maps (Kew Bull. Addit. Ser., 7). London: HMSO.

Descriptive flora, with keys to all taxa, synonymy, indication of local distribution, vernacular names, some taxonomic commentary, and notes on habitat, occurrence, phenology, origin, dispersal, and special features; list of mosses (by C. C. Townsend); indices to all names. The introductory section of this work, which essentially covers vascular plants, deals with physical features and geography, climate and other habitat factors, vegetation, and botanical exploration; statistics, references (pp. 7–8), a glossary, and a general key (pp. 17–31) are also included. No *exsiccatae* are cited (available upon separate application). See also WICKENS, G. E., 1974 (1975). *A field guide to the flora*

of Aldabra. xlii, 99 pp. London: The Royal Society. (Mimeographed.) [Checklist with keys; for field testing.]

Partial works

Included here are separate works on Cosmoledo and Astove. Other treatments, accounting also for Assumption, may be found in HEMSLEY (pp. 140–3) and GWYNNE and WOOD (for both, see **490**).

FOSBERG, F. R. and RENVOIZE, S. A., 1970. Plants of Cosmoledo Atoll. In STODDART (**490**), pp. 57–65.

FOSBERG, F. R. and RENVOIZE, S. A., 1970. Plants of Astove Island. In STODDART (**490**), pp. 101–11.

Annotated enumerations of non-vascular and vascular plants, with indication of *exsiccatae* and notes on distribution, ecology, dispersal, etc.

Notes

1 The term 'Malagassia' is here introduced as a uninomial for the southwest Indian Ocean islands, a geographically discrete and floristically interrelated group. It is derived from 'Malagasy' (Madagascar) and the ending is related to its archipelagic composition.

2 Another version of Sorenson's checklist, with some revisions, is found in Gressitt, J. L., 1964. *Insects of Campbell Island*, pp. 16–20 (Pacific Insects Monogr. 7). Honolulu.

3 On the life of Mueller and the relationship between him and the mid-century 'Kewites', see Kynaston, E., 1981. *A man on edge: a life of Baron Sir Ferdinand von Mueller*. Ringwood, Victoria: Allen Lane/Penguin Books. Mueller was to be one of the leaders in an Australian inter-colonial movement for scientific independence during the latter part of the nineteenth century.

4 Report of 1979 meeting of Committee of Heads of Australian Herbaria in *Australian Systematic Botany Society Newsletter* 21 (December 1979), pp. 7–9, and report from Bureau of Flora and Fauna, Department of Science (ABRS) *ibid.*, **23** (June 1980), pp. 5–6, as well as personal knowledge. The current plan for the new flora is for a series of revisionary treatments in 48 bound volumes, following – regrettably – the Cronquist system, with a 15-year target for completion; a standard format has already been agreed to and the scope of coverage is to reflect the 'current state of knowledge'. Writing of the new flora commenced in 1980 under the direction of the Commonwealth Bureau of Flora and Fauna and A. S. George has been named general editor. Publication (through the Australian Government Publishing Service) was expected to commence by August 1981. [The *ASBS Newsletter*, begun in 1973, has been chronicling many of the developments and discussions relating to the new Australian flora project.]

5 In many ways, the fragmentation criticized by Bentham a century ago – a factor contributing to his assuming sole responsibility for the execution of *Flora australiensis* – still exists, and the wish has been expressed that a 'Bentham redivivus', perhaps even located outside Australia, should 'take charge'. The existence of a separate Bureau of Flora

and Fauna at Canberra can here be seen to be advantageous as a 'seat' for a comprehensive national flora.

6 Most standard bibliographies lack complete bibliographic information on this work, even in concise form. See D. J. McGillivray, *Contr. New South Wales Natl. Herb.* 4(6): 366–8 (1973) for a full review.

7 Queensland is larger than Alaska or the southeastern United States (Regions 11 and 16 respectively). The centers of population (and users of a flora) are also rather widely spaced, although the majority live in the southeast (435).

8 For a recent review of the 'Torres Strait problem', with contributions from several disciplines, see Walker, D. (ed.) *Bridge and barrier.* (Australian National University, Department of Biogeography and Geomorphology, Publ. BG-3.) Canberra.

9 The original map depicted the status of floristic exploration as of 1963, and was utilized by E. Jäger for his world map of 1976 (reproduced here, with revisions as Map II).

10 Renvoize, S. A., 1979. The origins of Indian Ocean island floras. In *Plants and islands* (ed. D. Bramwell), pp. 107–29 (1979). London: Academic Press.

Division

5

Africa

The *Flora of Tropical Africa* and its twentieth century successors...are valuable tools for the developing nations and are therefore being supported by them. It is encouraging to see the publication of national floras by these independent countries, involving indigenous taxonomic research.

> Hepper in *Systematic Botany, Plant Utilization and Biosphere Conservation* (ed. Hedberg, 1979).

I can only, in conclusion, express the hope that this somewhat monumental and, at any rate, laborious work [*Flora of Tropical Africa*], may be found, as I believe certainly it will be, of real service to the material development of the resources of our African possessions. At the moment it perhaps is more appreciated in France and Germany than by our own countrymen.

> Portion of letter from W. T. Thistleton-Dyer to R. L. Antrobus, Colonial Office, December 8, 1905; quoted from Thistleton-Dyer, Botanical survey of the Empire; in *Bull. Misc. Inform.* (Kew), *1905*: 33 (1906).

Africa here includes the entire continent (bounded on the northeast by the Suez Canal) together with immediately adjacent islands, Socotra (with 'Abd-el-Kuri), and the Guinea Islands (Fernando Póo to Annobon). All Mediterranean islands are, however, incorporated in regions 61–63; Madeira, the Canaries and the Cape Verdes are part of Macaronesia (region 02); Ascension, St Helena, Tristan da Cunha, and Gough are all in region 03; and the Marion (Prince Edward) group is in region 08. All islands in the southwest Indian Ocean, including Madagascar, are referred to Division 4 as Superregion 46–49.

The largest of our ten divisions in terms of land area, Africa has been found to be geologically ancient but with a tropical flora less rich than that of the neotropics or the Indo–Pacific tropics, possibly associated with more erratic rainfall and greater influence of dry periods.[1] Associated with this is the more limited area of potential tall evergreen forest and the presence of fewer centers of local endemism. Woodland areas are also less rich than their counterparts in South America, although in both cases the *southerly* zones are more diverse floristically than the northern. Tropical Africa is estimated to have 30 000 species of vascular plants as opposed to at least 35 000 (more likely 40 000) for the Asia–Pacific tropics and about 90 000 for the neotropics. However, to tropical Africa must be added the estimated 17 000 species in southern Africa, one of the richest regions of its kind in the world, and the perhaps 10 000 species of temperate northern Africa, with its key center of endemism in the Atlas Mountains and others elsewhere.

Much of the history of African botany has been told in a series of contributions edited by Fernandes (1962), and some additional details on early botanists have been added by Hepper (1979). While numerous contributions were made from the mid-seventeenth to the mid-nineteenth century, they were at first directed largely to the northern fringe and to southern Africa, and tropical Africa yielded less information to Linnaeus in 1753 than the neotropics or the tropical Asia–Pacific region. The first major contribution from within the tropics was made by M. Adanson, resident in Senegal from 1749 to 1754, some of whose collections were accounted for by Linnaeus. More extensive inland botanical exploration began in the temperate fringes and in tropical west and northeast Africa in the late-eighteenth century but only spread to the rest of the continent in the nineteenth century as it became crisscrossed by numerous expeditions. Further stimuli to botanical exploration came with the development of the 'Kew floras' scheme of W. J. Hooker, the consequent foundation (in 1868) of *Flora of Tropical Africa* and revival (in 1877) of *Flora capensis* as Kew projects, the imposition of colonial rule and development of (largely rural) resources, and the gradual development (at first mainly in northern and southern Africa, but after 1900, with some stimulus from Germany, in the infratropical belt) of professional cadres with associate institutions including herbaria, gardens, and laboratories. The process was not uniform, with the result that the *Flora of Tropical Africa*, while by present standards largely obsolete, when written covered some regions better than others.

The early years of the present century witnessed other attempts at comprehensive 'primary' coverage. Engler directed much of his available resources at Berlin into energetic study of the African flora, beginning more or less with his own study of the montane flora (1892) and first account of the plants of German East Africa (1895). Numerous revisions and other papers were produced, culminating in *Die Pflanzenwelt Afrikas* (1908–25), a detailed, discursive account treating the whole continent not since emulated (and itself never completed). Associated with this was Thonner's *The Flowering Plants of Africa* (*Die Blütenpflanzen Afrikas*), published in German in 1908

and in English in 1915 and ever since much esteemed. Another key outlet for African studies was of course *Pflanzenreich*. Commendably, German efforts were more synthetic than floristic, and, in contrast to the British, comparatively little attention was given to the writing of colonial floras as an end in itself.

In the decades preceding World War I, more metropolitan powers became involved in African botany as they acquired colonial territories, but with fewer or less well-organized resources their contributions were more limited or fragmented. Nevertheless, rivalries ensued which have left their mark on the synonymy of many plant species. Among their generally more *ad hoc* contributions were a number of checklists as well as many 'contributions' and 'reports' but few major floristic works of lasting value.

As local needs developed, the period after 1919 became characterized by a greatly increased number of floras, checklists, and other floristic works of more limited scope, and a turn away from more spectacular, all-embracing comprehensive accounts, in part simply because the pool of information became too large but also due to the disappearance of an immediate justification for German contributions after 1918. With more collections, resources, and experience, the quantity and quality of Belgian, French, Italian and Portuguese contributions improved during this period. The concept of the African regional flora became instilled by *Flora of West Tropical Africa*, originally designed as a practical tool (albeit by herbarium botanists) but which soon gained stature in its own right among professionals and was to inspire other, usually more inflated, regional floras. Dendrofloras also began to make their appearance, through the efforts of the Imperial Forestry Institute and other organizations.

The more or less comprehensive regional flora – termed by Symington 'the next evolutionary step' – became a more general goal during the reassessments of work programmes forced upon metropolitan institutions by World War II, one which was also associated with changes in planned development strategies. In preparation of these works, a greater role was foreseen for resident workers and their institutions, especially in the infratropical zone with which most overseas-directed 'development' was concerned. French botanists already had regional flora programs under way for North Africa and Madagascar, and South Africa was to follow suit in the 1950s. The only African regions without such a project were the somewhat heterogeneous French Equatorial Africa and the Sudan, although in the former large-scale projects in the 'regional flora' genre were begun for the Cameroons and Gabon. In southern central Africa, the preexistence of *Conspectus florae angolensis*, prepared in Portugal, was recognized by the organizers of *Flora zambesiaca* who thereby limited the latter to the central and eastern parts of the region.

Since World War II, then, African floristic botany has to a large extent been dominated by the ponderous progress of the great institutional regional floras. Their styles vary considerably as do the resources available for their production, and their progress has been influenced by political and economic events. Lying behind all these is the size of the respective flora to be covered. Most of these works, however, are appearing at such a rate as to require many decades for their completion – and that may be not soon enough in the face of rapidly growing conservation and land-use problems, to which attention has, repeatedly, been drawn. For most of the regional floras are being run as semi-monographic outlets, in some respects an unhappy compromise between the earlier Englerian tradition of comprehensive systematic studies and practical needs at regional or national level. *Flora of West Tropical Africa* remains the only regional African flora with a concise manual-type format.

Many works of various kinds have also been produced at subregional or more local level, varying with needs, interests, and available resources, some emanating from metropolitan institutions but others from the growing number of African botanical institutions and herbaria, some of which have had a checkered history. The current pattern of progress involves an increasing amount of more specialized systematic and biological research as the general floristic picture has become better known, with a corresponding shift of emphasis in collecting to more critical, detailed work (though this, coupled with continuing 'primary' exploration, has continued to yield on average one new species daily since 1953, albeit on a somewhat declining curve). Associated with this has been the development of a pattern of cooperation between African institutions, where now there are more African botanists, and the metropolitan institutions in the northern hemisphere. This has included the appointment in some cases of 'liaison officers', including specialist collectors associated with flora projects. The shifts of emphases have tended to favor

the local institution, most clearly demonstrated where research into difficult-to-preserve plant parts or plant groups has been concerned. However, with the semi-monographic nature of most current African regional floras, the more critical approaches now demanded have also reduced quantitative taxonomic productivity (in numbers of species yearly). Nevertheless, the regional floras have already contributed to the creation of a sound taxonomic foundation in African botany and have stimulated much new research and the development of local cadres, although it is perhaps too early for many locally-produced 'national' floras.

Such 'national' works as have appeared in recent decades have largely taken the form of checklists; keyed descriptive floras have been written only for a few – Egypt, Libya, Algeria, Senegal, Nigeria, upland Kenya, Swaziland and Mozambique among them. Both resident and overseas botanists have been involved. Some 'dendrofloras' have also been produced, but relatively few local florulas for areas of special interest, a situation criticized by Aubréville (who in turn was criticized for his stand by 'opponents' of local florulas). More of the latter are surely needed – but their style, contents and format have to be carefully considered in relation to their potential audience, which in Africa is by no means uniform.

An important post-war development in African botany has been the formation of the 'Association pour l'Étude Taxonomique de la Flore de l'Afrique Tropicale' (AETFAT), which since 1952 has produced an annual index to new taxa as well as literature and which has also held periodic congresses every three (later four) years, the latest at Las Palmas (Canary Islands) in 1978. The published proceedings of these congresses give a good idea of ongoing developments, and have usually included 'progress reports'. The proceedings of the fourth congress, published in 1962, gave an extensive series of historical reviews which collectively constitute the best such resource available. More recently (1975) there has been formed the 'Organization for the Phyto-Taxonomic Investigation of the Mediterranean Area' (OPTIMA), which produces a semiannual newsletter and distributes collected separates as 'Cahiers OPTIMA'. This of course largely concerns only northern Africa.

J. Léonard, who for nearly three decades has been surveying the progress of African botany and has even contributed a map showing the degree of African floristic exploration (south of the Tropic of Cancer), has summarized future needs as follows (*Boissiera*, **24**: 19, 1975):

- Il convient donc d'intensifier la campagne d'exploration floristique de l'Afrique principalement en des régions jusqu'ici peu explorées;
- Il importe de poursuivre, et si possible d'intensifer, la publication des Flores régionales (de préférence du type de Flora of West Tropical Africa) se rapportant à de vastes contrées (grands pays ou groupe de pays);
- Il devient urgent d'envisager la préparation d'une Flora africana afin d'inaugurer la période synthétique de l'étude de la flore africaine.

While the third of these may be more academic, the first two are in accord with what is the greatest need: adequate documentation in the face of environmental destruction, for effective conservation recommendations and action plans, and for education. The last also calls for the production of more local field floras *in situ* – the third of Symington's evolutionary steps – and related scholastic works, as has also been cogently argued by Gentry for the neotropics and was done long ago by Corner in his *Wayside trees of Malaya* (**911**).

Progress

General surveys, including some historical background, are contained in the detailed zonal state-of-knowledge botanical reports by various specialists in CROSBY, M. (convenor), 1978. Systematic studies in Africa. *Ann. Missouri Bot. Gard.* **65**: 367–589. For Africa south of the Sahara, reference should also be made to AUBRÉVILLE, A., 1964. État actuel de la connaissance de la flore phanérogamique tropicale africaine et malgache. *Adansonia*, II, **4**: 4–7; LÉONARD, J., 1965. Carte du degré d'exploration floristique de l'Afrique au sud du Sahara. *Webbia*, **19**: 907–14, map; and HEPPER, F. N., 1979. The present stage of botanical exploration in Africa. In *Systematic botany, plant utilization, and biosphere conservation* (ed. I. Hedberg), pp. 41–6, map. Stockholm. [A revised version of the Léonard map, edited by Hepper, appears in KUNKEL, G. (ed.), 1979. *Taxonomic aspects of African economic botany*. Las Palmas.] Detailed country-by-country historical and current progress reviews have been published in most successive proceedings of the AETFAT, but particular mention is made here of the following anthologies: Histoire de l'exploration botanique de l'Afrique au sud du Sahara, 1962. In *Comptes rendus de la IV. réunion plénière de l'AETFAT* (Lisbonne, Coïmbre 1960)

(ed. A. FERNANDES), pp. 45–248, illus. Lisbon; Progress in the preparation of African floras, 1971. In *Proceedings of the VII. plenary meeting of the AETFAT* (München, 1970), (ed. H. MERXMÜLLER) pp. 13–90. Munich (publ. in Mitt. Bot. Staatssaml. München, **10**); and Progrès accomplis dans l'étude de la flore et de la végétation d'Afrique, 1976. In *Comptes rendus de la VIII. réunion plénière de l'AETFAT* (Genève, 1974) (ed. J. MIÈGE and A. STORK), pp. 519–639. Geneva (publ. in *Boissiera*, 24b).[2] Pertinent comments relating to tropical African floristic botany may also be found in the pantropical reviews of Prance (1978) and Symington (1943), both cited in Part I of this *Guide*. For South Africa, reference may be made to two successive historical surveys: PHILLIPS, E. P., 1930. A brief historical sketch of the development of botanical science in South Africa. *South African J. Sci.* **27**: 39–80; and his 1951 supplement in *South African Biol. Soc. Pamphlet*, **15**: 19–35. No overall review for northern Africa has been seen, and reference must be made to individual country reports published under the coordination of V. H. Heywood in CENTRE NATIONAL DE LA RECHERCHE SCIENTIFIQUE, 1975. *La flore du bassin méditerranéen*, pp. 15–142. Paris (full reference in Region 59, Introduction). Some progress reports on North African floras have appeared in AETFAT proceedings, but with the establishment of OPTIMA (Organization for Plant Taxonomy in the Mediterranean Area) in the mid-1970s most reports of current work have been appearing in the publications of that body (noted under **601**).

General bibliographies. Bay, 1910; Blake and Atwood, 1942; Frodin, 1964; Goodale, 1879; Holden and Wycoff, 1911–14; Jackson, 1881; Pritzel, 1871–7; Rehder, 1911. For bibliographies on particular regions, see the appropriate headings.

Divisional bibliographies

LÉONARD, J. 1965. Liste des flores africaines et malgaches récentes. *Webbia*, **19**: 865–7. [List of principal recent floristic works, arranged by countries. Nearly all of them are also included in this *Guide*.]

General indices. BA, 1926– ; BotA, 1918–26; BC, 1879–1939; BS, 1940– ; CSP, 1800–1900; EB, 1959– , FB, 1931– ; IBBT, 1963–9; ICSL, 1901–14; JBJ, 1873–1939; KR, 1971– ; NN, 1897–1944; RŽ, 1954– .

Divisional indices

ASSOCIATION POUR L'ETUDE TAXONOMIQUE DE LA FLORE DE L'AFRIQUE TROPICALE, 1952– . *AETFAT Index*, 1– . Brussels. [Annual bibliography recording new African taxa as well as floristic and systematic literature, with country index.]

Conspectus

500

Division in general

For tropical Africa, see **501**; for the high mountain and 'alpine' flora, see **503**; for the Sahara and other parts of dry northern Africa, see **505**. The comprehensive survey by Engler (not a flora in the strict sense) was never completed, but the contemporaneous keys by Thonner have been particularly commended and are still of much value.

Bibliography: distribution maps

LEBRUN, J.-P. and STORK, A. L., 1977. *Index des cartes de répartition: plantes vasculaires d'Afrique (1935–1976)* [Index of distribution maps: vascular plants of Africa (1935–1976)]. x, 138 pp. Geneva: Conservatoire et Jardin Botaniques. [Alphabetical species list, with citation of maps; bibliography, pp. 136–8.]

Keys to families and genera

THONNER, F., 1908. *Die Blütenpflanzen Afrikas.* xvi, 672 pp., 150 pls. (on 75l.), map. Berlin: Friedländer. English edn.: *idem*, 1915. *The flowering plants of Africa: an analytical key to the genera of African phanerogams.* xvi, 647 pp., 150 pls. (on 75l), map. London: Dulau. (Reprinted 1963, Weinheim, Germany: Cramer.) [Analytical keys to families and genera of seed plants in Africa, with synonymy, notes on range, and remarks on cultivation, other uses, etc.; statistical tables; index. An introductory section includes a bibliography of principal works on the African flora.] For supplement to the German edn., see *idem*, 1913. *Nachtrage und Verbesserungen....* 88 pp. Berlin.

ENGLER, A., 1908–25. *Die Pflanzenwelt Afrikas insbesondere seiner tropischen Gebiete. Grundzüge der Pflanzenverbreitung in Afrika und die Charakterpflanzen Afrikas.* Parts 1–3, 5(1). Illus., pls., maps (Die Vegetation der Erde, 9). Leipzig: Englemann.

Parts 2–3 of this work ('Charakterpflanzen Afrikas') contain a comprehensive, discursive, illustrated systematic account of the African flora, with emphasis on the families, genera, and more significant species and their role in the makeup of the vegetation; included are detailed accounts of local and regional distribution, and complete indices in each volume. Not completed; extends from the Pteridophyta to the end of the Archichlamydeae (Englerian system). Part 1 is entirely devoted to a general account of the vegetation and phytogeography of the continent, while part 5(1) provides supplementary material on vegetation in the tropical zone, based on further research. Part 4, to have covered the Sympetalae and the non-vascular plants, was never published.

Distribution maps

JARDIN BOTANIQUE NATIONAL DE BELGIQUE, 1969– . *Distributiones plantarum africanarum.* Fasc. 1– . Loose maps in portfolios. Brussels (later Meise).

Comprises a series of small folios of dot maps of individual species showing their distribution throughout the African continent, with indices. Each fascicle is devoted to a particular family. As of 1980, twenty fascicles, with more than 600 maps, have been published; present production is about two yearly, with some 30 maps in each.

501

Special geographical area: tropical Africa

The comprehensive *Flora of tropical Africa*, originally stimulated by Dr Livingstone, the African explorer, in 1864 upon his safe return to Great Britain, was begun the same year as a 'Kew flora' with four volumes planned. The publication of the first three volumes took place between 1868 and 1877, during which it became clear that additional volumes would be required (ultimately 12 physical parts in ten volumes). Preparation was then discontinued in disgust by Oliver because of loss of access to a key collection and was resumed only in 1891 after a nudge from Lord Salisbury – and then as an 'extra-official' project due to prior commitments to *Flora capensis* and other duties. Publication of the results of the revived programme began in 1897 and by the First World War had covered all spermatophytes save the Gramineae. Accounts for a considerable proportion of the latter were published up to 1937, but following termination of the work in the wake of World War II they were 'withdrawn', with official interest shifting to the preparation (or revision) of regional floras. The *Flora* was originally conceived in terms of west Africa ('Upper Guinea'), but because of important new collections coming in from there and from other parts of tropical Africa (notably those of Welwitsch and Livingstone/Kirk), expanding perceptions, and convenience the plan was rashly extended by 1863 to cover the *whole* of the infratropical zone – a truly vast area the size of North and Middle America together which then was little known even geographically, let alone botanically. For many areas it was decades before even a sketchily adequate basis for botanical knowledge was available, with the consequence that much of the *Flora* provides very incomplete coverage. With the passage of time – already in the 1920s it was considered outdated by those responsible for the *Flora of West Tropical Africa* – it has become mainly of historical interest, although in some regions it has yet to be wholly superseded.

OLIVER, D. *et al.*, 1868–77. *Flora of tropical Africa*. Vols. 1–3. London: Reeve. Continued as THISTLETON-DYER, W. T. *et al.* (eds.), 1897–1937. *Flora of tropical Africa*. Vols. 4–9, 10, part 1. London (later Ashford, Kent). (Vols. 1–8 partly reprinted, n.d., Ashford.)

Comprehensive, relatively concise descriptive flora of seed plants of infratropical Africa, with keys to genera and species, synonymy (with references), fairly detailed indication of distribution (including citations of *exsiccatae*), grouped into six broad regions (Upper Guinea, North Central, Nile Land, Lower Guinea, South Central, and Mozambique; not, however, corresponding to the regions adopted in this Guide), and notes on habitat, special features, etc., where known; indices to all taxa in each volume (no general index). A synopsis of families, together with Bentham's 'Outlines of Botany', appears in volume 1 and a definition of the six regions used is given in volume 7. The work, part of the 'Kew Series' of British colonial floras initiated by W. J. Hooker in the 1850s, was mostly written by herbarium botanists at Kew. When the work was finally suspended, the Gramineae (vols. 9–10) had not been completed. Along with other great floras of the day, the work ranks as 'primary' in the sense of Symington, a 'parent' to all the later regional floras.

503

Alpine and upper montane zones

The high mountain flora is well-documented in the accounts by Engler and by Hedberg; the latter is less broad in scope but is more critical and detailed, and includes keys.

ENGLER, A., 1892. *Über die Hochgebirgsflora des tropischen Afrika*. 461 pp. (Abh. Königl. Akad. Wiss. Berlin, Phys.-Math. Cl., 1891, 2). Berlin. (Reprinted 1975, Königstein/Ts.: Koeltz.)

Includes an enumeration of vascular plants recorded, with synonymy, references, citations of *exsiccatae* with localities and altitudes, phytogeographic categories, and overall distribution; index. The general enumeration is preceded by area lists of plants, with review and discussion for each, and by an account of botanical exploration, with references. Covers the mountains from Ethiopia through to Zimbabwe (Rhodesia) and Mozambique, the West African (Cameroon) mountains, the islands in the Gulf of Guinea, and the mountains of Angola.

HEDBERG, O., 1957. *Afroalpine vascular plants: a taxonomic revision.* 411 pp., 52 text-figs., 12 pls., tables (Symb. Bot. Upsal., 15(1)). Uppsala.

Critical, illustrated enumeration (pp. 21–253) of vascular plants of the alpine zone of central and east Africa, with descriptions of new taxa and keys to species (but not genera or families); full synonymy, with references; detailed indication of local range, with citations of *exsiccatae* or other records; notes and references on ecology, karyotypes, etc.; appendix (pp. 254–371) incorporating extensive taxonomic commentaries on individual species and genera; remarks on vicarious taxa; bibliography and index to all botanical names. The introductory section includes notes on prior botanical exploration, taxonomic concepts, and general notes on Afroalpine vegetation and ecology.

505

Deserts: the Sahara (and bordering areas)

Of the extensive desert areas of Africa, only the Sahara has had a discrete flora (Ozenda, 1977). The preparation of distribution maps of selected species for a wider area of dry northern Africa is, however, currently in progress under the direction of Lebrun; this particularly includes the Sahel, a critical drought area in the 1970s.

The Sahara

OZENDA, P., 1977. *Flore du Sahara.* 2nd edn. 622 pp., 177, 60 text-figs., 16 pls., 1 map (at end). Paris: Centre National de la Recherche Scientifique (CNRS). (First edn., 1958, as *Flore du Sahara septentrional et central.*)

Part II of this work comprises an illustrated manual-key to vascular plants, with limited synonymy, generalized indication of local and extralimital range, and brief notes on habitat, special features, etc.; marginal notations indicate references to the supplementary section (part III, pp. 465–589) added for this edition; while a glossary, notes on collecting and

preservation, extensive bibliography, and index to generic and family names appear as part IV. The introductory part I includes an extensive treatment of floristics, geobotany, and phytogeography of the region (also supplemented in part III with consideration of karyotypes and transitions in the flora included). No vernacular names are included. The limits of this revised edition encompass the whole of the Sahara from the Atlantic to 20 °E longitude (just east of the Tibesti massif) and from the Maghreb to the overlap with the Sahelian zone (about 18 °N latitude, with some parts to 20 °N, this being also the approximate northern limit of *Flora of west tropical Africa* (**580**)).

Distribution maps: northern Africa

LEBRUN, J.-P., 1977–9. *Éléments pour un atlas des plantes vasculaires de l'Afrique sèche,* 1–2. 50 + 40 maps (with overlays) (Études botaniques IEMVPT, 4, 6). Maisons-Alfort, Val-de-Marne: Institut d'Élevage et de Médecine Vétérinaire des Pays Tropicaux.

Comprises a series of distribution maps with accompanying descriptive text treating representative species of the drier parts of northern Africa, including the Sahel and the Sahara, with extensions to encompass southwestern Asia and nearby Macaronesian islands. The text contains synonymy (with references), brief descriptions of the plant concerned, habitat preferences, place in the ecosystem, and overall distribution, with some citations of *exsiccatae* and literature sources. No specified order is followed, but lists of the species covered appear at the beginning of each number.

Region

51

Southern Africa

Included here are all parts of the Republic of South Africa and its associated 'independent' states as well as Lesotho and Swaziland. For Namibia and Botswana, see **521** and **524** respectively.

The region is home to the rich and peculiar 'Cape' flora, highly esteemed horticulturally and collectively distinct enough to merit 'kingdom' status in the world-schemes of some phytogeographers, beginning with Diels. Interest in the flora began to develop from the time of the first Dutch settlements at the Cape in 1652 and records in the Sloane Herbarium show that some twenty people had collected in the area by 1753, the year of publication of Linnaeus' *Species plantarum*. For many, these collections were made merely in the course of stopovers, and serious botanical work dates only from the latter part of the eighteenth century. Thereafter, progress was rapid, especially after 1815, and by mid-century enough collections were available to justify preparation of a general flora as well as a 'genera plantarum'. The *Flora capensis* was launched by Harvey and Sonder in 1859 and initially covered Cape Colony, Kaffraria and Natal, the three English-dominated colonial entities of the period in southern Africa. However, the project lapsed in 1865 after the third volume and was not officially revived until 1877 as a result of increasing demand from the South African colonies, led by a developing local scientific community and caught up in northwards territorial and economic expansion. The revived *Flora capensis*, now taking in all of southern Africa to the Tropic of Capricorn and the 'great grey-green, greasy' Limpopo River, was all but completed between 1896 and 1925 (a final installment appeared in 1933), a period which also saw the formation of a number of the major Southern African herbaria, botanical gardens, and other institutions. A new capital, Pretoria, was established after formation of the Union of South Africa in 1911 and by the end of the decade it became the seat of the Botanical Survey of South Africa and its associated national herbarium. Since the 1920s the Survey (later the Botanical Research Institute of South Africa) has been responsible for botanical work at a national level, but important contributions have also been made by the large regional herbaria in Durban and Cape Town and (more locally) by the smaller, in part university, herbaria. The southern African region is now considered botanically moderately well known, with the Southern and Eastern Cape and patches elsewhere fairly well known and some remote areas still poorly collected.

In recent decades, the *Flora capensis* has become all but obsolete, and Sim's *Ferns of South Africa* (1915), covering the pteridophytes, is now much outdated, although a complementary modern checklist has been released by Schelpe (1969). Recognizing this, the Botanical Research Institute at Pretoria initiated in 1957 a project for a new general *Flora of Southern Africa* in 33 volumes, of which the first volume appeared in 1963. Progress to date, however, has been slow on account of the large size of the currently known vascular flora and the limited professional resources available; by 1974 only four per cent of the species had been revised and published.

Another production of the Botanical Survey is *The genera of Southern African flowering plants* (Phillips, 1926; revised in 1951 and again by Dyer in 1975–6), heir to the tradition of 'generic floras' as a medium of floristic communication for the region first established by Harvey in 1838 with his *Genera of South African Plants*. Phillips' *Genera* has long been regarded as a model of taxonomic work of its kind and has been indispensable to southern African botanists (along with a stablemate, Acocks' *Veld types of South Africa*, for vegetation and floristics).[3] To these works has now been added a classified bibliography (Bullock, 1978).

To these purely scientific works must be added a large range of more or less semipopular treatments of different aspects of the southern African flora. Premier among these is *The flora of South Africa* by R. Marloth (1913–32), a copiously illustrated, discursive work patterned upon Engler's *Die Pflanzenwelt Afrikas* and a forerunner of the many lavish monographs of different groups of 'Cape' plants published in the last three decades. The tree flora is especially well catered for, with four general works to suit different tastes and functions; the most comprehensive is *Trees of Southern Africa* by Palmer and Pitman (1972–3).

Coverage at the more local level, however, is still patchy, and what exists is usually in the form of checklists. Substantial descriptive manual-floras have been written only for the Cape Peninsula and Swaziland, both treating over 2000 species each, and a concise manual-key (not completed) for the Transvaal. Significant checklists have been published for parts of the Southern and Eastern Cape, Lesotho, Natal, Swaziland, the Kruger National Park, and Griqualand West, covering from 1500 to nearly 5000 species each. Some more or less semipopular works on the woody plants (or trees alone) have been written for the southern Cape, Natal, the Kruger National Park (an outstanding contribution), the Witwatersrand, and Orange Free State where particular interest has manifested itself. Many of these works have been produced privately or through local institutions.

It is thus to be hoped that additional effort can be put into prosecution of the *Flora of Southern Africa* so that its completion is not too far into the future.

– Vascular flora: 17000 species (D. B. Killick in

Boissiera, **24**: 633, 1976); 18500 species (P. Goldblatt in *Ann. Missouri Bot. Gard.* **65**: 369–436, 1978).

Bibliographies. General and divisional bibliographies as for Division 5.

Regional bibliographies

BULLOCK, A. A., 1978. In *Flora of Southern Africa: bibliography of South African botany (up to 1951)* (ed. O. A. Leistner), i, iii, 194 pp. Pretoria: South African Government Printer (for the Botanical Research Institute, Department of Agricultural Technical Services, Republic of South Africa). [Includes both author and classified (systematic) sections. This work is a revision of the bibliography on pp. 857–905 of Phillips, below.]

PHILLIPS, E. P., 1951. *The genera of South African flowering plants*, 2nd edn. [bibliography], pp. 857–905 (Mem. Bot. Surv. S. Africa, 25). Cape Town. (First edn. 1926; *ibid*., 10.)

TYRELL-GLYNN, W. and LEVYNS, M. R., 1963. *Flora africana: South African botanical books, 1600–1963*. [viii], 77 pp., plates. Cape Town: South African Public Library.

Indices. General and divisional indices as for Division 5.

510

Region in general

The classical *Flora capensis* begun by Harvey and Sonder in the mid-nineteenth century is now largely obsolete, with the three early volumes (1859–65) chiefly of historical interest. Publication of a successor work, the *Flora of Southern Africa*, has been under way since 1963 but after 16 years only six parts in five volumes (of a planned 33) have appeared. Partly filling this gap since 1926 have been successive editions of *The genera of Southern African flowering plants*, originally written by E. P. Phillips. Marloth's semipopular *Flora of South Africa* is also useful for much information not readily available in the more 'scientific' works. Several works are devoted to the trees, and one old work covers pteridophytes. For floristics and vegetation, a good background work is ACOCKS, J., 1975. *Veld types of South Africa*. [i], 128 pp., 104 photographs, 4 colored maps (Mem. Bot. Surv. S. Africa, 40). Pretoria. (First edn. 1952.) Also of interest is HUTCHINSON, J., 1946.

A botanist in Southern Africa. xii, 686 pp., illus., maps. London: Gawthorn. [Includes chapters on history and literature of Southern African botany, as well as floral regions.]

Keys to and surveys of families

DYER, R. A., 1977. *Flora of Southern Africa: key to families and index to the genera of Southern African flowering plants*. 60 pp. [Pretoria]: Botanical Research Institute, Department of Agricultural Technical Services, Republic of South Africa. [Based on *The genera of Southern African flowering plants*.]

RILEY, H. P., 1963. *Families of flowering plants of southern Africa*. xviii, 269 pp., 144 color photographs, 3 maps. Lexington, Kentucky: University of Kentucky Press. [Discursive introductory survey, with accounts of each family accompanied by (in most cases) a photograph of a representative species. Arrangement follows Hutchinson (1959). No keys provided.]

DYER, R. A., 1975–6. *The genera of Southern African flowering plants*. 2 vols. viii, 1040 pp., map. Pretoria: South African Government Printer (for the Botanical Research Institute, Department of Agricultural Technical Services). (First and second edns. see PHILLIPS, under regional bibliographies.)

Concise descriptive generic flora of seed plants, with analytical keys; includes synonymy (with references), citations of key floras or revisionary studies (for both genera and families), generalized indication of local and overall distribution of genera, and 1–2 lines of commentary on number of species, preferred habitat, and a few critical remarks; separate complete indices in each volume (for consolidated list, see Dyer, above) with addenda. An introductory section includes technical notes, statistics (nearly 2200 genera in the 3rd edn. as against 1645 recorded in 1926), and miscellaneous remarks. Vol. 2 was prepared with the assistance of Ms. A. Amelia Obermayer-Mauve and others. As in previous editions, the work covers the Republic of South Africa (with associated states), Namibia (Southwest Africa), Lesotho, and Swaziland.

DYER, R. A. *et al.* (eds.), 1963– . *Flora of Southern Africa*. Vol 1– . Illus. Pretoria: South African Government Printer (for Botanical Research Institute, Department of Agricultural Technical Services).

Large-scale descriptive 'research' flora, with family treatments mainly by specialists and staff botanists; in each family keys to genera and species, full synonymy (with references), literature citations, detailed indication of distribution, with citations of *exsiccatae* and other records, and summary of overall range, extensive taxonomic commentary, and notes on habitat, phenology, special features, etc.; figures of representative species and/or diagnostic attributes; indices to all botanical names at end of each volume. Vol. 1 includes a brief introductory section, with map. Designed to be used together with *The genera of Southern African flowering plants* (see above). Of the 33 volumes planned, to date (1981) vols. 1, 13, 16, parts 1–2, 22, 26 and 27, part 4 have been published. Volumes have been numbered, and families arranged, according to a modified Englerian sequence, but are issued only as accounts are ready; a volume may contain more than one family.

HARVEY, W. H. and SONDER, O. W., 1859 (1860)–65. *Flora capensis*. Vols. 1–3. Dublin: Hodges, Smith. Continued as THISTLETON-DYER, W. T. and HILL, A. W. (eds.), 1896–1933. *Flora capensis*. Vols. 4–7. London (later Ashford, Kent): Reeve. (Vols. 1–3 reprinted 1894, London; some volumes reprinted, n.d., Ashford; vols. 4 and 5 reprinted 1973, Lehre, Germany: Cramer.

Descriptive flora of seed plants, with analytical keys to genera and species (the latter sometimes represented only by synoptic devices, particularly in volumes 1–3), synonymy (with references), citation of *exsiccatae* with localities (more detailed in vols. 4–7) and generalized indication of overall distribution, taxonomic commentary, and notes on habitat, etc.; indices to all botanical names in each volume (no general index). Vols. 1–3 encompass only the Cape Colony, Kaffraria, and Natal; later volumes include the whole of southern Africa north to the Tropic of Capricorn and the Limpopo River. The work was treated as part of the 'Kew Series' of British colonial floras upon publication of volume 1, but not until 1877 did it become an official charge on Kew; the remaining volumes were prepared largely by herbarium botanists at that institution with some contributions by workers in southern Africa. By 1925 it was 'completed' (save for the Cycadaceae *sensu lato*, published as a supplement in 1933); no revision of the obsolescent early volumes was, however, undertaken. The volumes published to 1925 covered 11705 species and received colonial subventions totalling £3250.[4]

MARLOTH, R., 1913–32. *The flora of South Africa, with synoptical tables of the genera of the higher plants.* 4 vols. (in 6). Illus., plates (some in color). Cape Town: Darter.

Copiously illustrated, semi-popular descriptive account of the higher plants of southern Africa north to the Tropic of Capricorn and the Limpopo River (between Transvaal and Zimbabwe); includes keys to families and genera together with a consideration of the more important species and extensive remarks on their biology, adaptations, and special features; indices. Vernacular names and cross-references to *Flora capensis* are included. Not a flora in the strict sense, but valuable for its excellent illustrations and biological and other information not available in more conventional works. The overall style of this work bears comparison with Engler's *Die Pflanzenwelt Afrikas* and may have been inspired also by Warburg's three-volume *Pflanzenwelt* of 1911. A scientifically soundly based work, its influence on later works on the flora of southern Africa and its characteristic groups was considerable; it represents the first in a remarkable line of development in floristic writing which has few parallels elsewhere.

Special group – trees

The study of trees in southern Africa has attracted exceptional interest and for a comparatively small active population there is a wide and varied range of 'tree books' now available. All known tree species have been given a number in the semi-official *The national list of trees* [Die nasionale boomlys] by B. de Winter and J. Vahrmeijer (1972, Pretoria: van Schaik) and these numbers are used in the books by Palgrave, Palmer, and Palmer and Pitman described below.

BREITENBACH, F. VON, 1965. *The indigenous trees of southern Africa*. 5 vols. 345 text-figs. Pretoria: Government Printer (for the Department of Forestry, Republic of South Africa). (Mimeographed.)

Illustrated systematic descriptive account of the trees and larger shrubs of southern Africa, with keys to all taxa, limited synonymy, fairly detailed summary of internal range, vernacular names (Afrikaans, English, and the Bantu languages), and extensive notes on habitat, ecology, karyotypes, special features, pests and diseases, timber, various uses, and forestry practice; complete index in vol. 1 together with a key to families and bibliography. [This largely compiled work was regarded as an 'interim' publication, but has not itself been replaced although it has attracted some criticism (it is evidently not regarded as entirely reliable). Circulation has been comparatively limited due to withdrawal of many copies by the distributor following a copyright dispute.]

PALGRAVE, K. C., 1977. *Trees of southern Africa*. 1264 pp., 375 pls. (314 colored), numerous text-figs., maps. Cape Town: Struik.

Copiously illustrated, amply descriptive semipopular treatment of trees (including tree-ferns), with keys to all taxa, important synonymy, southern African and Rhodesian tree numbers, vernacular and standardized names (Afrikaans and English), general indication of distribution, with maps for all species, and notes on habitat, occurrence, phenology, properties and uses, and potential (with stress on arboriculture). The numerous fine illustrations include color plates, some photographic, and line drawings depicting technical details. An introductory section outlines the style and scope of the work and gives conservation rules, a list of southern African herbaria, and a general key to families, while an illustrated glossary, selected bibliography, and index conclude the work. The work was written in association with R. B. Drummond (see 525) and extends to include Namibia, Botswana, Zimbabwe (Rhodesia), and southern Mozambique (south of the Zambesi).

PALMER, E., 1977. *A field guide to the trees of southern Africa*. 352 pp., 395 text-figs., 32 color pls., map. London: Collins (Collins Field Guide Series).

Concise well-illustrated field manual-key to the more commonly encountered tree species (some 800 of the total of 1000) of southern Africa, with vernacular names (Afrikaans, English, Bantu languages), southern African tree numbers (full list, pp. 304–20), indication of distribution and habitat, and notes on properties, uses, potential, special (and diagnostic) features, etc.; select bibliography (pp. 321–2) and indices to Bantu-language names (mainly Zulu) and to botanical, Afrikaans, and English names (including synonyms). The numerous illustrations are compact, mainly stressing diagnostic technical details. In the introductory section is an illustrated glossary and vegetation-zone map as well as technical remarks and aids for the tyro. Based on the author's larger *Trees of southern Africa* (see next entry), this book is a remarkable attempt at compression; almost no other 'tree book' packs so many species into a limited compass while maintaining high professional standards.

PALMER, E. and PITMAN, N., 1972–3. *Trees of southern Africa*. 3 vols. 2000 halftones, 900 text-figs., 24 col. pl., map. Cape Town: Balkema.

Popularly oriented, discursive systematic and biological treatment of the native trees (and some large shrubs) of southern Africa south of the Kunene and Limpopo Rivers, with non-dichotomous keys to genera and species; fairly detailed synonymy, without references; vernacular names (Afrikaans and English); extensive remarks on taxonomic questions as well as habitat, life-form, phenology, biology, variability, special features, and local and extralimital distribution; indices to all botanical and vernacular names in each volume (general bibliography and indices in vol. 3). The introductory section (vol. 1) incorporates chapters on botanical exploration and research 'personalities', uses, properties, and special features of trees, and vegetation zones;

this volume also includes a chronological list of references and the 'National Check-List of Trees'.

Special groups – pteridophytes

Ferns and fern-allies are not covered in any general works on the southern African flora, and apart from Sim's elderly regional monograph no general works are available. A revised checklist has, however, been prepared by E. A. Schelpe.

SIM, T. R., 1915. *Ferns of South Africa*. 2nd edn., iv, 275 pp., 159 pls. Cape Town: Juta.

Briefly descriptive, illustrated treatment of ferns and 'fern-allies', with synoptic keys to all taxa; full synonymy, with references and citations; fairly detailed account of local range, with brief notes on extralimital distribution; remarks on habitat, ecology, special features, etc.; index to all botanical names. The introductory section includes chapters on fern morphology and biology, cultivation, identification and preservation, botanical exploration, and floristics and phytogeography together with a glossary. This work remains useful as none of the standard South African floras accounts for the lower vascular plants. See also SCHELPE, E. A., 1969. A revised checklist of the Pteridophyta of Southern Africa. *J. S. African Bot.* 35: 127–40.

511

Cape Province south of the Orange River (in general)

The Cape is here considered to include that part of the Cape Province south of the Orange River together with Ciskei and Transkei. Within this area are the former colonies of Cape Colony and Kaffraria. For Griqualand West and Bechuanaland, which lie north of the Orange River, see 519.

Apart from Sim's large-scale treatment of the woody flora, no overall treatment of Cape Province plants is available and reference must be made to general works on southern Africa or a variety of local florulas and checklists. A selection of these latter appears under 512.

SIM, T. R., 1907. *The forests and forest flora of the colony of the Cape of Good Hope*. vii, 361 pp., 161 pls., map. Aberdeen, Scotland: Taylor & Henderson.

The bulk (pp. 99–343) of this quarto work comprises an illustrated descriptive account of the trees and shrubs of the Cape Province south of the Orange River, with partially synoptic keys to all taxa together with synonymy (and references), fairly detailed account of local range, vernacular names, and notes on habitat, timber properties, history of utilization, etc.; complete index at end. The lengthy introductory section is mostly devoted to a discussion of the forests in the region and their composition, ecology, economics, utilization, and conservation.

512

Cape Province: regions

The many administrative divisions of Cape Province south of the Orange River, together with Ciskei and Transkei, are here grouped for convenience into six geographical regions. Only the first three, designated respectively as Cape Peninsula, Southern Cape and Eastern Cape in accordance with common usage, have separate works covering all or part of their areas; of these, the only descriptive manual-flora is that by Adamson and Salter (1950) covering the Cape Peninsula and Cape Town area.

I. Cape Peninsula

ADAMSON, R. S. and SALTER, T. M. (eds.), 1950. *Flora of the Cape Peninsula*. xix, 889 pp., maps. Cape Town: Juta.

Descriptive manual-flora of vascular plants of the Cape Peninsula from Cape Town southwards, with keys, limited synonymy, vernacular names, indication of local distribution (with many localities), taxonomic commentary, and notes on habitat, phenology, special features, etc.; complete indices. An introductory section gives accounts of physical features, vegetation, and human pressures as well as technical notes and a glossary. The Cape Peninsula, which includes Table Mountain, has perhaps the richest flora for an area of equivalent size anywhere in the world; this manual treats no less than 2622 species.[5]

II. Southern Cape (Olifants River to Humansdorp)

A useful introduction to this area, which includes an important belt of temperate rain forest, is the following: BREITENBACH, F. VON. 1974. *Southern Cape forests and trees*. Pretoria: South African Government Printer.

FOURCADE, H. G., 1941. *Check-list of flowering plants of the divisions of George, Knysna, Humansdorp and Uniondale*. 127 pp., map (Mem. Bot. Surv. S. Africa, 20). Pretoria.

Tabular systematic checklist of seed plants, with indication of occurrence by district as well as essential synonymy and cross references to *Flora capensis*; includes statistics (2969 native species and 166 aliens in 5429 square miles) and list of additions to *Flora capensis*; index to generic names. A brief introduction covers geography of the area and the history of collecting along with technical notes. (Seen by courtesy of J. H. Ross, Melbourne.)

III. Eastern Cape (Port Elizabeth to East London)

Within this area is the former territory of Kaffraria as used in *Flora Capensis* as well as by Sim (1894). Also included here is the present 'self-governing' territory of Ciskei.

MARTIN, A. R. H. and NOEL, A. R. A., 1960. *The flora of Albany and Bathurst*. xxiv, 128 pp., map. Grahamstown: Department of Botany, Rhodes University.

Annotated checklist of vascular plants, with vernacular names and notes on habitat, occurrence, variability and phenology, preceded by general notes and including an index. The introductory section comprises a survey of the vegetation. Covers 2390 species in an area centering on Grahamstown, between Port Elizabeth and East London.

SCHONLAND, S., 1919. *Phanerogamic flora of the divisions of Uitenhage and Port Elizabeth*. 118 pp., map (Mem. Bot. Surv. S. Africa, 1). Pretoria.

Briefly annotated enumeration of seed plants (2312 species), with indication of status, localities, flowering times, occurrence in the Cape Peninsula and/or Natal, and some critical notes; statistical summary. An introductory section covers geography, geology, climate, composition of the flora, vegetation formations, and phytogeography with also history of collecting. [This first publication of the Botanical Survey of South Africa (now the Botanical Research Institute) was seen as part of a 'regional floras' program.]

SIM, T. R., 1894. *Sketch and check-list of the flora of Kaffraria*. 92 pp. Cape Town: Argus Printing and

Publishing (for King William's Town Natural History Society).

Systematic list of vascular plants, with vernacular names and essential synonymy; no index. An introduction includes accounts of physical and biological features of the region, botanical exploration, vegetation formations, and aspects of utilization. Includes 2449 species for what is now the Eastern Cape, encompassing Port Elizabeth and East London along the southeastern coast. Now well out-of-date; partly superseded by the checklists of Martin and Noel and of Schonland (see above).

IV/V. Northwestern and Central Cape

No significant separate florulas or checklists appear to be available for any part of these areas, which together include the Great Karroo and Namaqualand as well as the dry central and western highlands south of the Orange River.

VI. Transkei

No significant separate floras or checklists are available for this nominally independent Black state, which lies between the Eastern Cape region of Cape Province (bordered by the Great Kei River), Lesotho (513), and Natal (514).

513

Lesotho

This state, under direct British rule before independence, was formerly known as Basutoland.

– Seed plants: 1537 species (Jacot Guillarmod).
JACOT GUILLARMOD, A., 1971. *Flora of Lesotho* (*Basutoland*). 474 pp., maps. Lehre: Cramer.

Systematic enumeration of cellular and vascular plants, with limited synonymy, detailed indication of local range (including *exsiccatae* and other records), and summary of extralimital distribution; lexicon of Sotho names, with equivalents; lists of plant uses, bibliography, and index to generic and family names. The introductory section includes accounts of the history, physical features, land utilization, and vegetation of Lesotho as well as a statistical analysis, a cyclopedia of

collectors, and a gazetteer. No keys or descriptions are included.

514

Natal and KwaZulu

This pair of polities, the South African province of Natal and the 'self-governing' territory of KwaZulu, is by southern African standards comparatively well-supplied with provincial-level floristic works.

– Vascular flora: 4818 seed-plant species (Ross, 1972), a figure now (1980) considered to be over 5000 (J. H. Ross, personal communication).

Ross, J. H., 1972(1973). *The flora of Natal*. 418 pp., 5 text-figs. (incl. 2 maps) (Mem. Bot. Surv. South Africa, 39). Pretoria.

Comprises a systematic enumeration of seed plants, with diagnoses of and keys to families and genera but having species only listed; includes essential synonymy, brief indication of local range with citation of representative, mostly recent *exsiccatae*, and taxonomic commentary; index to generic and family names at end. The introductory section includes a glossary and list of references. Supersedes BEWS, J. W., 1921. *An introduction to the flora of Natal and Zululand*. vi, 248 pp. Pietermaritzburg.

WOOD, J. M. (and EVANS, M. S.), 1898–1912. *Natal plants*. Vols. 1–6. 600 pls. Durban: Natal Government and Durban Botanic Society. (Reprinted 1970 in 1 vol., Lehre, Germany: Cramer.)

Comprises an atlas of seed plants, with plates and accompanying descriptive text; the latter includes references to *Flora Capensis*, generalized indication of local range (with some citations of *exsiccatae*), taxonomic commentary, and notes on habitat, uses, etc.; index to all taxa in each volume (no general index). No systematic order is followed. M. S. Evans was joint author of vol. 1 of this work, which was discontinued before completion.

Special group – woody plants

HENKEL, J. S., 1934. *A field book of the woody plants of Natal and Zululand*. xii, 252 pp., 2 pls. Pietermaritzburg: Natal University Development Fund Committee.

Artificial manual-key to woody plants, the keys mainly based upon vegetative attributes; includes general summaries of local range. A bibliography, glossary, and indices to botanical terms and all taxa are also incorporated.

MOLL, E. J. [1967.] *Forest trees of Natal.* 180 pp., illus., maps. [Pietermaritzburg]: Wildlife Protection and Conservation Society of South Africa (Natal Branch).

Popularly oriented pocket field manual of more important trees, with figures and dot maps; includes keys. Less complete than the manual of Henkel (see above), but more extensively illustrated.

515

Transvaal (including 'associated' polities)

For works on Kruger National Park, see **517**. Transvaal as delimited here includes the South African province of Transvaal, the states of Venda and (in part) Bophuthatswana, and the more or less 'self-governing' territories of KaNgwane, KwaNdebele, Gazankulu and Lebowa. The most recent complete work nominally covering the whole of this area is the checklist by Burtt-Davy and Pott-Leendertz (1912, with additions to 1920); Burtt-Davy's manual of 1926–32 was, alas, never completed. Both works also cover Swaziland (**516**) but for that area have now been superseded. Miscellaneous local checklists and other works exist, of which the most substantial is a 'tree book' for the Witwatersrand, the great gold-mining district centering on Johannesburg.

BURTT-DAVY, J., 1926–32. *A manual of the flowering plants and ferns of the Transvaal with Swaziland, South Africa.* Parts 1–2. pp. i–xxxv, 1–529, figs. 1–80, map. London: Longmans, Green.

Manual-key to vascular plants, with limited synonymy, indication of local range (including citation of *exsiccatae*) as well as a general summary of extralimital distribution, taxonomic commentary, and notes on uses, etc.; indices to generic and family names in each part. An appendix includes descriptions of new taxa. Not completed; covers families from pteridophytes through the end of the Umbelliferae (1926–34 Hutchinson system). [Parts 3 and 4, which would have concluded the work, were completed and in galley proof by 1941 but were never published (STAFLEU and COWAN, *Taxonomic Literature*, 2nd edn., vol. 1, p. 603).]

BURTT-DAVY, J. and POTT-LEENDERTZ, R., 1912. A first check-list of the flowering plants and ferns of the Transvaal and Swaziland. *Ann. Transvaal Mus.* 3: 119–82. (Reprinted separately.)

Unannotated systematic list of vascular plants,

with index to genera and families. The introductory part includes new combinations and a synopsis of families. For supplement, see POTT-LEENDERTZ, R., 1920. Addendum to the first check-list.... *Ibid.*, 6: 119–135.

Partial works

TREE SOCIETY OF SOUTHERN AFRICA, 1969. *Trees and shrubs of the Witwatersrand.* 2nd edn., illus. by B. Jeppe. xxi, 309 pp., 126 text-figs. (some in color). Johannesburg: Witwatersrand University Press.

Semipopular illustrated guide, with pictured keys to species, field descriptions, vernacular names; local range, and notes on habitats, special features, etc.; glossary, references, and complete index. The introductory section includes the circumscription of the area and propagation notes.

516

Swaziland

Prior to independence, this state was under direct British rule. In recent years, it has been supplied with two good general works, which supersede the partly incomplete treatments of BURTT-DAVY (see **515**).

– Seed plants: 2118 species (Compton).

COMPTON, R. H., 1966. *An annotated check list of the flora of Swaziland.* iii, 191 pp., 3 maps (J. S African Bot., Suppl. 6). Kirstenbosch, Mbabane.

Annotated list (pp. 25–79) of vascular plants, with notes on local range, habitat, vegetation formation-class, life-form, etc. An extensive appendix incorporates descriptive notes on individual plant families as represented in Swaziland, a lexicon of vernacular names, and an index to family and generic names. The introductory section contains descriptions of vegetation formations and a list of symbols.

COMPTON, R. H., 1976. *The flora of Swaziland.* 684 pp. (J. S African Bot., Suppl., 11). Kirstenbosch, Mbabane.

Concise descriptive manual-flora of seed plants with keys to genera and species, Swazi vernacular names, indication of distribution, with localities (list, pp. 7–8), representative *exsiccatae*, and altitudinal zonation (by three *veld* belts), and casual notes on habitat, phenology, etc.; appendix (pp. 675–6) of name changes from the checklist of 1966; index to family and generic names. An introductory section, with dedication to King Sobhuza II, includes an account of botanical exploration (beginning with Galpin in 1886), the background to the *Flora*, and technical notes. No family key is provided (see that of Dyer under **510**). [Although designed as a practical manual-flora, the format, imposed by the journal, is regrettably somewhat too large.]

517

Kruger National Park

This great wildlife reserve, one of the largest in Africa, is located in a strip of northeastern Transvaal, east of the Drakensberg divide from the Swazi border to the Limpopo, with Mozambique on the eastern boundary. Apart from a checklist of the flora (van der Schijff, 1969) there is a splendid modern work on the tree flora (van Wyk, 1972–4) supplanting Codd's sketchy treatment of 1951.

SCHIJFF, H. F. VAN DER, 1969. *A check-list of the vascular plants of the Kruger National Park*. [ii], 100 pp. (Publ. Univ. Pretoria, 53). Pretoria.

Includes an annotated checklist of 1838 vascular plant species, with limited synonymy, localities with citations of *exsiccatae*, and indication of habitat, etc. The introductory section covers plant communities and their significance in relation to park management, biogeography and the affinities of the flora, new records, statistics, and references.

Special group – woody plants

WYK, P. VAN, 1972–4. *Trees of the Kruger National Park*. 2 vols. xxii, 597 pp., 200 color pls., maps. Cape Town: Purnell.

Copiously illustrated semi-popular descriptive treatment of trees, with keys, vernacular names, many distribution maps as well as indication of localities, and notes on habitat, biology, special features, etc. Partly supplants CODD, L. E. W., 1951. *Trees and shrubs of the Kruger National Park*. 192 pp., 65 text-figs., 6 pls., map (Mem. Bot. Surv. S Africa, 26). Pretoria.

518

Orange Free State (Oranje-Vrystaat)

This area, which includes the territory of Qwaqwa and a small part of the Black state of Bophuthatswana, centers on Bloemfontein. It has had only a limited botanical literature; the field guide to woody plants by Venter is the first general floristic work as well as one of the few in southern Africa with Afrikaans text.

VENTER, H. J. T., 1976. *Bome en struike van die Oranje-Vrystaat*. [Trees and shrubs of the Orange Free State.] 240 pp., illus. Bloemfontein: de Villiers.

Artificially arranged students' field guide to erect woody plants, with small technical figures and distribution maps for each species; includes Afrikaans and English vernacular names. An illustrated glossary and list of references are provided at the end, and a general key in the introductory section. With an entirely bilingual text, this work encompasses both indigenous and naturalized species (154 in all).

519

Griqualand West, Bechuanaland and western Bophuthatswana

This area comprises those parts of the Cape Province north of the Orange River together with the western parts of the nominally independent Black state of Bophuthatswana. Griqualand West includes the famous diamond-mining center of Kimberley, while Bechuanaland takes in the southern part of the Kalahari Desert. Much of the area is imperfectly known; it is not covered by any works on Cape Province and only one local florula, described below, is available (Wilman, 1946).

Partial work

WILMAN, M., 1946. *Preliminary check list of the flowering plants and ferns of Griqualand West* (South Africa). vii, 381 pp., map. Cambridge, England: Deighton Bell; Kimberley: Alexander McGregor Memorial Museum.

Systematic enumeration of vascular plants of the Kimberley area and other parts of Griqualand West, with cross references to *Flora capensis*, vernacular names, local range (including citation of *exsicaatae*), and notes on habitat, ecology, phenology, uses, etc.; glossary and indices. No keys are included.

Region

52

South Central Africa

This region comprises the Caprivi Strip, Botswana, Namibia (Southwest Africa), Zimbabwe (Rhodesia), Zambia, Malawi, Angola, and Mozambique. Three serial floras (two still in progress) collectively cover the area: *Prodromus einer Flora von Südwestafrika*, *Conspectus florae angolensis*, and *Flora zambesiaca*.

While scattered collecting along the coasts was accomplished in the seventeenth and eighteenth centuries, and notable field work was done by Welwitsch in the mid-nineteenth century, serious botanical collecting dates mainly from the latter part of that century with penetration by European expeditions and subsequent imposition of colonial rule. The first overall checklist (for the new British territories) appeared in 1898, and collections of the day were worked into *Flora of tropical Africa*. The first regional herbarium was begun at Salisbury early in the present century, and currently is the largest (and most active) in this part of Africa. It was not until the 1950s, however, that a sufficient basis was available for a comprehensive flora, although some local checklists and other floristic contributions had appeared earlier and the Angolan *Conspectus* had commenced publication in 1937. As Namibia was preoccupied by another project, the organizers of *Flora zambesiaca* limited its scope to the central and eastern polities, i.e., the Zambesian subregion.

Production of *Flora zambesiaca* since 1960 has been relatively slow. While events of the last decade and a half in the region have contributed to delays in production, a major limiting factor is the continuing adherence to publication in volumes according to a predetermined sequence of families (in the tradition of the earlier Kew tropical floras) rather than in family fascicles or in volumes of randomly revised families, while at the same time depending upon contributions from a large number of specialists. This has been partly overcome by bypassing completion of volume 3, which is allocated entirely to the very numerous Leguminosae, and releasing volume 4 and the first part of volume 10 (the latter to be devoted entirely to Gramineae). A similar technical problem exists with the associated *Conspectus florae angolensis*. The present *Flore d'Afrique*

Centrale (560) originally appeared in like fashion but in 1967 commenced publication in family fascicles, like most other current serially issued tropical floras.

A notable modern contribution is Merxmüller's *Prodromus einer Flora von Südwestafrika* (1966–72), which was published over six years following 15 years of preparation. It represents an object lesson to flora writers in limitation of objectives, extensive preparation, thorough distillation, and rapid publication (but not so as to break one's pocket!). In many ways this work is reminiscent of Merrill's *Enumeration of Philippine flowering plants*, but it is improved over that work by the addition of keys and short descriptions.

Publication of *Flora de Moçambique* has also been revived after an apparent period of uncertainty following the events of 1974. However, because of *Flora zambesiaca*, no general floras are available for Malawi, Zambia, Zimbabwe (Rhodesia), and Botswana; the sole overall coverage is through checklists – available only for Malawi and Zimbabwe (Rhodesia).

With regard to the woody flora, an especially noteworthy book is *Trees of central Africa* (Coates Palgrave and Palgrave, 1957), with colored paintings and descriptive text. Several other works on the woody flora (or trees alone) are also available, covering all polities except Angola and Namibia, and the southernmost parts of the region are more or less encompassed by some Southern African tree books (510).

Botanical exploration has been considerable but somewhat patchy, and extensive remote areas are still poorly known.

– Vascular flora: about 6000 species (excluding Angola and Namibia; perhaps 10000 to 11000 for the whole of the region).

Bibliographies. General and divisional bibliographies as for Division 5.

Indices. General and divisional indices as for Division 5.

520

Region in general

See also **501** (*Flora of tropical Africa*). *Flora Zambesiaca* does not cover Angola or most of Namibia.

EXELL, A. W. *et al.* (eds., for Flora Zambesiaca Committee), 1960– . *Flora Zambesiaca*. Vols. 1– , Supplement. Illus. (some in color), maps. London: Crown Agents (distributed by HMSO).

Comprehensive illustrated descriptive 'research' flora of vascular plants, with keys to all taxa, full synonymy (with references and citations), symbolic summary of local range, with indication of *exsiccatae*, and generalized summary of overall distribution, taxonomic commentary, and notes on habitat, special features, etc.; colored frontispiece and indices to all botanical names in each volume. Volume 1 includes a general introduction in Portuguese and English, a historical account of botanical exploration, a glossary, and a list of principal references, along with a general key to families in English (Portuguese version in vol. 2). Covers the entire region except Angola (for which a parallel project is in progress) and Namibia. Publication of this flora, a joint project of the British Museum (Natural History) and the Botanical Institute of Coimbra University (Portugal) under control of a steering committee and also involving the herbaria at Kew and Salisbury, has been somewhat fitful, particularly since 1965; by 1975 only vols. 1, 2, 3, part 1, 10, part 1, and the Supplement (on the pteridophytes) had been published, but in spite of political and other difficulties preparation has been continuing and in 1978 volume 4 (Rosaceae through Cornaceae except Crassulaceae (families 62–93 on the modified Hutchinson system used)) appeared. Part of the very numerous Leguminosae (in vol. 3) and Gramineae (in vol. 10) remain to be completed, as well as the 'Sympetalae', 'Monochlamydeae', and the remainder of the Monocotyledoneae.

Partial work

BURKILL, I. H., 1898. List of the known plants occurring in British Central Africa, Nyasaland, and the British territory north of the Zambesi. In *British Central Africa* (ed. H. H. Johnston), 2nd edn., App. II, pp. 233–84, 284a–284l. London: Methuen.

List of vascular and non-vascular plants recorded from British Central African territories (corresponding to Zimbabwe, Zambia, and Malawi), with emphasis on the last two named; includes limited synonymy and local distribution (with citation of some *exsiccatae*); no index. A brief introductory section includes a short history of botanical exploration.

Special group – trees

The southern part of the region (Botswana, Zimbabwe, and southern Mozambique) is included in *Trees of southern Africa* by K. C. Palgrave (see **510**).

COATES PALGRAVE, O. H. and PALGRAVE, K. C., 1957. *Trees of Central Africa*. xxviii, 466 pp., 110 color pls.,

halftones, color frontispiece, map. [Salisbury]: National Publications Trust, Rhodesia and Nyasaland.

Semi-popular, richly illustrated 'coffee-table' guide to the more common and/or conspicuous trees of present-day Zimbabwe, Zambia and Malawi, with rather detailed descriptive text and notes on habitat, biology, special features, uses, folklore, etc.; index.

521

Namibia (Southwest Africa)

The Caprivi Strip is here excluded, being separately considered as **523**; it is the only part of Namibia included in *Flora Zambesiaca*. The rest of Namibia is, in addition to the recent flora by Merxmüller, variously covered by the floras and other manuals on southern Africa (see **510**); some older works only reach the Tropic of Capricorn, which passes not far south of Windhoek, while more recent works, including the latest edition of *Genera of southern African flowering plants* and the tree book of Palgrave (1977), extend right up to the Angolan boundary.

MERXMÜLLER, H. (ed.), 1966–72. *Prodromus einer Flora von Südwestafrika*. 35 fascicles. 2188 pp., maps. Lehre, Germany: Cramer.

Concisely descriptive flora of vascular plants, with analytical keys to genera and descriptive–diagnostic keys to species; full relevant synonymy, with references and citations; indication of *exsiccatae*, with localities, and concise summary of local range; taxonomic commentary and notes on special features; no individual indices. A map and index to families appear on the covers of each part. Fascicle 35 includes an introduction to the work, a list of family names, and a general key to families. For additions see ROESSLER, H. and MERXMÜLLER, H., 1976. Nachträge zum *Prodromus einer Flora von Südwestafrika. Mitt. Bot. Straatssamml. Munchen*, **12**: 361–73.

522

Angola

The major synthetic work on the area, the *Conspectus*, is complementary to *Flora Zambesiaca* and, like that work, is a product of long-standing collaboration between the Botanical Institute of Coimbra University (Portugal) and the British

Museum (Natural History). However, it is only some 40 per cent or so completed and the earlier parts are now somewhat outdated.

CARRISSO, L. *et al.* (eds.), 1937– . *Conspectus florae angolensis.* Vols. 1– . Illus. Lisbon: Junta de Investigações do Ultramar (later Junta de Investigações Cientificas do Ultramar).

Systematic enumeration of seed plants, with descriptions of new taxa and keys to genera and species; extensive synonymy, with references and principal citations; indication of *exsiccatae*, with local range, and general summary of extralimital range; notes on habitat, phenology, special features, etc.; figures of representative species (more abundantly provided from vol. 3 onwards); index to all taxa in each volume. A general introduction in both Portuguese and English appears in vol. 1. At this writing (1979), vols. 1–3 and 4, part 1 have appeared, the most recent part in 1970; family arrangement follows a modified Bentham and Hooker sequence. For pteridophytes, see SCHELPE, E. A., 1977. *Conspectus florae angolensis: Pteridophyta.* 197 pp., 33 pls. Lisbon. [Includes keys.]

523

Caprivi Strip

This portion of Namibia (South West Africa) is a narrow strip of land separating Botswana from Angola and Zambia (except that just to the east is a very short common boundary between Botswana and Zambia). It is the only part of Namibia included in *Flora Zambesiaca.* For separate coverage, see **521** (Merxmüller).

524

Botswana

See also **510** (all works except Dyer, although some do not cover the infratropical zone). The tree flora is also accounted for in the manuals of Palgrave (1977) and Palmer (1977). Much of the country is botanically still poorly known; no separate general flora or checklist is available.

MILLER, O. B., 1952. The woody plants of the Bechuanaland Protectorate. *J. S. African Bot.* **18**: 1–100, map.

Systematic enumeration of the woody plants of Botswana, with briefly descriptive notes; limited synonymy; vernacular names; generalized indication of local range, with citation of some *exsiccatae*; taxonomic commentary and notes on phenology, wood properties, etc.; glossary and lexicon of vernacular names; index to generic and family names. The introductory section includes lists of principal collectors and descriptive remarks on the physical features and vegetation formations of Botswana.

525

Zimbabwe (Rhodesia)

Much locally based effort has gone into contributions to the general *Flora Zambesiaca*, conceived as part of the externally-inspired scientific development of the former Federation of Rhodesia and Nyasaland, and not towards production of a 'national' flora. Apart from one aging checklist, the only substantial separate works deal with the woody flora; these latter are further supplemented by the semipopular works of Coates Palgrave and Palgrave (1957; see **520**) and Palgrave (1977; see **510**).

EYLES, F., 1916. A record of plants collected in Southern Rhodesia. *Trans. Roy. Soc. South Africa*, **5**(4): 273–564.

Systematic enumeration of cellular and vascular plants (2397 species), with limited synonymy (including references and citations), indication of *exsiccatae* with localities, and index to all botanical names. A list of basic references on the flora appears in the introduction.

DRUMMOND, R. B., 1975. A list of trees, shrubs, and woody climbers indigenous or naturalized in Rhodesia. *Kirkia*, **10**(1): 229–85, map.

Systematic checklist of woody plants, with standard numbers, limited synonymy, abbreviated indication of local and extralimital range, indication of habit, and citation of a representative *exsiccatae* (deposited in Salisbury); references to standard floras, revisions, etc., provided; index to family and generic names at end. Covers 1172 species. The map gives distribution units as used in the work. See also PARDY, A. A., 1951–6. Notes on indigenous trees and shrubs of Southern Rhodesia. *Rhodesian Agric. J.* **48**(3)–53(6) (variously). (Reprinted under separate cover, Salisbury.) [Halftone plates with accompanying descriptive text; includes vernacular names and notes on uses]; and STEEDMAN, E. C., 1933. *A description of some trees, shrubs and lianes of Southern Rhodesia.* xix, 191 pp., 92 half-tone pls. [Bulawayo]: The author. [Illustrated descriptive treatment of selected woody plants, with vernacular names and notes on uses.]

526

Zambia

Formerly known as Northern Rhodesia, and for a time after World War II included in the Federation of Rhodesia and Nyasaland. Except for the woody plants, no separate general works are available, save an introduction to the flora with keys to families prepared by N. C. Nath Nair for use at the University of Zambia (details not at present available).

WHITE, F., 1962. *Forest flora of Northern Rhodesia.* 482 pp., 72 text-figs., map. Oxford: Oxford University Press.

Briefly descriptive account of trees and shrubs, with keys to genera and species; limited synonymy, with citations of principal works; vernacular names; indication of *exsiccatae*, with fairly detailed summary of local range; taxonomic commentary and notes on habitat, associates, special features, uses, etc.; indices to vernacular, generic and family names. A bibliography appears in the introductory section. This work is partially updated by the following: FANSHAWE, D. B., 1973. Checklist of the woody plants of Zambia showing their distribution. iv, 48 pp., map (Zambia Forest Res. Bull., 22). Lusaka.

Special group – pteridophytes

KORNAŚ, J., 1979. *Distribution and ecology of the pteridophytes in Zambia.* 207 pp., 83 figs. (mostly maps), 6 pls. Warsaw/Krakow: Pánstwowe Wydawnictwo Naukowe.

Systematically arranged record and analysis of local distribution of Zambian pteridophytes (146 species), based on extensive fieldwork by the author; includes citation of *exsiccatae*, critical remarks (with revisions of Schelpe's *Flora zambesiaca* treatment), and notes on chorology, habitat, biology, phenology, etc., with distribution maps for each species and altitudinal zonation given. A general introductory section is also provided. [One of the most thorough treatments of African pteridophytes currently available.]

527

Malawi

Formerly known as Nyasaland, and for a time after World War II included in the Federation of Rhodesia and Nyasaland. Various checklists and other works covering parts of the flora are available, the best being that by Burtt-Davy and Hoyle (1958). A significant additional contribution to the flora, based on collections by L. J. Brass, is BRENAN, J. P. M. *et al.*, 1953–4. Plants of the Vernay Nyasaland expedition (1946), I–III. *Mem. New York Bot. Gard.* **8**: 191–256, 409–510; **9**: 1–132.

BURTT-DAVY, J. and HOYLE, A. C., 1958. *Check list of the trees and shrubs of the Nyasaland Protectorate.* 2nd edn., revised P. Topham. 137 pp., tables. Zomba: Government Printer. (First edn. 1936, Oxford, as *Check-lists of the forest trees and shrubs of the British Empire, 2: Nyasaland Protectorate*).

List of trees and shrubs, with limited synonymy, vernacular names, and brief notes (in tabular form) on habitat, life-form, etc.; lexicon of vernacular names, with botanical equivalents; index to genera. The introductory section includes an account of the forest types of the region, with indication of important species. For similar lists relating to the herbaceous flora, see BINNS, B., 1968. *A first check-list of the herbaceous flora of Malawi.* 113 pp., map. Zomba: Government Printer; also JACKSON, R. G., and WIEHE, P. O., 1958. *An annotated check-list of Nyasaland grasses.* 75 pp., illus. Zomba: Government Printer.

528

Mozambique

Much of the country north of the Zambezi River is also covered in *Die Pflanzenwelt Ostafrikas* by Engler (see **531**). For the tree flora in the area south of that river, see also PALGRAVE (1977; under **510**).

FERNANDES, A. and MENDES, E. J. (eds.), 1969– . *Flora de Moçambique.* Fasc. 1– . Illus. Lisbon: Junta de Investigações Cientificas do Ultramar.

Copiously illustrated, serially issued flora of vascular plants; each family fascicle includes keys to genera and families, full synonymy (with references and citations), vernacular names, localities with indication of *exsiccatae* and general summary of extralimital distribution, critical remarks, and notes on habitat, special features, occurrence, etc., and complete botanical index. Publication of fascicles is not in a set sequence of families. At this writing (early 1980), nearly 60 fasicles have appeared, the latest in 1979 after an interruption of five years occasioned through 'readjustment'. For ferns and fern-allies, see SCHELPE, E. A. C. L. E. and DINIZ, M. A., 1979. *Flora de Moçambique: Pteridophyta.* 257 pp. Lisbon.

GOMES E SOUSA, A., 1966–7. *Dendrologia de Moçambique.* 2 parts, 817 pp., 229 pls., text-figs., frontispiece, 2 color maps (Mem. Inst. Invest. Agron. Moçambique, Centr. Doc. Agrar., 1). Lourenço Marques [Maputo]: Imprensa Nacional de Moçambique.

Illustrated descriptive treatment *in extenso* of trees and shrubs, with keys only to genera; full synonymy, vernacular names, fairly detailed indication of local range with notation of formation-types and cursory summary of extralimital distribution, taxonomic commentary, and notes on habitat, phenology, wood structure, etc.; index only to illustrations. The introductory section includes general geographical notes and accounts of floristic zones, vegetation formations, and early and recent botanical exploration, with two accompanying maps.

Region

53

East tropical Africa

This is here considered to include Tanzania (with the offshore islands of Zanzibar and Pemba), Rwanda, Burundi, Uganda, and Kenya.

Because of Arab domination and distance from trade routes, early European contacts were limited until well into the nineteenth century, with the result that few old collections are known, and until 1860 most collections originated from Zanzibar and the area of Mombasa. The first major inland expedition was that of Speke and Grant in 1860, and afterwards contacts, aided greatly by the opening of the Suez Canal in 1869, increased rapidly, with many other expeditions being mounted. With these, serious collecting began in the region, and continued under colonial rule and afterwards until the greater part of the area, except for more remote parts of Kenya and Tanzania, is now considered moderately well to well known. Particularly energetic collecting programmes were prosecuted in present-day Tanzania by the Germans, who also established in 1902 in Amani the forerunner of the present East African Herbarium (now at Nairobi), today one of the largest herbaria in Africa and an important contributor to floristic work.

By 1950 collections had reached a level officially deemed sufficient for initiation of a regional successor to the abandoned *Flora of tropical Africa*, which had accounted for many of the earlier collections. Thus in 1952 there began to appear, as part of 'development' for the so-called East African Community,[6] the serial *Flora of Tropical East Africa* whose ponderous progress has dominated regional floristic botany ever since. As a 'Kew flora' it continues an unbroken tradition now 125 years old, seemingly undisturbed by changing sponsorship and means of publication. At this writing (1979) it covers about one-half of the vascular plant species (estimated at one-third in 1974), and may not be 'finished' until the end of the century.

Rwanda and Burundi, although geographically part of east Africa and at one time German territory, were mandated to Belgium after World War I and thus are covered in the present *Flore d'Afrique Centrale* (**560**) and in other works prepared by Belgian botanists, and not by *Flora of Tropical East Africa*.

Modern descriptive floras of more restricted scope are few; only upland Kenya (Agnew, 1974) and Rwanda (Troupin, 1971, 1978–   ) are so accounted for. Other coverage is very sketchy and some of it much outdated, such as Engler's *Verzeichnis* (Pflanzenwelt Ostafrikas, Teil C) of 1895 which is the latest 'separate' work for Tanzania. Better coverage exists for the woody plants, for which manuals, some lavishly illustrated, have been produced for each of the three 'Community' countries. The 'Afroalpine' flora has been covered in *Afroalpine vascular plants* (Hedberg, 1957; see **503**) as well as in a more general review of the African mountain flora by Engler (1892).

– Vascular flora: 10000–11000 species.

Bibliographies. General and divisional bibliographies as for Division 9.

Indices. General and divisional indices as for Division 9.

530

Region in general

See also **501** (*Flora of tropical Africa*); **503** (ENGLER; HEDBERG). The comprehensive *Flora of tropical East Africa* does not include Rwanda and Burundi.

TURRILL, W. B. *et al.* (eds.), 1952– . *Flora of tropical East Africa*. In fascicles. Illus., maps. London: Crown Agents (from 1980, Rotterdam: Balkema).

Comprehensive, detailed serial 'research' flora of vascular plants; each family fascicle includes keys to genera, species, and infraspecific taxa, full synonymy, with references and citations, typification, abbreviated designation of internal floristic districts, summary of internal range, with citation of (mainly representative) *exsiccatae*, and generalized indication of extralimital distribution, taxonomic commentary (often lengthy), and notes on habitat, special features, associates, etc., and complete botanical index. A general introduction to the work (1952) and a glossary appear in special fascicles. Publication of fascicles follows no set sequence, but tables for arrangement according to any one of three common systematic schemes (Bentham and Hooker, Engler, and the first version of Hutchinson) are provided. Progress on this work, prepared mainly at the Kew Herbarium, has been steady and at this writing (1979) about one-half of all species in something like 40 per cent of the families have been published (J. P. M. Brenan, personal communication).

531

Tanzania

The unit covers the mainland sector, formerly known as Tanganyika Territory and before that as German East Africa.

BRENAN, J. P. M. and GREENWAY, P. J., 1949. *Check-lists of the forest trees and shrubs of the British Empire*, 5: *Tanganyika Territory*, part II. xviii, 653, [ii] pp. Oxford: Imperial Forestry Institute.

Concise manual-checklist of woody plants, with keys only to species or groups thereof; references to key literature under family headings; limited synonymy; vernacular names; abbreviated indication of local range, with citation of representative *exsiccatae*; notes on habitat, phenology, associates, uses, etc.; index to genera. (Part I of this work, published in 1940, consists of a lexicon of botanical–vernacular and vernacular–botanical name equivalents, compiled by F. B. Hora.)

ENGLER, A. *et al.*, 1895. *Die Pflanzenwelt Ostafrikas und der Nachbargebiete*. Teil C: *Verzeichnis der bis jetzt aus Ost-Afrikas bekannt gewordenen Pflanzen*. [ii], ii, 433, 40 pp., 45 pls. Berlin: Reimer.

Systematic enumeration of vascular and non-vascular plants, with descriptions of new taxa and diagnoses of genera; synonymy, with references and principal citations; vernacular names (in several languages); symbolic indication of local range, with citation of some *exsiccatae*; general summary of overall range in Africa; notes on habitat, properties, uses, etc.; indices to all taxa and to vernacular names. The introduction includes a phytogeographic subdivision of Africa (39 regions) and a concise account of botanical exploration in former German East Africa and neighboring areas. The work covers the mainland part of Tanzania, Rwanda, Burundi, northern Mozambique, southern Uganda, and southern Kenya.

PETER, A., 1929–38. *Flora von Deutsch-Ostafrika*. Lfg. 1–3. 135 pls. (Repert. Spec. Nov. Regni Veg., Beih. 40(1): 540, 142 pp., 91 pls.; 40(2): 272, 36 pp., 44 pls.). Berlin: Fedde.

Keyed, rather uncritical enumeration of vascular plants based almost entirely upon the author's own as well as other German collections; includes descriptions of new taxa (in separately paged appendices), synonymy with references and citations, indication of *exsiccatae* with localities, general summary of overall distribution, and notes on phenology, abundance, uses, etc.; no indices. The introductory section includes a listing of Peter's itineraries of 1914–19 and 1925–6. Very incomplete; Lieferungen (issues) 1 and 2 cover pteridophytes, gymnosperms, and monocotyledons from Typhaceae through Cyperaceae, while Lieferung 3 covers dicotyledons from Casuarinaceae through Basellaceae (Englerian system). [Collections made by British workers were singularly neglected by this author, an ardent Germanophile. The work is regarded by Brenan (in BRENAN and GREENWAY, see above) as 'useful in providing a partial catalogue of Peter's very copious collections made in the [Tanganyika] Territory'.]

532

Zanzibar and Pemba

These islands were from the late nineteenth century under a separate British protectorate, but in 1964 became part of Tanzania. The only general work on the flora, not comprising a full descriptive treatment or checklist, is WILLIAMS, R. O., 1949. *The useful and ornamental plants in Zanzibar and Pemba.* ix, 497 pp., illus., 1 color pl. Zanzibar: [Protectorate Government]. [Descriptive account, without keys, of a wide cross-section of the flora, including the more common native and introduced plants.]

533

Kenya

For the southern part, see also **531** (ENGLER *et al.*). This and the upland zone are the most completely treated; coverage of the east and north is sketchy apart from the revisions in *Flora of tropical east Africa.* No separate general checklist is available.

DALE, I. R. and GREENWAY, P. J., 1961. *Kenya trees and shrubs.* xxvii, 654 pp., 110 text-figs., 81 halftones, 31 color pls. frontispiece, map. Nairobi: Buchanan's Kenya Estates; London: Hatchards.

Briefly descriptive manual of trees and larger shrubs, with keys to all taxa; limited synonymy; vernacular names; generalized indication of local range, with citation of representative *exsiccatae*; taxonomic commentary and notes on habitat, associates, biology, timber properties, and other uses, special features, etc.; glossary and indices to vernacular, trade, and botanical names. The introductory section includes a short general bibliography as well as a general key to families.

GILLETT, J. B. and McDONALD, P. G., 1970. *A numbered check-list of trees, shrubs and noteworthy lianes indigenous to Kenya.* iii, 67 pp. Nairobi: Government Printer.

Systematic checklist of woody plants, with each species numbered and provided with known trade and vernacular equivalents; indices. The introduction includes an explanation of the numbering system.

Partial work

AGNEW, A. D. Q., 1974. *Upland Kenya wild flowers: a flora of the ferns and herbaceous flowering plants of upland Kenya.* ix, 827 pp., illus. Oxford: Oxford University Press.

Concise, well-illustrated manual-flora of (mainly) non-woody plants (except Gramineae and Cyperaceae) of upland (southwestern) Kenya, with brief descriptions, keys to all taxa, very limited synonymy with a few key citations, symbolic indication of local range, representative *exsiccatae*, some critical remarks, and short notes on habitat, altitude, frequency, etc.; index to all botanical names. The introduction gives the background and scope of the work, glossaries, and key references. Treatments of some groups by specialists or collaborators. The Gramineae are separately covered by BOGDAN, A. V., 1958. *A revised list of Kenya grasses.* Nairobi.

534

Uganda

For the southern part, near Victoria Nyanza, see also **531** (ENGLER *et al.*). No separate general flora or checklist is available.

EGGELING, W. J. and DALE, I. R., 1951. *The indigenous trees of the Uganda Protectorate.* 2nd edn., xxx, 491 pp., 94 text-figs., 55 halftones, 20 color pls., frontispiece, map. Entebbe: Government Printer. (First edn. 1936.)

Briefly descriptive manual of trees (similar in layout to *Kenya trees and shrubs*), with keys to all taxa; limited synonymy; vernacular names; general indication of local range, with citation of representative *exsiccatae*; taxonomic commentary and notes on habitat, timber properties and other uses, special features, etc.; glossary and indices to botanical, trade, and vernacular names. The introductory section includes a list of principal references.

535

Rwanda

Formerly part of German East Africa and (later) the Belgian mandate of Ruanda-Urundi. Apart from the recent works of Troupin, coverage is also provided by *Pflanzenwelt Ostafrikas* (ENGLER *et al.*; see **531**) and

Flore d'Afrique Centrale and other works on Zaïre (see **560**).

TROUPIN, G., 1978. *Flore du Rwanda: Spermatophytes*. Vol. 1. xiii, 413 pp., 82 pls., map (Ann. Mus. Roy. Afrique Centr., Série-en-8°, Sci. Econ., 9). Butare, Rwanda; Tervuren, Belgium. (Also issued as Publ. Inst. Natl. Rech. Sci. Rwanda, 18.)

Keyed enumeration of seed plants, with limited synonymy, citation of key references, French and Kinyarwanda vernacular names, indication of local range (by districts, with citation of representative *exsiccatae*) and summary of overall distribution, critical remarks, and notes on occurrence, related species possibly to be found, biology, etc.; complete indices at end. An introductory section includes historical information, technical details, a list of major references, an extensive illustrated glossary, and a key to families. Covers the gymnosperms and 62 dicotyledon families from Ranunculaceae through Leguminosae in part (Caesalpinaceae), following the twelfth edn. of Engler's *Syllabus*, and represents an amplification of the author's own syllabus (see below). The illustrations depict representative members of the families included.

TROUPIN, G., 1971. *Syllabus de la flore du Rwanda (Spermatophytes)*. viii, 6, 76, 10, 24, 20, 31, 340, 5, 16 pp., illus. (Ann. Mus. Roy. Afrique Centr., Série-en-8°, Sci. Econ., 7). Butare, Tervuren. (Also issued as Publ. Inst. Natl. Rech. Sci. Rwanda, 8.)

Illustrated systematic treatment of a representative cross section of the seed-plant flora (about 500 species, i.e., one-quarter of the total), with descriptions of all families and the more important genera along with diagnostic keys; limited synonymy; vernacular names; brief indication of local range, habitat, etc.; lexicon of vernacular names, with botanical equivalents; index to all botanical names. Introductory chapters include an illustrated organography, glossary, explanations of scientific nomenclature and collecting techniques, and a key to families.

536

Burundi

Formerly part of German East Africa and (later) the Belgian mandate of Ruanda-Urundi. Limited local coverage has lately been provided by Lewalle but otherwise reference must be made to *Pflanzenwelt*

Ostafrikas (ENGLER *et al.*; see **531**) and *Flore d'Afrique Centrale* and other works on Zaïre (see **560**).

Partial work

LEWALLE, J. (compil.), 1970. *Liste floristique et repartition altitudinale de la flore du Burundi occidental*. 84 pp., maps, figures. Bujumbura: Université officielle de Bujumbura. (Mimeographed.)

Includes a tabular list of vascular plants, with indication of life-form, chorology, and altitudinal zonation; based on collections made by the author in western Burundi. See also *idem*, 1971. *Arbres du Burundi*, sér. 1 (essences autochtones). Fasc. 1 (with G. Gilbert). 61 pp. Bujumbura.

Region

54

Northeast tropical Africa

This region, the 'Horn of Africa', encompasses Ethiopia (including Eritrea), Somalia, Djibouti (former French Somaliland), and the islands of Socotra and 'Abd-al-Kuri to the east of the 'Horn'. The area is covered by two comprehensive works (one still in progress): *Adumbratio florae aethopicae* and *Enumeratio plantarum Aethopiae*.

Although little was known of the flora of the region until the latter part of the eighteenth century due to the paucity of contacts, knowledge and collections, notably from the Ethiopian highlands, increased fairly rapidly thereafter, giving European botanists their first real idea of the African tropical-montane flora. The first significant flora, regarded as only a preliminary attempt at coverage, was *Tentamen florae abyssinicae* by A. Richard (1847–51); a substantial checklist by G. Schweinfurth followed in 1867. Schweinfurth was to begin serious exploration in the Eritrean lowlands in the following decades and as colonial penetration proceeded collecting spread into Djibouti, the Ogaden, Somalia and Socotra (the last-named being of especial biological interest). By the 1940s it was considered that sufficient material was available as a basis for a new regional work.

This led to not one but *two* projects: one in Austria, where Cufodontis began work on his *Enumeratio* which appeared in 26 systematically arranged parts from 1953 to 1972 (later issued in book form), and the other in Italy where, because of colonial needs and interests, the 'Erbario coloniale' (later 'Erbario tropicale') within the state herbarium at Florence began work on a comprehensive *Adumbratio* with species distribution maps, of which the first fascicle appeared in 1953 and at this writing (January, 1980) has reached fascicle 32. Owing to limited available resources progress on this work has been slow and only some 5 per cent of the vascular species have been revised and published (3.3 per cent in 1974).

Other general works include a useful illustrated key to families by Burger (1967) as well as *Afroalpine vascular plants* by HEDBERG (1957; see **503**) and *Flora of tropical Africa* (**501**).

Coverage of individual units is patchy, with nearly all works dating from before 1940. The Italian involvement in the region resulted in floras of Eritrea (Pirotta, 1903–7, not completed) and Somalia (Chiovenda, 1929–36). The only significant official British contribution was a compiled checklist for the whole of present-day Somalia by Glover (1946), although important contributions on Socotra were made through non-official channels by Balfour (1888, 1903). French contributions have been few, without resulting in a flora or checklist for Djibouti.

The only works on woody plants are by Fiori (1909–12) for Eritrea and Breitenbach (1963) for Ethiopia, the latter limited to more common tree species and in large part compiled.

For a relatively detailed map and commentary on the present state of botanical exploration in the region, see MOGGI, G., 1976. Aperçu sur la connaissance floristique d'Ethiopie et de Somalie en vue d'une nouvelle édition de la carte du degré d'exploration floristique de l'Afrique. *Boissiera*, **24**: 593–6, maps. Recent work has rendered Cufodontis' *Enumeratio* noticeably incomplete, and a revision is currently in preparation (G. E. Wickens, personal communication).

– Vascular flora: about 7000 species (Moggi); 6323 spermatophytes are recorded by Cufodontis.

Bibliographies. General and divisional bibliographies as for Division 5.

Indices. General and divisional indices as for Division 5.

540

Region in general (including Ethiopia)

See also **501** (*Flora of tropical Africa*); **503** (ENGLER; HEDBERG). All works specifically relating to Ethiopia are included here for convenience.

Keys to families

BURGER, W., 1967. *Families of flowering plants in Ethiopia*. 236 pp., 74 pls. (Oklahoma Agric. Exp. Sta. Bull., 45). Stillwater: Oklahoma State University Press. [Comprises keys to and descriptions of families of seed plants of Ethiopia and adjacent regions, with references and representative illustrations; glossary, bibliography, and index at end.]

CUFODONTIS, G., 1953–72. *Enumeratio plantarum Aethopiae: Spermatophyta*. 26 parts (Bull. Jard. Bot. État, 23–42, *passim*, Suppl., pp. i–xxvi, 1–1657). Brussels. (Reprinted in 2 vols., 1975, Brussels.)

Systematic enumeration of seed plants of Ethiopia, Eritrea, Somalia, and Djibouti, with synonymy, references, and principal citations; includes also vernacular names, partly symbolic indication of local and extralimital distribution, with some *exsiccatae*; general index (in the book version). An introductory section (in German) in part 1 includes a map, bibliography, and list of abbreviations, and a revised map appeared in part 15 (1964). Originally issued as a series of supplements to successive volumes of *Bulletin du Jardin Botanique, Bruxelles*, and after completion reissued in book form (cf. F. G. Meyer in *Taxon*, **22**: 655–7, 1973). The work is now rather incomplete and outdated, and a revision, with the addition of the Sudan, is in preparation at the Kew Herbarium (G. E. Wickens, personal communication). For the pteridophytes, see *idem*, 1952. Enumeratio plantarum aethopiae, III: Pteridophyta. *Phyton* (Horn), **4**(1–3): 176–93. [Systematic enumeration in the same style as the main work. For additions, see PICHI-SERMOLLI, R. E. G., 1977. Novitates pteridologicae aethopiae. *Webbia*, **32**(1): 51–68, as well as his contributions to *Adumbratio florae aethopicae*.]

PICHI-SERMOLLI, R. E. G. *et al.*, 1953– . Adumbratio florae aethopicae. Parts 1– . Illus., maps. *Webbia*, 9, etc., *passim* (also issued separately).

Critical enumeration of the vascular plants of the entire region, including Socotra and 'Abd-el Kuri; each family fascicle includes synonymy, references and citations, vernacular names, detailed account of local distribution, with indication of *exsiccatae* and some distribution maps, taxonomic commentary, and notes on habitat, associates, etc. The area covered is defined in the introductory section (part 1). At this writing (mid-1979) 32 parts had been issued, the latest seen in *Webbia*, **33**(1) (1978). Each part usually covers one family, but no systematic sequence of publication is observed (in the separate reissue a plan of the work with indication of parts published appears in the covers). Both pteridophytes and seed plants are covered. [Preparation of this work is based in the 'Erbario tropicale' of the Florence Herbarium.]

Special group – forest trees

BREITENBACH, F. VON, 1963. *The indigenous trees of Ethiopia*, 2nd edn. 306 pp., 129 text-figs. Addis Ababa: Ethiopian Forestry Association (mimeographed). (First edn. 1960.)

Briefly descriptive, illustrated treatment of the more common trees of Ethiopia, with keys to families; index. The introductory section includes notes on physical features and vegetation formations together with a list of references. Based partly on EGGELING and DALE's *Indigenous trees of the Uganda Protectorate* (see **534**).

541

Eritrea

More recent floristic information on this area is accounted for in the general works on the region (see preceding entry).

FIORI, A., 1909–12. *Boschi e piante legnose dell'Eritrea*. 428 pp., 177 text-figs. (some halftones). Florence: Instituto Agricolo Coloniale Italiano.

This work comprises three sections: (I) a general survey of forests, deforestation, protection, and reforestation; (II) a descriptive treatment of arborescent vegetation (with illustrations), vegetation zones, and habitat factors, with references; (III) a systematic descriptive flora of woody plants, with keys, synonymy, references, citations (especially to Pirotta; see below), vernacular names, many illustrations, indication of localities with collectors as well as generalized summary of local and extralimital range and altitudinal zonation along with habitats, and some critical commentary; addenda, general key to genera (pp. 378–98), table of

specific gravities, lexicon of vernacular names with scientific equivalents, references (pp. 401–5), and index to scientific names. For more important tree species further notes on properties, wood, uses, etc., are given.

PIROTTA, R., 1903–7. *Flora della colonia Eritrea*. Parts 1–3. pp. 1–464, pls. 1–12 (Annuario Reale Ist. Bot. Roma, 8). Rome.

Non-systematic enumeration of vascular plants, with descriptions of new taxa and partial keys; includes synonymy (with references and citations), indication of *exsiccatae* with localities, critical commentary, and notes on habitat, diagnostic features, etc. The introductory section includes an account of botanical exploration. Not completed; the manuscript of a fourth and final part (by E. Chiovenda) was lost by the printing establishment to whom it had been sent (Blake and Atwood, 1942, p. 33).

542

Djibouti

This small state was previously known as the Territory of the Afars and Issas and before that as French Somaliland. No specific floras or lists have been published; however, a cursory introduction to the flora appears in CHEVALIER, A., 1938. La Somalie française: sa flore et ses productions végétales. *Rev. Int. Bot. Appl. Agric. Trop.* **19**: 663–87, 2 text-figs., pls. 17–19.

543

Somalia

Formed in 1960 from the union of the former territories of British and Italian Somaliland. One checklist of limited scope is available (Glover, 1947), supported (for the former Italian territory) by a series of collection reports (Chiovenda, 1929–36). Other records are accounted for in the general regional works of Cufodontis and Pichi-Sermolli *et al.*

GLOVER, P. E., 1947. *A provisional check-list of British and Italian Somaliland trees, shrubs and herbs including the reserved areas adjacent to Abyssinia*. xxviii, 446 pp., 13 pls., map. London: Crown Agents.

List of recorded vascular plants, with limited

synonymy and brief descriptive notes; lexicon of vernacular names, with botanical equivalents; bibliographical index to families and index to genera; general bibliography. The introductory section includes a description of physical features and the vegetation.

Partial work: former Italian Somaliland

CHIOVENDA, E., 1929. *Flora somala*, I. xvi, 436 pp., 50 pls. (on 26 ll.), map. Rome: Sindicato italiano arti grafiche (for the Ministero delle Colonie).

CHIOVENDA, E., 1932. *Flora somala*, II. xvi, 482 pp., 247 text-figs. (Lav. Ist. Bot. R. Univ. Modena, 3). Modena: Reale Orto Botanico.

CHIOVENDA, E., 1936. Flora somala, III. Raccolte somale dei G. Pollacci, L. Maffei, R. Ciferri e N. Puccioni fatte negli anni 1934 e 1935. *Atti Ist. Bot. 'Giovanni Briosi' R. Univ. Pavia*, IV, 7: 117–60, Pls. 1–12.

Comprises three self-contained annotated lists of separate collections of non-vascular and vascular plants, with synonymy, literature citations, indication of *exsiccatae* with localities, descriptions of new taxa, vernacular names, and indices. An introductory section in the first contribution (1929) includes chapters on vegetation, phytogeography, and floristic relationships. No cumulative index has been published. The 1929 volume is based largely on collections made by G. Stefanini and G. Paoli in both former British and Italian Somaliland and in the Ogaden of eastern Ethiopia. [Although nominally covering only the former Italian part of Somalia, the work is more detailed than that of Glover and moreover accounts also for some collections from former British Somaliland.]

545

Socotra and 'Abd-el-Kuri Islands

Little new has been published on these islands, which are of considerable biological and biogeographical interest, since the early twentieth century and Forbes's monograph. Having been attached to Aden, they are now part of the republic of South Yemen. A last British contribution to the flora, with many fine plates, is ROYAL BOTANIC GARDENS, KEW, 1971. New or noteworthy species from Socotra and Abd-al-Kuri. *Hooker's Icon. Pl.*, V, 7(4): Pls. 3673–700, map. [Includes descriptive text accompanying the plates.]

BALFOUR, I. B., 1888. *Botany of Socotra*. lxxv, 446 pp., 100 pls., map (Trans. Roy. Soc. Edinburgh, 31). Edinburgh.

Amply descriptive flora of non-vascular and vascular plants, without keys; synonymy, with references and citations (including references to standard treatments under generic headings); vernacular names; indication of *exsiccatae*, with localities, and summary of extralimital range (if applicable); extensive taxonomic commentary and notes on habitat, special features, etc.; figures of representative species; indices to all botanical and vernacular names. The introductory section includes accounts of physical features, history and geography, animal life, botanical exploration, major features of the flora and vegetation, endemic species, floristics, and phytogeography.

BALFOUR, I. B., 1903. Botany of Sokotra and Abd-el-Kuri. Angiospermae [and] Pteridophyta. In *The natural history of Sokotra and Abd-el-Kuri* (ed. H. O. FORBES), pp. 447–542, illus., pls. 26A–26B (color). Liverpool: The Free Public Museums; London: Porter (Liverpool Public Museums Special Bulletin).

Revised systematic enumerations of lower vascular and flowering plants, supplementary to the author's *Botany of Socotra* (see above) and incorporating the results of an 1898–9 expedition led by Forbes; includes indication of *exsiccatae* with localities and field data, some descriptions of new taxa as well as criticial commentaries, and indication of status; no separate indices. Other chapters in this natural history account for the non-vascular plants as well as animal groups. [Some recent additions to the flora are accounted for in the contribution from Kew noted above.]

Region

55

The Sudan (middle Nile basin)

This region here corresponds to the Sudan Republic, the largest country in Africa and one lying wholly within the tropics.

Serious botanical exploration dates from the mid-nineteenth century with the expeditions of Schweinfurth and of Speke and Grant, but with for long a nebulous political status, tardy development of general administrative control, and a paucity of resident botanists subsequent floristic progress was erratic and the first general checklist appeared only in 1929, with a revision by Andrews as *Flowering plants of the Anglo-Egyptian Sudan* (1950–6; additions by Wickens, 1969). However, much of the region, especially the long-troubled and remote south, remains poorly collected and documented; the white areas on Léonard's map of 1965 cover more than two-thirds of the country.

Floristic work, with new collections, is continuing and already by 1970 Andrews' *Flora* was considered very incomplete. For the southern parts, the standard works on northeast and east tropical Africa (region 53 and 54) have been considered more useful. Earlier collections are also cited in *Flora of tropical Africa.*

A significant recent contribution, relating to the Darfur district in the west, is *Flora of Jebel Marra* by Wickens (1976); this is a monograph on one of the great isolated Saharan massifs, from which no collections were made until 1920–1.

– Vascular flora: about 3500 species (Wickens); Andrews treats 3137 species.

Bibliographies. General and divisional bibliographies as for Division 5.

Indices. General and divisional indices as for Division 5.

550

Region in general

See also **501** (*Flora of tropical Africa*). The only nominal general flora (Andrews, 1950–6) is considered to be most useful in the central and northern parts of the Sudan, i.e., the Khartoum district and Nubia; elsewhere, especially in the south, its coverage is very poor and other floras, especially those of neighboring regions, should be used (E. A. Bari, personal communication).

ANDREWS, F. W., 1950–6. *Flowering plants of the Anglo-Egyptian Sudan.* 3 vols. Illus., map. Arbroath, Scotland: Buncle. (Vol. 3 entitled *Flowering plants of the Sudan.*)

Illustrated briefly descriptive flora of seed plants, with keys to genera and to some species; includes limited synonymy, generalized indication of local range (with some localities specified), miscellaneous notes, and index to all botanical names in each volume (no

general index). The introductory section includes an illustrated glossary–organography. For a supplement to this now very incomplete work, see WICKENS, G. E., 1969. Some additions and corrections to F. W. Andrews' *Flowering plants of the Sudan. Sudan Forests Bull.*, N.S. **14**: 1–49. Pteridophytes have been covered in a complementary enumeration, viz: MACLEAY, K. N. G. 1953. The ferns and fern-allies of the Sudan. *Sudan Notes & Rec.* **34**: 286–98.

Partial work

The phytogeographically significant, over 3000 m high isolated Sahelian massif of Jebel Marra in western Sudan is the subject of a recent analysis and enumeration by Wickens, thus forming a parallel to Quézel's study of the Tibesti massif to the northwest (see **593**). It forms a valuable supplement to the sketchy coverage by Andrews' flora.

WICKENS, G. E., 1976. *The flora of Jebel Marra (Sudan Republic) and its geographical affinities.* [ix], 368 pp., 34 text-figs., 8 pls., 208 distribution maps, frontispiece (area map), overlays (in pocket) (Kew Bull. Addit. Ser., 5). London: HMSO.

Includes (pp. 83–191) a systematic checklist of 982 species (966 vascular), with synonymy, references, citations of *exsiccatae*, and critical and ecological notes, without keys or descriptions; this is followed (pp. 233–337) by a long series of species distribution maps. The plant geography of the area and its affinities are discussed on pp. 35–59 and there is also a treatment of the vegetation and its history. An extensive list of references as well as indices appear at the end.

Region

56

Zaïre

This region coincides with the political limits of the Republic of Zaïre (until 1960 the Belgian Congo). For Rwanda and Burundi, formerly administered by Belgium under mandate but geographically part of east Africa, see **535** and **536** respectively.

Serious botanical exploration began only in the wake of the great expeditions of the latter part of the nineteenth century (including that of Stanley), and the subsequent establishment of the 'Congo Free State'. The first general checklist, *Sylloge florae congolanae*, was prepared by Durand and Durand after transfer of the 'Free State' from Léopold II to the

Belgian government, but at best it could represent only a fraction of the vascular flora. Sketchy coverage was also provided by *Flora of tropical Africa* (**501**).

By the 1940s, however, it was considered that the number of collections had become sufficient to justify the launch of a large-scale descriptive flora of the Belgian territories. The *Flore du Congo Belge et du Ruanda-Urundi* project was thus established in 1942, with publication commencing in 1948. At this writing (January, 1980) the work has covered between 40 and 50 per cent of the species after more than 31 years, and has continued production in spite of changes in sponsorship, format and name. It came to be regarded as a 'regional' flora in the sense of the first edition of *Flora of West Tropical Africa*, and set a trend soon to be emulated for most other regions of tropical Africa by the various 'official' botanical centers in the metropolitan countries responsible in Africa, particularly with the termination of the old *Flora of Tropical Africa* and the need for a successor work or works.

Belgian botanists also played a leading role in the creation of the present 'umbrella' organization for the coordination of African floristic and systematic botany, the 'Association pour l'Étude Taxonomique de la Flore de l'Afrique Tropicale' (AETFAT), which came into being about 1950.

Most organized floristic effort has in recent decades gone into production of *Flore d'Afrique Centrale*, as the *Flore du Congo Belge* is now known. Few local florulas have been written, and these largely before 1960; the leading ones are *Flore des Spermatophytes du Parc National Albert* (Robyns, 1948–55), covering what is now the Virunga Volcanoes National Park, and the incomplete *Flora des Spermatophytes du Parc National de la Garamba* (Troupin, 1956) for an area along the northern border. Some guides to forest trees have been published, the latest (Gillardin, 1959) never completed.

– Vascular flora: 10000–11000 species. The 3539 species revised in *Flore d'Afrique Centrale* to 1970 were thought to represent about 35 per cent of the total (seed plants).

Bibliographies. General and divisional bibliographies as for Division 5.

Indices. General and divisional indices as for Division 5.

560

Region in general

See also **501** (*Flora of tropical Africa*), **503** (Engler, Hedberg). Although *Flore d'Afrique centrale* is yet to be completed, with the change to publication in fascicles some early parts are being revised as circumstances permit.

Keys to families
Robyns, W., 1958. *Flore du Congo belge et du Ruanda-Urundi: tableau analytique des familles.* 67 pp. Brussels: Institut National pour l'étude agronomique du Congo. [Comprises a set of analytical keys.]

Durand, T. and Durand, H., 1909. *Sylloge florae congolanae* (Phanerogamae). iii, 716 pp. Brussels: A. de Boeck (for the Ministère des Colonies). (Also publ. as Bull. Jard. Bot. État., vol. 2.)

Systematic enumeration (in French) of the known seed plants of the then-Belgian Congo, with full synonymy, references, and citations; vernacular names; detailed indication of local range, with citation of *exsiccatae*; floristic statistics; indices to all vernacular and botanical names. The introductory section includes a listing of collectors. Rwanda and Burundi are not included, then being part of German East Africa (Region 53). [Published the year following transfer of the Congo from King Léopold II to the state.]

Robyns, W. *et al.* (eds.), 1948–63. *Flore du Congo belge et du Ruanda-Urundi* (from 1960 *Flore du Congo, du Rwanda et du Burundi*). Vols. 1–7, 8, part 1, 9–10. Text-figs., plates (some in color). Brussels: Institut National pour l'étude agronomique du Congo (INÉAC).

Jardin Botanique National de Belgique, 1967– . *Flore du Congo, du Rwanda et du Burundi* (from 1971 *Flore d'Afrique centrale* (Zaïre–Rwanda–Burundi)). Publ. in fascicles. Illus. Brussels.

Comprehensive illustrated descriptive flora of the present-day Republic of Zaïre together with Rwanda and Burundi (**535** and **536**), covering the vascular plants; includes keys to all taxa up to generic rank, full synonymy with references, literature citations, vernacular names, citation of *exsiccatae* with fairly detailed summary of local range and generalized indication of extralimital distribution, taxonomic commentary, and notes on habitat, biology, special features, etc.;

illustrations of representative species; indices in each volume (and later, each fascicle) to vernacular and botanical names. Volumes 1–10 followed a modified Englerian sequence, and cover the gymnosperms and about one-third of the known angiosperms; fascicles published from 1967 onwards each contain one family but do not follow a set sequence. The *Flora* project was begun in 1942; as of 1970, 3539 species had been accounted for – a little more than one-third of the vascular flora, leading to hints that the work would not be finished before 2020 (Léonard in MERXMÜLLER, 1971; see Division 5, Introduction, p. 253). [The change of publication method was made in the mid-1960s in the interests of flexibility. The fascicles are now arranged in two series, Ptéridophytes and Spermatophytes, and control of the work is vested in a managing editor.]

Partial works

An important 'florula' is available for the Virunga Volcanoes National Park (formerly Albert National Park) along the eastern border with Rwanda and Uganda, as well as an incomplete work for the Garamba National Park on the northern border.

ROBYNS, W. *et al.*, 1948–55. *Flore des spermatophytes du Parc National Albert*. 3 vols. Text-figs., plates (some in color), maps. Brussels: Institut des Parcs Nationaux du Congo Belge.

Keyed, illustrated descriptive treatment of seed plants, with synonymy, vernacular names, detailed indication of local range (including *exsiccatae*) and summary of extralimital distribution, taxonomic commentary, and notes on habitat, biology, etc.; full indices. The introductory section (in vol. 1) includes accounts of physical features, vegetation, phytogeography, and botanical exploration.

TROUPIN, G., 1956. *Flore des spermatophytes du Parc national de la Garamba*. Vol. 1. 349 pp., illus. (some in color), map. Brussels: Institut des Parcs Nationaux du Congo Belge.

Descriptive treatment of seed plants, without keys; includes synonymy, vernacular names, citation of *exsiccatae* with localities and general summary of extralimital range, and brief notes on habitat, biology, etc.; index. Not completed; encompasses the Gymnospermae and Monocotyledoneae. Garamba National Park is in the extreme northern part of the Zaïre Republic.

Special group – forest trees

The foresters' manual by Gillardin was evidently never completed and two earlier partial manuals have therefore been included here.

GILLARDIN, J. 1959. *Les essences forestières du Congo belge et du Ruanda-Urundi*. 378 pp., illus., map. Bruxelles: Ministère du Congo belge et du Ruanda-Urundi.

Foresters' checklist/manual of trees and important shrubs, vines, and herbs, without keys; limited synonymy; fairly detailed indication of local distribution; vernacular names; notes on habitat, special features, etc.; indices to vernacular, trade, and botanical names, the latter being cross-referenced to the *Flore du Congo* (see following entry). Not completed; extends up to the Leeaceae (modified Englerian system), not including Monocotyledoneae.

LEBRUN, J., 1935. *Les essences forestières des régions montagneuses du Congo oriental*. 263 pp., 28 text-figs., 18 pls. (Série scientifique, 1). Brussels: INEAC.

Foresters' manual of principal trees and other woody plants, with keys, references, vernacular names, indication of *exsiccatae* and summary of distribution, and notes on habitat, special features, properties, uses, etc.; indices. An account of floristic zones and a bibliography are also included. [Eastern Zaïre.]

VERMOESEN, C. 1931. *Manuel des essences forestières de la région équatoriale et du Mayombe*. Revised edn. xii, 282 pp., illus. (some in color). Brussels: Ministère des Colonies. (Original edn. 1923, Brussels, as *Les essences forestières du Congo Belge (région équatoriale et Mayombe)*.)

Pocket descriptive manual of principal forest trees and other woody elements of the moist high forest and seasonal monsoon forest/woodland zones, with vernacular names, indication of distribution, and notes on habitat, special features, uses, etc.; indices. [Central and Northern Zaïre.]

Region

57

Equatorial Central Africa and Chad

This somewhat heterogeneous region includes the states of Congo (Brazzaville), Gabon, Central African Republic, Chad, Cameroon, Equatorial Guinea, the territory of Cabinda (part of Angola), and the islands in the Gulf of Guinea (comprising the state of São Tomé and Príncipe and the insular possessions of Equatorial Guinea).

Some early collections were made along the coast and notably in the easily accessible Guinean Islands in the eighteenth and early-nineteenth centuries, with Cameroon Mountain following by the mid-nineteenth century, but systematic botanical exploration began only in the wake of colonial penetration and rule late in that century. Collections increased rapidly in succeeding decades, notably in the Cameroons, but in a region with some of the richest floras in tropical Africa a sufficient botanical basis even for 'national' floras was not available until after World War II. Some records were incorporated into *Flora of tropical Africa* (501).

Unlike other parts of tropical Africa, no 'regional' flora was launched in the post-war period, and with the lack of any earlier consolidated enumeration or flora this has led to a state of continuing fragmentation in floristic progress, reflecting also the geographical and political diversity of the region. Much of it still remains poorly known botanically and it may be that a comprehensive account was not practicable.

In the place of an overall work there have appeared a number of 'national' accounts of widely varying kinds. The most elaborate are the serial descriptive *Flore du Gabon* and *Flore du Cameroun*, both begun in the 1960s and by 1974 covering respectively one-quarter and one-eighth of their vascular flora. For the remaining units, only checklists are available, although serial descriptive floras were begun for the Guinea Islands. Congo (Brazzaville) was covered by Descoings (1961), mainland Equatorial Guinea (Rio Muni) by Guinea López (1947), Central African Republic by Tisserant (1950), southern Chad by Lebrun *et al.* (1972), and the Guinea Islands by Exell (1944–56, 1973). Some areas are more or less covered by general works on West Africa (580). The Guinea Islands serial floras are by Escarré (1968–9, not continued) for Fernando Póo and Jardim e Museu Agrícola do Ultramar (1972–) for São Tomé and Príncipe. The high mountain flora in the Cameroons was reviewed by ENGLER (1892; see 503).

The overall position at present is here regarded as unsatisfactory, with no general flora in progress or contemplated but at country level dominated by two large-scale projects running wastefully in parallel amidst a patchwork of checklists of variable quality and scope.

— Vascular flora: here estimated at between 12000 and 15000 species.
Bibliographies. General and divisional bibliographies as for Division 5.
Indices. General and divisional indices as for Division 5.

570

Region in general

See **501** (*Flora of tropical Africa*); **580** (AUBRÉVILLE; CHEVALIER; TARDIEU-BLOT). No separate regional works are available.

571

Cabinda

See **522** (CARRISSO *et al.*); **560** (ROBYNS *et al.*). This territory is politically part of Angola.

572

Congo Republic

Also known as Congo-Brazzaville, and formerly as the Middle Congo. The capital, Brazzaville, was before 1958 the main administrative and cultural center for this and other territories constituting French Equatorial Africa. On account of the limitations of the only available checklist, the floras of neighboring countries, especially Gabon (573) and Zaïre (560), should be consulted.

DESCOINGS, B., 1961. *Inventaire des plantes vasculaires de la République du Congo, déposées dans l'herbier de l'Institut d'Études Centre-Africaines* [de ORSTOM] *à Brazzaville*. Brazzaville: ORSTOM. (Mimeographed.)

Unannotated checklist of vascular plants, without synonymy or indication of local range. Essentially a herbarium catalogue.

573

Gabon

The ambitious *Flore du Gabon* project is still some little way from 'completion'. A useful precursor, dealing with the south-central uplands, is PELLEGRIN, F., 1924–38. *La flore du Mayombe d'après les récoltes de M. Georges Le Testu*. 3 parts. Illus. (Mém. Soc. Linn. Normandie, 26(2); N.S., Bot. 1(3, 4).) Caen.

– Vascular flora: about 6000 species (Floret in MIÈGE and STORK, 1976; see Division 5, Introduction, p. 253).

AUBRÉVILLE, A. *et al.* (eds.), 1961– . *Flore du Gabon.* Fasc. 1– . Illus. Paris: Muséum National d'Histoire Naturelle, Laboratorie de Phanérogamie.

Copiously descriptive, illustrated serial 'research' flora of vascular plants, with keys to genera and species in each family; treatments include synonymy, with references and citations, vernacular names, indication of local range with citations of *exsiccatae* and general summary of overall range, critical commentary, and notes on habitat, special features, etc.; indices to all botanical and vernacular names. No formal sequence of families is followed in publication; fascicles are numbered consecutively. At this writing (1979) 24 fascicles have appeared, each containing revisions of one or more families.

Special group – forest trees

SAINT AUBIN, G. DE, 1963. *La forêt du Gabon.* 208 pp., plates, maps (Publ. CTFT, 21). Nogent-sur-Marne: Centre Technique de Forestier Tropical.

Illustrated field manual of the more important forest trees, without keys; vernacular and trade names; generalized indication of local range; notes on habitat, timber properties, sylvical characteristics, etc.; indices to botanical and other names. The introductory section includes accounts of physical features, climate, vegetation, and forest composition (with maps).

574

Equatorial Guinea (Río Muni)

Only the mainland portion of Equatorial Guinea, formerly known as Río Muni, is accounted for here. Fernando Póo and Annobon are included with other Gulf of Guinea islands under **579**.

GUINEA LÓPEZ, E., 1946. *Ensayo geobotánico de la Guinea continental española.* 389 pp., illus. Madrid: Dirección General de Marruecos y Colonias.

Section 4 of this work (pp. 219–388) comprises an enumeration of the known vascular plants of Río Muni, with limited synonymy, concise general indication of local and extralimital range (with some citations and range maps), representative illustrations, and a bibliography; no index. The remaining sections deal with general aspects of tropical wet forest, its parameters and limiting factors, and vegetation classification.

575

Central African Republic

Formerly known as Ubangi-Shari and, for a time, the Central African Empire.

TISSERANT, C., Père, 1950. *Catalogue de la flore de l'Oubangi-Chari.* 166 pp., map (Mém. Inst. Études Centrafr., 2). Brazzaville. [? Repr. 1965.]

Briefly annotated enumeration of seed plants, with vernacular names, generalized indication of local range, taxonomic commentary, and notes on habitat, local ecology, biology, uses, etc.; no index. The introductory section includes a discussion of the value of vernacular names. [The author was a brother of the late Eugène Cardinal Tisserant, former dean of the Sacred College of Cardinals.]

576

Chad Republic

The northern part of Chad, largely high Saharan plateaux culminating in the large Tibesti massif, is here treated within region 59 (see **593**). For the remainder, somewhat better known botanically, the only floristic work of broad scope is the enumeration by Lebrun *et al.* (1972), bounded on the north by the sixteenth parallel.

LEBRUN, J.-P. *et al.*, 1972. *Catalogue des plantes vasculaires du Tchad méridional.* 289 pp., 18 maps (Études Botaniques IEMVPT, 1). Maisons-Alfort, Val-de-Marne: Institut d'Élevage et de Médecine Vétérinaire des Pays Tropicaux.

Annotated enumeration of vascular plants of the Chad Republic south of the sixteenth parallel, thus more or less excluding the Saharan zone; includes limited synonymy, indication of localities, and notes on habitat, special features, etc.; bibliography, statistics, and indices. Appendices cover floristics, phytogeography, useful plants, and other topics, while an account of botanical exploration appears in an introductory section.

577

Cameroon

This will here be taken to comprise that part of the republic mandated to France after World War I. The southern part of the formerly British-administered area, including the Bamenda Plateau and Cameroon Mountain (now also part of Cameroon), forms unit **578**. Like the *Flore du Gabon*, the *Flore du Cameroun* is a large-scale work which is still a long way from 'completion'; however, a cyclopaedia of collectors and a phytogeographical summary (both by R. Letouzey) have already been published.

– Vascular flora: about 8000 species (Letouzey in MIÈGE and STORK, 1976; see Division 5, Introduction, p. 253).

AUBRÉVILLE, A. *et al.* (eds.), 1963– . *Flore du Cameroun*. Fasc. 1– . Illus. Paris: Muséum National d'Histoire Naturelle, Laboratoire de Phanérogamie.

Copiously descriptive, illustrated serial 'research' flora of vascular plants, with keys to genera and species in each family; treatments include synonymy, with references and citations, vernacular names, indication of local range with citations of *exsiccatae* and general summary of overall distribution, critical commentary, and notes on habitat, ecology, uses, etc.; indices to all botanical and vernacular names. Fasc. 1, containing a phytogeographical summary, and fasc. 7, comprising a cyclopedia of collectors and a history of botanical exploration, are by R. Letouzey. No formal sequence of families is followed in publication; fascicles are numbered consecutively. At this writing (1979) 20 fascicles have appeared, each containing revisions of one or more families.

578

Bamenda Plateau and Cameroon Mountain

Since 1960, this formerly British-administered part of the Cameroon has been part of the republic of the same name. Special attention is called to it here as an area of great phytogeographical interest which includes the highest peak in all Equatorial Guinean Africa, Cameroon Mountain (4070 m). General coverage

is provided by the floras and other works on Cameroon (**577**), West Africa (**580**) and Nigeria (**581**). For separate coverage of the mountain flora, see **503** (ENGLER).

579

Islands in the Gulf of Guinea

See also **503** (ENGLER). The Guinea Islands comprise Fernando Póo, Principe, São Tomé, and Annobon. For further coverage of the first-named, which is nearest the mainland, see **580** (HUTCHINSON and DALZIEL). A close floristic affinity is shown with **578**, particularly with regard to the montane plants. The most useful works are the various lists by Exell and others, which include phytogeographical remarks.

– Angiosperm flora: Fernando Póo, 1105 species; Principe, 314 spp.; São Tomé, 601 spp.; Annobon, 208 spp. (Exell, 1973).

EXELL, A. W. *et al.* 1944. *Catalogue of the vascular plants of São Tomé (with Principe and Annobon).* xi, 428 pp., 26 text-figs., 3 maps. London: British Museum (Natural History).

Critical systematic enumeration (including descriptions of new taxa) of vascular plants, without keys; rather extensive synonymy, with references and citations; vernacular names; indication of *exsiccatae*, with localities, and general summaries of local and extralimital range; taxonomic commentary and miscellaneous other notes; index to all botanical names. The introductory section includes chapters on physical features, geology, climate, history and geography, and botanical exploration, together with remarks on the affinities of the flora and on phytogeography. For additions, corrections, etc., see *idem*, 1956. *Supplement* .... v, 58 pp., 3 text-figs. London.

EXELL, A. W., 1973. Angiosperms of the islands of the Gulf of Guinea (Fernando Póo, Principe, S. Tomé, and Annobon). *Bull. Brit. Mus.* [Nat. Hist.], *Bot.* **4**(8): 325–411.

Comprises a concise, systematically arranged tabular checklist of flowering plants of the Gulf of Guinea Islands, with numbers designating bibliographic references (post-1944) and letters designating the four islands as appropriate; index to genera and families at end. The introductory section gives a brief account of the main features of the combined flora and a phytogeographical analysis together with a numbered bibliography of relevant works published since the 1944 enumeration (see above). Note that this checklist also includes all records from Fernando Póo.

Partial works

ESCARRÉ, A., 1968–9. *Aportaciones al conocimiento de la flora de Fernando Póo*. Fasc. 1–2 (Acta Phytotax. Barcinonensia, 2, etc.). Barcelona.

Semi-revisionary serial flora, with descriptions, keys, synonymy, citations, indication of *exsiccatae*, taxonomic commentary, etc.; indices in each fascicle. At this writing (1979) only two fascicles have, apparently, been issued.

JARDIM E MUSEU AGRÍCOLA DO ULTRAMAR, PORTUGAL, 1972– . *Flora de São Tomé e Príncipe*. Fasc. 1– . Lisbon: Ministerio do Ultramar (later Ministerio da Cooperação).

Large-scale 'research' flora, issued serially; each fascicle with keys, descriptions, synonymy, citations, indication of *exsiccatae*, extralimital distribution, vernacular names, and some special notes; and separate index. All treatments by specialists, but without indication of editor. To date (1979) five fascicles have appeared, the latest in 1976.

Region

58

West tropical Africa

Within this region is all territory from the eastern boundaries of Nigeria and Niger Republic westward to the Atlantic at Senegal, with the northern limit approximately defined by the limits of the Sahara, ranging from 19 ° to 18 °N from west to east.[7] This corresponds with the northern limit of coverage (18 °N) of the current edition of the standard regional work, *Flora of West Tropical Africa*. Countries or areas thereof to the north, including western Sahara, are here considered part of northern Africa (region 59).

'West Africa' is botanically about the best-known part of tropical Africa. Early contacts in the seventeenth and eighteenth centuries were extensive, and much botanical material was collected around the many 'castles' (forts) and elsewhere by both visitors and residents. Extensive inland exploration and travel began in the late-eighteenth century, and by the 1860s a basis believed sufficient for a general flora had been attained. This became *Flora of tropical Africa* (1868–1937). Colonial penetration with, in time, the

French following the British, hastened collecting and floristic study, and after World War I local herbaria began to be established, of which the largest today is the Forest Herbarium at Ibadan (Nigeria).

The gradual obsolescence as well as unwieldiness of the *Flora of tropical Africa* gave rise to the first edition of the concise *Flora of West Tropical Africa* (Hutchinson and Dalziel, 1927–36). This much-esteemed work is one of the most concise but informative of tropical Floras. A second edition, largely in the same style but in three volumes to account for the great increase in known species, appeared from 1953 to 1972.

The existence of this 'staple' flora has, however, tended to deflect the publication of 'state' floras and (to a lesser extent) checklists and 'tree books'. The only current national floras are those for Senegal (Berhaut, 1967, 1971–), the illustrated work presently in progress representing an ambitious effort is, fortunately, making steady progress; Guinea-Bissau (*Flora da Guiné Portuguesa*, 1971– , but dormant since 1973); and Nigeria (*Flora of Nigeria*, 1970– , in continuation of *Nigerian Trees*). An introductory manual covering the major francophone centers is Roberty's *Petite flore de l'Ouest-Africain* (1954). Checklists of varying quality and scope exist for Liberia, Sierra Leone, Guinea-Bissau, Senegal, Mauretania, Gambia, and Niger.

The principal regional 'tree book' is Aubréville's *Flore forestière soudano-guinéenne* (1950), recently reprinted. Dendrofloras are also available for Nigeria, the Ivory Coast, Liberia (two), Sierra Leone, and Ghana, some of them illustrated. For pteridophytes, treatments in both French (Tardieu-Blot, 1953; includes also Region 57) and English (Alston, 1959, as part of *Flora of West Tropical Africa*) are available.

A descriptive flora (not yet completed) has also been written by Adam (1971) for the Nimba Massif in Liberia and Guinea, a phytogeographically significant area with a number of relicts.

For the Ivory Coast, an illustrated descriptive flora has been projected by L. Aké Assi and his colleagues. Elsewhere, there would appear to be some need for local florulas for areas of special interest. In general, the non-anglophone areas in this respect are better provided for, especially Senegal and the Ivory Coast. Added to these are the extensive contributions of the botanical branch (formed in the 1960s) of the French overseas cooperative, Institut d'Élevage et de Médecine Vétérinaire des Pays Tropicaux, as part of a programme of work covering much of northern dryland tropical Africa.[8]

West Africa is at present relatively well but rather unevenly explored, with considerable areas still comparatively neglected. Among lesser-known areas is the Guinea corner, especially Liberia, where the flora is the richest for the region. Some countries, such as Benin (Dahomey), Togo, Guinea, and Mali, lack any suitable checklist, and for Liberia and Sierra Leone the available checklists are of a poor standard. From a floristic point of view, the most valuable new contributions would be on Liberia and adjacent areas, where endemism is high, and on Togo/Benin, where extensive species transition reportedly occurs. Elsewhere, local endemism is limited, with distribution being constrained mainly by bioclimate.[9]

– Vascular flora: 7072 species recorded in the second edition of *Flora of West Tropical Africa* (Hepper).

Bibliographies. General and divisional bibliographies as for Division 5.

Indices. General and divisional indices as for Division 5.

580

Region in general

See also **501** (Flora of tropical Africa). In the first edition of *Flora of west tropical Africa*, coverage extended into northern Africa as far as the Tropic of Cancer.

HUTCHINSON, J. and DALZIEL, J. M., 1953–72. *Flora of west tropical Africa*. 2nd edn, revised R. W. J. Keay and F. N. Hepper. 3 vols. 462 text-figs., map. London: Crown Agents (distributed HMSO). (First edn. in 2 vols., 1927–36, London.)

Manual-flora of seed plants, with keys to all taxa, concise synonymy, principal citations, indication of representative *exsiccatae*, with general summaries of local and extralimital distribution, critical commentary, and brief notes on habitat, frequency, special features, etc.; index to all botanical names in each volume (cumulative index in vol. 3.) An introductory section includes an account of botanical exploration in the region together with a glossary and bibliography. For the pteridophytes, see below under Special groups.

Partial works

CHEVALIER, A., 1938. *Flore vivante de l'Afrique occidentale française*. Vol. 1. 360 pp., illus. Paris: Muséum National d'Histoire Naturelle.

Copiously descriptive flora of seed plants, with keys, synonymy, references and citations, indication of local and extralimital range, and notes on phenology, life-form, biology, etc.; indices to generic and family names. Along with French-administered territories in west Tropical Africa, this work also extends to the Sahara and the northern part of French-administered equatorial Africa. Not completed; discontinued after World War II.

ROBERTY, G., 1954. *Petite flore de l'Ouest-Africain*. 441 pp. Paris: Larose (for ORSTOM).

Concise manual of seed plants, with keys, limited synonymy, literature citations, local range, and notes on habitat, frequency, cultivation, etc.; lexicon of vernacular names; index. Only the more important species are described and keyed out. Covers Upper Volta, the Ivory Coast, Guinea, Mali (in part), and Senegal, i.e., the region south of 16 °N and west of the prime meridian.

Special groups – woody plants

AUBRÉVILLE, A., 1950. *Flore forestière soudano-guinéenne* [*AOF – Cameroun – AEF*]. 523 pp., 115 pls., 40 maps. Paris: Societe d'Éditions Geographiques, Maritimes, et Coloniales. (Reprinted 1975, Nogent-sur-Marne: Centre Technique de Forestier Tropical.)

Forest flora of the forest, woodland, and savanna–woodland zones of west (and central) Africa, with some keys to species, limited synonymy, vernacular names, local and extralimital distribution (with indication of some *exsiccatae* and range maps), and extensive notes on habitat, biology, special features, etc.; complete indices and numerous illustrations. An introductory section includes remarks on physical features, climate, features of the forest flora, and taxonomic problems, together with species lists indicating a given special feature, property, use, etc.

Special groups – pteridophytes

ALSTON, A. H. G., 1959. *The ferns and fern allies of West Tropical Africa*. 89 pp. London: Crown Agents (distributed by HMSO).

Intended as a companion to *Flora of west tropical Africa*, this treatment includes keys to all taxa, brief synonymy, literature citations, generalized indication of local and extralimital distribution (with some *exsiccatae*), critical commentary, and notes on habitat, frequency, etc.; index. Does not extend to equatorial Africa.

TARDIEU-BLOT, M. L., 1953. *Les ptéridophytes de l'Afrique intertropicale française*. 241 pp., illus. (Mém. Inst. Franç. Afrique Noire, 28). Dakar.

Descriptive treatment, with keys, full synonymy and

references, indication of *exsiccatae*, general summary of overall distribution, and notes on habitat, etc.; bibliography and index. An introductory section includes a list of principal collectors and remarks on floristic regions and fern ecology. Covers former French territories in west and central Africa, Cameroon Mountain, and islands in the Gulf of Guinea.

581

Nigeria

Of the *Flora of Nigeria* project, initiated by the late D. P. Stanfield as a continuation of *Nigerian trees*, only two fascicles have so far been published. Both works are semi-technical in nature.

KEAY, R. W. J., ONOCHIE, C., and STANFIELD, D. P., 1960–4. *Nigerian trees.* 2 vols. Illus., plates. Lagos: Nigerian Government Printer.

Briefly descriptive, semi-technical manual of trees, with keys to all taxa (including a special tabular key to genera in vol. 2), limited synonymy, principal citations, vernacular names, generalized indication of local and extralimital range (including representative *exsiccatae*), and notes on habitat, biology, special features, uses, etc.; general index to family and generic names in vol. 2. An introductory section contains an account of botanical exploration as well as an illustrated glossary. Continued by STANFIELD, D. P. and LOWE, J. (eds.), 1970– . *The flora of Nigeria.* Published in fascicles. Illus. Ibadan: Ibadan University Press. [Semi-technical revisionary treatments of various families, with concise descriptions, polyclave keys, limited synonymy, full-page figures and brief statements on habitat, distribution, special features, etc.; separately indexed. At this writing (mid-1979), only two fascicles (Gramineae and Cyperaceae respectively) have been published. The use of polyclave tables in place of the traditional Lamarckian analytical keys is a marked departure from usual practice.]

582

Benin (Dahomey) and Togo

No individual general floras or lists are available for either of these two sandwich states between Nigeria and Ghana. The area is characterized by the so-called biogeographic 'Dahomey Gap'.[10]

583

Ghana

The only extensive floristic work on this state, once known as the Gold Coast, is without keys.

IRVINE, F. R., 1961. *Woody plants of Ghana.* xcv, 868 pp., 142 text-figs., 34 pls. (some in color), frontispiece. Oxford: Oxford University Press.

Briefly descriptive forest flora, without keys; limited synonymy; vernacular names; general indication of local and extralimital range, with citations of representative *exsiccatae*; extensive notes on habitat, field features, properties, uses, etc.; lexicon of vernacular names; bibliography, illustrated glossary, and index to generic and family names. The introductory section includes lists of plants arranged by uses, with a tabular key. The work represents a revision of the author's *Plants of the Gold Coast* (1930, London).

584

Ivory Coast

Apart from Aubréville's major work on the forest flora, several local contributions have appeared, including a large paper by L. Aké Assi, but no complete flora or enumeration has yet been published.

– Vascular flora: about 5000 species (Aké Assi in MIEGE and STORK, 1976; see Division 5, Introduction, p. 253).

AUBRÉVILLE, A., 1959. *La flore forestière de la Côte d'Ivoire.* 2nd edn. Vols. 1–3. 351 pls., maps (Publ. CTFT, 15). Nogent-sur-Marne: Centre Technique Forestier Tropical. (First edn. 1936, Paris.)

Briefly descriptive forest flora, with analytical keys to genera and species and tabular keys to families; limited synonymy; vernacular names; generalized indication of local range, with citation of representative *exsiccatae*; taxonomic commentary and notes on habitat, biology, etc.; indices to generic and family names in vol. 3. The introductory section includes a short bibliography and a discussion of the forest zones of the Ivory Coast.

585

'Guinea'

Included here are **Liberia, Sierra Leone, Guinea Republic**, and **Guinea-Bissau** (former Portuguese Guinea). These diverse polities together form a somewhat natural unit, which incorporates most of the large, biogeographically significant Nimba massif. Apart from the general works on West Africa (**560**), the only coverage is provided by a miscellany of local works. An important recent contribution, not yet completed, is Adam's flora of the Nimbas. Relatively speaking, the flora is the richest in West Africa, aided by high and fairly consistent rainfall (especially in Liberia), and in part is still imperfectly known.

ADAM, J. G., 1971–5. *Flore descriptive des monts Nimba.* Parts 1–4. pp. 1–1586, Pls. 1–775 (Mém. Mus. Natl. Hist. Nat., sér. II/B, Bot., 20, 22, 24, 25). Paris.

Sect. 2 of this work (pp. 145 ff.) comprises an illustrated descriptive flora of this isolated mountain massif (which falls largely within Liberia and the Republic of Guinea), with limited synonymy, indication of *exsiccatae*, general summary of extralimital range, vernacular names, and notes on habitat, phenology, etc.; indices to all species provided. No keys are included. The lengthy general section includes an introduction to the region, the basis for the work, and chapters on climate, geology, and vegetation (principal zones 500–1200 m and over 1200 m), with special reference to the Liberian portion of the *massif*. The systematic part has not yet been completed as of this writing (1979).

I. Liberia

See also above (Adam). Apart from the aging, deficient compiled enumeration of Dinklage, uncoordinated works are also available for forest trees and pteridophytes. The country has never had a consistent history of botanical exploration and with the mountainous interior and difficulties of travel as noted by Harley it remains imperfectly known. It also has the highest and most consistent rainfall in West Africa giving rise to a relatively rich flora.

DINKLAGE, M., 1937. Verzeichnis der Flora von Liberia (ed. J. Mildbraed). *Repert. Spec. Nov. Regni Veg.* **41**: 235–71.

Unannotated list of known vascular plants, with very limited synonymy and citations of *exsiccatae*; no index. An account of botanical exploration appears in the introductory section. Now very incomplete.

HARLEY, W. J., 1955. The ferns of Liberia. *Contr. Gray Herb.*, N.S., **177**: 58–101, 1 pl., map.

Systematic enumeration of ferns, with synonymy, literature citations, critical remarks, and notes on habitat, frequency, etc.; list of references, index. A relatively lengthy introductory section covers physical features of Liberia, geography and travel problems, notes on prior collecting work, and discussion of phytogeography and vegetation with special reference to ferns. A phytogeographical map is included, as well as separate index.

KUNKEL, G., 1965. *The trees of Liberia.* 270 pp., text-figs., 80 halftones. Munich: Bayerischen Landwirtschafts-Verlag.

Copiously illustrated account of the more important trees, with tabular keys, synonymy, citations, vernacular names, notes on habitat, and indication of related species of lesser importance; lexica of vernacular and trade names; index. A glossary and an account of forest types appear in the introductory section. [An example of the kind, all too common, of 'results' of 'development contract' research, like Whitmore's guide to Solomon Islands trees (**938**).]

VOORHOEVE, A. G., 1965. *Liberian high forest trees.* 416 pp., 72 text-figs., 32 halftones. Wageningen, Holland: PUDOC.

Critical, well-illustrated account of more important rain forest trees; includes keys, extensive synonymy, literature citations, indication of local and extralimital range, vernacular and trade names, extensive critical remarks and copious notes on habitat, field attributes, utilization potential, etc., with mention of related species of lesser importance; glossary and general index. An introductory section includes accounts of physical features, climate, vegetation, forest types, and botanical exploration in Liberia. [Represents a thorough piece of work within its chosen compass, beside which Kunkel's simultaneously published work looks like a piker. Both works, however, take little heed of the user in the field, being more for the office.]

II. Sierra Leone

Sierra Leone has had a long history of botanical collecting, but useful publications on the flora are few; as in Liberia, forest trees are rather more fully documented. Gledhill's list is a very feeble effort, with even less documentation than the Liberian list of Dinklage – zero.

GLEDHILL, D. [1962]. *Check list of the flowering plants of Sierra Leone.* 38 pp. Sierra Leone: University College.

Unannotated, bare list of native and naturalized

flowering plants, based upon the *Flora of West Tropical Africa*; no index. Partially supplants LANE-POOLE, C. E., 1916. *The trees, shrubs, herbs, and climbers of Sierra Leone*. 159 pp. Freetown.

SAVILL, P. S. and FOX, J. E. D., 1971. *Trees of Sierra Leone*. 316 pp., illus., 1 color map. Freetown: Forestry Division, Sierra Leone. (Mimeographed.)

Descriptive manual of forest trees, with keys, local range, and notes on frequency, phenology, timber properties, uses, etc.; lexicon of vernacular names; index. An introductory section includes accounts of physical features, geology, climate, vegetation formations, and forest management.

III. Guinea Republic

For the Nimba massif, see also above under the main heading. The only separate general floristic work, strongly oriented towards useful plants and not to be considered a flora in the conventional sense, is POBÉGUIN, H., 1906. *Essai sur la flore de la Guinée française*. 392 pp., 80 pls., map. Paris. Contemporary conditions have not favored realization of any modern successor.

IV. Guinea-Bissau

For the Cape Verde Islands, see **025**. This state, the smallest in West Africa, was formerly known as Portuguese Guinea. A considerable but scattered floristic literature has built up since the 1930s, culminating firstly in a series of contributions from the Centro de Botânica of the Junta de Investigações do Ultramar (Pereira de Sousa, 1946–63) and secondly in the initiation of a serial flora by the downriver Jardim e Museu Agrícola do Ultramar (1971–), of which as of 1979 only a few fascicles had appeared, the last in 1973. [It would have been more beneficial if the latter project had been designed as a one-volume illustrated manual rather than as yet another 'flora-series'. Fortunately, reference can be made to the several works on Sénégal (**586**) lying just to the north.]

JARDIM E MUSEU AGRÍCOLA DO ULTRAMAR, 1971–3. *Flora da Guiné Portuguesa*. [Fasc. 1–5.] Lisbon: Ministerio do Ultramar.

A formal descriptive treatment, with each fascicle, comprising one family or a part thereof, containing keys, synonymy (with references and citations), vernacular names, indication of *exsiccatae* with localities, and notes on overall range, habitat, life-form, phenology, etc.; indices at end. At this writing (1979), only five fascicles have been published, covering Butomaceae and parts of the Leguminosae.

PEREIRA DE SOUSA, E., 1946–57. Contribuições para o conhecimento da flora da Guiné Portuguesa, I–VIII. *Anais Junta Invest. Colon.* **1**: 45–152; **3**, part 3(2): 7–85; **4**, part 3(1): 7–63; **5**, part 5: 7–64; **6**, part 3: 7–62; *Anais Junta Invest. Ultram.* **7**, part 2: 7–78; **11**, part 4(2): 7–38; **12**, part 3: 7–27.

PEREIRA DE SOUSA, E., 1960. *Contribuições para o conhecimento da flora da Guiné Portuguesa, IX.* 101 pp. (Estudos, Ensaios e Documentos, Junta Invest. Ultram., no. 77). Lisbon.

PEREIRA DE SOUSA, E., 1963. *Contribuições para o conhecimento da flora da Guiné Portuguesa, X.* 76 pp. (Mem. Junta Invest. Ultram., ser. II, no. 46). Lisbon.

Annotated enumeration of seed plants of Guinea-Bissau in the form of a series of ten contributions; the individual accounts include synonymy, references and citations of principal works, *exsiccatae* studied, summary of local and extralimital range, and notes on habitat, life-form, phenology, etc. along with an index. Part 10 includes a collation (p. 48) and a complete general index (pp. 49 ff.); this latter should be consulted in any use of this work. [The somewhat peculiar collation reflects administrative changes with regard to the publications of the Junta de Investigações do Ultramar.]

586

'Senegambia'

Includes the states of **Senegal** and **Gambia**. With a long history of botanical activity dating back to Adanson, Senegal is one of the best-known (and documented) countries in West Africa.

– Vascular flora: 2086 species (Lebrun).

BERHAUT, J., 1967. *Flore du Sénégal*. 2nd edn. [viii], 485 pp., text-figs., color pls. Dakar: Clairafrique. (First edn. 1954.)

Artificially arranged, analytical manual-key to seed plants, with vernacular names, brief indication of local range (with indication of some *exsiccatae*), and figures of representative species; glossary and list of families covered; indices to all botanical and vernacular names.

BERHAUT, J., 1971– . *Flore illustrée du Sénégal*. Vols. 1– . Illus. Dakar: Ministère du Développement

Rural, Direction des Eaux et Forêts (distributed by Clairafrique).

Spaciously planned atlas-flora of flowering plants, with ample descriptive text; includes synonymy, citations of *exsiccatae* with localities, vernacular names, and notes on habitat, occurrence, properties, uses, etc.; bibliography and indices to botanical names, vernacular names, and uses and properties. Of the 11 volumes projected, six have been published as of this writing (1979), covering families in alphabetical order from Acanthaceae through Nympheaceae. [The work is, however, in this author's view more prestigious than practical.]

LEBRUN, J.-P., 1973. *Énumeration des plantes vasculaires du Sénégal.* 209 pp., 6 pls., map (Études Botaniques IEMVPT, 2). Maisons-Alfort, Val-de-Marne: Institut d'Élevage et de Médecine Vétérinaire des Pays Tropicaux.

Briefly annotated, alphabetically arranged (by genera) checklist of vascular plants, with indication of family, synonymy, references and citations, and some critical commentary (in footnotes), as well as indication of range; bibliography of Senegal botany, 1930–72. An introductory section (with map) includes an account of botanical exploration and a survey of floristic regions. Additions to Berhaut's manual of 1972 (see above) are specially marked.

Partial work: Gambia

WILLIAMS, F. N., 1907. Florula gambica. *Bull. Herb. Boissier*, II, 7: 81–96, 193–208, 369–86.

Annotated enumeration of flowering plants, with some literature citations and descriptions of new or little-known species, indication of *exsiccatae* with localities and notes on distribution, critical remarks, and notes on uses, etc.; index. An introduction includes notes on collectors and local botanical history. See also JARVIS, A. E. C., 1980. *A checklist of Gambian plants.* 29 pp. Banjul, Gambia: Book Production and Material Resources Unit. (Mimeographed.) [530 species, with local vernacular names. Sources – not including Williams! – listed in the introduction on p. 1.]

587

Mauritania

The northern part of this state, lying within the Sahara, is here treated within region 59 (see **595**). An extensive study by Adam (1962) contains an overall checklist.

ADAM, J. G., 1962. Itinéraires botaniques en Afrique occidentale. *J. Agric. Trop. Bot. Appl.* 9: 97–200, 297–416, Pls. 1–18. (Reprinted separately, 236 pp., 18 pls., 1 fig.; Paris: Muséum National d'Histoire Naturelle.)

Pp. 345–416 comprise a systematic checklist of vascular plants of Mauritania, with synonymy and indication of climatic and soil preferences. The author's finds are asterisked. The remainder of the work is ecological and phytogeographical, with some attention to applied aspects.

588

Mali and Upper Volta

The northern part of Mali, lying within the Sahara, is here treated within region 59 (see **595**). No specific general floristic works are available for either state, and other information is limited and scattered.

589

Niger Republic

The northern part of Niger, lying within the Sahara, is here treated within region 59 (see **593**).

– Vascular flora: 1045 species (Peyre de Fabrègues and Lebrun).

PEYRE DE FABRÈGUES, B. and LEBRUN, J.-P., 1976. *Catalogue des plantes vasculaires du Niger.* 433 pp., map (Études Botaniques IEMVPT, 3). Maisons-Alfort, Val-de-Marne: Institut d'Élevage et de Médecine Vétérinaire des Pays Tropicaux.

Annotated systematic enumeration of vascular plants, with synonymy, references to principal works, localities and general range, with citations of *exsiccatae* for less common species (those not in *Flora of West Tropical Africa* specially marked), some critical remarks, and notes on alimentary and other uses; list of references and indices to family and generic names. An introductory section includes technical details as well as an account of botanical exploration (beginning with Mungo Park) and an introduction to the vegetation and phytogeography, with definition of three major chorological zones and indication of the practical limits of cultivation (about 15 °N.) on a map. Only nine of the 1045 species included are not in the current edition of *Flora of West Tropical Africa*; striking features are the almost total lack of endemism and orchid species. For additions, see BOUDOURESQUE, E., KAGHAN, S. and LEBRUN, J.-P., 1978. Premier supplement au 'Catalogue...'. *Adansonia*, sér. 2, **18**: 377–90.

Region

59

North Africa and Egypt

This region includes Egypt, Libya, Tunisia, Algeria and Morocco as well as the northern parts of Chad, Niger, Mali and Mauretania and the whole of the Western Sahara territory. It thus includes the 'Maghreb' which borders the Mediterranean Sea from Libya to Morocco as well as most of the Sahara. It is bounded on the east by the Red Sea and the Suez Canal and on the south by the Sudanian boundary and thence along the eighteenth to nineteenth parallel of latitude, gradually moving northwards from east to west.

The bulk of sustained botanical exploration in the region dates from the latter part of the eighteenth century to the present, although activity has lessened in some areas since decolonization, and since the nineteenth century there has been a resident botanical community with its institutions and herbaria.

However, for biohistorical reasons as well as the sheer size of the region, no comprehensive work has been achieved. For long, Egypt was considered within the 'Oriental' floristic sphere, and this may have induced Maire to limit his *Flore de l'Afrique du Nord* to areas not previously covered by a general flora, i.e., from Libya westwards. Moreover, that work appears to have its southern limit corresponding to the boundaries of Libya, Algeria, and Morocco as all areas south of 25 °N. were then included (in theory) within *Flora of west tropical Africa*. Further work has shown, however, that the Sahara is overall more closely allied floristically to the Mediterranean basin, and in the second edition of the west African flora a 'retreat' was made to 18 °N.

Another obstacle is that the original manuscript of *Flore de l'Afrique du Nord* was never completed by its author, and while what is available is planned for publication in full, no plans are in hand for completion of the work. In its place has been published a classified index to his *Contributions* (Lebrun and Stork, 1978),

and plans are being made for production of a 'Med-Check List' which would include North Africa, in accordance with a resolution of the European Science Research Councils.[11]

Separate works, however, are available for most countries and other units recognized here. Descriptive floras are available for Egypt (one in progress), Libya (in progress), Tunisia (incomplete), Algeria, and Morocco (incomplete). These vary widely in depth of coverage and quality, ranging from students' manual-keys through descriptive manual-floras to more detailed documentary floras (the most elaborate being Täckholm and Drar's *Flora of Egypt*). In addition, the Sahara west of the Nile is largely covered by Ozenda's *Flore du Sahara* (**505**). The most practical works are probably Täckholm's *Students' flora of Egypt* (2nd edn., 1974) and Quézel and Santa's *Nouvelle flore de l'Algérie* (1962–3). The suspended *Flore analytique et synoptique de la Tunisie* of Cuénod (1954) has been continued by G. Pottier-Alapetite as *Flore de la Tunisie*, with the first part of the treatment of dicotyledons appearing in 1979. The only other coverage is the elderly manual-key for Algeria and Tunisia by Battandier and Trabut (1902).

Checklists exist for some countries, notably Libya (two of recent date, of which one is in progress), Morocco, and Western Sahara. There are also older checklists for Libya and Tunisia, both published before 1914. The most detailed (and widely used) checklist is *Catalogue des plantes du Maroc* (Jahandiez et al., 1931–41, with later additions), but this is in need of considerable revision or replacement; the rather elaborate use of infraspecific categories in this work is now regarded as somewhat old-fashioned and obsolete.

A number of partial floras and enumerations exist for areas of special interest, both old and new. Noteworthy are Quézel's monograph on the Tibesti massif in northern Chad, one of the Saharan 'monadnocks', and the florulas of Nègre for the Marrakech district (in a wide sense) and Sauvage for the oak-forest of the Atlas and Rif, both in Morocco. On the other hand, no modern 'tree-books' are available for any except limited areas; the only work of broader scope is *Flore forestière de l'Algérie* by Lapie and Maige (1915).

Botanical knowledge for individual countries is uneven, with the best-known areas evidently being Morocco and northern Algeria and the Nile Valley and Delta of Egypt and from these east to the Sinai and west along the coast to Cyrenacia (Libya).

Progress

Reports for each country, with lists of recent literature, have been given in HEYWOOD, V. H. (coord.), 1975. Données disponibles et lacunes de la connaissance floristique des pays méditeranéens. In CENTRE NATIONAL DE LA RECHERCHE SCIENTIFIQUE [France]. *La flore du bassin méditerranéen. Essai de systematique synthétique*, pp. 15–142 (Colloques internationaux, CNRS, 235). Paris. An outcome of the meetings reported by this book (at Montpellier, 1974) was the formation in 1975 of the Organization for the Phyto-taxonomical Investigation of the Mediterranean Area (OPTIMA), an 'umbrella' organization also covering North Africa, whose semiannual *Newsletter* gives some indication of ongoing botanical work in the region (also reported through AETFAT congress proceedings). Additions to the country reports are contained in CASTROVIEJO, S., 1979. Synthèse des progrès dans le domaine de la recherche floristique et littérature sur la flore de la région méditerranéenne. *Webbia*, **34**: 117–31.

Bibliographies. General and divisional bibliographies as for Division 5.

Regional bibliography

Country bibliographies are given in the report by Heywood cited above.

Indices. General and divisional bibliographies as for Division 5.

590

Region in general

René Maire's legacy for posterity included manuscript sufficient for 20 volumes of his monumental *Flore de l'Afrique du Nord*, encompassing lower vascular plants, gymnosperms, monocotyledons, and dicotyledons through Leguminosae (on the Englerian system). At this writing (late 1979) 14 volumes have been published, the latest in 1977 following a break of nearly ten years due to technical and legal problems. At the present time, no plans are in hand for extension of the work beyond what manuscript is available; however, the considerable gap thus left is partly bridged by the recent publication of an index to Maire's numerous but scattered *Contributions à l'étude de la flore de l'Afrique du Nord* (Lebrun and Stork, 1978). In any

case, neither of these works covers Egypt, long considered within the *Flora Orientalis* floristic (and institutional) 'sphere' although it is in actual fact a phytogeographic 'bridge' between North Africa and the Middle East.

A successor work (at least in part) will now be the projected 'Med-Check List' (see **601**).

Bibliography (and index)

LEBRUN, J.-P. and STORK, A. L., 1978. *Index general des 'Contributions à l'étude de la flore de l'Afrique du Nord'* [par R. Maire]. 365 pp. (Études Bot. IEMVPT, 5). Maisons-Alfort, Val-de-Marne: Institut d'Élevage et de Médecine Vétérinaire des Pays Tropicaux. [Complete cross-index to 3697 precursory papers on North African botany written as a background to the *Flore de l'Afrique du Nord* from 1918 through 1949, with a bio-bibliography. Includes a systematic table of species and their citations in the various parts, with references; thus making it a kind of floristic checklist for the region, Maire having been by far the largest contributor.]

MAIRE, R. *et al.*, 1952– . *Flore de l'Afrique du Nord*. Vols. 1– . Illus., map (Encyclopédie Biol., 33, etc.). Paris: Lechevalier.

Comprehensive illustrated descriptive flora of vascular plants, with keys to genera and species; synonymy, with references and citations; detailed account of local range, with indication of *exsiccatae*; general summary of extralimital distribution; notes on karyotypes, habitat, life-form, special features, etc.; some taxonomic commentary; index to botanical names in each volume. Of the 20 volumes projected, at present (1980) 15 have been published, covering families from the pteridophytes through the Rosaceae (Englerian system) and will eventually go through the Leguminosae (the limit to which the original MS extends). The current editor is P. Quézel.

591

Egypt

See also **770–90** (BOISSIER). The Sinai Peninsula is considered to be geographically part of Asia and has thus been designated as **776**. The most useful and satisfactory modern work is Täckholm's *Students' flora of Egypt*, now in its second edition (1974) and accompanied by a book of supplementary notes. Unfortunately, the only completed descriptive flora

remains Muschler's unreliable *Manual flora of Egypt*.
– Vascular flora: 2085 species (Boulos, in Heywood, 1975).

MONTASIR, A. H. and HASSIB, M., 1956. *Illustrated manual flora of Egypt*. Vol. 1. 615 pp., 81 text-figs. [Cairo]: 'Misr'.

Descriptive manual of vascular plants, with keys to all taxa, essential synonymy, abbreviated indication of local distribution, and notes on habitat, abundance, etc.; figures of representative species; index to all botanical names. Offered as a successor to Muschler's *Manual flora of Egypt*, but encompasses only the Dicotyledoneae.

MUSCHLER, R., 1912. *A manual flora of Egypt*. 2 vols. xii, 1342 pp. Berlin: Friedländer. (Reprinted 1970, Lehre, Germany: Cramer.)

Briefly descriptive manual of vascular plants, with keys to genera and species, synonymy with references and citations, vernacular names, concise but relatively detailed indication of local range, summary of extralimital distribution, and notes on life-form, phenology, etc.; lexicon of vernacular names with botanical equivalents; index to all botanical names. Appendices include a glossary, a list of cultivated plants, and a lengthy tabular summary of local and regional distribution of all species in relation to Egypt, the Mediterranean basin, and elsewhere. For later additions and emendations, see SIMPSON, N. D., 1930. *Some supplementary records to Muschler's* Manual flora of Egypt. 59 pp. (Bull. Techn. Sci. Serv., Minist. Agric. (Egypt) 93). Cairo. [Muschler's *Manual flora* has been considered to be 'presumably as unreliable as his other publications' (BLAKE and ATWOOD, 1942, p. 9). For this and even more for bogus descriptive work published in the *Botanische Jahrbücher* and elsewhere, Muschler was dismissed from the Berlin Botanical Museum in 1913; a lawsuit against him was, however, disallowed on the grounds of psychological instability (STAFLEU and COWAN, *Taxonomic literature*, 2nd edn., vol. 3, p. 674). Curiously, his near-contemporary H. F. Wernham suffered a similar fate, although there was never any evidence of deliberate falsifications (W. T. Stearn in *Taxon*, **30**: 1–6, 1981.]

TÄCKHOLM, V., 1974. *Students' flora of Egypt*. 2nd edn. 888 pp., 292 pls., 16 color pls., map. Cairo: Cairo University Press. (First edn., 1956.) Accompanied by TÄCKHOLM, V. and BOULOS, L., 1974. *Supplementary notes* to Students' flora of Egypt, edn.

2. 136 pp., 16 pls. (Publ. Cairo Univ. Herbarium, 5). Cairo.

Concise manual of vascular plants, with keys to genera and species and a synoptic table of families; includes essential synonymy, concise indications of local range, and partly abbreviated notes on habitat, frequency, special features, etc.; illustrations of representative taxa; list of Arab names in script and in transliteration, with botanical equivalents; index to generic and family names at end. An introductory section includes among other matters remarks on plant collecting. The supplement includes more extensive notes on matters, including taxonomic problems, not readily included in the format of the main work; it might be considered a 'critical supplement'.

TÄCKHOLM, V., TÄCKHOLM, G. and DRAR, M., 1941–69. *Flora of Egypt*. Vols. 1–4 (Bulletin of the Faculty of Science, vols. 17, 28, 30, 36). Cairo: Cairo (*formerly* Fouad I) University. (Vols. 2–4 by V. Täckholm and M. Drar alone.)

Copiously descriptive flora of vascular plants, with keys to all taxa; includes extensive synonymy (with references and citations), vernacular names, detailed indication of local range with citation of *exsiccatae* along with generalized summary of extralimital distribution, and lengthy notes on habitat, ecology, biological features, properties and uses, historical and modern local 'lore', and (in each volume) a complete index. Not completed; coverage runs from the Pteridophyta through the Gymnospermae and Monocotyledoneae and the beginning (in reverse order) of the Dicotyledoneae (Casuarinaceae to Piperaceae). The Candollean system as used in Boissier's *Flora orientalis* has been the basis for the arrangement, and at publication of volume 4 some 30 per cent of the Egyptian vascular flora had been treated (V. Täckholm in MERXMÜLLER, 1971, pp. 17–18; see Division 5, Introduction, p. 253). [At that time, it was considered that a further 20 years would be needed to complete the work. As of 1980, nothing further had been published and with the deaths of M. Drar (1964) and V. Täckholm (1978) continuation of this work appears uncertain. It has been previously remarked upon for its perhaps excessive comprehensiveness, due to the inclusion of a considerable amount of non-floristic matter.]

592

Libya

A renewal of floristic studies followed the revolution of 1969, and since then a bibliography, two floras, and a checklist have appeared or are in progress. However, these have not as yet superseded the older standard floras of Durand and Barratte and of Pampanini. Much of the Saharan region in Libya is also covered in *Flore du Sahara* by OZENDA (see **505**).

– Vascular flora: about 1800 species (Boulos).

Bibliography

BOULOS, L., 1972. Our present knowledge on the flora and vegetation of Libya: bibliography. *Webbia*, **26**: 365–400.

ALI, S., EL-GADI, A. and JAFRI, S. M. H., 1976– . *Flora of Libya*. Parts 1– . Illus. Tripoli: Botany Department, Al-Faateh University of Tripoli and Arab Development Institute (distributed Koeltz, Königstein, W. Germany).

Illustrated, serially published full-scale flora with in each fascicle descriptions of taxa, keys, synonymy with references and literature citations, indication of local distribution by grid references (with some citations of *exsiccatae*), extralimital range, critical commentary, notes on phenology, and gridded map of Libya and index to taxa. An introduction to the project appears in fascicle 1, with the comment that a single-volume derivative work is also planned. The general plan of the work resembles that of the *Flora of Pakistan* (see **793**) with which the senior editor has also been associated; fascicles are numbered serially, but no systematic sequence is followed although all families published to date are listed in the latest fascicle. As of mid-1980, fascicles 1–86 had been published, and progress is likely to continue rapidly. [In this work, not much information is included on habitat, ecology, biology, or variability, although the citations of *exsiccatae* are indicative of there having been much new collecting since 1969.]

BOULOS, L., 1977. A check-list of the Libyan flora, 1. *Publ. Cairo Univ. Herb.* 7/8: 115–41.

BOULOS, L., 1979. A check-list of the Libyan flora, 2–3. *Candollea*, **34**: 21–48, 307–32. (Compositae by C. Jeffrey; corrections to that treatment in *ibid.*, **35**: 565–7, 1980.)

Comprises a systematic name list, including basionyms and selected synonymy with literature references. The first part covers pteridophytes, gymnosperms, and monocotyledons (Typhaceae through Orchidaceae), while part 2 covers 'early' dicotyledons (Salicaceae through Neuradaceae). Part 3 includes a treatment of the Compositae. In general, however, the same Englerian sequence as was followed in Pampanini's *Prodromo della flora cirenaica* (see below) is employed for this list. Part 1 is prefaced by a brief general review of studies on the Libyan flora.

DURAND, E. and BARRATTE, G., 1910. *Florae libycae prodromus: ou catalogue raisonné des plantes de Tripolitaine*. cxxvii, 330 pp., 20 pls., map. Geneva: Romet.

Systematic enumeration (in French) of vascular and non-vascular plants, with descriptions of new taxa; synonymy, with references and citations; detailed account of local range, with indication of *exsiccatae*, and general summary of extralimital range; taxonomic commentary and notes on phenology, life-form, etc.; addenda and index to generic and family names. The introductory section includes chapters on physical features, botanical exploration, phytogeography (including synoptic tables), endemic species, and geology together with a list of cultivated plants and a bibliography. The work covers the Egyptian coastal district of Mamara (west of the Nile Delta) together with the Libyan districts of Cyrenacia, Tripolitania, and Fezzan. Collaborators included P. Ascherson, W. Barbey and R. Muschler.

KEITH, H. G. [1973.] *A preliminary check list of Libyan flora*. 2 vols. 1047 pp., illus. n.p.: Published for Government of Libya. (Cover entitled *Libyan flora*.)

In this curious work, vol. 2 (pp. 173–1028) comprises a generically arranged alphabetical checklist of vascular plants, with synonymy, indication of status (and origin), vernacular names, local distribution, citation of some *exsiccatae*, notes on life-form, and miscellaneous casually presented notes; indication of family for each genus; photographs of characteristic plants; list of references (up to 1964). In Vol. 1 is a relatively brief introduction, with plan of the work and miscellaneous observations on climate, vegetation, and history (dated 1965), followed by a lengthy list of vernacular names with botanical equivalents and derivations. [The author, formerly in the British colonial service with a term in North Borneo (!), was FAO forestry advisor in Libya, 1955–60, and the present work is based to a large extent on his observations and collections as well as government herbarium holdings, and thus is not a fully retrospective work. Publication was evidently delayed until 1973 (1392 A.H.).]

Partial works

The following represent significant supplements to Durand and Barratte, but neither covers the whole of Libya.

PAMPANINI, R., 1914. *Plantae tripolitanae ab auctore anno 1913 ab lectae [et] repertorium florae vascularis tripolitanae.* xiv, 334 pp., 1 text-fig., 9 pls., map. Florence: 'Pellas'.

Systematic enumeration of vascular and non-vascular plants, based mainly upon collections by the author in 1913; includes synonymy and references, citation of *exsiccatae* with localities, and summary of overall range; bibliography and index to generic and family names.

PAMPANINI, R., 1931. *Prodromo della flora cirenaica.* xxxviii, 577 pp., 2 text-figs., 6 pls. Forlì: 'Valbonesi' (for Ministero delle Colonie).

Systematic enumeration of vascular and non-vascular plants, with descriptions of new or little-known taxa; includes synonymy (with references), citation of *exsiccatae* with localities, and miscellaneous notes; gazetteer, list of references, and complete index. An introductory section includes among other material an account of botanical exploration, notes on collectors, and tables of additions to the flora after 1910. For supplements, see *idem*, 1936. Aggiunte e correzione al 'Prodromo della flora cirenaica'. *Arch. Bot.* (Forlì) 12: 17–53; also *idem*, 1938. Aggiunte al 'Prodromo...' delle mie raccolte in Cirenacia negli anni 1933/1934. *Rendiconti Seminarii Fac. Sci. Reale Univ. Cagliari,* 8: 53–79.

593

Northern Chad and Niger

See also **589** (PEYRE DE FABRÈGUES and LEBRUN); **505** (OZENDA).

Partial work

Of key phytogeographic interest in this poorly known area is the large, isolated infra-Saharan Tibesti massif in the far north of Chad, with peaks rising to over 3400 m, thus forming, along with Jebel Marra (Sudan) and the Ahaggar (Algeria), an 'island' between the mountains of tropical Africa and the Maghreb. The study by Quézel fills a void covered by no other flora (except perhaps for Ozenda's *Flore du Sahara*) but is itself now equalled by a similar study for Jebel Marra by WICKENS (see **550**).

QUÉZEL, P., 1958. *Mission botanique au Tibesti.* 357 pp., plates, tables, maps (Mém. Inst. Rech. Saharien. Univ. Alger, 4). Algiers.

Includes an enumeration of seed plants of the great Tibesti massif with descriptions of novelties, general indication of local and extralimital range, and notes on habitat, frequency, etc.; statistical summary and index. Separate enumerations are given for the pteridophytes and non-vascular plants. An introductory section includes accounts of physical features, climate, etc., while the final chapters cover floristics, vegetation classification, and phytogeography. See also GILLET, H., 1968. *La végétation du massif de l'Ennedi (Nord-Tchad).* 206 pp., 21 text-figs., 33 halftone pls., tables, maps (Mém. Mus. Natl. Hist. Nat., sér. II/B, Bot., 17). Paris.

594

Algeria (Saharan zone)

For coverage of this vast area, see **505** (OZENDA) and **598** (QUÉZEL and SANTA). In the southeastern quarter is the large massif of Ahaggar, which reaches to more than 2900 m; with the Tibesti massif to the east (in Chad) it has been described as a phytogeographic 'stepping-stone' between the Mediterranean and the mountains of tropical Africa, but its affinities lie with the former (and during the Pleistocene were even more pronounced).

595

Northern Mali and Mauritania (Saharan zone)

See also **505** (OZENDA) and **587** (Adam).

MONOD, T., 1939. Phanérogames. In *Contributions à l'étude du Sahara occidental* (ed. T. Monod), vol. 2, pp. 53–211, Pls. 1–24. Paris: Larose. (Publications du Comité d'études historiques et scientifiques de l'Afrique occidentale française, sér. B, no. 5.)

Based largely upon 1934–5 field work by the author, this report, which forms part of a group of geographical, ethnographic, botanical and zoological studies, includes an annotated list of plants collected, with critical notes and indication of overall distribution within and beyond the area. A map of the region studied is included. The remainder of the report (parts 2 and 3) describe itineraries followed and give general remarks on floristic elements, divisions, etc., along with a bibliography. Treats vascular plants and charophytes.

596

Western Sahara

This former Spanish territory, at one time known as Rio de Oro, is at present (1979) politically disputed. In addition to the work described here, reference may also be made to GUINEA LÓPEZ, E., 1945. *La vegetación leñosa y los pastos del Sahara español.* 152 pp., illus., maps. Madrid: Instituto Forestal de Investigaciones y Experiencias. (Publication *hors serie*.)

GUINEA LÓPEZ, E., 1948. Catálogo razonado de las plantas del Sahara español. *Anales Jard. Bot. Madrid,* 8: 357–442, plates.

Systematic enumeration of seed plants, with limited synonymy, vernacular names, indication of *exsiccatae* with localities, generalized summary of overall range, taxonomic commentary, and notes on habitat, properties, uses, etc.; photographs of representative vegetation formations; bibliography and author's itinerary; no separate index.

597

Tunisia

Until publication in 1979 of the first volume of the treatment of dicotyledons (by Mme Pottier-Alapetite) in the revived *Flore de la Tunisie,* no substantial new floristic work had been produced for the country since the close of colonial rule in 1956. Other available standard works are now rather elderly. For practical use, the Algerian flora of QUÉZEL and SANTA (see **594**) is recommended.

BATTANDIER, J. A. and TRABUT, L., 1902 [1904]. *Flore analytique et synoptique de l'Algérie et de la Tunisie.* 406 pp. Algiers: Giralt.

Concise field manual-key to vascular plants, with limited synonymy, abbreviated indication of range, and notes on habitat, abundance, life-form, etc.; index to all botanical names. Not yet superseded for Tunisia.

BONNET, E. and BARRATTE, G., 1896. *Catalogue raisonné des plantes vasculaires de la Tunisie.* xlix, 519 pp. Paris: Imprimerie Nationale.

Systematic enumeration of vascular plants, with descriptions of new or little-known taxa; synonymy, with references and citations; fairly detailed indication of local range, with summary of extralimital distribution; taxonomic commentary and notes on habitat, life-form, phenology, associates, etc.; index to generic and family names. The introductory section includes chapters on physical features, geology, climate, botanical exploration, cultivated plants, vegetation, and phytogeography. [This work formed part of the results of French government scientific exploration in the 1890s.]

CUÉNOD, A., POTTIER-ALAPETITE, G. and LABBE, A. 1954. *Flore analytique et synoptique de la Tunisie.* Vol. 1. 39, 287 pp., 65 + text-figs., 2 maps. Tunis: Office de l'Expérimentation et de la Vulgarisation Agricoles, Tunisie.

POTTIER-ALAPETITE, G., 1979. *Flore de la Tunisie: Angiospermes – Dicotylédones* (Apetales – Dialypétales). 651 pp., illus. Tunis: Imprimerie Officielle.

Briefly descriptive flora of vascular plants, with keys to all taxa in volume 1, limited synonymy, French vernacular names, fairly detailed indication of local range, summary of extralimital distribution, and notes on habitat, life-form, phenology, frequency, etc.; figures of diagnostic features; index to all botanical names. The introductory section includes a brief historical review and a lengthy glossary. Volume 1 covers the pteridophytes, gymnosperms, and monocotyledons. The recently published continuation of the work by Mme Pottier-Alapetite, a long-awaited event, covers the apetalous and polypetalous dicotyledons; the gamopetalous families are expected to follow. [The 1979 volume has not been seen; it is cited from *Koeltz List,* **239** (1981).]

598

Algeria (Maghrebian zone)

The well-illustrated *Nouvelle Flore* of Quézel and Santa is now not so new, but no overall supplement has yet been published to this author's knowledge.

– Vascular flora: 3139 species (Quézel and Santa); (with Tunisia) 3300 species (Castroviejo, 1979).

QUÉZEL, P. and SANTA, S., 1962–63. *Nouvelle flore de l'Algérie et des regions désertiques meridionales.* Vols. 1–2. 1180 pp., numerous text-figs., pls. 1–112, A–J, 2 maps. Paris: Centre Nationale de la Recherche Scientifique.

Briefly descriptive, illustrated manual-flora of vascular plants, with keys to all taxa; limited synonymy; concise but relatively detailed indication of local range, with summary of extralimital distribution (by phytogeographic zones); French and local vernacular names; notes on infraspecific delimitation, habitat, frequency, associates, etc.; index to all botanical names (vol. 2). The introductory section contains aids to the use of the work, with maps. Covers all of Algeria except the extreme south.

Special group – woody plants

LAPIE, G. and MAIGE, A. [1915.] *Flore forestière de l'Algérie.* viii, 359 pp., 881 text-figs., 1 halftone, map. Paris: Orlhac.

Popularly oriented manual of woody plants, with artificial analytical illustrated keys to species (including also keys to woods and to winter features); generalized indication of local range; French and local vernacular names; discursive descriptive notes on habitat, biology, special features, uses, etc., with illustrations; glossary and indices to botanical and vernacular names. In addition to Algeria, the work also includes the more common woody plants of Tunisia, Morocco, and southern France.

599

Morocco

Included here are the territories of Ceuta and Ifni, but not Western Sahara. For additional coverage of woody plants, see **598** (LAPIE and MAIGE). The flora is the richest and most varied in the Maghreb, with many 'archaic' elements, and a considerable literature has resulted consequent to relatively detailed exploration since 1900. Continuing research and field work has rendered out-of-date the most important work on the flora, the *Catalogue* of Jahandiez and Maire, for which no supplements have appeared since 1956. Two significant partial floras are also in existence, covering the Marrakech basin and the oak forest zone respectively, and there is one general 'tree-book'.

– Vascular flora: about 4200 species (Castroviejo, 1979).

Bibliography

PELTIER, J. P., 1971. Bibliographie botanique marocaine. *Bull. Soc. Sci. Nat. (Phys.) Maroc,* **51**: 247–57.

JAHANDIEZ, É. and MAIRE, R., 1931–3. *Catalogue des plantes du Maroc (spermatophytes et ptéridophytes).* 3 vols. pp. i–lvii, 1–913, map. Algiers: 'Minerva' (distributed by Lechevalier, Paris). Continued as EMBERGER, L. and MAIRE, R., 1941. *Catalogue des plantes du Maroc,* 4: *Supplément...* pp. lix–lxxv, 915–1181 (Mém. Soc. Sci. Nat. Maroc, *hors série.*) Algiers.

Systematic enumeration of vascular plants, with rather detailed infraspecific treatment; includes relevant synonymy, references, abbreviated but detailed indication of local range, citation of representative *exsiccatae*, general summary of extralimital distribution, critical commentary, and notes on habitat, phenology, associates, etc.; index to all botanical names in vol. 3. A lengthy bibliography appears in the introductory section. For additions, see SAUVAGE, C. and VINDT, J., 1949–56. Notes botaniques marocaines: mise à jour du *Catalogue des plantes du Maroc.* Fasc. 1–4. *Bull. Soc. Sci. Nat. Maroc,* **29**: 131–62; **32**: 27–51; **34**: 217–34; **36**: 185–222.

SAUVAGE, C. and VINDT, J., 1952–4. *Flore du Maroc: analytique, descriptive et illustrée.* Parts 1–2. 102 text-figs. (Trav. Inst. Sci. Chérifien, Sér. Bot. 3). Tangier.

Illustrated descriptive flora, with keys to genera and species; synonymy, with references and citations; detailed account of local range; taxonomic commentary and notes on habitat, phenology, habit, etc.; indices in each part. In this work, only a few families have been produced, starting with Ericaceae and following the sequence of vol. 3 of Jahandiez and Maire's *Catalogue des plantes du Maroc.*

Partial works

NEGRE, R., 1962–3. *Petite flore des régions arides du Maroc occidental.* 2 vols. 982 pp., 150 pls. (2 in color), 2 maps. Paris: Centre National de la Recherche Scientifique.

Illustrated descriptive flora of vascular plants of the Marrakech basin north of the Grand Atlas, with keys to all taxa; limited synonymy; French, Arab, and local vernacular names; general indication of local and extralimital range; notes on habitat, life-form, associates, special features, etc., and taxonomic commentary; glossary, gazetteer, bibliography, and general index to all vernacular and botanical names.

SAUVAGE, C., 1961. *Flore des subéraires marocaines: catalogue des cryptogames vasculaires et des phanérogames.* xvi, 252 pp., map (Trav. Inst. Sci. Chérifien, Sér. Bot. 22). Rabat.

Systematic enumeration of vascular plants of the oak-forest zone, with symbolic indication of habitat, associates, life-form, and local range; more or less extensive taxonomic commentary and notes on variation and occurrence; indices to family and generic and to French vernacular names.

Special group – trees

EMBERGER, L. 1938. *Les arbres du Maroc et comment les reconnâitre.* 317 pp. Paris: Larose.

Semipopular field guide to trees and shrubs, with keys, descriptions, local range, French and local vernacular names, and miscellaneous notes on habitat, associates, etc.; index to genera and French vernacular names. Shrubs are dealt with in less detail, although included in keys.

Notes

1 Richards, P. W., 1973. Africa, the 'odd man out'. In *Tropical forest ecosystems in Africa and South America: a comparative review* (ed. B. S. Meggers *et al.*), pp. 21–6. Washington.

2 The first congress was held in 1951 at Brussels; successive congresses have been at three-, later four-yearly, intervals, with the ninth in 1978 at Las Palmas, Canary Islands. The AETFAT *Index* was begun in 1952; since its inception it has been used as a basis for statistical analyses of progress on the tropical and southern African (and southwestern Indian Ocean) floras compiled by J. Léonard and published in successive AETFAT proceedings.

3 On advantages of generic floras and synopses, see Just, T., 1953. Generic synopses and modern taxonomy: introductory essay. *Chron. Bot.* **14**(3): 103–14. Among others, the generic floras of southern Africa are considered.

4 For a short historical review of *Flora capensis*, see Thistleton-Dyer, W. T., 1925. Flora capensis. *Bull. Misc. Inform.* (Kew), *1925*: 289–93.

5 Referred to in Good, R. d'O., 1974. *The geography of the flowering plants* (4th edn., London), p. 224 as having, in proportion to its size, perhaps the richest of all floras.

6 An ambitious British-organized administrative scheme involving transport networks, utilities, research services, and many other government activities in Kenya, Uganda, and Tanzania, but now defunct.

7 Walter, H., 1973. *Die Vegetation der Erde*, 1, pp. 609–10, map. Jena.

8 Lebrun, J.-P., 1971. Les activités botaniques de l'Institut d'Elevage et de Médecine Vétérinaire des Pays Tropicaux. *Mitt. Bot. Staatssamml. München*, **10**: 86–90.

9 Richards, 1973; see note 1.

10 *Idem.*

11 European Science Foundation, *Report 1977*, pp. 81–2; *Report 1978*, pp. 95–7. In the latter reference, the plan and scope of the projected checklist are described.

Division

6

Europe

Flora URSS is thus completed. We remember all our colleagues, many of them long dead, who contributed to its achievement.

We have done what we could. We welcome the young botanists and wish them success.

Fecimus quod potuimus. Vivant sequentes.

E. G. Bobrov, *Nature*, **205**: 1049 (1965).

It is the hope of the editors [of *Flora Europaea*] that by wrestling with these problems they have, at least to some extent, made it unnecessary for the next generation to do so again, and have thus enabled them to devote more time to plants.

D. A. Webb, *Taxon*, **27**: 14 (1978).

The geographical limits of Europe adopted here are essentially the same as those worked out for *Flora Europaea*, with the omission of the tundra zone, the Arctic islands, and the Azores (included respectively in Regions 06, 05 and 02). The Caucasus, sometimes considered to be part of Europe, properly falls into Division 7 (northern, central, and southwestern Asia) as Region 74. Outside of the eight groups of polities here delimited, separate units have been designated for a number of major physiographic entities such as the Mediterranean, the Alps, and Carpathians, and the Urals.

Organized floristic study and the writing of floras as they are today internationally known began in Renaissance Europe, but had rudimentary parallels in ancient Greece and in China.[1] Major advances in methodology came through the work of such scholars as Ghini, Ray, Linnaeus, A.-P. de Candolle, Lamarck, W. J. Hooker, Martius, Bentham, J. D. Hooker, and Britton, and, in the modern era, the Flora Europaea Organization and the aborted Flora North America Program. Most of these developments took place in Europe, largely prior to 1860, influencing floras there and in other parts of the world. By then, European (and most other) floras had assumed their more or less 'modern' forms which since have changed little except in details.

The formative period of floristic botany in Europe may be taken as lasting from 1500 to 1623, the year of publication of the first general systematic survey, Bauhin's *Pinax theatri botanici*, which encompassed ancient and Renaissance writers. During this period, the field, which was largely synonymous with botany in general, gradually emerged from herbalism, with Luca Ghini in Bologna acting as a significant force for change through his technical innovations of the herbarium and botanical garden. Through his students and others, national floristic exploration began in Italy, Central Europe, France, the Low Countries, Spain and England and in succeeding centuries gradually spread through the rest of the continent and beyond. The last areas to be penetrated were the remoter parts of the Iberian peninsula, European Russia, and most notably southeastern Europe, where fierce natives, brigands and Turks had to be contended with.[2]

Prior to the mid-eighteenth century, floristic botany in Europe was largely in the hands of scholars and other professionals, of whom the most notable were L'Écluse, Tournefort, Ray, Haller, and of course Linnaeus. Ray and Linnaeus were both responsible for 'world floras', which were in effect pan-European floras as comparatively few plants from other parts of the world were then known. Indeed, Linnaeus's *Species plantarum* was arguably conceived to be, like Bauhin's *Pinax*, a codification of what was then known about plants in Europe; the tropical flora was thought to be comparatively 'uniform', and the plants of temperate Asia and North America were scarcely known. But *Species plantarum* and its successors were almost the last of their kind. With the spread of the Linnean reforms, floristic botany became, at least in the more 'civilized' countries, a respectable 'leisure' activity, aided by comparatively simple texts. This greatly increased the demand for 'local' floras and resulted in the development of amateur/professional 'national' botanical circles. In the hundred years from Waterloo to Sarajevo, extensive and detailed floristic exploration spread all through Europe, albeit more thinly in many parts of the south and distant east where amateurs were few and 'expeditions' correspondingly more important, as in most of the tropics, Australia, North America, and China and Japan at this period.

As the nineteenth century progressed, the growing professional community in many countries shifted their interests to a large extent to North Africa, the Orient, and further afield for scientific and other reasons, particularly from around 1840. In addition, there was a great professional growth in fields other than floristic or taxonomic botany. An increasing share of the conduct of floristic botany in Europe thus came into the hands of dedicated amateurs, schoolmasters and clergy, who for the latter part of the century and the first decades of that following were more or less to dominate the field, carrying on local (or more distant) exploration and observation and more or less blanketing the map with floras and other works. Many of them were to contribute major 'national' and even all-European works, especially in the period after 1870.

The years from the Franco-Prussian War to 1914 (and somewhat beyond into the 1920s) represented the zenith of this development (and of descriptive floristic and taxonomic botany in general, as Stafleu has noted). Especially characteristic of this era were the large-scale 'national' or supranational floras and other works, some very detailed, compiled usually by one or two (or more) authors; they corresponded in scope to the modern tropical 'research' flora in the sense of van Steenis, with detailed descriptions, elaborate synonymies and infraspecific treatments, and more or less extensive chorological information, although considerable variation in details (as well as in quality) existed. Examples included *Synopsis der mitteleuropäischen Flora* by Ascherson and Graebner, *Conspectus florae graecae* by Halácsy, *Flore de France* by Rouy and Foucauld, *The Cambridge British Flora* by Moss, *De flora van Nederland* by Heukels, *Prodrome de la flore belge* by de Wildeman and T. Durand, *Prodromus florae hispanicae* by Willkomm and Lange, and *Flore des Alpes Maritimes* by Burnat and Briquet. The most ambitious project, differing from the others by its greater engagement of specialists, was Hegi's *Illustrierte Flora von Mitteleuropa*, almost the last to be begun before 1914. While some of these works were the work, wholly or largely, of national herbaria, others, probably the majority, were sponsored privately or through non-governmental institutions. Some, like those for Greece and Spain, were 'colonial' floras, although within the latter major contemporary compilations were also produced. A similar publication development, mostly sponsored through state institutions, began in the Russian Empire in the 1890s; this was later to provide valuable source material for *Flora SSSR*.

The era also witnessed some attempts at pan-European works: *Plantae Europeae* (Richter,

1890–1903); *Conspectus florae Europaeae* (Nyman, 1878–90), and the bizarre *Flora Europae terrarumque adjacentium* (Gandoger, 1883–91) as well as the same author's *Novus conspectus florae Europae* (1910). Only that by Nyman has had lasting success, and ultimately constituted a major source for *Flora Europaea*.

The travel revolution of the mid-nineteenth century and the demands of an interested public for works in 'national' languages resulted in another significant development: the 'excursion' flora with analytical keys. Evidently inspired in the first instance by the *Flore française* of Lamarck and de Candolle, this generally condensed, purely practical type of work spread through Europe in train with Thomas Cook, although in some countries with large floras such works were sometimes abridged. The greatest number were produced in western, northern and central Europe. While in general only state-wide works are accounted for in this Guide, many such manuals, whose keys incorporate diagnoses but which lack separate descriptions, were also written for internal subdivisions or areas of special (e.g., touristic) interest, such as the Alps. Some of these works, such as Bentham's *Handbook of the British Flora* (first published 1858), were concise descriptive manuals with separate keys. The manual-key 'excursion flora' was to migrate to other parts of the world and its forms sometimes used for floras of a higher order, but has rarely been seen in North America. In Europe, these works have been more or less frequently reissued and/or revised and occasionally entirely rewritten, but by and large they do not represent 'creative' work and for that reason have been dubbed 'routine' floras.

The 1914–18 war and the changes brought by ecology, genetics and cytology as well as the relative decline of taxonomy and floristics within botany and general shifts in fashion brought the European floristic 'boom' to a close, although production of a few of the larger works lingered on afterwards for varying lengths of time, and one new project, *Flora polska* (with associated atlas) began to appear in 1919. However, after the dislocations caused by the war and the revolution, large-scale floristic work revived strongly in the Soviet Union during the inter-war period. At the end of the 1920s, the complete reorganization of social, cultural and scientific life and the formulation of national priorities led to the initiation of *Flora SSSR* in 1931, following preliminary planning. As *Flora*

SSSR began to appear, major republican and other 'regional' floras were commenced, continued or revised. Contemporaneously in Germany, the 'second edition' of Hegi's *Illustrierte Flora* was commenced, which at this writing (1980) is still in progress.

Almost all activity ceased or was redirected during World War II, although Rechinger's key *Flora aegaea* appeared in 1943 and heroic efforts were made to keep *Flora SSSR* going. Professional botanists who were preparing tropical floras became involved in assignments involving European plants, and it gradually became perceived how little was really known when a more comprehensive view was taken of the European flora. In the preceding decade or two, the needs of the developing subdisciplines of 'biosystematics' had also begun to have a 'recycling' effect with demands for more and better overall taxonomic accounts. The war more or less blew open the tight little 'national' schools of floristic botany which had been largely dominant from the late-eighteenth and early-nineteenth centuries; thinking was forced into new directions, which included consideration of the possibilities of collaborative floras covering large areas, like the tropical 'colonial' floras or *Flora SSSR* (with the latter exerting a very strong influence, whatever the merits or demerits of the Komarovian species concept).

The successes of the 'Third Reich' in the early part of the war increased German horizons, and in 1943 Rothmaler conceived a scheme for a collective *Flora Europaea* which would not only encompass the continent but also the Urals, the Caucasus, the Levant, North Africa, and Macaronesia. This was prevented by the ultimate outcome of the war and the disintegration of the Central European botanical profession along with the destruction of the Berlin and some other herbaria.

Nevertheless, horizons had been widened, and after the war there was a renewal in the preparation of floras as well as in other systematic studies. Major projects, some including distribution maps, were initiated in Scandinavia, Hungary, Romania, and (later) Serbia, Bulgaria, Switzerland and France, while others, such as the Soviet floras, Hegi's *Illustrierte Flora*, and *Flora polska* were continued. The first wholly new flora of the British Isles since the nineteenth century appeared in 1952, and there was a new round of new or revised 'routine' manuals, partly promoted by the expanding secondary and higher education market and afterwards by the increasing scientific and public interest in ecology and the environment.

For a large number of professional botanists, who now once more played a significant role in European floristic botany, this was not enough. Inspired by Rothmaler's proposals, which became more widely circulated after 1945, lobbying for further integration continued, and in 1954 there was mounted at the Eighth International Botanical Congress at Paris a session, 'Progress of Work on the European Flora', which is reported to have closed on a somber, inconclusive note apart from sonorous words about the great desirability of a general *Flora Europaea*. Like the challenge of *Flora SSSR* thrust upon the botanists in Leningrad, there was a feeling of doubt and fear in the face of such a project; it was still spoken of as a nebulous future ideal.

However, while big meetings may pass resolutions and appoint committees, it is often small, relatively uniform cadres which may actually initiate and push through major projects. The present *Flora Europaea* is said to have been born when a group of botanists, mainly British and furthermore once students at Cambridge University, met at a café opposite the Sorbonne after the inconclusive session and 'continued the discussion with the aid of appropriate refreshment'. This was in July 1954. By the end of 1955 the relatively radical basic principles had been evolved by this group and the first formal meeting of the Editorial Committee was held in January 1956 at Leicester. Twenty-one years later, at the final Flora Europaea Conference in Cambridge, the work was declared completed. A total of 187 authors had been involved from all over Europe and overseas, and upwards of £200 000 expended on a project which has dominated European botany for nearly a generation. Yet *Flora Europaea*, while taking twice as long as originally believed, was completed within one set of working lifetimes, unlike so many tropical floras. The first volume was published in 1964, and the fifth and final volume appeared in 1980.

Although some controversy has surrounded *Flora Europaea*, great benefits, both scientific and practical but too numerous to detail herein, have accrued from this vast undertaking. Already in 1962 Aymonin had hinted that more emphasis should be placed on southern Europe and the Mediterranean, and the work of the project clearly brought out the deficiencies in floristic knowledge of most parts of that area. Greater interest also developed in floristic mapping, and from experience in the British Isles,

Scandinavia and Finland, and Central Europe, the *Atlas Florae Europeae* project was mounted in the 1960s and has progressed gradually since then, with at this writing three parts published. As expositions of flora-methodology, the 'Green Books' (1958, 1960) published by the Flora Europaea Organisation have also been very influential.[3]

In an age when future planning has become almost a necessity, various ideas have been put forward as to the directions European floristic botany should pursue. Some authors, in recommending additional research, have suggested new directions including studies in groups of key scientific and practical importance, but also stressed the need for improvements in the quality and meaning of as well as access to old and new information. The need for keeping in motion the synthetic temper engendered by *Flora Europaea*, including the mapping of species, the increase in and promotion of integrated 'biosystematic' and 'ecosystematic' studies, investigations into the status of and changes to the flora, and the maintenance of control over the flow of information, has likewise been stressed. Others suggest a greater selectivity in priorities, with an increased emphasis on assistance to, and cooperation with, botanists concerned with the Mediterranean and tropical floras who all too frequently lack adequate support and may be too few to cope effectively with rich, threatened and often poorly known floras. The style and content of floras and manuals also are said to require review, with the additional point that in view of changes in the flora much of the stock of existing works must be viewed as largely obsolete in what they cover as well as in their method of treatment of the material. The determinism and 'finality' inherent in much floristic writing before 1914, including the detailed large-scale works, are philosophically and psychologically past us.

In recent years, one formal review has been made: by the 'Ad Hoc Group on Biological Recording, Systematics and Taxonomy' of the European Science Research Councils, acting through the European Science Foundation. This group was constituted in 1975 and made its first reports in 1977. In botany, four priorities for European floristics were recognized: (1) a European floristic information system (i.e., to carry on the work of *Flora Europaea*); (2) coordination in 'biosystematic' research and studies of critical groups; (3) extension of the work of the Threatened Plants Committee of IUCN; and (4) a flora of the Mediterranean Basin. Subsequent work has focused on amateur workers, the education of systematists and the extent of involvement of European botanists with tropical floras.

In part because of the wide agreement amongst the European botanical community on the need for more investigation of the flora of southern Europe and the Mediterranean, significant progress has already been made with the last-named of these four priorities. Existing knowledge, including literature surveys, and needs with respect to the Mediterranean flora were surveyed at a conference at Montpellier in 1974, and in 1975 the 'Organization for the Phyto-Taxonomic Investigation of the Mediterranean Area' (OPTIMA) came into being. Currently based in Berlin, the OPTIMA Secretariat publishes a semiannual news-letter along the lines of the *Flora Malesiana Bulletin* and circulates collections of separates. It has become recognized that realization of a descriptive flora for the whole of the Mediterranean basin is not at present practicable and attention has turned to the preparation of a critical enumeration, now known as the 'Med-Check List', the first formal meeting of which took place in December 1978. Discussions have also taken place on the European floristic information system concept, and in 1980 work on this (as the European Floristic, Taxonomic and Biosystematic Information System, an amalgam of (1) and (2) above) was initiated under the European Science Research Councils Taxonomy Group (S. Zobrist, personal communication).

Thus the trend towards interaction and coopera-tion amongst botanists in Europe, which began after World War II and was greatly accentuated by *Flora Europaea*, is likely to continue in so far as the existing political and ideological framework allows. Proper assessment of priorities within generally limited means will be a key feature, associated with greater formalization of supranational cooperation. Progress is likely to be relatively steady and unspectacular, with perhaps the major developments lying along the southern fringes and outside the continent.

Progress

No comprehensive or comparative historical accounts have been seen, but a sketchy general survey is given by HEYWOOD, V. H., 1978. European floristics: past, present and future. In *Essays in plant taxonomy* (ed. H. E. Street), pp. 275–89. London. British contribu-tions to overall study of the European flora are reviewed by STEARN, W. T., 1975. History of the British contribution to the study of the European flora. In

Floristic and taxonomic studies in Europe (ed. S. M. Walters and C. J. King), pp. 1–17 (*BSBI Conf. Rep.*, 15). Abingdon, England. Many national histories, not accounted for here, are available.

Flora Europaea. The genesis, development, and execution of this landmark have been described in many papers. The contemporary setting and early German proposals are given in ROTHMALER, W., 1944. Aufforderung zur Mitarbeit an einer Flora von Europa. *Repert. Spec. Nov. Regni Veg.* (Fedde), **53**: 254–70; GAMS, H., 1954. Flores européennes modernes: résultats acquis, objectifs à atteindre. In *Rapports et Communications avant le Congrès* (Huitième Congrès International de Botanique), sects. 2, 4, 5 & 6, p. 101. Paris; and VALENTINE, D. H., 1954. Progress of work on the European flora. *Ibid.*, pp. 87–92. Early developments are surveyed in HEYWOOD, V. H., 1957. A proposed flora of Europe. *Taxon*, **6**: 33–42; and *idem*, 1959. The Flora Europaea project: a report of progress. In *Ninth International Botanical Congress. Recent advances in botany*, vol. 1, pp. 941–4. Toronto. A 'mid-level' view is afforded by VALENTINE, D. H., 1971. Floristics in Europe. *BioScience*, **21**: 512–14. 'Valedictory' papers include Anonymous, 1977. Floras: [a] brief history of the Flora Europaea project. *Australian Systematic Botany Society Newsletter*, **13**: 9–12; and WEBB, D. A., 1978. Flora Europaea – a retrospect. *Taxon*, **27**: 3–14.

Other reviews of European floristic botany, which contain remarks on the content and style of floras, include AYMONIN, G., 1962. Où en sont les flores européennes? *Adansonia*, sér. 2, **2**: 159–71; European Science Foundation, 1977. ESF Annual Report 1977, pp. 78–82 and Annex 2. Strasbourg; HEYWOOD, 1978 (see above); and JONSELL, B., 1979. The present stage of botanical exploration: Europe. In *Systematic botany, plant utilization and biosphere conservation* (ed. I. Hedberg), pp. 34–40. Stockholm.

For the two sets of country-by-country reports associated with *Flora Europaea*, presented respectively at the second and seventh Flora Europaea symposia, and the Mediterranean area country reports of 1975, see immediately below under Divisional bibliographies.

The figures for the size of the vascular flora given in the headings for most countries are based on tables provided by Webb in his retrospective review. The larger pair of figures is inclusive of naturalized and adventive aliens.

General bibliographies. Bay, 1910; Blake, 1961 (with the omission of Germany, Austria, eastern

Europe and the USSR); Frodin, 1964; Goodale, 1879; Holden and Wycoff, 1911–14; Hultén, 1958; Jackson, 1881; Pritzel, 1871–7; Rehder, 1911; USDA, 1958. For national and regional bibliographies, see the appropriate headings.

Divisional bibliographies

In addition to Lawalrée's list (see below), unannotated lists of 'standard' floras are given in Hamann and Wagenitz, 1977 (see Region 64) as well as in each of the volumes of *Flora Europaea* (and its forerunner, the 'Green Books').

HEYWOOD, V. H. (ed.), 1963. A survey of taxonomic and floristic research in Europe since 1945. In *Proceedings of the II. Flora Europaea Symposium* (Florence, 1961) (ed. V. H. Heywood and R. E. G. Pichi-Sermolli), pp. 95–562 (in *Webbia*, 18). Florence. [Contains more or less complete bibliographies for every European country up to 1960 or 1961. For the period 1961–71, see Heywood 1974–5 below.]

HEYWOOD, V. H. (coord.), 1974–75. *VII. Flora Europaea Symposium* (Coimbra, 1971): *Floristic reports, 1961–71.* 834 pp. (as *Mem. Soc. Brot.* 24 (in 2 pts.)). Coimbra. [An extension to the reports in the preceding work; covers all countries except Iceland, Norway, Poland, Portugal, Turkey-in-Europe, the USSR, and Yugoslavia. For Spain see *Boissiera*, **19**: 23–60, 1971.]

HEYWOOD, V. H. (coord.), 1975. Données disponibles et lacunes de la connaissance floristique des pays méditerrannéens. See **601**/I (CENTRE NATIONAL DE LA RECHERCHE SCIENTIFIQUE, FRANCE).

LAWALRÉE, A., 1960. Indication des principaux ouvrages de floristique européenne (plantes vasculaires). *Natura Mosana*, **13**(2–3): 29–68. [Very useful, although in places uneven, list of 'standard' floras, with relatively complete bibliographic data. Some works on the adjacent regions of North Africa, the Caucasus, and Macaronesia are also included.]

General indices. BA, 1926– ; BotA, 1918–26; BC, 1879–1944; BS, 1940– ; CSP, 1800–1900; EB, 1959– ; FB, 1931– ; ICSL, 1901–14; JBJ, 1873–1939; KR, 1971– ; NN, 1879–1944; RŽ, 1954– .

Divisional indices

The question of the continued flow of information in European floristic botany consequent to the completion of the *Flora Europaea* project is, as already noted in the divisional introduction, a matter of current concern among the botanical profession in Europe.

BRUMMITT, R. K., *et al.* (comp.), 1966–71. *Index to European taxonomic literature for 1965* (and *1966, 1967, 1968, 1969*). (Regnum Veg. 45, 53, 61, 70, 80.) Utrecht; and BRUMMITT, R. K. and KENT, D. H. (comp.), 1977. *Index to European taxonomic literature for 1970.* Kew: Bentham-Moxon Trustees. [Classified bibliography, with a separate section for floras and related works. Now subsumed in *Kew Record* as to the years 1971 and following. Reviewed in BRUMMITT, R. K., 1973. A survey of the Index to European taxonomic literature 1965–1970. *Bol. Soc. Brot.*, II, **47**, Suppl: 41–55.]

KENT, D. H., 1954–70. Abstracts from literature. In *Bot. Soc. British Isles, Proc.* 1–17. [Summary accounts of new European floristic and taxonomic literature. From 1971 continued with a different emphasis as *BSBI Abstracts* (see Region 66).]

Conspectus

600

Division in general

With the completion of *Flora Europaea* in 1980, the conspectus of Nyman (with its supplements) has been superseded; however, it is retained here in lieu of a successor in its class. Also noticed here are Thonner's *Exkursionsflora von Europa* of 1901–18, recently reprinted, and the *Atlas florae europeae* edited by Jalas, of which as of 1981 five parts had been published.

– Vascular flora: 13 650 species and non-typical subspecies, of which 542 are in so-called 'critical' genera (*Alchemilla, Hieracium, Rubus, Sorbus*, and *Taraxacum*), are accounted for in *Flora Europaea*. Nyman listed 11 111 such taxa (exclusive of those in the above 'critical' genera) and Thonner gave a figure of 9876 seed plant species.

NYMAN, C. F., 1878–84. *Conspectus florae europaeae* (incl. *Supplementum I*). 1046 pp. Örebrö, Sweden: Bohlin. Continued as *idem*, 1889–90. *Conspectus florae europaeae: Supplementum II*. 404 pp. Örebrö.

Concise, partly compiled enumeration of the phanerogamic flora of Europe, with synoptic keys down to groups of species; includes the more important synonymy, generalized indication of distribution, brief taxonomic commentaries and miscellaneous notes, with index at end. The work has been an important source for the *Flora Europaea* project, and it has been noted that in making estimates of the number of species in the European flora, a useful rule of thumb was 'Nyman plus 20%.'[4] An unofficial supplement has also appeared, as follows: ROTH, E., 1886 (1885). *Additamenta ad 'Conspectum Florae Europaeae'*. 46 pp. Berlin.

THONNER, F., 1901. *Exkursionsflora von Europa*. x, 50, 356 pp. Berlin: Friedländer; and *idem*, 1918. *Exkursionsflora von Europa. Nachträge und Verbesserungen*. 55 pp. Berlin. (Reprinted in 1 vol. with added foreword, 1980, Leiden: Rijksherbarium.) French edn.: *idem*, 1903. *Flore analytique de l'Europe*. vi, 322 pp. Paris: Baillière.

Compact diagnostic and descriptive manual-key to the seed plant genera of Europe, with indication of geographical ranges of those genera of more limited distribution along with some synonymy; German vernacular names included. The introductory section includes a definition of geographical limits of the work along with its scope (which encompasses native, naturalized, adventive and/or extensively cultivated taxa), family and generic circumscriptions (after *Die natürlichen Pflanzenfamilien*), the method of key-construction, and sources, while at the end are a glossary, a list of author abbreviations, and a full index. [The new foreword in the 1980 reprint, by R. Geesink and R. van der Meijden, notes that this *Exkursionsflora*, though shrouded in obscurity like most of the author's other works, possesses considerable intrinsic merit. It has not been cited by Blake in his *Geographical guide*, vol. 2. The German and French editions are almost identical, save that no French version of the supplement appears to have been published. As of 1918 (*Nachträge*, pp. 52–3) Thonner recorded for Europe 9876 species of seed plants (9740 native) in 1264 genera (1164 native) and 143 families (31 native).]

TUTIN, T. G. *et al.* (eds.), 1964–80. *Flora europaea*. 5 vols. Maps. Cambridge: Cambridge University Press.

Briefly descriptive, synthetic flora of native, naturalized, and commonly cultivated vascular plants of Europe, with keys to all taxa down to subspecies; references given for accepted names, limited synonymy, concise indication of distribution (both in general and by country, with some extra-European ranges also given where relevant), indication of habitat and karyotypes (where known), critical taxonomic commentary on particular genera and families, complete index in each volume (with cumulative index in vol. 5). The introductory sections (in each volume) include a discussion of the scope and aims of the *Flora*, lists of basic and standard national and regional floras, citations of serials and other important literature, and a glossary of authors of plant names. The whole work covers 11 948 species and 2102 additional subspecies (in all, 13 650 taxa) in 1544 genera and 203 families, arranged on a modified Englerian system (*Syllabus*, 12th edn.) with some amendments. The best retrospective account of this landmark is by Webb.[5]

Distribution maps

JALAS, J. and SUOMINEN, J. (eds.), 1972– . *Atlas florae europeae*. Parts 1– . Maps [i–iv], 1– . Helsinki: Akateeminen Kirjakauppa (Academic Bookstore).

Comprises an atlas of distribution maps of the European flora, based on some 4400 50-kilometer squares. For each species, there is for each square an indication of occurrence, dates of records (before 1900, between 1900 and 1939, and since 1939), and status; the accompanying text gives the correct botanical name (usually that adopted for *Flora Europaea*), references to other published maps or relevant distribution reports, and occasional taxonomic and nomenclatural commentary (with references). The introductory section (part 1) gives remarks on the history of the project and the methodology pursued, as well as a list of collaborators. Circumscriptions of species, as with botanical names, generally follow *Flora Europaea*. At this writing (1980) five parts, comprising maps 1–668 with coverage from Psilotaceae through Basellaceae (sequence as in *Flora Europaea*, i.e., a modified Englerian system), have been published, but as yet the work is not far advanced.

Special group – woody plants (including trees)
The variety of literature on woody plants (and trees) of a popular and semi-popular nature prevailing in North America (see **100**) is now coming to be paralleled by a similar range of all-European works of varying quality, although it is more usual in Europe to include a substantial number of cultivated species. A small selection of works considered to be soundly based scientifically is given here, and some others are listed respectively under **601/I** and **601/II**. Many regional and national 'tree books' are also available, and certain of these have been accounted for in the *Guide*; however, no overall listing *per se* has been published to this reviewer's knowledge.

KRÜSSMANN, G., 1979. *Die Bäume Europas*. 2. Aufl. 172 pp., 405 figs. (on 88 pls., of which 8 are in color), 127 maps. Berlin: Parey. (First edn., 1968.)

Profusely illustrated, popularly oriented pocket-sized manual of native, naturalised, and commonly cultivated trees of Europe (154 angiospermous, 59 gymnospermous species), with tabular picture-keys to species, limited synonymy, vernacular names (in German, French, Italian, and English), generalized indication of range (with maps), brief notes on key distinguishing features, phenology, habitat, etc., and indices. The introductory section includes a terminological table and notes on use of the work. A list of key references is also included. The photographs, though numerous, are regrettably murky. [This second edition accounts more fully for Mediterranean trees than its predecessor.]

POLUNIN, O., 1976. *Trees and bushes of Europe*. xvi, 208 pp., colored illus. London: Oxford University Press.

Attractively produced, copiously illustrated popular guide to more commonly encountered woody plants 2 m in height or over, in atlas form, with colored photographs and/or drawings accompanied by descriptive and distributional notes; related, less common species are mentioned. An introductory section and index are also included, and vernacular names in principal western European languages given.

601

Major continental subdivisions

Included here are works for the Mediterranean Basin and for central and northern Europe.

I. Mediterranean Basin

The Mediterranean Basin, with its characteristic flora, occupies parts of three continents but as the largest part of the land area lies in Europe it is given here. No overall floras or enumeration have ever been published but in the last decade interest in the subregion among European botanists has greatly increased and with it have come attempts at greater coordination of activities. A general conference at Montpellier in 1974 led to the formation of OPTIMA (Organization for the Phytotaxonomic Investigation of the Mediterranean Area), a body resembling the Flora Malesiana Foundation (see Superregion 91–93) and Organization for Flora Neotropica (see Division 3) but differing from both in structure and activities; its secretariat is now in Berlin. Publications include a *Newsletter* (1975–) and a series of recirculated reprints, *Cahiers OPTIMA*. From 1975, biennial congresses have been held in different Mediterranean botanical centers (1975, Heraklion (Greece) and Crete; 1977, Florence; 1979, Madrid).

A project currently in progress under OPTIMA is the preparation of a checklist of the Mediterranean flora, considered to be a more feasible proposition than any form of more comprehensive work given available and likely resources; the most recent plan seen calls for publication by 1984.[6] At the Conservatoire et Jardin Botaniques, Geneva, one of three working 'bases' for the so-called 'Med-Check List', there has already been prepared and circulated (as Document 7) a *Liste codée des flores de base* (1979), containing 66 references (nearly all also treated in this *Guide*).

Bibliographies/progress reports

HEYWOOD, V. H. (coord.), 1975. Données disponibles et lacunes de la connaissance floristique des pays méditerranéens. In *La flore du bassin méditerranéen: essai de systématique synthétique* (Montpellier, 1974) (ed. CENTRE NATIONAL DE LA RECHERCHE SCIENTIFIQUE, France), pp. 15–142 (Colloques internationaux du CNRS, 235). Paris. [Country-by-country surveys, with lists of references. For extensions of these reports up to 1977 as contributed to the second OPTIMA congress in Florence, see CASTROVIEJO, S., 1979. Synthèse de progrès dans le domaine de la recherche floristique et littérature sur la flore de la région méditerranéenne. *Webbia*, 34: 117–31.][7]

Special group – woody plants

The following appears to be the first illustrated guide to woody plants of the Mediterranean area as a whole.

GÖTZ, E. 1975. *Die Gehölze der Mittelmeerländer: ein Bestimmungsbuch nach Blattmerkmalen.* 114 pp., 577 text-figs., map. Stuttgart: Ulmer.

Comprises an illustrated artificial key to woody plants of the Mediterranean Basin, based on vegetative features, especially leaves, with countries of occurrence indicated for each species and German vernacular names. A short list of references for individual countries and indices are also included.

II. Central and Northern Europe

Under this heading are given Hermann's *Flora von Nord- und Mitteleuropa* (1956) and (for trees) Mitchell's *Field guide* (1974), the latter now also available in German.

HERMANN, F., 1956. *Flora von Nord- und Mitteleuropa.* xi, 1134 pp. Stuttgart: Fischer. (Originally published 1912, Leipzig, as *Flora von Deutschland und Fennoskandinavien sowie von Island und Spitzbergen.*)

Concise manual-key to vascular plants, with limited synonymy, relatively extensive general indication of internal distribution (including altitudes), summary of extralimital range or source area, and symbolic designation of life-form, etc.; index to all botanical names. The introductory section contains *inter alia* a synopsis of orders and families. Does not include the Iberian Peninsula, western France, the British Isles, most of Italy, southeastern Europe, or the European part of the USSR; moreover, the nomenclature is not entirely in conformity with the International Code as prevailing at the time of publication.

Special group – trees

MITCHELL, A., 1974. *A field guide to the trees of Britain and northern Europe.* 415 pp., 640 text-figs., 40 color pls. London: Collins.

Illustrated descriptive semipopular treatment, with keys to all species, critical notes, vernacular names, and extensive commentary on special features, biology, phenology, properties, uses, occurrence, notable individual examples, etc.; index. Many naturalized or cultivated species are included, either as full entries or as notations, and forestry aspects are also discussed. Most useful in the British Isles and from northern France through the north European plain to Scandinavia.

603

Alpine and upper montane zones

It has been considered useful to include here a range of key works selected from the extensive though scattered and fragmented literature on the alpine and high-mountain floras of Europe, more often than not sundered by the lines between nations (and national floras). For convenience, they have been grouped under a range of subheadings, viz: I. Sierra Nevada (Spain); II. the Pyrenees (including Andorra); III. Alpes-Maritimes; IV. The Alps; V. The Apennines; VI. The Sudetens; VII. The Carpathians; VIII. Mountains of southeastern Europe; IX. Mountains of Norway and Sweden; and X. The Ural. For some of these, no suitable works appear to be available.

I. Sierra Nevada (Spain)

No works apart from 'wildflower books' appear to exist.

II. The Pyrenees (including Andorra)

The only modern account of the plants of this mountain chain, still incomplete though well along, is the serial *Catalogue-flore des Pyrénées* by Gaussen *et al.*, in progress since 1953. Earlier works, as listed in detail by Blake (1961) and other standard bibliographies, are too untrustworthy or very outdated. The account of Andorran plants by Losa and Montserrat (1951), however, provides a useful modern introduction to the Pyrenean flora to complement the *Catalogue-flore*.

GAUSSEN, H. *et al.* (comp.), 1953– . *Catalogue-flore des Pyrénées*. Parts 1– . Illus., map. *Monde Pl.* 48 (nos. 293–7), etc.

Systematic enumeration of vascular plants of the Pyrenees chain, with synonymy, notes on habitat, and detailed symbolic indication of local range arranged by districts; for map see *Monde Pl.* no. 333 (1961). Not yet completed though far advanced; at this writing (1980) covers pteridophytes, gymnosperms, monocotyledons, and dicotyledons through Labiatae (Englerian sequence); the latest part seen is in *Monde Pl.* 75 (nos. 403–5): 1–24 (1980) with further installments usually appearing quarterly. The best of the earlier general floras, although now very outdated, is PHILIPPE, X., 1859. *Flore des Pyrénées*. 2 vols. Bagnères-de-Bigorre, Hautes-Pyrénées.

Local work: Andorra

LOSA, M. and MONTSERRAT, P., 1951 (1950). *Aportación al conocimiento de la flora de Andorra*. 184 pp., illus., 6 pls. (on 3), maps (Monogr. Inst. Estud. Pirenaicos, 53, Botánica, 6). Zaragoza. (Also published 1950 as Actas del I. congreso internacional del Pirineo del Instituto de Estudios Pirenaicos, 1. San Sebastian.)

Systematic enumeration of vascular plants, bryophytes, and fungi, with limited synonymy, indication of local range, and notes on habitat; index at end. An introductory section deals with local geography, geology, botanical exploration, major features of the flora, vegetation associations, and floristics.

III. Alpes-Maritimes

See also FENAROLI (under IV. The Alps) as well as general floras of Italy (**620**) and France (**651**).

BURNAT, E. *et al.*, 1892–1931. *Flore des Alpes-Maritimes ou catalogue raisonné des plants qui croissent spontanément dans la chaîne des Alpes-Maritimes y compris le département français de ce nom et une partie de la Ligurie occidentale*. Vols. 1–7. Geneva: Georg (vols. 1–6); Conservatoire et Jardin Botaniques (vol. 7).

Annotated enumeration of vascular plants, with synonymy, indication of local range (with *exsiccatae* and other published records), and extensive taxonomic commentary; indices. Not completed; extends from Ranunculaceae through Compositae–Cynarioideae (Candollean system).

IV. The Alps

Of the considerable floristic' literature on the Alps, only two appear to be sufficiently comprehensive in coverage to merit full entries in this *Guide*. These are the five-volume *Atlas der Alpenflora* (1882–4; 2nd edn. 1897) published by the Deutscher und Österreichischer Alpenverein and the fine one-volume manual by Fenaroli, *Flora delle Alpi* (most recently revised in 1971 but with antecedents going back to 1902). Other current guides to Alpine plants, of which there is a considerable range, take the form of more or less popular and selective pocket-sized pictorial treatments with accompanying concise descriptive and ecological notes and thus strictly speaking fall outside the scope of the *Guide*. Some, however, are very well known and have had a wide influence in other parts of the world; a range of titles, cited in brief, is therefore presented following the description of the two main works.

DEUTSCHER UND ÖSTERREICHISCHER ALPENVEREIN, 1897. *Atlas der Alpenflora* (ed. E. Palla). 2nd edn. 5 vols. 500 color pls. Graz. Accompanied by DALLA TORRE, K. W. VON, 1899. *Die Alpenflora der österreichischen Alpenländer, Südbaierns und der Schweiz. Nach der analytischen Methode zugleich als Handbuch zu dem 'Atlas der Alpenflora'*. xvi, 270, [1] pp. Munich: Lindauer. (First edn. in combined form, 1882–4, Vienna, as DALLA TORRE, K. W. VON and HARTINGER, A. *Atlas der Alpenflora*. 5 vols.)

CLUB ALPEN ALLEMAND ET AUTRICHIEN, 1899. *Atlas de la flore alpine* (ed. H. Correvon). vii, 193 pp.; atlas of 500 color pls. in 5 vols. Geneva/Basle: Georg. (Reissued 1901, Paris; 6 vols.)

The text volume of this complementary pair of works comprises an analytical manual-key for field use, with on pp. xii–xvi of the introductory section a reference list of key works on Alpine floristics. The atlas consists of full-page plates, each of one species, termed by Blake (1961, p. 559) as 'fair to excellent...[but] without dissections'; in the German edition these are arranged on the Englerian system while in the French version the arrangement follows the Candollean system (the plates of the latter, however, bearing dual numbering). All the plates are by A. Hartinger. [Not seen; description and citations based on Blake and other sources.]

FENAROLI, L., 1971. *Flora delle Alpi*. 2nd revised edn. xi, 429 pp., 262 text-figs., 61 halftones, 44 color pls., 2 maps (in end papers). Milan: Martello. (First revised edn., 1955. Predecessor originally pub. 1902, Milan, as PENZIG, O. *Flora delle Alpi illustrata*; 3rd edn., 1932, as FENAROLI, L. *Flora delle Alpi e degli altri monti d'Italia*.)

Copiously illustrated, comprehensive manual-key to the vascular plants of the high country (above 2000 m) throughout the Alps as well as of the mountains of the Italian region (including Corsica, Sardinia, and Sicily); includes vernacular names (in several European languages), limited synonymy, generalized indication of internal and extralimital range, altitudinal zonation, and fairly detailed habitat and phytosociological notes together with some remarks on properties, uses, etc.; indices to vernacular and botanical names as well as associations. The introductory section, partly pedagogical, includes general remarks on physical, climatic, edaphic, and biotic factors as well as on vegetation associations and seres.

Popular field guides

FAVARGER, C., 1956–8. *Flore et végétation des Alpes*. 2 vols. 76 figs., 64 color pls. Neuchâtel: Delachaux et Niestlé. German edn.: *idem*, 1958–9. *Alpenflora*. 2 vols. Illus. Bern. [Vol. 1: Alpine zone; vol. 2: Subalpine zone and the Jura. Includes extensive descriptions of formations and communities.]

HEGI, G., MERXMÜLLER, H. and REISIGL, H. [1977]. *Alpenflora*. 25th edn., by H. Reisigl. 194 pp., 1, 40 pls. (32 in color), 1, 48 maps. Berlin: Parey. (First edn., 1905, Munich, by G. Hegi and G. Dunzinger.) [Pocket atlas-flora; comprises plates with accompanying text and distribution maps but no keys.]

LANDOLT, E., 1964. *Unsere Alpenflora*. 3rd edn. 223 pp., 25 text-figs., 72 color pls. Zurich-Zollikon: Schweizer Alpen-Club. French edn.: *idem*, 1963. *Notre flore alpine* (trans. R. Corbaz). 234 pp., 25 text-figs., 72 color pls. Italian edn.: *idem*, 1962. *La nostra flora alpina* (trans. G. Kauffman). 256 pp., 25 text-figs., 72 color pls. [Concise pocket manual, with keys; includes extensive introductory section and (on pp. 195–8) a list of key references on Alpine floristics.]

RAUH, W., 1951–3. *Alpenpflanzen*. 4 vols. Illus. (some in color). Heidelberg: Winter. (Revised edns. of vols. 1–2 later published by Quelle & Meyer, Heidelberg.) [Atlas-flora, with extensive introductory sections on ecology, biology, environment, etc.; no keys.]

SCHRÖTER, L. and SCHRÖTER, C., 1963. *Taschenflora des Alpenwanderers*. 29th edn., by W. Lüdi. 26 pls. (24 colored). Zurich: Schumann. (First edn., 1889; 25th edn., 1940.) [Pocket field atlas with short descriptive notes in German, French and English; no keys. Inspired *The mountain flora of Java* (see **903**).]

V. The Apennines (and mountains of Corsica and Sicily)

See FENAROLI (under IV. The Alps) for complete coverage of the high-mountain areas of the Italian peninsula and islands to the west.

VI. The Sudetens (Sudety)

KRUBER, P., n.d. *Exkursionsflora für das Riesen- und Isergebirge*. vii, 345 pp., 42 text-fig., 6 col. pl. Warmbrunn: Lepelt.

Manual-flora with synoptic keys, brief descriptions, German names, and notes on phenology, habit, habitat, frequency, special features, etc.; indices to family, generic, and local names. The introductory section includes keys to classes and to families based on the Linnean sexual system. No date of publication is given for this work, but it appears to have been in the 1920s or early 1930s. A later work on the area is FABISZEWSKI, J., 1971. *Rośliny Sudetów: atlas*. 154 pp., 60 color pls. Warsaw: PZWS. [Wildflower guide.]

VII. The Carpathians

No floras deal with the Carpathian chain as a whole; it is perhaps the longest mountain range in Europe, passing through several countries (at present Czechoslovakia, Poland, the Soviet Union, and Romania). Of these, the best-documented in modern terms is the Carpatho-Ukraine district. The various key works are here grouped under several subheadings. For an extensive bibliography, see Čopik (1976) below.

Local works: central Carpathians

Includes the Bieszczady (Beskids) and Tatry (Tatra); in the latter is the highest point in the Carpathians.

PAWŁOWSKI, B., 1956 *Flora Tatr/Flora tatrorum*. Vol. 1. 672 pp., 130 text-figs. Warsaw: Pánstwowe Wydawnictwo Naukowe.

Descriptive flora (in Polish) of vascular plants, with keys to all taxa, synonymy, vernacular names, local range and distributional remarks, and notes on variation, habitat, phenology, etc.; indices. An account of vegetation formations in the area appears in the introduction. Not completed; covers families from the Pteridophyta through the end of the Choripetalae (Englerian system). [The author died suddenly in 1971, while on Mount Olympus in Greece, so completion of this work appears uncertain.]

SAGORSKI, E. and SCHNEIDER, G., 1891. *Flora der Centralkarpathen/Flora carpatorum centralium.* 2 vols. 2 pls. Leipzig: Kummer.

Volume 2 consists of a concise manual of vascular plants in the central Carpathians, with keys, limited synonymy, local range, and notes on habitat, phenology, etc.; index at end. Volume 1 contains a general introduction, with accounts of geography, geology, climate, vegetation formations, and botanical exploration; bibliography, p. 112.

Ukrainian Carpathians

See also **695** (all works). A considerable amount of floristic and geobotanical work has been carried out in the Ukrainian Carpathians since World War II, and the area may now be considered as being relatively well-documented. Only the following, however, strictly speaking deals with the high-mountain flora; but its extensive bibliography is of value over a far greater area, including by and large the whole of the Carpathian chain.

COPIK, V. J., 1976. *Vysokogirna flora Ukrains'kikh Karpat.* 267 pp., 12 figs. Kiev: 'Naukova dumka'.

Systematic enumeration of the high-mountain vascular flora (475 species), with synonymy, local and extralimital distribution, karyotypes, some critical remarks, and notes on habit, ecology, phytosociological groups, etc.; references to *Flora SSSR* or other major works given under species headings. Pp. 167ff. comprise a detailed floristic/phytogeographic analysis, with emphasis also given to the biology and the development of the flora, while pp. 235–46 feature a rather detailed bibliography. In the introductory section are notes on physical features, climate, and floristic zones, with coverage extended to account for most of the Carpathian chain from Poland and Czechoslovakia to Romania and its geographical and botanical subdivisions. A complete index concludes the work.

Romanian Eastern Carpathians (Carpaţii Moldavi)

These divide Transylvania from Moldavia, and extend southwards to a point east of Braşov. No separate works are available so far as is known.

Southern Carpathians (Carpaţii Meridionali)

Also known as the Transylvanian Alps, this range lies on an east-west axis, dividing Transylvania from the plains of Wallachia, and links with the eastern Carpathians east of Braşov. No manuals appear to be available save for that by Beldie (1972) on the Bucegi massif southwest of Braşov, described below, but a number of local checklists compiled in conjunction with geobotanical studies exist, viz: BELDIE, A., 1967. *Flora şi vegetaţia munţilor Bucegi.* 578 pp., 68 halftones, tables, map. Bucharest: Academia RSR Press; BOŞCAIU, N., 1971. *Flora şi vegetaţia munţilor Ţarcu, Godeanu şi Cernei.* 494 pp., 39 figs., tables, maps. Bucharest; and NYÁRÁDY, E., 1958. *Flora şi vegetaţia munţilor Retezat.* 195 pp., 42 text-figs., 1 folding pl., tables. Bucharest. [The Bucegi massif, as noted, is southwest of Braşov and reaches 2507 m in Omu Peak; Ţarcu, Godeanu and Cernei form a group at the extreme west of the range towards the Danubian Iron Gates, with a maximum elevation of 2292 m; Retezat (Retyezat) is just east of the last three peaks named, in the western part of the range, and reaches 2506 m. No work has been seen covering Negoi, at 2543 m the highest peak in the range and the second highest in the whole of the Carpathians.]

BELDIE, A., 1972. *Plantele din Munţii Bucegi: determinator.* 409 pp., 62 text-figs. Bucharest. Academia RSR Press.

Manual-key to vascular plants, with indication of local distribution, notes on habitat, life-form, habit, phenology, etc., and representative illustrations; vernacular names; index. [Not seen by this writer; citation and description based on Koeltz Catalogue 262, p. 15 (1980) and other works of this author.]

604

The Ural

See also **710** (KRYLOV); **699** (KUČEROV); **064** (IGOSHINA). This geologically ancient range, the traditional dividing line between Europe and Asia, is more than a long chain of mountains: it also is, so to speak, a 'state of mind'. Recently it has also acquired status as an independent floristic region, as recognized by TZELEV in his 1976 revision of the grasses of *Flora SSSR* (*Prog. Bot.* **40**: 415 (1978); cited here under Komarov at **680**). It is therefore treated here as an independent unit. General floristic coverage is provided by the manual of Govoruchin (1937), now out of date, and two partial works: Vakar (1964) for more common plants of the central and southern Ural, and Mamaev (1965) for woody plants. All of these have keys and illustrations. The checklist of Igoshina for the high-mountain and arctalpine vascular flora is, as noted, listed at **064**.

GOVORUCHIN, V. S., 1937. *Flora Urala.* [Flora medio, boreali et polari uralensis.] 536 pp., 164 text-figs., color pl., portrait, folding map. Sverdlovsk: Oblast' Publishing Service.

Concise manual of vascular plants (1574 species) of the entire Ural (apart from the steppe region in the south), with keys to all taxa, limited synonymy, vernacular names, and brief notes on distribution, habitat, phenology, etc.; indices to generic, family, and vernacular names. The introductory section gives accounts of floristic regions and botanical exploration as well as aids for the use of the work.

Special group – herbaceous plants

VAKAR, B. A., 1964. *Opredelitel' rastenij Urala*. 2nd edn. 416 pp., 96 text-figs. Sverdlovsk: [Oblast'] Book Publishing Service. (First edn, 1961.)

Popularly oriented field guide to the vascular flora of the southern Ural, with vernacular and Russian-binomial names, diagnostic keys, brief general indication of local range, and notes on habitat, uses, special features, etc.; figures of representative species; indices to genera and families and to Russian-binomial names. The introductory section is pedagogical. Covers some 640 herbaceous and subwoody species which are more likely to be encountered; for trees and shrubs, the complementary work by Mamaev described below should be consulted. [Description in part based on information supplied by M. E. Kirpicznikov, Leningrad.]

Special group – woody plants

MAMAEV, S. A., 1965. *Opredelitel' derev'ev i kustarnikov Urala*. 116 pp., 35 text-figs., map (*Trudy Inst. Biol. Akad. Nauk SSSR, Ural'sk. Fil.*, no. 41). Sverdlovsk.

Illustrated manual of the spontaneous and commonly cultivated trees and shrubs of (mainly) the central and southern Ural (194 species), with diagnostic keys to all species; includes also vernacular names, general indication of local range or country of origin, and notes on habitat along with a list of references, aids for the use of the work, and an index to botanical names. The keys are wholly artificial, being arranged in six primary tables (compare Krüssman's *Die Bäume Europas* under 600).

608

Wetlands

In contrast to North America, comparatively few significant works have to date been published on the aquatic and marsh plants of Europe, although the treatment by H. Glück in A. Pascher's edition of *Die Süsswasser-Flora Mitteleuropas* is one of the earliest of its kind (predating the first edition of Fassett's *Manual* (108/I) by four years). A greatly expanded version of this work was published in 1980–1.

Northern Europe

CASPAR, S. J. and KRAUSCH, H.-D., 1980–1. Pteridophyta und Anthophyta, I–II. In *Süsswasserflora von Mitteleuropa* [von A. Pascher] (ed. H. Ettl *et al.*), 2nd edn., vols. 23–4. 2 vols. 109 + 119 pls. (with 2733 figs.), maps (end papers). Stuttgart: Fischer.

Descriptive manual-flora of vascular plants, with keys to all taxa, synonymy, indication of regional occurrence and overall distribution, karyotypes, and phenology; detailed discussion of habitat, preferences, associates, and special ecological and biological features; citations of published distribution maps; no index (in the first part). The work is opened by a concise introduction, with definition of ecological and phytosociological classes, and a general key. Covers hydrophytic and helophytic pteridophytes and monocotyledons; a second part, treating dicotyledons, is to follow. Supersedes GLÜCK, H., 1936. Pteridophyten und Phanerogamen. In *Die Süsswasser-Flora Mitteleuropas* (ed. A. Pascher), vol. 15. 486 pp., 258 figs. Jena: Fischer.

HASLAM, S. M., SINKER, C. and WOLSELEY, P., 1975. British water plants. *Field Studies*, 4: 243–351, illus. (incl. 28 pls.). (Reprinted separately, London.)

This work is in two parts: an illustrated key (with glossary), and a systematic enumeration of aquatic macrophytes with abbreviated but fairly detailed indication of habitat, substrates, habit, vernacular names, and geographical distribution. Illustrative plates are provided. In the introduction are given technical notes and tips for collecting as well as a brief indication of the scope of the work, while at the end is a complete index. Considered to be of good standard (C. D. K. Cook, personal communication).

Region

61

Iberian peninsula

In addition to Spain, Portugal, and Gibraltar, the region is also considered to include the Balearic Islands. The linguistic regions of Galicia and Catalonia are also accounted for individually, as each features a key separate flora of more recent date than Willkomm and

Lange's *Prodromus florae hispanicae* (1861–93), the leading 'regional' flora (which itself has yet to be superseded). Separate floras of more or less recent date, some still in progress, are also available for Portugal, Gibraltar, and the Balearic Islands.

No modern floras are, however, available for Spain. Present general works of Spanish origin include two students' manuals of varying scope (Lázaro é Ibiza, 1920–1; Caballero, 1940), a checklist (Guinea López and Ceballos Jimenez, 1974), and a dendroflora (Ruiz de la Torre, 1971). A second checklist, by B. E. Smythies, is in preparation. Heywood has noted that the evolution since the early-nineteenth century of two parallel floristic traditions – respectively by Spanish botanists and by foreigners (mainly from northern Europe) working in Spain – will make preparation of a modern critical flora of Spain a difficult task, not unlike that faced by a 'developing' tropical country which has chosen to prepare a flora. At the present time, it remains, along with parts of the Italian region and southeastern Europe, one of the least well documented areas in the continent, and parts are still botanically underexplored – a status comparable with the West Indies, India, and parts of west Africa or Australia.

Portuguese botanical effort has been more concerned with its *Ultramar* and, while the flora is relatively well known, it is not documented to the same standard as in northern Europe. *Nova Flora de Portugal*, based partly on *Flora Europaea*, has yet to be completed. For the Balearics, no modern general flora is available; Knoche's *Flora balearica* is outdated and keyless and Bonafé Barceló's illustrated work, though in the local Catalan, is more of a collectors' item as well as being limited to Mallorca. The most up-to-date listing is given by Duvigneaud (1979).

Bibliographies. General and divisional bibliographies as for Division 6. Very thorough coverage through 1959 is in particular provided by Blake 1961.

Indices. General and divisional indices as for Division 6.

611

Portugal

For the Azores, see **012**.

– Vascular flora: 2750–2950 species (2400–2600 native).

COUTINHO, A. X. P., 1939. *Flora de Portugal* (*plantas vasculares*). 2nd edn., revised R. T. Palhinha. ii, 938 pp. Lisbon: Bertrand (Irmãos). (First edn. 1913.)

Manual-key to native and naturalized vascular plants, with limited synonymy (no references), vernacular names, short descriptions, concise indication of local range (by provinces), and brief notes on phenology and ecology; glossary and indices to vernacular and botanical names at end. The introductory section includes a list of authors' names.

FRANCO, J. DO A., 1971. *Nova flora de Portugal*. Vol. 1. xxiv, 647 pp., 2 maps. Lisbon: Sociedade Astoria.

Concise descriptive account of vascular plants (based partly on *Flora Europaea*), with keys to all taxa, limited synonymy, and notes on habitats and local range; citations for standard treatments of particular genera under their headings; list of authors, guide to references, and index to botanical names. This first volume covers families from the pteridophytes through the Umbelliferae (as in volumes 1–2 of *Flora Europaea*); it includes the Azores as well as the Portuguese mainland.

SAMPAIO, G., 1946(1947). *Flora portuguesa*. 2nd edn., revised A. P. de Lima. xliii, 792 pp., 850 text-figs., 13 pls. Porto: Imprensa Moderno. (First edn., 1909–14, not completed.)

Illustrated manual-key to native and naturalized vascular plants, including vernacular names, brief general indication of local range, and phenological data; list of authors' names, glossary, and indices to vernacular and botanical names at end. No synonymy is included, and aliens and widely cultivated species are merely noted in smaller print.

612

Spain (in general)

Only works pertaining to Spain as a whole are given under this heading. The important separate works on Galicia, Catalonia, and the Balearic Islands are treated respectively under **613**, **615**, and **619**; Gibraltar appears as **618**. For Spanish settlements in Morocco (Ceuta, etc.), see **599**. [In addition to the works described here, a new checklist by B. E. Smythies is reportedly in preparation. Some remarks pertinent to the history of floristic studies in Spain and the preparation of floras of the country, applicable also to other parts of the Mediterranean, are given by Heywood (1978).[8]]

– Vascular flora: 5150–5250 species (4750–4900 native).

GUINEA LÓPEZ, E. and CEBALLOS JIMENEZ, A., 1974. *Elenco de la flora vascular española* (*Península y Baleares*). ii, 403 pp. Madrid: Instituto Nacional de Conservación de la Naturaleza (ICONA).

Systematic enumeration of vascular plants, with synonymy, citations, karyotypes (where known), habitat, altitudes, habit, and symbolic indication of local and overall distribution, but no critical remarks; indices to genera and families and addenda at end. A short introductory section contains a list of works consulted. Based in part on *Flora Europaea*.

HEYWOOD, V. H. (ed.), 1961. *Catalogus plantarum vascularum Hispaniae*. Part 1. xviii, 58, ii pp. Madrid: Instituto Botánico 'A. J. Cavanilles'.

List, with fairly detailed synonymy, of the native and naturalized vascular plants of Spain; index to families and genera. The introductory chapter includes a list of references for nomenclature. Not completed; includes Pteridophyta, Gymnospermae, and Dicotyledoneae from Ranunculaceae to Cruciferae (modified Bentham and Hooker system). Most of the work of compilation was done by P. W. Ball. The work is of no value for phytogeographic purposes. Text in English and Spanish.

LÁZARO É IBIZA, B., 1920–1. *Botánica descriptiva: compendio de la flora española*. 3rd edn. 3 vols. 1000 text-figs., map. Madrid. (First edn. 1896.)

Briefly descriptive flora of non-vascular and vascular plants, with synoptic keys to groups of species and to sections, genera, and families and incorporating brief indication of local range, vernacular names, notes on habitat, phenology, and special features, and floral diagrams and figures of representative species; appendix on vegetation and phytogeography; indices to all vernacular and botanical names. The introductory section (vol. 1) includes a glossary and several alternative general keys of various kinds, as well as a historical account of botanical work in Spain. Largely compiled from various sources, this work appears to have been designed mainly as a students' manual and textbook.

WILLKOMM, M. and LANGE, J., 1861–80. *Prodromus florae hispanicae, seu synopsis methodica omnium plantarum in Hispania sponte nascentium vel frequentius cultarum quae innotuerunt*. 3 vols. Stuttgart: Schweizerbart; also WILLKOMM, M., 1893. *Supplementum prodromi florae hispanicae 1862–93*. ix, 370 pp. Stuttgart. (Reprinted 1968, Lehre, Germany: Cramer.)

Briefly descriptive flora (in Latin) of native, naturalized, and commonly cultivated plants, with non-dichotomous analytical keys to genera (and synoptic keys to species in the larger genera); full synonymy, with references; vernacular names; detailed account of local range, with citations of *exsiccatae* and localities; general summary of extralimital range; notes and abbreviations on habitat, life-form, phenology, etc.; critical taxonomic commentary; addenda and corrigenda (in vol. 3); index to genera in vols. 1–2 and complete indices to all botanical and known vernacular names in vol. 3. The introductory section includes a synopsis of families as well as lists of abbreviations and authors' names.

Partial work

CABALLERO, A., 1940. *Flora analítica de España*. xiv, 617 pp., 268 text-figs. Madrid: Sociedad Anónima Española de Traductores y Autores.

Concise students' manual-key to the more common or significant native and naturalized plants, with concise descriptions, vernacular names, figures of representative species, illustrated glossary, and general index to botanical names (with vernacular equivalents and ecological notes). The work is in its scope limited to plants which are relatively common, widespread, or of economic or medicinal importance. No synonymy, except for a list of generic synonyms at the end, is included.

Special group – woody plants

RUIZ DE LA TORRE, J., 1971. *Arboles y arbustos de la España peninsular*. xxii, 512 pp., 133 pls. Madrid: Instituto Forestal de Investigaciones y Experiencias/ETS de Ingenieros de Montes.

Dendrological atlas-manual of trees and tree-like shrubs, with extensive descriptions of species, Spanish, Catalan, French, English and German vernacular names, and copious notes on ecology, phenology, distribution, diseases, and forestry aspects, and full-page drawings; illustrated glossary, periodical abbreviations, and indices at end. The introductory section includes a synoptic list (125 species covered), technical notes, and general and artificial keys to all species (pp. 1–35). An extensive bibliography appears on pp. 505–12.

613

Galicia

See also **612** (all works).

MERINO Y ROMÁN, B., 1905–9. *Flora descriptiva é ilustrada de Galicia.* 3 vols. Illus. Santiago de Compostela: 'Galaica'.

Descriptive flora of vascular plants, with partly synoptic keys to all taxa; vernacular names; fairly detailed indication of local range; notes on habitat, phenology, etc.; figures of representative species; addenda and corrigenda; remarks on floristics and phytogeography; indices to all botanical and vernacular names in vol. 3. The introductory section includes an illustrated glossary.

615

Catalonia

See also **612** (all works).

CADEVALL I DIARS, J., SALLENT I GOTÉS, A. and FONT QUER, P., 1913–37. *Flora de Catalunya.* 6 vols. Illus., 6 color pls. Barcelona: Institut de Ciencias.

Illustrated descriptive atlas-flora (in Catalan) of vascular plants, with keys to all taxa, limited synonymy, vernacular names, fairly detailed indication of local range, and notes on habitat, phenology, etc.; illustrated glossary (vol. 6); complete indices in each volume (general index in vol. 6). The introductory section (vol. 1) includes accounts of botanical exploration in the province and definitions of floristic zones. All species are illustrated in their work, which was patterned after Coste's *Flore de France.*

618

Gibraltar (and adjacent areas)

See also **612** (all works).

WOLLEY-DOD, A. H., 1914. *A flora of Gibraltar and the neighbourhood.* xxvi, 131 pp. (*J. Bot.* **50**, *Suppl.*) London.

Systematic enumeration of native and naturalized vascular plants of the Rock of Gibraltar and of neighboring parts of Andalucia, with some diagnoses, limited synonymy, indication of local range, and notes on habitat, phenology, etc.; addenda and index to genera. The introductory section includes accounts of local geography, geology, botanical exploration, and floristics.

619

Balearic Islands

See also **612** (all works). No suitable modern one-volume manual is available as a successor to the elderly work of Barceló y Combis (1879–81; in Spanish) or the multi-volume compendium of Knoche (1921–3; in French), the latter now considered out-of-date and moreover incomplete. The new illustrated flora in Balearic Catalan by Bonafè Barceló covers only Mallorca.

– Vascular flora: 1450–1550 species (1250–1450 native).

BARCELÓ Y COMBIS, F., 1879–81. *Flora de las Islas Baleares.* xlviii, 645 pp. Palma de Mallorca: Gelabart.

Briefly descriptive flora of the vascular (and some non-vascular) plants, with polychotomous keys to taxa above species level; limited synonymy; vernacular names; indication of range by island and locality (with citations); notes on habitat, phenology, uses, etc.; supplement (pp. 563–92); indices to generic, family, and vernacular names. The introductory section includes descriptions of the geography, geology, climate, and history of botanical work in the islands, as well as a list of authors. See also MARÈS, P. and VIGINEIX, G., 1880. *Catalogue raisonné des plantes vasculaires des îles Baléares.* xlvii, 370 pp., 9 pls. Paris. [Covers 1320 native species. All records are incorporated in Barceló's flora.]

DUVIGNEAUD, J., 1979. *Catalogue provisoire de la flore des Baléares.* 2nd edn. 43 pp. Liège: Département de Botanique, Université de Liège. (Mineographed.) (First edn. 1974.)

Systematic checklist of vascular plants, with indication by island based on available records and the author's collections and observations over several years; addenda and corrigenda (pp. 42–4); no index.

KNOCHE, H., 1921–3. *Flora balearica.* 4 vols. 47 pls., frontispiece, tables, maps. Montpellier: Roumégouis et Déhan. (Reprinted 1974, Königstein/Ts., Germany: Koeltz.)

Volumes 1–2 of this work contain an enumeration of non-vascular and vascular plants (covering all of the latter), with complete synonymy, references, fairly detailed indication of local range, general summary of extralimital distribution, and extensive notes on habitat, life-form, phenology, etc.; index. An introductory section (in vol. 1) includes a list of authors and tables of signs and abbreviations, as well as a gazetteer. Volume 3 is wholly devoted to floristics, phytogeography, and phytosociological analyses, while vol. 4 contains plates of illustrations and a map. The work is not of value for identification, being without keys.

Partial work – Mallorca

BONAFÈ BARCELÓ, F., 1977–80. *Flora de Mallorca.* 4 vols. Illus., maps. Palma de Mallorca: Moll.

Atlas-flora (in Mallorcan) of vascular plants, containing numerous photographic illustrations depicting all species with accompanying descriptive text; includes synonymy, vernacular names (Mallorcan, Catalan, Spanish) and indication of local occurrence, habitat, frequency, and external distribution; index in each volume (general index to work in volume 4). The introductory section contains a list of references (pp. xi–xviii) and an illustrated glossary, while in volume 4 are three appendices: an index to authors' names, a tabular list of vascular plants found throughout the Balearics (Ibiza and Formantera together here termed the Pityeuses), and a lexicon of vernacular names with their scientific equivalents. Arrangement of families follows the Englerian system.

Region

62

Italian peninsula and associated islands

Within this region are included the Italian mainland, Corsica, Sardinia, the Lipari Islands, Sicily, Malta, and Pantellaria (with the Isole Pelagie). Most, if not all, general works on Italy cover the whole of the region; however, as a fair literature exists for each of the main islands or island groups, these are accounted for separately. San Marino is here included with Italy.

Floristic exploration and documentation after four centuries are in general more advanced than in the Iberian Peninsula, but remain behind most of northern Europe. There is, however, a distinct gradient from north to south, islands excepted (among the latter, Corsica and Malta are perhaps the best known). General works of recent date include the illustrated manuals Fiori (1923–9, 1933) and Zangheri (1976), the former obsolescent, the latter more concise, and the abridged students' manual-key of Baroni (latest edition, 1969) and dendrofloras of Fenaroli (1967, 1976). The last-named author has also contributed *Flore delle Alpi* (latest edition, 1971; see **603**) which covers all montane and alpine areas in the region. [Another general flora, prepared by S. Pignatti and long-awaited, is reportedly in press at this writing, 1979.] The islands are covered through a variety of works, with the most up-to-date coverage in Corsica (Briquet *et al.*, 1910–61; Bouchard, 1977) and Malta (Borg, 1927; Haslam *et al.*, 1977). Cossu's Sardinian manual of 1968 lacks keys and descriptions and other treatments largely date from before 1914.

Bibliographies. General and divisional bibliographies as for Division 6. Very thorough coverage through 1959 is in particular provided by Blake 1961.

Indices. General and divisional indices as for Division 6.

620

Region in general

Works dealing with Italy as a whole (including San Marino) are treated here. For Fenaroli's *Flora delle Alpi*, see **603**/IV.

– Vascular flora: 6190 species in the whole of **620** and some adjacent areas (Zangheri); 5250–5350 species (4750–4900 native) on the Italian mainland (Webb).

ARCANGELI, G., 1894. *Compendio della flora italiana.* 2nd edn. xix, 836 pp. Turin: Loescher. (First edn. 1882.)

Concise manual of vascular plants, with non-dichotomous keys to families and genera along with synoptic keys to groups of species; limited synonymy; general indication of internal and extralimital range (the latter very abbreviated); notes on habitat, life-form, phenology, etc.; index to all botanical names. The work covers the Italian peninsula, Istria, Corsica, Sardinia, Sicily, and Pantelleria.

BARONI, E., 1969. *Guida botanica d'Italia.* 4th edn., revised and corrected S. Baroni Zanetti. xxxi, 545 pp., 16 color pls. (with 144 figs.). Bologna: Cappelli. (First edn. 1907, Rocca S. Casciano.)

Field manual-key to vascular plants, with very limited synonymy, concise summary of distribution in Italian region, and indication of habitat, life-form, phenology, etc.; colored figures of representative species; glossary and index to family and generic names. The brief introductory section is pedagogical. As with the second and third editions of 1932 and 1955, this compilation is largely based on Fiori's *Nuova flora analitica d'Italia*, with, however, the omission of many relatively uncommon and rare species. Covers the Italian peninsula, Corsica, Sardinia, Sicily, Istria, and the eastern French Riviera.

FIORI, A., 1923–9. *Nuova flora analitica d'Italia*. 2 vols. 22 text-figs. Florence: Ricci. (Reprinted 1969, 1974 in reduced format, Bologna: Edagricole.) Accompanied by FIORI, A. and PAOLETTI, G., 1933. *Iconographia florae italicae, ossia flora italiana illustrata*. 3rd edn. vii, 549 pp., 4419 text-figs. Florence. (Reprinted 1970, 1974 in reduced format, Bologna; 1st edn., 1896–1909, Padua.)

Generously scaled manual-flora of native, naturalized, and commonly cultivated vascular plants, with analytical keys to all taxa down to species level and synoptic keys to varieties, limited synonymy (with dates of publication), vernacular names, general indication of local and extralimital range, and brief notes on habitat, floristic zones, phenology, etc.; glossary of authors and complete index to botanical names in vol. 2. The species concept adopted is unusually broad, which has made statistics of the Italian flora difficult of comparison with those of neighboring areas. Although now joined by Zangheri's manual (see below), Fiori's work remains the standard 'modern' flora of Italy, encompassing the Italian mainland, Corsica, Sardinia, Sicily, Pantelleria, and nearby smaller islands; it also includes Istria and the Nice area. The accompanying atlas contains small, well-executed figures of every species and at least some varieties accounted for in the main work, with cross references, and includes a lexicon of vernacular names with botanical equivalents and separate index.

ZANGHERI, P. (assisted by BRILLI-CATTARINI, A. J. B.), 1976. *Flora italica (Pteridophyta–Spermatophyta)*. 2 vols. xxii, 1157 pp. (text); xxii pp., 7750 figs. on 210 pls. Padua: A. Milani (CEDAM).

In the text volume of this concise work are two parts: a general part covering life-forms, altitudinal zones, technical notes, symbols and abbreviations, and a glossary; and a special part comprising a manual-key to native, naturalized, and commonly cultivated vascular plants, with essential synonymy, vernacular names, karyotypes, indication of habit and life-form, phenology, altitudinal range, and internal and extra-limital distribution given in 1–2 lines for each; bibliography of key works, addenda and corrigenda, and complete indices at end. The accompanying atlas volume contains analytical figures for identification, with a separate skeleton index. Encompasses 8452 terminal taxa in 6190 species, of which 5692 species are keyed out and the remainder merely referred to in the text. The whole of the Italian region as defined here is included, along with the Nice area, Istria and Ticino.

Special group – trees

FENAROLI, L., 1967. *Gli alberi d'Italia*. 320 pp., 100 color pls., 102 maps. Milano: Martello.

Semipopular illustrated manual to the native, naturalized, and commonly cultivated trees of the Italian region (except Corsica), with vernacular names in several European languages; very limited synonymy; general indication of regional range (including maps for all species), notes on habitat, habit, phenology uses, etc.; indices to botanical and vernacular names. The introductory chapters deal with the function of forests (including pertinent remarks on the disastrous Florentine floods of 1966!), the uses of trees, and the importance of conservation measures; keys to major associations are also included. For a more elaborate treatment, with distribution maps, see FENAROLI, L. and GAMBI, G., 1976. *Alberi: dendroflora italica*. 717 pp., illus., maps. Trent: Museo Tridentino di Scienze Naturali.

621

Corsica

Apart from the two general floras described below, (neither of which provides complete coverage), this island, with a notably high endemic element, is covered by all standard floras and related works on Italy (620) as well as by French floras (651). The high-mountain zone is additionally covered by Fenaroli's *Flora delle Alpi* (603/IV).

– Vascular flora: 2250–2400 species (2150–2250 native).

BOUCHARD, J., 1977 (1978). *Flore pratique de la Corse*. 3rd edn. 405 pp., 50 pls. (incl. 2 maps) (Corse d'hier et de demain, 7). Bastia: Societe des Sciences

Historiques et Naturelles de la Corse. (First edn. 1968.)

Concise manual-key to vascular plants, with limited synonymy, French and local vernacular names, general indication of local range (with some citations), and brief, partly symbolic notes on habitat, phenology, life-form, frequency, uses, etc.; figures and photographs of representative species; list of endemic and phytogeographically interesting taxa, lexicon of Corsican local names, and index to generic and family names and their French equivalents. A brief introductory section includes a glossary, abbreviations of author's names, and a general key to families. Encompasses 1833 species.

BRIQUET, J. and LITARDIÈRE, R. DE, 1910–55. *Prodrome de la flore corse*. Vols. 1–3. Geneva: Georg [vols. 1, 2, part 1], Paris: Lechevalier [vols. 2, part 2, 3].

Systematic enumeration of vascular plants, with full synonymy (including references and citations), detailed indication of local range and *exsiccatae*; general summary of extralimital distribution, extensive taxonomic commentary and notes on habitat, life-form, phenology, ecology, etc.; index to generic and family names in each volume. The introductory section (vol. 1) includes bibliographic matter as well as notes on the plan of the work and records of field trips. Not completed; extends from the Pteridophyta up through the Solanaceae (Englerian system). For continuation, see BOUCHARD, J., 1961. *Materiaux pour un géographie botanique de la Corse*. 92 pp. N.p. [Concise enumeration with ecological notations; covers families from the Scrophulariaceae up to the beginning of the Ambrosiaceae (Compositae in part). Includes bibliography.]

622

Sardinia

See also **620** (all works). Additional records not accounted for in the works below appear in ATZEI, A. D. and PICCI, V., 1973. Note sulla nuova entità della flora Sarda non indicata in *Nuova flora analítica d'Italia* di A. Fiori per la Sardegna. *Arch. Bot. Biogeogr. Ital.* **49**(1–2): 1–70. [Scope of list; systematic enumeration of new records, with localities, citations of *exsiccatae* and critical notes. Includes references (pp. 63–70).]

– Vascular flora: 1900–2100 species (1900–2000 native).

BARBEY, W., 1884(1885). *Florae sardoae compendium. Catalogue raisonné des végétaux observés dans l'île de Sardaigne*. 263 pp., 7 pls., portrait. Lausanne: Bridel.

Systematic enumeration of the vascular and cellular plants with descriptions of new taxa, being mainly a supplement to Moris's *Flora sardoa* (see below); includes information on synonymy and geographical distribution (limited to data not found in the earlier work) as well as some notes on habitat, etc.; addenda (by P. Ascherson and E. Levier); index to generic and family names. The introductory section includes an account of the phytogeography of Sardinia in relation to neighboring lands as well as a list of endemic species, and pp. 123–69 describe G. Schweinfurth's journey of exploration within the island. In general, this work is something of a miscellany and reputedly not very reliable (Blake, 1961, p. 387).

COSSU, A., 1968. *Flora pratica sarda*. 365 pp., 22 pls. Sassari: Gallizzi.

Alphabetically arranged enumeration (by genera) of the more frequently encountered native, naturalized, and cultivated cellular and vascular plants, with vernacular names (in Sardinian dialect as well as in Italian), general indication of local range, and notes on habitat, life-form, cultivation, properties, uses, etc.; glossaries of vernacular names with botanical equivalents as well as of cultivars; short bibliography (pp. 263–4). The plates, of selected species, are reproduced from Moris's *Flora sardoa*.

MORIS, G. G., 1837–59. *Flora sardoa*. 3 vols., atlas. 114 pls., map. Turin: Piedmont Royal Printery.

Descriptive flora (in Latin) of the spontaneous and commonly cultivated seed plants of Sardinia, with synoptic keys to groups of species; full synonymy, with references and citations; indication of local range; vernacular names; critical taxonomic commentary; notes on habitat, life-form, phenology, uses, etc.; indices in each volume to generic and family names. The introductory section includes an account of the physical features of the island and its climate, as well as a synopsis of families. Not completed; covers the Dicotyledones and Gymnospermae. For continuation, see MARTELLI, U., 1896–1904. *Monocotyledones sardoae*. Fasc. 1–3. viii, 152 pp., 10 pls. Florence: Niccolai; Rocca S. Casciano: Cappelli. [Systematic treatment on similar lines to the parent work. Not completed; covers

Orchidaceae, Iridaceae, Amaryllidaceae, Dioscoreaceae, and part of the Liliaceae. Moris himself did not attempt the monocotyledons on account of his duties in later life as a political administrator (STAFLEU and COWAN, *Taxonomic literature*, 2nd edn., vol. 3, p. 586).]

624

Lipari Islands and Ustica

LOJACONO-POJERO, M., 1878. *Le isole eolie e la loro vegetazione, con enumerazione delle piante spontanee vascolari.* 140 pp. Palermo: Lorsnaider. (Also published as *Atti Soc. Acclim. Agric. Sicilia*, **17** (1878): 177–328.)

Includes an enumeration of vascular plants, with synonymy and citations (chiefly of regional literature), more or less detailed indication of local range, and notes on habitat and phenology. The lengthy introductory section gives accounts of the vegetation of the Lipari Islands and on the origin of individual stands, as well as notes on individual associations.

625

Sicily

See also **620** (all works). The alpine zone is also covered by Fenaroli's *Flora delle Alpi* (**603**/IV).
– Vascular flora: 2350–2600 species (2250–2450 native).

LOJACONO-POJERO, M., 1886–1909. *Flora sicula o descrizione delle piante vascolari spontanee o indigenate in Sicilia.* 3 vols. (in 5). 101 pls. Palermo: L. Pedone-Lauriel di Carlo Clausen.

Descriptive flora (partly in Latin) of vascular plants, with synoptic keys to genera and groups of species, fully synonymy, references and citations, detailed indication of local range, extensive taxonomic commentary, and notes on phenology, etc.; addenda and emendata in vol. 3 (pp. 412–47); index to all botanical names in each volume. The introductory section (vol. 1) includes material on the characters of Sicilian vegetation, the affinities of its flora, and on phytogeography. The plates depict representative species of the Sicilian flora. [The additions and corrections in vol. 3 have been reprinted from *Malpighia*, **20** (1906): 37–48, 95–119, 180–218, and 290–300.]

627

Pantelleria and Isole Pelagie

See also **620** (all works).

SOMMIER, S., 1922. *Flora dell'isola di Pantelleria.* 110 pp., portrait. Florence: Ricci (for Reale Istituto Botanico di Firenze).

Systematic enumeration of non-vascular (except algae) and vascular plants, with descriptions of new or less well-known taxa, synonymy, references and citations, taxonomic commentary, and notes on habitat, local occurrence, frequency, etc.; no index. Evidently based upon a manuscript left incomplete upon the death of the author.

629

Malta

Some works listed under **620** also cover these islands.

BORG, J., 1927. *Descriptive flora of the Maltese Islands, including the ferns and flowering plants.* 846 pp. Malta: Government Printing Office. (Reprinted 1976, Königstein/Ts., Germany: Koeltz.)

Briefly descriptive flora of vascular plants, with synoptic keys to families; some synonymy; general indication of local and extralimital range; vernacular names (Maltese and English); critical taxonomic remarks; notes on habitat, phenology, ecology, etc.; index to all botanical names. The introduction includes accounts of Maltese geography, geology, fossil records, climate, general floristics, phytogeography, and botanical exploration.

HASLAM, S. M., SELL, P. D., and WOLSELEY, P. A., 1977. *A flora of the Maltese islands.* lxxi, 560 pp., 29 figs., 66 pls. Msida, Malta: Malta University Press.

Concise manual-flora of vascular plants, with keys, very limited synonymy (where accepted name differs from that in Borg's flora), Maltese and English vernacular names, karyotypes, and notes on local and overall distribution, (with citations of earlier authorities), habitat, phenology, uses, and taxonomy; illustrations (on plates) of representative plant species; illustrated glossary, list of key works, and complete index at end. The introductory section gives notes on the genesis of the work, history of floristic studies in Malta, physical features, climate, soils, geology, and biotic influences including man, as well as descriptions of ecological areas and plant communities, a phytogeographical analysis, and technical notes; separate units on medicinal plants (by H. Micallef) and fruit tree growing (by J. Borg).

Region

63

Southeastern Europe

This region comprises the countries of the Balkan Peninsula (excluding Romania), Crete, and the western and central Aegean Islands. The eastern limit against Asia is as accepted for *Flora Europaea.*

The history and progress of botany in this region has been as balkanized as its politics. Separate traditions have tended to prevail in Greece, Bulgaria, and different parts of Yugoslavia, with the only unifying forces emanating from outside, mainly Central Europe.

In spite of the existence of a single sovereign state since 1921, floristic botany in Yugoslavia still largely revolves around the linguistico–cultural units represented by Slovenia, Croatia, and Serbia, all with different pathways in development of floras and other floristic literature. No 'all-union' flora has been completed although two such works were begun in the 1960s (*Analitička flora Jugoslavije*, a descriptive work with keys, and *Catalogus florae Jugoslaviae*, an enumeration). State floras, manuals, or checklists of varying formats exist for Slovenia, Croatia, Bosnia-Hercegovina (not yet completed), Montenegro, and Serbia, and there are some key partial works like *Flora velebitica* as well as (in the north) students' manuals, one of which (Domac, 1950) selectively covers the whole country. However, for some areas, floras of neighboring countries are used by preference or necessity.

The southern parts of the region, encompassing Greece, Albania and southern Yugoslavia, have also for long been treated as botanical *Kolonialgebieten* for students and other visitors from northern Europe. Mountainous, politically unstable, and to a large extent difficult of access and with but few (if any) resident botanists and institutions until fairly recent times, the

area centering on Macedonia and extending towards Bosnia, Albania, southern Greece and the Aegean, and Thrace has been botanically the most underexplored and documented part of Europe, and for long what field work was done there was carried out largely by outsiders from the Bulgarians northwards. Prominent among them have been Austrians and others from lands once part of Austria-Hungary.

Greece and the Aegean have, however, exercised the greatest attraction for foreign botanists ever since the onset of the 'Greek Revival', with the rich, diverse flora providing a major prop from its portrayal in *Flora graeca*. The intricacies of the plants, though, and the presence of separate traditions of floristic work (as with Spain) will make the preparation of a definitive critical flora of Greece a development for the future, particularly when more local resources are available. Notable earlier works, still widely used, include *Conspectus florae graecae* by Halácsy (1900–8, 1912) and *Flora aegaea* by Rechinger (1943; supplement 1949 and later papers). The development of Greek floristic botany is well surveyed in Greuter, W., 1975. Floristic studies in Greece. In Walters, S. M. *Floristic and taxonomic studies in Europe*, pp. 18–37, maps (BSBI Conference Report, 15). Abingdon, England.

No general floras are available for isolated Albania. The only items published of more than local interest are *Flora e Tiranës* (Paparisto *et al.*, 1962–5) and *Dendroflora e Shqipërisë* (Mitrushi, 1966), the one a district flora, the other, on woody plants, covering the whole country.

Several works cover Bulgaria, among them the multivolume *Flora na narodna republika Bălgarija* (1963–) and the manuals *Flora na Bălgarija* (Stojanov *et al.*, 1966–7) and *Ekskurzionna flora na Bălgarija* (Vălev *et al.*, 1960), the last-named abridged for student use.

Turkey-in-Europe is covered in recent terms by Davis' *Flora of Turkey* (**771**) as well as more particularly in a checklist by Webb (1966).

No work has fully replaced *Prodromus florae peninsulae balcanicae* (Hayek, 1924–33) as a regional flora. Written from outside the region, it has been fundamental to all later floristic work, and for some countries provides the only overall coverage.

In general, the standard of documentation is at a higher level in the north than in the south, where it has been influenced directly or indirectly by Hegi's *Illustrierte Flora von Mitteleuropa* and other Central

European works. In Greece, major props are Halácsy's and Rechinger's floras, in turn partly based on *Flora orientalis*, as well as Hayek's *Prodromus*.

Bibliographies. General and divisional bibliographies as for Division 6.

Indices. General and divisional indices as for Division 6.

630

Region in general

The otherwise comprehensive *Prodromus* by Hayek is limited on the northwest by the Sava River and the former Austrian boundary, thus excluding Vojvodina, Slavonia (north-eastern Croatia), Slovenia, and Istria.

HAYEK, A. VON, 1924–33. *Prodromus florae peninsulae balcanicae.* 3 vols. Map (Repert. Spec. Nov. Regni Veg., Beih., 30). Berlin. (Reprinted 1968, Königstein/Ts., W. Germany: Koeltz.)

Briefly descriptive flora of vascular plants, with analytical keys to families and genera and synoptic devices to species or groups thereof, extensive synonymy (with abbreviated references), concise symbolic indications and notes on local range, life-form, habitat, etc.; complete indices in each volume. The introductory section includes lists of symbols and abbreviations together with a methodological exposition concerning the work. The *Prodromus* is fundamental to floristic work in southeastern Europe, and was one of the five 'Basic Floras' constituting a primary source for the preparation of *Flora Europaea*.

631

Yugoslavia (in general)

Only general works dealing with the whole of Yugoslavia are listed here. Those covering individual states or parts thereof will be found under **632** to **634** (respectively northern, central, and southern Yugoslavia).

– Vascular flora: 5000–5150 species (4750– 4900 native); floristic knowledge is in general better in the north than in the south.

HORVATIĆ, S., 1954. *Ilustrirani bilinar. Priručnik za određivanje porodica i rodova višega bilja.* 767 pp., 174 text-figs. Zagreb: Institut za Botaniku Sveučilišta.

Illustrated handbook, with analytical keys, to genera and families, with descriptions, distinguishing features, critical remarks, and notes on distribution, special attributes, etc.

HORVATIĆ, S. (ed.), 1967– . *Analitička flora Jugoslavije.* [Flora analytica Jugoslaviae.] Vols. 1– . Maps. Zagreb: Institut za Botaniku Sveučilišta.

Descriptive flora of vascular plants, with some synonymy and keys to all taxa; detailed indication of local range and summary of extralimital distribution; symbolic notes on life-form, karyotypes, phenology, etc., and indication of habitat; extensive infraspecific treatment; no index. The introductory section contains chapters on physical features, geography, vegetation formations and plant communities (with references), and a summary of administrative subdivisions. At this writing (late 1979) only the pteridophytes, gymnosperms, and Ranunculaceae through Fumariaceae (together with family keys to dicotyledons) have been officially published as vol. 1, parts 1–2 (1967–73). However, three other parts (vols. 1, parts 3–4 and 2, part 1) and some 'supplements' were unofficially printed and distributed by others after Horvatić became seriously ill in 1974 and ultimately died (1976), but have been on ethical and other grounds disowned by the Yugoslav botanical community at a 1977 congress.[9] Formal continuation of this project in the near future is considered doubtful.[10]

MAYER, E. *et al.*, 1964– . *Catalogus florae Jugoslaviae.* Vols. 1– . Map. Ljubljana: Academia Scientiarum RP Socialistica Fœdrativae Jugoslaviae/ Academia Scientiarum et Artium Slovenica.

Comprehensive enumeration (in Serbo-Croat) of the non-vascular (to date only Musci) and vascular plants of Yugoslavia, with full synonymy, references, concise indication of local range and karyotypes, lists of references, and indices to generic names. No keys are provided. At this writing (1979) the project has for some time been dormant; all that has appeared (wholly before 1973) are vol. 1, parts 1–2 (Pteridophyta and Gymnospermae) and vol. 2, part 1 (Musci).

Partial works

DOMAC, R., 1950. *Flora: za određivanje i upoznavanje bilja.* 552 pp., 23 text-figs. Zagreb: Institut za Botaniku Sveučilišta.

Concise students' manual-key for identification of

vascular plants in Yugoslavia as a whole, with limited synonymy and brief indication of internal range; illustrated glossary and index to generic names. The introduction contains a list of more significant floristic references. [Apparently does not provide complete coverage of the flora.]

632

Northern Yugoslavia

Comprises the states of Slovenia and Hrvatska (Croatia). In addition to the works described below, the Istrian peninsula and adjacent mainland to the north (now divided between these two states and Italy but formerly the Austrian district of Küstenland) are covered by all standard Italian floras (620) and by Fritsch's *Exkursionsflora fur Österreich* (648), as well as by the following earlier local work: POSPICHAL, E., 1897–9. *Flora des österreichischen Küstenlandes*. 2 vols. 25 pls., map. Vienna: Deuticke.

I. Slovenia

See also 641 (all works); 648 (Fritsch). This state is not covered by Hayek's *Prodromus*.

MARTINČIČ, A. and SUŠNIK, F. (eds.), 1969. *Mala flora Slovenije*. 517 pp., 396 text-figs., maps (end papers). Ljubljana: Cankarjeva založba.

Concise manual of vascular plants, with keys to all species, very limited synonymy, vernacular names, and notes on habitat, associates, local range, phenology, etc.; indices to botanical and vernacular names of families and genera. The introductory section includes a chapter on descriptive morphology, a table of phytosociological units, and a key to families. Supersedes the following: PISKERNIK, A., 1951. *Ključ za dolocanje cvetnic in praprotnic*. 2nd edn. [Ljubljana.] (First edn. 1941.)

MAYER, E., 1952. *Seznam praprotnic in cvetnic Slovenskega ozemlja*. [Verzeichnis der Farn- und Blütenpflanzen des slowenischen Gebietes.] 427 pp. (Razpr. Slovensk. Akad., IV, 5; Inšt. Biol., 3.) Ljubljana.

Systematic enumeration of the vascular plants of Slovenia, with extensive synonymy (no references), concise general indication of local range, and index to generic and family names; full bibliography.

II. Hrvatska (Croatia)

Includes the regions of Slavonia, Croatia proper, and Dalmatia. Only the last two are covered by Hayek's *Prodromus*.

DOMAC, R., 1967. *Ekskurzijska flora Hrvatske i susjednik područja*. 543 pp. Zagreb: Institut za Botaniku Sveučilišta.

Concise manual-key to vascular plants (in Serbo-Croat), with limited synonymy, concise indication of local range, and notes on habitat, etc.; occasional technical commentary; index to generic and family names. A list of source works is also included. Subsequently reissued, with corrections, as *idem*, 1973. *Mala flora Hrvatske i susjednik područja*. 543 pp. Zagreb: 'Skolska Knjiga'.

Partial work

The one important modern work is Degen's *Flora velebitica*, covering the coastal Velebit Mountains in Dalmatia.

DEGEN, Á. VON, 1936–8. *Flora velebitica*. 4 vols. Illus. Budapest: Akadémiai Kiadó/Ungarischen Akademie der Wissenschaften.

Very detailed critical enumeration (in German) of the vascular and higher cellular plants of the limestone Velebit Range, without keys; includes full synonymy, citations of *exsiccatae*, taxonomic commentary, and copious supporting notes; full bibliography and complete index in vol. 3. The introductory section deals extensively with physical features, geography, vegetation, forestry, agriculture, botanical exploration, and other topics. Volume 4 is a condensed version of the work in Hungarian.

633

Central Yugoslavia

Comprises the states of Bosnia/Hercegovina and Crna gora (Montenegro). The small territory of Novipazar, an Ottoman administrative district occupied by Austria from 1878 to 1908/09 and attached to Bosnia/Hercegovina, is now mostly part of Serbia (634/II). *Flora Bosne*, begun by Beck in 1903, has yet to be completed after nearly 80 years although by 1980 it was very far advanced.

I. Bosnia and Hercegovina

BECK VON MANNAGETTA UND LERCHENAU, G., 1903–23. *Flora Bosne, Hercegovine i Novopazarskog sandžaka.* Vols. 1–2, 1[bis]. 484, 26 pp., illus. Sarajevo. (Reprinted in fascicles from *Glasn. Zemaljsk. Muz. Bosni Hercegovini* 16–35, *passim.*) Revised as *idem*, 1904–16. *Flora von Bosnien, der Herzegowina und des Sandžaks Novipazar.* Vols. 1, 2 (fasc. 1–3). Pp. 1–261. Vienna: Gerold. (Reprinted in fascicles from *Wiss. Mitt. Bosnien/Herzegowina* 9–13, *passim.*)

BECK-MANNAGETTA, G., 1927. *Flora Bosnae, Hercegovinae et regionis Novipazar.* Vol. 3. viii, 487 pp. (Posebna Izd. Srpska Kral. Akad., vol. 63). Belgrade.

BECK, G. and MALÝ, K., 1950–74. *Flora Bosnae et Hercegovinae.* Vol. 4, parts 1–3, revised and ed. by P. Fukarek (part 1) and Z. Bjelčić *et al.* (parts 2–3) (Posebna Izd. Biol. Inst. & Zemaljsk. Muz. Bosni Hercegovini, vols. 1–3.) Sarajevo.

Detailed systematic enumeration (in Serbo-Croat) of vascular plants of Bosnia, Hercegovina, and the former Ottoman district of Novipazar, with descriptions of new taxa; includes full synonymy, references and citations, indication of *exsiccatae* with localities, taxonomic commentary, and notes on habitat, phenology, etc.; indices to generic names (vols. 3 and 4 only). As of 1980 covers all families in Pteridophyta, Gymnospermae and Angiospermae up through the Dipsacaceae on the Englerian system; only Campanulaceae and Compositae remain to be published. The early parts, however, are now in need of revision. [A guide to the separate parts of vols. 1–2 and 1[bis] appears in vol. 3, pp. vii–viii, and a full collation of the whole work (save vol. 4, part 3) in STAFLEU and COWAN, *Taxonomic literature*, 2nd edn, vol. 1, pp. 158–9. The reprint of volumes 1 and 2 from *Glasnik Zemaljskog Muzeja u Bosni i Hercegovini*, which continued through to 1923 despite Princip's fatal shots on Appel Quay, is continuously paginated save for the section on Pteridophyta, which I have here designated as vol. 1[bis]. The German-language edition of this work, which contains many additions, was terminated after the Ranunculaceae, corresponding to p. 225 of the Serbo-Croat version.]

II. Crna gora (Montenegro)

ROHLENA, J., 1942. *Conspectus florae montenegrinae.* 506 pp. (Preslia, 20/21.) Prague.

Systematic enumeration (in Latin) of vascular plants (2623 spp.), with limited synonymy, citation of *exsiccatae* with localities, and brief indication of habitat; index to generic and family names at end. [In format the work resembles that of Mayer's 1952 enumeration of the Slovenian flora (632/I).]

634

Southeastern Yugoslavia

Included here are the states of Serbia (with the regions of Vojvodina and Kosovo) and Macedonia. The former district of Novipazar, at one time administered as part of Bosnia/Hercegovina, is now largely within Serbia.

Bibliography (Macedonia)

MICEVSKI, K., 1956. Bibliographie der Flora und der Vegetation Mazedoniens. *Posebni Izd. Filos. Fak. Univ. Skoplje*, 8: 99–118.

I. Vojvodina

See 643 (JAVORKA; JAVORKA and CSAPODY). This district, north of Belgrade, formed part of the Hungarian Kingdom, 'historical Hungary', before 1918. It is *not* included within the area covered by *Flora SR Srbije*.

II. Serbia (with Kosovo i Metonija)

JOSIFOVIĆ, M. (coordinator), 1970–7. *Flora SR Srbije.* [Flore de la Republique Socialiste de Serbie.] 9 vols. Illus. Belgrade: Srpska Akademija Nauka i Umetnosti.

Large-scale illustrated descriptive documentary flora of native, naturalized, and commonly cultivated vascular plants (in Serbo-Croat, with cyrillic script) of Serbia proper, Novipazar, and Kosovo, with keys to all taxa; includes full synonymy, references, vernacular names, citation of *exsiccatae* with localities and fairly detailed indication of local range, and notes on habitat, phenology, ecology, uses, etc.; indices to all botanical and vernacular names in each volume (general index to families and genera in vol. 9). The introductory section (vol. 1) includes an extensive historical account (with references) and an illustrated organography. Vol. 9 also contains a supplement (*dodenik*) to vols. 1–8. The work has the aspect and scale of Beck's flora of Bosnia (see **633**), but appears to have been to a greater extent compiled; by some authorities[11] regarded as uncritical.

PANČIĆ, J., 1874. *Flora kneževine Srbije*. [Flora principatus Serbiae.] xxxii, 798, [2] pp. Belgrade: Royal Serbian Printer; continued as *idem*, 1884. *Additamenta*.... 254, [2] pp. Belgrade. (Whole work reprinted in 1 vol. (as Posebna Izd. Srpska Akad., 492), 1976, Belgrade.)

Concise manual-key (in Serbo-Croat, with cyrillic script) of the vascular plants of Serbia proper (excluding Novipazar and Kosovo), with generalized indication of local range, notes on life-form, phenology, etc., and an index to family and generic names. The introductory section gives a general account of features of the flora and a history of botanical work in the area.

III. Macedonia

No properly consolidated general floras or enumerations are yet available for this most southerly of the Yugoslav states. Apart from the partial work given below, reference should be made to floras of adjacent states, especially those of Bulgaria (**639**). The area remains botanically inadequately known.

Partial work

BORNMÜLLER, J., 1925–8. Beiträge zur Flora Mazedoniens, I–III. 18 pls. *Bot. Jahrb. Syst.* **59**: 293–504; **60**, Beibl. 136: 1–125; **61**, Beibl. 140: 1–196.

Systematic enumeration of vascular plants, with synonymy, citation of *exsiccatae* and other records with localities, and extensive critical commentary; index to families in the third part. The introductory section includes a gazetteer and details of the author's itineraries.

635

Albania

The area of Albania as delimited in Hayek's *Prodromus* does not correspond to the limits of the modern state. The country remains botanically among the least known in Europe.

– Vascular flora: 3150–3300 species (3100–3300 native).

Partial work

PAPARISTO, K. *et al.*, 1961 (1962). *Flora e Tiranës*. 521 pp., 216 text-figs., 2 maps (end papers). Tirana: Univ. Shtetëvor i Tiranës. Accompanied by *idem.*, 1965. *Flora e Tiranës* (*Ikonografia*). 515 pp., 1300 text-figs. Tirana.

The text volume comprises a manual-key (for students) to vascular plants of the Tirana district, with limited synonymy, vernacular names, and notes on habitat, occurrences, phenology, etc.; illustrated glossary and lists of plants classified by special features, uses, properties, etc., in appendices; indices to botanical and to vernacular names at end. The atlas volume contains half-page figures, with cross references to the text.

Special group:– woody plants

MITRUSHI, I., 1966. *Dendroflora e Shqipërisë*. 520 pp., 617 text-figs. Tirana: Univ. Shtetëvor i Tiranës.

Illustrated manual-key to woody and sub-woody plants, with brief diagnoses, very limited synonymy, vernacular names, and notes on local range, habitat, phenology, status, origin, etc.; illustrated organography/glossary and indices to vernacular and botanical names. The introductory section includes general keys to families and genera. Partially supersedes the author's *Drurët dhe shkurret e Shqipërisë*. 604, vii pp., 289 text-figs., 9 maps (1 colored). Tirana, 1955. [Includes more descriptive details and commentary.]

636

Greece (mainland and western islands)

For Crete and the Aegean Islands, see **637**.

– Vascular flora: 4100–4250 species (3950–4100 native), excluding Crete.

Of the two general floras described here, the *Conspectus* by Halácsy is much the more important, although it does not include certain areas now within the modern Greek state which (except for the Dodecanese, Italian from 1912 through World War II) assumed its present form only in 1923. Numerous scattered contributions to the flora have been made since publication of the *Conspectus*, but the most substantial of these have concerned Euboea, Crete and the Aegean (**637**). For the present area, reference should also be made to Hayek's *Prodromus* (630). Much recent work is still unpublished.

A thorough historical survey and assessment of floristic progress in this botanically still inadequately known country, with some pertinent remarks concerning future work and inclusion of maps and a list of references, is given by Greuter.[12]

DIAPOULIS, K. A., 1939–49. *Ĕllēnikē khlōris.* Vols. A–B (in 3 vols.). Illus. Athens.

Manual-key to the vascular plants of Greece (chiefly the central and southern parts), with vernacular names, brief general indication of local range, numerous small text-figures, and indices to generic, family, and vernacular names. The introductory section (vol. 1) includes an illustrated organography/glossary, a Latin–Greek lexicon of terms, and lists of authors and references. To a large extent compiled from other sources, the work lacks a critical basis.

HALÁCSY, E. VON, 1900–8. *Conspectus florae graecae.* 3 vols. and supplement, Leipzig: Engelmann. (Reprinted 1968, Lehre, Germany: Cramer.) Continued as *idem*, 1912. Supplementum secundum Conspectus florae graecae. *Magyar Bot. Lapok*, **11**: 114–202.

Briefly descriptive flora (in Latin) of vascular plants, with synoptic keys to genera and species (or groups thereof), full synonymy, references and citations, indication of *exsiccatae* with localities, critical remarks, and notes on habitat, phenology, life-form, etc.; indices to all botanical names in each volume. Parts of northern and eastern Greece are not included. Despite its age and incompleteness, this work is still regarded as without peer.[13]

637

Crete and the Aegean Islands (including Euboea)

See also 636 (all works); 770–790 (Boissier). In addition to the general flora of Rechinger, a number of partial works are available, some published since 1943; a selection is given here.

RECHINGER, K. H., 1943. *Flora aegaea.* xx, 924 pp., 25 pls., 3 maps (Akad. Wiss. Wien, Math.-Naturwiss. Kl., Denkschr., 105/I). Vienna. Continued as *idem*, 1949. Florae aegaeae supplementum. *Phyton* (Horn), **1**: 194–228. (Both works reprinted in 1 vol., 1973, Königstein/Ts., W. Germany: Koeltz.)

Critical systematic enumeration of the vascular plants of the foreshores of the Aegean Sea and the Aegean Islands together with Crete, with analytical keys to species within genera and descriptions of new taxa; full synonymy, with references and citations; detailed indication of local range, with citation of *exsiccatae* and other records; notes on habitat, special features, etc.; annotated list of collectors; bibliography and index to all botanical names. The introductory section includes accounts of physical features, climate, etc., together with a gazetteer. The text is in German, the keys in Latin. Further additions are contained in GREUTER, W., 1973. Additions to the flora of Crete, 1938–72. *Ann. Mus. Goulandris*, **1**: 15–83.

Partial works: islands except Crete

CIFERRI, R., 1944. Flora e vegetazione delle Isole italianae dell'Egeo. 200 pp. (Atti Ist. Bot. Univ. Pavia, Lab. Crittogam., V, Suppl. A.) Pavia. (Mimeographed.)

Pages 21–136 comprise a systematic enumeration of vascular plants, with references and key citations of literature along with documented localities on the various islands (including Rhodes). The remainder of the work covers vegetation, chorology, and floristic plant geography as well as phytosociological classification, with references (pp. 183–92) and indices at the end, while in the introductory section is an account of the area and of previous botanical work therein. Covers 1186 species.

GREUTER, W. and RECHINGER, K. H., 1967. *Flora der Insel Kythera.* 206 pp., 4 full-page illus. (Boissiera, 13). Geneva.

Concise, partly descriptive account of vascular plants, with synonymy, indication of local range (with *exsiccatae*), taxonomic commentary, and notes on habitat, special features, etc.; phytogeographical considerations (in appendix); index to new names and combinations; generic index. The introductory section covers botanical exploration and previous literature. The work, ostensibly an account treating Kythera and Antikythera (SW Aegean Sea between the Peloponesus and Crete, but nearer the former), is also designed to serve as a first contribution to nomenclatural 'modernization' of the Greek vascular flora.

RECHINGER, K. H., 1961. Die Flora von Euboea. *Bot. Jahrb. Syst.* **80**: 294–465, pls. 4–10, 3 maps.

Consolidated report of collections made in the Greek Aegean island of Euboea subsequent to publication of *Flora aegaea* (see above); includes synonymy, localities with citations of *exsiccatae*, taxonomic commentary, and habitat and other data.

RUNEMARK, H. *et al.*, 1960– . Studies in the Aegean flora, I– . *Bot. Notis.* 113– and *Opera Bot.*, A, 13– , *passim*.

A series of contributions to the flora of the Aegean

Islands, particularly the smaller ones, based on field work by the senior author and his students from 1957 onwards. At this writing (1979) over 21 parts had been published, constituting a valuable supplement to Rechinger's *Flora aegaea*.

Partial works: Crete

– Vascular flora: 1700–1850 species (1600–1800 native).

GANDOGER, M., 1916. *Flora cretica*. 1818 pp. Paris: Hermann.

Uncritical systematic list (in French) of vascular plants and Characeae, with descriptions of some new taxa; citations of *exsiccatae* with localities; no index. An introductory section gives accounts of prior botanical work and the author's own trips (evidently made as a cover for clandestine activities) as well as a list of principal references; an appendix (pp. 120–81) deals with floristics. A gazetteer is also provided.

RECHINGER, K. H., 1943. *Neue Beiträge zur Flora von Kreta*. 184 pp., 14 text-figs., map (Akad. Wiss. Wien, Math.-Naturwiss. Kl., Denkschr., **105**/II, part 1.) Vienna.

Botanical report of an extensive collecting trip made by the author in 1942 during the German occupation; supplements Cretan records in the author's *Flora aegaea* as well as Gandoger's *Flora cretica*.

638

Turkey-in-Europe

For general works on Turkey, see **771**.

– Vascular flora: 2100–2250 species (2000–2100 native).

WEBB, D. A., 1966. The flora of European Turkey. *Proc. Roy. Irish Acad.* **65**–B(1): 1–100, 2 maps.

Systematic enumeration of vascular plants, with limited synonymy and concise indication of local range (with some localities); no index. The introductory section includes remarks on major floristic elements, while a gazetteer, references, and *desiderata* for further exploration conclude the work.

639

Bulgaria

Floristic knowledge in this country has greatly improved since early in the present century, and Bulgarian botanists have also contributed to the floristics of Macedonia and Thrace, parts of which have been claimed or from time to time administered by Bulgaria beyond present state boundaries.

– Vascular flora: 3550–3750 species (3500–3650 native).

Bibliographies

STOJANOV, N. and KITANOV, B., 1950. Literatur über die Flora und Pflanzengeographie Bulgariens im Jahrzehnte 1939–1948. *Izv. Bot. Inst.* (Sofia), **1**: 480–506.

KITANOV, B., 1960. *Literatur über die Flora und Pflanzengeographie Bulgariens 1949–1958*. Sofia.

JORDANOV, D. *et al.* (eds.), 1963– . *Flora na narodna republika Bălgarija*. [Flora reipublicae popularis bulgaricae.] Vol. 1– . Illus., map. Sofia: Bălgarska Naukite Akademija.

Comprehensive descriptive flora of vascular plants (along the lines of *Flora SSSR*), with keys to all taxa and fairly copious descriptions; includes full synonymy with references, Bulgarian equivalents and vernacular names, detailed exposition of local range and general summary of extralimital distribution, some taxonomic commentary, and notes on habitat, phenology, ecology, etc.; indices to all plant names in each volume. The introductory section (in vol. 1) contains chapters on botanical exploration and research in Bulgaria (this also in English), an illustrated organography, a glossary, and a list of authors (the latter three items also appearing in succeeding volumes). Of the projected total of 10 to 11 volumes, at this writing (late 1980) seven have been published, covering all families in the traditional Englerian sequence from the pteridophytes through Rosaceae (vols. 1–5), Leguminosae (vol. 6) and Oxalidaceae through Araliaceae (vol. 7).

STOJANOV, N., STEFANOV, B. and KITANOV, B., 1966–7. *Flora na Bălgarija*. [Flora bulgarica.] 4th edn. 2 vols. 1326 pp., 1549 text-figs. Sofia: Nauka i Izkustvo.

Illustrated manual-key to the vascular plants of Bulgaria, with limited synonymy, vernacular and transliterated names, fairly detailed indication of local range, and notes on habitat, phenology, ecology, etc.; indices to all botanical and other names (in vol. 2). The introductory section is pedagogic, while a list of author's names and an account of organography appear in the appendices.

Partial work

VĂLEV, S., GANČEV, I. and VELČEV, V., 1960. *Ekskurzionna flora na Bălgarija*. 735 pp., illus. Sofia: 'Narodna Prosveta'.

Illustrated manual-key to native, naturalized, and commonly cultivated vascular plants, with diagnoses, very limited synonymy, vernacular and transliterated names, indication of local range (or country of origin), and concise notes on ecology, habitat, life-form, special features, vertical zonation, etc.; lists of flowering times, ranges, vertical zonation, etc., of individual species; indices to botanical and vernacular names. The introductory section has a chapter on descriptive organography as well as remarks on the use of the work. Some 2250 species are covered in this work, which is intended mainly for student use; it thus tends to omit rare or local species. [Not seen by this author; citation and description based on data supplied by A. O. Chater, Leicester (now London).]

Region

64

Central Europe

This region encompasses Romania, Hungary, Czechoslovakia, Poland, East and West Germany, Austria, Liechtenstein, and Switzerland. It thus includes most of the territories of the pre-1918 German Reich and may be understood to encompass the Austro–Hungarian Empire, parts of which together comprise the traditional germanophone *Mitteleuropa* (here designated as **641**). It is roughly equivalent to historical usage, which designates a region 'lying between Emden and Geneva on the west and the Masurian Lakes and Transylvania on the east'.[14] With few exceptions, it is one of the floristically best-known and documented parts of Europe (and indeed the world), with four to five centuries of continuous study, the longest north of the Alps.[15] It is densely covered by manuals and other works of all kinds, only the most important of which can be listed here.

Dominating the regional floristic scene for most of the present century has been Hegi's *Illustrierte Flora von Mitteleuropa*, that ponderous but critical monument to Central European botanical scholarship which was in turn based on a long tradition of such thorough, sometimes pettifogging, but usually professional encyclopedic floras. It has, however, attained influence far and wide, and sets are widely dispersed through the botanical world. Yet its size has raised questions about the role and contents of floras.[16] The 'second edition' has yet to be 'completed', although the 'third edition'

was 'begun' some years ago as a revision of early parts of the 'second edition'. The work is accompanied by a critical checklist by Ehrendorfer *et al.* (1967; revised 1973), as well as a regional bibliography.

All countries in the region have more or less modern descriptive floras except for Austria, for which the major modern work is a very detailed critical checklist. Associated with these are field manual-keys, available for all countries and mostly recent save Fritsch's *Exkursionsflora für Österreich*, last revised in 1922. The most important of the modern contributions are *Flora der Schweiz und angrenzender Gebiete* (1967–73), *Flora Republicii Socialiste România* (1952–76) and Soó's *A Magyar flóra és vegetácio rendszertani-növeńyföldrajzi kezikönyve* (1964–80), the last a critical chorological–ecological synopsis of the Hungarian flora. *Flora polska* (1919–80) and *Atlas flory polskiej i ziem ósciennych* (1930–) should also be mentioned. The most useful book of illustrations is, however, Jávorka and Csapody's *Iconographia florae hungaricae* (1929–34; reissued 1975). Of the concise manuals, those by Rothmaler published as *Exkursionsflora für die Gebiete der DDR und der BRD*, II: *Gefässpflanzen*, and IV: *Kritischer Band* enjoy wide popularity.

Nevertheless, the majority of works published at present in the region belong to the 'routine' category of van Steenis, and the future trend is likely to be the same.

Bibliographies. General and divisional bibliographies as for Division 6 (except Blake, 1961, which covers only Switzerland).

Regional bibliographies

HAMANN, U. and WAGENITZ, G., 1977. *Bibliographie zur Flora von Mitteleuropa*. 2nd edn. 374 pp. Berlin: Parey. (First edn. 1970, Munich.) [Provides detailed coverage of floristic and other botanical works, mainly for 'Mitteleuropa' (**641**) but also includes a listing of 'standard' floras from other parts of the Holarctic zone. Designed as a companion to Hegi's flora. In the new edition, additions appear as a supplement, pp. 329–74.]

Indices. General and divisional indices as for Division 6.

640

Region in general

No general works cover the entire region as defined here, except to some extent *Vergleichende Chorologie der zentraleuropäischen Flora* by H. MEUSEL *et al.* (see 001). For the western subregion, or 'Mitteleuropa', see 641. For the Alps, see 603/IV; for the Carpathians, see 603/VII.

641

'Mitteleuropa'

Included here are German-language works relating to 'Mitteleuropa', or West Central Europe. This long-established botanical 'sphere' developed as a comprehensive working unit long before World War I; it encompassed the pre-1918 German Reich, the former Kingdom of Austria in a narrower sense (i.e., 'historical' Austria, including the present-day state, the Czech lands of Bohemia and Moravia, Slovenia, Istria and South Tyrol), Liechtenstein, and Switzerland. In effect, it is (or was) the region with the greatest concentration of German-speaking peoples. However, although circumscribed in this manner, the works listed below – especially Hegi's *Illustrierte Flora von Mittel-europa*, which is about the most comprehensive floristic work ever produced – are of value over a far wider area.

EHRENDORFER, F. (ed.), 1973. *Liste der Gefäss-pflanzen Mitteleuropas.* 2nd edn. xii, 318 pp., 2 figs., map. Stuttgart: Fischer. (First edn. 1967, Graz.)

List, with genera arranged alphabetically, of the native and naturalized vascular plants of 'Mitteleuropa' (defined as indicated in the heading above, but without much of former eastern Germany), giving accepted names of species and infraspecific taxa, standardized acronyms of names with numbers, symbolic indication of citations and of 'critical taxa', and geographical range within the region. The introductory section gives a list of abbreviations and of 'standard' floras utilized. The overall arrangement of this work resembles to some extent that of F. von Mueller's *Second systematic census of Australian plants* (see 420–50).

HEGI, G. (ed.), 1906–31. *Illustrierte Flora von Mittel-Europa* (later *Mitteleuropa*). 1st edn. 7 Bde. (in 13 vols.). 280 pls. (mostly in color), 3434 + 1273 figs. (incl. halftones, maps). Munich: Lehmann; Vienna: Pichler. (Bd. 6(2) reprinted 1954, Munich: Hanser; Bde. 4(3), 5(1), 5(2), 5(3), and 5(4) reprinted with addenda and emendata, 1964–6, Munich; pp. 580–1386 of Bd. 6(2) to be reprinted with new supplement as Bd. 6(4) of the 2nd edn, Berlin: Parey.) Addenda and emendata of 1964–6 collectively reissued in one volume as MERXMÜLLER, H. (comp.), 1968. *Nachträge, Berichtigungen und Ergänzungen zu den unveränderten Nachdrucken der Bände 4(3) und 5(1) bis 5(4) mit Verzeichnissen der lateinischen und deutschen Pflanzen-namen.* 168 pp. Munich: Hanser.

HEGI, G. (ed.) 1935– . *Illustrierte Flora von Mitteleuropa.* 2nd edn., revised and ed. by K. Suessen-guth *et al.* Bde. 1–6, partim (planned to occupy 7 Bde. in 18 vols.). Illus. (incl. color pls., halftones, maps). Munich: Lehmann (Bde. 1–2, 1935–9); Hanser (Bde. 3(1), 3(2) in part, 3(3), 4(1), 4(2A), 6(1), 6(2) in part, 6(3) in part, 1957–74); Berlin: Parey (remainder of Bde. 3(2), 6(2) and 6(3), including Lfg. A of Bd. 6(2), 1978–). (In progress. Bde. 4(2B) and 7 yet to commence publication; 6(2) not yet completed. Bd. 1 reprinted 1965, Munich. For Bd. 6(4), see above.)

HEGI, G. (ed.), 1966– . *Illustrierte Flora von Mitteleuropa.* 3rd edn. revised and ed. by H.-J. Conert, U. Hamann, W. Schultze-Motel and G. Wagenitz. Bd. 2, *partim* (planned to occupy 2 Bde. in 5 vols.). Illus. (incl. color pls., halftones, maps). Munich: Hanser (Bd. 2(1) in part, 1966–9); Berlin: Parey (remainder of Bd. 2(1), 1977–80, and subsequent issues). (In progress; Bde. 1(1A), 1(1B), 1(2) and 2(2) yet to commence publication as of the end of 1980.)

This work, perhaps the most compendious and copiously illustrated of all modern floras, covers in great detail the vascular plants of 'Mitteleuropa' (as defined in the unit heading above); includes keys to and lengthy descriptions of all taxa with principal synonymy and references, extensive critical discussion, detailed exposition of local and regional ranges (with many distribution maps), generalized account of extralimital range, and extensive notes on habitat, phenology, biology, ecology, karyotypes (in later editions), morphology, anatomy, palynology, chemistry, etymo-logy of names, properties, and uses as related to families, genera, and individual species; vernacular names in German, English, French and Italian; complete indices to botanical names in each volume (Band); general index (for first edition) in Bd. 7 incorporating all botanical and vernacular names in the entire work (as well as a key to families; guides to special categories of plants, a synopsis down to genera, a list of botanical authors, and an extensive glossary). The introductory section (in Bd. 1) includes a list of abbreviations and an additional glossary. The sequence of families follows the Englerian system. This is *the* standard detailed flora for Central Europe and was one of the five 'Basic Floras' employed as a primary source for the preparation of *Flora Europaea*. The second edition is yet to be completed; at this writing (1980) Bde. 4(2B), 6(4) and the index volume, Bd. 7 are unpublished and Bd. 6(2) is still incomplete, with two fascicles yet to be produced. Furthermore, a number of the other volumes (Bd. 4(3), the four sections of Bd. 5, and, eventually, Bd. 6(4)) are simply reprints (with supplements) of the corresponding sections of the first edition. The new so-called third edition (*Dritte Auflage*) is now planned as a five-volume revision of Bde. 1–2 of the second edition, published before World War II; to date (1980), all six fascicles of Bd. 2(1) and one of Bd. 1(3) have appeared. From July 1975, publication and distribution of this work has been in the hands of Verlag Paul Parey, Berlin. [Accompanying this work is an extensive source bibliography by U. Hamann and G. Wagenitz (1970, 2nd edn., 1977; see regional heading under Bibliographies).]

Distribution maps

For *Vergleichende Chorologie der zentraleuropäischen Flora* by H. MEUSEL *et al.* (1965–78, Jena), see 001.

Special group – woody plants

FITSCHEN, J., 1977. *Gehölzflora*. 6th edn., revised and ed. by F. H. Meyer *et al.* 396 pp., 651 text-figs. Heidelberg: Quelle & Meyer. (First edn. 1920; 5th edn. 1959.)

Concise pocket manual for identification of native, naturalized, and commonly cultivated trees and larger shrubs, with limited synonymy, vernacular names, indication of distribution and/or origin, and notes on habitat, occurrence, etc.; diagnostic figures; index.

642

Romania

The northwestern region of Transylvania (Siebenbürgen) was before 1918 a part of the Hungarian kingdom (in Austria–Hungary); it is, in addition to standard Romanian works, thus covered by the older Hungarian works of Jávorka, *Magyar flóra* (1925) and Jávorka and Csapody, *Iconographia florae hungaricae* (1929–34; reprinted 1975).

– Vascular flora: 3550–3750 species (3300–3400 native).

Bibliography

BORZA, A. and POP, E., 1921–47. *Bibliographia botanica Romaniae*. Parts 1–38 (Bull. Grăd. Mus. Bot. Univ. Cluj, 1–27, *passim*). Cluj.

BELDIE, A., 1977–9. *Flora României: determinator ilustrat al plantelor vasculare*. 2 vols. 1439 text-figs. Bucharest: Academia RSR.

Illustrated field manual-key to vascular plants, with essential synonymy, abbreviated indication of internal and extralimital distribution, and terse notes on habitat, life-form, habit, phenology, ecology, etc.; index to generic and family names. An introductory section gives technical details, main sources, and statistics (3350 species accredited to Romania); a separate key to woody plants is given together with the general family key.

BORZA, A., 1947. *Conspectus florae Romaniae regionumque affinum*. viii, 360 pp., 3 color pls. Cluj: 'Cartea Romaneasca'.

Systematic checklist of the vascular plants of Romania, intended to accompany the series 'Flora Romaniae Exsiccata', with concise indication of internal range (by regions); addenda and corrigenda, list of references, and index to generic names at end. Nomenclature follows Mansfeld's *Verzeichnis* (1940; see 647) where applicable. The labels for the series of *exsiccatae* include synonymy as well as habitat and other information.

PRODAN, I. and BUIA, A., 1966. *Flora mică ilustrată a RPR*. 5th edn. 676 pp., 699 text-figs., 43 pls. Bucharest: 'Agro-Silvica'. (First edn., 1928, Cluj.) [Based on the senior author's *Flora pentru determinarea şi descrierea plantelor ce cresc in România*. 2 vols., 1923, Cluj; 2nd edn., 1939.]

Illustrated manual-key to vascular plants, with limited synonymy, vernacular names, concise indication of local range, abbreviated notes on habitat, phenology, and phytosociology, and indices to vernacular names and to genera and families. A Hungarian version (not seen by this author) is also available (A. Borza, personal communication). [Now superseded by *Flora României* (see above), whose author, A. Beldie, notes in his introduction that the last serious revision of *Flora mică ilustrată* was probably in 1939.]

SĂVULESCU, T. *et al.* (ed.); NYÁRÁDY, E. J. *et al.* (comp.), 1952–76. *Flora Republicii Populare Romîne.* [Flora reipublicae popularis romanicae]. 13 vols. 1486 pls. Bucharest: Academia RPR. (Vols. 11–13 entitled *Flora Republicii Socialiste România* [Flora reipublicae socialisticae romaniae], and pub. by Academia RSR.)

Comprehensive descriptive documentary flora of vascular plants, including native, naturalized, and commonly cultivated plants, with keys to genera, species, and infraspecific taxa; includes extensive synonymy, references and citations (including Flora Romaniae Exsiccata), vernacular names in several languages, fairly detailed exposition of local range and summary of extralimital distribution, critical commentary, and extensive notes on habitat, life-forms, phenology, phytosociology, etc.; index to all botanical names in each volume. The introductory section in vol. 1 includes a historical account of Romanian botanical exploration and floristic and taxonomic research. In vol. 13 are a synopsis of the flora, general keys to families, addenda and corrigenda to vols. 1–12, notes on useful plants, endemics, rare species, etc., author abbreviations, and a general index to the whole work. [This Romanian flora is regarded as of a high standard, forming a baseline for further floristic studies especially in southeastern Europe.]

Special group – woody plants

BELDIE, A., 1953. *Plantele lemnosae din RPR.* 464 pp., 79 pls., 15 figs. Bucharest: 'Agro-Silvica'.

Descriptive treatment of woody plants, with keys and indication of distribution and/or origin, illustrations of representative features, and vernacular names. Indices and an alphabetical glossary are also included. Tree species are more thoroughly treated in NEGULESCU, E. G. and SĂVULESCU, A., 1965. *Dendrologie.* 2nd edn. 511 pp., 335 text-figs., folding map. Bucharest. (First edn. 1957.)

643

Hungary

Prior to 1918, the Hungarian kingdom, as a part of the Austro–Hungarian Empire, encompassed Hungary proper, Slovakia, Ruthenia, Transylvania (Siebenbürgen), the Banat, Slavonia, Croatia, and parts of modern Austria. These areas (except Croatia) are wholly covered by the complementary works of Jávorka (now superseded for Hungary proper) and Jávorka and Csapody. All other floras described here are limited to the territory of post-World War II Hungary.

– Vascular flora: 2550–2700 species (2250–2450 native).

Bibliography

GOMBOCZ, E., 1936–9. *A Magyar növénytani irodalom bibliográfiája* [Bibliographie der ungarischen botanischen Literatur,] [I–II]. 440, 360 pp. Budapest. [Volume 1 covers the years 1901 to 1925; vol. 2 is for the years 1578 to 1900.]

JÁVORKA, S., 1925. *Magyar flóra.* [Flora hungarica.] cii, 1307 pp., 13 pls., map. Budapest. 'Studium.'

Manual-key to vascular plants, with limited synonymy, vernacular names, concise indication of local range, brief summaries of overall distribution outside the Hungarian Basin, critical commentary, and notes on habitat, phenology, etc.; indices to botanical and vernacular names. The introductory section includes general chapters on biology, ecology, organography (with glossary), use of keys, etc. Although this work is now superseded for Hungary proper by Soó and Kárpáti's *Magyar flóra*, it is included here for its coverage of Croatia, Slavonia, Ruthenia, the Banat, Transylvania, and Slovakia. It also provides the textual basis for Jávorka and Csapody's *Iconographia*, described next.

JÁVORKA, S. and CSAPODY, V., 1979. *Iconographia florae partis austro-orientalis Europae centralis.* [Ikonographie der Flora des südöstlichen Mitteleuropas.] Revised edn. 704, 80 pp., 576 monotone and 40 color pls. (with 4090 figs.). Stuttgart: Fischer. (Originally pub. 1929–34 as *Iconographia florae hungaricae.*)

Systematically organized atlas of the vascular plants of Hungary and neighboring areas (corresponding to 'historical' Hungary), with each lithographed plate including separate figures for a number of species; captions include Hungarian vernacular and Latin names as well as concise notes on distribution (in Hungarian); the 80 pages added in the reissue comprise a German version of the captions (translated and augmented by Sz. Priszter). The introductory section in the reissue includes taxonomic changes (by R. Soó), bibliographic notes and a new introduction (by Sz. Priszter), a trilingual glossary, and a gazetteer; several indices conclude the work. All figures are the work of Vera Csapody. The extended area of coverage of this work, corresponding to that of Jávorka's *Magyar flóra* (see above), is a considerable asset, the more so as its geographical coverage is non-overlapping with Hegi's *Illustrierte Flora*. [Description based upon 1929–34 version and upon publisher's and other previews of the 1979 reissue, prior to seeing what is a most handsomely produced work.] Also issued in Hungarian as *idem*, 1975. *Közep-Európa délkeleti részének flórája képekben*. Budapest: Akadémiai Kiadó.

Soó, R. (ed.), 1964–80. *A Magyar flóra és vegetáció rendszertani-növényföldrajzi kézikönyve*. [Synopsis systematico-geobotanica flora vegetationsque Hungariae.] Vols. 1–6. Budapest: Akadémiai Kiadó.

Critical systematic–geobotanical enumeration and synopsis of vascular plants (and bryophytes) of Hungary, without keys (except to subspecific taxa); includes full relevant synonymy (with references and citations), detailed account of local range, general summary of extra-Hungarian distribution, taxonomic commentary, and notes on karyotypes, habitat, life-form, phenology, associates, properties, uses, etc.; indices to genera and to authors in each volume. The introductory section (in vol. 1) includes general chapters on taxonomy, geobotany, and the history of the flora, as well as a detailed phytosociological classification and an account of floristic regions. An appendix with addenda and corrigenda to vols. 1–2 appears in vol. 3 together with a key to plant societies. Volume 6 covers additions and corrections to vols. 1–5, together with a list of officinal drugs and an overall conspectus of the flora together with its geobotanical associations. [Vol. 7, when published under the editorship of Sz. Priszter, will contain a complete general index to the whole work.]

Soó, R., Jávorka, S. *et al.*, 1951. *A Magyar növényvilág kézikönyve*. 2 vols. xlvi, 1120 pp., 170 text-figs., map. Budapest: Akadémiai Kiadó.

Descriptive manual of native, naturalized, and commonly cultivated vascular and non-vascular plants, with keys to all taxa and citations of standard monographic and revisionary treatments; also includes synonymy, references, vernacular names, fairly detailed account of local range, generalized summary of extralimital distribution, and notes on habitat, phenology, habit, associations, karyology, properties, uses, etc.; lists of authors, medicinal plants, and errata; indices to botanical and vernacular names. The introductory section includes an explanation of the work, an extensive bibliography, and accounts of plant associations and floristic regions; illustrated glossaries are given at the beginning of each major plant group. Now partly superseded by Soó's *Synopsis* of 1964–73 (preceding entry) and the following work.

Soó, R. and Kárpáti, Z., 1968. Magyar flóra: harasztok [Pteridophyta] – virágos növények [Anthophyta]. In *Növényhatározó*, 2 (ed. T. Hortobágyi). 4th edn. 846 pp., 1990 text-figs. Budapest: Tankönyvkiadó. (First edn., 1952.)

Illustrated manual-key to native, naturalized, and commonly cultivated vascular plants, with abbreviated summary of local range, vernacular names, and notes on habitat, life-forms, associations, etc.; index to family, generic and vernacular names. The introductory section includes summaries of phytosociological units and floristic regions, technical notes, an illustrated glossary, lists of plants with particular features or properties, and a short reference list. [Based on preceding work; functionally succeeds Jávorka's *Magyar flóra* (1925).]

644

Czechoslovakia (in general)

Included here are modern works relating to Czechoslovakia as a whole. Floras for Czechy (Bohemia, Moravia, and Moravian Silesia) and Slovakia, and certain floras of neighboring areas with coverage extending to one or the other of these states, are described or cross-referenced under **645**. From formation of the republic until 1939 Czechoslovakia also encompassed Ruthenia (the Carpatho–Ukraine district), now part of the the Soviet Ukraine (**694**); for specific works on this area, however, see under the Carpathians (**603/VII**) and the SW Ukraine (**695**).

– Vascular flora: 3000–3150 species (2600–2750 native).

Bibliography

Futák, J. and Domin, K., 1960. *Bibliografia k flóre ČSR do r. 1952*. 883 pp. Bratislava: Slovenská akadémie vied. [20000 entries.]

Domin, K., 1935. *Plantarum Čechoslovakiae enumeratio, species vasculares indigenas et introductas exhibens*. 305 pp. (Preslia, 13–15). Prague.

Systematic census of vascular plants, with synonymy, references, and citations; descriptions and localities not included. Covers native, naturalized, and adventive species.

DOSTÁL, J., 1958. *Klíč k úplné květeně ČSR.* 2nd edn. 982 pp.; xviii, 408 text-figs. Prague: Československé akademie věd. (First edn. 1953.)

Illustrated, briefly descriptive manual of vascular plants, with keys to all taxa, limited synonymy, vernacular names, and symbolic indication of distribution and altitude, habitat, life-form, phenology, etc.; numerous diagnostic figures; list of authors and indices to all botanical and vernacular names. The introductory section includes a glossary. The work is essentially a revised and condensed version of the author's *Květena ČSR* (next entry).

DOSTÁL, J., 1949–50. *Květena ČSR.* 64, 2269 pp.; xviii, 711 text-figs. (Sbírka příruček československé botanické společnosti, sv. 2). Prague: Přírodovedecké nakladatelství.

Copiously illustrated descriptive flora of vascular plants, with keys to all taxa; synonymy; vernacular names; generalized indication of local range; taxonomic commentary and indication of habitat, phenology, life-form, associates, etc.; indices to botanical and vernacular names. The introductory section contains an illustrated glossary, a simplified general key to the more obvious families, a list of principal references, and addenda and corrigenda.

645

Czechoslovakia (individual states)

This unit gives listings of works relating respectively to the individual states of Czechy (Bohemia, Moravia, and Moravian Silesia) and Slovakia. In addition to floras of local authorship, both states are covered by works on adjacent regions: Czechy by various German-language works on Central Europe, Austria, and Germany; Slovakia by certain Hungarian-language works. For this, a historical basis exists; prior to 1918, Czechy was part of the old Austrian kingdom, while Slovakia formed part of 'historical Hungary,' i.e., the former Hungarian kingdom. Cross references to these works appear under the individual states.

Czechy (Ceské země)

See **644** (all works). German-language works include those by EHRENDORFER and HEGI (**641**), MANSFELD (**647**) and FRITSCH (**648**). As the two current all-republican works are in the first instance in Czech, no modern separate floras or manuals are available. Older works include POLÍKVA, F., 1901–04. *Názorná květena zemí koruny České.* 4 vols. Olomouc (Olmütz): Promberger; and *idem*, 1912. *Klíč k úplné květeně zemí koruny České.* 110, 864 pp., illus. Olomouc. [Several other floras for smaller areas, e.g., Bohemia, have also been published.]

Slovakia

See also **644** (all works). For Hungarian-language works, see **643** (JÁVORKA; JÁVORKA and CSAPODY).

FUTÁK, J. (ed.), 1966– . *Flora Slovenska.* Vols. 1– . Illus. Bratislava: Slovenská akadémie vied.

Comprehensive descriptive flora (vol. 2 onward) of vascular plants (in Slovak), with keys to all taxa; limited synonymy, with references; vernacular names; detailed exposition of local range, including citations; detailed taxonomic commentary and notes on habitat, ecology, biology, karyotypes, etc.; numerous figures and distributional maps (covering most species); indices to botanical and vernacular names. Volume 1 comprises a comprehensive illustrated Slovak botanical glossary and lexicon, compiled by J. Dostál and J. Futák. Not completed; at this writing (1979) only two volumes have appeared, the last in 1969.

NOVACKÝ, I. M., 1954. *Slovenská botanická nomenklatúra.* 227 pp. (Odborná terminológia, 11). Bratislava: Slovenská akadémie vied.

Systematic but unannotated checklist of Slovak vascular plants, with standardized vernacular equivalents. See also *idem*, 1943. Flora Slovenskj republiky. *Slovenská vlastiv.*, **1**: 333–99.

646

Poland

Those areas of modern Poland which formed part of the German Reich before 1918 are covered by all German-language floras and related works dealing with the region of 'Mitteleuropa', even to the present time;

this is also true of certain works on the flora of Germany in a narrower sense (i.e., the limits of 1937). As for Polish works, those sections of *Flora polska* and *Atlas flory polskiej* published before World War II also include large parts of Lithuania, White Russia, and the Ukraine (now in the USSR) but omit Pomerania, Silesia, and part of East Prussia (i.e., they cover the Poland of 1939). For the Sudetens and the Carpathians see also, respectively, 603/VI and 603/VII.

– Vascular flora: 2350–2600 species (2250–2450 native).

Bibliography
See also MANSFELD, R. (647).

SZYMKIEWICZ, D., 1925. *Bibliografia flory Polskiej*. 159 pp. (Prace Monogr. Komis. Fizjogr., 2). Cracow: Polska Akademji Umiejętności.

RACIBORSKI, M. *et al.* (eds.), 1919–80. *Flora polska*. 14 vols. Illus. Cracow: Polska Akademji Umiejętności (1919–47, vols. 1–6); Warsaw (later Warsaw/Cracow): Państwowe Wydawnictwo Naukowe (PWN) (1955, vols. 7–14).

Copiously descriptive documentary flora of vascular plants, with keys to genera and species; includes synonymy and vernacular names, fairly detailed indication of local range in Poland and adjacent areas, general summary of extralimital distribution, comments on taxonomic problems, and notes on habitat, phenology, special features, etc.; figures of diagnostic features; indices to all botanical names in each volume. At this writing (early 1981) the work has at last been completed, after much delay in completing the Compositae, especially with regard to finding specialists for such 'critical' genera as *Taraxacum* and *Hieracium* (K. Rostánski, personal communication).

For an accompanying folio atlas with descriptive commentary, see KULCZYŃSKI, S. and MĄDALSKI, J. (eds.), 1930– . *Atlas flory polskiej*. [Florae polonicae iconographia]. Vols. 1– (pub. in fascicles). Plates. Cracow: Polska Akademji Umiejętności (1930–6); Warsaw: Polska Akademia Nauk (later PWN) (1954–). (From 1954 entitled *Atlas flory polskiej i ziem ościennych*. [Florae polonicae terrarumque adiacientium iconographia.]) [Large-scale illustrations accompanied by text in Polish and Latin, arranged along the same sequence as *Flora polska*, with cross references. At this writing (1980) 33 fascicles in vols. 1–7, 9, 11, and 17 have been published, but only vols. 1, 5, and 9 have been completed. A total of 21 volumes is planned corresponding to the 14 volumes of the *Flora*. The plan

of the work as well as an indication of fascicles published appears in the covers of each new fascicle.]

SZAFER, W., KULCZYNSKI, S. and PAWŁOWSKI, B., 1953. *Rósliny Polskie*. xxviii, 1020 pp.; vii, 2187 text-figs., map. Warsaw: PWN.

Concise illustrated manual-key to native, natural-ized, and commonly cultivated vascular plants, with brief diagnoses, limited synonymy, vernacular names, indication of local range, and partly symbolic notes on habitat, life-form, phenology, etc.; indices to botanical and vernacular names. An introductory section, partly pedagogic, includes a list of authors and an illustrated glossary/organography. In addition to modern (post-1945) Poland, this work covers adjacent parts of neighboring countries. A second edition (not seen by this author) appeared at Warsaw in 1967.

Distribution maps
BIAŁOBOK, S., BROWICZ, K. *et al.*, 1963– . *Atlas rozmieszczenia drzew i krzewów w Polsce*. [Atlas of distribution of trees and shrubs in Poland.] Fasc. 1– . Maps. Warsaw: Pánstwowe Wydawnictwo Naukowe (for Zakład Dendrologii i Arboretum Kórnicke PAN).

Comprises a distributional atlas, with large-scale dot maps (on a grid) of ranges within modern Poland for individual species of trees and shrubs (accompanied by explanatory remarks in Polish, Russian and English), and summary maps of extralimital distribu-tion; many references provided. No systematic sequence is followed. As of 1980 the work was far advanced, with 29 fascicles published; additional fascicles are appearing at the rate of 2–3 per year.

Special group – woody plants
See above for *Atlas rozmieszczenia drzew i krzewów w Polsce*.

KOŚCIELNY, S. and SĘKOWSKI, B., 1971 (1972). *Drzew i krzewy: klucze do oznaczania*. 535 pp., illus. Warsaw: Pánstwowe Wydawnictwo Rolnicze i Leśne (PWRL).

Illustrated manual-key to native, naturalized and cultivated trees and shrubs, with indication of distribution and/or origin, limited synonymy, and notes on height, habitat (or other situation), phenology, uses, properties, special features, etc.; vernacular names included; index.

647

Germany

The post-1945 area of Germany is divided into two states, the German Democratic Republic (DDR) and the Federal Republic of Germany (BRD). However, all current florae, manuals, and lists (as well as all works relating to 'Mitteleuropa' in a wider sense) cover the whole of the German national territory, either within its present limits or within those of 1937. Only those manuals and other works relating to Germany proper are described here; those covering the wider area of 'Mitteleuropa' are accounted for under **641**. For the Alpine region, see also **603/IV**.

– Vascular flora: 3000–3150 species (2600–2750 native).

Bibliography

MANSFELD, R. (ed.), 1940 (1941). *Verzeichnis der Farn- und Blütenpflanzen des Deutschen Reiches.* [Bibliography, pp. 283–308.]

SUKOPP, H., 1960. Übersicht über die in der Zeit von 1945 bis 1959 erscheinen Gefässpflanzenflora Deutschlands. *Willdenowia* 2: 563–83.

GARCKE, A., 1972. *Illustrierte Flora von Deutschland und angrenzende Gebiete.* 23rd edn., revised and ed. K. von Weihe. xx, 1607 pp., 460 text-figs., 5 pls. Berlin: Parey. (First edn. 1849, Berlin, entitled *Flora von Nord- und Mittel-Deutschland.*)

Briefly descriptive manual-flora of vascular plants, with keys to all taxa; limited synonymy; vernacular names; brief summary of local and extralimital range, frequency, and occurrence; symbolic indication of habitat, life-form, phenology, karyotype, special features, etc.; remarks on variation and taxonomy; representative figures; illustrated glossary and index to all botanical and vernacular names. The introductory section includes aids to the use of the work, a synopsis, and key floristic references.

MANSFELD, R. (ed.), 1940 (1941). *Verzeichnis der Farn- und Blütenpflanzen des Deutschen Reiches.* 323 pp. (Ber. Deutsch. Bot. Ges., 58a). Jena.

Systematic census of the vascular plants of the Greater German Reich of 1940 (including Austria, the Czech lands and parts of Poland), with limited synonymy, vernacular names of genera, abbreviated indication of local range, and indices to vernacular and generic names and botanical synonyms. Pages 283–308 constitute a detailed bibliography of floristic literature on the region.

OBERDORFER, E., 1979. *Pflanzensoziologische Exkursionsflora.* 4th edn. (with T. Müller). 997 pp., 58 text-figs. Stuttgart: Ulmer. (Originally pub. 1949 as *Pflanzensoziologische Exkursionsflora für Süddeutschland*; 2nd edn., 1960, and 3rd edn., 1969 as *Pflanzensoziologische Exkursionsflora für Süddeutschland und die angrenzenden Gebiete.*)

This work, now extended to cover the whole of the two German states along with the Vosges and the Swiss and Austrian Alps, comprises a field manual-key to vascular plants with separate species entries containing non-descriptive information such as synonymy, karyotypes, ecology, phytosociological status, and chorological and other geographical data. Vernacular names are also included, and indices are given at the end of the work. Includes 3320 taxa (and 'bis' numbers).

ROTHMALER, W., 1976. *Exkursionsflora für die Gebiete der DDR und der BRD*, II: *Gefässpflanzen.* 8th edn., revised by H. Meusel and R. Schubert. 612 pp., 987 text-figs., map. Berlin [East]: Volk & Wissen Volkseigener Verlag. (Originally pub. 1959 as *Exkursionsflora für Deutschland*, II: *Gefässpflanzen*; title changed 1972.)

ROTHMALER, W., 1959. *Exkursionsflora für Deutschland*, III: *Atlas der Gefässpflanzen.* 2nd edn. 567 pp., 2550+ text-figs. Berlin. (Third edn. 1966.)

ROTHMALER, W., 1976. *Exkursionsflora für die Gebiete der DDR und der BRD*, IV: *Kritischer Band* [Gefässpflanzen]. Revised by W. Vent and R. Schubert. 4th edn. 811 pp., 743 text-figs. Berlin. (Originally pub. 1963 as *Exkursionsflora für Deutschland*, IV: *Kritischer Ergänzungsband – Gefässpflanzen.*)

Teil (part) II, the standard flora, comprises a manual-key to vascular plants of the two German states, with limited synonymy, vernacular names, general indication of local range, small text-figures depicting diagnostic features, brief notes on habitat, phenology, and association-type, and a combined index to generic, family, and vernacular names. The introductory section includes an illustrated glossary and chapters on the biology and distribution of plants. Treatment of 'critical groups' is relatively generalized. Teil III, the atlas, consists entirely of page-sized figures of nearly all recognized species and species aggregates of vascular plants, Teil IV, the 'Critical supplement', is actually an expanded manual of the flora, cross-referenced to Teil II, but with more detailed treatment and discussion (including diagnostic figures) of subspecies, aggregates, microspecies, varieties, hybrids, etc., with indication of habit, karyotypes, habitat, phenology, distribution, and chorological and phytosociological affinities, and including medicinal notes and vernacular names along with limited synonymy; list of more important literature, authors, and complete index at end. An introductory section features sections on organography (with illustrations), plant geography, and sociological divisions (pp. 20–43); general key to genera, etc. (pp. 44–73). [Both Teil II and IV are extremely compact works, representing much distillation from a vast body of information, and ought to set an example for others of their kind.]

SCHMEIL, O. and FITSCHEN, J., 1958. *Flora von Deutschland*. 67/68th edn., revised by H. Vörkel and G. Müller. xii, 515 pp., 698 text-figs. Jena: Fischer.

SCHMEIL, O. and FITSCHEN, J., 1976. *Flora von Deutschland und seiner angrenzenden Gebiete*. 86th edn., revised by W. Rauh and K. Senghas. 516 pp., 1103 text-figs. Heidelberg: Quelle & Meyer. (First edn., 1904.)

Concise illustrated manual-keys for students, with limited synonymy and brief indication of distribution, habitat, life-form, special features, etc. (partly symbolically); vernacular names included; indices. The brief introductory portions of these works, designed primarily for students at secondary and tertiary levels, include glossaries and notes on how to use the keys as well as explanations of abbreviations. The Heidelberg version is still being periodically reissued with continuing small amendments (with five reissues in an eight-year period), but for the Jena version no current information has been available.

648

Austria

Before 1918, the Austrian kingdom in a narrower sense (excluding Dalmatia, Bukovina, and Galicia) included, in addition to modern Austria, Istria, Slovenia, South Tyrol, and the Czech lands of Bohemia and Moravia. This portion of the old Austro–Hungarian Empire is the one encompassed by Fritsch's recently reprinted *Exkursionsflora* as well as by all German-language works relating to 'Mitteleuropa'. By contrast, Janchen's enumeration deals only with modern Austria. For works covering the Alpine region, see also **603**/IV.

– Vascular flora: 3300–3450 species (2900–3100 native).

Bibliography

JANCHEN, E., *Catalogus florae austriae*, Teil I. [Bibliography, pp. 1–50.]

FRITSCH, K., 1922. *Exkursionsflora für Österreich und die ehemals österreichischen Nachbargebiete*. 3rd edn. lxxx, 824 pp. Vienna: Gerold. (Reprinted 1973, Lehre, W. Germany, Cramer; 1st edn. 1897, Vienna.)

Manual-key to vascular plants, with limited synonymy, German vernacular names, symbolic indication of local range, and indices to generic and family names and to botanical synonyms. The introductory section includes a chapter on basic organography as well as a list (pp. xxi–xxiii) of the most important floras covering the region. In addition to present-day Austria, the work covers those areas of the former Austrian kingdom as indicated in the heading above.

JANCHEN, E., 1956–60. Pteridophyten und Anthophyten (Farne- und Blütenpflanzen). 4 parts. In *Catalogus florae austriae*, I (ed. K. HAFLER and F. KNOLL). 999 pp. Vienna: Österreichische Akademie der Wissenschaften (distributed by Springer, Vienna and New York). Continued as *idem*, 1963–7. *Ergänzungshefte* I–IV. 4 parts. Vienna.

Detailed systematic enumeration of native, naturalized, and commonly cultivated vascular plants, with numerous references to relevant revisions and monographs of individual genera and families as well as synopses of infrafamilial classifications; extensive synonymy, with references and citations; vernacular names; symbolic indication of local range; taxonomic commentary and notes on habitat, ecology, etc.; addenda (pp. 881–974); indices to vernacular, generic, and family names (for complete index to original work and all supplements, see *Ergänzungsheft* IV). The introductory section includes a very full bibliography.

649

Switzerland (with Liechtenstein)

See also **603**/IV (all works); **651** (BONNIER and DOUIN). The limits of *Flora der Schweiz* by Hess *et al.* extend well beyond actual Swiss frontiers.

– Vascular flora: 3000–3150 species (2600–2750 native).

BINZ, A., 1980. *Schul- und Exkursionsflora für die Schweiz mit Berücksichtigung der Grenzgebiete.* 7th edn., revised A. Becherer and C. Heitz. lix, 422 pp., 376 text-figs. Basel: Schwabe. (First edn. 1920.) French edn.: BINZ, A. and THOMMEN, E., 1976. *Flore de la Suisse y compris les parties limitrophes de l'Ain et de la Savoie.* 4th edn., revised P. Villaret. xxxiii, 398 pp., 434 text-figs., map. Basel: Schwabe; Neuchâtel: du Griffon. (First edn. 1941.)

Illustrated students' manual of vascular plants, with keys to all taxa; limited synonymy; German (or French) vernacular names; symbolic indication of local range; brief notes on occurrence, habitat, phenology, etc.; small figures of diagnostic features; list of toxic plants; indices to generic and family names and to German (or French) vernacular names. The introductory section has an illustrated glossary and a list of references. Only slight differences exist between the German and French editions.

HESS, H. E., LANDOLT, E. and HIRZEL, R., 1967–73. *Flora der Schweiz und angrenzender Gebiete.* 3 vols. Illus., maps (some in color), diagrams. Basel: Birkhäuser.

Copiously illustrated, rather bulky descriptive treatment of native, naturalized, and commonly cultivated vascular plants, with keys to all taxa; complete synonymy, without references; fairly detailed account of local and extra-Swiss range; taxonomic commentary and notes on habitat, ecology, karyotypes, variability, etc.; indices to all taxa in each volume. The introductory section includes chapters on the Tertiary and Quaternary history of the flora, vegetation formations, phytosociology, floristics, and phytogeography. A corrected second edition of vol. 1 was published in 1976. Accompanied by *idem*, 1976. *Bestimmungsschlüssel zur Flora der Schweiz.* iii, 657 pp., *c.* 1500 figs. Basel. [Illustrated manual-key based on the larger *Flora*, with small marginal analytical figures and notes on occurrence; glossary and indices.]

SCHINZ, H. and KELLER, R., 1914–23. *Flora der Schweiz*, I: *Exkursionsflora.* 4th edn., revised by H. Schinz and A. Thellung. xxxvi, 792 pp., 172 text-figs. Zurich: Raustein. (First edn. 1900; 3rd edn. 1909.) French edition: *Idem*, 1909 (1908). *Flore de la Suisse*, I. 3rd edn., transl. E. Wilczek and H. Schinz. xxii, 690 pp., 128 text-figs. Lausanne.

SCHINZ, H. and KELLER, R., 1914. *Flora der Schweiz*, II: *Kritische Flora.* 3rd edn., revised by H. Schinz and A. Thellung. xviii, 582 pp. Zurich. (First edn. 1900, together with the *Exkursionsflora*.)

The first section of this work (Teil I) comprises a general manual-key to native, naturalized and commonly cultivated vascular plant species (and some subspecies), with limited synonymy, vernacular names, symbolic indication of distribution, and notes on taxonomy, habitat, phenology, etc.; indices to all botanical and vernacular names at end. The introductory section includes a glossary as well as other aids to the use of the manual. The second section (Teil II) is a critical supplement, with detailed treatments of subspecies, varieties, 'microspecies', hybrids, etc., together with extra-Swiss distribution of all taxa listed in both parts; a list of adventive plants is also given. [No French version of the *Kritische Flora* is available. The work as a whole is still valued in spite of the appearance of the more encyclopedic *Flora der Schweiz* by Hess, Landolt, and Hirzel (see above).]

THOMMEN, É., 1967. *Taschenatlas der schweizer Flora mit Berücksichtigung der ausländischen Nachbarschaft.* 4th edn., revised A. Becherer. xvi, 303 pp., 3055 text-figs. Basel: Birkhäuser. (First edn. 1945.) French edn.: THOMMEN, É., 1961. *Atlas de poche de la flore suisse comprenant les régions étrangères limitrophes.* 2nd edn. revised A. Becherer. xv, 303 pp., 3055 text-figs. Basel: Birkhäuser. (First edn. 1945.)

Small pocket atlas of vascular plants, with limited synonymy, vernacular names, a small figure (or figures) of each species, and indices to genera and families as well as to vernacular names. An appendix gives an account of species found around the borders of Switzerland, especially in the vicinity of Geneva. The German and French editions are virtually identical.

Region

65

Western Europe (Mainland)

This region consists of France, Belgium, Luxembourg, and the Netherlands. Like Central Europe, it is by and large a well-collected and well-documented region, densely covered by manuals and other works, although in France a lot of the provincial works are comparatively old and the general approach to floristics is less sophisticated.

France has a number of general floras, the most outstanding being Coste's famous *Flore descriptive et illustrée de la France* (1901–6), which was directly influenced by Britton and Brown's contemporary *Illustrated Flora* (**140–50**). This has in recent years (1972–7) been acquiring a supplement, which will constitute a fourth volume. Other floras are those by Bonnier: the *Flore complète portative* (with de Layens) and the thirteen-volume *Flore complète illustrée* (with Douin) with 721 colored plates, published early this century; the *Quatre flores de France* by Fournier, recently reissued but not revised; the fourteen-volume *Flore de France* by Rouy *et al.* (1893–1913), with detailed infraspecific treatments in the style of the day; and, lastly, *Flore de France* by Guinochet and de Vilmorin (1973–), a manual (eventually to be in five volumes) following the polyphyletic, 'stachyosporic'/ 'phyllosporic' Emberger system. However, the view is taken here that it should be possible to write a good one-volume descriptive work, as the French flora is of the same order, though slightly larger, as that of Texas (**171**) and is somewhat smaller than that of California (**195**) – both states with highly regarded, one-volume manual-floras.

The Low Countries are provided with a plethora of floras of several kinds. Both Belgium and Holland have had large-scale documentary floras, of which the current representatives are respectively *Flore générale de Belgique* by Lawalrée (1952–66) and *Flora neerlandica* of the Royal Netherlands Botanical Society (1952–), both incomplete as of this writing (1981). In addition, they have had many 'routine' manual-keys, of which the leaders are *Nouvelle flore de la Belgique* by de Langhe *et al.* (latest version, 1978) and *Flora van*

Nederland by Heukels and van Ooststroom (latest version, 1977), both illustrated. The Belgian works conventionally include Luxembourg.

Bibliographies. General and divisional bibliographies as for Division 6. Very thorough coverage through 1959 is in particular provided by Blake 1961.

Indices. General and divisional indices as for Division 6.

651

France (including Monaco and the Channel Islands)

The region of Alsace-Lorraine, under German rule from 1871 to 1918, is considered to be part of the botanists' 'Mitteleuropa' and is thus covered by a number of German-language works listed under **641** and **647**. Northern France is covered in *Nouvelle flore de la Belgique* by DE LANGHE *et al.* (**656**), while the Channel Islands are part and parcel of general works on the British Isles (**660**). The areas bordering on Switzerland (**649**), especially Geneva, are covered in some Swiss works, especially *Flora der Schweiz* by Hess *et al.* The Pyrenean zone (with Andorra) is dealt with by works listed under **603/II**; for the Alpes-Maritimes, see **603/III**; for the Alps, see **603/IV** (all works, but especially FENAROLI, *Flora delle alpi*).

– Vascular flora: 5000–5150 species (4300–4450 native), excluding Corsica.

BONNIER, G. and DOUIN, R., 1911–35. *Flore complète illustrée en couleurs de France, Suisse et Belgique* (*comprenant la plupart des plantes d'Europe*). 13 vols. 721 color pls. (La végétation de la France, Suisse et Belgique, II). Paris, Brussels, Neuchâtel (vols. 1–8); Paris, Brussels (vol. 9); Paris (vols. 10–13): Ministère de l'Instruction Publique, France (distributed by Orlhac, Paris; Lebègue, Brussels; and Delachaux & Niestlé, Neuchâtel). Accompanied by BONNIER, G. and LAYENS, G. DE, [1909]. *Flore complète portative de la France et de la Suisse*. xxvii, 426 pp., 5338 text-figs., 2 maps (La végétation de la France, Suisse, et Belgique, I). Paris: Orlhac. (Many reissues.)

Comprises (in vols. 1–12) a folio atlas of colored plates, with accompanying descriptive text; the latter includes synonymy, vernacular names, local and extralimital range, and notes on habitat, properties, uses, etc.; general commentary on characteristics and relationships of families and genera; indices to all plant names. Volume 13 (Tableau générale) is devoted to a general index to the entire work as well as addenda and corrigenda. The accompanying *Flore complète portative*, in small octavo format, comprises tabular illustrated analytical keys to all vascular plant species, with vernacular names and brief indication of distribution and frequency, and includes an index.

COSTE, H., l'abbé, 1901–06. *Flore descriptive et illustrée de la France, de la Corse et des contrées limitrophes.* 3 vols. 4354 text-figs., colored map. Paris: Klincksieck. (Reprinted 1937, 1950.)

Illustrated descriptive account of the vascular plants of France, with keys to all taxa, synonymy, vernacular names, general indication of local range, and concise notes on overall distribution, habitat, phenology, etc.; indices to all botanical and vernacular names in vol. 3. The introductory section in volume 1 contains an illustrated glossary and a survey of the general features of the flora together with an essay on French floristics and vegetation by C. Flahault. The plan of this work, like several others of its genre, follows that of Britton and Brown's *Illustrated flora of the northern [United] States* (see **140–50**).[17] The *Flore descriptive et illustree* is by general consent considered to be a classic among European floras, and was designated as a 'Basic Flora' of primary importance in the preparation of *Flora Europaea*. For additions, see *idem*, 1972–77. *Flore descriptive et illustrée de la France.* Suppléments 1–5, by P. Jovet and R. de Vilmorin. Pp. 1–589, illus. Paris: Blanchard. [At present, 1979, covers all families in the pteridophytes, gymnosperms, and dicotyledons (species 1–4263).]

FOURNIER, P., 1977. *Les quatre flores de France, Corse comprise.* 2nd revised edn. 2 vols. xlviii, 1106 pp., 48 text-figs., maps (endpapers); 304 pls. on 308 pp. (with 4216 figs.). Paris: Lechevalier. (Original edn. 1934–40; revised edn. 1961.)

This work, lately reorganized and reissued in a new format, now comprises separate volumes of text and atlas. The text-volume (unaltered from the previous edition) is a manual-key to vascular plants, with brief diagnoses, limited synonymy, vernacular names, and symbolic indication of floristic zone(s), local range, frequency, special features, etc.; general index to all botanical and vernacular names. An introductory section gives a glossary, translations of the commonest specific epithets, and a general key to families. The atlas volume includes all the figures of the previous edition but in a greatly enlarged form, rendering them rather clearer. However, the convenience of the previous editions has been to some extent lost in this new version.

GUINOCHET, M. and VILMORIN, R. DE, 1973–8. *Flore de France.* Parts 1–3. Illus. Paris: Centre National de la Recherche Scientifique (distributed by Doin).

Manual-key to vascular plants of mainland France and Corsica, with very limited synonymy, brief indication of local and extralimital range, and notes on habitat, ecology, phytosociological affinities, special features, etc.; figures of critical details; no index. The introductory section includes chapters on theoretical considerations, the plan of the work, and a phytosociological key and synopsis. At this writing (1979) three of the five parts planned have been published, covering pteridophytes, gymnosperms, and Santalaceae through Orchidaceae (families 1–135 on the Emberger system as adapted from his *Traité de Botanique* (*Systématique*), II: *Les végétaux vasculaires.* Paris, 1960).

ROUY, G. *et al.*, 1893–1913. *Flore de France ou description des plantes qui croissent spontanément en France, en Corse et en Alsace-Lorraine.* 14 vols. Asnières, Rochefort: the authors; Paris: Deyrolle. (Also pub. as supplements to Ann. Soc. Sci. Nat. Charente-Infér.)

Detailed comprehensive descriptive flora of vascular plants, with keys to all taxa; full synonymy, with references and citations; indication of localities, with records, and detailed summary of local and extralimital range; extensive taxonomic commentary; notes on habitat; index to all botanical names in each volume (general index to family and generic names in vol. 14). The introductory section includes a bibliography. The species concept adopted is rather broad, but concomitant with this is the very detailed treatment of infra-specific taxa. Collaborators included J. Foucaud (vols. 1–3), E. G. Camus (vols. 6–7), and N. Boulay (vol. 6). For a summary companion work, see ROUY, G., 1927. *Conspectus de la flore de France.* xvi, 319 pp., portrait. Paris: Lechevalier. [Catalog of vascular plant taxa as accepted in the larger work.]

Special group – woody plants

ROL, R. and JACAMON, M. 1963–9. *Flore des arbres, arbustes, et arbrisseaux.* 4 parts, index. Illus. (covers in color). Paris: Agricole (Maison Rustique).

Artificially arranged, illustrated treatment of woody

plants (somewhat in the style of HOUGH's 1907 *Handbook of the trees of the northern states and Canada*; see **140**), with photographs of leaves, buds, flowering parts, fruits, habit, etc., and botanical and dendrological descriptions with extensive notes. Climbers are also included. The four parts deal respectively with: (1) 'plains and hills', (2) 'mountains', (3) 'Mediterranean region', and (4) 'introduced species'.

656

Belgium

See also **647** (SCHMEIL and FITSCHEN, 1968); **651** (BONNIER *et al.*).

– Vascular flora (including Luxembourg): 1900–2100 species (1600–1800 native).

HAUMAN, L. and BALLE, S., 1934. *Catalogue des ptéridophytes et des phanérogames de la flore belge.* 126 pp. (Bull. Soc. Roy. Bot. Belgique, 66, Suppl.) Gembloux.

Systematic list of vascular plants, with synonymy and indication of range by districts; statistics of the flora; bibliography of relevant literature from 1903 to 1934; index.

DE LANGHE, J.-E. *et al.*, 1978. *Nouvelle flore de la Belgique, du Grand-Duché de Luxembourg, du nord de la France et des régions voisines.* 2nd edn. cv, 899 pp., text-figs., map. Meise (near Brussels): Patrimoine du Jardin Botanique National de Belgique. (First edn. 1973.)

Illustrated manual-key to vascular plants, with limited synonymy, indication of local and extralimital range, vernacular names, and notes on habitat, phenology, habit, special features, etc.; list of new combinations and illustrated glossary; indices to all botanical and vernacular names. The introductory part includes sections on the geographical limits of the work and its plan as well as a brief bibliography. Also incorporated is a 35 page key to woody plants based on vegetative features. This critical work, which incorporates much commentary by specialists, also extends its coverage to the southern Netherlands (i.e., the southern phytogeographic district). Succeeds Crepin's *Manuel de la flore de Belgique*, Goffart's *Nouveau manuel de la flore de Belgique et des régions limitrophes* (1934), and Mullenders *et al.*, *Flore de la Belgique, du nord de la France et des régions voisines* (1967).

LAWALRÉE, A., 1950. Ptéridophytes. In *Flore générale de Belgique* (ed. W. ROBYNS *et al.*), sér. III. iv, 194 pp., illus., maps. Brussels: Jardin Botanique de l'Etat.

LAWALRÉE, A., 1952–6. Spermatophytes. *Ibid.*, sér. IV, vols. 1–4, and 5, part 1. Illus., map. Brussels.

Copiously descriptive illustrated flora of pteridophytes (sér. III) and spermatophytes with keys to genera, species, and infraspecific taxa; the extensive documentation includes full synonymy (with references and citations), detailed, documented indication of local range, summary of extralimital distribution, critical commentary, and extensive notes on karyotypes, habitat, ecology, biology, phenology, etc.; distribution maps and illustrations of representative species; index to all botanical names at end of each volume. Not completed; in sér. IV only gymnosperms and dicotyledons (Salicaceae through Thymeleaceae) are covered. The other series (I and II) were designated for non-vascular plants.[18] Partly supersedes WILDEMAN, É. DE and DURAND, T., 1898–1907. *Prodrome de la flore belge.* 3 vols. Brussels: Castaigne. [Covers both vascular and non-vascular plants.]

Distribution maps

ROMPAEY, E. VAN, DELVOSALLE, L., and collaborators, 1978–9. *Atlas de la flore belge et luxembourgeoise: ptéridophytes et spermatophytes.* [Atlas van de Belgische en Luxemburgse flora: pteridofyten en spermatofyten.] 2 vols.: text, 116 pp.; atlas, 293 pp. 1452 maps. Brussels: Jardin Botanique National de Belgique. (First edn. 1972.)

Comprises distribution maps of most Belgian vascular plant species, save those of extreme ubiquity, with indication of era of collection and of presence or absence; usually only one species depicted per map. Six maps appear on a page and the grid used is composed of 4-kilometer squares. In the second edition (not seen by this author) a volume of explanatory text has been added.

657

Luxembourg

No separate flora are listed here, as the Grand Duchy is wholly covered by most floras of adjacent states, especially those of Belgium.

658

The Netherlands

See also **647** (SCHMEIL and FITSCHEN, 1968).
– Vascular flora: 1750–1900 species (1400–1600 native).

HEIMANS, E., HEINSIUS, H. W., and THIJSSE, J. P., 1965. *Geillustreerde flora van Nederland*. 21st edn., revised by J. Heimans *et al.* viii, 1182 pp., 6000 + text-figs., maps. Amsterdam: Versluys. (First edn., 1899.)

Copiously illustrated manual-key to spontaneous and commonly cultivated vascular (and selected non-vascular) plants, with limited synonymy, vernacular names, notes on habitat and on local and extralimital range, and symbolic indication of life-form and phenology; index. At the end of the introductory section are pedagogical chapters on taxonomy and phytogeography, a key to phytosociological units, a synopsis of families, and remarks on pollinators and galls.

HEUKELS, H. and OOSTSTROOM, S. J. VAN, 1977. *Flora van Nederland*. 19th edn. v, 925 pp., lxxi + 1038 text-figs., map. Groningen: Wolters-Noordhoff. (First edn., 1900, entitled *Geillustreerde schoolflora voor Nederland*.)

Illustrated manual-key to vascular plants, with limited synonymy, vernacular names, brief notes on habitat, distribution, and frequency, and symbolic indication of life-form and phenology, indices to all botanical and vernacular names. The general section includes chapters on the history of botanical work in the Netherlands and on matters geobotanical, a list of phytogeographical districts, an illustrated glossary, and general keys of several sorts to families. A shortened pocket version is: HEUKELS, H. and OOSTSTROOM, S. J. VAN, 1968. *Beknopte school- en excursie-flora voor Nederland*. 12th edn., revised by S. J. van Ooststroom. Groningen. (First edn., 1932, as *Beknopte schoolflora voor Nederland*.)

KONINKLIJKE NEDERLANDSE BOTANISCHE VER-ENIGING (Weevers, T. *et al.*), 1948– . *Flora neerlandica*. [Flora van Nederland.] Vols. 1– . Illus. Amsterdam.

Documentary descriptive flora in the form of a series of revisions, with keys to genera and species, extensive synonymy, references and citations, detailed indication of local range, taxonomic commentary, and notes on ecology, biology, etc.; figures of diagnostic or critical attributes; index to all taxa in each fascicle. At this writing (1979), nine fascicles have been published, constituting vol. 1, parts 1–6 and vol. 4, parts 1–2, 9; no systematic sequence is followed in publication but the arrangement follows the Englerian system; presently available are the pteridophytes, gymnosperms, monocotyledons, and a few families (and genera) of the dicotyledons. Since the early 1960s only one fascicle has been produced (on part of *Hieracium*), and currently, very little time is available for preparation of this work.[19] Partly supersedes HEUKELS, H., 1909–11. *De flora van Nederland*. 3 vols. 2047 text-figs., map. Leiden: Brill. [Illustrated descriptive flora, with keys, synonymy, detailed local range, habitat notes, vernacular names, etc.; complete indices.]

Distribution maps

MENNEMA, J. *et al.* (eds.), 1980. *Atlas of the Netherlands flora*, vol. 1: Extinct and very rare species. 266 pp., 333 maps. The Hague: Junk.

Comprises a distributional atlas of the native and introduced Dutch vascular flora, featuring gridded maps with associated descriptive text (in Dutch, with summary in English) covering the pattern of records, current status, and distribution elsewhere; cross references to *Flora Europaea* are given as well as probable reasons for decline (if appropriate). Volume 1 includes an extensive introductory section, with list of more general literature, while at the end are a comprehensive source bibliography and general index. Prepared at the Netherlands and European Flora section of the Rijksherbarium, Leiden; three volumes in all are planned (see *Blumea*, 25: 118 (1979)). [The decline of much of the Dutch vascular flora in recent decades is starkly illustrated in this fine work, a model of its kind.]

Region

66

The British Isles

This is construed to encompass the large group of islands off the coast of Western Europe which includes Great Britain and Ireland, the Isle of Man, the

Hebrides, Orkney and Shetland, and all associated smaller islands, with the addition of isolated Rockall in the North Atlantic. The Channel Islands have been included with France (651).

Floristic work in the British Isles has a history of nearly 450 years, beginning with William Turner in 1538–48,[20] and the vascular plants are now by world standards supremely well-documented, rivalled perhaps only by Central Europe (region 64), the Low Countries (region 65 in part), and Scandinavia and Finland (region 67). There is a vast number of general and local floras and related works, and the recording of new information is highly organized through such units as the Biological Records Centre and the Botanical Society of the British Isles. Only a small selection from the literature can be recorded here.

The leading modern descriptive manual-flora is *Flora of the British Isles* by Clapham, Tutin, and Warburg (1952; revised 1962). Associated with it is *Excursion flora of the British Isles* (1959; revised 1967). However, there are always those who do otherwise; some still rely on the nineteenth-century classic manuals of Bentham and of Hooker while others have gone out with picture-books, of which the best-known modern example is Keble Martin's *Concise British Flora in Colour* (1965). With respect to collections of illustrations, of which many exist beginning with *English Botany* in the late eighteenth century, the best of the contemporary sets is Ross-Craig's *Drawings of British Plants* (1948–74); another series, by S. J. Roles, accompanies *Flora of the British Isles*. Nevertheless, there is less of a choice among 'scientific' manuals than in France or Germany, and entirely lacking is a modern 'critical' flora (P. D. Sell, personal communication). Partly because of the marked overseas orientation of much of the botanical profession during the age of elaborate pure phytography, but also because Sowerby and Smith's illustrated *English Botany* (1790–1814 and later supplement; 2nd edn. by J. T. B. Syme, 1863–86, with supplement 1891–2) was regarded as *the* definitive descriptive 'regional' flora (upon which the smaller manuals were largely based), there is no large-scale work on the British Isles comparable with Rouy and Foucaud's *Flore de France* or, for that matter, Hegi's *Illustrierte Flora*. J. D. Hooker had such a work in mind, but the only contemporary effort of its kind was the *Cambridge British Flora* by Moss (1914–20), of which but a small part was ever completed.

Apart from the many county floras, not accounted for here, separate floras are available for Wales (*Welsh flowering plants*, 1934; revised 1957) and Ireland (*An Irish flora*, 1943; latest revision, 1977) but not for Scotland.

Bibliographies. General and divisional bibliographies as for Division 6. Very thorough coverage through 1959 is in particular provided by Blake, 1961.

Regional bibliographies

KERRICH, G. J. *et al.*, 1978. *Key works to the fauna and flora of the British Isles and northwestern Europe*. 4th edn. xii, 180 pp. London: Academic Press (for the Systematics Association). (First pub. 1953 as *Bibliography of key works for the identification of the British fauna and flora*.)

SIMPSON, N. D., 1960. *A bibliographical index of the British flora*. xix, 429 pp. Bournemouth: The author. [This most exhaustive of botanical bibliographies has 35 000 entries (Besterman in *Biological sciences: a bibliography of bibliographies*, 1971, Totowa, N.J.).]

Indices. General and divisional indices as for Division 6.

Regional indices

BOTANICAL SOCIETY OF THE BRITISH ISLES, 1971– . *BSBI Abstracts*, 1– , comp. D. H. Kent. London. [Classified taxonomic, biosystematic and floristic abstracts relating to British and Irish vascular plants. Issued more or less annually; part 8 pub. 1978.]

660

Region in general

The titles described below represent only a proportion of all those available. Special mention should be made, however, of the following: BENTHAM, G., 1858. *Handbook of the British flora*, subsequently revised by J. D. Hooker and finally in 1924 by A. B. Rendle; and HOOKER, J. D., 1870. *The student's flora of the British Islands*, last revised in 1884, both still sworn to by many for their method and conciseness; BABINGTON, C. C., 1843. *Manual of British botany*, last revised by A. J. Wilmott, 1922, for infraspecific taxa; and KEBLE MARTIN, W., 1965. *The concise British flora in colour*, perhaps the outstanding recent popular work and enjoying wide current use.

– Vascular flora: Great Britain, 2350–2600 species (1700–1850 native); Ireland, 1350–1450 (1000–1150 native).

BUTCHER, R. W., 1961. *A new illustrated British flora*. 2 vols. viii, 1016; viii, 1080 pp. London: Leonard Hill [Books].

Atlas-flora of vascular plants, with each species occupying one page and comprising a full-page figure with modest descriptive text; the latter includes scientific and vernacular names, diagnostic features, some notes on the habit, habitat, occurrence, and phenology of the plant concerned, and origin of the material used for the drawing (most are from life). Synonymy appears only where the name used differs from that in *Flora of the British Isles* (see below). Short generic descriptions are also included. The introductory portion includes a section on 'descriptive botany' akin to that incorporated by Bentham in his colonial floras as well as the plan of the work (pp. 29–30) and an extensive artificial key, while at the end of vol. 2 is a complete general index. The arrangement of the work follows Dandy's *List* (see below). Based upon the author's *Further illustrations of the British flora* and its predecessors which had been prepared to accompany later editions of Bentham's *Handbook of the British flora*.

CLAPHAM, A. R., TUTIN, T. G., and WARBURG, E. F., 1962. *Flora of the British Isles*. 2nd edn. xlviii, 1269 pp., 87 text-figs. Cambridge: Cambridge University Press. (First edn., 1952.) Accompanied by ROLES, S. J., 1957–64. *Flora of the British Isles: Illustrations*. 4 vols. 1910 figs. Cambridge.

Briefly descriptive flora of native, naturalized, and commonly cultivated vascular plants, with keys to all taxa; limited synonymy; vernacular names; generalized indication of local and extralimital range; symbolic indication of karyotypes, phenology, life-form, and number of vice-counties in which present; taxonomic commentary and notes on habitat and special features; figures of critical characteristics; glossary and index to all botanical and vernacular names. The introductory section includes a synopsis of families.

CLAPHAM, A. R., TUTIN, T. G., and WARBURG, E. F., 1981. *Excursion flora of the British Isles*. 3rd edn. xxxiii, 499 pp., 10 text-figs. Cambridge: Cambridge University Press. (First edn. 1959.)

Concise manual-key to vascular plants, with very limited synonymy, vernacular names, notes on habitat, and local range (no extralimital distribution), and symbolic indication of phenology and life-form; glossary and index to family, generic, and vernacular names.

DANDY, J. E., 1958. *List of British vascular plants*. xvi, 176 pp. London: British Museum (Natural History).

Systematic list of native and naturalized vascular plants, including infraspecific taxa, with synonymy and miscellaneous notes; index. Alien, extinct, and unconfirmed species as well as those known only from the Channel Islands are especially indicated.

ROSS-CRAIG, S., 1948–74. *Drawings of British plants*. 31 parts, index. Plates (index, 39 pp.). London: Bell.

Comprises octavo-sized, systematically arranged (on a modified Bentham and Hooker system) plates of all native and naturalized species of British vascular plants (except Cyperaceae and Gramineae, available in semi-popular form elsewhere). This series is considered to be among the best of its kind in the present century. For Gramineae, see HUBBARD, C. E., 1954. *Grasses*. xii, 428 pp. Harmondsworth: Penguin.

Distribution maps

PERRING, F. H. and WALTERS, S. M. (eds.), 1962. *Atlas of the British flora*. xxiv, 432 pp., numerous dot maps. London: Nelson. (Reprinted 1976, East Ardsley, Yorkshire: EP Publishing.)

PERRING, F. H. (ed.) assisted by SELL, P. D., 1968. *Critical supplement to the* Atlas of the British flora. vii, 159 pp., numerous dot maps. London. (Reprinted 1978, East Ardsley, Yorkshire.)

These two works together comprise a distribution atlas of native and naturalized vascular plant species of the British Isles, with the 'critical groups' mapped in the Supplement. Mapping is by presence or absence in a square, with distinction according to the era of collection (sometimes this is not made, however) and occasional other qualifications. Four maps appear on a page. The base map is founded on the British National Grid, subdivided into 10×10 kilometer squares. The introductory section provides an explanation of the work (which was the first of its kind to be produced, at least in part, with the aid of mechanized means of data processing and plotting), while at the end is an index of names. The Critical Supplement treats such genera as *Sorbus, Alchemilla, Taraxacum* and *Hieracium* in more detail, and is separately indexed. [The Atlas, with its supplement, is now regarded as one of the classics in its genre, and has inspired similar efforts elsewhere.]

662

Wales

HYDE, H. A. and WADE, A. E., 1957. *Welsh flowering plants*. 2nd edn., revised by A. E. Wade. xi, 209 pp., 4 pls., 7 maps. Cardiff: National Museum of Wales. (First edn. 1934.)

HYDE, H. A. and WADE, A. E., 1962. *Welsh ferns: a descriptive handbook*. 4th edn. 122 pp., illus. Cardiff.

The treatment of flowering plants comprises a systematic enumeration, with vernacular names (English and Welsh), limited synonymy, brief indication of distribution-class, life-form, etc., and more extensive summary of local range (with some citations of *exsiccatae*), with list of references, index to collectors, and general index to family, generic and vernacular names. An introductory section gives accounts of plant life-forms, phytogeography, the composition of the Welsh flora, and the (old) counties of Wales. The companion handbook on ferns is a descriptive treatment, with keys and illustrations for field use, but otherwise broadly similar.

663

Scotland

No separate general floras or enumerations of recent date are available. The latest work is evidently HOOKER, W. J., 1821. *Flora scotica, or a description of Scottish plants arranged both according to the artificial and natural methods*. 2 vols. London: Hurst-Robinson.

665

Ireland

SCANNELL, M. J. P. and SYNOTT, D. M., 1972. *Census catalogue of the flora of Ireland*. 127 pp., map. Dublin: Stationery Office.

Systematic list of native and established alien vascular plants, with relevant synonymy, equivalent English and Gaelic vernacular names as known, and indication of internal range (by vice-counties); index to generic, family, and vernacular names. The introductory part gives brief accounts of Irish geography, floristic districts, plant habitats, etc., as well as a 'bibliography of vice-county distribution'. [Not seen by this author; prepared from notes supplied by A. O. Chater.]

WEBB, D. A., 1977. *An Irish flora*. 6th edn. xlii, 277 pp., 160 text-figs. Dundalk: W. Tempest (Dundalgan Press). (First edn. 1943.)

Concise manual-flora of native, naturalized, and commonly cultivated vascular plants, with keys to all taxa, limited synonymy, Irish (Gaelic) and English vernacular names, generalized indication of local range, and notes on habitat, phenology, etc.; index to all vernacular and botanical names. The introductory section includes a glossary, with figures separate from the text. In this (sixth) edition keys have been rewritten and units metricated.

669

Rockall Island

Only algae and lichens have been recorded for this desolate rock in the North Atlantic, a British possession about 250 km northwest of Ireland.

Region

67

Northwestern Europe (Scandinavia and Finland)

This region consists of Iceland, the Faeroes, Denmark, Norway, Sweden, and Finland. However, floras and other works dealing exclusively with Lapland or the Arctic islands to the north are not described here but rather in Regions 05 and 06. Some works given here also cover East Fennoscandia, which corresponds to the Soviet territories of Karelia and the Kola Peninsula.

Like other parts of central and northern Europe, the region has enjoyed a long history of floristic work, dominated in the eighteenth century by the work of

Linnaeus and his students, and today is one of the best-documented areas in the world, a status favored also by the comparatively small vascular flora. A wide range of regional, national, and more local floras and other works is available.

The leading descriptive regional flora is Hylander's *Nordisk kärlväxtflora* (1953–66), alas, not completed; associated with this critical work is the same author's regional checklist, *Förteckning över Nordens växter*, I: *Kärlväxter* (1955; supplement, 1959). It is to be hoped that at some future time Hylander's flora can be continued to completion. At state and territory level, manual-keys are available for all five countries and the Faeroes in more or less recent productions, with one covering two countries (*Norsk og Svensk Flora* by J. Lid). A special feature of many, if not all, of these works is the presentation of family keys employing the 24 Linnean classes as part of their organization. Among works of more local scope, the only one accounted for here is *Skånes flora* (Weimarck, 1963) which covers Denmark as well as the floristically distinctive southernmost province of Sweden. Finland has a major critical flora in Latin (*Conspectus florae fennicae*, 1888–1926, a basis for all later work), two leading works in Finnish, and an older manual-key in Swedish, reflecting the links of botany with contemporary social and educational developments.

Bibliographies. General and divisional bibliographies as for Division 6. Very thorough coverage through 1959 is in particular provided by Blake 1961.

Indices. General and divisional indices as for Division 6.

670

Region in general

For *Atlas över växternas utbredning i Norden* (Atlas of the distribution of vascular plants in NW Europe) by E. Hultén, see below.

HYLANDER, N., 1955. *Förteckning över Nordens växter*, I: *Kärlväxter*. [List of the plants of NW Europe, I: Vascular plants.] x, 175 pp. Lund: Gleerup.

Systematic list of the vascular plants of northwestern Europe (including also east Fennoscandia, Iceland, and the Faeroes), with symbolic indication of distribution and/or origin; table of protected plants; indices to genera and to synonymous names. The

explanation of signs and abbreviations appears both in Swedish and English. For additions, corrections, and deletions, see *idem*, 1959. Tillägg och rattelser.... *Bot. Not.* **112**: 90–100.

HYLANDER, N., 1953–66. *Nordisk kärlväxtflora*. Vols. 1–2. Illus., maps. Stockholm: Almqvist & Wiksell.

Comprehensive critical descriptive flora of the vascular plants of northwestern Europe, with geographical limits as for the author's *Förteckning över Nordens växter*, incorporating keys to genera, species, and infra-specific taxa, concise synonymy (with dates), vernacular names, detailed exposition of internal range together with summaries of extralimital distribution, taxonomic commentary, notes on karyotypes, habitat, phenology, life-form, etc., and figures of diagnostic and critical features; short lists of references; no indices (except to genera in inside covers). This important work is one of the five 'Basic Floras' constituting a major source for *Flora Europaea*; however, it only covers families from the Pteridophyta up through the Polygonaceae (Englerian system), and with the author's death in 1970 its future progress is in doubt (R. Santesson, personal communication).

Distribution maps

HULTÉN, E., 1971. *Atlas över växternas utbredning i Norden*. [Atlas of the distribution of vascular plants in Northwestern Europe]. 2nd edn. 56, 531 pp., 1984 distribution maps on 496 pls. Stockholm: Generalstabens Litografiska Anstalt (distributed by Almqvist & Wiksell). (First edn. 1950.)

Distribution atlas of vascular plants based upon Scandinavia and Finland, with individual colored maps for each species (four to a page); includes limited synonymy, Swedish vernacular names, and phenological data (for some species); indices to vernacular and botanical names. The general part preceding the plates includes sections on general questions of plant distribution, historical phytogeography, physical features of the region, geology, climate, etc. (with special extra maps). Apart from Scandinavia and Finland, maps also include East Fennoscandia and the Soviet Baltic republics.

671

The Faeroes

See also **672** (OSTENFIELD and GRÖNTVED).
– Vascular flora: 450–500 species (400–500 native).
HANSEN, K., 1966. *Vascular plants of the Faeroes: horizontal and vertical distribution*. 141 pp., maps (Dansk Bot. Ark., vol. 23, part 3). Copenhagen.

Atlas-account of vascular plant records, with mapped treatments of 363 species (based on field work in 1960–1 and earlier records) including indication of vertical distribution (three life zones); records of adventives and index. An introductory section describes physical features, geological and vegetation history, climate, and biotic influences, along with botanical exploration.

RASMUSSEN, R., 1952. *Forøya flora*. 2nd edn. xxviii, 232 pp., 108 text-figs. Tórshavn: Jacobsens Bókhandils. (First cdn. 1936.)

School- and excursion-manual (in Faeroese) of vascular plants, with keys to all taxa; essential synonymy; concise indication of local range; vernacular names; notes on habitat, life-form, etc.; figures of diagnostic attributes; illustrated glossary/organography and index to vernacular and botanical names. The introductory section is pedagogic.

672

Iceland

This land, on the fringe of the Arctic, was comparatively poorly known botanically until the present century.
– Vascular flora: 500–600 species (450–550 native).
GRÖNTVED, J., 1942. The Pteridophyta and Spermatophyta of Iceland. In *The botany of Iceland* (ed. J. GRÖNTVED, O. PAULSEN, and T. SØRENSEN), vol. 4, part 1. 427 pp., 177 text-figs. (incl. maps). (Copenhagen), London: Munksgaard.

Comprehensive, critical enumeration of vascular plants (including full accounts of *Taraxacum* and *Hieracium*); extensive synonymy, with references and citations; vernacular names; detailed accounts of local range, with distribution maps, and general summaries of extralimital distribution; taxonomic commentary

and notes on ecology, life-form, phenology, etc.; bibliography and indices to vernacular and botanical names. The introductory section includes accounts of physical features, climate, soils, vegetation, phytogeography, and botanical exploration and research.

LÖVE, Á., 1977. *Íslenzk ferðaflóra*. 2nd edn. 429 pp., 576 text-figs., 25 pls. (some in color), map. Reykjavik: Almenna Bókafélagið. (First edn., 1970.)

Illustrated manual of vascular plants (in Icelandic) with keys to all taxa; very limited synonymy; vernacular names; symbolic indication of local range; notes on karyotypes, ecology, life-form, phenology, etc.; indices to vernacular and botanical names. The introductory section includes a glossary, notes on the use of the work, and a list of protected plants. Replaces the same author's *Íslenzkar jurtir* (Kaupmannahöfn, 1945).

OSTENFIELD, C. H. and GRÖNTVED, J., 1934. *The flora of Iceland and the Faeroes*. cciv, 195 pp., 2 maps. Copenhagen: Levin & Munksgaard.

Manual-key to vascular plants, with limited synonymy, generalized indication of local range, and notes on habitat, life-form, phenology, etc.; glossary and indices to family and generic names as well as to Icelandic and Faeroese vernacular names. The introductory section includes a short bibliography. This work, which remains the standard English-language manual on the Icelandic flora, was originally prepared for visitors.

STEFÁNSSON, S., 1948. *Flora Íslands*. 3rd edn., revised by S. Steindórsson. lviii, 407 pp., 253 text-figs., map. Akureyri: Íslenzka Náttúrufraeðífélag (distributed by Norðri, Reykjavik). (First edn. 1901.)

Illustrated descriptive manual (in Icelandic) of vascular plants, with relatively limited treatment of 'microspecies'; limited synonymy; vernacular names; generalized indication of local range; notes on habitat, phenology, life-form, etc.; illustrated glossary and map of districts; indices to all botanical and vernacular names.

674

Denmark

See also **676** under Partial works (WEIMARCK). The part of southern Denmark under German rule from 1866 to 1918 is also covered by all works listed under **641**.

– Vascular flora: 1750–1900 species (1400–1600 native).

HAGERUP, O. and PETERSSON, V., 1956. *Botanisk atlas: Danmarks dækfrøede planter*. 550 pp., [515] pls. Copenhagen: Munksgaard. English edn.: HAGERUP, O. and PETERSSON, V., 1959. *A botanical atlas, I: Angiosperms*. Trans. H. Gilbert-Carter. xvi, 550 pp., [515] pls. Copenhagen: Munksgaard.

Atlas of illustrations of the flowering plants of Denmark, often with details (in some cases, the details only); no text except for some family or specific descriptions and botanical and Danish (or English) vernacular names. Essentially all native and some adventive species are included, but microspecies are not treated in detail. A supplementary volume covers the gymnosperms, ferns, fern-allies, and the genera of mosses.

RAUNKIAER, C. 1950. *Dansk ekskursions-flora*. 7th edn., revised by K. Wiinstedt. xxxi, 380 pp. Copenhagen: Gyldendal (Nordisk Forlag). (First edn. 1890.)

Manual-key to vascular plants, with limited synonymy, concise indication of local range, vernacular names, and symbolic notes on ecology, life-form, phenology, etc.; glossary and index to vernacular, generic, and family names. Included also is an explanation of the author's well-known life-form system.

ROSTRUP, E. and JØRGENSEN, C. A., 1973. *Den danske flora*. 20th edn., revised by A. Hansen. 664 pp., 141 text-figs., map (in pocket). Copenhagen: Gyldendal. (First edn. 1860, as *Vijledning i den danske flora*.)

Illustrated manual-flora of vascular plants, with keys to all taxa, limited synonymy, vernacular names, concise indication of local range, and notes on habitat, phenology, etc.; index to vernacular, generic, and family names. The introductory section includes an illustrated glossary/organography and an explanation of 'botanical districts'. Vernacular names are used in all keys and taxon headings together with scientific names.

675

Norway

For Lapland, see also **061**. For Jan Mayen and Svalbard (Spitsbergen and Bear Island), see also **051** to **053**.

– Vascular flora: 1700–1850 species (1400–1600 native).

LID, J., 1974. *Norsk og svensk flora*. 2nd edn. 808 pp., 439 text-figs. (incl. 2 maps). Oslo: Norske Samlaget. (Originally published 1944 as *Norsk flora*; 1st edn. of *Norsk og svensk flora*, 1963.)

Illustrated manual of the vascular plants of Norway and Sweden, with keys to all taxa, limited synonymy, general indication of local range for the two countries, Norwegian and Swedish vernacular names, and notes on karyotypes, ecology, phenology, etc.; glossary and etymological lexicon; index to vernacular, generic, and family names.

NORDHAGEN, R., 1940. *Norsk flora*. xxiii, 766 pp. Oslo: Aschehoug (W. Nygaard). Accompanied by *idem*, 1970. *Norsk flora: Illustrasjonsbind*, I. xxxvi pp., pls. 1–638 (incl. 772 text-figs.). Oslo.

Descriptive manual of vascular plants, with keys to all taxa, limited synonymy, brief generalized indication of local range, vernacular names, and notes on ecology, phenology, etc.; indices to vernacular and to botanical names. The introductory section includes a glossary as well as a general key to families. The accompanying atlas, with figures by M. Bodther, comprises three parts, of which parts 1–2 originally appeared in 1944; it covers lower vascular plants, gymnosperms, monocotyledons, and Salicaceae through Fumariaceae.

676

Sweden

See also **675** (LID). For Lapland, see also **061**.

– Vascular flora: 1900–2000 species (1600–1800 native).

KROK, T. O. B. N. and ALMQVIST, S., 1960. *Svensk flora*, I: *Fanerogamer och ormbunkväxter*. 25th edn., revised by E. Almqvist. iv, 403 pp., 195 text-figs., 96 color pls. Stockholm: Svensk Bokförlaget. (First edn. 1883.)

Illustrated manual of vascular plants, with keys to all taxa (including a Linnaean general key to families), concise indication of local range, vernacular names, and symbolic notes on life-form, phenology, etc.; lexicon of Latin terms; indices to vernacular, generic, and family names as well as to synonymous names. The introductory section includes a glossary, a keyed Linnaean synopsis, and a key to plant groups with small or difficult flowers based on vegetative attributes.

LINDMAN, C. A. M. 1926. *Svensk fanerogamflora.* 2nd edn., x, 644 pp., 329 text-figs. Stockholm: Norstedt & Söners. (First edn. 1918.)

Manual of vascular plants, with keys to all taxa (including a general key to families and genera based on the Linnaean method), limited synonymy, vernacular names, concise indication of local and extralimital range, and notes on ecology, life-form, phenology, etc.; indices to vernacular, generic, and family names. The introductory section includes a glossary and a key to Linnaean classes.

Partial works
This outstanding manual has of late been produced for the southernmost province of Scania; it also covers all Danish plants.

WEIMARCK, H., 1963. *Skånes flora.* xxiv, 720 pp., 1 color pl., maps (end papers). Lund: Corona.

Descriptive manual of vascular plants, with keys, limited synonymy, vernacular names, fairly detailed indication of local range, summary of overall range in southern Sweden and Denmark, and notes on karyotypes, ecology, phenology, etc.; complete indices. A glossary is also included.

678

Finland

Some works described below also cover East Fennoscandia, which largely corresponds to the Soviet regions of Karelia (**686**) and the Kola Peninsula (**687**).
 – Vascular flora: 1450–1550 species (1250–1450 native).

ALCENIUS, O., 1953. *Finlands kärlväxter: de vilt växande och allmännast odlade.* 12th edn., revised Å. Nordström. 428 pp. Helsinki: Söderström. (First edn. 1863.)

The standard Swedish-language manual of the vascular plants of Finland, with keys to all taxa

(including a Linnaean general key); limited synonymy; vernacular names; symbolic indication of local range; notes on habitat, phenology, life-form, etc.; index to generic names. Apparently last revised in 1930 and merely reprinted since then.

HIITONEN, I., 1933. *Suomen kasvio.* 771 pp., 437 text-figs., map. Helsinki: Kustannusosakeyhtiö Otava.

Illustrated descriptive manual of vascular plants, with keys to all taxa (synopsis and general key organized on the Linnean method), limited synonymy, vernacular names, concise indication of local range, and notes on ecology, life-form, etc.; indices to vernacular and to generic and family names. The introductory section includes a glossary together with the general key. Partly based upon, and covering the same area as, Hjelt's *Conspectus florae fennicae* (see below).

HIITONEN, I., 1934. *Suomen putkilokasvit.* 158 pp., map, Helsinki: Kustannusosakeyhtiö Otava.

Systematic list of the vascular plants of Finland and East Fennoscandia, paralleling the author's *Suomen kasvio*; includes limited synonymy, symbolic indication of local range, gazetteer, and index to generic names; a map, with abbreviations of botanical districts, is also incorporated.

HIITONEN, I. and POIJÄRVI, A., 1958. *Koulu- ja retkeilykasvio.* 9th edn. 472 pp., 311 text-figs., 7 pls., map. Helsinki: Kustannusosakeyhtiö Otava. (First edn. 1932.)

Copiously illustrated students' manual-flora of wild and commonly cultivated vascular plants, with keys to all taxa, essential synonymy, vernacular names, symbolic indication of local (by botanical districts) and extralimital range, life-form, and phenology, and concise notes on habitat and special features; occasional taxonomic commentary; map of botanical districts and indices to generic, family, and vernacular names. The general key to families (and some genera) is based on the Linnaean system (as in Swedish floras), and is preceded by a 10-page glossary. Based on the senior author's *Suomen kasvio*, but with updated distribution data over the area of pre-1939 Finland (as covered in this work); Russian Karelia and Lapland are omitted. [Not seen by this reviewer; citation and description based mainly on information supplied by A. O. Chater, Leicester (now London).]

HJELT, H., 1888–1926. *Conspectus florae fennicae.* 7 parts in 2 vols. Illus. (Acta Soc. Fauna Fl. Fenn., vols. 5, part 1; 21, part 1; 30, part 1; 35, part 1; 41, part 1; 51, part 1; 54). Helsinki.

Detailed critical enumeration of vascular plants of Finland and East Fennoscandia, with synonymy, detailed indication of local range including localities, taxonomic commentary, and other notes; bibliography and index in each part. Part 1 (1888–95, in 3 fascicles) treats vascular plants other than dicotyledons; Parts 2–7 (1902–26) cover dicotyledons. Geographical information in Latin; taxonomic and other notes in Swedish. Represents a series of family revisions for Finland and East Fennoscandia, and is fundamental for all later floristic work in the area. Symbols are explained in *idem*, 1888. Notae 'Conspectus florae fennicae'. 20, [4] pp. Helsinki.

Region

68/69

USSR-in-Europe (and all-Union works)

This, the largest European region, comprises the European part of the Soviet Union, from its western borders to the limits of Asia (the latter approximately as defined for *Flora Europaea*). However, the Arctic tundra zone and the islands of the Arctic Ocean are excluded; for these, see Regions 05 and 06.

The fascinating, many-sided history of botanical exploration and floristic recording and analysis, which includes some distinctive features, has been recorded in various ways by many authors. Early centers included St Petersburg (Leningrad) and (later) Moscow, Kiev, and Dorpat (Jurjew, Tartu), all dating from the eighteenth century. It was from these centers, then largely staffed by foreign botanists, that botanical exploration and the compilation of floras began. Effective local coverage was, in European Russia, first attained in the Baltic territories and the St Petersburg region (Ingria), followed by Middle Russia and the Ukraine. The first general flora, *Flora rossica* by Pallas, appeared in 1784–8 but unfortunately was never completed. This was followed by the Dorpat professor Ledebour's highly regarded and complete *Flora rossica* (1842–53).

St Petersburg as a botanical center, however, only began to rise to its present importance from the 1820s. In 1823, two years before the Decembrists' revolt, the

Botanical Garden was reorganized as a scientific institution, and 12 years later the Botanical Museum became a separate unit within the Academy of Sciences. Together these institutions were to contribute mightily to the systematic exploration of Russian territories in Europe and Asia as well as other countries, notably the 'Chinese Border' territories of Central Asia, although their greatest impact before the October Revolution was in the preceding three or four decades. It was then that a considerable number of floras and related works were added to the array of classical mid-nineteenth-century works. From the 1890s, both Garden and Museum organized systematic programmes for floristic exploration and flora preparation, most likely under the stimulus of S. I. Korshinsky who held senior posts at both institutions from 1892 to 1900, and, aided by increases in staff and the impact of government survey programmes under the Transmigration Bureau (which yielded vast collections), large parts of these programmes began to be realized. Several, partly overlapping, flora projects were begun; for European Russia this resulted in the preparation at the Garden of *Flora Evropejskoj Rossii* in three volumes (Fedtschenko and Flerow, 1908–10). Other notable contributions of this period included *Flora srednej Rossi* by Majevski (First edn., 1892; 9th edn., 1964), the standard manual-key for the Russian heartland; *Flora srednej i južnoj Rossii, Kryma, i Severnogo Kavkaza* by Schmal'hausen (1895–7), for the Ukraine, Lower Don region, the Crimea, and Ciscaucasia; *Tentamen florae Rossiae orientalis* by Korshinsky himself (1898); and an illustrated flora for the Moscow region by Syreishchikov (1906–14). In 1913 there appeared the first separate manual-key for Northern Russia, *Opredelitel' rastenij lesnoj polosy Severo-Vostoka Evropejskoj casti Rossii*. Routine floras appeared in the Baltic provinces and Lithuania.

Thus, by 1917 the modern regional flora tradition had become well-established, perhaps in part because the prospect of a new comprehensive flora as a successor to Ledebour's work seemed too daunting. However, detailed *local* coverage was comparatively limited outside of Finland, Ingria, the Baltic provinces, Poland, and the Ukraine.

For a decade or so after the Revolution, except for the disruptive years between 1917 and 1922, the pattern of floristic work and publication remained much as it had been in earlier decades. Some pre-war flora projects were continued, and under the new Soviet regime strong support for biological surveys resumed

Table 1. *Lexicon of Russian words used in the titles of Soviet floras**

ASSR	Autonomous Soviet Socialist Republic
čast'	portion
Dal'nyj Vostok (Dal'nego Vostoka)	the [Soviet] Far East (of the Far East)
derevo (derevja)	tree (trees)
dikorastuščij	wild, spontaneous
flora	flora *or* a Flora
izdanie	edition, publication
južnyj	southern
Kavkaz	the Caucasus
konspekt	conspectus or enumeration (without extensive descriptions; may or may not have keys)
krai	an administrative 'territory' constitutionally under direct administration by a union republic
kustarnik	shrub
lesnoj	forested
oblast'	province (or region)
okrestnost'	environs, vicinity
opredelitel'	'manual-key': a work with descriptive keys and concise indications of habit, range, phenology, etc. Often (but in my opinion clumsily) translated as 'the keys', 'determinator', or 'determination'
osennij	autumnal
ostrov	island
ozero	lake
polosa	zone, belt
poluostrov	peninsula
Predbajkal'ja	the Cis-Baikal region, west of the lake and including Bratsk and Irkutsk
Prijenisej	the Jenisei (Yenisei) River basin
RSFSR	Russian Soviet Federative Socialist Republic. The largest of the fifteen union republics, with numerous internal subdivisions
ranne-letnij	early summer
rastenie (rastenij)	plant (of plants)
SSR	Soviet Socialist Republic. One of the fourteen 'national' republics
SSSR	Union of Soviet Socialist Republics
severnyj	southern

(*cont.*)

spisok	list (or 'elenchus'): a *concise* annotated enumeration, but partly synonymous with *konspekt*
srednij	middle, central; see also *tsentral'nyj*
tsentral'nyj	central
vesennij	spring, vernal
vostočnyj	eastern
vysokogornyj	alpine, high-mountain
vysšij (vysšikh)	higher (of higher ...s). Now used as equivalent to 'vascular' when referring to plants
Zabajkal'ja	the Trans-Baikal region, east of the lake; includes most of Buryat-Mongolia as well as Chita Oblast'
zapadnyj	western

* Nominative singular forms are used, with other forms given in parentheses where significantly different.

after the civil war. A new regional flora, *Flora Rossiae austro-orientalis*, written by several authors under the editorship first of B. A. Fedtschenko and later B. K. Shishkin, was begun in 1927 and completed in 1936; it is both the last of the 'primary' regional floras of European Russia and a forerunner of the collective method of preparation which characterized *Flora SSSR*; both Fedtschenko and Shishkin were to be heavily involved in that project.

The conception of *Flora SSSR* was influenced by the thorough reorganization of science conducted under Party direction beginning from the late 1920s, and it featured as a primary project of the new Botanical Institute (formed from a merger of the Botanic Garden and Museum in 1931) in the first Five Year Plan. Initiated in 1929 and formally begun in 1931, the project dominated Soviet floristic botany for more than three decades. Its origin, progress, and completion have been described by several authors. Contributing greatly to the success of the work was the authority wielded by Komarov, further enhanced between 1936 and 1945 in his capacity as president of the Academy of Sciences. Ninety-two collaborators in all were involved in the work.

The concentration of efforts upon *Flora SSSR* did not, however, break the regional flora tradition. In addition, more botanical centers came into being in all parts of the Soviet Union. Many of these have contributed considerably to the large number of regional, republican, and local floras which appeared after World War II, promoted by the success of *Flora SSSR*. In the European part of the Soviet Union there had already been produced *Flora Severnogo Kraja* (Perfil'ev, 1934–6) and the early volumes of *Flora URSR* (Bordzilovsky *et al.*, 1937–65), respectively for the northern region and the Ukraine. Afterwards, regional floras were written for the Kola Peninsula, Karelia, the Leningrad region, the Baltic Republics (two of these not yet completed), Moldavia, White Russia, the Crimea, and parts of northern, middle and eastern Russia, including a new manual for the Moscow region. The Soviet Arctic from 1960 was also being covered by a critical regional flora under the editorship of Tolmatchev, who was also responsible for the production of the most recent regional work, again on the northern region, *Flora Severo-Vostoka Evropejskoj časti SSSR* (1974–7). Coverage by regional works, however, is not uniform and there are some areas for which no recent works in this genre are available. Moreover, they vary considerably in scope from manual-keys to full-scale documentary works like those for many other European countries.

For the European part of the Soviet Union as a whole, however, there was no work to replace the flora of Fedtschenko and Flerow, although since 1907 there had been a one-volume manual-key by Taliev, *Opredelitel' vysšikh Evropejskoj časti Rossii*, a Moscow-school work last revised in 1957 by Stankov as *Opredelitel' vysšikh rastenij Evropejskoj časti SSSR* and long a standard students' work. This, and a desire to introduce new taxonomic principles and methods of data presentation, led An. A. Fëdorov and others at Leningrad to begin a new work, *Flora evropejskoj časti SSSR*. Publication began in 1974 and at this writing (1980) four volumes (of eleven planned) have appeared. The standards set by *Flora Europaea* have been used as guidelines. In addition, a volume of additions and corrections to *Flora SSSR* has been produced by Czerepanov (1973) and a revision of the grasses by Tzelev was published as *Zlaki SSSR* in 1976. A total of some 21 000 species has now been recorded for the Soviet Union according to Czerepanov.

The woody plants are covered in several works of varying scope, the most comprehensive being Sokolov's *Derevja i kustarniki SSSR* (1949–62; supplement, 1965). To these has now been added a series of maps, *Arealy derev'ev i kustarnikov SSSR*, prepared under Sokolov's direction and issued commencing in 1977.

– Vascular flora (Soviet Union as a whole): 17 520 species (*Flora SSSR*); 21 000 species (Czerepanov, 1973). For figures for the European part of the Soviet Union alone, see below under **680**.

Progress

The corpus of Russian literature on the history of botany in the USSR is considerable, and only a small selection most useful in the floristic context can be given here. Two general source works are: Asmous, V. C., 1947. *Fontes historiae botanicae rossicae* (Chron. Bot., vol. 11, no. 2.) Waltham, Mass.; and Lipschitz, S. Ju., 1947. Sistematika, floristika i geografija rastenij. In *Očerki po istorii russkoj botaniki* (ed. Breslavets, L. P. *et al.*), pp. 9–114. Moscow: Moscow Society of Naturalists Press. A number of histories for particular regions, institutions, societies or other special categories have also been published. For English-speaking readers, the best introductory account, though centering mainly on the Komarov Botanical Institute and its antecedents in Leningrad, is Shetler, S. G., 1967. *The Komarov Botanical Institute: 250 years of Russian research*. Washington, who gives a bibliography including many Russian references. Recent developments and trends are reviewed in that work and in Bobrov, E. G., 1963. Sostojanie i perspktivy izučenia otečestvennoj flory. *Bot. Žurn. SSSR*, **48**: 1729–40; Takhtajan, A. L., Tolmatchev, A. I., and Fëdorov, An. A., 1965. Izučenie flory SSSR, dostiženija i perspektivy. *Ibid.*, **50**: 1365–73 (see also Tolmatchev, A. I., 1966. *J. Gen. Biol.* (USSR) **27**: 411–22); and Fëdorov, An. A., 1971. Floristics in the USSR. *BioScience* (USA), **21**: 514–21. Under Regional bibliographies (see below) is listed the unfinished historico–bibliographic review by Kirpicznikov of post-1917 floras and manuals.

Flora SSSR. The secondary references on this work are numerous, and only a selection is essayed here. In Russian: Bobrov, E. G., 1965. 'Flora SSSR', rabota nad nej i značenie etogo izdanija. *Bot. Žurn. SSSR*, **50**: 1374–83; Linczevski, I. A., 1966. 'Flora SSSR' (notula bibliographica). *Novost. Sist. Vysš. Rast.* (Leningrad), **3**: 316–30; and Kirpicznikov, M. E., 1967. 'Flora SSSR' – krupnejšee dostiženie sovetskikh sistematikov. *Bot. Žurn. SSSR*, **52**: 1503–30 (with extensive bibliography). In English: Heywood, V. H. and Bobrov, E. G., 1965. Preparation of 'Flora

URSS'. *Nature*, 205: 1046–9 (abr. and trans. from Bobrov 1965); Kirpicznikov, M. E., 1969. The *Flora of the USSR. Taxon*, 18: 685–708 (abr. and trans. from *idem.*, 1967; includes bibliography).[21]

Bibliographies. General and divisional bibliographies as for Division 6. (except Blake, 1961).

Regional bibliographies

Kirpicznikov, M. E., 1968–9. Übersicht über die wichtigsten Floren und Bestimmungsbücher, die in der Sowjet-union während der letzten 50 Jahre erscheinen sind, I–II. *Bot. Žurn. SSSR*, 53: 845–55; 54: 121–36. [Annotated review; to date covering only the European part of the USSR.]

Lebedev, D. V., 1956. *Vvedenie v botaničeskuju literaturu SSSR*. 382 pp. Moscow/Leningrad: AN SSSR Press. [Selected, briefly annotated guide to Russian systematic, floristic, and geobotanical literature, arranged by subject and region; indices. 3000 entries.]

Lipschitz, S. Ju., 1975. *Literaturnye istočniki po flore SSSR*. [Florae URSS fontes]. 232 pp. Leningrad: 'Nauka'. [Bibliography of 2368 items, arranged by author within five major geographical divisions. A short section on 'school-floras' is also included.]

Indices. General and divisional indices as for Division 6.

680

Region in general

The works described here fall into two general categories, respectively for All-Union works and for works on the European part of the Soviet Union alone. Each category includes some key works on the woody flora (including 'tree books').

All-Union (*Vsesojuz*) works

Komarov, V. L., Shishkin, B. K. *et al.* (eds.); Bobrov, E. G. *et al.* (comp.), 1934–64. *Flora SSSR* [Flora URSS]. 30 vols., index. Illus., maps. Moscow/Leningrad: AN SSSR Press (index: 'Nauka'). (Vol. 11 reprinted 1945, Moscow/Leningrad; vols. 1–13 reprinted 1964, Lehre, Germany: Cramer.) English edn.: *idem*, 1960– . *Flora of the USSR. Trans.*

N. Landau and P. Lavoott. Vols. 1–21, 24. Illus., maps. Jerusalem: Israel Program for Scientific Translations.

Comprehensive descriptive flora of the vascular plants of the entire Soviet Union, with keys to all taxa; extensive synonymy, with references and citations; vernacular and binomial Russian names; rather detailed account of distribution within and outside the Union; taxonomic commentary and notes on habitat, life-form, phenology, associates, paleobotany, uses, etc.; references to published illustrations; indices to all botanical and other names in each volume; separate general index to complete work. Volume 1 includes a brief historical account and an exposition of taxonomic principles employed, while the index volume includes maps showing floristic regions (the basis for individual distribution accounts). The English translation, in progress for more than ten years, is now more than half completed. The work represents one of the five 'Basic Floras' fundamental to the preparation of *Flora Europaea*. Among modern floras, *Flora SSSR* represents a *tour de force*: a landmark of the twentieth century in the same way as was von Martius' *Flora brasiliensis* in the nineteenth century. A considerable historico-bibliographic and interpretative literature (see above) has grown up around this work, which contains some 22000 pages of text and 1250 plates.[22] Of the English version, the most recent volume was published in 1977, but it is understood that the translation project has now been discontinued. For a comprehensive index of additions to the *Flora*, see Czerepanov, S. K., 1973. *Svod dopolnenij i izmenenij k 'Flore SSSR'* (tt. I–XXX) [Additamenta et corrigenda ad 'Floram URSS' (vols. I–XXX)]. 667 pp. Leningrad: 'Nauka'. [Alphabetically arranged list of additions and changes to *Flora SSSR* from 1934 to 1971, with synonymy, references, citations, and indication of types. Includes an extensive introduction (in English and Russian) and a bibliography (pp. 20–3).] Volume 2, covering the grasses, has been revised as Tzelev, N. N., 1976. *Zlaki SSSR*. 788 pp., 9 text-figs., 16 pls. Leningrad: 'Nauka'.

Special group – woody plants (including trees)
The major contribution in this class is the six-volume work edited by Sokolov, which includes an extensive range of cultivated as well as spontaneous species. A selection of students' keys and a popular guide are also given here.

Lapin, P. I. (ed.); Borodina, N. A. *et al.*

(comp.), 1966. *Derevja i kustarniki SSSR*. 637 pp., illus. (some in color), maps. Moscow: 'Mysl''.

Semi-popular illustrated manual to native and principal introduced woody plants of the USSR, divided into two major sections for these respective groups. The first section (pp. 210–540) includes descriptions of individual native species as well as notes on their internal and extralimital range, altitudinal zonation, special features, uses, potential, etc; there are also short associated notes on closely related but rarer species. The second section (pp. 543–606) provides a similar treatment for exotic species. The introductory section includes general remarks, a definition of terms (pp. 7–24), a map of geobotanical regions, and a series of polyclaves based on regions and morphological characters (pp. 30–209), while at the end is a list of references (pp. 609–11) and an index to Russian names.

Sokolov, S. Ja., 1949–62. *Derevja i kustarniki SSSR*. 6 vols. Illus., maps. Moscow/Leningrad: AN SSSR Press. Accompanied by Sokolov, S. Ja. and Svjazeva, O. A., 1965. *Geografija drevesnykh rastenij SSSR* (vol. 7 of *Derevja i kustarniki SSSR*). 265 pp., map. Moscow/Leningrad: 'Nauka'.

Comprehensive descriptive account of the native, naturalized, and more or less commonly cultivated trees and shrubs of the USSR, with keys to all taxa; includes extensive synonymy (without references), vernacular and Russian binomial names, fairly detailed indication of internal distribution (with some maps) and summary of extra-Soviet distribution or country or region of origin, and notes on habitat, phenology, biology, dendrological features, cultivation, hybrids, uses, etc.; indices to all botanical and other names in each volume (general index to work in vol. 6). The supplementary volume of 1965 is devoted to statistics, lists, and a summary and general analysis of the woody flora (with extensive bibliography).

Sokolov, S. Ja., Svjazeva, O. A. and Kubli, V. A., 1977– . *Arealy derev'ev i kustarnikov SSSR*. [Areographia arborum fruticumque URSS.] Vol. 1– (each in 2 parts). Maps (in portfolios). Leningrad: 'Nauka'.

The textual parts of this distributional atlas comprise analyses of the distribution of each species of tree, shrub and woody vine mapped, while the distributions themselves are presented on loose maps in portfolios. Each textual part includes a bibliography and Latin name index. As of 1981 two volumes, each with accompanying map portfolios, had been issued,

covering on 189 maps families from Taxaceae to Aristolochiaceae and Polygonaceae to Rosaceae (Englerian system). [Not seen by this reviewer; prepared with the assistance of information supplied by Mrs J. Diment, British Museum (Natural History). The work is a counterpart, albeit of broader scope, to the atlas of United States trees by Little (**100**).]

European part (Evropejskoj časti) of the USSR

This largely corresponds with that used in *Flora Europaea* and for the present *Guide*.

– Vascular flora: 4450–4600 species (4100–4300 native).

Fedorov, An. A. *et al.*, 1974– . *Flora evropejskoj časti SSSR* [Flora partis europaeae URSS]. Vol. 1– . Illus., map, portraits. Leningrad: 'Nauka'.

Partially descriptive flora of vascular plants, with keys to all taxa, diagnoses (of families and genera), full synonymy (with references and principal citations), indication of internal range (by geobotanical districts) and extralimital distribution, localities, karyotypes, taxonomic commentary, and notes on habitat, special features, etc. Diagnoses of subspecies are provided (this signaling a change from the philosophy of *Flora SSSR*, in which the concept of polytypic species was not recognized). Indices to all Latin and Russian names appear at the end of each volume. An introductory section (in volume 1) includes historical and philosophical chapters, a synopsis of orders and families, the plan and scope of the work, a map of floristic districts, a bibliography of key works (pp. 22–7) and general key to families. A description of the project, whose ideology has been strongly influenced by *Flora Europaea*, has been given by its senior author.[23] At this writing (1979), four volumes (of eleven planned) have been published, covering Pteridophyta, Gymnospermae, and various families of Monocotyledoneae and 'Sympetalae'. Although families have been pre-numbered on a modified Englerian sequence, actual publication, as in *Conspectus florae asiae mediae* (**750**) is as families or groups thereof are ready.

Stankov, S. S. and Taliev, V. I., 1957. *Opredelitel' vysšikh rastenij Evropejskoj časti SSSR*. 2nd edn. 740 pp., 645 text-figs. Moscow: 'Sovetskaja Nauka'. (First edn. 1949.) (Originally pub. 1907 as *Opredelitel' vysšikh rastenij Evropejskoj časti Rossii*, by V. I. Taliev; 9th edn. 1941.)

Manual-key to the vascular plants of the European part of the USSR, with limited synonymy, vernacular and binomial Russian names, very concise indication of internal distribution, and notes on habitat, life-form, phenology, etc.; figures of representative species; indices to all botanical and other names. The introductory section includes an illustrated organography/glossary, a synopsis of the system adopted, and a list of references.

Special group – woody plants (including trees)

VANIN, A. I., 1967. *Opredelitel' derev'ev i kustarnikov.* 2nd edn. 236 pp., illus. Moscow: 'Lesnaja Promyslennost''. (First edn., 1956.)

Students' manual to 187 more common and significant species of indigenous and most common introduced woody plants, consisting mainly of four chapters of keys using different kinds of morphological and field characters as well as a chapter with species descriptions; distribution only very sketchily indicated. Intended for use all over the Soviet Union. See also ANDRONOV, N. M. and BOGDANOV, P. L. *Opredelitel' drevesnykh rastenij po list'jam.* 127 pp., 107 text-figs. Leningrad: Leningrad State University Press. [Illustrated manual-key, based on leaves and other vegetative features.]

681

Kaliningradskaja Oblast'

This territory represents the northern half of the former German province of East Prussia, and includes Kaliningrad (Königsberg or Królewiec), now a Soviet naval base. The area, in addition to the two works described below, is covered by works listed under **641** (HEGI) as well as by some German works still recognizing the boundaries of 1937 (see **647**).

POBEDIMOVA, E. G., 1956(1955). Sostav, rasprostranenie po rajonam i khozjajstvennoe značenie flory Kaliningradskoj oblasti. *Trudy Bot. Inst. AN SSSR*, III, **10**: 225–329.

Includes a concise systematic enumeration of vascular plants, with notes on local distribution and habitat, life-form and habit, phenology, special features, potential, etc.; list of non-vascular plants; bibliography. The introduction includes itinerary details and a map, while the remainder of the work deals with geobotany, forests, etc. Based on results of a 1949–51 expedition.

STEFFEN, H., 1940. *Flora von Ostpreussen.* iii, 319 pp. Königsberg: Gräfe und Unzer.

Concise field manual of vascular plants, with analytical keys, brief descriptions, German vernacular names, and partly symbolic notes on habitat, life-form, phenology, and extralimital range; local range usually not specifically indicated. An introductory section includes remarks on plant geography and vegetation.

682

Litovskaja SSR (Lithuania)

The eastern part of Lithuania, centering about the present capital city of Vilnius, is also covered by parts of Polish works published up to 1939. The remainder of the country, which formed an independent state between the wars, is additionally covered by the following systematic enumeration: HYRNIEWIECKI, B., 1933. *Tentamen florae Lithuaniae.* xv, 369 pp., 61 text-figs., 2 pls. (Arch. Nauk Biol. Towarz. Nauk. Warszawsk., 4). Warsaw.

Bibliography

ŠAPIRAITÉ, S. (ed.), 1971. *Lietuvos botanikos bibliografija 1800–1965.* 528 pp. Vilnius: Lietuvos TSR Mokslų Akademija, Centrine Biblioteka.

NATKEVIČAITÉ-IVANAUSKIENÉ, M. P. (ed.), 1959–. *Lietuvos TSR flora.* Vols. 1– . Illus. (some in color), maps. Vilnius: Valstybine Politines ir Mokslines Literaturos Leidykla (later 'Mokslas').

Illustrated descriptive flora (in Lithuanian) of vascular plants, with keys to all taxa; includes full synonymy, references and citations, Lithuanian and Russian vernacular (and Russian binomial) names, rather detailed indication of local range, summary of extralimital distribution, and notes on habitat, phenology, life-form, associates, uses, etc., as well as some critical remarks; Russian- and German-language summaries as well as complete indices at end of each volume. The introductory section (in vol. 1) provides a historical account of botanical work in Lithuania. At this writing (1979), five volumes have been published (the latest in 1976), accounting for pteridophytes, gymnosperms, and Typhaceae through Dipsacaceae (Englerian system); one further volume, on Compositae, will complete the work.

SNARSKIS, P., 1968. *Vadovas Lietuvos augalams pažinti*. Revised edn. 502 pp., illus. Vilnius: 'Mintis'. (First edn. 1954, under title *Vadovas Lietuvos TSR augalams pažinti*.).

Manual-key to vascular plants, with very limited synonymy, vernacular names, concise indication of local range, and brief notes on habitat, phenology, etc.; figures of some more common species; indices to generic, family, and Russian and Lithuanian vernacular names. [Not seen by this reviewer; description based on information supplied by A. O. Chater, Leicester (now London).]

683

Latvijskaja SSR (Latvia)

GALENIEKS, P. (ed.), 1953–9. *Latvijas PSR flora*. 4 vols. Illus. (some in color). Riga: Latvijas Valsts Izdevniecība.

Illustrated descriptive flora (in Lettish) of vascular plants, with keys to all taxa; includes relatively limited synonymy, references, Lettish and Russian vernacular names (and Russian binomials), concise indication of local and extralimital distribution, and notes on habitat, life-form, phenology, etc.; summaries in Russian as well as complete indices at end of each volume.

PETERSONE, A. and BIRKMANE, K., 1958. *Latvijas PSR augu noteicējs*. 762 pp., illus. Riga: Latvijas Valsts Izdevniecība.

Manual-key to native, naturalized, and commonly cultivated vascular plants, with limited synonymy, vernacular names, generalized indication of local range, and concise notes on habitat, life-form, phenology, floral formulae, etc.; figures of diagnostic features; illustrated glossary; list of authors and indices to all botanical and other names.

684

Estonskaja SSR (Estonia)

In the eastern part of this present-day Soviet republic lies Tartu (Dorpat or Jurjew), the oldest botanical center in the Baltic republics.

– Vascular flora (including also **682** and **683**): 1600–1750 species (1400–1600 native).

EESTI NSV TEADUSTE AKADEEMIA, ZOOLOGIA JA BOTANIKA INSTITUUT. 1953– . *Eesti NSV floora* [Flora Estonskoj SSR]. Vols. 1– . Illus., maps. Tallinn: Eesti Riiklik Kirjastus ('Valgus').

Illustrated descriptive flora (in Estonian) of vascular plants, with keys to genera, species, and infra-specific taxa; full synonymy, with references; Estonian and Russian vernacular (and binomial) names; generalized indication of local and extralimital range, with numerous dot maps; taxonomic commentary and notes on habitat, phenology, uses, special features, etc.; complete indices (together with a summary in Russian) in each volume. Of the eleven volumes projected, all but volume 9 had been issued by 1979. Revised editions of vols. 1 and 2 have also appeared. The sequence followed is that of Grossheim (a modified ranalean system with affinities to that of Takhtajan).

KASK, M. and VAGA, A., 1966. *Eesti taimede määraja*. 1188 pp., 1149 text-figs. Tallinn: 'Valgus'.

Concise descriptive manual of vascular plants, with keys to all taxa, limited synonymy, Estonian and Russian vernacular (and binomial) names, brief indication of local range, and notes on habitat, phenology, life-form, associates, etc.; figures of diagnostic features; indices to all botanical and other names.

KASK, M. *et al.*, 1972. *Taimede välimääraja*. 526 pp., 662 text-figs., 5 pls. Tallinn: 'Valgus'.

Concise illustrated manual-key to vascular plants, with brief diagnoses, limited synonymy, vernacular names, and partly symbolic indication of habitat, life-form, habit, frequency, and phenology; naturalized and widely cultivated plants also included. The introductory section gives *inter alia* a synopsis of families and genera and an illustrated glossary; at the end are found a table of protected plants and indices to all Latin and Estonian names. In part condensed from *Eesti taimedemääraja*.

685

Greater Leningrad Region

See also **678** (HIITONEN 1933, 1934). Part of this area, which includes the former Ingria, was under Czarist rule administered separately from Middle Russia and today corresponds approximately to the Leningradskaja, Pskovskaja, and Novgorodskaja *oblasti*. It is *not* encompassed by Majevski's *Flora srednej polosy Evropejskoj SSSR* (**691**).

Partial works

SHISHKIN, B. K., 1955–65. *Flora Leningradskoj oblasti.* 4 vols. Illus. Leningrad: Leningrad State University Press.

Briefly descriptive, critical flora of the Leningrad region and adjacent districts, with keys to all taxa; synonymy, with some references; generalized indication of local and extralimital range; vernacular and binomial Russian names; taxonomic commentary and notes on habitat, associates, biology, uses, etc.; full indices in each volume; general index to generic, family, and widely used vernacular names in vol. 4.

MINJAEV, I. A., SHMIDT, V. M., and SOKOLOV, M. V. (eds.), 1970. *Konspekt flory Pskovskoj oblasti.* 176 pp., 2 maps. Leningrad: Leningrad State University Press.

Concise enumeration of vascular plants, with essential synonymy, references and citations, Russian binomial names, local range, and many notes on habitat, ecology, special features, status, life-form, and phenology; bibliography and indices. The introductory section gives *inter alia* accounts of botanical progress, the general features of the flora, and floristic regions. [Not seen by this author; description prepared from data supplied by M. E. Kirpicznikov, Leningrad.]

686

Karel'skaja ASSR

See also **678** (HIITONEN 1933, 1934); **688** (PERFIL'EV).

RAMENSKAJA, M. L. 1960. *Opredelitel' vysšikh rastenij Karelii.* 485 pp., 94 text-figs. (incl. 2 maps), 12 color pls. Petrozavodsk: Karel'skoj ASSR Government Publishing Service.

Manual-key to vascular plants, with limited synonymy, generalized indication of local range, vernacular and Russian-binomial names, and notes on habitat, phenology, biology, etc.; figures of diagnostic features; index to all botanical and other names. The introductory section includes a brief account of vegetation associations and floristic regions. The spring flora is also accounted for in CHERNOV, V. N., 1955. *Vesennjaja flora Karelo-Finskoj SSR: opredelitel' vesennikh rastenij.* 154 pp., 90 text-figs. Petrozavodsk. [Includes glossary and bibliography; covers plants flowering before 15 June.]

687

Kola Peninsula

See also **678** (HIITONEN, 1933, 1934); **688** (PERFIL'EV). The area corresponds approximately to that of Murmanskaja Oblast'. The arctic zone has also been designated as **062**.

GORODKOV, B. N. and POJARKOVA, A. J. (eds.), 1953–66. *Flora Murmanskoj oblasti.* 5 vols. Illus., maps. Moscow/Leningrad: AN SSSR Press (later 'Nauka').

Illustrated descriptive flora of vascular plants, with keys to all taxa; synonymy, with references and citations; vernacular and binomial Russian names; generalized indication of local and extralimital range, with dot maps for each species; taxonomic commentary and notes on habitat, phenology, associates, etc.; addenda and corrigenda (vol. 5); indices to botanical and other names in each volume. The introductory section includes a short list of important floristic references.

688

Northern Russia

This area is here considered to correspond to a combination of Arkhangel'skaja oblast' (except for the tundra zone) and the Komi ASSR. For Vologda *oblast'*, see **689**. Vascular flora: 1300–1450 species (1250–1450 native), including **686** and **687**.

PERFIL'EV, I. A. 1934–6. *Flora Severnogo Kraja.* Vyp. 1, 2–3 (in 2 vols.). 160, 407 pp.; 111 text-figs. (in vol. 2). Archangel: Severnoj Kraj Press.

Partially illustrated manual-key to vascular plants, with essential synonymy, Russian vernacular and binomial names, fairly copious diagnoses, and concise notes on local range, habitat, life-form, and phenology, indices to Russian and botanical names of genera and families (vol. 2); list of references (vol. 2). No extralimital distribution is indicated. Volume 2 also includes a list of new taxa discovered in the area from 1500 onwards. Coverage is from the Gorkovskaja and Ivanovskaja *oblasti* in the south to the islands in the Arctic Ocean and from Karelia to the Urals. [Description originally prepared from notes and facsimile pages supplied by M. E. Kirpicznikov,

Leningrad, and E. Hultén, Stockholm, and modified following examination of copies in Leningrad and Geneva. Very few copies appear to be available outside the Soviet Union.]

SNYATKOV, A. A., SIRJAEV, G. I., and PERFIL'EV, I. A., 1922. *Opredelitel' rastenij lesnoj polosy Severo-Vostoka Evropejskoj časti Rossii.* 2nd edn., revised and ed. I. A. Perfil'ev. 215 pp., illus. Vologda: State Press, Vologda Oblast' Branch. (First edn. 1913.)

Manual-key to vascular plants of the forest region of northern and northeastern Russia, with limited synonymy, vernacular and binomial Russian names, concise indication of local range, and notes on habitat, life-form, phenology, etc.; indices to generic, family, and Russian names. The introductory section is mostly pedagogical. [Although superseded for the whole of 688 by later works, it has been included here for the record and for its coverage of 689 and the northern half of 698, i.e., down to the limits of the zone of potential continuous forest vegetation.]

TOLMATCHEV, A. I. (ed.), 1974–7. *Flora Severo-Vostoka Evropejskoj časti SSSR* [Flora regionis boreali-orientalis territoriae europaeae URSS]. 4 vols. Maps. Leningrad: 'Nauka'.

Non-descriptive flora of vascular plants, with analytical keys, full synonymy (with references and relevant citations), Russian binomials, indication of localities (with citations) and summary of internal and extralimital distribution, and notes on phenology, special features, etc. At the end of each volume are dot maps for all species and subspecies as well as indices for all Komi, Russian, and Latin names; no consolidated index, however, appears to be available. Covers the entire area as defined above, with the addition of the tundra zone to the north (063).

Partial work

TOLMATCHEV, A. I. *et al.*, 1962. *Opredelitel' vysšikh rastenij Komi ASSR.* 359 pp., 15 text-figs. Moscow/Leningrad: AN SSSR Press.

Manual-key to vascular plants, with very limited synonymy, standardized and vernacular Russian and Komi names, and notes on local range, habitat, phenology, etc.; indices to all botanical and other names. A brief introduction to the work is also provided.

689

Northern Middle Russia

See 688 (PERFIL'EV; SNYATKOV *et al.*). This area is here considered to center on the present-day Vologda *oblast'* but additionally includes the northeastern part of Kostroma *oblast'* and the northern lobe of Kirov *oblast'* (both formerly included in Vologda *oblast'*). It is *not* encompassed by Tolmachev's recent *Flora Severo-Vostoka Evropejskoj časti SSSR*, although wholly included within the northern forest zone accounted for in the earlier subregional floras.

691

Middle Russia

This area, which encloses the Russian heartland, is bounded on the west and south by the Ukraine and Belorussia, on the north and east by the Volga Valley and the Volga-Don Canal, and on the northwest by the Greater Leningrad Region (685).

MAJEVSKI, P. F., 1964. *Flora srednej polosy Evropejskoj časti SSSR.* 9th edn., revised and ed. B. K. Shishkin. 880 pp., 325 text-figs. Leningrad: 'Kolos'. (First edn. 1892, Moscow, under title *Flora srednej Rossii.*)

Manual-key to vascular plants, with limited synonymy, Russian vernacular and binomial names, concise indication of local range, and partly symbolic notes on habitat, phenology, life-form, associates, uses, etc.; small text-figures of diagnostic features; list of authors and complete indices to botanical and vernacular names.

Partial work: Moscow region

A number of the administrative subdivisions within the area of Middle Russia are covered by separate works; however, special notice should be taken of the manual for the greater Moscow region described below.

VOROSHILOV, V. N., SKVORTSOV, A. K., and TIKHOMIROV, V. N., 1966. *Opredelitel' rastenij Moskovskoj oblasti.* 367 pp., 313 text-figs., 1 map (in end papers). Moscow: 'Nauka'.

Field manual-key to vascular plants of the greater Moscow region, with diagnoses, indication of local range, and notes on habitat, phenology, etc.; index. [Prepared under the auspices of the Moscow Society of Naturalists.]

692

Belorusskaja SSR (White Russia)

The western part of this Soviet republic, up to a line running some way west of Minsk, was from 1919 to 1939 part of Poland (646) and thus is covered by Polish works or parts thereof published during that period.

SHISHKIN, B. K. *et al.* (eds.), 1948–59. *Flora BSSR.* 5 vols. Illus. Moscow: Otiz-Sel'khozgiz (vol. 1); Minsk: AN BSSR Press (vols. 2–5).

Illustrated descriptive flora of vascular plants, with keys to all taxa; brief indication of synonymy, with references; generalized summary of local and extra-limital range; vernacular and binomial Russian names; partly symbolic notes on habitat, life-form, phenology, etc.; complete indices in each volume; general index to family and generic names as well as a list of principal references in vol. 5. The introductory section in vol. 1 contains a historical account of floristic work in the area.

SHISHKIN, B. K. *et al.* (eds.), 1967. *Opredelitel' rastenij Belorussii.* 871 pp., 325 figs., maps. Minsk: 'Vyšejšaja škola'.

Manual-key to vascular plants (in White Russian), with limited synonymy, vernacular and binomial names, concise indication of local range (by botanical districts), and notes on habitat, life-form, phenology, etc.; figures of representative species; indices to all botanical and other names. The introductory section includes an organography/glossary. Several distribution maps of species are also included.

693

Moldavskaja SSR (Bessarabia)

Between the two World Wars, most of this present-day Soviet republic was part of Romania. It has also included Bukovina, but this is now part of the Ukraine.

HEIDEMANN (GEIDEMANN), T. S., 1975. *Opredelitel' vysšikh rastenij Moldavskoj SSR.* 2nd edn. 576 pp.,

100 text-figs. (incl. map). Kishinev: 'Štiintsa'. (First edn. 1954.)

Illustrated manual-key to vascular plants, with very limited synonymy, short diagnoses, vernacular names (in Russian and Moldavian), occasional indication of local range, and partly symbolic notes on habitat, life-form, phenology, special features, etc.; bibliography (p. 534), glossary, and indices to all Russian, Moldavian, and Latin names.

694

Ukrainskaja SSR (in general)

For the Southwestern Ukraine see also 695. Between 1919 and 1939 the western part of this large and populous Soviet republic was included in Poland, while the Carpatho-Ukraine district (or Ruthenia) formed part of Czechoslovakia (after 1939 the latter came under Hungary until the close of World War II). Within the Western Ukraine also is the eastern part of Galicia, which with Ruthenia and adjacent Bukovina came within the Austro–Hungarian Empire.

– Vascular flora: 3100–3250 species (2900–3100 native), including 693 and part of 692 but not 696.

BORDZILOVSKY, E. I., FOMIN, A. V. *et al.* (eds.), 1938–65. *Flora URSR.* [Flora RSS Ucr.] 12 vols. Illus., 2 maps. Kiev: AN URSR Press (vols. 1–11); 'Naukova dumka' (vol. 12). (A preliminary edition of vol. 1 was issued in 1937.)

Comprehensive illustrated descriptive flora (in Ukrainian) of vascular plants (including the Crimea), with keys to all taxa (a general key to families appearing in vol. 3); extensive synonymy, with references and citations (together with illustration references); vernacular and binomial names; extensive indication of local and USSR-wide distribution, with citation of some *exsiccatae*; summary of extra-USSR range; taxonomic commentary and notes on habitat, life-form, phenology, associates, etc.; Latin diagnoses of new taxa and complete indices to botanical and other names at end of each volume (no comprehensive index). The introductory section (vol. 1) includes an account, with maps, of the floristic regions of the Ukraine. The grasses have recently been revised as PROKUDIN, YU. M. *et al.*, 1977. *Zlaki Ukraini.* 518 pp., 74 figs., 165 maps. Kiev.

ZEROV, D. K. *et al.* (eds.), 1965. *Viznačnik roslin*

Ukraini. 2nd edn. 877 pp., 605 text-figs., map. Kiev: 'Urožai'. (First edn. 1950, under title *Viznačnik roslin URSR.*)

Manual-key (in Ukrainian) to vascular plants of the Ukraine (including the Crimea), with limited synonymy, vernacular and binomial names, concise indication of local range, taxonomic remarks, and brief notes on habitat, life-form, phenology, etc.; many figures of diagnostic features; appendices on terminology, collecting methods, use of keys, etc.; indices to all botanical and other names.

Special group – woody plants

LYPA, A. L., 1955–7. *Opredelitel' derev'ev i kustarnikov (dikorastuščikh i kul'tiviruemykh) v Uk.SSR.* 2 vols. Illus. Kiev: Kiev State University Press.

Descriptive account (in Russian) of wild and cultivated Ukrainian trees and shrubs with keys and illustrations; includes synonymy, vernacular and binomial names, and extensive notes on local and extralimital distribution, habitat, phenology, variation, cultivation, etc.; indices to all botanical and other names. [Not seen by this author; description based on information supplied by A. O. Chater, London.]

695

Southwestern Ukraine (Ruthenia)

See also **603**/VII and **694** (all works). This area is defined as that part of the western Ukraine southwest of the Dniester and Prut Rivers, encompassing the Ukrainian Carpathians and including much of Ruthenia and Bukovina. The part lying south of the mountain divide is the area once called Carpathian Russia or Carpatho-Ukraine but now comprising Zakarpatskaja Oblast', whose sovereignty has changed several times in the present century. Across the range, in the foothills southwest of Czernowitz, Vavilov in August 1940 made his last field trip.[24] The vascular flora is, at over 2000 species, rich for a smallish area.

ČOPIK, V. J., KOTOV, M. J. and PROTOPOPOVA, V. V. (eds.), 1977. *Vizačnik roslin Ukrains'kikh Karpat.* 435 pp., 338 text-figs., map. Kiev: 'Naukova dumka'.

Manual-key (in Ukrainian) to vascular plants (2012 species, including those native, naturalized, adventive, and commonly cultivated), with species-leads including indication of habitat phenology occurrence (by floristic districts), and habit along with limited synonymy and Ukrainian-binomial names; some localities given as well as taxonomic notes; complete indices at end. The introductory section gives the background to the work, technical details, a description of the ten floristic districts in the area (with map), and a general key to families. For separate works limited to the Carpatho-Ukraine, see FODOR, S. S., 1974. *Flora Zakarpattja.* 208 pp. L'vov (L'viv): 'Vyšča Skola'; JAROŠENKO, P. D., 1947. *Korotkij vizačnik roslin Zakarpattja.* 99 pp. Užgorod: Užgorod State University; and POPOV, M. G., 1949. Konspekt flory Zakarpatskoj oblasti. In *idem, Očerk rastitel'nosti i flory Karpat*, pp. 176–300 (Mater. Pozn. Fauny Fl. SSSR, N.S., vyp. 13 [Otd. Bot., vyp. 5]). Moscow.

696

The Crimea

See also **695** (all works). The Crimea is characterized by considerable local endemism, due to its mountainous character. Vascular flora: 2250–2400 species (2150–2250 native).

RUBTSOV, N. I. (ed.), 1972. *Opredelitel' vysšikh rastenij Kryma.* 550 pp., 504 text-figs. Leningrad: 'Nauka'.

Copiously illustrated manual-key to vascular plants, with concise ciagnoses, limited synonymy, Russian vernacular and binomial names, local range, and (partly symbolic) short notes on habitat, life-form, phenology, ecology, and frequency; indices to all botanical and other names. An introductory section gives aids for the use of the work.

WULFF, E. W. *et al.*, 1927–69. *Flora Kryma* [Flora taurica]. 3 vols. portrait, maps. Yalta: Nikitskij Bot. Sad (vol. 1, parts 1–3); Moscow/Leningrad: Otiz-Sel'khozgiz (vol. 1, part 4, vol. 2, part 1); Moscow: Sel'khozgiz (vol. 2, part 2, vol. 3, part 1), 'Sovetskaja Nauka' (vol. 2, part 3), 'Kolos' (vol. 3, part 2); Yalta: Nikitskij Botaničeskij Sad (vol. 3, part 3).

STANKOV, S. S. and RUBTSOV, N. I. (eds.), 1959. *Addenda et corrigenda ad Vol. 1 'Florae tauricae'.* 127 pp. (Trudy Gosud. Nikitsk. Bot. Sada, 31). Yalta.

Comprehensive critical enumeration of vascular plants, with diagnostic keys to all taxa and full descriptions of novelties; also includes extensive synonymy and references, literature citations, detailed indication of local range, with many *exsiccatae* and other records, and generalized statement of extralimital distribution, taxonomic commentary, and notes on habitat, associates, etc.; tables of contents and literature lists in each part (but only vol. 3 fully indexed). The introductory section (in vol. 1, part 1) contains a historical account of botanical work in the Crimea and a brief description of its floristic districts, while vol. 1, part 4 includes a map of these districts. The entire work was produced under the auspices of the Nikitskij Botanical Garden in Yalta, close to the site of the historic Three Power Talks of 1945.

697

Lower Don region

This area, lying between the eastern limits of the Ukraine, the North Caucasian lowlands, and the lower Volga basin, is with one exception not covered by any separate floras or enumerations. It centers on Rostov Oblast', but also includes a part of Stavropol Krai, the southern part of Volgograd Oblast' (south of the Volga-Don Canal), and the western lobe of the Kalmytsk ASSR. The eastern limit is considered to be the same as the western limit of coverage by *Flora jugo-vostoka Evropejskoj časti SSSR*, i.e., along the Jergeni Rise at 44 °E. The only regional work actually covering the area (now otherwise superseded) is Schmal'hausen, I., 1895–7. *Flora srednej i južnoj Rossii, Kryma i Severnogo Kavkaza.* 2 vols. Portrait. Kiev: Kusnerev. [Briefly descriptive flora of vascular plants, with partial keys, synonymy, and fairly detailed indication of local range.]

Bibliography

Novopokrovsky, I., 1938. Literatura po botaničeskoj geografii Rostovskoj oblasti, Krasnodarskogo kraja, Ordžoni-kidzevskogo kraja i Dagestana. *Trudy Rostov obl. biol. obšč*, 2: 5–45. [442 titles.]

Pereselenkova, L. M., 1950. *Derevja i kustarniki Rostovskoj oblasti.* 107 pp., illus., map. Rostov-na-Donu: Rostovs Oblast' Press.

Semitechnical illustrated descriptive account of native and commonly cultivated woody species, with notes on habitat, local range, properties, uses, and potential; artifical general key to all species; no index. The introductory section gives aids to the use of the work, basic organography, and an account of the area and its vegetation (with map).

698

Southeastern Russia

This generally imperfectly known subregion comprises the lower Volga Basin from the Caspian Sea at 45 °N, upstream through Volgograd to a point north of Saratov and ranges eastwards into extreme western Kazakhstan; the eastern limit follows the Ural River, part of the traditionally recognized boundary between Europe and Asia.

– Vascular flora (including parts of **697** and **699**): 2300–2500 species (2200–2450 native).

Fedtschenko, B. A. and Shishkin, B. K., 1927–38. *Flora jugo-vostoka Evropejskoj časti SSSR.* [Flora Rossiae austro-orientalis.] 3 vols. (in 6 parts), index. Illus., maps. Leningrad: Glavnyj Botaničeskij Sad (parts 1–5, comprising vols. 1–2); Moscow/Leningrad: AN SSSR Press (part 6, corresponding to vol. 3, and index). (Parts 1–5 also pub. as *Trudy Glavn. Bot. Sada SSSR*, vol. 40, parts 1–3, and vol. 43, parts 1–2.)

Full descriptive flora of vascular plants, with keys to all taxa, limited synonymy, Russian-binomial and vernacular names, limited synonymy, detailed account of local range (by botanical regions; see maps in parts 1 and 2), with citations and some range maps, taxonomic commentary, and notes on habitat, associates, etc.; figures of representative species; complete index (in separate volume). In addition to the area as circumscribed in **698**, this work also includes much of Eastern Russia as delimited in **699**, with a northern limit at the Kama and Bjeloja Rivers and an eastern limit in the southern Ural.

699

Eastern Russia

In this work Eastern Russia is considered to represent the area between the Volga and Ural Rivers and the Ural Mountains, extending from the Ural River and a line just south of Kujbyšev north to the southern boundary of Komi ASSR, thus including among other areas most of Kirov Oblast' (except its northern lobe), the whole of Perm' and Sverdlov *oblasti*, the Udmurt ASSR, the Bashkir ASSR, and large parts of Kujbyšev and Orenburg *oblasti*. The circumscription of this area has been based on the geographical limits of Korshinsky's *Tentamen florae Rossiae orientalis*, which includes a clear map. The northern and southern parts (separately considered below) are also covered respectively by other, more recent works, noted under the relevant subheadings. On the western fringe, the area is to some extent overlapped by Majevski's *Flora* (see **691**). For the Ural Mountains, see also **604**.

KORSHINSKY, S., 1898. *Tentamen florae Rossiae orientalis*. xix, 566 pp., 2 maps (Zap. Imp. Akad. Nauk, Fiz.-Mat. Otd., VIII, 7(1)). St Petersburg.

Systematic enumeration of vascular plants (in Latin), without keys; synonymy, with references; detailed indication of local range, with citations; generalized summary of extralimital range; taxonomic commentary and notes on habitat, etc.; index to all botanical names. The introductory section includes a description of the region (with maps), an account of botanical districts, and an extensive literature survey, together with an index to collectors. Over much of its goegraphical range, this work has now been superseded. For a trial essay for a descriptive flora, containing also an important introductory essay on taxonomic philosophy and floristic and vegetation zones, see *idem*, 1892. *Flora vostoka Evropejskoj Rossii*. Vyp. 1. 227 pp., 3 pl. Tomsk.

Partial works – non-forest zone

This zone, except for the Ural (**604**), extends north to about 54–55 °N. and falls in large part within the mixed forest-and-steppe vegetation formation. In addition to Korshinsky's *Tentamen*, Fedtschenko and Shishkin's *Flora jugo-vostoka Evropejskoj časti SSSR* (see **698**) covers this area almost entirely. However, two significant recent local works are given below; one (Terekhov) is a 'spring flora'.

KUČEROV, E. V. (ed.), 1966. *Opredelitel' rastenij Baškirskoj ASSR*. 493 pp., 95 text-figs. Moscow: 'Nauka'.

Illustrated manual-key to vascular plants of Bashkiria, centering around Ufa; includes limited synonymy and brief notes on local range, habitat, form, etc., and index.

TEREKHOV, A. F., 1969. *Opredelitel' vesennikh i osennikh rastenij Srednogo Povolžija i Zavolžija*. 3rd edn. 464 pp., 100 text-figs., 4 pls. Kujbyšev: Kniž. Izd. (First edn. 1930.)

Illustrated students' manual-key to the spring flora of the Kujbysev district in the central Volga Basin, with brief notes on phenology, habitat, local range, habit and life-form, etc.; glossary and index.

Partial works – forest zone

This zone, with the exception of the Ural (**604**), has its southern limit between 54–55 °N, i.e., the southern limit of potentially continuous forest vegetation. Besides being covered by Korshinsky's *Tentamen*, it is entirely (or almost entirely) included in SNYATKOV *et al.*, *Opredelitel' rastenij lesnoj polosy severo-vostoka Evropejskoj časti Rossii* (see **688**). Modern manuals, however, have followed current administrative boundaries, as in the works of anonymous author (1975) for the Kirov Oblast' (comprising in large part the former Vyatka *gubernija*) and Efimova (1972) for the Udmurt ASSR.

ANON., 1975. *Opredelitel' rastenij Kirovskoj oblasti*. Vol. 2. Kirov.

[Manual-key to higher plants. Not seen by this author; cited from *Prog. Bot.* **40**: 415, 427 (1978).]

EFIMOVA, T. P., 1972 (1971). *Opredelitel' rastenij Udmurtii*. 224 pp., illus., 8 pls. Iževsk: 'Udmurtija'.

Concise field manual-key to vascular plants, with brief diagnoses and partly symbolic notes on habitat, life-form, phenology, local and extralimital range, and special features; includes general key to families, Russian nomenclatural equivalents, and indices. The introductory section includes an illustrated glossary and notes on the use of the work.

Notes

1 The earliest known 'regional' flora is in fact the fourth-century Chinese *Nan-fang ts'ao-mu chuang*, covering southern China. An English translation by H. L. Li was announced for publication in 1979.

2 Stearn, W. T., 1958. Botanical exploration to the time of Linnaeus. *Proc. Linn. Soc. London*, **169**: 173–96.

3 The 'Green Books', first seen by the present author in 1963, directly influenced the original conception of this *Guide* and its 1964 version.

4 Webb, D. A., 1978. *Flora Europaea* – a retrospect. *Taxon*, **27**: 3–14.

5 *Idem.*

6 European Science Foundation, *Report 1977*, pp. 78–82; *Report 1978*, pp. 95–7. Strasbourg. [In the latter reference the plan and scope of the projected checklist are described.]

7 This latter report is a product of the 'Commission pour l'encouragement et la coordination de l'exploration floristique, la préparation des flores et de travaux monographiques'.

8 Heywood, V. H., 1978. European floristics: past, present and future. In *Essays in plant taxonomy* (ed. H. E. Street), pp. 275–89. London.

9 Lovrić, A.-Ž., *OPTIMA Newsletter*, **6**: 38–40 (1978).

10 Jäger, E. *Fortschr. Bot.* **40**: 413 ff. (1978).

11 *Idem.*

12 Greuter, W., 1975. Floristic studies in Greece. In *Floristic and taxonomic studies in Europe* (ed. S. M. Walters and C. J. King), pp. 18–37, 4 maps (BSBI Conference Reports, 15). Abingdon, Berks.

13 *Idem.*

14 Webb, D. A., 1978. *Flora Europaea* – a retrospect. *Taxon*, **27**: 13.

15 Stearn, W. T., 1975. History of the British contribution to the study of the European flora. In *Floristic and taxonomic studies in Europe* (ed. S. M. Walters and C. J. King), pp. 1–17; p. 2, map (BSBI Conference Report, 15). Abingdon, Berks.

16 Heywood, V. H., 1970. The new Hegi Compositae [review]. *Taxon*, **19**: 937–8.

17 It was the first edition of 'Britton and Brown', completed in 1898, which stimulated the Paris publisher Klincksieck to realize a similar semi-popular work for France. Connections in the French botanical community put him into contact with Coste in 1899. The value of this flora is largely due to Coste's insistence on forming descriptions from fresh plants (or specimens) and to Klincksieck's choice of artists. Coste's voucher collection is now preserved at Montpellier (MPU). See Aymonin, G. G., 1974. La naissance de la 'Flore descriptive et illustrée de la France' de l'Abbé Hippolyte Coste. *Taxon*, **23**: 607–11.

18 Lawalrée's work has been regarded as similar in format and style to the several tropical floras undergoing elaboration at the Laboratoire de Phanérogamie, Paris. See Aymonin, G. G., 1962. Où en sont les flores européennes? *Adansonia*, II, **2**: 159–71. At present the work is dormant (V. Demoulin, personal communication).

19 Mennema, J., 1979. The Rijksherbarium and its contribution to the research on the Netherlands and European flora. *Blumea*, **25**: 115–19.

20 Stearn, 1975 (see note 15).

21 Fëdorov has stated that the reviews of Kirpicznikov and Linczewski give the course of events regarding the *Flora SSSR* 'most accurately' [and dispassionately]. Bobrov's account, even in its edited English version, is more personal, as is only to be expected from one associated with the project 'for the duration'. Shetler's review is detached but detailed and well informed. Kirpicznikov's review additionally gives a consideration of competing taxonomic philosophies in relation to treatments in the *Flora*, including the advantages and disadvantages of the Komarovian method.

22 For statistics see Kirpicznikov, 1967, 1969 in the introduction to Region 68/69, under Progress, Flora SSSR.

23 Fëdorov, An. A., 1971. Floristics in the USSR. *BioScience* (USA), **21**: 514–21.

24 Medvedev, Zh. A., 1969. *The rise and fall of T. D. Lysenko* (trans. I. M. Lerner), pp. 67–70. New York: Columbia University Press.

Division

7

Northern, central, and southwestern (extra-monsoonal) Asia

The methodological basis of the work on the *Flora URSS* was the morphological-geographical method, the concept of the race as an actual biogeographical unit. That [this method] was generally adopted...is by no means fortuitous. Those who live and work in the USSR...cannot but think in terms of geography... This ideological trend and the progressive character of the *Flora URSS* are frequently ignored, and are possibly not sufficiently well understood by foreign botanists.

E. G. Bobrov, *Nature*, **205**: 1048 (1965).

On a publié des catalogues des flores locales, mais tous [les] riches matériaux étaient épars, sans liaison entr'eux, souvent difficiles à consulter; il était [donc] indispensable de les réunir, de les comparer, de les relier ensemble, et c'est le travail que j'aborde aujourd'hui.

Boissier, *Flora orientalis*, **1**: i (1867).

Generally speaking, Division 7 comprises that half of the Asiatic continent which is for the most part boreal, cool-temperate, semiarid, or arid; in other words, it is largely beyond the influence of the summer monsoon. The sub-Arctic and Arctic zones are excluded; for these, see regions 05 and 06. The western limits correspond to those of Division 6 (Europe), thus taking in the Caucasus (as region 74), while the eastern and southern limits border for their entire length on Division 8 (southern, eastern, and southeastern Asia). Within the division, nine geopolitical regions have been delimited; those constituting the Asiatic part of the USSR are additionally grouped as Superregion 71–75, while those in southwestern Asia (except the Caucasus) are grouped as Superregion 77–79. That part of central Asia which is outside the USSR remains separate as Region (Superregion) 76. Additional units have been allocated for key physiographic areas, at present the various high mountain chains along the southern border such as the Karakoram, Pamirs, and Altai–Sayan system.

The history of floristic work and current programs in flora writing in extra-monsoonal Asia cannot be readily reviewed as a whole, and is therefore accounted for in the introductions to the superregions. It may be noted here, however, that a significant field of development in the floristics of the semi-continent has been in studies of the genesis, spread and present distribution of the high-mountain flora of northern and central Asia, as evidenced by the considerable body of published work now available which includes a number of floras worthy of inclusion in this Guide.

General bibliographies. Bay, 1910; Frodin, 1964; Goodale, 1879; Holden and Wycoff, 1911–14; Hultén, 1958, Jackson, 1881; Pritzel, 1871–7; Rehder, 1911; USDA, 1958. See also superregional headings.

General indices. BA, 1926– ; BotA, 1918–26; BC, 1879–1944; BS, 1940– ; CSP, 1800–1900; EB, 1959– ; FB, 1931– ; ICSL, 1901–14; JBJ, 1873–1939; KR, 1971– ; NN, 1879–1943; RŽ, 1954– .

Conspectus

700

Division in general

For comprehensive floras, see supraregional headings.

703

Mountains of extra-monsoonal Asia

The works accounted for within this unit all relate to the high-mountain or 'alpine' flora of various mountain ranges or systems in extra-monsoonal Asia from the Caucasus to the mountains of northeast Siberia. One work on forest flora has also been included here. Only one treatment, described below, provides something approaching an overall treatment of part of the high-mountain flora; all other works selected relate only to a single range or system of ranges, and are listed under specific subheadings arranged roughly from west to east. The overall output of floras and related works on the high-mountain flora of extra-monsoonal Asia has been considerable, greatly exceeding even North America (**103**) though not of course Europe (**603**). Much of this is the work of Russian and Soviet botanists; for useful background material on the progress and range of high-mountain floristic, geobotanical and chorological studies in the Soviet Union, see SUKACHEV, V. N. (ed.), 1960. *Materialy po izučeniju flory i rastitel'nosti vysokogorij.** 304 pp., illus., maps. (I. *Vsesojuznoe soveščanie po problemam izučenija i osvoenija flory i rastitel'nosti vysokogorij*, Leningrad, 1958; [Proceedings].) Moscow/ Leningrad: AN SSSR Press, trans. into English as *idem*, 1965. *Studies on the flora and vegetation of high-mountain areas.* viii, 293 pp., illus., maps. Jerusalem: Israel Program for Scientific Translations; and TOLMATCHEV, A. I. (ed.), 1974. *Rastitel'nyj mir vysokogorij i ego osvoenie.* 339 pp., illus. Leningrad: 'Nauka'. The first-named of these collections also represents the work of the first in the series of All-Union specialist meetings on high-mountain flora and vegetation, since 1958 held at approximately three-yearly intervals in various Soviet centers. [These meetings have been organized under the auspices of the All-Union Botanical Society of the USSR and affiliated bodies; and the papers (*Tezisy dokladov*) of at least the fifth (Baku, 1971) and sixth (Stavropol', 1974) have been published.]

TOLMATCHEV, A. I. (ed.), 1974. *Endemičnye vysokogornye rastenija Severnoj Azii. Atlas.* 335 pp., 228 maps. Novosibirsk: 'Nauka'.

* For a lexicon of Russian words used in titles see Table 1, p. 347.

Atlas of distribution maps of 228 high-mountain endemic species depicting their spread throughout northern Asia, with signed commentaries in a separate section (by a sizeable number of contributors), containing notes on chorology and taxonomy, karyotypes, and references; bibliography (pp. 95–8). A brief preface is also given (pp. 3–4). [Nearly all species treated are herbaceous and the work can thus be regarded as a chorological treatment of 'alpines'. Control of preparation of this work was in the hands of a working committee under the direction of the late Professor Tolmatchev.]

I–II. High Caucasus, Transcaucasus and Elburz Ranges

No separate florulas have been seen.

III. Hindu Kush

No separate florulas have been seen.

IV. The Karakoram

For other works, see 798. This highest of ranges in central Asia forms, with the Pamir (see below), the celebrated 'Roof of the World'; large parts are under permanent ice and snow and its highest point reaches 8611 m in K2 (Mt Chogoru or Godwin-Austen). No work has yet appeared to succeed the relatively comprehensive treatment by Pampanini (1930–4), although the range has been visited by a number of expeditions in succeeding decades.

PAMPANINI, R., 1930. La flora del Caracorùm. In *Relazione scientifiche della Spedizione italiana De Filippi nell'Himàlaia, Caracorùm e Turchestàn Cinese, 1913–14* (ed. SPEDIZIONE ITALIANA DE FILIPPI NELL'HIMÀLAIA, CARACORÙM E TURCHESTÀN CINESE), ser. II: Geologici e geografici (dir. G. Dainelli), vol. 10, pp. 1–290, 32 text-figs. (incl. maps), Pls. 1–7 (incl. 1 map). Bologna: Zanichelli.

PAMPANINI, R., 1934. Aggiunte alla flora del Caracorùm. In *ibid.*, vol. 11, pp. 143–78, Pls. 7–10. Bologna.

Critical systematic enumeration of the vascular plants, bryophytes, and fungi of the Karakoram north of the upper Indus River from Gilgit southeastwards to Lake Pancong, with descriptions of new taxa; includes full synonymy, references, citations, indication of *exsiccatae* and other records with localities, some taxonomic commentary, and notes on habitat, special features, etc.; index to genera at end. The introductory section includes chapters on the history of botanical work in the area (with provision of individual route maps), the itinerary of the De Filippi expedition, and a list of maximum altitudes for various species, while following the enumeration proper is a full bibliography. All known records have been incorporated, so this work is retrospective and thus fundamental to all later floristic study on the Karakoram. The 1934 supplement covers additions arising from a return expedition by Dainelli in 1930, and a corresponding revision of phytogeographical tables.

V. Kunlun Shan

No separate florulas are available for any part of this remote northern bordering range of the Tibetan Plateau, whose highest peak is Ulugh Muztagh or Mutzu T'a-ko (7723 m) in the central part of the range.

VI. The Pamir

Like the Karakoram, the Pamir form part of the so-called 'Roof of the World'. Within it is the highest peak in the Soviet Union, Pik Kommunizma (Mt Communism; 7495 m). The central plateau lies mainly to its south within the Tadzhik Union Republic, with extensions into northeastern Afghanistan and extreme western Sinkian. The early compilations of O. A. Fedtschenko (1903–14, 1907) have been supplanted by an 'alpine flora' by Ikonnikov (1963).

IKONNIKOV, S. S., 1963. *Opredelitel' rastenij Pamira.* 281 pp., 31 pls., map (Trudy Bot. Inst. Akad. Nauk Tadžiksk. SSR, 20). Dushanbe.

Manual-key to vascular plants of the Pamirs, with limited synonymy, Russian and local vernacular names, fairly detailed general indication of local range, brief summary of extralimital distribution, taxonomic commentary, and notes on habitat, ecology, and phenology; list of references and indices to all botanical and other plant names. The introductory section incorporates chapters on physical features, climate, floristics, phytogeography, and botanical exploration. Succeeds FEDTSCHENKO, O. A., 1903. *Flora Pamira.* iv, 239, 10 pp., 8 pls., 1 folding map (Trudy Imp. S.-Peterburgsk. Bot. Sada, 21, pp. 233–471.) St

Petersburg; *idem*, 1904–14. *Dopolnenie k 'Flore Pamira'*, nos. 1–5 (*Ibid.*, 24: 123–54, 313–55; 28: 97–126, 455–514; 31: 441–90); and *idem*, 1907. *Opredelitel' Pamirskikh rastenij*. 64 pp. (*Ibid.*, 28: 129–90.)

VII. Tian Shan

This extensive complex of mountains, with several more or less parallel and overlapping ranges of different geological ages and with altitudes ranging up to 7439 m (Pik Pobedy), runs north of the Syr Darya from just east of Tashkent across Soviet Middle Asia (largely within Kirghizstan to the south of Frunze) and into western Sinkiang (Xinjiang) in China, therein (with a maximum altitude of 5445 m in Mt Pokota near Urumchi) separating the Great Tarim Basin to its south from Dzungaria to its north. The most recent independent treatment, with keys and descriptions, is evidently the following: FEDTSCHENKO, B. A., 1904–5. *Flora zapadnogo Tjan-Sanja*, [pts. I–II]. (Trudy Imp. S.-Peterburgsk. Bot. Sada, vol. 23, pp. 249–532; vol. 24, pp. 155–260). St Petersburg. However, by far the greater part of the range within Soviet territory is now encompassed by the more recent *Flora Kirghizskoj SSR* (see **755**) and other relevant floras of Central Asia, as well as GRUBOV (see **760**).

VIII. Altai-Sayan mountain system (in general)

This group of ranges in southern Siberia consists of three major divisions, each here treated separately: the Altai Mountains, the Western Sayans, and the Eastern Sayans. Separate works of relatively recent vintage are available for the latter two ranges, along with the forest flora of Koropachinsky (1975) and 'alpine flora' of Malyschev (1968). The latter work in particular resembles and recalls *Rocky Mountain flora* by W. A. Weber (see **103**).

KOROPACHINSKY, I. JU., 1975. *Dendroflora Altajsko-Sajanskoj gornoj oblasti*. 290 pp., 141 illus. (incl. halftones, maps). Novosibirsk: 'Nauka'.

A descriptive treatment of the woody plants of the Southern Siberian mountains (including the whole of Tuva), with keys, synonymy and references, regional distribution (including dot maps for each species), and extensive remarks on external features and geographical patterns as well as uses and properties; this is followed

(pp. 226 ff.) by a floristic analysis of the region covered, characteristic species in each 'type', and history and dynamics of the flora and vegetation. A table of altitudinal zones, bibliography, and index to botanical names appear at the end. In the introductory section are illustrated accounts of physical features, climate, vegetation and some altitudinal and aspect zonation, vegetation zones and formations, forest composition with size ranges, and statistics of the woody flora. [The work covers substantial parts of southern Siberia and adjacent middle Asia and Mongolia, and is an expansion of the author's earlier work on Tuva (see **724**).]

MALYSCHEV, L. I., 1968. *Opredelitel' vysokogornykh rastenij južnoj Sibiri*. 284 pp., 37 text-figs., map. Leningrad: 'Nauka'.

Manual-key to vascular plants of the alpine regions of southern Siberia, with limited synonymy, concise indication of local range, Russian-binomial and other vernacular names, and notes on habitat, life-form, phenology, etc.; list of references and indices to all botanical and other names at end. [Includes the Altai and the Sayans.]

Partial works: The Altai

This range, with its highest peak (Belukha) reaching 4506 m, arises in the extreme southeastern part of Western Siberia and extends far into Sinkiang (Xinjiang) in China, where it forms the boundary between that territory and the Mongolian People's Republic. To its south lies the Dzungarian Basin. No specific treatment of the mountain flora, which is said to exhibit affinities with that of the central Rocky Mountains in Colorado (see **103**), has evidently been published since before 1840. The classic work is LEDEBOUR, K. F., 1829–34. *Flora altaica*. 4 vols. Berlin: Reimer, to which supplements were added in 1835 and 1841. Of a later date, but covering a much larger area of Western Siberia and thus not a high-mountain flora in the sense adopted here, is KRYLOV, P. N., 1901–14. *Flora Altaja i Tomskoj gubernii*. 7 vols. Tomsk: Tomsk University. Modern works for identification include Krylov's monumental *Flora Zapadnoj Sibiri* (see **710**) and the manual-key by Malyschev (see above). The Altai have the distinction of being one of the earliest 'alpine' areas in extra-monsoonal Asia to have been botanically explored.

Partial work: western Sayans

The western Sayans form the boundary between Tuva and Krasnoyarsk Krai, and are somewhat lower than their eastern and western neighbors.

KRASNOBOROV, I. M., 1976. *Vysokogornaja flora Zapadnogo Sajana*. 379 pp., 32 figs., 96 maps, tables. Novosibirsk: 'Nauka'.

Concise keyed enumeration of high-mountain vascular plants, with synonymy, references, local distribution (with dot maps for all species), and notes on habitat, special features, potential, etc.; tabular floristic summary with species karyotypes, range type, soil preferences and presence/absence recorded; bibliography (pp. 289–98) and complete indices. An extensive introductory section accounts for botanical exploration (with portraits and itineraries), physical features, climate, vegetation formations with *Charakterpflanzen*, and statement and review of species concepts.

Partial work: eastern Sayans

Situated along the southern fringe of Cisbaikalia, with a maximum altitude of 2922 m, the eastern Sayans arise to the south of Krasnoyarsk and run southeastwards, following the boundaries of Tuva and Mongolian People's Republic until petering out to the south of Lake Baikal.

MALYSCHEV, L. I., 1965. *Vysokogornaja flora vostočnogo Sajana* [Flora alpina montium Sajanensium orientalium]. 368 pp., illus., maps. Moscow/Leningrad: Nauka.

Includes a briefly descriptive flora, with diagnostic keys, synonymy, citations, Russian binomials, fairly detailed indication of local range (with some dot maps), and notes on habitat, special features, ecology, etc.; list of references and indices. The remainder of the work is primarily on geobotany and ecology.

IX. The Stanovoy Uplands

This area lies northeast of Lake Baikal, but generally is just south of the line of the Baikal–Amur Magistral railway; the maximum elevation is 2999 m. For a recent analysis of the high-mountain flora, with many distribution maps, see MALYSCHEV, L. I. (ed.), 1972. *Vysokogornaja flora Stanovogo Nagorja. Sostav, osobennosti i genezis.* 272 pp., 7 figs., 318 maps. Novosibirsk: 'Nauka'.

704

The Ural

See **604** (all works).

705

Steppe and desert areas

Much of Division 7 is a land of deserts and steppes, but as there has been a natural tendency to work up floras for politically delimited areas for the most part before those of physiographic and/or vegetational entities few, if any, specific works of suitable scope exist. One such key work relating to the steppe and desert areas of Mongolia is given below.

Mongolian steppe and desert areas (including the Gobi)

Comprises an area centering upon the Gobi Desert and falling largely in the Mongolian People's Republic and the Inner Mongolian Autonomous Region of China. The flora begun by Norlindh in 1949 has yet to be completed. For alternative coverage, see **760**, **761**, and **763**.

NORDLINDH, T., 1949. *Flora of the Mongolian steppe and desert areas.* Fasc. 1. 155 pp., 18 text-figs., 16 pls., maps. (Reports from the scientific expedition to the northwestern provinces of China under the leadership of Dr Sven Hedin, 11 (Botany, 4).) Stockholm.

Systematic enumeration of vascular plants, with keys to species within genera, full synonymy, references and citations, detailed indication of local range, with several dot maps, summary of overall distribution, and extensive critical commentary and biological and horticultural notes. An introductory section includes an account of prior botanical exploration in the region and the work of the expedition. Covers pteridophytes, gymnosperms, and monocotyledons (Typhaceae through Gramineae) on the Englerian system.

707

Lake Baikal and environs

One of the outstanding natural features in Division 7 is Lake Baikal, for which on account of the existence of some floras relating to the surroundings, a special subheading has been thought desirable. This unique and spectacular lake is the deepest body of fresh water in the world and for the most part is surrounded by high mountains. The littoral flora was documented by Popov and Busik (1966). For other coverage see **725**, **726**, and **727** (all works, but especially *Flora tsentral' noj Sibiri* by MALYSCHEV and PESCHKOVA, 1979).

POPOV, M. G. and BUSIK, V. V., 1966. *Konspekt flory poberežij ozera Bajkal* [Conspectus florae litorum laci Baical]. 214 pp., 15 text-figs. (incl. maps). Leningrad: 'Nauka'.

Concise enumeration of the vascular plants along the foreshores of Lake Baikal, with essential synonymy, citation of *exsiccatae*, and indication of localities; bibliography and diagnoses of new taxa; no index. The introductory section incorporates chapters on the physical features of the lake and its environs, floral regions, and botanical exploration past and present. Partly supplanted by MALYSCHEV, L. I. and PESCHKOVA, G. A. (eds.), 1978. *Flora Pribajkal'ja*. 320 pp., 6 tables. Novosibirsk: 'Nauka'. [Contains a series of floristic and other contributions by various authors (including lists, according to E. E. Jäger in *Prog. Bot.* **42**: 333 (1980)) but not properly a flora in the sense normally accepted in this *Guide*.]

Superregion
71–75
USSR-in-Asia*

The whole of this superregion corresponds to the Asiatic part of the USSR, including Siberia (regions 71–72), the Soviet Far East (region 73), the Caucasus (region 74), and Soviet Middle Asia (region 75). All 'standard' regional and other floras for this area are accounted for herein, except for those few pertaining entirely to the sub-Arctic and Arctic zone (regions 05 and 06). The remainder of central Asia, not forming part of the Soviet Union, constitutes Region 76.

Gmelin's pioneering *Flora sibirica* appeared in St Petersburg in 1747–69, thus partly preceding the advent of binomial nomenclature. The first coverage of the Caucasus was given in Bieberstein's *Flora taurico-caucasica* (1808–19). The next work to cover the whole of Asiatic Russia was Ledebour's *Flora rossica* (1842–53), written when botanical exploration in the superregion was still not very far advanced, and extending to include Alaska. In the same general period other works covering particular regions appeared, among them *Flora altaica* by Ledebour and others, *Flora baicalensi-dahurica* by Turczaninow, *Primitiae florae amurensis* by Maximowicz, and the unfinished *Flora caucasi* by Ruprecht, each respectively a source work for Regions 71 through 74 (Region 75, middle Asia, then was largely outside the Russian Empire). After these mid-century developments, however, no further efforts at comprehensive coverage took place until the 1890s. At that time, there began a general rise in development in Russia, which had a positive effect on botanical (and floristic) research, including the appointment of additional staff to the two large botanical institutions in St Petersburg, the Imperial Botanic Garden and the Botanical Museum of the Academy of Sciences. With greater opportunities, both institutions launched major projects on the Russian flora. The Botanic Garden set up a program of critical regional works, covering (a) the Caucasus, under Kuznetsov and others (*Flora caucasica critica*, 1901–18); (b) Middle Asia, under O. A. and B. A. Fedtschenko

---

* For a lexicon of Russian words used in titles see Table 1, p. 347.

(*Plantae asiae mediae*, 1906–16); (c) European Russia, under B. A. Fedtschenko and Flerow (*Flora Evropejskoj Rossii*, 1908–10); and, lastly (d) Asiatic Russia in general, also under B. A. Fedtschenko and Flerow (*Flora asiatskoj Rossii*, 18 fascicles, 1912–24). The Museum, with more limited resources, established a series covering Siberia and the Far East under the direction of Borodin and Litwinow (*Flora Sibiri i Dal'nego Vostoka*, 6 vols., 1913–31). As a precursor to the latter series, Litwinow prepared a floristic bibliography for Siberia (1909) and a historical review (1908). Neither of these Asian series covered more than ten per cent of the flora, although both continued to appear for several years after the 1917 October Revolution.

By 1930, however, both works had been terminated in the wake of widespread administrative changes and critical reassessment in the latter part of the 1920s, which also featured the gradual imposition of central control over the sciences. Transfer of the Botanic Garden to the Academy of Sciences took place in 1925, and in 1931 came the merger of the Botanical Museum with the Botanic Garden to form what is now the Komarov Botanical Institute. The critique for this rationalization made special mention of wasteful duplication of effort, notably the parallel flora projects on Asiatic Russia! In 1929, as part of the program for the First Five-Year Plan (1929–34), planning for an All-Union flora was initiated, and in 1931–2, concomitant with the merger, work began on *Flora SSSR*, ultimately completed in 1964. The two earlier Asiatic flora projects, however, provided a useful basis for the new flora.

The *Flora SSSR* finally achieved a tremendous task: the effective consolidation of floristic knowledge for Soviet Asia in modern terms. The project had in turn built upon large new collections resulting from systematic biological surveys carried out in many parts of the Soviet Union after 1920. For the woody plants, the *Flora* has been paralleled by Sokolov's *Derevja i kustarniki SSSR*.

These achievements have provided the major basis for most other modern regional, republican, territorial and more local floras, enumerations and manuals in the Asiatic parts of the USSR, although regional works in areas such as western Siberia and the Caucasus where local botanical work had been long established have also provided an important foundation. Many of the later floras were written as part of planned scientific development and the survey and inventory of natural and other resources in a given area, at first under guidance from Leningrad (and Moscow) and later from (or together with) republican or regional institutions. The present pattern of research and writing appears to be one of noticeable decentralization – but also some stagnation.

Progress

For history, see BORODIN, I., 1908. *Kollektory i kollektsii po flore Sibiri*. 245 pp. (Trudy Bot. Muz. Imp. Akad. Nauk, 4). St Petersburg; KRYLOV, G. V. and SALATOVA, N. G., 1969. *Istorija botaničeskikh i lesnykh issledovanij v Sibiri i na Dal'nem Vostoke*. Novosibirsk; and MALYSCHEV, L. I., 1979. Razvitie botaniki v Sibiri v Sovetskij period. *Bot. Žurn. SSSR*, **64**: 112–20. Other accounts are cited under Region 68/69. Early work in the Caucasus is reviewed by V. I. LIPSKY in his *Flora Kavkaza* (1899), and in middle Asia likewise by Lipsky in his pioneering work *Flora Srednej Azii* [Flora Asiae mediae] (1903), pp. 249–337. Recent developments have been surveyed briefly in the cited paper by Malyschev (not, however, for the Caucasus and Middle Asia).

General bibliographies. Bay, 1910; Frodin, 1964; Goodale, 1879; Holden and Wycoff, 1911–14; Hultén, 1958; Jackson, 1881; Pritzel, 1871–7; Rehder, 1911; USDA, 1958.

Supraregional bibliographies

LEBEDEV, D. V., 1956. *Vvedenie v botaniceskuju literaturu SSSR*. 382 pp. Moscow/Leningrad: AN SSSR. [Described in Division 6, Region 68/69 under Regional bibliographies.]

LIPSCHITZ, S. JU., 1975. *Literaturnye istočniki po flore SSSR*. 232 pp., table. Leningrad: 'Nauka'. [Also described in Division 6, region 68/69 under Regional bibliographies. Both this and the preceding have separate sections for the major parts of Soviet Asia: the Caucasus, Siberia and the Far East, and middle Asia.]

LITWINOW, D. I., 1909. *Bibliografija flory Sibiri*. ix, 458 pp. (Trudy Bot. Muz. Imp. Akad. Nauk, 5). St Petersburg. [Relates mainly to Regions 71, 72, 73.]

MERRILL, E. D. and WALKER, E. H., 1938. *A bibliography of Eastern Asiatic botany*. xlii, 719 pp., maps. Jamaica Plain, Mass.; Arnold Arboretum of Harvard University.

WALKER, E. H., 1960. *A bibliography of Eastern Asiatic botany: Supplement 1*. xl, 552 pp. Washington:

American Institute of Biological Sciences. [These two latter works relate mainly to Regions 72 and 73. For description, see Superregion 86–88.]

General indices. BA, 1926– ; BotA, 1918–26; BC, 1879–1939; BS, 1940– ; CSP, 1800–1900; EB, 1959– ; FB, 1931– ; ICSL, 1901–14; JBJ, 1873–1939; KR, 1971– ; NN, 1879–1943; RŽ, 1954– .

710–50

Superregion in general

In addition to the two great All-Union works which comprehensively cover this vast area, *Flora SSSR* (1934–64) and *Derevja i kustarniki SSSR* (1949–62), brief mention is here made of two earlier, more or less competing flora-series on the Asiatic part of the USSR referred to in the introduction above: BORODIN, I. P. and LITWINOW, D. I. (eds.), 1913–31. *Flora Sibiri i Dal'nego Vostoka*. Fasc. 1–6 (in 9 parts). St Petersburg: Botaničeskij Muzej, (I)AN; and FEDTSCHENKO, B. A. and FLEROW, A. F. (eds.), 1912–24. *Flora asiatskoj Rossii*. Fasc. 1–15; N.S., fasc. 1–3. St Petersburg: Glavnyj (Imperatorskij) botaničeskij sad. [Both series comprise revisions of various families, extending also into the Filicales and part of the Musci. The first-named series covers only Siberia and the Far East, without middle Asia or the Caucasus. Both projects were terminated in favor of *Flora SSSR* following the ideological revolution of the late 1920s, but have provided a basis for subsequent floristic studies in so far as their limited extent of coverage (probably no more than 10–15 per cent of the flora) allows.]

All-Union works

KOMAROV, V. L. *et al.* (eds.), 1934–64. *Flora SSSR* [Flora URSS]. 30 vols., index. Illus., maps. Moscow/Leningrad: AN SSSR Press. [For full description, see **680**.]

SOKOLOV, S. JA., 1949–62. *Derevja i kustarniki SSSR*. 6 vols., supplement. Illus., maps. Moscow/Leningrad: AN SSSR. [For full description, see **680**.]

Region

71

Western Siberia

This region includes most of the territory of the Ob'-Irtysh river system and its tributaries; it extends from the Ural Mountains in the west to the boundary of Krasnoyarsk Krai in the east, and from the Altai Mountains and the Kazakhstan boundary in the south to the 'tree line' in the north (for the Arctic zone, see **064**). The following are its principal political/administrative subdivisions: the *oblasti* of Sverdlovsk, Kurgan, Chelyabinsk, Orenburg (in part), Omsk, Tyumen (in part), Tomsk, Novosibirsk, and Kemerov and Altai Krai.

Western Siberia, which formerly on an administrative basis extended well into the steppes and semi-deserts of the Kazakh, had something of a regional flora already by the 1830s (Ledebour's *Flora altaica*) but the formation of a botanical school at Tomsk University in 1885 and the work there of P. N. Krylov and (later) B. K. Shishkin led to the development of more intensive botanical exploration, aided by increasing settlement, development of communications in the wake of the Great Siberian Railway and, especially after 1917, planned resource surveys. The first significant modern flora was Krylov's *Flora Altaja i Tomskoj gubernyi* (1901–14), later superseded by the first great regional flora in the Soviet Union, *Flora zapadnoj Sibiri* (12 vols., 1927–49 and 1961–4), begun by Krylov and continued after his death by Shishkin and by L. P. Sergievskaja. Based to a large extent on critical studies, this latter flora has long been regarded as a model of its kind.[1] The wide and unifying influence of this work and the limited development (at least until recently) of botanical centers other than that at Tomsk University have, in contrast to central Siberia, led to a relative absence of more local manuals or checklists. The first significant work of this kind to appear in modern times is *Opredelitel' rastenij Novosibirskoj oblasti*, published in 1973 and covering the Novosibirsk district, an important population and scientific center encompassing among other things the institutes of the Siberian branch of the Academy of Sciences.

– Vascular flora: 2805 species (Krylov, 1927–49).

Bibliographies. General and supraregional bibliographies as for Superregion 71–75.

Regional bibliography

 AKSENOVA, N. N., 1965. *Flora zapadnoj Sibiri: bibliografičeskij ukazatel'*. 105 pp. Tomsk: Tomsk State University Press. [Unannotated; titles arranged by subject, with author index.]

Indices. General indices as for Superregion 71–75.

710

Region in general

 In addition to the massive, well-known regional flora of Krylov, a practical manual is now available for the Novosibirsk *oblast'* in the eastern part of the region.

 KRYLOV, P. N. *et al.*, 1927–64. *Flora zapadnoj Sibiri* [Flora sibiriae occidentalis]. 12 vols. Tomsk: Russkoe botaničeskoe obščestvo, Tomskoe otdelenie (and successors); Tomsk State University Press (vol. 12). (Vol. 1 reprinted 1955; vol. 5 reprinted 1958, Tomsk; originally published 1901–14 as *Flora Altaja i Tomskoj gubernyi*.)

 Comprehensive, large-scale descriptive flora of vascular plants of western Siberia (and parts of northern Kazakhstan), with keys to all taxa, synonymy, references and citations, rather detailed indication of local and regional distribution (with summary of extralimital range), Russian binomials and vernacular names, taxonomic commentary, and notes on habitat, special features, etc.; indices to all botanical and other names in each volume (complete general index in vol. 12). The introductory section (vol. 1) includes chapters on administrative subdivisions (the old *gubernija* system of Czarist Russia has been used in this work) and botanical districts as well as an artificial general key to families. The limits of the *Flora* range from longitude 59° to 90°E and from latitude 49° to 73°N. B. K. Shishkin and L. P. Sergievskaja were largely responsible for completing this work; the latter was also editor of the two-part supplementary volume 12, which in addition to the amendments to vols. 1–11 and the general index contains an account of botanical work in the region from 1927 onwards.

Partial works: southeastern subregion

For woody plants, reference may also be made to a recently published manual [not yet seen at the time of writing; cited in *Prog. Bot.* 42: 332, 342 (1980)] for the whole of the southeastern part of Western Siberia, with detailed distribution maps: KHLONOV, JU. P., 1979. *Derev'ja i kustarniki jugo-vostočnoj časti Zapadnoj Sibiri*. 126 pp., 50 maps. Novosibirsk: 'Nauka'.

 KOROLEVA, A. S., KRASNOBOROV, I. M. and PEN'KOVSKAJA, E. F., 1973. *Opredelitel' rastenij Novosibirskoj oblasti*. 368 pp., 281 text-figs. Novosibirsk: 'Nauka'.

 Illustrated manual-key to vascular plants, with diagnoses, very limited synonymy, Russian binomials and vernacular names, and brief notes on habitat, phenology, and special features; local range not specifically indicated. The introductory section includes definitions of morphological terms. Complete indices are also provided.

Region

72

Central and eastern Siberia (with Yakutia)

 As here delimited, this vast expanse consists of Siberia from the western limits of the Yenisei basin (approximately 84° to 88°E) east through Cisbaikalia to the limits of Transbaikal, Dahuria, and Yakutia, and from the southern boundary of the USSR north to the 'tree line' (the Arctic and sub-Arctic zones form part of Regions 05 and 06). It thus includes Krasnoyarsk Krai, Khakassk Autonomous Oblast', Tuva ASSR, Irkutsk Oblast', the Buryat-Mongol ASSR, Yakutsk ASSR, and Chitinsk Oblast'.

 Botanical exploration proceeded spottily, mainly along the rivers and from centres along the old trans-Siberian road, especially Irkutsk; the first regional work, Turczaninow's *Flora baicalensi-dahurica*, appeared in the 1850s. Later, exploration was organized from a base at Tomsk University. For several decades, the pattern of publication was in the form of floristic reports of expeditions, local surveys, etc., in an *ad hoc* fashion. Only in the 1920s and afterwards, and then somewhat fitfully, did modern checklists and floras begin to appear; these have generally been based on administrative units or parts thereof. Some of these works are still incomplete. Put simply, the region as defined here has been too large, in great part too remote and until recent decades difficult of access, and as a whole too under-developed and decentralized ever to have had a separate 'regional' flora, in contrast to nearly all other parts of the USSR. *Overall* coverage has been provided only in All-Union works such as *Flora SSSR*.

In addition to Popov's *Flora srednej Sibiri* of the 1950s, which covers essentially the same southern subregion as *Flora baicalensi-dahurica*, a new subregional work, *Flora tsentral'noj Sibiri*, has appeared: the 'third generation'.[2] This will extend to the subregion of the upper Lena and its tributaries, northeast of Lake Baikal, an area through which the Baikal–Amur Magistral (BAM) railway is being constructed. Other coverage of the region is found in a miscellany of largely uncoordinated works of more limited scope, some not completed, covering Krasnojarsk Krai, or the Jenisei basin (Reverdatto *et al.* 1937, 1960– ; Čerepnin, 1957–67);[3] Cisbaikalia (Djagilev, 1938, a 'spring flora'); Tuva (Sobolevskaja, 1953), Yakutia (Petrov, 1930; Karavajev, 1958; Tolmatchev, 1974); and Transbaikalia (Fedtschenko *et al.*, 1929– ; Sergievskaja, 1966–72). Floras or checklists are also available for the high-mountain areas along the southern border for the foreshores of Lake Baikal and (in Tuva) for the woody plants, as well as for more restricted areas not accounted for here.

Bibliographies. General and supraregional bibliographies as for Superregion 71–75; see also Superregion 86–88 (Merrill and Walker).

Indices. General indices as for Superregion 71–75.

720

Region in general

For separate works on the southern alpine zone, see 703/VIII; for the Stanovoy Uplands to the northeast, see 703/IX; for special coverage of the Lake Baikal littoral, see 707. The most comprehensive current regional works are *Flora srednej Sibiri* by M. G. Popov (1957–9), described below, and the recent *Flora tsentral'noj Sibiri* by MALYSCHEV and PESCHKOVA (1979), see 725. Neither, however, covers the complete region as circumscribed here. Both works omit Tuva (724) as well as most areas north of 60 °–61 °N., thus limiting themselves to what is traversed by the railway lines with even there some parts imperfectly covered.

The more recent of the two works additionally omits southern Krasnoyarsk Krai, but gives more extended as well as deeper coverage to the northeastern sector, e.g., in the Stanovoy Uplands.

POPOV, M. G., 1957–9. *Flora srednej Sibiri*. 2 vols. 918 pp., 104 text-figs., 2 portraits. Moscow/Leningrad: AN SSSR Press.

Briefly descriptive flora of vascular plants, with synoptic keys to genera and groups of species; limited synonymy, with citations; vernacular and binomial Russian names; generalized indication of local and extralimital range; taxonomic commentary and notes on habitat, etc.; figures of representative species; general indices to all botanical and other names in vol. 2. The introductory section includes a list of principal references, a synopsis of families, and a survey of local floristics (with maps). Covers the region from the upper Jenisei through Cis- and Transbaikalia to Dahuria and from the Sajan Mountains on the southern frontier to approximately 61 °N.

721

Krasnoyarsk (Krasnojarsk) Krai (in general)

For more coverage of the southern part of this vast territory, which includes by far the greater part of the Yenisei River basin, see 722. The general flora begun by Reverdatto in 1937 has still to be completed, although as of 1980 it was very far advanced with nine of the ten or eleven parts published.

REVERDATTO, V. V. and SERGIEVSKAJA, L. P., 1937. *Konspekt Prijenisejskoj flory* [Conspectus florae jenisseensis]. Fasc. 1. 46 pp. Tomsk: Biologičeskij naučno-issledovatel'skij institut, Tomsk State University.

REVERDATTO, V. V. *et al.*, 1960– . *Flora Krasnojarskogo kraja*. Fasc. 2– . Maps. Tomsk: Tomsk University Press. (fasc. 2, 5(2), 5(4), 6, 7/8, 9(2)); Novosibirsk: 'Nauka' (fasc. 3/9, 4/5(1)).

These two evidently complementary works together make up a serially published, systematic treatment of the vascular plants of the Yenisei Basin, or Krasnoyarsk Krai (before 1935 known as Prijenisejsk Krai), with descriptions of species (except in fasc. 1), diagnostic keys to genera and species, Russian binomials and vernacular names, synonymy, citations of pertinent literature, generalized indication of local range (from fasc. 2 onward expanded to include *exsiccatae*), northern limits, and miscellaneous notes; each part fully indexed (except in fasc. 1). An introductory section (fasc. 1) provides accounts of the history of botanical exploration and sources for the flora as well as family accounts for Polypodiaceae through the Hydrocharitaceae (Englerian system), while in the continuation work begun in 1960, the coverage available (as of 1980) extends over the remaining mono-cotyledons (fasc. 2–4), the Salicaceae through Amaranthaceae (fasc. 5(1) and 5(2)), the Papaveraceae through Rosaceae (fasc. 5(4)), the Leguminosae (fasc. 6), the Geraniaceae through Cornaceae and Pyrolaceae through Boraginaceae (fasc. 7/8), the Labiatae (fasc. 9), and the Solanaceae, Scrophulariaceae, and Campanul-aceae (fasc. 9(2)). [Fascicles published from Tomsk University have appeared as separate entities save for nos. 7 and 8 which were issued jointly under one cover, like those fascicles published by 'Nauka' at Novosibirsk. Preparation of this flora, however, has been largely a long-term project of the Krylov Herbarium at Tomsk University (*Index Herbariorum*, part I, 7th edn., p. 286), and in recent years has been under the leadership of A. V. Polozhij. It is expected that the work will be complete in 10 fascicles (each in one or more sections).]

722

Krasnoyarsk (Krasnojarsk) Krai (southern part)

This unit covers the whole territory south of the Angara River, and centers on Krasnoyarsk; also included is the Khakassk Autonomous Oblast'. Within this area are most centers of population and scientific activity in the *krai*. To the six-part documentary flora of Čerepnin (1957–67) has now been added a manual-key by Begljanova *et al.* (1979).

BEGLJANOVA, M. I. *et al.*, 1979. *Opredelitel' rastenij juga Krasnojarskogo kraja*. 669 pp., 324 text-figs. Novosibirsk: 'Nauka'.

Illustrated manual-key to vascular plants, with abbreviated indication of habit, phenology, habitat, distribution, properties, uses and potential, etc., Russian nomenclatural equivalents, and limited synonymy; general indices to all names at end. The introductory section contains an illustrated glossary as well as a general key to families. Distribution units within the southern Krasnojarsk Krai are as used in Čerepnin's flora (see below), which this work partially supplants.

ČEREPNIN, L. M., 1957–67. *Flora južnoj časti Krasnojarskogo kraja*. 6 parts. 122 text-figs., map. Krasnoyarsk: Krasnojarsk State Pedagogical Institute.

Documentary, non-descriptive manual of vascular plants, with diagnostic keys to all taxa, synonymy, references and citations, Russian vernacular names and binomials, localities with indication of *exsiccatae*, and many notes on habitat, occurrence, properties, uses, habit, phenology, etc.; maps of area in fascicles 1 and 6; general indices to all vernacular, Russian binomial, and Latin names in fascicle 6 (fasc. 5 separately indexed). A general key to families appears in fascicles 1 and 5. The 58th parallel (just north of the lower Angara) forms the northern limit of this flora, which was evidently written as a source work for middle schools.

723

Krasnoyarsk (Krasnojarsk) Krai (northern part)

This is the territory from the Angara River north to the limits of tree growth, and includes the middle Yenisei and the central Siberian Plateau. For the sub-Arctic north Siberian plain, including the Taimyr Peninsula, see 066; for Severnaja Zemlja, see 056. Local coverage here is fragmentary, and reference must be made to the general works of REVERDATTO and others (721, all works). For *Flora Putorana*, see 066.

724

Tuva (Tuvinsk) ASSR

See also 703/IX (all works). This territory, at one time loosely associated with the Manchu Empire and later known as 'Tannu Tuva', was incorporated into the Soviet Union after World War II.

KOROPACHINSKY, I. JU. and SKVORTSOVA, A. V., 1966. *Derevja i kustarniki Tuvinskoj ASSR*. 184 pp., 76 text-figs., map. Novosibirsk: 'Nauka'.

Illustrated forest flora, with analytical keys, synonymy, general indication of local and extralimital range (with local distribution maps), and extensive notes on habitat, biology, special features, properties, uses, etc.; index to botanical names. The introductory section includes chapters on floristics, vegetation, ecology, and botanical exploration.

SOBOLEVSKAJA, K. A., 1953. *Konspekt flory Tuvy.* 245 pp., portrait, map. Novosibirsk: AN SSSR Press (West-Siberian Branch).

Systematic enumeration of vascular plants, with limited synonymy and citations of relevant literature, Russian vernacular names and binomials, fairly discursive indication of local range (with some citations) but no extralimital distribution, and some taxonomic commentary and notes on uses; indices to all Russian and botanical names. The prefatory section contains a short account of botanical exploration in Tuva. [Description in part prepared from notes supplied by M. E. Kirpicznikov, Leningrad. The press run for this flora was exceptionally small – 300 copies.]

725

Eastern Central Siberia (in general)

This subregion of Central Siberia, which includes areas usually considered as geographically part of Eastern Siberia, is here delimited as essentially comprising Irkutsk and Chita Oblasti together with the Buryat-Mongol ASSR, with Lake Baikal in the center (for other coverage, see **707**), the Mongolian border on the south (see also **703/VIII**) and the Stanovoy Uplands in the northeast (see also **703/IX**). This corresponds with the area covered by *Flora tsentral'noj Sibiri* by Malyschev and Peschkova, described below.

MALYSCHEV, L. I. and PESCHKOVA, G. A., 1979. *Flora tsentral'noj Sibiri* [Flora Sibiriae centralis]. 2 vols. 1048 pp., 1+1284 maps, frontispiece. Novosibirsk: 'Nauka'.

Briefly descriptive flora of vascular plants (2311 native and naturalized species in 596 genera), with keys to all taxa, limited synonymy, Russian nomenclatural equivalents, karyotypes, some critical remarks, and indication of habitat, occurrence, and distribution (individual dot maps, covering 55 per cent of the species, grouped together at end of each volume) along

with special features; complete general indices in volume 2. The introductory section (volume 1) gives information on the background of the work, earlier floras, definition of the limits of the area covered and of its subdivisions; main references, p. 10. [Despite the title, the limits of the work, which are more or less the same as defined for **725**, differ from those adopted by M. G. Popov for his *Flora srednej Sibiri* (**720**) in that the latter also includes the southern part of the Krasnoyarsk Krai (north to about 61 °N.). Related works include a progress report by Malyschev (with source bibliography)[2] and the following 'precursor': MALYSCHEV, L. I. and PESCHKOVA, G. A., 1978. *Flora Pribajkalja*. 320 pp., 6 tables. Novosibirsk: 'Nauka'. [Not seen by this reviewer; reported to contain floristic lists by E. E. Jäger in *Prog. Bot.* 42: 333 (1980) but possibly not a monographic flora or enumeration in the sense normally accepted in this *Guide*.]

726

Cisbaikalia

This area is here considered to correspond for the most part with the limits of Irkutsk Oblast'. No specific general flora or enumeration is available; the most nearly suitable work is the 'spring flora' of Djagilev (1938). Full modern coverage is provided by MALYSCHEV and PESCHKOVA (**725**).

DJAGILEV, V. F., 1938. *Opredelitel' vesennikh i ranne-letnikh rastenij Predbajkalja.* 223 pp., illus. Irkutsk: Oblast' Publishing Service.

Illustrated, semipopular students' manual-key to the spring and early summer vascular plants of Irkutsk Oblast' and the Lake Baikal littoral, with Russian binomial and vernacular names and notes on habitat, phenology, local distribution, uses, biological features etc.; list of references and index to Russian binomial names. The introductory section includes chapters on collecting methods and terminology as well as on general features of the vegetation. Arrangement of families is on the Englerian system. [Description prepared in part from notes supplied by M. E. Kirpicznikov, Leningrad. No 'alpines' are included in this work.]

727

Transbaikalia (Zabajkalja)

This includes the region formerly known as Dahuria. As delimited here, Transbaikalia corresponds approximately to the area of the Buryat-Mongol ASSR and Chitinsk Oblast', with the eastern limit formed by a line from the Silka-Amur river junction north to the Olyomka River (120° to 122 °E). The western limit is largely formed by Lake Baikal, and in the north are the Stanovoy uplands. See also **725** (MALYSCHEV and PESCHKOVA).

FEDTSCHENKO, B. A. *et al.* (eds.), 1929– . *Flora Zabajkalja*. [Flora transbaicalica.] Fasc. 1–7. Pp. 1–760, figs. 1–356, maps. Leningrad: Geografičeskoe obščestvo (fasc. 1–2, 1929–31); Moscow/Leningrad: AN SSSR Press (fasc. 3–6, 1937–54); Leningrad: 'Nauka' (fasc. 7, 1975). (Published on behalf of the Vsesoyuznoe Botaničeskoe Obščestvo, Leningrad.)

SERGIEVSKAJA, L. P. *et al.*, 1966–72. *Flora Zabajkalja*. 2nd edn. Vyp. 1–4. portrait, map. Tomsk: Tomsk University Press.

The first edition of this work, in manual-key format, comprises a briefly descriptive account of vascular plants, with limited synonymy, Russian binomial and vernacular names, fairly detailed indication of local range (by subdivisions as delimited on an area map, supplied in fascicles 1, 6 and 7), and notes on habitat, phenology, special features, biology, etc.; representative illustrations; no indices. Fascicle 1 includes a consideration of floristic districts as adopted for the work. Not yet completed; as of fascicle 7 complete through the Cornaceae (Englerian system). Progress has been intermittent due perhaps to its [semi-official] status and the availability and interest of particular individuals as editors; the project has been a charge of the All-Union Botanical Society of the USSR. [In 1980 fascicle 8 appeared.]

The revised edition of 1966–72, a project of the herbarium at Tomsk University and the Tomsk branch of the Botanical Society, has the format of a briefly descriptive flora, with synonymy, references, citations of key literature, detailed indication of local range (including localities and authorities), extralimital distribution, some taxonomic commentary, and notes on habitat; no illustrations. The first fascicle (1966) includes in its introductory section an account of floristic districts as used in the work (with map) and a concise encyclopedia of collectors in Transbaikalia. Evidently not continued, perhaps in part due to the death of the senior author in 1975; covers pteridophytes, gymnosperms and monocotyledons, corresponding to the first two fascicles of the first edition. [Descriptions of both editions prepared with the assistance of information supplied by M. E. Kirpicznikov and V. I. Grubov, Leningrad; the latter has from fascicle 7 acted as editor of the work.]

728

Yakutsk (Jakutsk) ASSR

POPOV'S *Flora srednej Sibiri* and (to a somewhat greater extent) MALYSCHEV and PESCHKOVA's recent *Flora tsentral'noj Sibiri* (**720**) take in only the southwestern corner of this huge area. The zone north of the 'tree line' has also been designated as **067**. The incomplete flora of Petrov, with a very expansive format, is complemented by a checklist by Karavaev (1958) and a manual-key by Tolmatchev (1974).

KARAVAEV, M. N., 1958. *Konspekt flory Jakutii.* 190 pp., 12 text-figs. Moscow/Leningrad: AN SSSR Press.

Systematic enumeration of vascular plants (1523 species), with limited synonymy, concise indication of local range, and brief notes on habitat, special features, etc.; bibliography; no index. The introductory section includes chapters on the major features of the flora, floristic regions, and past and present botanical exploration.

PETROV, V. A., 1930. *Flora Jakutii.* Vol. 1. xii, 221 pp., 75 text-figs. (incl. maps). Leningrad: AN SSSR Press.

Detailed descriptive flora of vascular plants, with keys to all taxa; full synonymy, with references; citation of *exsiccatae*, with indication of local range (incl. some dot maps), and general summary of extralimital range; notes on habitat, etc.; index to all botanical names. Not completed; extends from the Pteridophyta through the Poaceae [Gramineae] (Englerian system).

TOLMATCHEV, A. I. (ed.), 1974. *Opredelitel' vysšikh rastenij Jakutii.* 543 pp., 70 pls. Novosibirsk: 'Nauka'.

Concise illustrated manual-key to vascular plants

of this vast territory, with brief diagnoses, Russian and Yakut vernacular names, and short notes on habitat and internal range. A very brief introduction is given, and there are indices to Russian and Yakut family and generic names as well as to all Latin names. An interesting feature of this work is the clear indication in range summaries for individual species of presence in the tundra or sub-Arctic zone.

Region

73

Soviet Far East

This region encompasses all the areas of the Soviet Union associated with the north Pacific Ocean (except for Anadyr-Chukotsk which is wholly in the sub-Arctic zone – see **068**) together with the Amur Basin west to the Silka River junction. Principal political/administrative units include Primorsk Krai; Khabarovsk Krai; Amursk Oblast'; Magadansk Oblast' (except sub-Arctic zone); Kamchatsk Oblast' (including the Commander Islands); and Sakhalin Oblast' (including the Kurile Islands). For convenience, however, this region has been subdivided into seven geographical units (**731–737**) corresponding to the 'districts' of Voroshilov's *Flora sovetskogo Dal'nego Vostoka* (1966), the latest overall checklist.

Early botanical exploration was sporadic and localized and the first regional flora appeared only in the 1860s. A major impetus to more detailed collecting work came only with the construction of the Great Siberian Railway system and settlement activities from the 1880s onwards; in the vanguard was V. L. Komarov, whose field work in the region laid a basis for his floras of Manchuria (1901–7; see **860/I**), the Kamchatka Peninsula (prepared around 1910, but not published until the late 1920s), and the Soviet Far East generally (1931–2, with E. Klobukova-Alisova). He also compiled a bibliography on the flora and vegetation of the region (1928; extended by others to 1969). The last two works were doubtless stimulated by the creation of the Far East Branch of the Soviet Academy of Sciences in the 1920s. Further collecting, practical needs, and the influence of the Flora SSSR resulted in a new cycle of works for the region (or parts thereof) appearing from the 1950s. If the Japanese contributions on Sakhalin

and the Kuriles as well as the critical work of Hultén on Kamchatka (1927–30) and elsewhere are included, floristic coverage is now quite substantial, although still sketchy for some remote areas. Good coverage is also available for the diverse woody flora.

Bibliographies. General and supraregional bibliographies as for Superregion 71–75; see also Superregion 86–88 (Merrill and Walker).

Regional bibliographies

KOMAROV, V. L., 1928. *Bibliografija k flore i opisaniju rastitel'nosti Dal'nego Vostoka* [Bibliography of the flora and of the vegetation of the Far East]. 279 pp. (Zap. Južno-Ussurijsk. Otd. Gosud. Russk. Geogr. Obšč., vyp. 2). Vladivostok. [Comprehensive list, with subject and author indices; 1202 entries.]

GOROVOJ, P. G. and SOPOVA, M. S. (eds.), 1973. *Flora, rastitel'nost' i rastitel'nye resursy Dal'nego Vostoka. Ukazatel' literatury* (1928–1969 gg.) [Flora, vegetation and plant resources of the Far East. Bibliography (1928–1969)]. 552 pp. Vladivostok. [An extension of the Komarov bibliography, arranged by subject headings.]

Indices. General indices as for Division 7.

730

Region in general

The only overall work including keys remains the manual by Komarov and Klobukova-Alisova (1931–2); the flora by Voroshilov (1966) is merely an enumeration, with only partial keys. A separate manual-key is now available for Primorja and Priamurja, the areas with largest demand (see **731, 732**). Two general works are also available for the woody plants.

KOMAROV, V. L. and KLOBUKOVA-ALISOVA, E. N., 1931–2. *Opredelitel' rastenij Dal'nevostočnogo kraja.* [Key for the plants of the Far Eastern Region of the USSR.] 2 parts, x, 1175 pp., 330 pls., color frontispiece. Leningrad: AN SSSR Press.

Concise illustrated manual-key to vascular plants, with limited synonymy, Russian vernacular and binomial names, brief indication of internal range, and short notes on special features of particular genera; separate keys to genera of aquatic and woody plants; abbreviations; indices to all Russian and botanical names. Encompasses the entire Soviet Far East together with the Kamchatka Peninsula and Sakhalin, but now superseded in the southern part of this range (for Amur Basin southward see **731**, **732**).

VOROSHILOV, V. N., 1966. *Flora sovetskogo Dal'nego Vostoka*. 477 pp. Moscow: 'Nauka'.

Gives a concise enumeration of vascular plants of the entire region, with keys only to species within genera; synonymy, with references; vernacular and 'standardized' Russian names; fairly detailed indication of local range (by districts); some taxonomic commentary; bibliography; indices to family, generic, and vernacular names. The introductory section includes statistical details together with an outline of the features of the flora.

Special groups – woody plants

USENKO, N. V., 1969. *Derevja, kustarniki, i liany Dal'nego Vostoka*. 416 pp., 58 pls., maps (end papers). Khabarovsk: Khabarovsk Book Publishing Service.

Dendrological flora of woody plants, with descriptions, illustrations, regional distribution, notes on special features, and silvicultural data; ranges of 32 leading species shown on endpaper maps; references (p. 384); index to Russian names with Latin equivalents. No keys are provided, and less common species are treated in small type. Covers 498 species.

VOROBIEV, D. P., 1968. *Dikorastuščije derevja i kustarniki Dal'nego Vostoka: opredelitel'*. 277 pp., 40 text-figs. Leningrad: 'Nauka'.

Descriptive manual of the indigenous woody plants of the Soviet Far East, with keys to genera and species; Russian vernacular and binomial names; general indication of internal and extralimital range; some taxonomic commentary; lists of references and indices to all Russian and botanical names. No synonymy is included. Covers 475 species.

731

Primorja

This is considered to correspond with the Vladivostok district, or Primorsk Krai. A general manual and a 'tree-book', both fairly recent, are available specifically for this area and Priamurja (732).

VOROBIEV, D. P., *et al.*, 1966. *Opredelitel' rastenij Primorja i Priamurja*. 491 pp., 192 text-figs. Moscow: 'Nauka'.

Concise manual-key to the vascular plants of Primorja and the Amur Basin (including the Khabarovsk district), with limited synonymy, vernacular and 'standardized' Russian names, brief indication of

habitat and local range, and occasional taxonomic remarks; special key to genera of woody plants; indices to all botanical and other names. The introductory section includes statistical details as well as a short chapter on floristics.

Special groups – woody plants

VOROBIEV, D. P., 1958. *Opredelitel' derev'ev i kustarnikov Primorja i Priamurja*. 184 pp., text-figs., 50 pls. Blagoveščensk: Amur Book Publishing Service.

Illustrated artificial analytical manual-key to woody plants for field use, with separate keys based on different field and other characters; systematic synopsis; no index. Covers 353 species.

732

Priamurja

Comprises the Amur Basin (including the Khabarovsk district) as far upstream as the Šilka River junction at the boundary of Chita Oblast'. The area is covered by the same manuals as Primorja (see **731** above).

733

Sakhalin

The southern half of this island, also known as Karafuto, was under Japanese administration from 1905 to 1945. General floras are available in both Russian (Vorobiev *et al.*, 1974) and Japanese (Sugawara, 1937, 1937–40). Also accounted for here are two partial works by Japanese workers and a 'tree book' by Tolmatchev.

SUGAWARA, S., 1937. *Karafuto-no shokubutsu* [Plants of Saghalien]. [iv], 490 pp., map. Toyohara (Sakhalin).

Systematic enumeration (pp. 50–138) of vascular plants, without keys; includes synonymy and references, Japanese vernacular and scientific names, generalized indication of local range throughout the island, and indices to all botanical and Japanese names. The introductory section has a chronological table of local botanical research and exploration as well as a descriptive historical account, while extensive appendices (in Japanese) provide descriptive remarks on woody, naturalized, and useful plants (separately indexed).

SUGAWARA, S., 1937–40. *Karafuto shokubutsu zushi* [Illustrated flora of Saghalien]. 4 vols. 42 text-figs., 892 pls., 2 maps. Tokyo: The author. (Reprinted, 1975.)

Large-scale atlas-flora of vascular plants, without keys; the plate of each species is accompanied by a text in Japanese containing a short description, synonymy (with references and citations), Japanese vernacular and scientific names, citations of *exsiccatae* and literature record, generalized indication of local and extralimital distribution, and notes on habitat, special features, biology, etc.; complete indices to Japanese and botanical names in each volume. The introductory section (vol. 1) includes an account (with itinerary maps) of botanical research and exploration in the island as well as a bibliography. [This well-produced and documented work originally was produced in a limited edition (500 sets) and evidently was uncommon outside Japan. However, it was reprinted in 1975.]

VOROBIEV, D. P. *et al.*, 1974. *Opredelitel' vysšikh rastenij Sakhalina i Kuril'skikh ostrovov* [Key for the vascular plants of Sakhalin and Kurile Islands]. 372 pp., 64 pls. Leningrad: 'Nauka'.

Concise illustrated manual-key to vascular plants (in Russian), with diagnoses, limited synonymy, Russian names, and indications of local range, habitat, and endemicity; some critical remarks also included. The introductory section gives the plan of and basis for the work, while at the end is a separate key to genera with woody species, a list of literature (p. 362), and complete indices.

Partial works

KUDO, Y., 1923. A contribution to our knowledge of the flora of northern Saghalien. *J. Fac. Agric. Hokkaido Univ.* 12(1): 1–68, pls. 1–12.

Enumeration of vascular plants (based mainly on collections by the author), with synonymy, relevant citations, notation of *exsiccatae* with localities, some taxonomic commentary, and remarks on habitat, etc.; index. An introductory section (in English) includes descriptive remarks on the vegetation.

MIYABE, K. and MIYAKE, T., 1915. *Flora of Saghalin.* 26, 648, 19, and 10 pp.; 13 pls. Toyohara: Saghalin Government. (In Japanese.)

Concise, keyed descriptive flora of the vascular plants of southern Sakhalin, with limited synonymy, essential citations, generalized indication of local and extralimital distribution, and brief taxonomic and other notes; complete indices. For an incomplete enumeration in English, see MIYABE, K. and KUDO, Y. 1930–4. *Flora of Hokkaido and Saghalien.* Parts 1–4. pp. 1–528 (J. Fac. Agric. Hokkaido Univ., 26). Sapporo. [Covers families from the pteridophytes through Polygonaceae on the Englerian system.]

Special groups – woody plants

TOLMATCHEV, A. I., 1956. *Derevja, kustarniki, i derevjanistye liany ostrova Sakhalina: kratkij opredelitel'.* 171 pp., 83 halftones. Moscow/Leningrad: AN SSSR Press.

Illustrated semipopular manual-key to woody plants, with general indication of local range and more or less extensive remarks on habitat, ecology, phenology, biology, etc.; Russian binomial and vernacular names; no index.

734

Ochotia

Includes the western part of Magadan Oblast' and the northern part of Khabarovsk Krai. No recent separate general floras or enumerations are available.

735

Kamchatka Peninsula

Includes the Peninsula and the region immediately to the north of it. In spite of the similarity in publication dates of the two general floras, Hultén's work, as well as being more critical, is the more recent by 15 years.

HULTÉN, E., 1927–30. *Flora of Kamchatka and adjacent islands.* 4 vols. Illus., maps (Kongl. Svenska Vetenskapsakad. Handl., ser. III, vol. 5, parts 1–2, vol. 8, parts 1–2). Stockholm.

Detailed descriptive flora of vascular plants, without keys; extensive synonymy, with references and literature citations; notation of *exsiccatae*, with indication of local range (including distribution maps), and general summary of extralimital range; taxonomic commentary and extensive notes on habitat, ecology, etc.; index to all botanical names in vol. 4. The introductory section (vol. 1) incorporates accounts of prior botanical exploration and the author's Kamchatka itinerary together with descriptions of the various plant communities and formations. In addition to the Peninsula, this work also covers the Commander Islands and the northern Kuriles.

KOMAROV, V. L., 1927–30. *Flora poluostrova Kamčatki*. [Flora peninsulae Kamtschatka]. 3 vols. Illus., plates, maps. Leningrad: AN SSSR. (Reprinted 1951, Moscow, Leningrad: AN SSSR, as vols. 7–8 of *V. L. Komarov: Opera selecta*.)

Copiously descriptive flora of vascular plants, with keys to all taxa; synonymy, with references; some Russian names; citation of *exsiccatae*, with localities, and general summary of extralimital range; taxonomic commentary and notes on habitat, occurrence, special features, etc. (habitat and occurrences also in Latin); indices to botanical names in each volume. The introductory section includes an account of botanical exploration, while in vol. 3 is found a summary discussion (in very defective English) of the general features of the flora and vegetation.[4]

736

Commander Islands

See also **110** (HULTÉN, 1941–50, 1968); **111** (Hultén, 1960); **735** (HULTÉN, 1927–30).

VASIL'EV, V. N., 1957. *Flora i palaeogeografija Komandorskikh ostrovov*. 259 pp. Moscow/Leningrad: AN SSSR Press.

Briefly descriptive flora of vascular plants, with keys to all taxa; full synonymy, with references and literature citations; notation of *exsiccatae*, with localities, and general summary of extralimital range (including distribution classes); 'standardized' Russian names; taxonomic commentary and notes on habitat, life-form, etc.; annotated bibliography; index to all botanical and other names. The introductory section contains chapters on physical features, climate, floristics, vegetation, and botanical exploration and research; a postscript (pp. 201–36) comprises a detailed floristic, phytogeographic, and historical analysis.

737

Kurile Islands

See also **733** (VOROBIEV); **735** (HULTÉN). From 1905 to 1945 these islands were wholly under Japanese administration, but since 1945 they have been incorporated into the Soviet Union. Two works, one rather outdated, covering the group as a whole have appeared (see below); for local literature of more recent date, including the numerous papers of Tatewaki, general and supraregional bibliographies should be consulted, as well as the regional bibliography of Gorovoj and Sopova (1973).

MIYABE, K., 1890. The flora of the Kurile Islands. *Mem. Boston Soc. Nat. Hist.* **4**(7): 203–75, map.

Systematic enumeration of the known vascular plants, without keys or descriptions (except for new taxa); synonymy, with references; general indication of local range; taxonomic commentary; notes on habitat and special features; no index. The introductory section has chapters on physical features and floristics together with a statistical summary.

VOROBIEV, D. P., 1956. *Materialy k flore Kuril'skikh ostrovov*. 79 pp. (Trudy Dal'nevostočnogo fil. 'Komarova', ser. Bot., 3). Vladivostok: AN SSSR.

Systematic enumeration (pp. 6–78) of known vascular plants of the archipelago, including all earlier reports, with indication of local range (by islands), Russian names, and habitat; bibliography (p. 78–9). The introductory section gives an account of botanical exploration and research in the area and statistics of the flora (994 species, as against 317 species recorded by Miyabe in 1890).

Region

74

The Caucasus

This region is here considered to include the Soviet Republics of Armenia, Georgia, and Azerbaijan, along with the High Caucasus, Ciscaucasia (the north slope) and most of the region lying between the Sea of Azov to the west and the Caspian Sea. The northern boundary is approximately equivalent to that adopted for Flora Europaea, i.e., along the Kuma River in the east from its mouth as far as Priumsk, then on a line running north of Stavropol as far as the Jeja River, thence along this to its mouth in the Sea of Azov northwest of Novorossiysk.

With a long history of botanical exploration and the first regional flora published in the early nineteenth century, the rich, varied Caucasian flora, containing many relict Eurasian elements and including some unusual subtropical pockets, is relatively well-known and an extensive range of floras and other works is available, some still in progress. Unfortunately, among these latter is the second edition of Grossheim's *Flora Kavkaza*, begun 40 years ago. In addition, the one-volume general manual by the same author, *Opredelitel' rastenij Kavkaza.* (1949), is now in need of a revision.

Among the numerous republican and district floras are two complete works on Georgia (*Flora Gružii*, 1941–52, 2nd edn., 1971– ; and *Opredelitel' rastenij Gruzii*, 1964–9), both in Georgian, and one on Azerbaijan (*Flora Azerbajdžana*, 1950–61). The *Flora Armenii* of Taktajan (1954– ; six volumes now published), arranged according to his system, has yet to be completed. For the north slope, or Ciscaucasia, there is a concise checklist, *Spisok rastenij Severnogo Kavkaza i Dagestana*, by Flerow (1938). An incomplete general work on trees and shrubs is *Dendroflora Kavkaza* (1959–), which is supplanting an earlier work by Medwedew (1919) but in less concise form. Complete floras, or works on woody plants, at district level are available for Krasnodarsk Krai, Daghestan, parts of western Georgia including the Black Sea coast, the environs of Tbilisi and Jerevan, and the Apsheron Peninsula near the oil city of Baku. However, Fedorov has criticized some of these works for excessive imitation of *Flora SSSR*; to him, only the *Flora Armenii* among the large Caucasian works has the most suitable design for a regional flora, with lesser emphasis on descriptions and more on geographical distribution and critical commentary.

Bibliographies. General and supraregional bibliographies as for Superregion 71–75; see also Superregion 77–79 (Field).

Regional bibliography

LIPSKY, V. I., 1899. *Flora Kavkaza* [Flora caucasi]. pp. 1–116 (bibliography). Trudy Tiflis. Bot. Sada, 4). St Petersburg.

Indices. General indices as for Superregion 71–75.

740

Region in general

The second edition of Grossheim's *Flora Kavkaza*, begun in 1939, has yet to be completed although it has reached its seventh volume and is far advanced in the system. *Dendroflora Kavkaza* is also not yet completed. These and the other regional works are all in Russian.

GROSSHEIM, A. A., 1949. *Opredelitel' rastenij Kavkaza.* 747 pp., map. Moscow: 'Sovetskaja Nauka'.

Concise manual-key to vascular plants of the whole Caucasus region, with limited synonymy, vernacular and binomial Russian names, brief indication of local and extralimital range, and notes on habitat, phenology, life-form, etc.; special keys to families represented only by cultivated species; indices to botanical and Russian names of families and genera. The brief introductory section includes a map with floristic regions.

GROSSHEIM, A. A. *et al.* (eds.), 1939– . *Flora Kavkaza.* 2nd edn. Vols. 1– . Illus., maps. Baku: AN Azerbajdžanskoj SSR (vols. 1–3); Moscow/Leningrad: AN SSSR Press (later Leningrad: 'Nauka') (vols. 4–). (First edn., 1928–34, Tiflis (later Baku).)

Comprehensive descriptive 'research' flora of vascular plants, with keys to all taxa, full synonymy, references and citations, Russian and local vernacular names (but no Russian 'standarized' binomials), detailed indication of local range, with distribution maps for every species, taxonomic commentary, and notes on habitat, life-form, special features, etc.; diagnoses of new taxa; indices to generic names in Russian as well as all scientific names in each volume. An introductory section (vol. 1) includes a map of floral regions (repeated in later volumes). Not yet completed; at this writing (1979) seven volumes have been published, covering families through the Scrophulariaceae (Englerian system).

Special groups – woody plants

GULISASHVILI, V. Z. *et al.*, 1959– . *Dendroflora Kavkaza.* Vols. 1– . Illus. Tbilisi: AN Gruzinskoj SSR Press.

Copiously illustrated descriptive dendrological treatment of the woody plants of the Caucasus (including native and introduced species), with partial keys, limited synonymy, references, 'standardized' Russian binomials and Russian and local vernacular names, general indication of local and extralimital distribution of native species, with distribution maps, origin of exotic species, and notes on habitat, special features, ecology, biology, adaptation, silvicultural aspects, management, etc.; lists of references and complete indices in each volume. At this writing (1979) five volumes have appeared, covering families through the Punicaceae (Englerian system).

MEDWEDEW, JA. S., 1919. *Derevja i kustarniki Kavkaza.* 3rd edn. iv, 485 pp. Tiflis. (First edn., 1883.)

Descriptive account of known woody plants, with keys to species, limited synonymy, Russian names, relatively detailed indication of local range, and notes on habitat, dendrology, etc.; artificial keys to genera; bibliography and indices to Latin and Russian names and to local vernacular names. Now largely superseded by *Dendroflora kavkaza* (see above).

741

Ciscaucasia (Predkavkaz) in general

This area consists of the north slope and foothills of the main Caucasus range, extending north to the mutual boundary of Asia with Europe as defined in this *Guide* (see p. 297). It includes the Krasnodar and Stavropol' Krai in the north and west, the Daghestan ASSR in the east, and several smaller administrative units. The checklist of Flerow (1938) and the two manuals of Galushko (1967, 1978–) more or less cover the area as a whole; some substantial partial works are also available, here listed under 742–744.

Bibliography

NOVOPOKROVSKY, I., 1938. Literatura po botaničeskoj geografii Rostovskoj oblasti, Krasnodarskogo kraja, Oržoni-kidzevskogo kraja i Dagestana. *Trudy Rostov obl. biol. obšč.,* 2: 5–45. [442 titles.]

FLEROW, A. F., 1938. *Spisok rastenij Severnogo Kavkaza i Dagestana.* [Elenchus plantarum in Caucaso septentrionale nec non Daghestana sponte crescentium.] 693 pp., portrait. Rostov-on-Don: Rostovizdat.

Concise systematic enumeration of vascular plants, with essential synonymy, detailed indication of local and altitudinal range, and brief notes on

taxonomy, habitat, special features, etc.; addenda; extensive bibliography (755 titles) and index to orders and families.

GALUSHKO, A. I., 1978–80. *Flora Severnogo Kavkaza.* Vols. 1–2. Illus., maps. Rostov-on-Don: Rostov University Press.

Field manual-key to vascular plants, with abbreviated indication of internal distribution, altitudinal limits, habitat, and occurrence, Russian botanical equivalents, and limited synonymy; indices (in each volume) cover only Latin scientific names. The brief introductory section contains a general key to families as well as brief remarks on the background for the work (but without mention of the other standard works on the area). Coverage presently (end of 1980) extends from the pteridophytes through Verbenaceae (Englerian system); the third volume will complete the work. A total of some 3900 species is projected for inclusion.[5]

Special group – woody plants

GALUSHKO, A. I. (ed.), 1967. *Derevja i kustarniki Severnogo Kavkaza.* 536 pp., 118 text-figs. Nal'čik.

Illustrated flora of woody plants, with moderately long descriptions, Russian names, relatively detailed indication of local range (for native species) or region of origin and places of cultivation (for introduced species), and notes on habitat, special preferences, biological forms, cultivation, uses, and potential; bibliography and indices to all Russian and Latin plant names at end. The introductory section gives sources for the work, its plan, and a general key to families. Encompasses the entire North Caucasian Slope, including Daghestan, and extends into parts of Rostov Oblast'.

742

Western Ciscaucasia

Corresponds to Krasnodarsk Krai and other administrative units formerly included within it.

KOSENKO, I. S., 1970. *Opredelitel' vysšikh rastenij Severo-Zapadnogo Kavkaza i Predkavkaza.* 613 pp., 9 text-figs. Moskva: 'Kolos'.

Concise manual-key to all vascular plants of this area (3150 species), with diagnoses, Russian names, and brief (partly symbolic) indication of local range, habitat, phenology, life-form, uses, and potential; essential synonymy; complete indices at end. The introductory section includes an illustrated glossary. This work essentially covers Krasnodarsk Krai and neighboring areas.

743

Central Ciscaucasia

Corresponds to Ordžonikidzevsk Krai and other administrative units formerly included within it. No separate floras or related works are known to this reviewer.

744

Daghestan

Corresponds to the limits of Daghestan ASSR. The only available separate work is limited to woody plants.

LEPEKHINA, A. A., 1971. *Opredelitel' derev'ev i kustarnikov Dagestana.* 243 pp., 31 text-figs., map. Makhačkala: Dagestanskoe Učebno-Pedagogičeskogo Press.

Descriptive manual of native and introduced woody plants, with keys to all taxa, diagnoses, Russian names, and notes on habitat, local range, phenology, habit, special features, uses, potential, etc.; short glossary, list of literature (pp. 223–6) and complete indices at end. The introductory section includes *inter alia* an account (with map) of dendrofloristic areas and a general key to families.

746

Georgian (Gruzinskaja) SSR

The homeland of Djugashvili features two all-republic floras, a multi-volume flora (currently under revision) and a manual-key both in Georgian, and a number of notable local florulas in Russian of which two are for the Black Sea autonomous republics respectively centering on Sukhumi and Batumi and one (Kolakovskij, 1961) is on subtropical Colchidia (Colchis), an area of considerable botanical interest and also the mythical abode of Medea.

BOTANIČESKIJ INSTITUT, AKADEMIA NAUK GRU-ZINSKOJ SSR, 1964–9. *Opredelitel' rastenij Gruzii.* 2 vols. 2 maps. Tbilisi: 'Metsniereba'. (Also a title-page in Georgian.)

Concise manual-key to vascular plants (in Georgian), based mainly on *Flora Gruzii*; includes limited synonymy, nomenclatural equivalents, local vernacular names, concise indication of local range, and notes on habitat, life-form, phenology, etc.; indices to all botanical and other names in each volume (combined index to families in vol. 2). A map of the region with floristic districts, is provided in the introductory section.

MAKASHVILI, A. K., SOSNOVSKIJ, D. I. and KHARADZE, A. L. (eds.), 1941–52. *Flora Gruzii.* [Flora Georgiae.] 8 vols. Illus. Tbilisi: AN Gruzinskoj SSR Press. (Also a title-page in Georgian.)

KETSKHOVELI, N. N. *et al.* (eds.), 1971– . *Flora Gruzii.* 2nd edn. Vols. 1– . Illus., maps. Tbilisi: 'Metsniereba'. (Also a title-page in Georgian.)

The first edition comprises a detailed descriptive flora (in Georgian) of vascular plants, with keys to genera and species, significant synonymy with references and literature citations, Georgian nomenclatural equivalents and local vernacular names, concise generalized indication of local and extralimital range, and notes on habitat, phenology, special features, etc.; figures of representative species; indices to all botanical and other names in each volume (no cumulative index available).

The second edition is similar in format and style to its predecessor but additionally features species distribution maps. At this writing (late 1981) seven volumes, covering pteridophytes, gymnosperms, and angiosperms (on the Grossheim system) through Leguminosae, have been published.

Partial works: Black Sea district
Florulas for the Abkhazian Autonomous Republic (centering on Sukhumi) and the Adzharian Autonomous Republic (centering on Batumi) as well as for Colchidia are listed here.

DIMITRIEVA, A. A., 1959. *Opredelitel' rastenij Adžarii.* 447 pp. Tbilisi: AN Gruzinskoj SSR Press (in association with the Botaničeskij Sad, Batumi).

Concise manual-key to the vascular plants of the Adzharian ASSR in the southwestern part of the Georgian Republic, centering on Batumi with its botanical garden; the species leads include brief notes on habitat, local range, etc.; index.

KOLAKOVSKY, A. A., 1938–49. *Flora Abkhazii.* 4 vols. Illus. Sukhumi: Institut dlja Abkhazskoj Kulturu, AN SSSR (later AN Gruzinskoj SSR).

Keyed, briefly descriptive flora of vascular plants, with synonymy, Russian and Georgian names, local range, and notes on habitat, phenology, etc.; indices. [The Abkhazian ASSR is in the northwestern part of the Georgian Republic, centering on Sukhumi with its well-known botanical gardens.]

KOLAKOVSKY, A. A., 1961. *Rastitel'nyj mir Kolkhidy.* 460 pp., 231 illus., map (Mater. Pozn. Fauny Fl. SSSR, Mosk. Obšč. Ispytat. Prirod., N.S., Otd. Bot., 10). Moscow.

Pages 121–459 constitute a very concise keyed manual of vascular plants of subtropical Colchidia, with brief notes on ecology, phenology, and local range; glossary. [Colchidia (Colchis) lies along the central Georgian Black Sea coast.]

Partial work: Tbilisi district

MAKASHVILI, A. K., 1952–3. *Flora okrestnostej Tbilisi.* 2 vols. Illus. Tbilisi: Stalin State Institute.

Descriptive manual-flora of the environs of Tbilisi (formerly Tiflis), capital of the Georgian Republic. [Not seen by this reviewer; cited from Lebedev 1956.]

747

Armenian (Armjanskaja) SSR

In addition to the incomplete multivolume flora begun by Takhtajan in 1954, a concise manual for woody plants is also available (Sosnovskij and Makhatadze, 1950) and a florula for the Yerevan district (Takhtajan and Fëdorov, 1972). All these are in Russian.

TAKHTAJAN, A. L. (ed.), 1954– . *Flora Armenii.* Vols. 1– . Illus., map. Jerevan: AN Armjanskoj SSR Press.

Copiously illustrated flora of spontaneous and commonly cultivated vascular plants, with descriptive keys to all taxa, full synonymy, references, 'standardized' Russian binomials and vernacular names, concise indication of local and extralimital distribution, taxonomic commentary, and notes on habitat, special features, biology, cultivation, etc.; complete indices in each volume. At this writing (1980), six of the 12 volumes projected have appeared. The author's system of families, now widely known internationally, is used.

Partial work: Jerevan district

TAKHTAJAN, A. L. and FËDOROV, AN. A., 1972. *Flora Jerevana: opredelitel' dikorastuščikh rastenij Araratskoj kotloviny.* 394 pp., 118 text-figs. Leningrad: 'Nauka'.

Illustrated manual-key to vascular plants of the Ararat basin around Yerevan, with limited synonymy, vernacular names, and notes on habitat, phenology, local distribution,

etc.; index. Supplants two earlier works by the same authors: *Flora Jerevana: opredelitel' rastenij okrestnostej Jerevana* and *Atlas risunkov k 'Flore Jerevana'* (both published 1945, Jerevan).

Special group – woody plants

SOSNOVSKIJ, D. I. and MAKHATADZE, L. B., 1950. *Kratkij opredelitel' derev'ev i kustarnikov Armjanskoj SSR.* 103 pp. Jerevan: AN Armjanskoj SSR Press.

Systematic manual-key to woody plants, with short diagnoses and notes on habit, distribution, special features, uses, potential, etc.; Russian and Armenian names; artificial key to genera and complete indices. Also includes a summary in Armenian (pp. 76–94).

748

Azerbaijan (Azerbajdzanskaja) SSR

See also **790** (Rechinger). A portion of the Azerbaijan Republic, especially the Talysh lowland and parts of the Kara-Araks Basin near the Caspian Sea, is encompassed within the limits of *Flora Iranica*. The oil-bearing Apsheron Peninsula is also botanically distinctive as being of somewhat insular character.

KARJAGIN, I. I., *et al.* (eds). 1950–61. *Flora Azerbajdžana.* 8 vols. 368 pl., maps. Baku: AN Azerbajdžanskoj SSR Press.

Detailed descriptive flora of vascular plants, with keys to all taxa, extensive synonymy, references, 'standardized' Russian binomials and local vernacular names, generalized indication of local and extralimital distribution, taxonomic commentary, and notes on habitat, special features, phenology, etc.; figures of representative species; complete indices in each volume (cumulative generic index in volume 8). An introductory section (vol. 1) has a list of principal references for the area as well as a map of local floristic regions.

Partial work: Baku and the Apsheron Peninsula

KARJAGIN, I. I., 1953. *Flora Apšerona.* 439 pp., illus. Baku: AN Azerbajdžanskoj SSR Press.

Manual-key to vascular plants of the Apsheron Peninsula (which projects into the Caspian Sea, and on whose southern side Baku is situated). [Not seen by this reviewer; cited from Lebedev, 1956.]

Region

75

Soviet Middle Asia

This comprises the Soviet republics of Kazakhstan, Kirghizstan, Uzbekistan (including the Kara-Kalpakian ASSR), Tadzhikstan, and Turkmenistan. The limits of the region correspond to political-administrative boundaries except in the extreme west, where the boundary (along the Ural River from Orsk downstream) is the same as that adopted for *Flora Europaea*. A small portion of Kazakhstan thus is excluded from Region 75.

Extensive botanical exploration, as in much of Africa, began only in the latter part of the nineteenth century with the spread of Russian influence into the region, but by 1917 much had already been accomplished through the central institutions in Leningrad via several expeditions and surveys and two floristic works: Lipsky's *Materialy dlja flory Srednej Azii* (1901–9) and the Fedtschenkos' *Conspectus florae Turkestanicae* (1906–16). After the Revolution and the establishment in 1918 of a 'base' at Tashkent – ever since the scientific 'capital' of the region – widespread collecting took place, with much effort focused along the foothills of the mountain ranges from Ashkhabad and Dushanbe (Stalinabad) to Alma-Ata (where all the republic capitals are located). This exploration was in the 1950s and 1960s, but in some areas even before World War II, followed by the publication of full 'republic' floras in the wake (and sometimes in imitation) of *Flora SSSR*. A number of local works have also been written for centers of demand. At present, this network of coverage is extensive though of uneven quality and completeness; revision is currently being effected through a concise comprehensive regional manual-flora, *Opredelitel' rastenij Srednej Azii* [Conspectus florae Asiae mediae] (1968–), a project of the Botanical Institute at Tashkent; at present (1980) five of the ten volumes planned have been published.

Coverage by partial or district floras in this vast region of mountains, desert and steppe is, however, still comparatively limited; it is in this respect rather less thoroughly covered than the Caucasus. A separate work covers the Kara-Kalpak ASSR near the Aral Sea (Bondarenko, 1964), and local floras have appeared for the environs of Tashkent, Ashkabad, Leninabad, and Dushanbe as well as for the high Pamir (703/VI). A number of works on woody plants, all relatively technical, are also available.

Nevertheless, the publication achievements in 60 years of this collective fraternal botanical activity are impressive, when it is considered that the region is about two-thirds the size of Australia. The standard of some of the work, however, is considered by Fëdorov to be less than satisfactory.

Bibliographies. General and supraregional bibliographies as for Superregion 71–75; see also Superregion 77–79 (FIELD).

Indices. General indices as for Superregion 71–75.

750

Region in general

The early regional enumeration, *Conspectus florae Turkestanicae*, by O. A. and B. A. Fedtschenko (6 parts, 1906–16; German edn., 1905–13, 1923) has since 1968 been progressively replaced by a new work, *Conspectus florae Asiae mediae*, now relatively far advanced.

KOVALEVSKAJA, S. S. (ed.), 1968– . *Opredelitel' rastenij Srednej Azii* [Conspectus florae Asiae mediae]. Vols. 1– . Illus. Tashkent: 'FAN'.

Concise manual of the vascular plants of Soviet middle Asia, with descriptive keys to all taxa, full synonymy, references and citations, Russian binomial as well as local vernacular names, generalized indication of internal distribution, occasional taxonomic commentary, and brief notes on habitat, life-form, phenology, special features, etc.; indices to all botanical and other names in each volume. Based principally on the herbaria of the Tashkent State University and the Uzbek Academy of Sciences, the most substantial resource in middle Asia. Families published in groups in volumes as ready, without regard to predetermined systematic sequence. At this writing, five volumes (of a projected ten) have been published.

751

Kazakh (Kazakhskaja) SSR

This largest of Soviet republics (apart from the RSFSR), with vast stretches of steppe and desert, has an illustrated manual and a 'tree book' as well as a multivolume general flora, all in Russian.

Bibliography

PAVLOV, N. V., 1940. *Literaturnyje istočniki po flore i rastitel'nosti Kazakhstana.* 182 pp. (Trudy Kazakhstansk. Fil. Akad. Nauk SSSR, 19). Moscow/Leningrad.

INSTITUT BOTANIKI, AKADEMIJA NAUK KAZAKHS-KAJA SSR, 1969–72. *Illjustrirovannyj opredelitel' rastenij Kazakhstana.* 2 vols. Illus., map. Alma-Ata: 'Nauka', Kazakhskoj SSR.

Copiously illustrated, succinct manual-key to the vascular plants of Kazakhstan, with limited synonymy, Russian binomial and local vernacular names, brief indication of local range, and notes on habitat, life-form, phenology, etc.; general index (in vol. 2). A brief introduction to the use of the work appears in vol. 1.

PAVLOV, N. V. (ed.), 1956–66. *Flora Kazakhstana.* 9 vols. Illus., maps. Alma-Ata: AN Kazakhsk. SSR Press (later 'Nauka', Kazakhsk. SSR).

Briefly descriptive flora of vascular plants, with keys to all taxa; full synonymy, with references and citations; Russian binomial and local vernacular names; detailed indication of local range and generalized summary of extralimital distribution; some taxonomic commentary; notes on habitat, life-form, phenology, uses, etc.; diagnoses of novelties; indices to all botanical and other names in each volume (cumulative index to families and genera in vol. 9). The introductory section (vol. 1) contains an illustrated organography and a map with floristic regions.

Special groups – woody plants

MUŠEGJAN, A. M., 1962–6. *Derevja i kustarniki Kazakhstana: dikorastuščije i introdutsirovannyje.* 2 vols. 20 pls. Alma-Ata: Kazsel'khozgiz (vol. 1); 'Kainar' (vol. 2).

The main part (section III) of this work comprises a descriptive account of the native and exotic trees and shrubs of Kazakhstan, with analytical keys, synonymy, principal citations, Kazakh and Russian names, general indication of local and extralimital range (as well as countries of origin), and notes on occurrence, habitat, cultivation, properties, uses, etc., both in the republic as well as (where relevant) elsewhere

in the USSR; extensive bibliographies and full indices in both volumes. The introductory sections (I and II) include a short account of sources for the work and a scheme for dendrofloristic regions, while two concluding sections (IV and V) deal with the origin and distribution (respectively) of the native and introduced woody species. Based in part of Sokolov's *Derevja i kustarniki SSSR.* [Not seen by this author; description based on data supplied by M. E. Kir-picznikov, Leningrad.]

752

Turkmen (Turkmenskaja) SSR

See also **790** (RECHINGER) with regard to the southern (Iranian/Afghan) fringe. In addition to the 'research' flora, the first such initiated in the region, there is a recent manual-key by Čopanov and others (1978–) and the following illustrated florula for the environs of Ashkhabad: NIKITIN, V. V., 1965. *Illjustrirovannyj opredelitel' rastenij okrestnostej Aškhabada.* 458 pp., 116 text-figs. Moscow/Leningrad: 'Nauka'.

ČOPANOV, P. C. *et al.*, 1978. *Opredelitel' khvošče-obraznykh, paporotnikoobraznykh, golosemjannykh i odnodol'nykh rastenij Turkmenistana.* 330 pp., 33 text-figs. Ashkhabad: 'Ylym'.

Manual-key to lower vascular plants, gymnosperms, and monocotyledons (519 species), with annotated key-leads, limited synonymy, and vernacular names; index. Of this volume, only 500 copies were printed. Reviewed by V. Botchantzev in *Bot. Žurn. SSSR,* **64**: 277–8 (1979). Volume 2, covering part of the dicotyledons, was published in 1980 as noted in Royal Botanic Gardens, Kew. Library: *Current Awareness List* (1980, no. 1), p. 56.

FEDTSCHENKO, B. A., POPOV, M. G. and SHISH-KIN, B. K. (eds.), 1932–60. *Flora Turkmenii.* 7 vols. Illus., map. Leningrad: AN SSSR (vol. 1); Askhabad: Turkmensk. Gosizdat (vol. 2); Turkmensk. fil. AN SSSR (later AN Turkmensk. SSR) Press (vols. 3–7).

Full-scale descriptive flora of vascular plants, with keys to all taxa; includes limited synonymy, localities with citations of *exsiccatae*, generalized indication of internal and extralimital range, and notes on habitat, life-form, phenology, etc.; indices to botanical names in each volume (except vols. 1–2). Volume 1 contains a brief preface (with map of floristic regions) and a general discussion of the significant botanical features of all vascular families in Turkmenia.

753

Uzbek (Uzbekskaja) SSR

The multi-volume republic flora has been supplemented for the Kara-Kalpak ASSR (by the Aral Sea) by a separate work (see **754**).

Bibliography

DEVJATKINA, A. V., 1966. *Rastitel'nyj i životnyj mir Uzbekistana: bibliografičeskij ukazatel'* (*1917–52*). 468 pp. Tashkent: 'FAN' (AN Uzbekskoj SSR).

SCHREDER, R. R., KOROVIN, E. P. and VVEDENSKY, A. I. (eds.), 1953–62. *Flora Uzbekistana*. 6 vols. Illus. Tashkent: Uzbekskij filial AN SSSR (later AN Uzbekskoj SSR) Press.

Full-scale descriptive flora of vascular plants, with keys to all taxa; full synonymy, with references; 'standardized' Russian and some Uzbek vernacular names; general indication of local and extralimital range; taxonomic commentary and notes on habitat, life-form, phenology, uses, etc.; diagnoses of new taxa; indices to all botanical and other names in each volume (cumulative indices in vol. 6). The introductory section (vol. 1) contains an illustrated organography, an account of Uzbekistan botanical regions, and a short history of botanical exploration and research.

754

Kara-Kalpak (Kara-Kalpakskaja) ASSR

See also **753** (Schreder, *et al.*). This autonomous republic is linked administratively with the Uzbek SSR.

BONDARENKO, O. N., 1964. *Opredelitel' vysšikh rastenij Karakalpakij*. 303 pp., 94 text-figs. Tashkent: 'Nauka' (Uzbekskoj SSR).

Concise manual-key to vascular plants, with essential synonymy, brief indication of local range, notes on life-form, phenology, etc., and index to botanical names. [The general character of this desert flora, lying in the delta of the Amu Darya (Oxus) and around the Aral Sea, is indicated by the presence of representatives of 40 genera of Chenopodiaceae (described in 41 pages), but only one fern species!]

755

Kirghiz (Kirghizskaja) SSR

See also **703**/VII (FEDTSCHENKO) and **760** (GRUBOV) for other treatments of the Tian Shan, much of the Soviet part of which lies within the Kirghiz Republic.

SHISHKIN, B. K. and VVEDENSKY, A. I. (eds.), 1950–62. *Flora Kirghizskoj SSR: opredelitel' rastenij Kirghizskoj SSR*. 11 vols. Illus. Frunze: Kirghizskij filial AN SSSR (later AN Kirghizskoj SSR) Press.

VYKHOTSEV, Y. V., 1967–70. *Flora Kirghizskoj SSR: Dopolnenie*. Vyp. 1–2. Illus. Frunze: 'Ilim'.

Briefly descriptive flora of vascular plants, with keys to all taxa; includes limited synonymy (without references), Russian botanical binomials and local vernacular names, generalized indication of local and extralimital distribution, and notes on habitat, life-form, phenology, uses, etc.; diagnoses of new taxa; indices to all botanical and other names in each volume (general index in volume 11). In volume 1 there is an illustrated organography along with a general key to families, while vol. 2 includes a general descriptive introduction to the flora and its phytogeographic relationships. The supplement of 1967–70 includes corrections and additions in all families as well as some revised keys.

756

Tadzhik (Tadžikskaja) SSR

For the Pamirs, see also **703**/VI (IKONNIKOV). The large-scale republic flora, covering some of the highest country in the Soviet Union, has still to be completed after more than 40 years (20 years since resumption in 1957), although progress has been continuing. Local florulas are also available respectively for the Dushanbe and Leninabad districts (Grigorëv, 1953; Komarov, 1967) and for Badakshan (Ikonnikov, 1979).

Bibliography

MARGOLINA, D. L., 1941. *Flora i rastitel'nost' Tadžikistana: bibliografija*. 346 pp., portrait. Moscow/Leningrad: AN SSSR Press.

KOMAROV, V. L. and GONCHAROV, N. F. (eds.),

1937. *Flora Tadžikistana.* Vol. 5. Illus. Moscow/Leningrad: AN SSSR Press. [Leguminosae only.]

OVCHINNIKOV, P. N. (ed.), 1957– . *Flora Tadžikskoj SSR.* Vols. 1–4. Moscow/Leningrad: AN SSSR Press (later Leningrad: 'Nauka').

Full-scale semi-revisionary descriptive flora of vascular plants, with keys to all taxa, detailed synonymy, references, localities with citations of *exsiccatae* and general indication of local and extralimital distribution, notes on habitat, life-form, phenology, special features, uses, etc., and some taxonomic commentary; detailed lists of references and complete indices at end of each volume. The introductory section, volume 1, contains an account of botanical districts (with map) and a historical survey, while in the early volume 5 is found an extensive descriptive account of floristics and vegetation. With the appearance of volume 4 in 1975 the gap prior to volume 5 was filled, completing accounts of families from the pteridophytes through the Leguminosae (Englerian system), but at this writing (mid-1979) nothing further has been published. [A total of ten volumes has been planned for this series, which in contrast to the other regional and republican floras in Middle Asia (save some early volumes) has from the beginning been a Leningrad-based project.]

Partial works

GRIGORËV, JU. S., 1953. *Opredelitel' rastenij okrestnostej Stalinabada.* 299 pp., 56 text-figs. Moscow/Leningrad: AN SSSR Press.

Concise manual-key to vascular plants of the Dushanbe (formerly Stalinabad) district, with Russian binomials and vernacular names, occasional taxonomic notes, and brief indication of local range, life-form, habitat, phenology, special features, uses, etc.; indices.

KOMAROV, B. M., 1967. *Opredelitel' rastenij Severnogo Tadžikistana.* 495 pp., illus. Dushanbe: 'Doniš'.

Manual-key to the vascular plants in the northern lobe of Tadzhikistan (centering on Leninabad and the area south of Tashkent), with diagnoses, limited synonymy, Russian binomial and local vernacular names, local distribution, and notes on habitat, life-form, phenology, cultivation, uses, etc.; indices.

IKONNIKOV, S. S., 1979. *Opredelitel' vysšikh rastenij Badakhšana.* 400 pp., 25 pls. Leningrad: 'Nauka'.

Manual-key to vascular plants, with partly symbolic indications of habit, phenology, habitat, local range and altitudinal zonation; includes limited synonymy and Russian nomenclatural equivalents; addenda, references (pp. 372–4) and index at end. The introductory section includes background information and a brief conspectus of the vegetation as well as of the history of floristic work in the area (which is bordered to the west by the Amu Darya and centers on Khorog, with altitudes ranging from 1700 to 7000 m). Covers 1567 species.

Region (Superregion)

76

Central Asia

This region, formerly constituting the greatest part of the 'Chinese Border', essentially consists of all the territory between Soviet Asia to the north and west, 'China proper' to the south and east, and the Indian subcontinent to the south, except that northeast China (Manchuria) has had to be referred to Region 86 for practical reasons. With the exception of the Mongolian People's Republic, the entire region forms part of the People's Republic of China. The Chinese portion includes Sitsang (Tibet), Qinghai (Chinghai), Sinkiang (including 'Chinese Turkestan', i.e., the Takla Makan basin and Dzungaria), Inner Mongolia, and the western part of Kansu.

Botanical exploration in the region has had a relatively long but, until well into the twentieth century, sporadic history. Much early work was carried out by Russian exploring expeditions with scientific parties, whose botanical collections are now largely in Leningrad. Contributions were also made by the British (mainly in Tibet) and others. Difficulties of travel, extremes of climate, and sparse population along with an absence of scientific centers restricted more detailed collecting, and while many 'reports' and 'contributions' were published few consolidated floristic treatments appeared, the most notable being the various works (some not completed) of Maximowicz. From World War I until 1949, several expeditions from Europe and the United States, as well as China and the Soviet Union, penetrated different parts of the region and more sizeable collections were made; settlement also increased, a process continued since 1949 (in Tibet, 1951) when botanical work became exclusively a Chinese and Soviet activity. In recent decades, scientific centers have been established in Ulan Bator, Sining, Urumchi and Lhasa, and more detailed botanical work has been in progress, but only sketchy information has been available to the writer.

The first synthetic floristic work, the serial *Plantae asiae centralis* led by V. I. Grubov, began to appear in 1963 and at this writing has had seven parts published (of 15 planned). No such previous coverage has been available; the region was excluded from *Index florae sinensis* although it is being included in the modern general floras of China (see 860–80). The work is based primarily on collections available in Leningrad. Of unit floras, the only modern works are the unfinished *Flora of the Mongolian steppe and desert areas* by Norlindh (1949), *Konspekt flory Mongol'skoj Narodnoj Respubliki* by Grubov (1955; supplement 1972), and the very recent *Flora intramongolica* by Fu and others (volumes 2–4, 1977–9). Plans evidently have also existed for a flora and vegetation survey of Tibet (Si-tsang) to be carried out by Chinese botanists, and most recently this has been done through a biological center in Tsinghai Province; a first publication on central Tibetan vegetation appeared in 1967 and a second survey, on the flora and fauna of the A-li region in the west, was issued in 1979. No consolidated flora, however, has yet been published. Likewise, no separate flora is available for the vast Sinkiang Uighur Autonomous Region, once known as Chinese Turkestan.

The region, sketchily documented and but unevenly collected, is thus overall the least known area of its size in Asia. A useful alternative introduction to the flora and vegetation is however, the illustrated vegetatiological work by WANG, C. W., 1961. *The forests of China: with a survey of grassland and desert vegetation.* xiv, 313 pp., 78 text-figs., map (Maria Moors Cabot Foundation, Harvard Univ., Spec. Publ., 5). Petersham, Mass.

Progress. Early and recent activities are usually treated as part of Chinese botanical history (see Superregion 86–88). The most useful sources are the works by Bretschneider and Cox. Expeditions from British India are reviewed in South Asian sources (see Superregion 81–84).

General bibliographies. Bay, 1910; Frodin, 1964; Goodale, 1879; Holden and Wycoff, 1911–14; Hultén, 1958; Jackson, 1881; Pritzel, 1871–7; Rehder, 1911; USDA, 1958.

Regional bibliographies

See also MERRILL and WALKER, 1938 and WALKER, 1960 (Superregion 71–75). [Both cover the entire region except for Sinkiang.]

LEBEDEV, D. V., 1963. Materialy k bibliografii po flore i rastitel'nosti Tsentral'noj Azii. In *Rastenija Tsentral'noj Azii* (ed. V. I. Grubov), fasc. 1: 99–166.

Moscow. [Unannotated, alphabetically arranged list. A supplement has appeared in fasc. 6 (1971) of the same work.]

General indices. BA, 1926– ; BotA, 1918–26; BC, 1879–1939; BS, 1940– ; CSP, 1800–1900; EB, 1959– ; FB, 1931– ; ICSL, 1901–14; JBJ, 1873–1939; KR, 1971– ; RŽ, 1954– .

760

Region in general

See also general works on China (**860–80**; all references *except* Forbes and Hemsley); however, coverage of central Asia in at least some of these works is apt to be sketchy. Virtually the whole of the region is now being covered in the critical *Plantae asiae centralis* produced under the direction of V. I. Grubov.

GRUBOV, V. I. (ed.), 1963– . *Rastenija Tsentral'noj Azii.* [Plantae asiae centralis.] Vyp. 1– . Illus., maps. Moscow/Leningrad: AN SSSR Press (later Leningrad: 'Nauka'). English edn.: *idem*, 1965. *Plants of Central Asia.* Fasc. 1. Jerusalem: Israel Program for Scientific Translations.

Detailed regional conspectus of vascular plants, with diagnostic keys to genera, species, and infraspecific taxa and descriptions of novelties; extensive synonymy, with references and citations of relevant literature; detailed indication of internal distribution, with citation of *exsiccatae* and incorporation of some range maps; general summary of extralimital range; notes on habitat, special features, etc., together with taxonomic commentary; index to all botanical names in each fascicle. Fascicle 1 includes an extensive account on floristics and geobotanical regions, together with a large regional bibliography (the latter supplemented in fasc. 6). The work, based upon the central Asian collections at the Komarov Botanical Institute, is expected to encompass 15 fascicles; at this writing (1979) seven have been published, the latest in 1977. Fascicles may contain one or more families; in the latter case they may be systematically related although overall no preset sequence is followed in publication. Some treatments are by specialists.

761

Mongolian People's Republic

Also known as Outer Mongolia, this former territory of the Manchu (Ch'ing) Empire has had since the early 1920s a close alliance with the Soviet Union. Through this link has come much scientific and technical assistance. Coordinating these efforts to a large extent has been the Mongolian Commission of the Academy of Sciences of the USSR, in whose *Trudy* Grubov's conspectus of 1955 was published. As of 1972 Grubov had recorded 2008 vascular plant species. Earlier Russian exploration is reflected in the incomplete enumeration of Maximowicz, which contains analytical keys. Latterly there has been added the first local flora in Mongolian: ŽAMSRAN, C., ÖLSI-KHUTAG, N. and SANČIR, C., 1972. *Ulaanbataar ortschmyn urgamal tanykh bitschig.* Ulan Bator. [Not seen; cited from *Prog. Bot.* **35**: 309, 320 (1973).]

GRUBOV, V. I., 1955. *Konspekt flory Mongol'skoj Narodnoj Respubliki.* 308 pp., 1 color pl., 4 portraits, 3 maps (Trudy Mongol'sk. Komiss. Akad. Nauk SSSR, vol. 67). Moscow: AN SSSR Press.

Systematic enumeration of vascular plants, of the Mongolian People's Republic, without keys, synonymy, or references; species headings include vernacular names, a few critical remarks, and notes on habitat, biology, special features, etc. Indices to family and generic, and to vernacular names are also provided. An introductory section incorporates chapters on local botanical exploration and research, geobotanical and floristic regions, and statistics of the flora together with a glossary, bibliography, and list of collectors. Additions and corrections are contained in *idem*, 1972. Dopolnenija i ispravlenija k 'Konspektu flory Mongol'skoj Narodnoy Respubliki'. *Nov. Sist. Vysš. Rast.* **9**: 270–98. [Further changes have been published in a series of papers published in *Botaničeskij Žurnal SSSR* from volume 56 (1971) onwards; as of 1978 seven such contributions had appeared.]

MAXIMOWICZ, C. J., 1889. *Enumeratio plantarum hucusque in Mongolia nec non adjacente parte Turkestaniae sinensis lectarum.* Fasc. 1. iv, 138, [8] pp., 14 pls. (Historia naturalis itinerum N. M. Przewalskii per Asiam centralem, pars botanica, vol. 2.) St Petersburg: Press of the Imperial Academy of Sciences.

Systematic enumeration of the seed plants of Mongolia and adjacent parts of the Chinese Sinkiang

(Xinjiang) Autonomous Region, with keys to genera and species, descriptions of new taxa, synonymy, citation of *exsiccatae* with indication of localities, summary of overall range, and taxonomic commentary; index at end. Text in Latin and Russian. Not completed; covers only Thalamiflorae and Disciflorae (Candollean system).

763

Inner Mongolia (north of the Great Wall)

See also 705 (NORLINDH); 761 (both works); 860/I (KITAGAWA; KOMAROV; NODA). As delimited here, Inner Mongolia is considered to correspond to the greater part of the former provinces of Ningsia, Suiyuan and Chahar and part of the former district of Jehol. This includes the present Inner Mongolian Autonomous Region, Ningsia Hui Autonomous Region north of the Huang Ho, and the northern desertic part of Kansu (north of the Sinkiang railway line). However, Inner Mongolia from 1949 to 1969 administratively encompassed the whole of 763 along with the western part of former Manchuria (North-East China; 860/I) and Cis-Ningsia (south of the Huang Ho), except that both Cis- and Trans-Ningsia were reabsorbed by Kansu in 1954 (Cis-Ningsia (869) again acquired separate status in 1958). In 1969, the northern extension of Inner Mongolia was returned to the three reconstituted provinces of North-East China and Trans-Ningsia was transferred to Cis-Ningsia, the combined polity retaining quasi-provincial status.

FU, H.-C. *et al.* [Inner Mongolian Botanical Records Compiling Group], 1977–9. *Flora intramongolica*. Vols. 2–4. Illus. Huhehaot'e, Inner Mongolia: Inner Mongolian People's Press.

Illustrated descriptive flora, with keys, Chinese and local nomenclatural equivalents, limited synonymy, indication of internal and extralimital distribution, and notes on habitat, special features, properties and uses, etc.; indices in each volume. A brief introduction to each volume is also provided. Coverage to date (end of 1980) extends from Polygonaceae through Umbelliferae (modified Englerian system; monocots will evidently follow dicots). Citation of authorship follows usage in Royal Botanic Gardens, Kew. Library: *Current Awareness List* (1980, no. 2), p. 51.

764

Sinkiang Uighur (Xinjiang Weiwuer) Autonomous Region

The only important partial account for this area to date is Maximowicz's incomplete enumeration covering Mongolia and eastern Sinkiang ('Chinese Turkestan'; see 761). Otherwise, no general coverage is available and reference must be made to general works on China (860–80) or to *Plantae asiae centralis* by GRUBOV (760).

765

Kansu (Gansu): western part

Western Kansu, as here defined, includes all that part of the province lying northwest of the Huang Ho (with Lanchou City just outside, being on the south bank of the river) and south of the deserts and steppes of Inner Mongolia in an extended sense (763). To the south of Western Kansu lies Tsinghai (767). Part of the area is covered by Maximowicz's incomplete *Flora tangutica*, which was centered in northeastern Tsinghai. No provincial floras or other separate works treating western Kansu are available; reference must be made to general works on China (860–80) or to *Plantae asiae centralis* by GRUBOV (760).

767

Tsinghai (Qinghai)

See also 768 (HEMSLEY). The only important partial account on this province, which includes keys, is that by Maximowicz described below. Plants collected from this area by J. F. Rock in 1925 have been accounted for by REHDER and WILSON, and by REHDER and KOBUSKI (see 860–80 under Partial works), where also general works on China may be found.

MAXIMOWICZ, C. J., 1889. *Flora tangutica; sive Enumeratio plantarum regionis Tangut (Amdo), provinciae Kansu, nec non Tibetiae praesertim orientali-borealis atque Tsaidam, ex collectionibus N. M.*

Przewalskii atque G. M. Potanin. Fasc. 1. xviii, 114 pp., 31 pls. (Historia naturalis itinerum N. M. Przewalskii per Asiam Centralem. Pars botanica, 1). St Petersburg: Press of the Imperial Academy of Sciences.

Systematic enumeration of seed plants recorded from northern and northeastern Tsinghai and adjacent southwestern Kansu, with keys to genera and species, descriptions of novelties, synonymy, citation of *exsiccatae* with localities, extralimital range, and taxonomic commentary; index. Text in Latin and Russian. Not completed; covers dicotyledon families in Thalamiflorae and Disciflorae (Candollean system). [The Tangut district centers on the great lake of Tsinghai (Koko Nor), while the area around Hsi-ning (Xining), the present provincial capital, was at the time within Kansu. Under the Ch'ing dynasty, Tsinghai constituted a district of Tibet.]

768

Tibet/Sitsang (Xizang)

In addition to the single work described below, a useful general introduction to the botany and vegetation of Tibet is WARD, F. KINGDON, 1935. A sketch of the geography and botany of Tibet, being materials for a flora of that country. *J. Linn. Soc. Bot.*, **50**: 239–65. A number of recent Chinese publications, representing the results of botanical surveys made since 1951, are now also available, but none constitutes a flora in the strict sense. Other information may be found in general works on China (860–80).

HEMSLEY, W. B. (with PEARSON, H. H. W.), 1902. The flora of Tibet or High Asia. *J. Linn. Soc., Bot.* **35**: 124–265, map.

Systematic enumeration of vascular plants known from Tibet and Tsinghai Province [the latter then still part of Tibet], with descriptions of new taxa, concise synonymy (with references and relevant citations), indication of localities and *exsiccatae*, taxonomic commentary, and notes on habitat, special features, etc.; no separate index. The introductory section incorporates accounts of physical features, climate, and botanical exploration (with itineraries), while an appendix gives a discussion of vegetation, biological features, and phytogeography. [This treatment was based upon Tibetan plants housed in the Kew Herbarium.]

Superregion

77–79

Southwestern Asia

This large region includes most of the traditional 'Orient', now expanded and called 'the Middle East'; as delimited here it extends from the Bosporus and Dardanelles and from the Suez Canal eastward to the long escarpment bordering the Indus Basin, and from the Soviet border in the north to the Indian Ocean, including the whole of the Arabian Peninsula. Except for infratropical Arabia, this substantially corresponds to the area in Asia covered by Boissier's *Flora orientalis*.

Early botanical exploration and reporting in the superregion, very fragmented and frequently beset with hazards, was effectively consolidated by Boissier in his comprehensive *Flora orientalis*, begun in the 1860s and completed by Buser in 1888. It has ever since represented a foundation for regional floristic work and flora writing; for the sake of conformity, the Candollean classification system used by Boissier has been adopted in a number of the current floras. Most user needs were so well satisfied that there was little demand for more up-to-date floras until recent decades; the only substantial works to appear from the late-nineteenth century until World War II were the Beirut-based *Flore du Liban et de la Syrie* by Bouloumoy (1930) and *Flora of Syria, Palestine and Sinai* by Post and Dinsmore (1932–3), the latter originally having appeared in 1896, Holmboe's *Studies on the vegetation of Cyprus* (1914), and the gap-filling, critical *Flora des tropisches Arabien* (Schwartz, 1939) as well as Blatter's more sketchy *Flora arabica* (1919–23, 1936). A miscellany of more local works was also produced for such scattered areas as the North-West Frontier, Baluchistan, Aden, Transjordan, the Sinai, and parts of Turkey.

However, by 1951 it was true to say, as did G. H. M. Lawrence in his *Taxonomy of Vascular Plants*, that 'most of the literature...is very old', let alone obsolete, with little evidence of much ongoing work. Not perceived was the great burst of new floristic activity which took place almost throughout the superregion, both by outside botanists (mainly from Europe) but also by nationals within the area, either in their own countries or in others. This was in a sense a continuation of developments starting in about 1920 which included the growth of tertiary institutions, the foundation of a number of herbaria, and the development of a botanical profession within the superregion. Widespread flora-writing, however, was limited by the constraints of local resources and by then-current fashions elsewhere. Much collecting was accomplished between the wars, an activity accelerated after World War II. From the 1950s and notably in the 1960s, as interest revived and the oil-enriched means became available, new floras were published or begun for every country in the superregion. Leading works include the *Flora of Turkey, Flora of Cyprus, Nouvelle flore du Liban et de la Syrie, Flora palaestina, Flora of Iraq, Flora iranica*, and *Flora of West Pakistan*. A flora of the Arabian peninsula is reportedly in preparation, and compilation of a dendrological distribution atlas was in 1975 progressing at Kórnik, Poland under US PL-480 assistance. Several checklists as well as floras of more limited scope have also been produced. Some assistance to flora writers concerned with the superregion has been given by the publication of *Flora SSSR* and regional floras in the Caucasus and Soviet middle Asia. Thus, the current situation is very different from 1939, when, with the composition of *Flora orientalis* and other great nineteenth-century floras not yet too remote in spirit, floristic work was still thought to have 'run its course', despite the standard of documentation in reality being very deficient; Boissier never regarded his work as being 'final' in any sense, but mainly as a consolidation of what was then known.

Nonetheless, even with their more imposing size, many of the floras now in progress are in varying degree still little more than preliminary surveys or *prodromi*, covering areas more or less sketchily known. The coverage they provide is about at the level of much of Middle America, south Asia, or some parts of Africa; more detailed documentation has for the most part yet to come, including the results of 'biosystematic' and other more sophisticated studies. Other problems are continuing political turbulence, which may interrupt the progress of current flora projects although the majority of the large ones are based outside the superregion, and the apparent lack of coordination between these projects (except informally). However, the establishment of the 'umbrella' organization, OPTIMA, in 1975 may serve to some extent as a force for collective progress; already it has 'rescued' one flora project accounted for here, and inspired another – the Middle East checklist of Zohary et al. (1980–).

Progress
Early history has been surveyed by Boissier in volume 1 of *Flora orientalis* (1867). No overall accounts of later history have been seen, although reviews for individual countries, either in the introductions to the various floras or in separate articles, are available. For a fairly personal but pragmatic view of current developments, see MEIKLE, R. D., 1971. Co-ordination of floristic work and how we might improve the situation. In *Plant life of South-West Asia* (ed. P. H. DAVIS et al.), pp. 313–31. Edinburgh. Recent progress in the countries on and near the Mediterranean (Region 77), with lists of references, has been surveyed in separate articles in HEYWOOD, V. H. (coord.), 1975. Données disponibles et lacunes de la connaissance floristique des pays méditerranéens. In *La flore du bassin méditerranéen: essai de systématique synthétique* (CENTRE NATIONAL DE LA RECHERCHE SCIENTIFIQUE, France), pp. 15–142 (Colloques internationaux du CNRS, 235). Paris (see also **590, 601**).

General bibliographies. Bay, 1910; Frodin, 1964; Goodale, 1879; Holden and Wycoff, 1911–14; Hultén, 1958, Jackson, 1881; Pritzel, 1871–7; Rehder, 1911; USDA, 1958.

Supraregional bibliographies
FIELD, H. (comp.), 1953–64. *Bibliography on Southwestern Asia.* 7 vols. Coral Gables, Fla.: University of Miami Press; and FIELD, H. and LAIRD, E. M., 1969–72. *Bibliography of Southwestern Asia: Supplements* I–VIII. Coconut Grove, Fla: Field Research Projects. [Encompasses the Caucasus, Egypt, and Soviet middle Asia as well as the whole of southwestern Asia as here delimited; coverage in the original work extends up through 1959. Cumulative botanical subject indices have been provided by R. C. Forster.]

ZOHARY, M., 1966. Selected standard floras. In *Flora palaestina*, 1 (ed. M. Zohary and N. Feinbrun-Dothan), pp. xii–xiv. Jerusalem. [A short list, with a depth of coverage very similar to the present *Guide*.]

General indices. BA, 1926– ; BotA, 1918–26; BC, 1879–1944; BS, 1940– ; CSP, 1800–1900; EB, 1959– ; FB, 1931– ; ICSL, 1901–14; JBJ, 1873–1939; KR, 1971– ; NN, 1879–1943; RŽ, 1954– .

Supraregional indices
Limited coverage is provided in the biannual OPTIMA Newsletter (OPTIMA Secretariat, Berlin).

770–90

Superregion in general

The whole of the superregion has long been dominated by the comprehensive *Flora orientalis* of E. Boissier, written in the style of the contemporaneous 'Kew floras' then being prepared for different parts of the British Empire.

BOISSIER, E., 1867–88. *Flora orientalis, sive enumeratio plantarum in Oriente a Graecia et Aegypto ad Indiae fines hucusque observatarum.* 5 vols. and supplement. 6 pls., portrait. Basel, Geneva: Georg. (Vol. 1 reprinted 1936, Geneva, Herbier Boissier; whole work reprinted 1963–4, Amsterdam, Asher.)

Descriptive flora of vascular plants (in Latin), with synoptic keys to genera and species; includes essential synonymy, references and citations, vernacular names, generalized indication of internal and extra-limital range (with citation of some *exsiccatae* and localities), taxonomic commentary, and miscellaneous observations; indices to all botanical names in each volume (cumulative index to family, generic, and vernacular names in vol. 5). The introductory section in volume 1 incorporates chapters on botanical exploration, floristic region, limits of the Flora, and philosophical and methodological considerations. The *Supplementum* (or vol. 6), edited by R. Buser following Boissier's death in 1885, consists wholly of additions and corrections to the five volumes of the main work. [The plan of *Flora orientalis* was based upon that of the 'Kew Series' of British colonial floras; furthermore, the eastern limits of the work, by gentleman's agreement with J. D. Hooker, were matched – along a natural physical and botanical boundary – by the western limits of the *Flora of British India*. The overall geographical limits also recall the Alexandrine Empire and the sphere of Hellenistic civilization (the extent of which was then being rediscovered by travellers and archeologists). The southern limit, however, was set at the Tropic of Cancer, which almost bisects the Arabian peninsula and runs not far south of Aswan.]

ZOHARY, M., HEYN, C. C., and HELLER, D., 1980. *Conspectus florae orientalis*, 1. xiv, 107, 2 pp., 2 maps. Jerusalem: Israel Academy of Sciences and Humanities.

Subtitled *An annotated catalogue of the flora of the Middle East*, this work, which covers countries from Turkey (Asia Minor) and the east Aegean islands of Greece along with Egypt to Iran and southward to the foot of the Arabian peninsula, comprises an annotated regional checklist, featuring basionyms (where appropriate), name used in *Flora orientalis* (see above) where appropriate, and more common synonyms along with 'chorotypes' (chorological classes) representing overall distribution together with summary of regional distribution to species level in fairly detailed fashion. To date (end of 1980) covers Papaverales and Rosales (following the system of the Englerian *Syllabus*, edn. 12). [Aimed at presenting an 'overview' accounting for recent floristic and other taxonomic work, not attempted since the time of Boissier. In contrast to the forthcoming *Med-Check List* (see **601**/I), this Hebrew University-based work emphasizes distribution and chorology rather than critical taxonomy and nomenclature. Not seen for this commentary; based on review by W. Greuter in *OPTIMA Newsletter*, **10/11**: 45–6 (1980).]

Region

77

The Levant

Also known loosely as 'The Orient', the region here called the Levant consists of Turkey-in-Asia (Asia Minor), the adjacent islands of the eastern Aegean, Cyprus, Syria, Lebanon, Israel and the 'West Bank' (in Palestine), the Sinai peninsula, Jordan, and Iraq. Within this region is the subregion known as the 'Holy Land'. The overall limits correspond almost precisely with those of the Asiatic possessions of the former Ottoman Empire (except for the Hejaz, now part of Saudi Arabia (region 78)).

The first subregional or unit manual-flora within the Levant as here defined was the first edition of Post's *Flora of Syria, Palestine and Sinai* (1896), written from a base at the American University at Beirut, where the first herbarium in the region had been established. This flora of the 'Holy Land' was followed by a cycle of collecting and miscellaneous publications, including studies of Biblical plants, but the modern cycle of floras did not begin until about 1930. Modern floras for Turkey and Iraq were not begun until the 1960s, and at this writing are still in progress, the latter but slowly. For Syria and Lebanon, additional floras have been published by Thiébaut (1936–53) and Mouterde (1966–), the latter with atlases, and for Israel, the West Bank and Jordan – Palestine in the wide sense – *Flora palaestina* (Zohary and Feinbrun-Dothan, 1966–) is providing fairly detailed coverage, with three (of four) volumes (each with atlas) having been published by 1978. The Sinai peninsula is covered by floras of Egypt (**591**) as well as by *Flora of Syria, Palestine and Sinai,* but few separate works are available, except Zohary's checklist and phytogeographic analysis of 1935. Publication of a modern *Flora of Cyprus* (Meikle, 1977–) has only recently begun although preliminary work has long been in progress. All the major floras have been largely prepared and published outside the region, except Post's flora and the various French-language works and also *Flora palaestina*, based respectively in Lebanon and Israel. Students' manuals have been published for Israel and Iraq, as well as on a more local scale elsewhere, and there are some more or less complete checklists. Botanical knowledge is thus relatively far advanced for southwestern Asia, save for less accessible desert or mountainous areas such as the eastern part of Turkey and eastern Syria and Jordan.

Bibliographies. General and supraregional bibliographies as for Division 7 and Superregion 77–79.

Indices. General indices as for Superregion 77–79.

770

Region in general

See 770–90 (*Flora orientalis*).

771

Turkey-in-Asia (Asia Minor)

General works for the whole of modern Turkey are given under this heading. For the European part, see **638**.

Bibliographies
AYTÜG, N. and ÇAKMAN, A., 1972. *Türkiye flora bibliografyasi,* 1843–1968. vii, 38 pp. (TÜRDOK, Bibliografya Ser., 5). Ankara. (Mimeographed.)

DAVIS, P. H. and EDMONDSON, J. R., 1979. Flora of Turkey: a floristic bibliography. *Notes Roy. Bot. Gard. Edinb.*

37: 273–83. [Literature base for *Flora of Turkey* project, including precursory papers.]

KRAUSE, K., 1927–31. Die botanische Literatur über die Türkei. *Repert. Spec. Nov. Regni Veg.* **24**: 113–26; **28**: 136–41.

ZEYBEK, N., 1972. *A bibliography of the papers on the taxonomy and ecology of Turkish flora, 1841–1971.* Izmir-Bornova: University of Ege. (Mimeographed.)

DAVIS, P. H. (ed.), 1965– . *Flora of Turkey and the east Aegean islands.* Vols. 1– . Illus., maps. Edinburgh: Edinburgh University Press.

Comprehensive descriptive flora of vascular plants, with keys to all taxa; extensive synonymy, with references and principal citations; vernacular names; localities, with indication of *exsiccatae*, and general summary of internal and extralimital range (together with a number of distributional maps); taxonomic commentary and notes on habitat, phenology, associates, etc.; indices to all botanical names in each volume. The introductory section (in vol. 1) provides chapters on geography, climate, floristics, and phytogeography as well as a list of references. Aided by a cadre of writers in Edinburgh, production of this flora has been steady and at this writing (mid-1979), six volumes of the eight planned have been published; a supplement is projected for additions and corrections.

Partial work
BIRAND, H., 1952. *Türkiye bitkileri* [Plantae turcicae]. viii, 330 pp. (T. C. Ankara Üniv., Fen Fakültesi, Yayinlari, um. 58; Botanik, um. 1). Ankara.

Enumeration of vascular plants of Turkey, based mainly on herbarium collections of Ankara University; includes localities, citation of *exsiccatae*, and brief notes on habitat; index. By no means a complete catalogue of the flora.

772

Cyprus

The works by Chapman and Holmboe and the new flora by Meikle are to some extent interrelated. The last-named represents a splendid addition to botanical literature on the Middle East.

CHAPMAN, E. F., 1949. *Cyprus trees and shrubs.* 88 pp., illus. Nicosia: Government Printer. (Reprinted 1967.)

Briefly descriptive, somewhat nontechnical treatment of Cyprus woody plants, with keys to all taxa

Greek vernacular names; indication of local range, with citation of some *exsiccatae*; notes on habitat, phenology, local uses, and wood properties; indices to all botanical and vernacular names. An illustrated glossary is included together with a general key to genera.

HOLMBOE, J., 1914. *Studies on the vegetation of Cyprus.* 344 pp., 143 text-figs., 7 pls. (Bergens Mus. Skrifter, N.S. 1 (2).) Bergen.

Incorporates a systematic enumeration of known vascular plants (based partly upon the author's collections), with synonymy and references; indication of *exsiccatae*, with localities; brief descriptive notes, with emphasis on special features, and extra details of new or interesting plants; list of references; index to generic names. The introductory section includes accounts of the physical features and geography of the island as well as a history of botanical exploration and the author's itinerary, while at the end of the work is a discussion of vegetation formations and plant communities along with an account of historical phytogeography.

MEIKLE, R. D., 1977. *Flora of Cyprus.* Vol. 1. xii, 832 pp., 52 pls., map. Kew: Bentham-Moxon Trustees (Royal Botanic Gardens).

Concise descriptive flora of seed plants, with keys to genera and species (no key to families), full synonymy with references, citations of principal regional works, typifications, localities with citations of (selected) *exsiccatae*, generalized indication of local (with altitudes) and overall distribution, critical remarks, and notes on habitat, special features, etc.; appendices with abbreviations, list of collectors, and list of new taxa and names; complete index. A precursory part covers physical features, climate, botanical subdivisions (with map and remarks on *Charakter-pflanzen*; eight units recognized), and history of botanical exploration (mainly since 1905, the year of Holmboe's visit), and technical details. Partly complements Chapman's work on the woody plants. [One of the best of all floras to appear in recent years, this work is written and published very much in the tradition of the famous 'Kew Series' of the nineteenth and early-twentieth centuries, striking a compromise between practicality and documentation. This first of two volumes covers gymnosperms and dicotyledons (Ranunculaceae to Theligonaceae) on a modified Bentham and Hooker system.]

OSORIO-TAFALL, B. F. and SERAPHIM, G. M., 1973. *List of the vascular plants of Cyprus.* v, 137 pp.,

Nicosia: Ministry of Agriculture and Natural Resources, Cyprus.

Systematic name list, with synonymy and indication of endemics and introduced species as well as those of uncertain status; summary of the flora with statistics (1810 species in 668 genera) and index to families and genera. A brief introduction is also given, detailing the basis for the work.

773

'Holy Land' (in general)

This is here defined as a purely geographical subregion, extending approximately from the Turkish border to the Gulf of Suez and from the Mediterranean Sea to the Syrian-Jordanian desert ('Arabia deserta'). It overlaps parts of 774, 775, and 776.

POST, G. E., 1932–3. *Flora of Syria, Palestine and Sinai.* 2nd edn., revised by J. E. Dinsmore. 2 vols. 774 text-figs., map. Beirut: American University of Beirut. (First edn. 1896.)

Briefly descriptive flora of vascular plants, with analytical keys to families, genera, and species in larger genera as well as synoptic keys to species; essential synonymy, with principal citations (including Löw's *Die Flora der Juden*); indication of internal range by districts, with some citations and localities; Arabic and English vernacular names; notes on habitat, life-form, and phenology; complete indices to all botanical and vernacular names at end of vol. 2. The introductory section gives details of sources for the work as well as a glossary and bibliography. The *Flora* covers a region extending from the Taurus Mountains in the north to Ras Muhammad (near Sharm-el-Sheikh) in the south and from the Mediterranean to the Syrian–Jordanian desert. The arrangement of families in this work follows the Candollean system used in *Flora orientalis.*

774

Syria and Lebanon

See also 773 (POST). Partly for historical reasons (both were French mandates after World War I), these two countries have always been grouped together for floristic purposes and are accordingly treated as one unit.

MOUTERDE, P., 1966– . *Nouvelle flore du Liban et de la Syrie*. Vols. 1– (each in 2 parts). Plates. Beirut: Dar el-Machreq (Imprimerie Catholique) (distributed by Orientale; vol. 3 by Orientale and by OPTIMA Secretariat, Berlin-Dahlem).

Briefly descriptive flora of vascular plants, with keys to all taxa; limited synonymy; fairly detailed indication of internal range, with localities and citations, and brief summary of extralimital range; extensive taxonomic commentary and notes on habitat, phenology, etc.; bibliography (in vol. 1); indices to all botanical names. Each text volume is accompanied by an atlas with full-page plates of nearly all species. The introductory section (in vol. 1) covers floristic zones, general features of the flora, life-form categories, and the history of botanical exploration; a gazetteer is also provided. At this writing (mid-1979), vols. 1–2 (with atlases) and the text of vol. 3, parts 1–3 have been published, covering all families up through the Campanulaceae (Englerian system). Owing to the author's death in 1972, the subsequent severe political troubles, and editorial problems, publication of vol. 3 has been delayed, although the MS was originally considered virtually ready for press. Considerable revision was, however, deemed necessary by the present editors, A. Charpin and W. Greuter, and it is now planned to bring out the volume in fascicles; the first appeared in 1978. A precursory checklist for the whole work appeared as *idem*, 1965. *Flore du Liban et de la Syrie*. i, 101 pp. Beirut: The author. (Mimeographed.) [Systematic checklist, with indication of local range and frequency.]

THIÉBAUT, J., 1936–53. *Flore libano-syrienne*. 3 vols. 24 pls. Cairo: Institut d'Égypte (vols. 1–2); Paris: Centre Nationale de la Recherche Scientifique (vol. 3). (Vols. 1–2 originally publ. as *Mém. Inst. Égypte*, **31** (1936), **40** (1940); reprint of vol. 2, 1947, Paris, CNRS.)

Comprises a manual-key to vascular plants (in French), with limited synonymy, generalized indication of local range, and notes on habitat, phenology, etc.; addenda; indices in each volume (no general index). The introductory section includes a glossary, while the plates depict 200 representative species.

775

Palestine (including Israel, the 'West Bank', and Jordan)

See also 773 (POST). For floristic purposes, these polities have generally been considered as one unit and are accordingly treated together. They incorporate the former mandated territories of Cis- and Trans-jordanian Palestine.

ZOHARY, M., 1976. *A new analytical flora of Israel*. iii, 540, iv pp., 957 text-figs. Tel Aviv: Am Oved.

Manual-key to vascular plants (in Hebrew), with descriptions of families; includes Latin and Hebrew names, limited synonymy, generalized indication of local range, phenology, habitat, etc., and numerous small analytical figures; indices to Hebrew names and to families and genera, followed by corrigenda. The general part contains information for the student and the tyro, the plan of the work (pp. 14–15), a glossary, notes on botanical history, vegetation, phytosociology, etc., and a list of main references (pp. 45–6) and general key to families (pp. 48–65). This standard Hebrew-language work, now in its third and much altered version, contains many more figures than its predecessors. Succeeds EIG, A. *et al.*, 1960. *Analytical flora of Palestine*. 2nd edn. 66, 515 pp., 29 pls., map. Jerusalem. (First edn. 1948.)

ZOHARY, M. and FEINBRUN-DOTHAN, N., 1966–78. *Flora palaestina*. Vols. 1–3 (each in 2 parts). Plates. Jerusalem: Israel Academy of Sciences and Humanities.

Briefly descriptive flora of vascular plants, with keys to all taxa, synonymy with references and principal citations, generalized indication of local and extralimital distribution, taxonomic commentary, and notes on habitat, phenology, special features, ecology, uses, etc.; maps of botanical districts; diagnoses of new taxa and index to all botanical names in each volume. Each text volume is, as also in Mouterde's *Nouvelle flore*, accompanied by an atlas with illustrations of most species captioned with botanical and Hebrew vernacular names. The work, like the *Analytical flora* (see above), covers former Cis- and Trans-jordanian Palestine, with a further extension into the desert to the east. Of the four volumes projected, three, each with text and atlas, have appeared at this writing (mid-1980), covering all families up to the end of the Dicotyledoneae (12th edn.

of the Engler *Syllabus*). A precursory atlas of selected flowering plants is FEINBRUN, N., ZOHARY, M., and KOPPEL, R., 1949–58. *Iconographia florae terrae Israëlis* (later *Flora of the land of Israel: iconography*). 3 parts, 150 pls. (15 colored). Jerusalem: Palestine Journal of Botany (parts 1–2); Weizmann Science Press (part 3).

776

Sinai Peninsula

See also **591** (all works); **773** (Post). The peninsula is normally part of Egypt politically but geographically belongs to Asia; the Suez Canal acts as a continental boundary.

ZOHARY, M., 1935. Die phytogeographische Gliederung der Flora der Halbinsel Sinai. *Beih. Bot. Centralbl.*, Abt. B, **52**: 549–621, 3 figs. (incl. map).

Includes a tabular checklist of 942 species with indication of life-form, distribution, and phytogeographical affinities, with list of references, as well as general considerations on physical features, exploration, characteristics of the flora, and affinities.

778

Iraq

For the northern highland region, see also **790** (*Flora iranica*). A non-technical description of the *Flora of Iraq* appears in TOWNSEND, C. C., 1980. The 'Flora of Iraq' project. *Ur*, **2**: 19–23.

Bibliography

GUEST, E. and BLAKELOCK, R. A., 1954. Bibliography of Iraq. *Kew Bull.* 9: 243–249.

AL-RAWI, A., 1964. *Wild plants of Iraq, with their distribution*. 232, 18 pp., 2 maps (Dir. Gen. Agric. Res. (Iraq), Tech. Bull., 14). Baghdad: Government Press of Iraq.

Briefly annotated systematic list of vascular plants of Iraq, with concise indication of local range (arranged by districts); indices to generic and family names. The introductory section incorporates accounts of physical features, climate, vegetation, floristic zones, and progress in botanical exploration; a list of references appears at the end of the work. Essentially supersedes

ZOHARY, M., 1950. *The flora of Iraq and its phytogeographical subdivision*. 201 pp. (Dir. Gen. Agric. (Iraq), Tech. Bull., 31). Baghdad.

GUEST, E., TOWNSEND, C. C. and AL-RAWI, A. (eds.), 1966– . *Flora of Iraq*. Vols. 1– . Illus., maps. Baghdad: Ministry of Agriculture, Republic of Iraq.

Comprehensive descriptive flora of vascular plants, with keys to all taxa; extensive synonymy, with references and relevant citations; vernacular names; generalized indication of local range, with citation of *exsiccatae*, and summary of extralimital range; taxonomic commentary; notes on habitat, karyology, phenology, ecology, uses, special features, etc.; indices in each volume. In the introductory section in vol. 1 there are chapters on physical features, climate, vegetation, floristic zones, local names and languages, and the history of botanical exploration; there is also a gazetteer, a glossary, and a bibliography. A summary in Arabic is included. Of this detailed although concise work, nine volumes have been projected, with the flowering plants according to the 1959 version of the Hutchinson system; at this writing (mid-1979) vols. 1 (Introduction), 2 (lower vascular plants, gymnosperms, and Ranunculaceae through Rosaceae), 3 (Leguminosae), and 9 (Gramineae) have been published. The publication mode is the same as for *Flora Zambesiaca*: volume by volume on a preset systematic arrangement.

RECHINGER, K. H., 1964. *Flora of lowland Iraq*. 746 pp. Weinheim: Cramer.

Briefly descriptive manual-flora of vascular plants, with keys to all taxa; limited synonymy; fairly detailed indication of local range; occasional taxonomic commentary; notes on habitat, etc.; index to all botanical names. The work also includes notes on floristic zones, a list of frequently cited collectors, a glossary, addenda, and a list of references. [Intended as a 'students' flora', but at the price asked largely out of reach of its potential clientele.]

Partial work

For the Baghdad district, reference may be made to the following incomplete work: AGNEW, A. D. Q., 1962. *Flora of the Baghdad district*, I. *Monocotyledons*. ii, 170 pp. Baghdad. [Concise illustrated descriptive treatment, with keys, intended for student use.]

Region

78

Arabian peninsula

As here delimited, the Arabian peninsula includes all the land (and associated islands) south of the southern boundaries of Jordan and Iraq. The bulk of the area is occupied by Saudi Arabia, with the remainder comprising the Arab Republic of Yemen, South Yemen (Aden), Oman (with Muscat), the United Arab Emirates (Trucial States), Qatar, Bahrain, and Kuwait. Works dealing with Saudi Arabia alone are included under the general heading (**780**). Because of the very small number of works dealing purely with the smaller states, no subdivision of the region into units is here attempted.

General coverage is provided by *Flora orientalis* (north of the Tropic of Cancer) and by the partly compiled *Flora arabica* of Blatter (1919–23, 1936) and (for the infratropical zone) the critical *Flora des tropischen Arabien* of Schwartz (1939). With much collecting in more recent years, these works soon became notably outdated, based as they were on a very limited foundation. A new general flora is now in preparation, and a students' manual, *Flora of Saudi Arabia* (Mighadid and Hammouda, 1974) has been published, as well as two bibliographies. Some older florulas covering areas formerly under British rule are also available, such as Blatter's *Flora of Aden* (1914–16) and Dickson's *The wild flowers of Kuwait and Bahrain* (1955), now joined by a checklist for Qatar.

Progress

The earlier history of work in the Peninsula is covered by BLATTER, E., 1933. *The botanical exploration of Arabia.* 51 pp. Reprinted from *Rec. Bot. Surv. India*, 8(5): 451 [bis]–501 [bis], Delhi, as an addition to his *Flora arabica* (see below).

Bibliographies. General and supraregional bibliographies as for Division 7 and Superregion 77–79.

Regional bibliographies

BATANOUNY, K. H., 1978 (1398 A.H.). *Natural history of Saudi Arabia: a bibliography.* xii, 113, 8 pp. (Publ. Biol., 1). Jeddah: King Abdulaziz University. [Botany, pp. 21–31; 77 references, in part on non-vascular plants.]

RAHIM, M. A., 1979. *Biology of the Arabian Peninsula: a bibliographic study from 1557–1978.* xxiv, 180 pp. (Saudi Biol. Soc. Publ., 3). N.p.

Indices. General and supraregional indices as for Superregion 77–79.

780

Region in general

For the extra-tropical zone, see also **770–90** (*Flora orientalis*). General works on Saudi Arabia appear here.

BLATTER, E., 1919–23, 1936. *Flora arabica.* Fasc. 1–5. ii, 519, xlix pp. (Rec. Bot. Surv. India, 8 (1–4, 6).) Calcutta (later Delhi).

Systematic enumeration, without keys, of the seed plants of the Arabian peninsula (including parts of the Sinai Peninsula, Negev, etc., north to 31 °N), with synonymy, references, and fairly extensive literature citations: vernacular names; generalized indication of internal range (based on four large quadrants), with localities and indication of *exsiccatae*; summary of extralimital distribution; general index to all botanical names. An addition to this volume (as fasc. 6) is Blatter's historical account of botanical exploration in Arabia (see above), which appeared both separately and as *Records of the Botanical Survey of India*, vol. 8, part 5, pp. 451 [bis]–501 [bis].

MIGAHID, M. and HAMMOUDA, M. A., 1974. *Flora of Saudi Arabia.* 573 pp., text-figs., halftones, map, Riyadh: Riyadh University.

Students' manual-flora of vascular plants, with keys to all taxa, diagnostic descriptions, and indication of distribution and life-form; list of scientific names with Arab vernacular equivalents and vice versa; glossary and index to family and generic names. An introductory section includes an account of floristic regions, a list of orders and families in the Saudi Arabian flora (94 families) with general key, a list of references, and a rationale for the work. Patterned after Täckholm's *Students' flora of Egypt* (see **591**). For revised version of this work, with a greatly increased number of illustrations including many color photographs see MIGAHID, A. M., 1978. *Migahid and Hammouda's 'Flora of Saudi Arabia'.* 2nd edn. 2 vols., 940 pp., Illus. (some in color), map. Riyadh. [Volume 1, dicotyledons; vol. 2, monocotyledons.]

SCHWARTZ, O. 1939. *Flora des tropischen Arabien.*

393 pp. (Mitt. Inst. Allg. Bot. Hamburg, 10). Hamburg.

Systematic enumeration of vascular plants of the Arabian peninsula south of the Tropic of Cancer, with descriptions of new taxa but without keys; extensive synonymy, with references and citations; detailed indication of *exsiccatae* and other records, with localities, and general indication of extralimital distribution; short list of references; index to all botanical names. The introductory section includes remarks on phytogeographical zones and on botanical exploration in the region. [The region encompassed by this work was never covered by Boissier's *Flora orientalis*.]

Local work: Gulf of Aden/Red Sea states

BLATTER, E., 1914–16. *Flora of Aden*. 3 parts, 418, xix pp., plates, maps (Rec. Bot. Surv. India, 7). Calcutta.

Briefly descriptive flora of Aden and environs, with keys, synonymy, indication of local and extralimital range (including citation of *exsiccatae*), vernacular names, and index. An introductory section incorporates remarks on physical features, vegetation, and botanical exploration.

Local works: Persian Gulf states

BOULOS, L., 1978. Materials for a flora of Qatar. *Webbia*, **32**(2): 369–396, 10 figs., map.

Alphabetically arranged (by families) enumeration, with citation of *exsiccatae* and some critical notes; plants of special interest illustrated. Based on the collections of Obeid and Boulos, the first to be made in Qatar (1975–7); includes 260 species. A map appears in the introduction.

DICKSON, V., 1955. *The wild flowers of Kuwait and Bahrain*. 144 pp., 7 pls. (some in color), text-figs., 6 maps. London: Allen & Unwin.

Semipopular work, with lists of the Kuwait flora (by V. Dickson) and that of Bahrain (by R. d'O. Good), each incorporating vernacular names, local range, and short descriptive notes (the latter only in the Kuwait list). Remarks on land vegetation, marine algae, and plant collectors are also included. A singularly miscellaneous book.

Local work: Oman

The best available account on the plants of this sultanate in the southeastern part of the Arabian Peninsula, although not here considered a proper flora or checklist, is MANDAVILLE, J. P., JR., n.d. Plants. In *Scientific results of the Oman flora and fauna survey 1975* (ed. D. L. HARRISON *et al.*), pp. 229–67, pls. 1–24 (some in color). [Muscat]: Ministry of Education and Culture, Oman. (Journal of Oman Studies, Special Report.) [Includes checklist of plants collected in support of an account of the vegetation; emphasizes montane areas

scarcely explored in the past. References, p. 267. Noted in *Royal Botanic Gardens, Kew. Library: Current awareness list* (1980, no. 5), p. 33.]

Region

79

Iranian Highland

This region comprises the 'highland' area from northeastern Iraq through Iran and Afghanistan to the upland western and northern parts of Pakistan, finding its eastern limit along the Indus escarpment and in the Hindu Kush and Pamirs. However, for convenience, Hazara and Azad Kashmir (east of the Indus) and northern Kashmir (Gilgit, Baltistand and Ladakh) are also included here although their floras are essentially of Himalayan (region 84) and central Asiatic (region 76) affinities respectively. The region thus includes all of present-day Pakistan except for the lowland areas along the Indus (**811, 812**). The Iranian Highland additionally extends into the Azerbaijan Soviet Republic (**748**) and the southern fringe of Soviet middle Asia (region 75).

The whole region, with the exception of Hazara, Azad Kashmir, Gilgit, Baltistan and Ladakh (**796–799**) and the addition of those parts of the Soviet Union mentioned above, is covered by the modern comprehensive *Flora iranica*, the first general flora for a century and one accounting for the extensive collecting and other floristic and systematic work accomplished since World War II. For the three individual countries within the region, however, few practically useful floras are available. In Iran, there is only Parsa's voluminous but largely compiled *Flore de l'Iran* (1943–52), now being revised as *Flora of Iran* (1978–) in twelve volumes, only one of which had appeared at the time of writing. The French version also nominally covered Afghanistan, otherwise provided with only two other general works, neither of them complete: *Flora of Afghanistan* (Kitamura, 1960; supplement 1966) and *Symbolae afghanicae* (Køie and Rechinger, 1954–65). For Pakistan, however, there is the serial *Flora of West Pakistan* (1970–), now far advanced, and *An annotated catalogue of the vascular plants of West Pakistan and Kashmir* by Stewart (1972). The latter country features also a number of floras and checklists of more limited scope for the highland provinces and territories, both old and modern, covering Baluchistan (Burkill, 1909), the North-West Frontier and associated tribal districts (several works), and 'Pothohar', a geographical area centering on Rawalpindi (Bhopal and Chaudhri, 1977–). A number of miscellaneous works exist for northern Kashmir, the most important being *La flora del Caracorùm* (Pampanini, 1930; supplement 1934) (see **703**). The only major dendrological work is Sabeti's *Forests, trees and shrubs of Iran*, a coffee-table work published in 1976; however, Brandis' *Indian trees* can be used in the eastern part of the region.

While collecting in all countries has been extensive since World War II, coverage is still uneven, with Baluchistan perhaps the least known; northern Pakistan and northern and western Iran, where a number of scientific centers and tertiary institutions exist – mostly established in the last four decades or so – have been the most thoroughly examined. Present documentation is barely adequate overall, although helped by a number of recent bibliographies.

Progress

Separate historical accounts are available for Iran and Pakistan, viz: FEDTSCHENKO, B. A., 1945. [The investigators of the flora of Iran.] *Bot. Zurn. SSSR*, 30: 31–43; and STEWART, R. R., 1967. Plant collectors in West Pakistan and Kashmir. *Pakistan J. Forest.* 17: 337–63. Some recent developments in Pakistan are described in CHAUDHRI, M. N., 1977. The Pakistan herbarium. *Pakistan Syst.* 1(2): 100–5.

Bibliographies. General and supraregional bibliographies as for Division 7 and Superregion 77–79.

Indices. General indices as for Division 7 and Superregion 77–79.

790

Region in general

Since 1963, the region has been the beneficiary of the critical *Flora iranica*, produced under the direction of K. H. Rechinger at Vienna and now (1979) far advanced. Hitherto, it has been covered only by *Flora orientalis* (**770–90**), now long outdated, and imperfectly by a variety of 'national' and local floras.

RECHINGER, K. H. (ed.), 1963– . *Flora iranica.*

Fasc. 1– . Illus., map (on covers). Graz, Austria: Akademische Druck- und Verlagsanstalt.

Concise comprehensive descriptive flora of vascular plants, published in family fascicles; each fascicle includes keys to genera and species, more important synonymy, references and principal citations, detailed indication of internal range, with citations of *exsiccatae* and other records, summary of extralimital distribution, critical commentary, notes on habitat, biology, special features, etc., and index to all botanical names. No systematic sequence is followed in publication, but tables of five possible systems are given with every fascicle, with families published specially designated. The area covered is given in detail under the regional heading; except for the Karakoram, the work encompasses the entire region defined here. Progress on this outstanding flora has been steady, and at this writing (late 1980) 146 of the projected 170 fascicles have been published.

791

Iran

The two standard works on Iran proper are both somewhat theatrical, especially in their more recent versions. Furthermore, the new edition of Parsa's flora, commenced in 1978, exhibits little improvement over its uncritical predecessor of 1943–52 despite its more imposing format, and its continuation is moreover uncertain in view of recent political changes in Iran.

Bibliography

FREY, W. and MEYER, H.-J., 1971. Botanische literatur über den Iran. *Bot. Jahrb. Syst.* 91: 348–82.

PARSA, A., 1943–52. *Flore de l'Iran.* 5 vols. and supplement (in 6). Illus. Teheran: Muséum d'Histoire Naturelle.

PARSA, A., 1959. *Flore de l'Iran.* Vol. 7, parts 1–2 (Publications of the University of Tehran, 504). Teheran.

PARSA, A., 1960. *Flore de l'Iran.* Vol. 8 (*ibid.*, 613). Teheran.

PARSA, A., 1966. *Flore de l'Iran.* Vol. 9 (*ibid.*, 613/9). Teheran.

PARSA, A., 1974. *Flore de l'Iran.* Vol. 10. Teheran. (Reported to exist as indicated in *Acta Ecologia Iranica*, 1: 85 (1976), but, so far as is known, not available outside Iran.)

Largely compiled, briefly descriptive flora (in French) of vascular plants of Iran and Afghanistan, with synoptic keys to genera and groups of species; includes synonymy (with references and principal citations), detailed indication of local range with citation of *exsiccatae*, general summary of extralimital distribution, Persian vernacular names, and notes on phenology, life-form, etc.; indices to vernacular and botanical names in each volume. The introductory section (in vol. 1) contains a general key to families as well as chapters on physical features and botanical exploration, a gazetteer, a glossary, and a list of references. Further supplements to this work include a complete general index as volume 7 (in two parts, 1979) and the rarely seen volumes 8 and 9 (1960, 1966) with additions and changes to volumes 1 and 2. No information on the purported tenth volume has been available. [For review and bibliographic data, see LAMOND, J. M., 1978. Notes on Parsa's 'Flore de l'Iran', volumes 8 and 9. *Notes Roy. Bot. Gard. Edinburgh*, **35**: 349–64. Lamond comments that the work is 'badly written and ill-presented', with plentiful errors and questionable typification practices.] A revised version in English, with twelve volumes projected, has begun to appear: *Idem*, 1978. *Flora of Iran*. Vol. 1. Illus. (some in color) (Publ. Natl. Sci. Res. Council, Iran, 21). Teheran: Ministry of Science and Higher Education.

Special group – woody plants

SABETI, H., 1976 (2535 A.F.). *Forests, trees and shrubs of Iran*. 810 pp. (Persian text), 64 pp. (English text), illus. (some in color), portrait, maps. Teheran: National Agriculture and Natural Resources Research Organization, Iran.

This generously formatted work, in Persian with a summary section in English, is a descriptive atlas of 986 species of woody plants, without keys and in a dictionary arrangement by scientific names (including synonyms). The text for accepted species includes synonymy, vernacular names (in Persian, English, French, and German), botanical description, and commentary, also for native species photographs (partly in color) of habit, technical details, etc., and distribution maps with summary (the latter also in English); index to Persian names. An introductory section deals with floristic regions and phytogeography. The English section, separately paged at the 'back', contains the same introductory matter with, in

addition, an index of families, indices to English, French, and German vernacular names, and a selected bibliography (pp. 57–62). English descriptions of two new species (invalidly published) are also included. Now of coffee-table dimensions, the work is a revised and expanded version of *idem*, 1966. [*Native and exotic trees and shrubs of Iran*.] xii, 430, [ii] pp., illus. Teheran.

792

Afghanistan

See also **791** (Parsa). The two works described below are the most comprehensive available, although neither provides complete coverage of the flora. That by Kitamura is essentially an amplified expedition report, while the series of Køie and Rechinger is a partial precursor to *Flora iranica*. No separate floras have been hitherto available.

Bibliography

BRECKLE, S. W. *et al.*, 1969. Botanical literature of Afghanistan. *Notes Roy. Bot. Gard. Edinburgh*, **29**: 257–371.

KITAMURA, S., 1960. Flora of Afghanistan. In *Results of the Kyoto University scientific expedition to the Karakoram and Hindukush, 1955* (ed. COMMITTEE OF THE KYOTO UNIVERSITY SCIENTIFIC EXPEDITION TO THE KARAKORAM AND HINDUKUSH), vol. 2. ix, 486 pp., 105 text-figs., color frontispiece, 2 maps. Kyoto.

KITAMURA, S., 1966. Additions and corrections to 'Flora of Afghanistan'. *Ibid.*, vol. 8, pp. 67–154.

Systematic enumeration of seed plants, with synonymy, references and citations, vernacular names, localities with *exsiccatae* and general summary of local and extralimital range; list of references and index to botanical names. An introductory section includes chapters on prior botanical exploration and research, the Kyoto expedition itinerary, and definition of floristic and phytogeographic zones. The supplement of 1966 is based on later field work in the region.

KØIE, M. and RECHINGER, K. H., 1954–65. *Symbolae afghanicae*. 6 parts. Illus. (Kongel. Danske Vidensk. Selsk., Biol. Skr., 8(1), etc.). Copenhagen.

Comprises a series of critical enumerations of Afghan representatives in various families of vascular plants, with keys for some genera and descriptions of novelties, limited synonymy, indication of *exsiccatae* with localities, and extensive taxonomic commentary; indices in each part. Not a flora in the strict sense, but covers a very wide range of families and may be regarded as complementary to Kitamura's flora.

793

Pakistan (in general)

Under this heading are included bibliographies and floristic works pertaining to Pakistan as a whole or to substantial portions thereof not corresponding to a given province or territory. These latter appear as separate units: **794**, Baluchistan (with Quetta); **795**, North-West Frontier Province (except Hazara); **796**, Hazara; **797**, Azad Kashmir; **798**, Gilgit Agency and Baltistan; **799**, Ladakh; **811**, Sind; **812**, (West) Punjab (with Bahawalpur). For Indian-occupied Jammu and southern Kashmir, east of the so-called 'Cease-Fire Line', see **841**. Ladakh, is, in part, also under Indian administration.

The area of present-day Pakistan was prior to 1947 never covered as a whole by a single flora. The country lies abreast of the Indus escarpment zone and the deep dry upper Indus valley of Kohistan, a major biogeographic disjunction used in the nineteenth century as the mutual boundary between *Flora orientalis* and *Flora of British India*, the two standard works of the day respectively for southwest and south Asia. This disjunction was later utilized as the eastern limit of *Flora iranica*. Later local coverage never amounted to more than a smattering of scattered florulas and enumerations (except in the Punjab), and few centers for detailed floristic work existed. Large areas of the mountainous districts remained, botanically, very poorly known.

Since independence, much new botanical exploration has been accomplished and some local floras published. From 1970, available information has been consolidated in two nation-wide works: the *Flora of West Pakistan* edited by E. Nasir and S. I. Ali (1970–) and the *Annotated catalogue* by R. R. Stewart (1972). Much of the northern hill and mountain country centering on Rawalpindi and Islamabad and extending to the northern and western frontiers is being treated additionally in *Flora of Pothohar and the adjoining areas* by F. G. Bhopal and Md. N. Chaudhri (1977–). However, it is still argued that many areas, particularly in the mountainous northern districts, are in need of further collecting and study.[6]

Pakistan here corresponds to the former West Pakistan. East Pakistan separated in 1971 as Bangladesh (**835**).

Bibliography

KAZMI, S. M. A., 1970–7. *Bibliography on the botany of West Pakistan and Kashmir and adjacent regions*, I–V. Coconut Grove, Fla.: Field Research Projects. [The 'adjacent regions' include nearly all of Afghanistan and parts of Iran, Soviet central Asia, China, and India.]

STEWART, R. R., 1956. A bibliography of the flowering plants of West Pakistan and Kashmir. *Biologia* (Lahore), 2: 221–30. [Somewhat selective.]

NASIR, E. and ALI, S. I. (eds.), 1970– . *Flora of West Pakistan*. Fasc. 1– . Illus., maps. Karachi: The editors (under sponsorship from the US Department of Agriculture; distributed from Department of Botany, University of Karachi).

Detailed descriptive flora of seed plants, published in family fascicles; each fascicles contains keys to genera and species, synonymy, references and citations, localities with indication of respresentative *exsiccatae* (referred to a grid system), general indication of overall range (internally and externally), notes on habitat, frequency, phenology, special features, uses, etc., and index to all botanical names. The work, produced with the aid of surplus agricultural funds available under USA PL-480, has been appearing fairly rapidly but not in systematic order and without numbering in a preset scheme. At this writing (mid-1979) part 123 has been reached. For some families, interim mimeographed reports under Ali's editorship have also been produced.

STEWART, R. R., 1972. *An annotated catalogue of the vascular plants of West Pakistan and Kashmir*. xviii, 1028 pp., portrait, map. (Part of NASIR and ALI, *Flora of West Pakistan*; see previous entry.)

Systematic enumeration of vascular plants known from Pakistan and the whole of Kashmir, with essential synonymy, references and citations, localities with indication of selected *exsiccatae*, critical remarks, and notes on habitat, local range, etc.; list of authors and index of collaborators; index to all botanical names. An introductory section incorporates a general description of the region together with accounts of floristics, vegetation, botanical exploration and analysis, and future research needs. The work is now considered to be in need of revision, with perhaps ten per cent of Pakistan species yet to be included (Chaudhri, 1977).[6]

Partial work: Pothohar

BHOPAL, F. G. and CHAUDHRI, MD. N. 1977– . Flora of Pothohar and the adjoining areas, I–II. *Pakistan Syst.* 1(1): 38–128; 1(2): 1–98, illus. (some colored in part II).

Concise descriptive floristic treatment, with keys to all taxa, limited synonymy, references and citations (usually major floristic works or revisions), local distribution, overall range, and some critical notes; no habitat or biological data. At this writing (late 1979) the Centrospermae as well as Casuarinaceae through Polygonaceae (Englerian system) have been published. Centering on Rawalpindi (and Islamabad), this work covers the Margalla Hills (northern Punjab), Hazara, and Azad Kashmir as well as adjoining parts of North-West Frontier Province (from Peshawar to Chitral), Gilgit Agency, and Baltistan (an area not yet defined in the text, however).[7]

794

Baluchistan (with Quetta)

See also **793** (all works). This Pakistani province is made up of Quetta and the former Baluchistan Agency.

BURKILL, I. H., 1909. *A working list of the flowering plants of Baluchistan*. 136 pp. Calcutta: Superintendent of Government Printing, India. (Reprinted after 1950, Baluchistan Forestry Department.)

Systematic enumeration of known seed plants, with limited synonymy, citations of *exsiccatae* or other sources, localities, vernacular names, and notes on economics, uses, etc.; index. An introductory section describes local botanical exploration among miscellaneous remarks. For additions, see BLATTER, E., HALLBERG, F. and McCANN, C. C., 1919–20. Contributions towards a flora of Baluchistan. *J. Indian Bot.* 1: 56–61, 84–91, 128–38, 169–78, 226–36, 263–70, 344–52.

795

North-West Frontier (except Hazara)

See also **793** (all works). Hazara appears as **796**, while **795** here comprises the remainder of the present North-West Frontier Province from Waziristan and Dera Ismail Khan north through Malakand and Mardan to Swat, Kalam, Dir and Chitral (the last four districts named were at one time administered centrally as 'tribal trust territories'). No general flora for this area is available, and for convenience the various local florulas have been grouped under two subheadings: the North-West Frontier proper (Waziristan to Malakand and Mardan) and the former tribal states (Swat to Chitral).

Local works: North-West Frontier

BLATTER, E. and FERNANDEZ, J., 1933–4. Flora of Waziristan, I–V. *J. Bombay Nat. Hist. Soc.* 36: 665–87, 950–77, pls. 1–3, map; 37: 150–71, 391–424, 604–19, pls. 4–8.

Concise enumeration of higher plants, with descriptions of new taxa, essential synonymy, references and citations of key literature, vernacular names, localities with indication of *exsiccatae* and summary of overall (extralimital) range, and brief notes on habitat, phenology, etc.; short list of references. A prefatory section includes remarks on botanical exploration and a definition of the limits of the area. The few plates depict a range of vegetation formations, with notable *Charakter-pflanzen*. [Waziristan is in the southern part of the province, from the Baluchi border north to about 33° 30′N; it thus stops short of the Peshawar district.]

QURAISHI, M. A. and KHAN, S. A., 1965–72. Flora of Peshawar district and Khyber agency, 1–2. *Pakistan J. Forest.* 15 (1965): 364–93, 41 figs.; 17 (1967): 203–54; 22 (1972): 153–219, 323–83.

Copiously illustrated, briefly descriptive flora of seed plants, with keys, limited synonymy, general indication of local and extralimital range, and brief notes on habitat, etc.; no indices. An introduction in part 1 discusses physical features and geography of the area, which centers on Peshawar and extends westward to include Khyber Pass, the famous gateway to South Asia. Not yet completed; extends from Ranunculaceae through Crassulaceae on the Bentham and Hooker system.

Local works: Swat, Kalam, Dir and Chitral

DUTHIE, J. F., 1898. The botany of the Chitral relief expedition. *Rec. Bot. Surv. India*, 1: 139–81, map.

Systematic enumeration of vascular plants and bryophytes collected and/or observed by the writer, with descriptions of some new taxa; citation of *exsiccatae* with localities and brief notes are included, but no index. An introductory section gives a description of physical and botanical features of Chitral as well as a map of the expedition's routes.

STEWART, R. R., 1967. Check list of the plants of Swat State, NW Pakistan. *Pakistan J. Forest.* 17: 457–528.

Systematic list of vascular plants, with limited synonymy, local range (including some *exsiccatae*), and habitat notes; no index. The introductory section gives a general description of physical and botanical features of the region, a list of collectors, and statistical tables.

796

Hazara

See 793 (all works, but especially BHOPAL and CHAUDHRI); 812 (BAMBER; PARKER; STEWART). This part of the North-West Frontier Province lies east of the Indus and centers on Abbotabad, north of Rawalpindi.

797

Azad Kashmir

See 793 (all works, but especially BHOPAL and CHAUDHRI). This area includes the Mirpur, Poonch (Punch) and Muzaffarabad districts of Kashmir west of the 'cease-fire line' and coming under Pakistani administration. For the remainder of Pakistani-administered Kashmir, including Gilgit, Baltistan, and Ladakh, see 798–799.

798

Gilgit and Baltistan

See also 703 (PAMPANINI); 760 (GRUBOV *et al.*); 793 (all works). This area corresponds to the part of northern Kashmir north of the upper Indus River and west of the 'cease-fire line' between Pakistan and India. Dominating the area is the greater part of the Karakoram, much of it under permanent ice and snow and rising to a height of 8611 m at K2 (Mt Godwin Austen). The impoverished flora is of central Asiatic affinity and thus is outside the scope of *Flora iranica*. The best coverage is provided in the works cross-referenced above, but some further details can be found in Kitamura's enumeration (1964), the only 'separate' florula.

KITAMURA, S., 1964. Flowering plants of West Pakistan. In *Results of the Kyoto University scientific expedition to the Karakoram and Hindukush, 1955* (ed. COMMITTEE OF THE KYOTO UNIVERSITY SCIENTIFIC EXPEDITION TO THE KARAKORAM AND HINDUKUSH), vol. 3, pp. 1–161, figs. 1–60. Kyoto.

Partial enumeration of the seed plants in the mountainous areas of Gilgit and Baltistan, with descriptions of novelties, synonymy, references and literature citations, indication of *exsiccatae* with localities, general summary of overall range, and notes on habitat, etc.; no separate index. An introductory section gives accounts of prior botanical exploration, the itinerary of the Kyoto expedition, and phytogeographical relationships.

799

Ladakh

See also 703 (PAMPANINI); 760 (GRUBOV *et al.*); 793 (all works). This area, centering on Leh and including the Ladakh Range and the eastern Karakoram, is part of northern Kashmir north of the Indus but falls abreast of the 'cease-fire line' and thus is divided between Pakistani and Indian administration, with the northeastern Ladakh Plateau (Aksai Chin) effectively under Chinese control (cf. 768). As in Gilgit and Baltistan, the flora is basically of central Asian (and Tibetan) affinity, and falls outside the scope of *Flora iranica*. Apart from the coverage provided in the general works, two separate floras are available (Stewart, 1916; Kachroo *et al.*, 1977).

KACHROO, P., SAPRU, B. L., and DHAR, U., 1977. *Flora of Ladakh: an ecological and taxonomical appraisal.* x, 172 pp., figs. 1–6, 4–24, tables (incl. maps). Dehra Dun: Bishen Singh Mahendra Pal Singh.

Annotated systematic enumeration of seed plants, with diagnoses, synonymy, references, localities with indication of representative *exsiccatae*, and notes on habitat, phenology, etc.; index to all botanical names; references (p. 166). The introductory section gives generalities on physical features, climate, biota, people and land use, the economy of the area and planned development, and finally on vegetation, floristics and phytogeography, and life-forms.

STEWART, R. R., 1916. The flora of Ladak, western Tibet, I–II. *Bull. Torrey Bot. Club*, 43: 571–90, 625–50. (Reprinted separately, 1973, Dehra Dun: Bishen Singh Mahendra Pal Singh.)

Includes a systematic list of vascular plants, based mainly on collections by the author, with occasional synonymy, citation of *exsiccatae* with localities, published records, taxonomic commentary, and miscellaneous observations; list of references. An introductory section covers geography, climate, geology, and botanical exploration as well as details of the author's itinerary. The area covered falls wholly within eastern Kashmir, west of the Chinese boundary. All records are now incorporated in the author's *Annotated catalogue* (1972; see **793**), and the work has been at least partly superseded by the flora by Kachroo *et al.* (see above).

Notes

1 Shetler, *Komarov Botanical Institute* (1967), p. 74.
2 Malyschev, L. I., 1978. Svodka 'Flora tsentral'noj Sibiri' i rabota nad nej. *Bot. Žurn. SSSR*, **63**: 1358–63.
3 A new, one-volume manual of southern Krasnojarsk Krai appeared in 1979.
4 E. Hultén (personal communication) has noted that Komarov's flora was written some time after 1908–9, when its author travelled extensively in Kamchatka, but almost certainly well before 1917. Unfortunately, the manuscript was lost following submission to the publishing house and did not turn up for nearly 10 years. However, following its rediscovery in the mid-1920s, evidently no attempt at revision was made, despite a copy of all Hultén's notes having been made available to Komarov. Consequently, when finally published the work was already some 15 or more years out of date.
5 See also Galushko, A. I., 1975. Rukovodstvo dlja avtorov 'Flory Severnogo Kavkaz'. 22 pp. Rostov-on-Don, Stavropol. These 'instructions to contributors' led to an ideological exchange, similar to but, at least in print, more subdued than the great Backer-Koorders controversy in Java (**918**): KIRPICZNIKOV, M. E., 1977. Tak li nužno pisat' rukovodstva? [Is this how manuals should be written?] *Bot. Žurn. SSSR*, **62**: 136–8; and GALUSHKO, A. I., 1979. Po povodu retsenzii M. E. Kirpicznikova 'Tak li nužno pisat' rukovodstva?' *Ibid.*, **64**: 1354–9.
6 Chaudhri, M. N., 1977 (1978). The Pakistan herbarium. *Pakistan Syst.* **1**(2): 100–5.
7 I wish to thank Mr Abid beg Mirza, formerly of the Wau Ecology Institute, Wau, Papua New Guinea, for assistance with the definition of 'Pothohar'.

Division

8

Southern, eastern, and southeastern (monsoonal) Asia

In these matters (of style and phraseology) my *Flora of the British Islands* has been followed; the style there adopted having been suggested by the requirements of the Professors of Botany in the Scotch universities and approved by them, seemed to me to be equally applicable to a more extended [work]...It is as a hand-book to what is already known, and a pioneer to more complete works, that the present is put forward.

> J. D. Hooker, Preface to *Flora of British India* (1872).

It should be remembered that outside the small herbarium and botanical library in Hong Kong herbaria and botanical libraries were non-existent in China previous to 1916.

> H.-L. Li, *Proc. Linn. Soc. London*, **156**: 39 (1944).

Beyond the Great Wall [of China] there is not enough rainfall to support agriculture; below it there is.

> R. M. Nixon, *The Real War* (1980).

This division essentially consists of that half of the Asiatic continent under the influence of the summer monsoon. Its western and northern limits, which impinge upon Division 7, originate at the Arabian Sea coast, west of Karachi and run along the western edge of the Indus basin and then along the upper course of the Indus River and the 'cease-fire line' to the Kashmiri-Chinese border. From there it runs along the Himalayan frontier of China as far as the Burmese–Indian border and from there northward along the Tibet–Szechwan provincial boundary (west of the 'Tibetan Marches'), the eastern limit of Qinghai (Chinghai) Province, and along the Huang Ho through Kansu (passing by the city of Lanzhou) and the Ninghsia Hui Autonomous Region to the boundary of the Inner Mongolian Autonomous Region; and thence eastward approximately along the Great Wall, the southeastern boundary of the Region, and the Chinese-Mongolian and Chinese-Soviet frontiers to the Sea of Japan southwest of Vladivostok. Also included in Division 8 are Korea, Japan, the Ryukyus, the Bonin and Volcano Islands, Parece Vela, the Daito Islands, Taiwan, the South China Sea islands (Paracels and Spratlys), and the Coco, Andaman, and Nicobar Islands. However, Peninsular Malaysia is excluded, being referred to West Malesia (region 91) in Division 9, and the Laccadive and Maldive Islands are referred to region 04 in Division 0. Nine politico–geographical regions have been delimited, additionally grouped into

four superregions (South Asia, 81–84; Japan and
Korea, 85; China 'proper' and Manchuria, 86–88; and
Southeastern Asia, 89). These latter by and large also
represent the biohistorical foci of the hemi-continent,
and under their headings are therefore presented the
appropriate background surveys.

General bibliographies. Bay, 1910; Frodin, 1964;
Goodale, 1879; Holden and Wycoff, 1911–14; Jackson,
1881; Pritzel 1871–7; Rehder, 1911; Sachet and
Fosberg, 1955, 1971; USDA, 1958. See also supraregional headings.

General indices. BA, 1926– ; BotA, 1918–26; BC,
1879–1944; BS, 1940– ; CSP, 1800–1900; EB, 1959– ;
FB, 1931– ; IBBT, 1963–9; ICSL, 1901–14; JBJ,
1873–1939; KR, 1971– ; NN, 1879–1943; RŽ, 1954– .
See also supraregional headings, notably for *Flora
Malesiana Bulletin.*

Conspectus

800

Division in general

No flora or related work covers the whole of the division. What comprehensive floras exist have been written for the four superregions, notably 81–84 and 86–88, and reference should be made there.

803

Alpine and upper montane zones

Substantial portions of the hemi-continent are at comparatively high elevations, but apart from the Himalaya – here delimited in more or less political terms as Region 84, as most floras within it are so bounded and/or cover much more than the alpine and upper montane zones – there exists little specific literature falling within the scope of this *Guide*. However, floras of the high elevation areas of the Himalaya alone will be considered under this heading.

Three floras are given below: one by Rau (1975) on the western Himalaya and two by Takeda (1960, 1961; the latter with K. Tanabe) on Japan.

I. The Himalaya

For other works, see Region 84.

RAU, M. A., 1975. *High altitude flowering plants of West Himalaya.* x, 234 pp., 25 figs., 5 photographs. Howrah, West Bengal: Botanical Survey of India.

Concise systematic account, with descriptions of and general notes on families, keys to genera, and lists of species in each genus with localities (arranged by districts) and altitudinal zonation and occasional synonymy; addenda, glossary, references (pp. 207–10), and index to botanical names. The introductory section covers physical features, climate, insolation, the extent of the alpine flora, habitats and general features of the flora, noteworthy plants, the origin of the Himalaya, and phytographic history. Covers the mountain tracts of Uttar Pradesh and Himachal Pradesh in northwest India above 3300–3600 m, and encompasses over 1000 species.

II. Japan

TAKEDA, H., 1959–62. *Genshoku Nihon kōzan shokubutsu zukan.* [Alpine flora of Japan in color.] 2 vols. 109, 114 pp., pls. (some in color), table (vol. 2) (Flora and fauna of Japan, vols. 12, 28). Osaka: Hoikusha. (Reprinted subsequently.)

Comprises an atlas of habit photographs of alpine wild flowers with associated descriptive text and notes and botanical names in Latin with Japanese transliterations and vernacular equivalents; illustrated glossary and complete indices at end. No keys are provided. The introductory section includes a description of alpine vegetation. This book, of which the first volume (all seen by this reviewer) treats 281 species, strongly resembles in impact the Hegi-Merxmüller *Alpenflora* (603/IV) and represents a distillation from a considerable amount of other published work. See also *idem*, 1957. *Kōzan shokubutsu zu-i* [Illustrations of alpine plants]. 4, 310, 17 pp., 442 figs., 100 halftones, 8 color pls. Tokyo. (First pub. 1933.)

TAKEDA, H. and TANABE, K., 1961. *Nihon kōzan shokubutsu zukan* [Illustrated manual of alpine plants of Japan]. Revised edn. 15, x, 347 pp., illus. (some in color). Tokyo: Hokuryukan. (First edn., by H. Takeda alone, 1950.)

Pocket manual of alpine plants, for field use (in contrast to the preceding, which is for the home or office): [Not seen by this reviewer; cited from *National Union Catalog* 1963–7, vol. 52, p. 499. Earlier editions listed in MERRILL and WALKER, Supplement, p. 347 (Superregion 71–75).]

804

Highlands: the South Indian hills

FYSON, P. F., 1932. *The flora of the South Indian hill stations:* Ootacamund, Coonor, Kotagiri, Kodaicanal, Yercaud, and the country round. 2 vols. xxix, 697 pp.; atlas of 611 pls. Madras: Government Press. (Originally pub. 1915–20 in 3 vols. as *The flora of the Nilghiri and Pulney hill-tops*; this edn. reprinted 1974, Dehra Dun: Bishen Singh, Mahendra Pal Singh.)

Copiously illustrated, semipopular descriptive flora of the seed plants of the high country on the Nilghiris, Palnis, and Shevaloys above about 4000 ft (some 1200 m); includes limited synonymy, English common names, fairly detailed indication of local distribution, general summary of extralimital range, and notes on habitat, phenology, biology, special features, etc.; complete index at end. [Designed and written for above all the annual 'migration' to the well-known and popular 'hill-stations' of peninsular India, as named in the title. The volume of plates moreover was representative of another peninsular tradition: the botanical atlas as originated by Rheede and continued by Wight and by BEDDOME (820).]

808

Wetlands

The only available works in this genre are for the South Asian superregion; that by Biswas and Calder (1937) is a comparatively early example which for India has been partially supplanted by the manual of Subramanyam (1962). A revision of the Biswas and Calder work, however, appeared in 1955.

South Asia (including India) and Burma

BISWAS, K. and CALDER, C. C., 1955. *Handbook of common water and marsh plants of India and Burma.* 2nd edn., revised by K. Biswas. xvi, 216 pp. (incl. 32 pls.), 6 halftones (Health Bulletin (India), no. 24). Delhi: Manager of Publications, Government of India. (First edn., 1937.)

Amply descriptive account of the more common aquatic macrophytes (including 178 vascular plant species) with accompanying figures, with keys (to genera and species only), indication of habitat and distribution, and special features; plates (pp. 179–216) and index to scientific names at end. A glossary is also included. The introductory section deals with general limnological processes, general features of aquatic vegetation, periodicity, control methods, and surveys of coastal halophytes and mangroves as well as other topics. [The second edition has been oriented to a greater extent towards botanical users, but it retains the conservative style of its predecessor; as an account it is more conventional than that of Subramanyam described below.]

SUBRAMANYAM, K., 1962. *Aquatic angiosperms.* viii, 190 pp., 63-text-figs., color frontispiece (Botanical Monographs, Council of Scientific and Industrial Research, India, no. 3). New Delhi.

Illustrated descriptive account of more common flowering hydrophytes, with keys, synonymy and references, literature citations, generalized indication of range, habitat, phenology, etc., and notes on specialized morphological features. Illustrations separate from text; references (pp. 177–80), list of karyotypes, and general index at end. The introduction includes a general key to families along with a consideration of degree of coverage.

Superregion

81–84

South Asia (Indian subcontinent)

The limits of South Asia (or the Indian subcontinent) are here considered to be the western edge of the Indus Basin (the mutual boundary between Boissier's *Flora orientalis* and Hooker's Flora of *British India*), the upper course of the Indus River in northern Kashmir; the Chinese border in the Himalaya (in the east as recognized by India); and the Burmese frontier in the east. It thus includes the southern and eastern parts of Pakistan, almost all of India, Nepal, Sikkim, Bhutan, Bangladesh, and Sri Lanka (Ceylon).

Floristic work and publication in the modern sense may be said to begin with the work of Rheede tot Draakestein and his assistants over three centuries ago; their *Hortus malabaricus* (1678–1703), based on freshly gathered material as well as centuries-old local botanical knowledge and manuscripts, laid a foundation for all future work in the superregion. These activities early spread to Sri Lanka through one of the assistants, Hermann, whose collections were written up by others, notably Linnaeus in *Flora zeylanica* (1747). A new cycle of activity began with the extension of British control in the latter part of the eighteenth century after the Seven Years' War through the work of the 'United Brotherhood' in southern India and Sri Lanka; one of its members was W. Roxburgh who later (1793) moved on to Calcutta as superintendent of the then recently-founded Indian Botanic Garden. His pioneering *Flora indica*, however, appeared only after his death (1820–4; 2nd edn., 1832).

The gradual extension of the British Raj during the course of the nineteenth century witnessed a rapid and widespread expansion of botanical work, especially into Baluchistan, the Northwest frontier, the Himalaya, Kashmir, Assam, and other mountainous interior country, aided after 1850 by the spread of railways. Additional botanical centres came into existence and a great number of floristic contributions, large and small, were published. A first attempt at a modern *Flora indica* (1855) by J. D. Hooker and T. Thomson was not continued, but after dissolution of the East India Company following the 1857 mutiny and the establishment of formal government, botanical resources could be reorganized and increased and Hooker, with the aid of specialists, finally summarized most available information in his *Flora of British India* (1872–97), a part of the 'Kew Series'. In Sri Lanka, the work of Thwaites, Trimen and others was consolidated in the *Handbook of the flora of Ceylon* by Trimen and Hooker (1893–1900, with a supplement by Alston in 1931). Forest botany began serious development in the 1860s, and the regional forest floras of Beddome (1869–73) and Stewart and Brandis (1874) as well as other contributions laid the basis for Brandis's *Indian trees* (1906). Beddome also contributed a regional manual on pteridophytes (1883; 2nd edn. 1892).

In 1890, the first Botanical Survey of India was established, which with administrative changes and other developments ushered in a period of greater decentralization of botanical work. From bases in Howrah, Madras, Poona and Saharanpur (later Dehra Dun), there were produced under the Survey's auspices in the first four decades of the present century all the well-known presidency and other regional floras, and support given to several 'forest floras'. Many floristic works for western India and Pakistan were in addition contributed by Fr E. Blatter and others (among them H. Santapau) from St Xavier's College, Bombay. It was this group, along with interested individuals at some other tertiary institutions and in forest botany circles (especially at Dehra Dun), who were largely responsible for pursuing organized floristic work and education after the reduction of the original Botanical Survey of India subsequent to the 1900s and its virtual abolition by the 1930s in that stagnant but ominous inter-war period so well characterized by E. M. Forster and the historian J. Morris, a development likely also influenced by changing botanical fashions after 1900.

After the political changes of 1947–50, the obsolescence of the various available floristic works and the need for a renewal of floristic and systematic botany was gradually perceived in the new nations of South Asia. In India, the Botanical Survey of India was revived in the 1950s, becoming a statutory body later encompassing the Indian (formerly Royal) Botanic Garden and seven regional centers. A large amount of floristic contributions and several floras, mainly at state or local level, have been published. A new enumeration of Indian plants by S. K. Mukerjee was begun in the BSI *Bulletin* in 1959, but lapsed in 1962. In Pakistan,

recent progress has been more evident; R. R. Stewart's checklist was published in 1972 and the *Flora of West Pakistan* (see **793**), begun in 1970, is now far advanced, with over 125 parts now available. In Sri Lanka, the *Flora of Ceylon* project was organized in the late 1960s; publication of the results as *A revised handbook to the flora of Ceylon* begun in 1973 but due to a change of arrangements is now scheduled to appear from 1980–83 in six volumes as the *Flora of Ceylon*. A *Flora of Bangladesh* began publication the year after independence. The latest development is the initiation of a serial *Flora of India* by the Botanical Survey of India as from 1978.

All these new productions, however, have not replaced the basic superregional treatises: Beddome's *Handbook to the ferns of British India, Ceylon, and the Malay Peninsula* (1892), Brandis' *Indian trees* (1906), and, most importantly, Hooker's *Flora of British India* (1872–97), one of the greatest floras of all time. Additions and corrections to Hooker's flora were published by Calder *et al.* (1926), Razi (1959) and Nayar and Ramamurthy (1976), and the grasses (except for the Bambuseae) were wholly revised by Bor (1960). A 'companion' to Beddome's *Handbook* by Nayar and Kaur appeared in 1972 (reissued 1974). In spite of their now-obsolete nomenclature, based on the 'Kew Rule' and other nineteenth-century practices, these works continue to be used widely and have all been reprinted within the last decade or so.

Since World War II, a considerable amount of new field work has been carried out through much of the superregion, much of it under the Botanical Survey of India, although to date little specialist collecting has been done and large forest trees are said to have been neglected. Notable field projects have been mounted in Sri Lanka, the Himalaya, Karnataka, northeastern India, and Pakistan among others. But 'many densely vegetated areas are yet to be explored', and more effort in this and in the writing of floras and related works is required. The considerable collections now to hand will serve as a partial basis for a much-needed new round of subregional, state, and district as well as national floras to supplant the many works published from about 1890 to 1940; more details appear under the regional headings.

In 1977 a general symposium of the Botanical Survey of India was held at Howrah for assessment of progress and needs. At the same time, an outline of the Survey's *Flora of India* program was given; this entails

1. a national flora, to appear serially; 2. state flora 'analyses' (enumerations); 3. district (or local) floras, with keys, descriptions, etc. (some already published); and 4. works on non-vascular cryptogams and special topics as well as monographs.[1] In general, however, South Asian work is perhaps less well-coordinated than that on Greater Malesia or Africa.

The long period over which the present stock of standard floras in South Asia has appeared has also been one of many boundary changes, especially within India as reorganization of internal boundaries took place according to linguistic, ethnic and other criteria. The units adopted here represent a compromise between the mosaic as represented in the last decade of the British Raj (*circa* 1937) and the modern states, provinces, and territories in the successor nations.[2]

Progress

For early and recent history, see particularly BURKILL, I. H., 1965. *Chapters on the history of botany in India.* xi, 245 pp., maps. Delhi; and SANTAPAU, H., 1958. *History of botanical researches in India, Burma and Ceylon*, 2: *Systematic botany of angiosperms.* vi, 77 pp. Bangalore (for the Indian Botanical Society). Among earlier progress reports may be cited AGHARKAR, S. P., 1938. Progress of botany during the last twenty-five years. In *Progress of science in India* (ed. B. Prasad), pp. 742–67. Calcutta; BISWAS, K., 1943. Systematic and taxonomic studies on the flora of India and Burma. In *Proceedings 30th Indian Science Congress* (Calcutta), vol. 2, pp. 101–52, figs. 1–11. Calcutta; SANTAPAU, H., 1962. The present state of taxonomy and floristics in India after independence. *Bull. Bot. Surv. India*, 4: 209–16; and MAHESHWARI, P. and KAPIL, R. N., 1963. *Fifty years of science in India: progress of botany.* viii, 178 pp., illus., map. Calcutta. Recent progress is surveyed sketchily by LEGRIS, P., 1974. Vegetation and floristic composition of humid tropical continental Asia. In *Natural resources of humid tropical Asia* (ed. UNESCO), pp. 217–38. (Natural resources research, 12). Paris; and rather extensively although unevenly in the many contributions in BOTANICAL SURVEY OF INDIA, 1977. *All-India symposium on floristic studies in India: present status and future strategies* (Howrah, 1977): Abstracts, 52 pp. Howrah, West Bengal; and in the *Annual Reports, Botanical Survey of India* and *Flora Malesiana Bulletin*. Some remarks respecting the subregion are also given in the general survey of tropical floristics by Prance (1978).

General bibliographies. Bay, 1910; Frodin, 1964; Goodale, 1879; Holden and Wycoff, 1911–14; Jackson, 1881; Pritzel, 1871–7; Rehder, 1911; USDA, 1958.

Supraregional bibliographies

BLATTER, E., 1909. A bibliography of the botany of British India and Ceylon. *J. Bombay Nat. Hist. Soc.* **20**: lxxix–clxxvi.

NARAYANASWAMI, V., 1961–5. Indian botany. Parts 1–2. In *A bibliography of Indology* (National Library of India), vol. 2, parts 1–2. Calcutta. [A projected part 3 has not appeared.]

SANTAPAU, H., 1952. Contributions to the bibliography of Indian botany. *J. Bombay Nat. Hist. Soc.* **50**: 520–48; 51: 205–259.

STEWART, R. R. 1956. A bibliography of the flowering plants of West Pakistan and Kashmir. *Biologia* (*Lahore*), 2(2): 221–30.

General indices. BA, 1926– ; BotA, 1918–26; BC, 1879–1944; BS, 1940– ; CSP, 1800–1900; EB, 1959– ; FB, 1931– ; IBBT, 1963–9; ICSL, 1901–14; JBJ, 1873–1939; KR, 1971– ; NN, 1879–1943; RŽ, 1954– .

Supraregional indices

STEENIS, C. G. G. J. VAN and JACOBS, M. (eds.), 1948– . *Flora malesiana bulletin*, 1– . Leiden: Flora Malesiana Foundation (later Rijksherbarium). (Mimeographed.) [Issued nowadays annually, this contains a substantial bibliographic section also very thoroughly covering South Asian literature. A fuller description is given under Superregion 91–93.]

810–40

Superregion in general (including India)

The superregion is still dominated by Hooker's *Flora of British India*, although time has rendered it incomplete and keys and nomenclature are now obsolete. Three sets of additions have been made, the latest in 1976 (the latter two, however, relate only to modern India), and a modern treatment of grasses (Bor, 1960) has replaced that in volume 7 of Hooker's *Flora*. General manuals are also available for the ferns (Beddome, 1892, with supplement in 1972) and trees (Brandis, 1906). To these classics have been added a dictionary of genera for India (Santapau and Henry, 1973) and a forest botany manual (Bor, 1953).

At the end of 1978 there commenced publication of a serial *Flora of India*, to be issued, like the *Flora of West Pakistan* and *Flora of Bangladesh*, in fascicles.[3] Floras are also in progress for Sri Lanka and Nepal.

– Overall vascular flora: 13000–15000 species (Hooker).

Dictionary

SANTAPAU, H. and HENRY, A. N. (assisted by B. Roy and P. Basu), 1973. *A dictionary of the flowering plants in India.* vii, 198 pp. New Delhi: CSIR of India. [Annotated alphabetical checklist of 2900 generic names (both accepted and synonymous) with essential references, indication of families, brief descriptions, range, and approximate number of species, with the number represented in India given as well as any outstanding forms; index to vernacular and other common names.]

BOTANICAL SURVEY OF INDIA, 1978– . *Flora of India.* Pub. in fascicles. Illus. Howrah, West Bengal: Botanical Survey of India.

Comprises a documentary descriptive flora, published serially; each fascicle, with one or more families, contains keys to all taxa, six- to nine-line descriptions, synonymy (with references and citations), typification, general distribution and altitudinal zonation, notes on karyotypes and phenology (but not ecology), illustrations of representative species, and index. [Not seen; description and citation based on review in *Flora Malesiana Bull.* 32: 3202 (1979). Preparation of the work is under the direction of S. K. Jain. Publication commenced in December 1978, and as of this writing (late 1980) five fascicles have appeared.]

HOOKER, J. D., 1872–97. *Flora of British India.* 7 vols. London: Reeve. (Reprinted in covers entitled 'Flora of India' at various dates to 1961, Brook nr. Ashford, Kent: Reeve; 1973, 1978, Dehra Dun: Bishen Singh Mahendra Pal Singh. A Peking reprint of about 1956 is also known.)

Concise, briefly descriptive flora of seed plants, with modified synoptic keys to genera and species; includes synonymy (with references and citation of key literature), generalized indication of internal range (with some localities and *exsiccatae* recorded), summary of extralimital distribution, taxonomic commentary, and notes on habitat, special features, etc.; complete general index to botanical names in vol. 7 (vols. 1–6 also individually indexed). Treatments of some families by specialists. The preface to vol. 1 is very brief. [The geographical spread of this great work, the 'star of India' in the pleiad of 'Kew floras' initiated under W. J. Hooker, includes the entire superregion as here delimited along with Burma, peninsular Thailand (at the isthmus of Kra), and the then-Straits Settlements and Malay States (although coverage of the last three is rather sketchy, and for Burma reasonably adequate only for the southern part).]

For a more complete (and still useful) general introduction to the Indian flora (as well as a philosophical chapter on taxonomy which is almost without peer), see HOOKER, J. D. and THOMSON, T., 1855. *Flora indica.* Vol. 1 (all pub.). xvi, 280, 285 pp., map. London. [Contains also a folding map and an explanation of the divisions used in the *Flora of British India*.]

Since 1897, three lists of additions and corrections have been produced, viz: CALDER, C. C. *et al.*, 1926. *List of species and genera of Indian phanerogams not included in Sir J. D. Hooker's 'Flora of British India'.* ii, 157 pp. (Rec. Bot. Surv. India, vol. 11, part 1). Calcutta. (repr. 1978, Dehra Dun); RAZI, B. A., 1959. *A second list of species and genera of Indian phanerogams not included in J. D. Hooker's 'Flora of British India'.* ii, 56 pp. (*Ibid.*, vol. 18, part 1); and NAYAR, M. P. and RAMAMURTHY, K., 1976. Third list of species and genera of Indian phanerogams not included in J. D. Hooker's 'Flora of British India' (excluding Bangladesh, Burma, Ceylon, Malayan Peninsula and Pakistan). *Bull. Bot. Surv. India*, vol. 15, pp. 204–34. [These three lists bring coverage up to 1975.] For a modern treatment of Gramineae (except Bambuseae), with analytical keys, see BOR, N. L., 1960. *The grasses of India, Burma, Ceylon and Pakistan, excluding Bambuseae.* xviii, 767 pp., illus. Oxford: Pergamon. (Reprinted with additions and corrections, 1973, Königstein/Taunus, W. Germany: Koeltz.)

MUKERJEE, S. K., 1959–62. Enumeration of Indian flowering plants. *Bull. Bot. Surv. India*, 1: 138–41; 2: 99–107, 293–7; 3: 99–101, 351–5; 4: 39–47.

Systematic enumeration, with synonymy, references and citations; generalized indication of internal and extralimital range. Some treatments provided by collaborators. The geographical range of this work includes Nepal, Sikkim, and Bhutan but not Pakistan, Bangladesh, Burma, or Ceylon. Not completed; only a few families (Ranunculaceae to Annonaceae on the Bentham and Hooker system) have been published, the last in 1962.

Special groups – woody plants

BOR, N. L., 1953. *Manual of Indian forest botany*, xv, 441 pp., 31 pls. Bombay: Oxford University Press (India).

Elementary descriptive treatment of the more important trees, shrubs, and other woody plants of India, with keys to families, genera, and (in larger genera) species; limited synonymy, without references or citations; notes on local range, habitat, life-form, special botanical features, wood properties, and uses; index to all botanical names. Intended as a students' manual.

BRANDIS, D., 1906. *Indian trees.* xxxii, 767 pp., 201 text-figs. London: Archibald, Constable. (Reprinted 1907, 1908, 1911, London; 1971, 1978, Dehra Dun, Bishen Singh Mahendra Pal Singh.)

Descriptive treatment of trees, shrubs, woody climbers, bamboos, and palms of the Indian Empire, Burma, and Ceylon, with synonymy, references and citations, English and local vernacular names, generalized indication of internal and extralimital distribution, taxonomic commentary and notes on habitat, phenology, special features, properties, and uses; addenda and index to all botanical and vernacular names. Less important and lesser-known species are featured in smaller type. The introductory section includes chapters on forest districts and forest vegetation, sources for the work, and special botanical features, together with a synopsis of families.

Special group – pteridophytes

BEDDOME, R. H., 1892. *Handbook to the ferns of British India, Ceylon and the Malay Peninsula.* 2nd edn., with supplement. xiv, 500, 110 pp., 299 text-figs., frontispiece. Calcutta: Thacker, Spink. (Reprinted 1969, Delhi: 'Today and Tomorrow's'.) (First edn., 1883.)

Briefly descriptive, illustrated fern flora, with synoptic keys to genera and species, citations of major sources, generalized indication of internal range and elevations, more or less extensive taxonomic commentary, and index to all botanical names. The supplementary section in the 1892 edition comprises additions and alterations to the original text (unchanged from the first edition). For further additions and changes, and revised nomenclature, see NAYAR, B. K. and KAUR, S., 1972. *Companion to R. H. Beddome's* 'Handbook to the ferns of British India, Ceylon, and the Malay

Peninsula'. 196, vii pp. New Delhi: Pama Primlane. (Reissued in larger format with additional introductory matter, 1974, New Delhi: Pama Primlane, and Königstein/Taunus; W. Germany: Koeltz. iv, 244 pp.)

Region

81

Indus basin and valley; northwest and central India

Within this region are the Pakistani provinces of Sind and West Punjab (with Bahawalpur) along with the Indian states (and territories) of East Punjab, Haryana, Delhi, Uttar Pradesh (excluding the Himalayan hill and mountain tracts), Rajasthan, and Madhya Pradesh (the last-named comprising parts of the former Central Provinces together with the Central India Agency). Generally speaking, it corresponds to 'North-West and Central India' of the former British Raj, except that all the hill and mountain tracts in the Himalaya and beyond the Indus have been excluded (for these, see regions 79 and 83).

Botanical exploration in the region has been fairly extensive but varying in thoroughness; lesser known areas include the arid lands centering on the Indian desert and parts of the central plateau and other hilly upland tracts. Comparatively few state or provincial floras or forest manuals were published from the 1930s to the 1960s, and the basic works are now at least 50 years old, including the subregional *Plants of the Punjab* (Bamber, 1916) and *Flora of the upper Gangetic Plain* (Duthie, 1903–22, but lacking Gramineae). However, in the last two decades a revival of floristic work has taken place, and floras, of varying quality, have, in addition to Pakistan in general (see **793**: *Flora of West Pakistan*, commenced in 1969, has made rapid progress), been published for West Punjab (in part; see Bhopal and Chaudhuri, 1977– , under **793**), the (East) Punjab plains (Nair, 1978), Delhi (Maheshwari, 1963–6), Uttar Pradesh (Raizada, 1976, as supplement to Duthie's flora), Western Rajasthan (Puri, 1964), and the Indian Desert (Bhandari, 1978), as well as a number of districts. There is also a series of floristic contributions by Panigrahi (1965–6) for Madhya Pradesh, hitherto without a regional flora and covered only by two old forest floras by Haines (1916) and Witt (1916) having different limits. Papers on the grasses and the pteridophytes of the upper Gangetic plain, never covered by Duthie, have also been published. In general, the region is fairly well covered, except for Madhya Pradesh and adjacent areas where a new state flora is very much needed.

Bibliographies. General and supraregional bibliographies as for Divison 8 and Superregion 81–84.

Indices. General and supraregional indices as for Division 8 and Superregion 81–84.

810

Region in general

See also **793** (all works); **810–40** (all works).

STEWART, J. L. and BRANDIS, D., 1874. *The forest flora of North-West and Central India*. 2 vols. xxxi, 608 pp.; atlas of 70 pls. London: Allen. (Reprinted 1972, Dehra Dun: Bishen Singh Mahendra Pal Singh.)

Briefly descriptive treatment of trees and shrubs, with non-dichotomous keys to genera and species; synonymy, with references; vernacular names; generalized indication of local and extralimital range; extensive notes on habitat, life-form, phenology, associates, timber, uses, etc., as well as taxonomic commentary; indices to all botanical, English, and vernacular names. The plates are all by W. H. Fitch. The introductory section includes a general description of the woody flora of the region and its composition, a short glossary, and a list of references.

811

Sind (Pakistan)

See also **793** (all works); **822** (COOKE; TALBOT). Prior to 1947 Sind formed part of the Bombay Presidency. Karachi and Hyderabad are the main population centers, and for these and the Indus delta between them local florulas are available as well as the provincial flora by Sabnis (1923–4).

SABNIS, T. S., 1923–4. The flora of Sind. *J. Indian Bot. Soc.* **3**: 151–3, 178–80, 204–6, 227–32, 277–84; **4**: 25–7, 50–70, 101–15, 134–48.

Systematic enumeration of seed plants, with

references and detailed indication of localities and *exsiccatae*; general summary of overall range; list of references (in introductory section). Forms an areal supplement to Cooke's *Flora of the Presidency of Bombay*, which covers this province.

Partial works

BLATTER, E., MCCANN, C. C., and SABNIS, T. S., 1929. *The flora of the Indus Delta*. vi, 172 pp., 140 text figs, 50 halftones. Bombay: Indian Botanical Society.

This mainly ecological and phytogeographical work includes (pp. 1–38) a systematic enumeration of vascular plants, with synonymy, localities, and citations.

JAFRI, S. M. H., 1966. *The flora of Karachi (coastal West Pakistan)*. iv, 375 pp., 345 text-figs., 3 pls. Karachi: The Book Corporation.

Students' manual-flora of vascular plants, with keys, descriptions, illustrations, fairly extensive synonymy, relatively detailed indication of local range, summary of extralimital distribution, taxonomic commentary, and notes on habitat, uses, etc.; list of references and index to all botanical names.

812

(West) Punjab and Bahawalpur (Pakistan)

See also **793** (all works). Floras pertaining to the whole of the Punjab and adjacent tracts published before 1947 are described under this heading. Of the several population centers, florulas are available for the Lahore and Rawalpindi districts.

BAMBER, C. J., 1916. *Plants of the Punjab*. iii, 652, xxviii pp., 6 pls. Lahore: Superintendent of Government Printing, Punjab. (Reprinted 1976, Dehra Dun: Bishen Singh Mahendra Pal Singh.)

Descriptive manual of seed plants, with artificial keys to genera and groups of species; generalized indication of local range, including some citations of records; English and Urdu vernacular names; occasional notes on uses, special features, etc.; complete index to botanical and vernacular names. A brief glossary is incorporated in the introductory section. Nomenclature follows for the most part that of the *Flora of British India*. Coverage encompasses Jammu, Kashmir, and the North-West Frontier Province (of Pakistan) as well as Bahawalpur and the East and West Punjab. For supplement, see SABNIS, T. S., 1940–1. A contribution to the flora of the Punjab plains and the associated hill regions. *J. Bombay Nat. Hist. Soc.* **42**: 124–49, 342–79, 533–86.

PARKER, R. N., 1924. *A forest flora of the Punjab with Hazara and Delhi.* 2nd edn. 591 pp. Lahore: Superintendent of Government Printing. (Reprinted as '3rd edn.', 1956; 2nd reprint, 1973, Dehra Dun: Bishen Singh Mahendra Pal Singh; 1st ed., 1918.)

Briefly descriptive manual of trees and shrubs, with keys to genera and species, limited synonymy, references, vernacular names, generalized indication of local range and elevation, taxonomic commentary, and notes on habitat, phenology, special features, uses, etc.; complete index. An introductory section has a glossary and an account of the general features of the flora.

STEWART, J. L., 1869. *Punjab plants.* [iv], xiv, 269, 106 pp., map. Lahore: Government Press. (Reprinted 1977, Dehra Dun: Bishen Singh.)

Systematic enumeration of the known vascular plants of the Punjab and adjacent hill regions with notes on habitat, special features, natural products, and uses; limited synonymy; English and local vernacular names; complete indices. Introductory chapters include a discussion of the flora and vegetation and its phytogeographical relationships, along with notes on vernacular names, natural products, and uses. The area covered by this work extends beyond the Punjab to Garwhal and Kumaon in the Himalaya and to Jammu, Kashmir, and the North-West Frontier Province, and there is a certain bias toward useful plants.

Partial works

KASHYAP, S. R. and JOSHI, A. C., 1936. *Lahore district flora.* 285 pp., 218 text-figs. Lahore: University of the Punjab.

Illustrated, briefly descriptive students' manual-flora, with keys to genera and species, summary of local range, vernacular names, and notes on habitat, phenology, etc.; glossary and index to generic, family, and vernacular names.

STEWART, R. R. [1958]. *The flora of Rawalpindi district, West Pakistan,* x, 163 pp. Rawalpindi. (Originally pub. 1957–8 in parts in *Pakistan J. Forest.*)

Systematic enumeration of vascular plants, with synonymy, vernacular names, summary of local range (with inclusion of some localities), critical commentary, and notes on habitat, biology, special features, cultivation, uses, etc.; index to family and generic names. For supplements, see *idem*, 1961. Additions and corrections to the flora of Rawalpindi district, West Pakistan. *Pakistan J. Forest.* **11**: 51–63, 222–95.

813

(East) Punjab and Haryana (India)

For general works, see also **812**. The Himalaya hill and mountain tracts are excluded from this unit, which is thus almost entirely covered in the new state flora by Nair (1978). For Delhi see **814**.

NAIR, N. C., 1978. *Flora of the Punjab Plains.* xx, 326 pp. (Rec. Bot. Surv. India, 21(1)). Howrah, West Bengal.

Briefly descriptive manual-flora with keys to all taxa, synonymy, references and citations, vernacular names, indication of local range, with representative *exsiccatae*, some taxonomic remarks, and notes on phenology, habitats, associates, frequency, uses, etc.; sources (for naturalized species); indices to botanical and vernacular names. An introductory section deals with physical features, geology, soils, climate, earlier botanical work and basis for the present flora, and floristics and vegetation, with list of principal references. Covers 807 native or naturalized species in an area from 28° 50′ N to 32° 50′ N and 73° 30′ E to 78° E., i.e., from Delhi northwards.

814

Delhi

Delhi lies between the Punjab Plains and the upper Ganges Plain. It has seemed convenient to treat it here as a separate unit.

MAHESHWARI, J. K., 1963. *The flora of Delhi,* viii, 447 pp. New Delhi: Council for Scientific and Industrial Research.

MAHESHWARI, J. K., 1966. *Illustrations of the flora of Delhi.* xx, 282 pp., 278 text-figs. New Delhi: Council for Scientific and Industrial Research.

Keyed, briefly descriptive flora of seed plants, with accompanying atlas; includes synonymy, references, citations, some indication of *exsiccatae* and general summary of local range, vernacular names, and notes on habitat, phenology, uses, special features, etc.; bibliography and complete index. The introductory section covers physical features, climate, soils, earlier botanical work, floristics, and vegetation.

815

Uttar Pradesh (upper Gangetic plain)

The Himalayan hill and mountain tracts, including the Dehra Dun district, are here excluded. The standard flora of Duthie (1903–22), which, however, omits grasses and pteridophytes, just reaches that area, but to the south extends across the hilly parts of northern Madhya Pradesh and into the eastern part of Rajasthan (816–817). A revised checklist was published by M. A. Rau in 1968, and a consolidated supplement has also appeared (Raizada, 1976). Companion works have been produced for the pteridophytes (Chowdhury, 1973) and the grasses (Raizada *et al.*, 1961–6).

DUTHIE, J. F., 1903–22. *Flora of the upper Gangetic Plain and of the adjacent Siwalik and sub-Himalayan tracts.* 3 vols. Map. Calcutta: Indian Government Printer. (Reprinted in 2 vols., 1960, Calcutta: Botanical Survey of India.)

Briefly descriptive manual-flora of seed plants (except Gramineae), with keys to genera and species, synonymy, references and citations, generalized indication of internal range, taxonomic commentary, and notes on phenology, biology, uses, etc.; no index (evidently terminated unfinished). Designed as a subregional flora for the former United Provinces (and Oudh), i.e., Uttar Pradesh (except for the Himalayan tracts), the former Central India Agency of princely states, now part of Madhya Pradesh (but from 1950 to 1956 the separate states of Madhya Bharat and Vindhya Pradesh), and the eastern part of the former Rajputana Agency of princely states (now Rajasthan). The Narbada River forms the approximate southern limit for the work. A general index, compiled by S. K. Jain, was published in 1952 by the Council for Scientific and Industrial Research of India. [Duthie was based at the old Company garden at Saharanpur, north of Delhi towards Dehra Dun in the northern plains, and in charge of a herbarium subsequently (1908) transferred to Dehra Dun.] For a substantial supplement, arranged on the same sequence as the original work and similar in style, see RAIZADA, M. B., 1976. *Supplement to Duthie's 'Flora of the upper Gangetic Plain and of the adjacent Siwalik and sub-Himalayan tracts'.* vii, 355 pp. Dehra Dun: Bishen Singh Mahendra Pal Singh.

For Gramineae, see RAIZADA, M. B. *et al.*, 1961–6. Grasses of the upper Gangetic Plain. *Indian For. Rec.*, N.S., Bot. **4**: (iv), 171–277; **5**: 151–220, 10 pls. (Panicoideae, I–II); *Indian For.* **92**: 637–42 (Pooideae). [The Panicoideae are fully treated, with keys, distributions, records, and remarks, but the Pooideae are given only as a checklist.] For pteridophytes, see CHOWDHURY, N. P., 1973. *The pteridophyte flora of the upper Gangetic Plain.* vii, 91 pp., 10 pls. New Delhi: Navayug Traders. [Includes keys and descriptions.]

RAU, M. A., 1968. *Flora of the upper Gangetic Plain and of the adjacent Siwalik and sub-Himalayan tracts: check list.* 87 pp. (Bull. Bot. Surv. India, Suppl. 2). Calcutta.

Brief introduction; systematic checklist of vascular plants (except Gramineae), incorporating additions (275 species), nomenclatural changes, and revised synonymy resulting from the progress of botany since publication of Duthie's flora.

816

Madhya Bharat and Vindhya Pradesh

See 815 (all works). These states, originally formed in 1950 from the princely states of the erstwhile Central India Agency, were absorbed in 1956 into the state of Madhya Pradesh (819).

Local work: Bhopal district

OOMMACHAN, M., 1977. *The flora of Bhopal.* xi, 475 pp., 15 pls., tables, 2 maps. Bhopal, Madhya Pradesh: Jain Brothers.

Descriptive flora of seed plants, with keys to taxa from generic level downwards; includes synonymy and references, literature citations, local and overall distribution (including some specimen citations), vernacular names, and notes on variability, habitat, occurrence, phenology, etc.; some critical remarks in footnotes; lists of plants in particular situations, references, lexicon, and index to families and genera at end. The introductory section covers physical and biotic features, vegetation and forests, human influence, plant life-forms, and floristic statistics (836 species). [Bhopal, now in the state of Madhya Pradesh, was formerly a princely state in the Central India Agency.]

817

Eastern Rajasthan

See **815** (all works). The unit covers all that part of the state (before 1947 the Rajputana Agency) from the Aravalli Hills eastwards, and includes Jaipur and Ajmer. [Mt Abu in the geologically old Aravallis, rising to 1724 m, has an isolated patch of evergreen forest around its summit, unusual for an otherwise dry area.]

818

Western Rajasthan

This area, which includes the Thar (or Indian) Desert, constitutes all that part of the state west of the Aravalli Hills. It is not covered in any regional flora, but existing knowledge has been consolidated in a separate flora by Puri *et al.* (1964), supplanted now for the southwestern desert quarter by Bhandari (1978).

PURI, G. S. *et al.*, 1964. *Flora of Rajasthan* (*west of the Aravallis*). 159 pp., 1 text-fig., 1 pl. (Rec. Bot. Surv. India, 19(1)). Calcutta.

Systematic enumeration of seed plants recorded from the western part of Rajasthan, floristically the poorest in India; includes keys to all taxa, synonymy (with references and citations of more important literature), indication of local range with citations of *exsiccatae*, etc., and notes on habitat, special features, etc.; list of references and index to generic and family names. The introductory section provides chapters on physical features, geography, other biota, vegetation, and prior botanical exploration and research. This work, to a certain extent compiled, is marred by many inconsistencies (see review of Gupta and Bhandari in *Ann. Arid Zone*, 4: 236–8 (1965)).

Partial works: Indian Desert

BHANDARI, M. M., 1978. *Flora of the Indian Desert.* viii, 472 pp., 133 text-figs., 13 pls., map. Jodhpur: Scientific Publishers (distributed by United Book Traders).

Illustrated descriptive flora of seed plants, with keys to all taxa, synonymy, references, vernacular names, indication of local range with citation of *exsiccatae*, overall distribution, critical remarks, and notes on habitat, frequency, occurrence, phenology, uses, etc.; list of references and indices to all vernacular and scientific names.

An introductory section includes descriptions of physical features, climate, soils, biotic factors, history of botanical work in the area and the basis for the present work, and accounts of floristics, vegetation, and plant geography. This good district flora treats 592 species in 319 genera in the SW half (Jodhpur, Jaisalmer and Barmer) of **818**. Supplants BLATTER, E. and HALLBERG, F., 1918–21. The flora of the Indian desert (Jodhpur and Jaisalmer). *J. Bombay Nat. Hist. Soc.* 26: 218–46, 525–51, 811–18, pls. 1–31; 27: 40–7, 270–9, 506–12, pls. 32–7. [Systematic enumeration of vascular plants, with extensive remarks on the vegetation attached.]

819

Madhya Pradesh and adjacent areas

Modern Madhya Pradesh, the largest state in India, does not correspond to the former Central Provinces of the British Raj. Between 1947 and 1956 Berar and neighboring districts were transferred to Maharashtra (Bombay) and Madhya Bharat and Vindhya Pradesh (the one-time Central India Agency) were annexed; other boundary adjustments were also made. Much earlier, in the 1910s, part of Chota Nagpur, formerly entirely in the Presidency of Bengal, had been transferred to the former Central Provinces. For convenience, Berar and adjacent areas now in Maharashtra are treated as part of **819**, as is also Chota Nagpur, but the districts of the former Central India Agency are separately designated as **816**.

Botanical development in this inland plateau area, part of which lies in the Deccan, has been comparatively slow; no general flora has ever been published nor has it been included in any of the standard regional floras. The woody plants were covered in a series of manuals by Haines and Witt (1910–16). A resumption of botanical work took place with the development of the Botanical Survey of India and organization of its Central Circle; some floristic contributions have been made by Panigrahi (1965–6) and others.

HAINES, H. H., 1916. *Descriptive list of trees, shrubs, and economic herbs of the Southern Circle, Central Provinces.* xxviii, 384 pp. Allahabad: Pioneer Press. [Reprinted, with revisions, from a series in *Indian Forester* 38–40 (1912–14).]

Descriptive manual-flora of woody plants (and herbs of economic significance), with non-dichotomou keys to genera and species (including also an artificial key at end of work); essential synonymy; vernacular names; generalized indication of local range; notes on habitat, phenology, properties, uses, etc.; glossary and index to generic and vernacular names. The introductory section includes a description of the region and sections on geology and soils as well as a synopsis of families.

WITT, D. O., 1916. *Descriptive list of trees, shrubs, climbers and economic herbs of the Northern and Berar Forest Circles, Central Provinces.* xiii, 79, 247, viii, xviii, ii, xvi, ii, ii pp. Allahabad: Pioneer Press.

Forms a companion work to that of Haines on the Southern Circle, featuring a similar scope and layout but differing in the presence of analytical keys, accounts of forest formations, and a delineation of botanical sub-regions. Both this and its companion encompass native, naturalized, and commonly cultivated species.

Partial and local works

HAINES, H. H., 1910. *A forest flora of Chota Nagpur.* vii, 634, xxxvii pp., map. Calcutta: Superintendent of Government Printing. (Reprinted 1974, Dehra Dun, Bishen Singh Mahendra Pal Singh.)

Keyed, briefly descriptive manual-flora of trees, shrubs, vines, and more important herbs of this then-extensively forested large area in northeastern Madhya Pradesh and adjacent Bihar, with vernacular names, indication of local range, and notes on habitat, special features, etc.; indices. At the time of writing, the area was a constituent division of the old Presidency of Bengal, which covered much of eastern India; soon after publication, however, it was divided between the Central Provinces and the newly organized province of Bihar and Orissa (the latter now two states).

PANIGRAHI, G. *et al.*, 1965–6. Contributions to the botany of Madhya Pradesh. Parts 1–3. *Bull. Bot. Surv. India*, **8** (1966): 117–25 (part 1); *Proc. Natl. Acad. Sci. India*, B (Biol. Sci.), **35** (1965): 87–98, 99–109 (parts 2–3).

A systematic enumeration of flowering plants, with synonymy, references, local range, and miscellaneous notes. Intended as additions to existing knowledge on the flora of the state. Not yet completed; extends from Dilleniaceae through Convolvulaceae on the Bentham and Hooker system.

PATEL, R. I., 1968. *Forest flora of Melghat.* xlviii, 380 pp. Dehra Dun: Bishen Singh Mahendra Pal Singh.

Keyed descriptive manual-flora of woody plants of the Melghat district in the Berar circle (now in Maharashtra), with limited synonymy, indication of local range, and notes on occurrence, habitat, special features, etc.; index.

Region

82

Peninsular India and Sri Lanka

This region comprises all of India south of the Narbada River, the southern limits of the former Central Provinces, the northeastern limits of the former Madras Presidency, together with Sri Lanka (Ceylon).

With a long history of botanical exploration, beginning in the sixteenth century, peninsular India and Sri Lanka may currently be considered to be floristically relatively well-known. However, the basic stock of subregional floras is mostly 50 or more years old; these include *Flora of the Presidency of Bombay* by Cooke (1901–9), *Flora of the Presidency of Madras* by Gamble and Fischer (1915–38), and *Handbook of the flora of Ceylon* by Trimen and Hooker (1893–1900, with supplement by Alston, 1931). Coverage of many areas by these works is moreover sketchy, and for the former Nizam's Dominions (Hyderabad-Deccan) no separate flora has ever been published. Revisions of these have not materialized or have been slow in coming. On the other hand, active collecting projects have taken place in Karnataka, the Peninsula proper, Sri Lanka, and elsewhere, thereby laying the basis for new district and state floras. The largest amount of modern work has come from Karnataka, in the interior uplands (where active systematics centers have existed at Mysore and Bangalore), and in Sri Lanka (in which the Ceylon Flora project stimulated considerable new collecting and reinvestigation of the island's flora and, after a false start in publication, has finally since 1980 begun to emerge as the *Flora of Ceylon*). A number of district floras have appeared in other states and territories since 1950, and some forest flora-manuals have also been published (e.g., Patel, 1971, for Gujarat; Soma-sundaram, 1967, for the southern states of India). Nevertheless, more exploration (before it is too late in many areas) and documentation, particularly from the eastern part of the region, is greatly needed in terms of currently accepted standards in floristics.

Bibliographies. General and supraregional bibliographies as for Division 8 and Superregion 81–84.

Indices. General and supraregional indices as for Division 8 and Superregion 81–84.

820

Region in general

See also **810–40** (all works). For the illustrated work by Fyson (1932) on the distinctive montane flora of the Southern Indian hill-tops, see **804**.

BEDDOME, R. H., 1869–73. *Flora sylvatica for South India*. 2 vols. 330 + 28 pls. Madras: Gantz (for the Madras Government). (Reprinted 1978, Dehra Dun, Bishen Singh Mahendra Pal Singh.)

Comprises two main sections: (1) a 'Foresters' Manual of Botany for Southern India (pp. i–ccxxxviii)', including Bentham's 'Outlines of Botany' (for colonial floras), a synopsis and index of genera and families, and a descriptive treatment of genera and species, with keys to genera and extensive notes on botanical attributes, wood, properties, uses, etc.; and (2) an iconography of 330 plates (not in systematic order) with accompanying descriptive text, featuring every species treated in the first volume; in this text are given formal details of synonymy, local distribution, and occurrence. The additional 28 plates in vol. 2 feature morphological analyses of representatives of a number of different South Indian genera. A synoptic index to plates appears in vol. 1, which, along with the 'Foresters' Manual', follows the Bentham and Hooker system.

SOMASUNDARAM, T. R., SRI, 1967. *A handbook on the identification and description of trees, shrubs and some important herbs of the forests of the southern states for the use of the Southern Forest Rangers' College, Coimbatore*. iv, 563, 9 pp. Delhi: Manager of Publications, Government of India.

Concise field manual of native and principal introduced woody plants (as well as the larger and/or more important herbs), with keys to nearly all taxa, concise descriptions, limited synonymy and key references, vernacular names, generalized indication of local range, remarks on size, habitat, phenology, special features, properties, and uses; errata, glossary and complete indices at end. The introductory section includes a general key to families.

821

Gujarat

The state of Gujarat took its present shape by 1960 after a variety of changes from 1947; it comprises parts of the old Bombay Presidency and the Western India and Gujarat Agency of princely states. In 1961 the former Portuguese enclave of Diu became Indian and, although it is (with Goa and Daman) a separate territory, may be included here. The state is floristically diverse, but beyond coverage in the Bombay Presidency floras of Cooke and Talbot (**822**), no general flora is available. A state-wide work for the woody plants is, however, now available as a pocket-sized manual, but all other floras are more or less local.

– Vascular flora: estimated at 1800 species in 803 genera.[4]

PATEL, R. I., 1971. *Forest flora of Gujarat State*. ii, 381 pp. Baroda: Forest Department, Gujarat (distributed by Gujarat Government Press, Ahmadabad).

Briefly descriptive flora of woody plants and larger herbs, with analytical keys to all taxa; includes synonymy with citations of key literature, vernacular and English names, general indication of local range, and notes habitat, life-form, phenology, etc.; indices provided respectively for botanical, vernacular, and English names. The introductory section includes a description of the region and remarks on its climate, geology, soils, vegetation, and history of botanical work. [Not seen by this reviewer; account prepared from notes supplied by the author.]

Partial works

The decentralized pattern of Gujarati life and activities, paralleling its physical and floristic diversity, has given rise to a goodly number of local florulas. A representative selection is given below.

BLATTER, E., 1908–9. On the flora of Cutch, [I–II]. *J. Bombay Nat. Hist. Soc.* **18**: 756–77; **19**: 157–76.

Systematic enumeration of seed plants of this arid, salty, seasonally marshy area near the Sind (Pakistan) border, with generalized indication of local and extralimital range and brief notes on habitat, phenology, and frequency; appendices on the features of the flora, vegetation, cultivated plants, and species biology are also included.

CHAVAN, A. R. and OZA, G. M., 1966. *The flora of Pavagadh (Gujarat State, India)*. vii, 296 pp., 2 pls. (Botanical Memoirs, 1). Baroda: University of Baroda.

Briefly descriptive, keyed flora of seed plants of Pavagadh Hill, a monadnock in eastern Gujarat; includes vernacular names, synonymy (with references), indication of local range (with some *exsiccatae*), and notes on habitat and phenology; list of references and index to all botanical names. Partially designed as a students' manual.

SANTAPAU, H., 1962. *The flora of Saurashtra.* Part 1. x, 270 pp. Rajkot: Saurashtra Research Society.

Keyed, briefly descriptive flora of seed plants, with synonymy, vernacular names, generalized indication of local range (with many localities), and notes on special features, phenology, uses, etc.; no index. Not completed; extends from Ranunculaceae through Rubiaceae on the standard Bentham and Hooker system. See also SANTAPAU, H. and JANARDHANAN, K. P., 1966 (1967). *The flora of Saurashtra: check list.* 58 pp. (Bull. Bot. Surv. India, Suppl. 1). Calcutta. [Unannotated list of seed plants, with essential synonymy.]

SAXTON, W. T. and SEDGWICK, L. J., 1918. Plants of northern Gujarat. *Rec. Bot. Surv. India*, 6: 207–323, i–xiii, map.

Annotated systematic checklist of seed plants (614 species), with citations of major works, vernacular names, localities, critical remarks, and notes on uses, occurrence, behaviour, etc.; index. An introductory section covers physical features, climate, soil, features of the vegetation, floristics and ecology, and statistics, and also points out the inadequacy of coverage of the area by Cooke's Bombay Presidency flora (822). In a short appendix are notes on the cryptogamic flora.

822

Bombay (Maharashtra in part)

This unit is here defined as all that part of the former Bombay Presidency south of the Narbada River together with the former Portuguese enclaves of Daman and the Dadra and Nagar Haveli Territory. For Goa, see 823; for Kurg (Coorg), see 826 (with Mysore). Since 1950, the limits of Bombay in the strict sense have changed markedly; apart from Kurg, parts of the south have been ceded to Kerala and Karnataka (Mysore) and to the east part of the former Central Provinces was incorporated. In 1960 the districts north of Narbada were ceded to Gujarat and the state was renamed Maharashtra. For convenience and for understanding the existing literature, a limited modification of the pre-1950 boundaries has been adopted as already indicated, and the resulting area thus includes North and South Kanara, the northern part of the Western Ghats, and well-known centers such as Surat, Pune (Poona), and Bombay. Works dealing with the whole of the old Presidency appear under this heading. The long history of botanical endeavor in 'Bombay' is partly embodied in the now-elderly floras of Cooke (1901–6) and Talbot (1909–11) and the supplements of Blatter and McCann (1926–35) and some florulas. Among these latter should be mentioned NAIRNE, A. K. [1894]. *The flowering plants of Western India.* xlvii, 401 pp. London: Allen (reprinted 1976), an introductory work for Bombay and Gujarat.

COOKE, T., 1901–9. *The flora of the Presidency of Bombay.* 2 vols. (in 3). ix, 2, 645, 1083 pp. London: Taylor and Francis. (Reprinted 1958, 1967 in 3 vols., Calcutta: Botanical Survey of India.)

Descriptive flora of vascular plants, with keys to genera and species, synonymy, references and citations, vernacular names, indication of local range (with citation of some *exsiccatae*), general summary of extralimital distribution, taxonomic commentary, and notes on phenology, uses, etc.; complete index at end of vol. 2. Less concise in general than the other contemporary regional Indian floras. For a long series of additions and corrections, see BLATTER, E. and McCANN, C. C., 1926–35. Revision of the flora of the Bombay Presidency, I–XXVII. *J. Bombay Nat. Hist. Soc.* 31–38 (*passim*), illus. [Serially published systematic enumeration, with descriptions of novelties, revised nomenclature, and miscellaneous notes. Not completed; covers nearly all flowering plant families from Menispermaceae to Cyperaceae on the first Hutchinson system.]

TALBOT, W. A., 1909–11. *Forest flora of the Bombay Presidency and Sind.* 2 vols. 541 text-figs. Poona: Bombay Presidency Government. (Reprinted with corrections by M. B. Raizada, 1975, Dehra Dun: Bishen Singh Mahendra Pal Singh.)

Illustrated descriptive forest flora of coffee-table dimensions, with keys to genera and species, extensive synonymy, references and citations, vernacular names, generalized indication of local and extralimital range, and notes on habitat, phenology, special features, wood details, properties, uses, etc.; index to all botanical and vernacular names in each volume. The introductory section incorporates descriptive treatments of the vegetation and the forest lands of the Presidency. Represents a much enlarged version of the author's *Trees, shrubs, and woody climbers of the Bombay Presidency* (2nd edn., 1902, Bombay).

Partial work

SANTAPAU, H., 1967. *The flora of Khandala on the Western Ghats of India*. 3rd edn. xxv, 372 pp., map (Rec. Bot. Surv. India, 16(1).) Calcutta. (First edn. 1953.)

Systematic enumeration of seed plants in the vicinity of this 'hill-station' near Bombay, with synonymy and references, vernacular names, some citations of *exsiccatae*, taxonomic commentary, and notes on occurrence, habitat, phenology, special features, etc.; list of references and complete index.

823

Goa

This ancient Portuguese territory, the scene of Garcia da Orta's botanical work four and a half centuries ago, was incorporated into India in 1961 and is presently a Union Territory under direct rule from Delhi.

VARTAK, V. D., 1966. *Enumeration of plants from Gomantak, India, with a note on botanical excursions around Castle Rock*. 5, 167 pp., frontispiece, map. Poona: Maharashtra Association for Cultivation of Science.

Tabular systematic list of vascular plants, with limited synonymy, vernacular names, local range, and indication of habitat and frequency; list of references. Partly supplants DALGADO, D. G., 1898. *Flora de Goa e Savantvadi*. xvi, 290 pp. Lisbon.

824

Hyderabad/Deccan (Nizam's Dominions)

The Nizam's Dominions, with their capital at Hyderabad in the Deccan, formerly constituted the largest princely state in the former Indian Empire, with the ruler having precedence over all his more than 500 peers. In 1950, however, the Dominions were for political reasons dispersed amongst the new states of Karnataka (Mysore), Maharashtra, and Andhra Pradesh, with Hyderabad City coming into the latter. For convenience, though, they have been retained here as a separate unit. They were largely bypassed in the general expansion of Indian botany of the nineteenth and early twentieth centuries and were never included in one of the regional floras or provided with a separate general work (except for the incomplete forest flora of Partridge given below). A large part of the dry Deccan falls within the area, which is still imperfectly known.

PARTRIDGE, E. A., 1911. *Forest flora of HH the Nizam's Dominions (Hyderabad-Deccan)*. 2, 3, 60, viii, 433, iii, xi, xi pp. Hyderabad: Pillai.

Descriptive field manual of more representative woody plants, with key literature citations, vernacular names, and copious notes on range, frequency, localities, timber, other products, uses, special features, and phenology (but not much on habitats); analytical keys to genera; indices to all English, local and scientific names. The introductory section includes a descriptive organography (separately indexed) and a synoptic key to families.

825

Andhra Pradesh (in restricted sense)

A large part of modern Andhra Pradesh was part of the old Madras Presidency before 1947, with other areas comprising various princely states, especially in the northeast on the Orissa border. However, major boundary changes and the creation of Andhra Pradesh followed Indian independence in 1947 and the area of 825 (the northern part of the old Presidency) now includes parts of Karnataka and Orissa. A major part of Hyderabad, here kept separate (824), has also been incorporated into Andhra Pradesh after dismemberment of the former princely state in 1950. The southern limit of 825 is here taken as the state boundary with Tamil Nadu. Also included here is the former French enclave of Yamam (now part of Pondicherry Union Territory). Botanical exploration, especially in less accessible areas, has been uneven and apart from *Flora of the Presidency of Madras* by Gamble and Fischer (828), no general coverage is available.

826

Karnataka in part (old Mysore and Kurg)

The former princely state of Mysore, centering on Bangalore and Mysore and also including the Hassan district in the northwest, was reconstituted after 1950 in a much enlarged form, with inclusion of parts of Hyderabad (821), the northern part of the former Madras Presidency (825), and the southernmost part of the erstwhile Bombay Presidency (823). The latter included the former state of Kurg (Coorg), at one time associated with Mysore. The modern state is now known as Karnataka. For convenience, **826** is here defined in terms of pre-1950 Mysore with the inclusion of Kurg, thus corresponding to the limits of Cameron's forest manual (1894) and Razi's checklist (1950). This same area also falls within the limits of *Flora of the Presidency of Madras* by Gamble and Fischer (**828**). The imperfect coverage of these works has now been supplemented by two substantial district florulas, with others in preparation;[5] however, compilation of a general state flora is seen to be something for the future in view of limited resources and the uneven level of botanical knowledge in the enlarged state. In addition to the works described below, mention is made of two new district floras, not yet seen: RAO, R. R., (in press). *A synoptic flora of Mysore district c.* 400 pp. New Delhi: Today and Tomorrow's; and RAZI, B. A., in press. *Flora of Chikkamagalur district.* Dehra Dun: Bishen Singh Mahendra Pal Singh.

CAMERON, J., 1894. *The forest trees of Mysore and Coorg.* 3rd edn. viii, 334, xxxvi pp., maps. Bangalore: Mysore Government Central Press. (Reprinted 1978, Dehra Dun: Bishen Singh; 1st edn. 1880.)

Systematic descriptive treatment of more common native and introduced woody plants, with emphasis on trees; includes citations of key literature and figures, vernacular names, local range, and notes on uses, special features, sylviculture, etc.; indices to all English and local vernacular names as well as scientific names. Appendices include classified tables of uses, potential, special features, etc. No keys are provided, and the work does not constitute a complete woody flora.

RAZI, B. A., 1946. A list of Mysore plants. *J. Mysore Univ., B* 7(4): 39–81.

Systematic list of seed plants recorded within the limits of the former princely state, with indication of Raunkiaeran life-forms, list of references. To a large extent it is based on the records in Gamble and Fischer's *Flora of the Presidency of Madras*.

Partial works

RAMASWAMY, S. V. and RAZI, B. A., 1973. *Flora of Bangalore district.* 1, 739 pp., map. Mysore: Prasaranga, University of Mysore.

Descriptive flora of higher plants, with keys, synonymy, references and citations, localities, and notes on habitat, phenology, frequency, etc.; indices. An introductory section covers sources for the work, geography, geology, vegetation, history of botanical exploration, and agriculture and includes a general key to families and a statistical synopsis of the flora. Bangalore, the largest city in modern Karnataka, lies in the eastern part of the old state of Mysore.

SALDANHA, C. J. and NICHOLSON, D. H., 1976. *Flora of Hassan District, Karnataka, India.* viii, 915 pp., 132 text-figs., 19 color pls., map. New Delhi: Amerind Publishing Co. (for the Smithsonian Institution and US National Science Foundation, Washington). (Reprinted with addenda, 1978.)

Concise descriptive flora of vascular plants, with keys to all taxa, synonymy, references, indication of local range and citations of selected *exsiccatae*, overall distributions, critical remarks, and notes on phenology, occurrence, habitat, uses, and desiderata for future work; representative illustrations (usually one per family), and complete index. No vernacular names are included. An introductory section gives the background for the work, physical features of the district, climate, and man's influence, and an account of the vegetation along with botanical history and the work of the flora project; remarks on economic botany and a general key to families are also included. This flora, prepared with the aid of surplus agricultural funds made available under US PL-480[6] is thought to account for some 75 per cent of the vascular flora of Karnataka (in its modern sense).

827

Kerala

The modern state of Kerala, along the old Malabar Coast, has a long botanical history, having been the main area covered by Rheede's *Hortus malabaricus* late in the seventeenth century (but for which no good modern analysis is yet fully available). During the British Raj the area was for the most part in the Madras Presidency, with the rest constituting princely states such as Travancore. The present state was organized in 1950. The entire area is included within *Flora of the Presidency of Madras* (828) and other works on south India (820). However, no separate general account is available and the state is still considered underexplored, with many poorly-known species in need of further investigation.[7] In particular, little has been done until recently to follow up the work carried out by Bourdillon on the forest flora very early this century.[8] The most significant local coverage available is for the former Travancore State (see below).

Local works: Travancore

RAMA RAO, M., 1914. *Flowering plants of Travancore.* xiv, 495 pp. Trivandrum, Kerala: Travancore Government Press. (Reprinted 1976, Dehra Dun: Bishen Singh Mahendra Pal Singh.)

Systematic enumeration of seed plants from this former princely state on the Malabar Coast, with variously extended notes on habitat, distribution, habitat, uses, and properties as well as English and local vernacular names and references to illustrations; indices to English and to local vernacular names as well as genera and families at end. A brief introductory section supplies the background to the work. Covers 3535 species; intended only as a preliminary account. Forest trees are covered in greater depth in BOURDILLON, T. F., 1908. *Forest trees of Travancore.* xxxii, 456 pp., plates. Trivandrum. (Reprinted 1976, Dehra Dun.) [Includes keys.]

828

Tamil Nadu (Madras)

The historic state of Madras, now known as Tamil Nadu, was in the latter part of the eighteenth century the center of activities of the botanical 'United Brotherhood' and in the nineteenth century of R. Wight, R. H. Beddome and others who contributed the basis of our knowledge of the flora of southern India. The first local herbarium was established in 1853, and a distinctive botanical 'identity' has been maintained to the present time. 828 is here defined as the approximate limits of the present state, with the inclusion of Pondicherry enclaves (French until 1954, now a Union Territory); under the British Raj, however, it was part of the Madras Presidency which also took in most of Kerala (827), a large part of Andhra Pradesh (825) and other areas. The standard flora of Gamble and Fischer, *Flora of the Presidency of Madras*, includes all of these as well as contiguous princely

states, especially Mysore and Coorg (826). However, Fyson's useful flora of the hill stations appears under 804.

GAMBLE, J. S. and FISCHER, C. E. C., 1915–38. *Flora of the Presidency of Madras.* 3 vols. 2017 pp. London: Adland. (Reprinted 1957, 1967, Calcutta, Botanical Survey of India.)

Concise descriptive manual-flora of seed plants, with diagnostic keys to all taxa; under species headings are incorporated synonymy (with references and citations), fairly detailed general summaries of local range, vernacular names, and notes on habitat, special botanical features, uses, etc. The introductory section includes a glossary and a description of the general features of the flora, while the last volume contains complete indices to botanical and vernacular names. The plants of the former princely state of Mysore are also included.

829

Sri Lanka (Ceylon)

After more than three centuries of investigation, the flora of what has been called 'Serendip'[9] is relatively well-known and a stock of good floristic works is available. In the 1950s, however, it became evident that Trimen's *Handbook* was becoming obsolete as well as scarce; fortunately, in 1967–8 it become possible to launch a new Flora of Ceylon project, with the aid of surplus United States funds, and much new collecting and specialized investigations were carried out. The first fascicle of the *Revised Handbook to the Flora of Ceylon* was published in 1973, but progress was slow and by 1979 only one more had been produced, using newsprint. It became necessary to seek an alternative method of publication, and at this writing the work is appearing as *Flora of Ceylon* in six volumes, the first in 1980. Many precursory papers have already appeared in *Ceylon Journal of Science* and elsewhere.[10]

Bibliographies

ALWIS, N. A., 1978. *Bibliography of scientific publications relating to Sri Lanka.* v. 244 pp. Colombo: Social Science Research Centre, National Science Council of Sri Lanka. (Mimeographed.) [Botany, pp. 121–50; 304 main and cross-references.]

GOONETILEKE, H. A. I., 1970–6. *A bibliography of Ceylon.* 3 vols. (Bibliotheca asiatica, 5 covers vols. 1–2; 14

covers vol. 3.) Zug, Switzerland: Inter-Documentation Co. [Botany and horticulture, 222 items; vegetation, forests, and soils, 180 items.]

ABEYWICKRAMA, B. A., 1959. A provisional check list of the flowering plants of Ceylon. *Ceylon J. Sci., Biol. Sci.* **2**(2): 119–240.

Systematic census of flowering plants recorded from Ceylon, with limited synonymy and cross references to Trimen and Hooker's *Handbook* (see below); index to families. Partly supersedes WILLIS, J. C. and WILLIS, M., 1911. *A revised catalogue of the flowering plants and ferns of Ceylon, native and introduced.* 188 pp. Colombo.

DASSANAYAKE, M. D. and FOSBERG, F. R. (eds.), 1980– . *A revised handbook to the flora of Ceylon.* 2nd edn., Vol. 1– . Maps. New Delhi: Amerind (for the Smithsonian Institution and the National Science Foundation, Washington, DC).

Amply descriptive flora of higher plants, with keys to genera and species, synonymy (with reference and citations, in some cases rather extensive), vernacular names, indication of distribution (and origin), taxonomic commentary, and notes on habitat, vegetation type, frequency, phenology, elevations, etc.; indication of *exsiccatae* with localities. A few species distribution maps and an end-paper district map appear in vol. 1, but no index or illustrations. The brief, partly historical introduction in vol. 1 precedes ten family accounts – rather variable in treatment – in no predetermined sequence. Similar groups of families will be published in subsequent volumes as accounts are ready; a total of six volumes is presently projected. Developed as a successor to Trimen's *Handbook* (see next entry) and somewhat resembling it in style, this version of the *Revised Handbook* represents the results of the joint Smithsonian Institution/University of Ceylon/Royal Botanic Gardens, Peradeniya 'Ceylon Flora Project' (originally begun in 1968, and largely financed from US PL-480 surplus funds). See also ABEYWICKRAMA, B. A. and DASSANAYAKE, M. D. (eds.), 1973–7. *A revised handbook to the flora of Ceylon.* 1st edn. Vol. 1, parts 1–2. Illus. Peradeniya: University of Ceylon. [Illustrated descriptive treatment with keys, synonymy, *exsiccatae*, distribution, vernacular names, critical remarks, and other notes; modelled on *Flora Malesiana*. Publication suspended consequent to printing and other difficulties; family accounts therein to reappear in Amerind edition after revision.]

TRIMEN, H. and HOOKER, J. D., 1893–1900. *Handbook of the flora of Ceylon.* 5 vols., atlas of 100 color pls. London: Dulau. (Reprinted 1974, Dehra Dun, Bishen Singh Mahendra Pal Singh.)

ALSTON, A. H. G., 1931. *Supplement to the Handbook of the flora of Ceylon* (vol. 6). vi. 350 pp. London: Dulau.

Briefly descriptive flora of vascular plants, with keys to genera and species, synonymy, references and pertinent citations, vernacular names (English, Sinhalese, and Tamil), fairly detailed indication of local range (with some *exsiccatae*), generalized summary of extralimital distribution, taxonomic commentary, and notes on habitat, uses, etc. Volume 1 incorporates a general sketch of the physical features and climate of the island, while vol. 5 contains a synopsis and general key to orders and families, together with accounts of the vegetation and the history of botanical exploration. Complete indices are also given in vol. 5. Volume 6 contains additions and corrections to the original work. The fine atlas (in larger format) contains 100 hand-colored plates of characteristic species. The work officially forms part of the 'Kew Series' of British colonial floras, like the *Flora of British India* (**810**–**40**). The original *Handbook* was completed by Hooker after the death of Trimen in 1896.

Partial work: Kandy area

The following is useful as a students' flora in the central region, especially around Kandy and Peradeniya.

ALSTON, A. H. G., 1938. *The Kandy flora.* xvii, 109 pp., 404 text-figs. Colombo: Ceylon Government Press.

Keyed, briefly descriptive illustrated manual-flora of seed plants of the Kandy-Peradeniya district, with limited synonymy, vernacular names, and notes on phenology, life-form, etc.; complete index. The introductory section includes a description of the general features of the flora and vegetation, while an appendix contains a glossary of terms as well as a 'List of grasses, rushes, and flowerless plants'.

Special groups – woody plants

ABEYESUNDERE, L. A. J. and ROSAYRO, R. A. DE, 1939. *Draft of first descriptive check-list for Ceylon.* 115 pp. (Check-lists of the forest trees and shrubs of the British Empire, 4). Oxford: Imperial Forestry Institute.

Enumeration of trees and shrubs recorded from Ceylon, with synonymy, citations of pertinent literature, vernacular names, generalized indication of local range, and notes on habitat, phenology, special botanical and dendrological features, and uses; list of references and indices to generic and vernacular names. An appendix lists records of exotic Coniferae.

WORTHINGTON, T. B., 1959. *Ceylon trees.* 429 pls., map. Colombo: Colombo Apothecaries' Co.

Illustrated atlas of 429 photographic plates with accompanying text, depicting the more commonly encountered native and exotic trees of Ceylon; the text includes descriptive notes on habitat, local range, life-form, biology, distinguishing features, wood, properties, uses, etc., as well as limited synonymy (with cross references to Trimen and Hooker's *Handbook*), vernacular names, and extralimital range; index to all vernacular, generic, and family names. The introductory section includes addenda, corrigenda, a glossary, and a list of references. Most of the photographs were made from freshly collected specimens.

Region

83

Eastern India and Bangladesh

Included herein are all the states and territories from Bihar and Orissa eastward to the Burmese frontier, as well as Bangladesh. However, the hill and mountain zones of the Himalaya, including the Darjeeling district, are excluded (for these, see Region 84).

Until 1910, this entire area was under two administrative units: the great Presidency of Bengal, stretching from Bihar and Orissa to East Bengal, and Assam. From then onwards there has been a sporadic process of devolution, leaving by the 1970s ten separate polities (exclusive of Arunachal Pradesh), of which one, Bangladesh, is an independent state.

From the time of Roxburgh to the early-twentieth century, substantial botanical work was done in the region and a number of basic floras written, notably *Botany of Bihar and Orissa* (Haines, 1921–5, with supplement by Mooney, 1950), *Bengal plants* (Prain, 1903), and the incomplete *Flora of Assam* (U. N. Kanjilal *et al.*, 1934–40). Comparatively few substantial floras have appeared since 1950, but exploration in more remote areas has continued and some source materials, recently joined by a few district floras in better-known areas such as Meghalaya and West Bengal, have been published. To these must be added *Flora of Bangladesh*, begun in 1972 as a serial work but appearing less rapidly than the *Flora of West Pakistan*. *The Flora of Eastern India* (Mitra, 1958), a students'

manual, ceased to appear after one volume and moreover suffers from flawed and tortuous keys. Generally speaking, therefore, much work has been done in the region, both before and after 1950, but can not be considered well-consolidated by modern standards.
Bibliographies. General and supraregional bibliographies as for Division 8 and Superregion 81–84.
Indices. General and supraregional indices as for Division 8 and Superregion 81–84.

830

Region in general

See also **810–40** (all works). The whole region (with adjacent parts of the Himalaya) is nominally covered in Mitra's manual but of this largely compiled work the only volume published has been that on the monocotyledons.

MITRA, J. N., 1958. *Flowering plants of eastern India.* Vol. 1: *Monocotyledons.* xx, 389 pp., map. Calcutta: World Press (Private).

Manual-key to flowering plants (to date covering only the Monocotyledoneae), with brief diagnoses, limited synonymy, vernacular names, and generalized indication of internal range (with some localities mentioned); glossary and index to all botanical names. The introductory section includes a synopsis of orders and families. Covers the entire region from Bihar and Orissa to Assam; however, the rather tortured layout of the keys (which are in part nonanalytical) may render this book somewhat difficult to use.

831

Bihar

See also 833 (PRAIN). The present state of Bihar represents the major part of the province of Bihar and Orissa originally detached from the Presidency of Bengal in 1911. Orissa, however, became a separate entity (832) in 1936 and after independence of India other areas became part of West Bengal. In addition, much of the Chota Nagpur forest district, entirely associated with the Bengal Presidency before 1911, is now part of Madhya Pradesh and Haines's forest flora has therefore been included with other works on that state (see **819**). The 'regional' flora of Haines (1921–5) and its supplement by Mooney (1950) both deal with the former British Raj province and associated princely states. Recent work, however, suggests that some areas of the state are insufficiently collected.

HAINES, H. H., 1921–5. *The botany of Bihar and Orissa.* 2 vols. x, 199, 1350 pp., 2 maps. London: West, Newman. (Reprinted 1961 in 3 vols., Calcutta, Botanical Survey of India.)

Briefly descriptive flora of vascular plants, with keys to genera and species; synonymy; vernacular names; fairly detailed indication of local range, with citation of *exsiccatae*, and general summary of extralimital distribution; taxonomic commentary and notes on habitat, uses, etc.; complete index to botanical and vernacular names. The introductory section includes chapters on physical features, climate, vegetation, ecology, and botanical exploration. For additions and corrections, see MOONEY, H. 1950. *Supplement to* 'The botany of Bihar and Orissa'. iii, 294 pp. Ranchi: Catholic Press. [Descriptive list of additions, together with accounts of the physical features and floristics of Bihar and Orissa.]

832

Orissa

See **831** (HAINES); **833** (PRAIN). Until 1936, the British-administered parts of Orissa were included with Bihar (and before 1911 in the Presidency of Bengal); they then became a separate administrative unit. After 1947 the princely states of the Eastern States Agency and the districts of Koraput and Ganjam (the two latter part of or associated with Madras Presidency (see **828**)) were included in the new state of Orissa. As with Andhra Pradesh (**825**) and Bihar, many parts in the hilly interior districts are considered to be undercollected. No separate flora is yet available.

833

Bengal (in general)

This is here understood to refer to the whole of the old Presidency of Bengal as it existed up to 1911 and for which Prain's *Bengal plants* (1903), still in part

current, was written. It then included most of modern Bihar and Orissa as well as the state of West Bengal, the now-independent nation of Bangladesh, and the state of Tripura (formerly part of East Bengal). Associated with it was the Eastern States Agency now mostly in Orissa. Prain's flora, however, does not cover the northern hill tracts around Darjeeling (which are here listed as **845**).

PRAIN, D., 1903. *Bengal plants.* 2 vols. 1319 pp., map. Calcutta: Government of Bengal. (Reprinted 1963 in 2 vols., Calcutta: Botanical Survey of India.)

Descriptive manual of vascular plants, with keys to genera and species, an artificial general key to genera, limited synonymy, principal literature citations, vernacular names, generalized indication of local range (with some localities), and notes on taxonomy, habitat, occurrence, diagnostic features, life-form, properties, uses, etc.; complete general index. Extensive descriptions are provided only for genera and families. An introductory section includes a synopsis of families and a general description of Bengal (in the pre-1911 sense) with notes on botanical subdivisions and floristics. The work includes all the area defined under the heading, minus the hill tracts of North Bengal as noted.

834

West Bengal (India)

See **833** (PRAIN). The northern hill tracts, not covered in *Bengal plants*, comprise **845**. Modern West Bengal includes parts of the former province of Bihar and Orissa and some erstwhile foreign enclaves, among them Serampore and Chandernagore. Separately published florulas have been few; the best work is the 1979 treatment of Bennet for Howrah district (see below).

Local work: Howrah district

BENNET, S. S. R., 1979. *Flora of Howrah district.* viii, 406 pp. New Delhi: Periodical Export Book Agency; Dehra Dun: International Book Distributors.

Descriptive flora, with keys, synonymy and references, literature citations (key works), brief indication of local range along with the author's *exsiccatae*, and notes on habitat, frequency, occurrence, phenology, uses, special features, etc.; index to scientific names at end. In the introductory section are brief surveys of physical and biotic features, land use, vegetation (nearly all anthropogenic) and its structural elements and synusiae, earlier botanical work in the area, and floristic statistics (605 species); a general key to families is also included. [Provides the first consolidated coverage of the area since *Bengal plants* by Prain (833); viewed as precursory to a state flora.]

835

Bangladesh

See 833 (PRAIN). Formerly comprising most of the administrative unit of East Bengal, Bangladesh was between 1947 and 1971 part of Pakistan; since then it has been an independent nation. It includes the former Assamese district of Sylhet, but not the former East Bengal district of Tripura, which remained with India and is now a state (836). A serial documentary work, *Flora of Bangladesh*, commenced publication in 1972.

KHAN, M. S. and HUQ, A. M. (eds.), 1972– . Flora of Bangladesh. Fasc. 1– . Illus.; map in each fascicle. Dacca: Agricultural Research Council, Bangladesh (for the Bangladesh National Herbarium). (Parts 1–3 reprinted 1975.)

Documentary descriptive flora of higher plants, published in fascicles with one or more families; each family treatment contains keys to genera and species, synonymy, references and citations, detailed indication of local distribution in 19 districts, with representative *exsiccatae*, summary of extralimital distribution, vernacular (including English) names, notes on origin (if naturalized), habitat, occurrence, karyotypes, special features, etc.; with list of references consulted and index at end. As of 1980 11 fascicles each with one or more families, had been published; no predetermined systematic order has been established.

836

Tripura (India)

See 833 (PRAIN). Before 1947 this state was part of East Bengal, but it then opted to remain with India. Until the 1950s, it was botanically poorly known, but in 1956–61 and again in 1962 considerable collections were made by D. B. Deb. His results and observations on the flora and vegetation and a review of past work appear in DEB, D. B., 1963. Bibliographical review on the botanical studies in Tripura. *Bull. Bot. Surv. India*, 5: 49–58.

837

Old Assam Region (except Arunachal Pradesh)

This is here defined as encompassing the states of Assam and Meghalaya (the latter created in 1972) and the territory of Mizoram (between Tripura and Manipur). For Arunachal Pradesh, see 847. Before 1947 the old province of Assam also included Manipur (838) and Nagaland (839) as well as the Sylhet district south of the Khasi Hills (the latter now part of Bangladesh). The general flora begun by Kanjilal in 1934 was never completed (of the monocotyledons only the grasses were ever published) and its earlier volumes have a bias towards the woody flora. The remainder of the monocotyledons have been covered in Mitra's manual (830) and (in part) in scattered papers under the auspices of the Botanical Survey of India.[11]

KANJILAL, U. N. *et al.*, 1934–40. *Flora of Assam.* Vols. 1–5 (in 6). Shillong: Assam Government.

Briefly descriptive flora of seed plants (with a bias toward woody species in earlier volumes), including keys to genera and species, vernacular names, relatively detailed indication of local range, and notes on phenology, etc.; indices to all botanical and vernacular names in each volume. The introductory section (vol. 1) includes descriptive accounts of physical features, climate, ecology, etc., a glossary, a synopsis of families (following the Bentham and Hooker system), and a list of references. Not completed; lacks any treatment of monocotyledons except for the Gramineae (by N. L. Bor) in vol. 5, although this gap is partly bridged by Mitra's *Flowering plants of eastern India* (830). Supplants KANJILAL, P. C., 1925. *Descriptive list of trees and shrubs of the Eastern Circle*. Calcutta: Superintendent of Government Printing.

838

Manipur

See also 837 (KANJILAL *et al.*). In recent years Manipur has become a separate state. A special survey by D. B. Deb in the late 1950s has resulted in a two-part enumeration of flowering plants (1961–2); prior to this, botanical exploration of this tribal border area had been spasmodic.

DEB, D. B., 1961. Monocotyledonous plants of Manipur Territory. *Bull. Bot. Surv. India*, **3**: 115–38.

DEB, D. B., 1962. Dicotyledonous plants of Manipur Territory. *Ibid.*, **3**: 253–350.

These papers together constitute an enumeration of the known flowering plants of the area, with brief diagnoses, synonymy (with references), and generalized indication of local range (including citations of *exsiccatae*). The introductory sections of both parts contain accounts of physical features, climate, and botanical exploration and long lists of references, together with statistics of the flora.

839

Nagaland

See 837 (KANJILAL *et al.*). Although some lists of collections from this area (e.g., from around the town of Kohima) have been published, botanical penetration of this mountainous land on the Burmese border has been slow and to this reviewer's knowledge no separate flora or checklist has been published. Formerly associated with Assam, Nagaland is now a separate state after a period as a Union Territory.

Region

84

The Himalaya: foothill and mountain zones

This region extends along the Himalaya from the upper Indus River (in northern Kashmir) to the North-East Frontier Agency north of Assam, passing through Kashmir (with Jammu), the Punjabi Himalaya, the hill and mountain tracts of Uttar Pradesh, Nepal, Sikkim and northern Bengal, and Bhutan. Its southern limit is along the base of the foothills, and it is bounded on the north by China.

The Himalaya, a natural phytogeographic unit of great significance, has a history of botanical exploration dating from early in the nineteenth century, but until the 1950s at least collecting was comparatively patchy, with much emphasis on a small number of areas; large tracts, including Nepal, Bhutan, and what is now Arunachal Pradesh remained poorly known for a variety of reasons. The multiplicity of polities has also hindered any overall floristic synthesis of the region. In the last two decades or so, though, several field trips and systematic/floristic analyses by Japanese botanists under the overall direction of H. Hara have compiled some synthetic works on the east-central zone (eastern Nepal to Bhutan), and Hara's group has combined with botanists at the British Museum (Natural History), where major collections from Nepal were available, to put together the first modern systematic account for that country. Much collecting has reportedly also been carried out in the North-West Himalaya and Arunachal Pradesh under the Botanical Survey of India and other agencies.

The outstanding modern floras in a region otherwise noted for a hodgepodge of scattered, technically uneven works of relatively local scope comprise the results of the two groups mentioned above: *Flora of eastern Himalaya*, of which three parts have appeared (1966–75) from Tokyo, and *Enumeration of the flowering plants of Nepal* from London, to appear in three volumes (1978–). District and state florulas and other works have since the 1960s also appeared for north Bengal and Sikkim, Bhutan (also long without any separate flora), Nepal, and parts of Uttar Pradesh (Naini Tal, Dehra Dun), Himachal Pradesh, and Kashmir (Jammu, Srinagar). There is also a recent 'Alpenflora' for the North-West Himalaya (RAU, see 803). A desirable future step, with the material now available, would be the compilation of a critical synthetic enumeration for the whole region, which would serve to eliminate, as in Mexico and Central America, much superfluous synonymy.

Bibliographies. General and supraregional bibliographies as for Divison 8 and Superregion 81–84.

Indices. General and supraregional indices as for Division 8 and Superregion 81–84.

840

Region in general

See also **810–40** (all works). Apart from the general floras for South Asia, no single work covers the whole of the Himalaya and its southern foothills. Since the 1950s, however, Japanese expeditions have been involved in fairly widespread botanical exploration, especially in the central and eastern Himalaya, in pursuance of a long-term programme of study of biogeographical links between this region and Japan. The results of this work have been made available in a number of significant floristic publications, notably *The flora of eastern Himalaya*, a critical series begun by H. Hara in 1966 and now in its third volume.

HARA, H. (ed.), 1966. *The flora of Eastern Himalaya*. 744 pp., 68 text-figs, 40 pls. (some in color). Tokyo: University of Tokyo Press.

HARA, H. (ed.), 1971. *The flora of Eastern Himalaya: second report*. 393 pp., 24 pls. (some in color) (Bull. Univ. Tokyo Mus., 2). Tokyo: University of Tokyo Press.

OHASHI, H. (ed.), 1975. *The flora of Eastern Himalaya: third report*. xv, 458 pp., 33 pls., other illus., maps (Bull. Univ. Tokyo Mus., 8). Tokyo: University of Tokyo Press.

The main work comprises a systematic enumeration, without keys of the cellular (except algae) and vascular plants of eastern Nepal, Sikkim, and northern Bengal; it includes rather extensive synonymy (with references and citations), detailed indication of local ranges, general summaries of extralimital distribution, some taxonomic commentary, and notes on habitat, uses, etc.; gazetteer and index to all botanical names. The introductory section includes chapters on karyology and phytogeography, together with a selected list of references. The Second Report contains additional records based upon further collecting in Nepal, Sikkim, and Bhutan as well as three revisionary and cytological studies; there is also an additional list of references. The Third Report is based on the fifth expedition in 1972 and includes an additional list of vascular plants (pp. 13–205) as well as a list of bryophytes, a number of special studies including revisions, an additional bibliography, and indices to localities and to botanical names. These well-executed volumes have been based on the results of five successive Tokyo University-sponsored expeditions up to 1972, but also account for older literature.

841

Kashmir (with Jammu)

See also **793** (all works); **812** (BAMBER; PARKER). As circumscribed here, Kashmir (with Jammu) is bounded on the north and northwest by the upper course of the Indus River and on the west by the 'cease-fire line' which forms the *de facto* boundary between Pakistan and India (Jammu and Kashmir being claimed by both countries). The remaining areas of Kashmir not included here, among them Gilgit, Baltistan and Ladakh, are treated with Region 79. Separate works on this area, which includes the famous Vale of Kashmir, are few and uncoordinated, although one or two new works have appeared lately.

BLATTER, E., [1927–9]. *Beautiful flowers of Kashmir*. 2 vols. 2 color frontispieces, 62 pls. (mostly in color). London: Staples and Staples.

Popular descriptive treatment, with keys only to species; includes notes on diagnostic features, local occurrence, habitat, altitudinal zonation, phenology, and general distribution, some English names, and indices to scientific and English names in both volumes. A glossary is also included. Only a selection of the more conspicuous species is treated but the work is lavishly illustrated by watercolors. Nomenclature follows *Flora simlensis* (**842**).

LAMBERT, W. J., 1933. *List of trees and shrubs for Kashmir and Jammu forest circles*. ii, 40 pp. (Forest Bull., India, 80). Calcutta.

Annotated list of known trees and shrubs, with concise indication of localities (including observations by the author or literature citations); index to generic names. Supplements Parker's *Forest flora of the Punjab* (1924; see **812**), which also covers part of this area.

Local works

In addition to the title given below, the following has been announced: SHARMA, B. M. and KACHROO, P. *Flora of Jammu and plants of neighbourhood*. Dehra Dun.

SINGH, G. and KACHROO, P., 1976 (1977). *Forest flora of Srinagar and plants of neighbourhood*. x, 278 pp., 11 pls., tables, map. Dehra Dun: Bishen Singh Mahendra Pal Singh.

Part of this work comprises a systematic enumeration of the vascular plants of the Srinagar district – the Vale of

Kashmir – with synonymy, references, vernacular names, local distribution, and indication of habitat, phenology, and altitudinal zonation; occasional critical remarks; bibliography (pp. 233–8), extensive appendix on ethnobotany, and index to all names. The lengthy preliminary section includes descriptions of physical and biotic features, history of botanical work, floristic statistics (661 species in the district; 2928 in the whole of Kashmir) and units, an account of the outstanding features of the flora in different localities, with lists (special emphasis is given to the Srinagar woodland), and a description of the vegetation and floristics of the Srinagar area; a list of new plant records and species concludes the section.

842

Himachal Pradesh

See also **810** (Stewart and Brandis); **812** (Bamber; Parker). Prior to 1947 this area consisted of the Punjab Hill States, Chamba, and the Punjabi hill district of Kangra. Himachal Pradesh was formed from the erstwhile princely states after independence of India and in 1966 the Kangra district was incorporated with this state. Within this area is the great 'hill station' of Simla, for long a summer seat of Indian governments. Two significant florulas are currently available.

Partial works

Collett, H., 1921. *Flora simlensis*. 2nd edn. lxviii, 652 pp., 200 text-figs., map. Calcutta: Thacker, Spink. (Reprinted 1971, Dehra Dun: Bishen Singh; 1st edn., 1902.)

Illustrated, briefly descriptive semipopular manual-flora of seed plants (about 1300 species) of the Simla district in the southern part of Himachal Pradesh, with keys to genera and species, English names, local and extralimital range, and notes on habitat, phenology, biology, special features, etc.; index to all botanical and English names. An introductory section (by the Kew botanists W. B. Hemsley and W. T. Thistleton-Dyer) includes accounts of botanical exploration, vegetation, phytogeography, and notes on particular life-form classes together with a synopsis of families, a glossary, and list of references. [The work reminds this reviewer of a contemporaneous English county flora, and, indeed, the local flora has a temperate facies, with Simla itself lying at over 2000 m in altitude.]

Nair, N. C., 1977. *Flora of Bashahar Himalayas*. xxxii, 360 pp., 24 pls. Hissar, Haryana: International Bioscience Publishers.

Descriptive enumeration of seed plants, with synonymy, references, literature citations, brief diagnosis, localities with *exsiccatae* (mainly from Dehra Dun Forest Herbarium), indication of phenology, and other notes; index to all scientific names at end. The introductory section covers physical and biotic features, previous work on the area, and floristic statistics (1629 species) with list of references (pp. xxx–xxxi). [Covers the former Bashahar State, an area of some 10 000 km² on the southern fringe of Himachal Pradesh ranging from 650 to 6930 m in altitude and including Simla.]

843

Uttar Pradesh: Himalayan tracts

This mountanous area includes the Chakrata-Dehra Dun foothill district (bordering the plains) as well as the districts of Tehri Garhwal (a former princely state) and Kumaon. A number of 'hill stations' are located here, notably Dehra Dun, Mussorie, and Naini Tal. The area largely falls outside the limits of Duthie's *Flora of the upper Gangetic Plain*, which just reaches the level of Dehra Dun (for this work, see **815**). The broken terrain and previously fragmented administrative history has resulted in a variety of local florulas but no general work, an unfortunate state of affairs. The most useful handbooks are accounted for below under four subheadings: I. Dehra Dun (including Chakrata), II. Tehri Garhwal, III. Naini Tal and IV). Kumaon.

I. Dehra Dun district (including Chakrata)

The works by Babu and Kanjilal for the Dehra Dun area are more or less complementary. To these has now been added a flora for the Mussorie area north of Dehra Dun.

Babu, C. R., 1977. *Herbaceous flora of Dehra Dun*. vii, 721 pp., map. Delhi: Publications and Information Directorate, CSIR, India.

Descriptive florula of seed plants (1230 species) of a herbaceous or subwoody character; includes keys to all taxa, synonymy, references and citations, indication of localities with selected *exsiccatae*, and notes on origin (for naturalized and other alien plants), habitat, phenology, etc., as well as critical remarks; index to all botanical names at end. An introductory section includes accounts of physical features, geology, soils, climate, vegetation and biotic factors, and previous botanical exploration along with a floristic and phytogeographic analysis (with statistical table). No vernacular names are included. Complements Kanjilal's long-standard manual for woody plants (see below).

KANJILAL, U. N., 1928. *Forest flora of the Chakrata, Dehra Dun and Saharanpur forest divisions, United Provinces.* 3rd edn., revised by B. L. Gupta. xxiii, 558 pp. Calcutta: Central Publications Branch, Government of India. (First edn. 1901, as *Forest flora of the school circle, NWP*; 2nd edn., 1911, as *Forest flora of the Siwalik and Jaunsar divisions of the United Provinces of Agra and Oudh.*)

Keyed, briefly descriptive manual-flora of woody plants, with limited synonymy, vernacular names, indication of local range, and notes on habitat, special features, etc.; complete index. A glossary is also included. In addition to the foothill zone around Dehra Dun and northwestwards, coverage of this work also extends to Saharanpur in the adjacent plains. The work has long served as a students' manual for classes at the Forestry College, Dehra Dun but, although several times reprinted (most recently in 1969) it has not yet been revised.

RAIZADA, M. B. and SAXENA, H. O., 1978 (1979). *Flora of Mussorie.* Vol. 1 (of 2). lvi, 645 pp. Dehra Dun: Bishen Singh Mahendra Pal Singh.

Briefly descriptive flora of seed plants, with keys to all taxa, synonymy with references, literature citations, local range including indication of *exsiccatae* seen as well as overall distribution, vernacular names, and notes on habitat, phenology, properties, and uses, etc.; indices to both scientific and vernacular names at end of volume. The introductory section covers physical and biotic features, floristics and vegetation, phytogeography, plants particular to given habitats, adventive and naturalized species, and history, followed by a more detailed consideration of forest vegetation and a general key. A comparison of the area with Simla and vicinity (1219 species v. 1326 at the former viceroyal station) is presented. Coverage in the first volume is of native, naturalized, adventive, and commonly cultivated species from Ranunculaceae through Labiatae (Bentham and Hooker system). [Mussorie, in the Himalaya foothills north of Dehra Dun, is a past and present 'hill station'.]

II. Tehri Garhwal district

No suitable separate florulas are available.

III. Naini Tal district

GUPTA, R. K., 1968. *Flora nainitalensis: a handbook of the flowering plants of Naini Tal.* xxix, 489 pp., 40 pls. New Delhi: Navayug Traders.

Keyed, briefly descriptive manual-flora of seed plants, with synonymy and citations, indication of local and extralimital range, taxonomic commentary, and notes on habitat, phenology, uses, etc.; glossary, lexica of vernacular names, and botanical index. The introductory section covers botanical exploration, general features of the area and its

vegetation, floristics, and statistics of the flora. The work covers among other areas the environs of the 'hill station' of Naini Tal, long the summer capital of Uttar Pradesh (and, before it, the United Provinces). In style and format it resembles Collett's *Flora simlensis* (841).

IV. Kumaon districts (Garhwal and Almora)

This Himalayan division of the former United Provinces comprises the districts of Almora (Tarai), just west of the Nepal boundary, and Garhwal to its northwest. It lies to the north of Naini Tal.

OSMASTON, A. E., 1927. *A forest flora of Kumaon.* xxxiv, 605 pp., map. Allahabad: Superintendent of Government Press, United Provinces. (Reprinted 1978, Dehra Dun: Bishen Singh Mahendra Pal Singh.)

Briefly descriptive flora of woody plants of the Kumaon district, with keys, essential synonymy, vernacular names, local range, and notes on taxonomy, habitat, phenology, etc.; glossary, list of references, and complete indices. An introductory section covers physical features, climate, and vegetation of the area.

STRACHEY, R., 1906. *Catalogue of the plants of Kumaon and of the adjacent parts of Garhwal and Tibet.* Revised by J. F. Duthie. 269 pp. London: Reeve. (Reprinted 1974, Dehra Dun: Bishen Singh Mahendra Pal Singh.)

Annotated list of known vascular plants, bryophytes, and lichens recorded from the given area, with tabular indication of habitat, diagnostic features, phenology, local and extralimital range (with some localities), and elevations; index to generic and family names at end. An introductory section includes floral statistics and a list of references. The original version of this list appeared in Atkinson's *Gazetteer of the Himalayan districts of the NW Provinces and Oudh* (1882).

844

Nepal

The rich flora (*c.* 7000 species of vascular plants) of this Himalayan kingdom remained relatively little explored until some three decades ago, and very few works had been published although important collections had been made there from early in the nineteenth century. The present greater ease of access to Nepal has resulted in a great deal of collecting, including visiting British and Japanese expeditions, and substantial materials are now available. Several publications have appeared in recent years, and results are now being summarized in *An enumeration of the flowering plants of Nepal* by H. Hara and others (1978–). In addition to the floristic works accounted for here, the following work on Nepal's forests provides a useful general introduction: STAINTON, J. D. A., 1972. *Forests of Nepal.* xvi, 181 pp., 32 color pls., 5 maps. London: Murray.

Bibliography

DOBREMEZ, J. F., VIGNY, F. and WILLIAMS, L. H. J., 1972. *Bibliographie du Népal.* Vol. 3, Sciences naturelles, part 2: Botanique. 126 pp., 4 pls., 9 maps. Paris: Centre National de la Recherche Scientifique. [732 references, arranged by author with detailed subject index; notes on herbaria, botanical gardens, and collectors.]

Keys to families and genera

PANDE, P. R. *et al.,* 1967–8. *Keys to the dicot genera in Nepal.* 2 parts. Kathmandu: HM Government of Nepal. [Comprises, in two small volumes, analytical keys to the families and genera of Nepalese dicotyledons. Sources for this work are given in the introductory section.]

HARA, H., STEARN, W. T., and WILLIAMS, L. H. J. 1978–9. *An enumeration of the flowering plants of Nepal.* Vols. 1–2 (vol. 2 (1979) ed. H. Hara and L. H. J. Williams). Illus., maps. London: British Museum (Natural History).

Systematic enumeration of seed plants, with synonymy, detailed references and citations, local distribution and altitudinal zonation (with representative *exsiccatae*) and overall range outside Nepal, critical remarks (sometimes extensive), and miscellaneous notes; list of new names and index to family and generic names. Some keys are included, as well as citation of key literature under family and generic headings, but no descriptions are given. The introductory section includes sections on the genesis of the work, the history of botanical work in Nepal, geography and physical features, vegetation (with map), horticultural contributions, and technical notes; lists of depository herbaria, collectors (by S. Sutton), and route maps of recent field trips are also included.

MALLA, S. B. *et al.* (eds.), 1976. *Catalogue of Nepalese vascular plants.* ix, 211, 12, 40 pp. (Bull. Dept. Medicinal Pl. Nepal, 7). Kathmandu: Ministry of Forests, Nepal.

Unannotated name list of vascular plants, with synonymy, covering 3453 species; based on herbarium holdings in Kathmandu. An introductory section gives the basis for the work, collecting history (with map of routes and areas), and a brief account of the vegetation

with synoptic table. Selected references and an index to genera and families are also provided. The *Bulletin* series contains other floristic works but these by and large are outside our scope.

Partial work

KITAMURA, S., 1955. [Flowering plants and ferns.] In *Flora and fauna of Nepal Himalaya* (ed. H. Kihara), pp. 73–290, illus. (Scientific results of the Japanese expedition to the Nepal Himalaya, 1952–3, vol. 1.) Kyoto: Kyoto University.

Includes a systematic enumeration of vascular and non-vascular plants (excluding Fungi), with descriptions of new taxa, synonymy (with citations), and indication of local range; figures and photographs of representative species; index at end of volume (no separate index). An introductory section includes among other material accounts of phytogeography and ecology (with illustrations), and a map of areas visited (in central Nepal).

845

Sikkim and northern Bengal

The best available modern work, which more or less centers on this area, is *Flora of Eastern Himalaya* by H. HARA and others (see **840**). However, two local works specifically accounting for this area (one of which is incomplete) have been included here, and a local bibliography is also available. Northern Bengal, including the famous 'hill station' of Darjeeling, is not covered in Prain's *Bengal plants* (**833**). In 1975, Sikkim, previously a self-governing princely state, was made a state of India following internal political troubles.

Bibliography

MATTHEW, K. M., 1970. A bibliography of the botany of Sikkim. *Bull. Bot. Soc. Bengal,* **24**: 57–9. [Short, not exhaustive list.]

BISWAS, K., 1966. *Plants of Darjeeling and the Sikkim Himalayas.* Vol. 1. 8, vii, 540, xxviii pp., *c.* 100 halftones, 47 color pls. Alipore: West Bengal Government Press.

This generously proportioned work comprises a copiously illustrated descriptive treatment, without keys but including synonymy, vernacular names, local range, notes on habitat, special features, etc., and an index. Not completed; extends from Ranunculaceae through Ericaceae (Bentham and Hooker system).

COWAN, A. M. and COWAN, J. M., 1929. *The*

trees of Northern Bengal, including shrubs, woody climbers, bamboos, palms and tree ferns, 178 pp. Calcutta: Government of Bengal.

Systematic enumeration, without keys, of the woody plants of northern Bengal (and Sikkim); includes limited synonymy, vernacular names, general indication of local range, altitudinal distribution, and notes on taxonomy, habitat, life-form, phenology, associates, wood, uses, etc.; index to all botanical and vernacular names. Essentially a revision of J. S. Gamble's *List of trees, shrubs and large climbers found in the Darjeeling district, Bengal* (1878, Calcutta; 2nd edn. 1898).

846

Bhutan

See also **840** (HARA *et al.*). As with Nepal, some early collecting trips were made into this remote feudal state in the nineteenth century, but the first general botanical survey was made only in 1963–5 consequent to improved access and fewer restrictions. Collecting trips have also made by the Japanese expeditions involved with the Himalaya floristic project (see **840**). The results of the Indian botanical survey were published in 1973 and the first results of the Edinburgh-based Bhutan Flora project appeared in 1980.

GRIERSON, A. J. C. and LONG, D. G. (compil.), 1980. *A provisional checklist of the trees and major shrubs (excluding woody climbers) of Bhutan and Sikkim*. 51, 2 pp. Edinburgh: Royal Botanic Garden. (Mimeographed.)

Interim nomenclatural checklist, arranged according to the Englerian system. Produced in connection with the Bhutan Flora project, initiated in 1975 and described by the above authors in *Notes Roy. Bot. Gard. Edinburgh*, 36: 139–41 (1978). [Partially succeeds the treatment for northern Bengal and Sikkim by COWAN and COWAN (**845**).]

SUBRAMANYAM, K., (ed.), 1973. *Materials for the flora of Bhutan*. xii, 278 pp. (*Rec. Bot. Surv. India*, 20(2)). Calcutta.

Annotated enumeration of vascular plants, without keys; includes principal synonymy (with references) and citations, short diagnoses (of key features), and localities (with indication of *exsiccatae*);

a few other notes are given as required. An appendix gives a list of more abundant and important medicinal plants, and a selected bibliography and index to all botanical names conclude the work. The introductory section gives a summary of places visited, previous botanical exploration, and general remarks on physical features, climate, and vegetation. Based principally on collections made by the Botanical Survey of India in 1963–5, this account is the first consolidated modern account of the higher plants of Bhutan.

847

Arunachal Pradesh

See **837** (KANJILAL *et al.*). This area, also claimed by China, was formerly part of Assam, but after the independence of India it became a territory known as the North-East Frontier Agency and in 1972 achieved statehood as Arunachal Pradesh. The land is mountainous and difficult of access for the most part; botanical collecting was sporadic and the area remained very inadequately known until after World War II. Since 1950, considerable collections have been made and the flora is now estimated as having some 3000 species of seed plants.[12] However, the new information has not yet been synthesized into a separate state flora.

Region (Superregion)

85

Japan, Korea, and associated islands

Included under this heading are the main islands of Japan proper, the Ryukyu Islands, the Nanpô Shotô (south of Honshu), the Bonin (and Volcano) Islands (also known as the Ogasawaras and Kazans respectively), the Daitō (Borodino) group, Parece Vela, and Korea.

The interaction of Japanese with Western botany began early in the seventeenth century, with the establishment of the Dutch East India Company in Nagasaki. For more than two centuries, this channel remained virtually the only means by which knowledge of the Japanese flora flowed out to the West. Notable early collectors included Thunberg in 1773 and von Siebold in 1823, both of whom later prepared floristic accounts (the latter jointly with Zuccarini) in 1784 and 1836–70 respectively. After the 'opening' of Japan in 1853, some collecting prior to the Meiji restoration of 1868 was done by Wright, Maximowicz and Savatier, the latter continuing until 1876. Savatier also collaborated with Franchet (in Paris) on *Enumeratio plantarum japonicarum* (1873–9), the fundamental early synthesis of Western and Japanese knowledge of the flora of Japan and a basis for all later research and writing.

The Japanese botanical profession grew rapidly from the 1870s, along with the nation in general and its scientific community. Some collecting trips continued to be made by Western botanists but the bulk of the detailed botanical survey of Japan has been carried out by Japanese. The first manual appeared in 1906 and enumeration in 1904–12. By the Second World War the flora was comparatively well-known and today is covered at a standard corresponding to California, the eastern United States or parts of Europe; all current standard works date from after 1950, and one's choice is mainly a matter of taste. The standard technical floras are those by Ohwi, *Nihon shokubutsu* (1965) and its English version, *Flora of Japan* (1965), an American-style manual-flora with keys, and *Makino's new illustrated flora of Japan* (1961), an illustrated atlas-flora in a style akin to Britton and Brown's *Illustrated flora*, but without keys. Of the semi-popular illustrated works, the best are those in the *Flora and Fauna of Japan* series of the Hoikusha publishing house in Osaka, written under the direction of S. Kitamura (1954–64; partial revision, 1971–). An incomplete *Enumeratio spermatophytarum japonicarum* by Hara (1948–54) attempted to provide a critical nomenclatural basis for modern work. A large range of regional and local florulas is also available for the islands of Japan proper.

In other parts of the region, the level of documentation has generally improved markedly since 1950. Some works on the Bonin (Ogasawara) and Volcano (Kazan) Islands have been published since return of these islands to Japan in 1969, notably *The nature in the Bonin Islands* whose senior author, T. Tuyama, was a significant contributor to the islands' floristics before World War II. No proper flora, however, has been published. With respect to the Ryukyu Islands, two large floras have been published

in recent years: *Flora of the Ryukyu Islands* (Hatusima, 1971, in Japanese; revised 1975) and *Flora of Okinawa and the southern Ryukyu Islands* (Walker, 1976, in English). These complement Masamune's critical enumeration of 1951–64. Some 'tree books' have also been published. Hatusima's flora also covers the Daito Islands, for which no separate account has appeared since 1941.

Botanical exploration in Korea was very sporadic before 1905, and the first general flora did not appear until Palibin's enumeration of 1898–1901, an outcome of its 'Russian' era. At the same time the country was covered in *Index florae sinensis* (1886–1905). From then onwards, ending only with World War II, most collecting and flora writing was done by Japanese, with notable contributions being made by Nakai in his *Flora koreana* (1909–11) and *Flora sylvatica koreana* (1915–39). The major Korean contribution is the two-volume *Korean flora* by Chŏng (1956–7; revamped 1965), but this is regarded as less than satisfactory and a new critical flora is much needed. The country is evidently less well explored botanically than Japan and certainly less well documented at the present time. An illustrated guide to woody plants is also available.

Progress
A useful review of history and current knowledge, mainly for Japan, is KOYAMA, T., 1961. Available knowledge on the flora of Japan and its neighbouring regions. In *Ninth International Botanical Congress. Recent advances in botany*, vol. 1, pp. 937–40. Toronto. The introduction in the English version of Ohwi's *Flora of Japan* should also be consulted. No similar reports on progress for other areas have been seen; reference should be made to the floras concerned.

Bibliographies. General bibliographies as for Division 8.

Regional bibliographies
MERRILL, E. D. and WALKER, E. H. *A bibliography of Eastern Asiatic botany.* See Superregion 86–88.

WALKER, E. H. *A bibliography of Eastern Asiatic botany: Supplement.* See Superregion 86–88.

Indices. General indices as for Division 8.

Regional index
STEENIS, C. G. G. J. VAN and JACOBS, M. (eds.), 1948– . *Flora malesiana bulletin*, 1– . Leiden: Flora Malesiana Foundation (later Rijksherbarium). (Mimeo-

graphed.) [Includes a substantial bibliography. For fuller description, see Superregion 91–93.]

KANAI, H., 1975– . *List of literature related to plant taxonomy and phytogeography of Japan*. Tokyo. [Coverage as of 1979 available for the years 1971 through 1976; issued serially.]

851

Japan

This unit is here defined as encompassing the main Japanese islands of Hokkaido (Yezo), Honshu (Hondo), Shikoku, and Kyushu, along with a number of nearby small groups including Osumi Gunto, Goto Retto, Tsushima, Oki Gunto, Sado, and Izu Shichito; this corresponds to the limits of the standard modern flora of Japan (Ohwi, 1965) available in both Japanese and English. The flora of Japan is at present one of the best known in Asia, at a level corresponding to that of California, the northeastern United States, or France, and a large number of general and partial floristic works is available. Only a limited selection can be described here.

Dictionary

HONDA, M., 1957. *Nomina plantarum japonicarum*. 2nd edn. [4], 389, 126 pp. Tokyo: Koseisya-Koseikaku. (First edn., 1939.) [Systematic list of vascular plants, with standard equivalents in Japanese. A list of necessary new combinations, with required documentation, and indices are also provided.]

HARA, H., 1948–54. *Enumeratio spermatophytarum japonicarum*. Vols. 1–3. Tokyo: Iwanami Shoten. (Reprinted in 1 vol., 1972, Königstein/Taunus, W. Germany: Koeltz).

Critically compiled, systematic enumeration of native, naturalized, and commonly cultivated seed plants of Japan and adjacent islands, with detailed synonymy, references and pertinent citations; general indication of local range; standardized Japanese names; index to genera. Except for the Latin-language references and botanical names, the text is in Japanese. Not completed; covers only the Metachlamydeae and part of the Archichlamydeae (Geraniaceae through Cornaceae) in the Englerian system.

KITAMURA, S. *et al.*, 1957–64. *Coloured illustrations of herbaceous plants of Japan*. 3 vols. Text-figs., 198 color pls. Osaka: Hoikusha. (Revised edn. of vol. 1

(Sympetalae), 1958; all vols. subsequently reprinted more or less annually.) (Flora and fauna of Japan, 15, 16, 17.)

KITAMURA, S. and OKAMOTO, S., 1959. *Coloured illustrations of trees and shrubs of Japan*. viii, 306 pp., 72 text-figs., 68 color pls. Osaka: Hoikusha. (Reprinted subsequently.) (Flora and fauna of Japan, 21.)

TAGAWA, M., 1959. *Coloured illustrations of the Pteridophyta of Japan*. iv, 270 pp., 8 text-figs., 72 color pls. Osaka: Hoikusha. (Reprinted subsequently.) (Flora and fauna of Japan, 24.)

These related works together constitute an outstanding semipopular series of copiously illustrated manuals covering all (or most) of the vascular plants in Japan; with slight variations they include keys to all taxa, limited synonymy, standard Japanese names, vernacular names in English, French and German (for genera), general summaries of local and extralimital distribution, and notes on habitat, phenology, karyotypes, special features, etc. Each volume has a list of references and complete indices. The three volumes on herbaceous plants respectively cover Sympetalae, Choripetalae, and Monocotyledoneae. The volume on pteridophytes also contains an annotated enumeration of all taxa. Collaborators have included G. Murata, M. Hori, and T. Koyama. A revised edition of the treatment of woody plants, in two volumes and featuring a new set of colored plates, is KITAMURA, S. and MURATA, G., 1971–9. *Coloured illustrations of woody plants of Japan*. Revised edn. 2 vols. 438 text-figs., 144 color pls. Osaka: Hoikusha. (Flora and fauna of Japan, 49, 50.)

MAKINO, T., 1961. *Makino's new illustrated flora of Japan*. 3rd edn., revised under supervision of F. Maekawa, H. Hara, and T. Tuyama. vii, 12, 1060, 77 pp., 3894 text-figs., 4 color pls., portrait. Tokyo: Hokuryukan. (First edn. 1925, as *Illustrated flora of Japan*; 2nd edn. 1940, as *New illustrated flora of Japan*. All editions reprinted several times.)

Illustrated descriptive atlas-flora of most vascular and the more important non-vascular plants, *without keys*; includes scientific names, standardized and vernacular Japanese names, general summaries of local range, and notes on habitat, special features, uses, etc.; etymological and terminological glossaries, a lexicon of genera, and indices to botanical and other names at end. This work, modeled to some extent on Coste's *Flore de France* (651) or the earlier versions of Britton and Brown's *Illustrated flora of the northern United States*

(140–50), has for many years been the standard traditional-style flora of Japan, with origins rooted in classical Japanese (and Chinese) herbals but written in a manner compatible with international conventions. The plan of this work has evidently had a considerable influence elsewhere in eastern Asia for it has been followed (with variations) by standard floras in Korea, China, and Vietnam.

OHWI, J., 1965. *Nihon shokubutsu-shi*. 2nd edn. [iv], 7, 1560 pp., 10 text-figs., 10 pls., 66 color illus. Tokyo: Shibundo. (First edn. 1953–7.) English edn. OHWI, J., 1965. *Flora of Japan* (ed. F. G. Meyer and E. H. Walker). vii, 1067 pp., 17 text-figs., 16 pls., maps (partly in end papers). Washington: Smithsonian Institution.

Briefly descriptive flora of vascular plants, with keys to all taxa, limited synonymy, Japanese equivalents, generalized indication of local range, and notes on habitat, special features, etc.; list of authors and indices to all botanical and Japanese names. The introductory section includes chapters on botanical exploration, important students of the flora, floristic zones, and a general key to families. The Japanese names in the English edition are in the Roman alphabet. The work is the standard modern 'Western-style' flora of Japan, contrasting with Makino's flora (see preceding entry).

Distribution maps

HORIKAWA, Y., 1972–6. *Atlas of the Japanese flora. An introduction to the plant sociology of East Asia.* Vols. 1–2. 862 maps. Tokyo: Gakken.

Systematically arranged folio atlas of distribution maps of the flora of Japan and neighboring regions (in English), based on records from grid squares 20×20 km and incorporating associated descriptive text; the latter incorporates principal synonymy and notes on life-form, phenology, and uses. A general index is given at the end of each volume. Each map also has 'satellite' maps depicting altitudinal zonation of the species in question. The introduction to the work incorporates a concise sketch of the vegetation of Japan. A further three volumes are projected. [Citation and description based partly on a review by C. G. G. J. van Steenis in *Flora Malesiana Bull.* 27: 2214–16 (1974).] See also HARA, H. and KANAI, H., 1958–9. *Distribution maps of flowering plants in Japan.* 200 maps. Tokyo.

852

Nanpô Shotô

This archipelago, lying between the Izu Shichito just off central Honshu (851) and the Bonin Islands (853), includes Hachijo-jima, Aoga-shima, Urania, Sumisu-jima, Tori-shima, and So-fu Gan (Lot's Wife). No general floras or enumerations specifically covering these islands appear to be available; any knowledge is scattered through miscellaneous papers. Aoga-shima has been treated in three papers by M. Mizushima in *Misc. Rept. Res. Inst. Nat. Resources (Japan)*, 38: 106–26 (1955); 41/42: 76–80 (1956); 45: 64–8 (1957).

853

Bonin (Ogasawara) and Volcano (Kazan) Islands

This distinctive pair of island groups, lying just outside the tropics, is by some authorities considered to be within biogeographical Oceania, but Hosokawa in the 1930s demonstrated that the vascular flora as a whole was most closely related to that of mainland Asia, the Ryukyus and Japan. This view, upheld by Balgooy,[13] is maintained here with the inclusion of these groups in Region 85. However, following the pioneer studies of Wilson, Nakai and Tuyama before 1940, no consolidated flora or enumeration appears to have been published; from World War II until 1969 the islands were under US occupation and the first new natural history work, *The nature in the Bonin Islands*, appeared only in 1970. This color-plate work provides a good introduction to the island biota, but does not contain a full enumeration of vascular plants. The woody plants are also reviewed in WILSON, E. H. 1919. The Bonin Islands and their ligneous vegetation. *J. Arnold Arbor.* 1: 97–115.

KOBAYASI, S., 1978. A list of plants occurring on the Ogasawara (Bonin) Islands. *Ogasawara Research*, 1: 1–33, 3 maps.

This contribution from the Ogasawara Research Committee of the Tokyo Metropolitan University and its Makino Herbarium represents an annotated systematic checklist of vascular plants, with romanized Japanese names, indication of presence (or not) in the

Volcano Islands group as well as distribution outside the Ogasawaras and/or origin (if applicable), and notes on habit and habitat. No synonymy or specimen citations are provided. A total of 483 species is now recorded from the Ogasawaras proper, with 152 endemic and 114 introduced or commonly cultivated. [A copy of this work was made available by courtesy of K. Woolliams, Waimea Arboretum, Haleiwa, Oahu, Hawaii.]

NAKAI, T., 1930. [Plants of the Bonin Islands.] *Bull. Biogeogr. Soc. Japan*, **1**: 249–78, pls. 16–18.

Comprises an unannotated systematic list of vascular plants, with botanical names and their Japanese equivalents, together with a detailed phytogeographical analysis of the flora and discussion of distribution patterns. The illustrations depict a selection of endemic species. The significant disjunction between the flora of these islands and the Marianas (subsequently known as 'Hosokawa's Line') is pointed out.

TUYAMA, T., 1935–9. Plantae boninenses novae vel criticae, I–XII. *Bot. Mag. Tokyo* **49** (1935): 367–74, 445–52, 505–12, pls. 1–17; **50** (1936): 25–32, 129–34, 374–9, 425–30, pls. 18–36; **51** (1937): 22–4, 125–32, pls. 37–43; **52** (1938): 463–7, 567–72, 6 pls.; **53** (1939): 1–7, 3 pls.

Gives descriptions of, and notes upon, new or little-known species of vascular plants, with citations of *exsiccatae* and remarks on taxonomy, habitat, special features, etc. Illustrations are provided for the more interesting taxa.

TUYAMA, T. and ASAMI, S., 1970. *The nature in the Bonin Islands*. 2 vols. Illus. (mostly in color), maps. Tokyo: Hirokawa Shoten.

Volume 1 (text) includes a section on the flora (pp. 109–41), with tabular checklists of vascular plants but no single consolidated enumeration for the archipelago. Vegetation is also discussed. The rest of the text volume deals with physical features, geology, climate, fauna, human influences, conservation, etc. Volume 2 (plates) is an attractive color atlas, with pp. 43–130 including 213 photographs of the islands' vegetation and characteristic plants. The work, though not containing a proper 'flora', represents the first consolidated general account for these islands since before World War II.

854

Daito (Oagari or Borodino) Islands

See also **855** (HATUSIMA; MASAMUNE). One separate account is available for these Ryukyuan outliers in the Philippine Sea.

MASAMUNE, G. and YANAGIHARA, M., 1941. (On the flora of Ōagari-zima (Daitō Islands), Ryukyu Archipelago), I–III. *Trans. Nat. Hist. Soc. Taiwan*, **31**: 237–50, 268–74, 317–30.

Parts II–III of this work contain an enumeration (in Japanese) of all known vascular plants from this island group, with botanical and equivalent Japanese names and miscellaneous notes on local range, habitat, etc.

855

Ryukyu Islands

This group is here considered to include the four sets of islands (Tokara Gunto, Amami Gunto, Okinawa Gunto and Sakishima Gunto) between the Osumi Islands just off Kyushu (**851**) and Taiwan (**886**). The two southern groups were under United States civil administration from 1951 to 1969. Two overall works are available: a descriptive flora in Japanese (Hatusima, 1971) and an enumeration in English (Masamune, 1951–64). These are accompanied by an illustrated 'tree book' (Walker, 1954). For the two southern groups, a substantial descriptive flora in English (Walker, 1976) has recently been published, supplanting two local florulas. The general works by Hatusima and Masamune also cover the more remote Daito Islands (**854**) to the east.

HATUSIMA, S., 1971. [*Flora of the Ryukyu Islands*.] x, 940 pp., 18 text-figs. (incl. 6 maps), 30 pls. (some in color), portraits. Okinawa: Okinawa Association for Biology Education.

Descriptive flora of vascular plants, with keys to all taxa, synonymy (with dates), standardized Japanese and local vernacular names, general summary of local and extralimital range, taxonomic commentary, and notes on habitat, special features, uses, etc.; indices to all botanical and other names at end. An introductory section includes chapters on physical features, geography, geology, climate, special features of the flora, history of botanical exploration and research, and an illustrated organography. Text in Japanese. A revised, somewhat expanded and more copiously illustrated edition (not seen by this reviewer) appeared as *idem*. 1975. *Ryukyu shokubutsu-shi* [Flora of the Ryukyus]. 16, 27, 1002 (actually 1023) pp., 27 half-tone pl., 62 col. photogr., 6 maps. Okinawa. – A predecessor of this flora is HATUSIMA, S., and AMANO, T. 1967. [Flora of Okinawa.] Rev. ed. iii, 218 pp. Okinawa. (1st ed., 1958.)

MASAMUNE, G., 1951–64. Enumeratio tracheophytarum Ryukyu insularum. 10 parts. *Sci. Rep. Kanazawa Univ., Biol.* 1 (1951): 33–54, 167–99; 2 (1953): 87–114; 2A (1954): 59–117; 3 (1955): 101–82, 253–338, figs. 1–7; 4 (1956): 45–134, 201–80, figs. 8–23; 5 (1957): 85–121, figs. 24–31; 9 (1964): 119–56, 1 color pl.

Systematic enumeration of vascular plants, with synonymy, references and citations, generalized indication of local and extralimital range, taxonomic commentary, and notes on habitat, special features, etc.; figures of plants of special interest; index to families and genera in part 10. Text in English. The part containing Orchidaceae was published rather later than its predecessors. Synonymy and literature citations have been relative thoroughly worked out in this compilation, the last of several made by this author (see also Taiwan, Hainan, and Borneo).

Partial work: Southern Ryukyus

The earlier floras of Okinawa by Hatusima and Amano (1967, in Japanese) and Sonohara and others (1952, in English) have now been supplanted by the large new flora by Walker (1976). This covers Okinawa Gunto and Sakishima Gunto, the areas under American administration until 1969.

WALKER, E. H., 1976. *Flora of Okinawa and the southern Ryukyu Islands*. 1159 pp., 209 text-figs. (incl. 1 color pl.). Washington: Smithsonian Institution Press.

Descriptive manual of vascular plants, with keys to all taxa, synonymy and key references, fairly detailed indication of local distribution, with mention of representative *exsiccatae* arranged by district, standard Japanese names (transliterated, with derivation) and local Ryukyu names, critical commentary, and notes on habitat, occurrence, behaviour, associates, etc.; glossary, author abbreviations, cyclopedia of collectors, selective bibliography (pp. 1099–1104), and complete indices at end. The illustrations are largely of woody plants. An introductory section accounts for the genesis of the work and gives technical notes as well as the key to families. This conservatively planned work is similar in style and format to the English edition of Ohwi's *Flora of Japan* (851), and is an outgrowth of SONOHARA, S., TANADA, S., and AMANO, T., 1952. *Flora of Okinawa*. 237, 28, 50 pp., 1 pl. Okinawa. (Mimeographed.) [It is a moot point, however, whether such a large and bulky work is appropriate for a group of relatively small islands, and then only an administratively determined portion thereof. It is unfortunate that the work was not extended to include the whole of the Ryukyus.]

Special group – forest trees

WALKER, E. H., 1954. *Important trees of the Ryukyu Islands*. v, 350 pp., 209 text-figs., frontispiece, 2 maps. Okinawa: United States Civil Administration.

Illustrated descriptive treatment of the more significant tree species, with keys to all taxa (including an artificial general key); generalized indication of local range; Okinawa and standard Japanese names; notes on habitat, wood structure, uses, etc.; references and indices to all botanical and other names.

858

Korea

Korea here includes both the Democratic People's Republic of Korea (North) and the Republic of Korea (South), the latter extending to Cheju Do (Quelpart) in Korea Strait. Although a number of floras and related works have been published, the level of botanical exploration is evidently rather less than in Japan and the basis for floras correspondingly weaker. A good critical modern flora is much needed; Chŏng's works are not highly regarded (K. C. Oh, personal communication).

Bibliography

GOODE, A. M., 1956. An annotated bibliography of the flora of Korea. *Trans. Kentucky Acad. Sci.* 17: 1–32. (Provisional edn. 1955, Washington, as separate work.) [Includes 305 entries.]

Dictionaries

MORI, T., 1922. *An enumeration of plants hitherto known from Corea*. 10, 3, 3, vii, vii, 372, 174 pp., 5 pls. Seoul: Chosen Government. [Systematic list of all vascular plants, with equivalents in Japanese, Korean, and Chinese; tabular summary of the flora and list of references; complete indices.]

TO, PONG-SŎP and IM, NOK-CHAE, 1955. *Chosŏn singmul myŏngchip* [Sbornik nazvanij rastenij Korei/ Nomina plantarum koreanum]. 10, 364 pp. Pyongyang: Academy of Sciences, Democratic People's Republic of Korea). [Systematic list of vascular plants, with Korean equivalents, covering all families, genera and species; complete indices.]

CHŎNG, TAI HYUN [KAWAMOTO, TAIGEN], 1957. *Hanguk singmul togam* [Korean flora]. Vol. 1: [Trees and shrubs]. 2nd edn. [7], 13, 7, 507, 82 pp., 1013 text-figs., frontispiece, map. Seoul: Synji Sa. (First edn. 1943, in Japanese, as *Chosen sinrin shokubutsu zusetsu*.)

CHŎNG, TAI HYUN [CHONG T'AE-HYŎN], 1956. *Hanguk singmul togam* [Korean flora]. Vol. 2: [Herbaceous plants]. [2], 2, 3, 11, 1025, 129 pp., 2050 text-figs., 2 pls., map. Seoul: Synji Sa. (Reprinted 1957, Seoul.)

Complementary atlas-floras covering all Korean vascular plants, without keys; they include large figures with accompanying descriptive text and (in vol. 1 only) dot distribution maps. The text accounts (in Korean) feature occasional synonymy, Korean and Japanese scientific and vernacular equivalents (with transliterations in the Roman alphabet), general indication of local (by province) and extralimital range, and notes on habitat, phenology, special features, uses, etc.; indices to all botanical and other names. The layout of vol. 1 features three species (with figures) and one range map for each two pages; in vol. 2, there are two species (with figures) per page. To a large extent this flora has been based on Nakai's various works (see below).

CHŎNG, TAI HYUN (T'AE-HYŎN), 1965. *Tracheophyta*. 1824 pp., 340 pls. (some in color) with 3051 figs., 12 plain pl. (in glossary), 2 color pls., 2 color maps (frontispiece), portrait (Illustrated encyclopedia of fauna and flora of Korea, vol. 5). Seoul: Samhwa (for Ministry of Education).

CHŎNG, TAI HYUN, 1970. *Tracheophyta: appendix*. 232 pp., 40 pls. with 355 figs. (Illustrated encyclopedia of fauna and flora of Korea, vol. 5a). Seoul.

Atlas-flora of vascular plants, with text and figures divided, the latter on separate plates; the text includes description, Korean and romanized Japanese nomenclatural equivalents, distributional data, altitudinal zonation (where necessary), and notes on habitat, origin, occurrence, uses, etc., while the plates each have nine small figures (these partly in color). The introductory section begins with an explanatory section and continues with glossary-tables, a list of references (p. 26), and synoptic tables of contents (in Korean and in Latin), while at the end of the work are indices to

all scientific names and their Korean and romanized Japanese equivalents. [Nomenclature largely follows the 1952 checklist by Nakai, itself a revision of his flora of 1909–11, described below.] The supplement of 1970 includes additional species in the same format, but with all illustrations in monochrome. On pp. 6–7 therein is a statistical table, indicating a total of 3406 specific and infraspecific entities known for Korea.

LEE, TCHANG BOK. 1979. *Illustrated flora of Korea*. 990, 2 pp., illus., 2 color pls., color signature. Seoul: Hyangmunsa.

Amply-dimensioned atlas-flora of vascular plants (3160 mainly specific entities), systematically arranged with descriptions and small figures (four to a page, in the style of Makino's flora of Japan (**851**)) and Korean nomenclatural equivalents. The several appendices comprise tables of illustrations of botanical features, a glossary of derivations of botanical names, an explanation of author abbreviations, Latin-Korean, Korean-Japanese, and Japanese-Korean nomenclatural lexica, and indices to specific and infraspecific taxa in Korean and Latin. A short introduction and two synoptic tables of contents (respectively in Korean and Latin) precede the main text. [The work is very well produced, but contains virtually no information on ecology, distribution, etc., as usually provided in floras.]

NAKAI, T., 1909–11. *Flora koreana*. 2 vols. Illus. (J. Coll. Sci. Imp. Univ. Tokyo, vol. 26, pp. 1–304, i–ii, Pls. 1–15; vol. 31, pp. 1–573, Pls. 1–20). Tokyo.

Systematic enumeration (in Latin) of known vascular plants, with detailed synonymy, references and citations; citation of records and representative *exsiccatae*, with localities, and general indication of extralimital range; keys to genera and species; descriptions of new taxa (with illustrations); addenda and complete index to botanical names in vol. 2. This work represents the standard modern basis for later floristic treatments. For a revision and updating of this inventory, see *idem*, 1952. A synoptical sketch of Korean flora. *Bull. Natl. Sci. Mus. Tokyo*, **31**: 1–152. [Systematic list of 3176 species of vascular plants, with summary.]

Partial work

LEE, YEONG NO, 1976. [Flowering plants.] 895 pp., 186 color pls. (with 889 photographs) (Illustrated encyclopedia of fauna and flora of Korea, 18). Seoul: Samhwa (for the Ministry of Education, Korea).

Copiously illustrated large-scale semipopular guide to wild flowers and flowering trees of Korea, with four main sections corresponding to the seasons (each with separate table of contents); within each section species are described and annotated, with accompanying photographs, and a general summary of the special features of each season is given. A synopsis in English (pp. 779–826) covering all species, a bibliography (pp. 827–30) and complete indices (Latin, Korean, Japanese and Chinese) conclude the work. The introductory sections (both in Korean and English) give the basis of the work, its plan, and an introduction to climate, floristics, life-forms, and phenological patterns; table of families (pp. 25–9). 889 species are covered, none in more than one of the four main sections. Some species keys are provided, but none for genera and families.

Special group – woody plants

FOREST EXPERIMENTAL STATION, KOREA, 1973. *Illustrated woody plants of Korea*. Revised edn. iv, 237, 15, 10 pp., 755 text-figs. Seoul. (Original edn. 1966.)

Atlas-manual of woody plants, with small figures accompanied by descriptive text and scientific names with Korean transliterations; keys to all species (pp. 190–237); complete indices to Latin and other names at end.

NAKAI, T., 1915–39. *Flora sylvatica koreana*. Parts 1–22. Plates. Seoul: Forest Experiment Station, Chosen Government.

Large-scale atlas of woody plants, with engraved plates accompanied by copious descriptions; accounts of species include synonymy, references and citations, Japanese and Korean names, generalized indication of local and extralimital distribution, and taxonomic commentary, all arranged family by family (the latter not in systematic order). Preceding each family treatment is a conspectus of *all* known species (both woody and non-woody) recorded for Korea, with references, literature citations, vernacular names, and notes on distribution. Text in Japanese and Latin. Apart from the arrangement of families, this work compares somewhat with Sargent's *Silva of North America* (**100**).

Special group – pteridophytes

PARK, MAN KYU [MAN-KYU], 1975. *Pteridophyta*. 549 pp., 43 text-figs., 60 pls. with 258 figs. (some in color), 7 plates (some in color), incl. map, portrait (Illustrated encyclopedia of fauna and flora of Korea, 16). Seoul: Samhwa (for Ministry of Education, Republic of Korea).

The text of this semi-monographic treatment is in two sections: (*a*) a descriptive portion, with indication of distribution, some taxonomic commentary, and notes on habitat, etc., along with critical figures and keys to all taxa; and (*b*) an appendix with a synoptic list, a key to rhizome sections, a floristic summary and statistical table (272 species admitted), an account of life-forms, distributional and

chorological tables, and (in sect. 6) formal nomenclature, with synonymy, references, citations, life-form, romanized Japanese names, and Korean nomenclatural equivalents. The introductory section recounts the plan of the work and the history of fern studies concerning Korea, followed by a synoptic table of contents, while at the end are the list of references (pp. 529–31) and general indices to Japanese, Korean and Latin names. [The work is in part a companion to the treatment of flowering plants by Yeong No Lee (see above).]

Table 2. *Chinese provincial names*

Conventional – 'Post Office' (as used in text)	Pinyin (current Chinese usage)
Anhwei	Anhui
Chekiang	Zhejiang
Fukien	Fujian
Hainan Tao	Hainan Dao
Heilungkiang	Heilongjiang
Honan	Henan
Hopeh	Hebei
Hunan	Hunan
Hupeh	Hubei
Inner Mongolian A.R.	Neimenggu A.R.
Kansu	Gansu
Kiangsi	Jiangxi
Kiangsu	Jiangsu
Kirin	Jilin
Kwangsi Chuang A.R.	Guangxi Zhuang A.R.
Kwangtung	Guangdong
Kweichow	Guizhou
Liaoning	Liaoning
Ningsia Hui A.R.	Ningxia Huizu A.R.
Shansi	Shanxi
Shantung	Shandong
Shensi	Shaanxi
Sinkiang Uighur A.R.	Xinjiang Weiwuer A.R.
Szechwan	Sichuan
Tibet/Sitsang A.R.	Xizang A.R.
Tsinghai	Qinghai
Yunnan	Yunnan

No Pinyin equivalent is to be used for Taiwan, Macao, or Hong Kong. Hainan Tao is part of Kwangtung Province. Not shown here are the South China Sea Islands.

A.R., Autonomous Region.

Superregion

86–88

China (except Chinese central Asia)

This superregion corresponds approximately with the traditional eighteen provinces of 'China proper' together with the three provinces of North-East China (Manchuria), i.e., that part of the People's Republic of China bounded on the south by Burma and Viet Nam, on the west by Tibet (Sitsang) and Chinghai Province, on the north by Inner Mongolia and the eastern part of the Soviet Union, and on the east by Korea and the North and South China Seas – with the inclusion of Taiwan, Hainan, and the Paracels (Hsi-sha) and Spratlys (Nan-sha). In effect, it is the China of the summer monsoon; those tracts excluded here constitute the dry interior, once known as the 'Chinese Border'. With Mongolia, the provinces and territories of the dry Chinese interior make up Region 76.

The vascular flora of China has been of great interest to the botanical world from the mid-nineteenth century when its richness and diversity became generally appreciated. An added feature was the existence of a large body of 'traditional' phytographic literature, with many copiously illustrated works, dating back in some cases for nearly two millenia. However, botany as a modern profession began only after the revolution of 1911, and thus the publication of two major national works, *Chung kuo chung tz'u chih wu t'u chih* [Flora reipublicae popularis sinicae] (1959–), now making progress, and *Iconographia cormophytorum sinicorum* (1972–6), complete in five volumes with two supplements reportedly projected, are testimonials to the remarkable if at times interrupted progress of floristic botany in China since early in the century, when local resources for such work were virtually non-existent.

The development of botany in China has been treated by Chun as having had (by 1930) three phases, to which must now be added a fourth (and possibly fifth) covering the period since 1949. The first of these is that of ancient traditional 'encyclopedic' research and the writing of herbals, dictionaries and other compilations. Many of these works are still of value today, and the task of systematization in modern terms of the material involved is immense and largely unfinished: only a beginning was made by Bretschneider.

The second phase was that of collecting and research by foreigners, which was concurrent with the last developments of the first phase and in the twentieth century overlapped with the third phase. For mercantile and other reasons, the Ch'ing (Qing) rulers until 1840 kept tight controls on foreign contacts, which were limited geographically to Canton (Guangzhou) and the Portuguese colony of Macao. It is thus from these places that most of the earliest collections from China were made; until 1771 these remained a trickle, with only some 260 species being recorded by J. R. Forster in his *Flora sinensis*, published in Osbeck's *A voyage to China and the East Indies* of that year. With an increasing number of collectors and recorders, however, a considerable increase in the number of known Chinese species took place in the years before the First Opium War.

The first of the 'unequal treaties' which followed that war included cession of the island of Hong Kong, more treaty ports, and more freedom of movement. The collections of Champion, Hinds, Wright, Fortune, and others from Hong Kong enabled Bentham to write the first florula of any part of China, *Flora hong-kongiensis* (1861; supplemented in 1872). Collections were also made in other treaty ports and in accessible parts of the interior, in which travel was still restricted, notably by Fortune in the 1840s and 1850s and by Hance afterwards.

A second war in 1860 was followed by further treaties, in which most of the interior became open to outside travel and the customs and some other services were mandated to foreigners, because at that time they were better to be trusted. From then onward came the great surge in botanical exploration in China, which also extended to Taiwan, the Ryukyus and Korea, although contact with the latter was very limited. In the three decades or so before the end of the nineteenth century, collecting was dominated by French Catholic missionaries, among them A. David who first made the biological riches of mountainous western China known to the world, several Russians, including Potanin, Przewalski, Maximowicz, and (at a later date) Bretschneider and the young Komarov, some British, among them the customs official A. Henry and (at Hong Kong) the gardens superintendent C. Ford, and

miscellaneous others. However, it was less active as a period for garden introductions. A thorough review has been presented by Bretschneider in his classic history (1898).

The turn of the century was fittingly marked by the compilation of *Index florae sinensis* (1886–1905; supplement 1911), sponsored in part by the Linnean Society of London. This was the first modern catalogue of Chinese seed plants, covering (to 1905) over 8200 species and including the mainland, Taiwan, Hong Kong, Korea and the Ryukyu Islands, and has provided, like its counterpart *Biologia centrali-americana: Botany* in Mexico and Central America, a key foundation for future research.

From 1900 began the last parts of the 'Western' phase. The first of these was 'the Wilson era', dominated by collectors sponsored by gardening interests, particularly in Britain and the United States where a marked change in gardening fashion had come on at this time. Notable collectors of the period to about 1940 included Wilson, Farrer, Purdom, Forrest, Kingdon War, Silvestri, Handel-Mazzetti, and Rock. Western China and adjacent parts of Tibet were the main focus of activities, although other parts were visited as well, notably the province of Hupeh, in the search for 'hardy' plants. Although the main emphasis was on seeds and other material for gardens, most collectors obtained large quantities of herbarium material which was energetically studied in the great botanical centers of Europe, Japan, and the United States. As in the latter part of the nineteenth century, large, autarchic 'prestige' reports, sometimes with outside sponsorship, were prepared on particular collections, perhaps, it may be said, without too much regard for the consequences, a state of affairs also affecting New Guinea and some other botanical 'landmark' areas for which well-established single centers were lacking. Apart from the nineteenth-century reports by Franchet and Maximowicz, examples are *Plantae chinenses forrestianae* (1911–30), compiled at Edinburgh, *Plantae wilsonianae* (1911–17) from Boston, Mass., *Die Flora von Central-China* (1900–1) from Berlin, *Symbolae sinicae* (1929–37) from Vienna, and *Plantae sinenses* (1924–47) from Uppsala. Large contributions were also made from Paris, where work on the Indochinese flora was being prosecuted concurrently by Finet, Gagnepain, and Lecomte, and Le Mans, France, where the uncritical Léveillé worked up many collections from western China, Guizhou and parts of eastern China where French missionaries were still active. With World War II and the subsequent change of government in China, this pattern of research and publication largely lapsed. Because of the great scattering of efforts among many nations, however, no serious effort at synthesis in the 'Western' sphere could be made; a *Flora of China* project initiated at the Arnold Arboretum in the 1950s was abortive, resulting in but a limited amount of publishable work (e.g., Malvaceae, Compositae). Merrill has noted that by 1930 some 21 centers outside China had been involved in Chinese floristics; in many cases, little of the material in their possession has been duplicated by specimens in Chinese herbaria. The location of so many of the types of Chinese plants outside China – nearly 18000 by 1930, and far more than is the case with North America – has created a serious and lasting problem for Chinese floristic botanists; the plea of Chun for assistance in overcoming this unfortunate hiatus is being made once again by a new generation as foreign contacts are resumed (Bartholomew *et al.*, 1979).

However, 'the Wilson era' and its aftermath were also to see the development of Chun's third phase of Chinese botany, namely the entry of Chinese into the profession and their emergence as collectors and researchers, accompanied by the formation of herbaria, libraries, laboratories, botanical gardens and field stations. This had been strongly influenced by the great interest of foreign botanists in China, but was given impetus in the cultural reforms which followed the 1911 revolution which ended the Qing dynasty. From the 1920s floristic publications by Chinese authors began to appear in considerable numbers, and at the same time a strong tradition of biological survey developed which over a period of 30 years yielded some hundreds of thousands of collections for herbaria. By 1930 six significant herbaria were in existence and since then the number has increased further, despite some amalgamations and transfers after 1949. Among the first generation of professional botanists were S. S. Chien and W. Y. Chun, later to become the founders of the *Flora reipublicae popularis sinicae* in the 1950s.

Closely associated with the growth of the Chinese profession were Merrill, acting as a 'consultant', and Levine and McClure (Lingnan University) and Steward (Nanking University), along with Metcalf (Lingnan and Fukien Christian Universities). Apart from teaching, these American botanical missionaries collected more or less extensively, established herbaria

and living collections, and began or contributed to floristic projects. Some standard floras or enumerations were written by them. In the temperate north, considerable assistance was received from Rehder at the Arnold Arboretum on accounts of woody plants. Many early collections from the Chinese biological surveys were determined by Handel-Mazzetti, Smith, Diels and other overseas authorities as well as Merrill. In northeastern China, floristic work progressed rapidly after entry of the Japanese, and the foundation was laid for a number of works, aided by the earlier publication of *Flora Man'čžurii* (1901–07) by Komarov, the first 'provincial' Chinese flora. The Japanese presence in Taiwan also led quickly to the publication of floristic works, so that by the 1940s a substantial basis had been laid there as well as in the northeast for future modern floras.

Elsewhere in China, more emphasis was laid on education, biological survey, and institution-building; national works were mainly in the form of students' manuals and dictionaries while those at provincial level mainly took the form of checklists. Some showpiece atlas-floras were also published, as well as a number of works on woody plants such as *Chinese economic trees* by Chun (1922), *A catalogue of the trees and shrubs of China* by Chung (1924), still regarded as a key work, *Forest botany of China* by Lee (1935; supplement 1973), and *Chung kuo shu mu fên lei hsüeh* by Ch'en (1937; revised 1957). A large number of works outside our scope were also produced. However, collections, facilities and experience probably were, even more than in India at the same period, quite inadequate for the production of descriptive floras; this was a task for another generation.

The 1937–45 war with Japan was marked by great disruption of scientific activities but, because of the move of many institutions to western China, resulted in improved knowledge of the flora and contributed to the establishment of regional herbaria in Shensi, Szechwan and Yunnan. Only a short time was available after the war for reorganization before the revolution of 1949 brought profound structural changes, ending Chun's third phase of Chinese botanical history.

Accounts of developments in floristic botany since 1949, regrettably, are few. Science and higher education in China were entirely reorganized, including the formation of an all-embracing Academy of Sciences under central control. The Institute of Botany at Peking was organized in 1950 by amalgamation of a number of units, including several herbaria. The South China Institute of Botany (now the Kwangtung Institute of Botany) and some other currently provincial botanical institutions similarly came into being through fusion of existing units. Among these were some college and university herbaria, notably those built up at Christian institutions such as Lingnan University. Shortly after these physical and administrative changes came the establishment of *Acta Phytotaxonomica Sinica*, which superseded virtually all other outlets for floristic contributions in China and has ever since remained the officially sanctioned medium for Chinese systematists (though suspended in 1966–74). The existence of such a separate taxonomic journal in the Chinese Academy of Sciences 'stable' is, however, indicative of a certain special significance of the field in the development of China.

These post-revolutionary developments very largely followed Soviet models; and for most of the 1950s there was extensive interaction between Chinese and Soviet (and East European) scientists. It is quite possible that a desire to emulate *Flora SSSR* and the network of regional Soviet floras inspired at this time the establishment of the *Flora reipublicae popularis sinicae* project at Peking and Canton and the later initiation of some provincial and subregional flora projects at these and other centers. Certainly the format of the national flora bears similarities to *Flora SSSR*.

Apart from the imposition of central direction, the events of 1957–8, including the 'great leap forward', gave a strong impetus to the production of floras and related works at both national and regional level. A *Flora illustrata plantarum primarum sinicarum* was begun in this period, and in 1958 there was initiated the project which was finally published in the 1970s as *Iconographia cormophytorum sinicorum*. A regional flora for northeastern China (Manchuria) commenced publication in 1958 and *Flora hainanica* began to appear in 1964. However, by 1966 only three volumes of the national flora had been issued.

The next six years were dominated by the Cultural Revolution, during which almost all scientific work was disrupted. A semblance of normal conditions began to return in 1972, and from the mid-1970s publication of floras and related works has been rapid. Two hundred botanists are said to be engaged on *Flora reipublicae popularis sinicae* alone, with many families being revised through teamwork at a number of locations. It is at present the most ambitious flora project in the world. Several provincial floras have been commenced or revived, although some have not at this writing reached publication. *Flora hainanica* was completed in 1977, and work is underway on floras of northeastern China, Kiangsu, Chekiang, Kwangtung, Hupeh, Tibet, and the Tsinling Range so far as is known. It has been suggested that, ultimately, all provinces and subregions will be provided with floras. A similar comprehensive work is projected for woody plants. On the fringes, a *Flora of Taiwan* has been largely completed and work is underway on a Hong Kong flora.

At the present time, Chinese floristic botany might be considered to be still in its fourth (post-1949) phase, but with the increased general emphasis on science and technology, the dominance of the national flora project, and renewed interaction with foreign botanists, the period from the mid-1970s should perhaps be considered as a fifth phase. It is foreseen as a period during which floristic documentation will improve rapidly, reaching perhaps the level of the West Indies, parts of Africa, Central America or the Soviet Union. A hindrance, however, will be the limited knowledge of Chinese among most foreign workers. Negotiations for an English version of *Flora reipublicae popularis sinicae* are, however, reportedly in progress. The present stock of general works is surveyed under **860–80** below.

Progress

For foreign involvement, see BRETSCHNEIDER, E., 1898. *History of European botanical discoveries in China*. 2 vols. St Petersburg; and COX, E. H. M., 1945. *Plant hunting in China*. London. (The former is renowned for its thoroughness.) The period from about 1915 to 1940 or so, when Chinese entered the profession on a large scale, is treated by LI, HUI-LIN, 1944. Botanical exploration in China during the last twenty-five years. *Proc. Linn. Soc. London*, **156**: 25–44. Sources for the post-1949 era include BARTHOLOMEW, B. *et al.*, 1979. Phytotaxonomy in the People's Republic of China. *Brittonia*, **31**: 1–25; CHING, REN-CHANG, 1979. [Twenty years of Chinese systematic botany.] *Acta Phytotax. Sin.* **17**: 1–6; GOULD, S. H. (ed.), 1961. *Sciences in Communist China* (American Association for the Advancement of Science, Publ. 68). Washington; and HU, SHIU-YING, 1975. The tour of a botanist in China. *Arnoldia*, **35**: 265–95. Useful papers by Chun and by Merrill (1931) on botanical progress appear in Symposium on the flora of China. In *Proceedings of the Fifth International Botanical Congress*. pp. 513–36.

Bibliographies. General bibliographies as for Division 8.

Supraregional bibliographies

LIU JU-CH'IANG, 1930. Important bibliography on the taxonomy of Chinese plants. *Bull. Peking Soc. Nat. Hist.* 4(3): 17–32.

MERRILL, E. D. and WALKER, E. H., 1938. *A bibliography of Eastern Asiatic botany.* xlii, 719 pp., maps. Jamaica Plain, Mass.: Arnold Arboretum of Harvard University. [Provides detailed coverage of floristic and geobotanical literature, with regional and subject indices, for China proper as well as Korea and Japan (Region 85), the 'Chinese Border' (Region 76), and central and eastern Siberia and the Soviet Far East (Regions 72, 73). 35,000 entries.]

WALKER, E. H., 1960. *A bibliography of Eastern Asiatic botany: Supplement 1.* xl, 552 pp. Washington: American Institute of Biological Sciences. [Additions to the foregoing work, with regional and subject indices.]

Indices. General indices as for Division 8.

Supraregional index

STEENIS, C. G. G. J. VAN and JACOBS, M. (eds.), 1948– . *Flora malesiana bulletin*, 1– . Leiden: Flora Malesiana Foundation (later Rijksherbarium). (Mimeographed.) [Issued nowadays annually, this contains a substantial bibliographic section also very thoroughly covering much East Asian literature. A fuller description is given under Superregion 91–93.]

860–80

Superregion in general

The publication of *Flora reipublicae popularis sinicae*, begun in 1959 and projected to comprise 80 volumes, has greatly accelerated since the mid-1970s. When and if it is completed, it will provide the first modern general account of the Chinese vascular flora. Forbes and Hemsley's *Index florae sinensis* (1886–1905; supplement 1911) has long been a standard reference but covers only part of the mainland – corresponding to Superregion 86–88 – and is now long out of date,

accounting for little more than a quarter of the known vascular flora (and omitting pteridophytes). There are also three abridged manuals, of which the most comprehensive by far, with some 8000 vascular plant species, is *Iconographia cormophytorum sinicorum* (1972–6), now complete in five volumes with two supplementary volumes projected. The other, older manuals are *Chung kuo chih wu t'u chien* (Chia and Chia, 1937; revised 1955), similar to the *Iconographia* but much more selective, and *Chung kuo chung tz'u chih wu fên lei hsüeh* (Chêng, 1954–8), a three-volume descriptive account with partial keys. For the woody flora, two descriptive works, neither very satisfactory, and an aging checklist are available (which may be superseded by a projected seven-volume general account), and in the unfinished *Flora illustrata plantarum primarum sinicarum* are useful introductory accounts to pteridophytes, grasses, and legumes. These are supported by two standard dictionaries and two sets of keys to families. Most works are in Chinese, except *Index florae sinensis* and Lee's *Forest botany of China* which are in English (although in all works botanical names and many references appear in roman script). An excellent general survey of the flora and vegetation for English-speaking readers is provided by WANG, CHI-WU, 1961. *The forests of China: with a survey of grassland and desert vegetation.* xiv, 313 pp., 78 text-figs., map (Maria Moors Cabot Foundation, Harvard Univ., Spec. Publ., 5). Petersham, Mass.

Apart from the general works accounted for here, which appear under four subheadings (Comprehensive works, Abridged general works, Special groups – woody plants, and Special groups – pteridophytes), a representative selection of botanical reports and related literature relating to the vast collections made in central and western China from the 1870s to the 1940s is given under a further subheading, Partial works. However, for detailed references to the immense literature, reference should be made to the standard bibliographies.

– Vascular flora (including Region 76): 8271 seed-plant species were recorded in *Index florae sinensis* (which did not cover Region 76). By 1930, when Chun reported to the Fifth International Botanical Congress, 18 000 had been recorded, and the latest estimate, from Bartholomew *et al.* in their 1979 trip report, suggests a figure of around 32 000 species.

Dictionaries

How Foon-chew [Hou K'uan-chao] (ed.), 1958. *Chung kuo chung tz'u chih wu ko shu tz'u tien.* i, 553 pp. Peking (Beijing): K'o hsüeh chu pan she (Academia Sinica Press). [This 'Dictionary of Chinese seed plants' includes all families and genera as well as some species, accompanied by Chinese transliterations of the names and brief notes for each taxon (with citations of pertinent monographs and revisions). Definitions of descriptive botanical terms are incorporated, and there are also lists of genera under their respective families.]

K'ung Ch'ing-lai (and 12 collaborators), 1918. *Chih wu hsüeh ta tz'u tien* [Botanical nomenclature: a complete dictionary of botanical terms]. 1726 pp., illus. Shanghai: Shang Wu Yin Shu Kuan (Commercial Press). (Fifth edn. 1923; reprinted 1933.) [Illustrated dictionary of vascular plants, with short descriptive text, some synonymy, and generalized indication of internal range; indices to botanical, common European, and Japanese nomenclatural equivalents. Botanical terms are also defined, so altogether the work is something of a Chinese 'Willis'. (Not seen by this reviewer; description and citation prepared in part from notes supplied by J. Needham, Cambridge.)]

Keys to families and genera

Hu Hsien-hsu *et al.*, 1954–5. Chung kuo chih wu k'o shu chien so piao [Claves familiarum generumque plantarum sinicarum]. *Acta Phytotax. Sin.* **2**: 173–586. [Analytical keys to the genera and families of Chinese vascular plants; systematic synopsis of families and genera; index. Based largely on Engler's *Syllabus der Pflanzenfamilien* and other standard works.]

Institute of Botany, Academica Sinica, 1979. *Claves familiarum generumque cormophytorum sinicorum.* 733 pp., illus. Peking: K'o hsüeh chu pan she (Academia Sinica Press). [Succeeds Hu *et al.*, 1954–5.]

Kêng Yi-li and Kêng Pai-chieh, 1958. *Chung kuo chung tz'u chih wu fên ko chien so piao.* [3rd edn.] ii, 108 pp. Peking. (First edn. 1948.) [This 'key to the families of phanerogams in China' also includes a synopsis of the families recorded. (Not seen by this reviewer; cited from Walker, 1960.)]

Comprensive works

Forbes, F. B. and Hemsley, W. B., (1886–) 1888–1905. *Index florae sinensis. An enumeration of all plants known from China proper, Formosa, Hainan, Corea, the Luchu Archipelago and the island of Hongkong, together with their distribution and synonymy.* 3 vols. 14 pls., map (*J. Linn. Soc., Bot.*, **23** (1886–8): 1–521, Pls. 1–14, map; **26** (1889–1902): 1–592; **36** (1903–05): 1–686). London. (Originally issued both as journal and as separate work, only the latter carrying the title *Index florae sinensis*; reprinted 1980, Königstein/Taunus, W. Germany: Koeltz.)

Systematic enumeration, in English, of seed plants recorded from the territories indicated in the title, with descriptions of new taxa, full synonymy, references and citations, fairly detailed indication of internal distribution, brief summary of extralimital range, and miscellaneous taxonomic and other notes. Volume 3 includes at its end some remarks on the history of botanical exploration in the area, a 'first supplement' (by M. Smith) for the period 1886–1904, and a complete index to botanical names. [The work was actually carried out by Hemsley at the Kew Herbarium with the assistance of a dozen collaborators and special financial grants; Forbes merely acted as a zealous promoter, although he himself had collected extensively in China in 1857–74 and 1877–82.] For the 'second supplement', see Dunn, S. T., 1911. *A supplementary list of Chinese flowering plants.* (*Ibid.*, **39**: 411–581.) London. [Alphabetically arranged list of species under their respective genera, with references.]

Institute of Botany, Academia Sinica, 1959– *Chung kuo chung tz'u chih wu t'u chih* [Flora reipublicae popularis sinicae]. In several volumes. Illus. Peking (Beijing): K'o hsüeh chu pan she (Academia Sinica Press).

This great work, the first comprehensive flora of all China, is a multivolume documentary descriptive flora of vascular plants; each family treatment includes keys to genera and species, full synonymy, references and citations, generalized indication of local and extralimital distribution, vernacular names as well as Chinese transliterations of scientific names, and notes on habitat, special features, and indices to all Chinese and Latin names. The style of the work reflects that of *Flora SSSR*, taken as a model at the time the Chinese flora project was launched in the 1950s, and the sequence of the 80 volumes planned follows the standard Englerian system as used in the Soviet flora. However, the issue of volumes (or parts thereof) has been made only as family treatments have become ready (and political circumstances have permitted). Some volumes are in two parts. Large families will occupy two (or more) parts, while in some cases one volume will contain two or more closely related families. Until 1974–5, progress on this work was comparatively slow, with only three parts published before 1966 when the Cultural Revolution caused a suspension (at least outwardly) of all activity. Since the fall of the 'Gang of Four' in 1976, publication has been greatly accelerated, with five parts alone appearing in 1977–78. A large number (up to 200) of botanists is engaged in one way or another in this project, organized in several cadres under the control of senior workers and guided by a ten-man editorial committee. At the outset of the work the leaders were Sung-shu Chien and Woon-yung Chun (both now deceased). Notable contributors of family treatments so far have included R. C. Ching, Y. Tsiang, and T. T. Yü. External reports on the progress of the work have been given by Stafleu (1974),[14] Hu (1975), and most thoroughly (including a detailed list of families with actual or probable collaborators) by Bartholomew (1979) who, in a letter to the writer (31 August 1979), noted that the work 'will probably be completed within the next ten years or so' although not by 1985 as targeted by its sponsors. At this writing (end of 1980) the number of volumes and half-volumes published was approaching 30, ranging from vol. 2 (Pteridophyta, 1) to vol. 68 (Scrophulariaceae, 2). The latest reviews are YÜ, TE-TSUN, 1979 (1980). Special report: status of the 'Flora of China'. *Syst. Bot.* **4**: 257–60, which covers the organization and progress of the project, and HEDGE, I. C. *et al.*, 1979. The 'Flora of China'. *Notes Roy. Bot. Gard. Edinburgh*, **37**: 467–468, in which defects such as a narrow species concept and too much of an adherence to Komarovian methodology and style are singled out for comment.

For a short time in the early 1950s, a parallel, privately supported Chinese flora project existed in the United States. A large card file was developed, but only one 'pilot' fascicle of the planned flora was ever published: ARNOLD ARBORETUM OF HARVARD UNIVERSITY, 1955. *Flora of China* (Family 153, Malvaceae, by Shiu-ying Hu). Illus. Cambridge, Mass. [Systematic treatment, with keys, brief descriptions, synonymy, references and citations, indication of internal distribution, and miscellaneous notes; index.] See also HU, SHIU-YING, 1965–9. *The Compositae of China*. 704 pp. Taipei. (Repr. from *Quart. J. Taiwan Mus.* **18**(1)–**22**(2) *passim*.)

Abridged general works

Under this subheading are listed *Iconographia cormophytorum sinicorum* and two students' manuals.

CHÊNG MIEN, 1954–8. *Chung kuo chung tz'u chih wu fên lei hsüeh*. 3 vols. Illus. [Shanghai]: Hsin Ya Shu Tien (New Asia Book Co.).

This 'Manual of Chinese seed plants' is a concise, illustrated systematic account (on the Englerian

system) designed for students, with analytical keys to families and genera and synoptic keys for species or groups thereof, limited synonymy, diagnostic descriptions, Chinese transliterations of scientific names, notes on distribution, habitat, special features, etc., and many small figures of representative species. Descriptions of genera and families are also given. Appendices include a synopsis of families and a glossary, and (presumably) indices to names in Chinese and Latin. Now in part supplanted by *Iconographia cormophytorum sinicorum* (see below) but forming a useful model for an 'amplified' generic flora suitable for developing countries in the tropics with large floras and limited scientific resources. [Not seen by this reviewer; description compiled from information and sample pages of vol. 1 sent by courtesy of Messrs Eugene Wu, Cambridge, Mass., and R. Frodin, now of Thetford Center, Vermont.]

CHIA TSU-CHANG [Kia, Tchou-tsang] and CHIA TSU-SHAN [Kia, Tchou-shan], 1955. *Chung kuo chih wu t'u chien*. Revised edn. 4, 4, 16, 1529, 62, 70 pp., 2602 text-figs. Peking: Chung-hua shu chü. (Reprinted 1958; 1st edn., 1937.)

This 'Atlas of Chinese plants' is an illustrated descriptive atlas-flora of a selected range of non-vascular and vascular plants, with analytical keys, limited synonymy, Chinese transliterations of scientific names, generalized indication of internal distribution, and notes on habitat, life-form, special features, properties, uses, etc.; indices to Latin and Chinese botanical names and some European vernacular names. Line drawings (6 × 4 cm) are provided for each species included. Now largely superseded by the more comprehensive *Iconographia cormophytorum sinicorum* (see below). [Not seen by this author; citation and description based on Walker, 1960 as well as on notes supplied by Dr J. Needham, Cambridge.]

INSTITUTE OF BOTANY, ACADEMIA SINICA, 1972–6. *Iconographia cormophytorum sinicorum*. 5 vols. 8374 figs. Peking (Beijing): K'o hsüeh chu pan she (Academia Sinica Press).

Semitechnical atlas-flora (in simplified characters) of a substantial selection of Chinese vascular plants and bryophytes (some 10000 species in all), with accompanying descriptive text; the latter includes for each species a botanical description, Chinese transliteration of the scientific name, and general notes on internal distribution, occurrence, habitat, properties, uses, special features, etc., but omits botanical synonymy and references. The figures, of almost uniformly excellent quality, contain habit and botanical details of the plants depicted. Volume 1 also includes an illustrated glossary and organography, while all volumes contain analytical keys to genera and families (and, in many cases, species) as well as complete indices (abridged general index to the work in vol. 5). The work, now complete, follows the Englerian system (in the vascular plants) and was compiled largely by more junior botanists in Peking and other centers (Bartholomew, 1979); a supplementary volume or volumes are, however, planned in the wake of the larger *Flora* project. [It evidently provided a significant and strongly practically oriented focus for taxonomic botany in China in the later years of the Cultural Revolution. Coverage seems to be more complete for northern and eastern provinces than for other regions of China. In format, the *Iconographia* resembles Coste's *Flore de France* (**651**), but with (usually) two species per page.]

Special groups – woody plants (including trees)

The principal modern manual (though not complete) used in China is that by Ch'en Yung (1957). The checklist of Chung (1924) was important in its day but is now long outdated. Lee's work of 1935 is useful for English-speaking users, but regrettably his supplement of 1973 is uncritical and is without illustrations.

CH'EN YUNG, 1957. *Chung kuo shu mu fēn lei hsüeh*. 2nd edn. 2, 2, 6, 14, 2, xxxxiv, 6, 106, 13, 1191, 4, 64, 67, 6, 18, 13 pp. (main text); 11, 60 pp. (appendix); 1159 text-fig. Shanghai: Science and Technology Press. (First edn., 1937, Shanghai: Agricultural Association of China.)

This 'Illustrated manual of Chinese trees and shrubs' comprises a briefly descriptive, extensively illustrated but largely compiled account of about 2100 selected native and introduced species, with keys to all taxa, limited synonymy, Chinese transliterations of scientific names, generalized indication of internal range, and notes on habitat, special features, uses, etc.; glossary, bibliography, summary of the work, and indices to all Latin, Chinese and English names. Introductory chapters include a list of sources, an illustrated organography, detailed synopses of families and genera, and a general key to families. The present work is essentially a reprint of the 1937 edition with subsequent additions and corrections covered in the appendix. [Now in need of extensive revision.]

CHUNG HSIN-HSUAN, 1924. *A catalogue of the trees and shrubs of China*. 271 pp. (Mem. Sci. Soc. China, 1(1)). Shanghai.

Compiled systematic checklist of known Chinese trees and shrubs (4840 species), with limited synonymy and indication of distribution by provinces; tabular summary and index to family and generic names. No literature citations are provided. The texts is mainly in English, with a summary in Chinese; the cut-off year for coverage is 1920. Not yet superseded, although by now quite outdated.

LEE, SHUN-CH'ING, 1935. *Forest botany of China*. xlvii, 991 pp., 272 text-figs. Shanghai: Commercial Press.

Illustrated descriptive treatment (in English) of the more common trees (and some shrubs) of China, with keys to families and genera, limited synonymy, generalized indication of local range and altitudinal zonation, and notes on habitat, special features, etc.; family summaries, glossary, and index to botanical names at end. An introductory section includes a synoptic list of orders and families. In general, this work is more useful for the northern and eastern parts of old China; coverage of other areas seems weak. For an unillustrated, uncritical and poorly documented supplement through 1963, comprising mainly additions and corrections (including some new keys) but accounting also for all species in the original work (thus covering in all 4073 species), see *idem*, 1973. *Forest botany of China: supplement*. xi, 477 pp. Taipei: Chinese Forestry Association. [The 10-year delay in publication is unexplained.]

Special groups – pteridophytes

For a recent overall review of this group in China, see CHING REN-CHANG, 1978. The Chinese fern families and genera: systematic arrangement and historical origin. *Acta Phytotax. Sin.* **16**(3): 1–19; **16**(4): 16–37.

FU SHU-HSIA, 1957. *Flora illustrata plantarum primarium sinicarum: Pteridophyta*. 6, ii, 280 pp., 346 text-figs. Peking (Beijing): K'o hsüeh chu pan she (Academia Sinica Press).

This work comprises an illustrated descriptive manual of more important or widespread pteridophytes in China, covering 437 species in 130 genera with limited synonymy, internal and extralimital distribution, habitat and other notes, and keys; index at end. (Other volumes in this series, not accounted for here, cover grasses and legumes.) Based on *idem*, 1954. [Genera of Chinese ferns and fern allies.] viii, 203 pp., 102 text-figs. Peking. [Not seen by this reviewer; reference from Koeltz Book List No. 205.]

Partial works

As noted under the main heading, there is here included a selection of the more important botanical reports on the many collections from (mainly) central and western China made by visiting explorers from 1870 to the 1930s. The background of these field trips, some of long duration, has been ably presented by Bretscheider in his *History of European botanical discoveries in China* and Cox in his *Plant hunting in China* (see 'Progress' under Superregion 86–88, p. 444; the latter is recommended as an introduction to the subject. Together, the reports provide a scattered but useful coverage of the rich flora of a part of China until recently somewhat sketchily covered by the various general works.

DIELS, L., 1900–1. Die Flora von Central-China. *Bot. Jahrb. Syst.* **29**: 169–659, figs. 1–5, pls. 2–5.

Report of collections made chiefly in western Hupeh, eastern Szechwan, and southern Shensi, accompanied by a phytogeographic summary. For additions, see *idem*, 1905. Beiträge zur Flora des Tsin Ling Shan und andere Zusätze zur Flora von Central-China. *Ibid.*, **36**, Beibl. 82: 1–138. For a Chinese version of the original paper, see *idem*, 1932–4. [The flora of Central China] (trans. Tong Koe-yang). *J. Coll. Sci. Sun Yat-sen Univ.* **3** (1932): 684–723; **4** (1932–3): 85–132, 221–81, 427–79, 619–71; **5** (1933–4): 267–316, 455–506; **6** (1934): 25–60, 277–316.

FRANCHET, A., (1883–)1884–8. *Plantae davidianae ex Sinarum imperio.* 2 vols. 44 pls. (Nouv. Arch. Mus. Hist. Nat. Paris, II, **5** (1883): 153–272, pls. 7–16; **6** (1883): 1–126, pls. 11–18; **7** (1884): 55–200, pls. 6–14; **8** (1885): 183–254, pls. 2–10; **10** (1888): 33–198, pls. 10–17.) Paris. (Reprinted separately, Paris: Masson; 2nd reprint (in 1 vol.), 1970, Lehre, Germany: Cramer.)

Enumeration (in two parts) of collections made by Père A. David, the renowned pioneer explorer of western China and the Tibetan Marches, from 1866 to 1874 in many parts of China and Mongolia (notably during his second trip of 1868–70 in western Szechwan and adjacent areas). As a pioneer general sketch of the Chinese flora, this work is of great importance as it gives one of the first outlines of the overall floristics and phytogeography of China and what were then the 'borderlands' in Central Asia.

FRANCHET, A., 1889–90. *Plantae delavayanae, sive enumeratio plantarum quas in provincia chinensi Yun-nan collegit J.-M. Delavay.* Fasc. 1–3. pp. 1–240, pls. 1–48. Paris: Klincksieck.

Systematic enumeration, with descriptions of new taxa, of the collections of Père J.-M. Delavay from northwest Yunnan. Not completed; extends from Ranunculaceae through Saxifragaceae (Candollean system).

HANDEL-MAZZETTI, H. (ed.), 1929–37. *Symbolae sinicae. Botanische Ergebnisse der Expedition der Akademie der Wissenschaften in Wien nach Südwest-China, 1914–18.* 7 vols. (in 11 parts). Illus. Vienna: Springer.

Systematic enumeration (with descriptions of new taxa) of non-vascular (vols. 1–5) and vascular plants based mainly on the collections of H. Handel-Mazzetti and C. Schneider made in central and western China between 1914 and 1918, largely in northern Yunnan, southwestern Szechwan, and the adjacent 'Tibetan Marches'. Treatments of many groups contributed by specialists.

LAUENER, L. A. *et al.*, 1961– . Catalogue of the names published by Hector Léveillé, I– . *Notes Roy. Bot. Gard. Edinburgh*, **23**: 573–96, etc.

Systematic evaluation of all taxa from eastern Asia published by H. Léveillé, including synonymy, reductions, and literature references and citations; in some cases indication of holotypes, paratypes, and syntypes is made. At this writing (late 1980), 13 parts have appeared, the latest (*ibid.*, **38**: 453–85 (1980)) covering the Loganiaceae through Verbenaceae and Phrymaceae. By now, the 'Polypetalae' and most of the 'Sympetalae' in the Bentham and Hooker system have been reviewed. The work represents a revision and amplification of the following: REHDER, A., 1929–37. Notes on the ligneous plants described by H. Léveillé from eastern China. [14 pts.] *J. Arnold Arbor.* **10–18**, *passim*.

MAXIMOWICZ, C. J., 1890–2. Plantae chinenses Potaninianae nec non Piaszekianae. *Trudy Imp. S.-Peterburgsk. Bot. Sada*, **11**: 1–112.

Systematic enumeration of higher plants from the collections of G. N. Potanin and others, mainly from the subregion of 'northwestern China' (Shensi and Kansu) and adjacent parts of Mongolia. [Not seen by this reviewer; cited from Merrill and Walker, 1938.]

PAX, F., 1922. Aufzählung der von Dr Limpricht in Ostasien gesammelten Pflanzen. In *Botanische Reisen in den Hochgebirgen Chinas und Ost-Tibets* (ed. W. Limpricht), pp. 298–515, illus. (Repert. Spec. Nov. Regni Veg., Beih. vol. 12). Berlin: Fedde.

Comprises a systematic enumeration, with descriptions of new taxa, of Limpricht's collections from western Szechwan and the adjacent 'Tibetan Marches' made in 1913–14. Treatments of several groups by specialists.

REHDER, A., 1929–37. See under LAUENER, L. A. *et al.* above.

REHDER, A. and WILSON, E. H., 1928–32. Enumeration of the ligneous plants collected by J. F. Rock on the Arnold Arboretum expedition to northwestern China and northeastern Tibet. *J. Arnold Arbor.* **9**: 4–27, 37–125, pls. 12–23; **13**: 385–409; and REHDER, A. and KOBUSKI, C. E., 1933. An enumeration of the herbaceous plants collected by J. F. Rock for the Arnold Arboretum. *Ibid.*, **14**: 1–52.

Systematic enumerations of J. F. Rock's vascular plants from northwestern Yunnan, western Szechwan,

Chinghai Province, and southern Kansu; includes descriptions of new taxa, citations of *exsiccatae*, and descriptive commentary.

ROYAL BOTANIC GARDEN, EDINBURGH. 1911–12. Plantae chinenses forrestianae. *Notes Roy. Bot. Gard. Edinburgh*, **5**: 65–148, 161–308.

Systematic enumeration, with descriptions of new taxa, of collections from Forrest's first Yunnan expedition, 1904–7. For numerically arranged catalogs of collections from this and later expeditions of Forrest's, see DIELS, L., 1912–3. Plantae chinenses forrestianae. *Ibid.*, **7**: 1–411 (first expedition, 1904–7, nos. 1–5099); also ROYAL BOTANIC GARDEN, EDINBURGH, 1925–30. Plantae chinenses forrestianae. *Ibid.*, **14**: 75–393 (fifth expedition, 1921–3, nos. 19334–23258); **17**: 1–406 (fourth expedition, 1917–20, nos. 13599–19333).

SARGENT, C. S. (ed.), 1911–17. *Plantae wilsonianae*. 3 vols. (in 9 parts) (Publ. Arnold Arbor., 4). Jamaica Plain, Mass.

An elaborately produced systematic enumeration of the woody plants collected by E. H. Wilson (and others) during the years 1907–09 and 1910, principally in western Hupeh, eastern and western Szechwan, and Kiangsi; includes several partial revisions of genera, descriptions of new taxa, extensive synonymy (with references and citations), indication of *exsiccatae* with localities, and keys. Treatment of many groups contributed by specialists.

SMITH, H. *et al.*, 1924–47. Plantae sinenses a Dre. H. Smith annis 1921–2 lectae [1924 et 1934]. *Acta Horti Gothob.* 1(1924): 1–187; 2(1926): 83–121, 143–184, 285–328; 3(1927): 1–10, 65–71, 151–5; 5(1930): 1–54; 6(1931): 67–78; 8(1933): 77–81, 127–46; 9(1934): 67–145, 167–83; 12(1938): 203–359; 13(1940): 37–235; 15(1944): 1–30, figs. 1–195, pls. 1–6; 17(1947): 113–64.

Enumeration, with descriptions of new taxa, of the author's collections from western and northwestern Szechwan and the adjacent 'Tibetan Marches' as well as from Shansi. Not completed; contributions not in systematic sequence. Treatments of most groups by specialists.

Region

86

North China (including north-east China)

As delimited here, North China includes the northeastern provinces of Heilungkiang, Kirin and Liaoning (formerly the North-East Province, and long known as Manchuria), the northern provinces of Shantung, Hopeh, Honan, and Shansi, and the northwestern provinces and territories of Shensi, Kansu (eastern part), and Ningsia Hui Autonomous Region. For convenience, the former territory of Jehol is here considered part of Hopeh. The western part of Kansu, beyond the Hwang Ho and Tao Ho, is floristically part of central Asia and therefore is treated as part of Region 76.

A considerable amount of exploration was accomplished in North China by foreign botanists in the nineteenth and early-twentieth centuries, with later extensions to Manchuria and Northwestern China. Additional extensive work was carried out during the biological surveys mounted after 1920, and in Manchuria collecting was assiduously pursued by the Japanese after 1931. Regional botanical centers have been established at Mukden (Shenyang) and Wukung (near Sian) for the northeast and northwest respectively, and in recent years floras have been compiled at both of them.

At present, there is no separate flora covering the whole region, which has in any case for botanical purposes been treated as three subregions. However, it is relatively well-covered in general floras of China, the most thoroughly of course in the new national *Flora*. The available separate accounts are largely subregional, and few provincial floras have been written at all. Northeastern China was long politically separate and at times even under foreign rule or domination; several floras are now available, of which two are in Chinese (one is still to be completed), one in Japanese, one (now elderly) in Russian, and one in English (the most recent, published in 1979). Northwestern China has two non-overlapping works: one on the Shensi-Kansu-Ningsia upland loess basin (Lo and Hsu, 1957) and one on the Tsinling Range (*Flora tsinlingensis*, 1974–) along the southern perimeter. Little, however, is available on the northern heartland; an enumeration of woody plants by Rehder (1923–6) was never completed.

Bibliographies. General and supraregional bibliographies as for Division 8 and Superregion 86–88.

Indices. General and supraregional indices as for Division 8 and Superregion 86–88.

860

Region in general

See also **860–80** (all general works). No flora covers the entire region as here defined. However, most of it is covered by partial works centering on one of three subregions: North-east China, North China, and North-west China. Each of these is here given a subheading under which the appropriate works are listed. North-east China corresponds almost entirely to the former Manchuria (now Heilungkiang, Kirin, and Liaoning); North China centers on Peking and includes the provinces of Hopei (with Jehol), Shantung, Honan, and Shansi; while North-west China here includes Shensi, southern Kansu, and Ninghsia Hui Autonomous Region.

Keys to families

LIU JU-CH'IANG, 1934. *Systematic botany of the flowering families in North China.* 2nd edn. xvi, 218 pp., 300 text-figs. Peking: Vetch (The French Bookstore). (First edn. 1931.) [Students' handbook, with keys to and descriptions of families as well as illustrations of representative species, vernacular names, taxonomic commentary, and notes on cultivation, uses, etc.; glossary and descriptive organography.]

I. North-east China (Manchuria)

This area is currently (1979) administered as three provinces of China: Heilungkiang (**861**), Kirin (**862**) and Liaoning (**863**). From the late-nineteenth century it has had a checkered history of political and administrative changes. The northern portion, approximately corresponding to Heilungkiang province, was at the beginning of the twentieth century a Russian concession related to the construction and operation of the 'Chinese Eastern Railway'. From 1931 to 1945 the whole area, with the addition of Jehol, was the Japanese puppet state of Manchukuo. After 1949 it became a single North-East Province; later the western hill zone (Great Khingan Mountains) was attached to Inner Mongolia and the remainder divided into three provinces. More recently, the western hill zone was again returned to the respective provinces.

Five different floras are available: one in English, one in Japanese (with parts also in English), one in Russian (with some Latin), and two in Chinese. Three are copiously illustrated, and only the English work lacks keys.

KITAGAWA, M., 1979. *Neo-lineamenta florae manshuricae.* viii, 716 pp., 12 pls., map. Vaduz, Liechtenstein: Cramer (Flora et vegetatio mundi, 6).

Systematic enumeration of vascular plants (2708 species), with synonymy, references, Japanese names in transliteration, and indication of local and extralimital distribution; list of principal literature and index to genera at end. The introductory section includes notes on the preparation of the work and accounts of the vegetation, floristics and phytogeography of the subregion. The few illustrations depict species of special interest. Only spontaneously occurring species are accounted for in this work. Succeeds *idem*, 1939. *Lineamenta florae manshuricae.* 488 pp., 12 pls., map (Rep. Inst. Sci. Res. Manchukuo, 3, App. 1(1)). Hsinking.

KOMAROV, V. L., 1901–7. *Flora Man'čžurii.* 3 vols. Illus. (Trudy Imp. S.-Peterburgsk. Bot. Sada, 20, 22, 25). St Petersburg. (Reprinted 1949–50, Moscow/Leningrad: AN SSSR Press, as vols. 3–5 of *V. L. Komarov: Opera Selecta.*)

Comprehensive descriptive flora of vascular plants, without keys (except to species in the larger genera), full synonymy, references and citations, indication of *exsiccatae* with localities, general summary of local range, taxonomic commentary, and extensive notes on habitat, life-form, special features, etc.; indices to all botanical names in each volume. An introductory section (vol. 1) incorporates a discussion of species concepts as well as accounts of floristic regions and botanical exploration. Text partly in Russian, partly in Latin.

LIU SHEN-O [Liou Tchen-ngo], 1955. *Tung-pei mu pen chih wu t'u shih* [Illustrated flora of woody plants of the North-East Province]. [ii], ii, 568 pp., 26 text-figs., 175 pls., 2 maps. Mukden: Institutum Silviculturae et Soli, Academia Sinica.

LIU SHEN-O, 1958– . *Tung-pei ts'ao pen chih wu shih* [Flora plantarum herbacearum Chinae borealiorientalis]. Fasc. 1– . Illus. Peking: K'o hsüeh chu pan she (Academia Sinica Press).

These complementary works comprise extensively illustrated descriptive accounts respectively of the woody and herbaceous vascular plants of the Northeastern subregion, with keys to all taxa, synonymy, references and citations (more copious in the latter work), indication of local and extralimital range, Chinese transliterations of scientific names, critical remarks, and notes on habitat, special features, etc.; Latin diagnoses of new taxa in appendices (in the latter work); indices to all names. Lists of references are also given. In the introductory section of the former work appear an account of general features of the flora, a floristic analysis, and an illustrated glossary. Publication of the latter work was suspended between 1959 and 1975; at present, parts 1–2 (1958–9) and 3, 5–6, 9 and 11 (1975–77) are available, and thus it is more than half completed. [In part not seen by this reviewer; description prepared with the aid of material kindly supplied by Eugene Wu, Cambridge, Mass.]

LIU SHEN-O, 1959. *Claves plantarum Chinae boreali-orientalis*. vii, 655 pp., 224 pls. Peking (Beijing): K'o hsüeh chu pan she (Academia Sinica Press).

Illustrated manual-key (in Chinese) to vascular plants, with essential synonymy and Chinese transliterations of scientific names; list of references and complete index. Separate descriptions, as well as a synopsis, of families are also provided.

NODA, M., 1971. *Flora of the North-East Province (Manchuria) of China*. 10, 17, 3, 1613 pp., 237 pls., text-figs., portrait, maps. Tokyo: Kazama Bookshop.

Illustrated, briefly descriptive flora of non-vascular (chiefly algae) and vascular plants, with keys to genera, species, and infraspecific taxa, synonymy, references and citations, Japanese transliterations of scientific names, local names, generalized indication of local and extralimital distribution, and miscellaneous notes; complete indices. Text in Japanese. The introductory section (in Japanese and English) includes accounts of Manchurian floristics and vegetation formations as well as botanical exploration and research, with a separate list of references. The limits of this work are those of Manchuria (Manchukuo) before World War II; the author was an 'old Manchurian hand'.

II. North China

This area is traditional 'north China', from Peking to Shantung and the borders of Shensi. However, the work by Rehder on the ligneous flora also covers much of **III. North-west China** (see below).

LIU SHEN-O [Liou Tchen-ngo] (ed.), 1931–6. *Flore illustrée du nord de la Chine: Hopei (Chih-li) et ses provinces voisines*. Fasc. 1–4. plates. Peking: Académie Nationale de Peiping.

Large-scale atlas of flowering plants, with descriptive text (in Chinese and French) including synonymy, general indication of local range, taxonomic commentary, and keys to genera and species for each family. Not completed; covers only four families.

REHDER, A., 1923–6. Enumeration of the ligneous plants of northern China, I–III. *J. Arnold Arbor*. 4: 117–92; 5: 137–224; 7: 151–227, 1 pl.

Systematic enumeration of the known woody plants of North China, with full synonymy (including references and citations), notation of *exsiccatae* with localities, general indication of local and extralimital distribution, taxonomic commentary, and miscellaneous other notes; no separate index. Not completed; covers families from Ginkgoaceae through Sapindaceae (Englerian system). The work, whose limits include North-West China (see below) as well as North China, was based primarily on collections available at the Arnold Arboretum.

III. North-west China

This area encompasses much of the upland 'loess region' of Kansu and Shensi as well as the large Tsinling (Qinling) Range in the southern part. The two main works, *Flora tsinlingensis* (1974–) and *Shen Kan Ning p'ang ti chih wu shih* (1957) are somewhat complementary, but together do not cover the whole area.

LO TIEN-YU and HSU WEI-YING, 1957. *Shen Kan Ning p'ang ti chih wu shih*. 2, 274 pp., 266 illus. (incl. maps). Peking (Beijing): K'o hsüeh chu pan she (Academia Sinica Press).

This account of the vascular plants of the Shensi-Kansu-Ningsia basin comprises a briefly descriptive manual-flora, with keys to all genera and species, essential synonymy, indication of local range, and notes on special features, uses, etc.; indices to all names at end. Includes Chinese transliterations of botanical names and photographic illustrations of pressed specimens. An introductory section includes chapters on geography, climate, and vegetation (with map), notes on particular woody plants, a synoptic table of contents (pp. 26–49), and a key to families (pp. 50–7). [The Shensi-Kansu-Ningsia basin is a loess upland north of Sian and west of the Hwang Ho which encompasses central Shensi and parts of Kansu and

Ningsia and is bounded on the west by the Liu-p'an Shan and on the north by the Pai-yü Shan. Within this area is Yenan, the World War II stronghold of Mao Tse-tung and his army.]

NORTHWESTERN INSTITUTE OF BOTANY, ACADEMIA SINICA, 1974– . *Flora tsinlingensis.* Vols. 1, parts 1–2, 2. Illus. Peking (Beijing): K'o hsüeh chu pan she (Academia Sinica Press).

Illustrated descriptive manual-flora of vascular plants, with analytical keys to all taxa, extensive synonymy, references and citations, vernacular names, Chinese transliterations of scientific names, indication of local range and altitudinal zonation as well as extralimital distribution, and notes on special features, properties, uses, etc. An appendix (in each part) includes Latin diagnoses of new taxa, and complete indices to Chinese and Latin plant names are also provided. At this writing (1979) parts 1 and 2 of volume 1 as well as volume 2 have been published, covering (in vol. 1) gymnosperms, monocotyledons and dicotyledons (Saururaceae through Rosaceae on the Englerian system) and (in vol. 2) pteridophytes.

861

Heilungkiang (Heilongjiang)

See 860/I (all works). No separate provincial floras are available.

862

Kirin (Jilin)

See 860/I (all works). No separate provincial floras are available.

863

Liaoning

See 860/I (all works). No separate provincial floras are available.

864

Hopeh (Hebei, Hopei, Chih-li)

This province, within which is the capital district of Peking (Beijing), was formerly known as Chih-li. As now delimited, it also includes the greater part of the formerly separate Jehol. The only general works on the province proper remain those by Léveillé (1917a, b); these, like his other works, must be used together with the reviews by Lauener and by Rehder (see 860–80 under Partial works). The more common trees are additionally covered in the following: CHOU HANG-FAN, 1934. *Ho-pei hsi chien shu mu t'u shuo* [The familiar trees of Hopei]. xii, 370 pp., 43 text-figs. (Handb. Peking Soc. Nat. Hist., 4). Peking. (Parallel editions in Chinese and English.)

LÉVEILLÉ, H., L'abbé, 1917a. Catalogue des plantes de Pékin et du Tché-Li. *Bull. Acad. Int. Géogr. Bot.* **27**: 70–87.

Unannotated list of vascular plants, with families alphabetically arranged; phytogeographic summary at end.

LÉVEILLÉ, H., L'abbé, 1917b. *Flore de Pékin et du Tché-Li.* 5th edn. 122 pp. [Le Mans]: The author. (Limited, holographed edition from author's MS.)

Similar in format and content to the foregoing work, but with the addition of keys to families. For additions, as well as keys to genera, see *idem*, 1917c. *Flore de Pékin et de Chang Hai et des provinces du Tché-Li et du Kiang-Sou.* 2nd edn. 112 pp. [Le Mans]: The author. (Limited, holographed edition from author's MS.)

Partial work: Jehol districts
This former province was after 1911 first carved out from the northern part of Hopeh and parts of adjacent Manchuria and Inner Mongolia. Between 1931 and 1945 it was under Japanese control and formed part of Manchukuo. After 1949 it was broken up once more, the greater portion being united with Hopeh.

NAKAI, T., HONDA, M., SATAKE, Y., and KITAGAWA, M., 1936. *Index florae jeholensis.* [ii], 108 pp., 3 pls. (Report of the first scientific expedition to Manchukuo, sect. IV, 4). Tokyo.

Systematic list of vascular plants (924 species), with Japanese standard names; localities, with citations of *exsiccatae* and literature records; descriptions of new or little-known taxa; no index. A brief introduction (in English) precedes the main text.

865

Shantung (Shandong)

No separate floras or enumerations for the province as a whole are available. Of some use, however, is a florula for the Tsingtao (Ch'ingdao) district, before World War I the German 'concession' of Kiautschou.

Local work: Tsingtao district

LOESENER, T., 1919. Prodromus florae tsingtauensis. Die Pflanzenwelt des Kiautschou-Gebietes. *Beih. Bot. Centralbl.* 37(2): 1–206, pls. 1–10.

Systematic enumeration of all known vascular and non-vascular plants, with descriptions of new taxa, synonymy, citations of *exsiccatae* with localities, etc.; no keys. Certain family treatments contributed by specialists. [Tsingtao is situated on the southern side of the Shantung peninsula.]

866

Honan (Henan)

See 860/II (all works). No separate provincial floras or enumerations are available. The southern part is also covered by Steward's manual-flora for the lower Yangtze Basin (see 870).

867

Shansi (Shanxi)

See 860/II (all works). No separate provincial floras or enumerations are available.

868

Shensi (Shaanxi)

See 860/III (all works). There is reportedly a general flora of the province by Tsi-an Peh (1959), but no details have been available as of this writing (1979).

869

Ningsia Hui (Ningxia Hui) AR and southern Kansu (Gansu)

See 860/III North-west China (all works). The area as delimited here comprises the eastern and southern parts of Kansu (i.e., east of Lanchow) and Ningsia Hui Autonomous Region (the latter separated from Kansu in 1949). The drier western parts of Kansu, reaching towards Sinkiang, comprise 765; their flora has more marked central Asiatic affinities. The more mountainous parts of southern Kansu are also partially covered in some of the botanical reports listed under 860–80 (Partial works), especially those by Maximowicz and by Rehder and Wilson. For Ningsia, reference should also be made to *Flora intramongolica* (see 763) and to Norlindh's *Flora of the Mongolian steppe and desert areas* (see 705). [Modern Ningsia does not correspond in limits to the former Inner Mongolian province.]

Region

87

Central China

This region comprises the provinces of Kiangsu, Chekiang, Anhwei, Kiangsi, Hunan, Hupeh, and Szechwan. The so-called 'Tibetan Marches' are also designated separately, although this area, at one time a separate province of Sikang, is now divided between Tibet (Sitsang) and Szechwan.

Except for the treaty ports and their hinterland in the east, botanical penetration of the region began only after 1860. Much of the outside interest was directed towards the mountainous western subregion, then hardly known. A large number of the collections were described in imposing reports, and in the early-twentieth century much material for gardens was also obtained. Later work was directed at general biological survey as Chinese botanists became more numerous, with comprehensive floristic knowledge and flora writing among the aims. A number of regional botanical centers now exist, both in the eastern (at

Nanking, Hangchow, and Wuhan) and western (at Chengtu) subregions.

The eastern subregion has for the most part been treated in *Vascular plants of the lower Yangtze* (Steward, 1958), a work based on long field experience by the author and his students. This work, in English, is now being joined by a two-volume flora of Kiangsu, *Chiang-su chih wu chih* (1977–) and a similar flora for Chekiang is projected. Further to the west, a four-part *Flora hupehensis* commenced publication in 1978. Information on the plants of the western subregion remains very scattered; the only provincial flora is an uncritical compilation by Léveillé (1918) on Szechwan, based mainly on the French missionaries' collections. However, a small portion falls within Hupeh, with its current flora project. The southern fringe, in Kiangsi and Hunan, is also inadequately documented, and with no significant local centers separate floras are some time away.

Bibliographies. General and supraregional bibliographies as for Division 8 and Superregion 86–88.

Indices. General and supraregional indices as for Division 8 and Superregion 86–88.

870

Region in general

See also **860–80** (all general works). For convenience, two subregions can be recognized: east central China, from I-ch'ang on the Yangtze River eastwards, and west central China, from mountainous western Hupeh and Hunan through Szechwan to the Tibetan Marches. The eastern subregion is largely covered by Steward's manual of 1958, while coverage of the western subregion is very scattered although fairly abundant. For the latter, a selection of the more important botanical reports, some of which also extend to Hunan and Kiangsi, is given under **860–80** (Partial works).

I. East central China

STEWARD, A. N., 1958. *Manual of vascular plants of the lower Yangtze Valley, China.* vii, [6], 621 pp., 510 text-figs., 2 maps. Corvallis: Oregon State College.

Detailed illustrated descriptive flora, with keys to all taxa; synonymy, with references; Chinese and English vernacular names; fairly detailed indication of local distribution and summary of extralimital range; miscellaneous notes on habitat, special features, etc.; glossary (in English, with Chinese equivalents); index to botanical and English vernacular names. The introductory section contains remarks on physical features, geography, and vegetation together with a list of references. The geographical limits of this work include Kiangsu, Chekiang, Anhwei, the northern parts of Hunan and Kiangsi, eastern Hupeh, and southern Honan.

II. West central China

See **860–80** (Partial works) and provincial headings.

871

Kiangsu (Jiangsu)

See also **870** (STEWARD). With regard to the works by Léveillé described below, see the nomenclatural reviews of LAUENER and REHDER (under **860–80**, Partial works). His obsolete provincial floras are currently in train of replacement by a new provincial flora from the Kiangsu Institute of Botany.

KIANGSU INSTITUTE OF BOTANY, 1977. *Chiang-su chih wu chih* [Flora of Kiangsu]. Vol. 1. [iii], 502 pp., 742 figs. (incl. 12 color pls.), 23 pls. (in glossary). Nanking.

Illustrated descriptive flora of vascular plants, with keys to all taxa, very limited synonymy, indication of local distribution and altitudinal range, and notes on habitat, occurrence, uses and properties, etc.; Chinese nomenclatural equivalents; illustrated organography-glossary, summary of new taxa, and general indices at end. In the introductory section are accounts of physical features, climate, vegetation, land use, etc. (pp. 1–6). For references, see p. 464. [All published to date (late 1980); covers all groups from the pteridophytes through the end of the monocotyledons (Englerian system). It is likely that four volumes in all will complete the work.]

LÉVEILLÉ, H., L'abbé, 1916. Catalogus plantarum provinciae chinensis Kiang-Sou hucusque cognitarum. *Mem. Real Acad. Ci. Barcelona*, III, **12**: 543–65. (Reprinted separately, 25 pp.)

Unannotated list of vascular plants, with descriptions of some new taxa; families alphabetically arranged. For keys to families and genera, see *idem*, 1917. *Flores de Chang-Hai et du Kiang-Sou.* 5th edn. 132 pp. [Le Mans, France]: The author. (Lithographed MS.) Further additions appear in the author's *Flore de Pékin et de Chang Hai* (864).

Local work: Shanghai district

Sui Bin-shen, 1959. *Enumeratio plantarum civitatis Shanghai.* vi, 138 pp., 80 text-figs. Shanghai.

Briefly annotated systematic list, with limited synonymy and symbolic indication of abundance and local range; illustrations of representative species; indices. An introductory section gives a summary account of the flora (1450 species).

872

Chekiang (Zhejiang)

See also **870** (Steward). No successor to the incomplete provincial flora of Cheng and Chien has yet materialized.

Cheng Wan-chun and Chien Sung-shu, 1933–6. An enumeration of vascular plants from Chekiang. Parts 1–4. *Contr. Biol. Lab. Chin. Assoc. Advancem. Sci.*, Sect. Bot., 8(1933): 298–306; 9(1933–4): 58–91, 240–304, pls. 4–6, 23–28; 10(1936): 93–155, pls. 13–18.

Systematic enumeration of seed plants, with descriptions of new taxa, synonymy, references and citations, notation of *exsiccatae*, with indication of localities, taxonomic commentary and miscellaneous notes. Not completed; includes families from Ginkgoaceae through Rosaceae (Englerian system).

Special groups – woody plants

Lin Kang, 1936. Enumeration of woody plants in Chekiang Province, China. *J. Forest.* (*Tsinan*), 6: 1–32.

Systematic checklist. [Not seen by this reviewer; cited from Walker, 1960.]

873

Anhwei (Anhui)

See also **870** (Steward). Attention should also be called to one modern florula: Chen Pang-chieh *et al.*, 1965. *Observationes ad floram Hwangshanicum.* [iv], 2,

335 pp., text-fig., maps. Shanghai. [Summary of botanical results of field studies in the Huang Shan, a mountain chain in the southern part of the province; includes systematic checklist.] The only general provincial work is limited to the woody plants.

Rehder, A. and Wilson, E. H., 1927. An enumeration of the ligneous plants of Anhwei. *J. Arnold Arbor.* 8: 87–129, 150–99, 238–40.

Systematic enumeration of woody plants, based mainly on collections made 1922–5, with relevant synonymy, references, and citations; indication of *exsiccatae*, with localities; some taxonomic commentary.

874

Kiangsi (Jiangxi)

The northern part of this province is covered by Steward's manual for the lower Yangtze basin (870). For the remainder, only scattered coverage is available; the most extensive references can be found in the botanical reports on the collections of Handel-Mazzetti and of Wilson (the latter by Sargent); for these, see **860–80** (Partial works).

875

Hunan

For the northeastern part, see **870** (Steward). Apart from the obscure catalog of Chou described below, little save scattered papers is available for this province to date.

Chou Hang-fan, 1924. Hu-nan chih-wu t'u chih. *Hu-nan chiao-yü tsa-chih* [Special Number Educ. Misc. Hunan], 4(7): 1–2, 1–77, pls. 1–4.

This checklist of the plants of Hunan comprises a systematic enumeration of 737 species; text in Chinese with Latin botanical names. Incorporates many references to traditional Chinese botanical literature, but now very incomplete. [Not seen by this reviewer; data from Merrill and Walker, 1938.]

876

Hupeh (Hubei)

Coverage of the low eastern part of this province is provided by Steward's flora of the lower Yangtze Basin (870). Until recently, coverage of the hilly western part has been available only in very scattered publications, with the most useful being the botanical reports of SARGENT (on Wilson's collections) and DIELS; for these, see 860–80 (Partial works). However, 1978 saw the commencement of a new provincial flora, *Flora hupehensis*, a work of the Hupeh Institute of Botany, Wuhan.

[INSTITUTE OF BOTANY, PROVINCE OF HUPEH], 1978–80. *Flora hupehensis*. Parts 1–2. Illus. Wuhan.

Copiously illustrated descriptive manual-flora of seed plants, with keys to all taxa, synonymy, Chinese transcriptions of botanical names, indication of local range, and notes on habitat, status, ecological preferences, etc.; indices to botanical names in Chinese and Latin. All species treated are illustrated by small figures in boxes. Part 1 covers Gymnospermae and Sauruaceae through Lauraceae on the Wettstein (?) system; part 2, Papaveraceae through Sabiaceae. Four parts in all are projected.

877

Szechwan (Sichuan)

This large and diverse province, which includes the Chengtu Plain, possesses a relatively copious, albeit very scattered, literature, but lacks any general flora apart from the obsolete compilation of Léveillé (1918). Much additional and more accurate information is to be found in the general botanical reports of DIELS, SARGENT, HANDEL-MAZZETTI, FRANCHET and others; these and the reviews of Léveillé's work by LAUENER and REHDER are given under 860–80 (Partial works).

LÉVEILLÉ, H., L'abbé, 1918. *Catalogue illustré et alphabétique des plantes du Seu-tchouen*. [2], 221 pp., 66 pls. (some in color). Le Mans: The author. (Limited, holographed edition of 10 copies from author's MS.)

Systematic enumeration of vascular plants, with limited synonymy and references; no index. A tabular summary of the flora and some phytogeographical remarks are appended. Now very incomplete and moreover as uncritical as his other floras (see 881, Yunnan; 882, Kweichow).

878

'Tibetan Marches'

This mountainous but horticulturally important source area, formerly consisting of petty Tibetan states and at one time constituting a separate province of Sikang or Changtu (Chamdo), is now divided between Tibet (Sitsang) and Szechwan, with the present boundary falling along the upper Yangtze. However, its biogeographical links are with monsoon Asia, and as defined here it extends from the Ta-hsüeh Shan (in which is situated Minya Konka) west to approximately 96 °E. or the former western boundary, with Tsinghai (Chinghai) to the north and Yunnan, Burma and northeastern India to the south (here bounded by the so-called McMahon line). Much early collecting in the Marches was accomplished by French missionary-priests beginning with Père A. David, and in the present century they were joined by many other botanists from Europe and the United States. From this work there has grown up a considerable but scattered literature; of this, the general reports of FRANCHET (*Plantae davidianae*), HANDEL-MAZZETTI, PAX, REHDER and WILSON, and SMITH *et al.* (see 860–80 under Partial works) are perhaps the most useful. [See also 768.]

Region
88

South China

South China, as here circumscribed, includes the provinces of Yünnan, Kweichow, Kwangtung, and Fukien, the Kwangsi Autonomous Region, the territories of Hong Kong and Macao (here included with Kwangtung), the large islands of Taiwan and Hainan, and the low island groups of the Paracels (Hsi-sha) and Spratlys (Nan-sha) in the South China Sea.

Extensive botanical exploration in the modern sense began only after 1860. Foreign collectors had previously been restricted closely to Guangzhou (Canton), with the addition after 1842 of Hong Kong and the environs of the additional treaty ports made available. However, with the exception of parts of Yunnan with potential for good garden plants, collecting remained relatively localized until well after 1900, although some individual contributions were considerable. Diseases (including malaria), wild animals, aborigines and lamas, and travel difficulties all presented problems, as in west-central China at the same period. Nevertheless, much South China material is recorded in *Index florae sinensis*, and some pioneer provincial floras were written, certain of them yet to be superseded. These latter include *Flora hong-kongiensis* (Bentham, 1861), *Flora of Kwangtung and Hong Kong* (Dunn and Tutcher, 1912), *Flore de Kouy-Tchéou* and *Catalogue des plantes de Yun-nan* (Léveillé, 1914–15, 1915–17), and a number of works on Taiwan culminating in *Icones plantarum formosanarum* (Hayata, 1911–21; supplement, 1925–32). After 1900 a goodly number of foreign collectors worked in northwestern Yunnan.

With the establishment of Chinese biological surveys after 1920 came systematic collecting through most of South China and Hainan, and the warm-temperate, subtropical and tropical flora became much better known. Botanical centers became established at Guangzhou (now the second most important in China), Guilin, and Kunming as well as in Taipei, over and above the small herbarium in Hong Kong. Few synthetic floristic works appeared, however, apart from checklists and a never-completed *Flora of Fukien* (Metcalf, 1942), until well after the revolution of 1949. Since 1956, floras have been published for the environs of Guangzhou (1956, 1957), Hainan (*Flora hainanica*, 1964–77), Yunnan (*Flora yunnanica*, 1977–) and Taiwan (*Flora of Taiwan*, 1975–8), and a flora of Kwangtung is projected. Preparation of a new florula for Hong Kong is also in progress at this writing (1979). Floristic accounts also exist for the Paracels and Spratlys. No modern floras are, however, to hand for the three remaining large units. Two woody floras are available for Taiwan (Liu, 1960–1; Li, 1963).

Due to the larger size of the flora as well as other factors, in South China the province has been the working unit for floristic studies and flora writing (with Hainan obtaining separate recognition), in contrast to North China where the few separate works are basically subregional. Consequently, no overall works on South China have come into being independently from the nation-wide floras, enumerations and other accounts.

Bibliographies. General and supraregional bibliographies as for Division 8 and Superregion 86–88.

Indices. General and supraregional indices as for Division 8 and Superregion 86–88.

881

Yunnan

The compiled list of Léveillé is now wholly obsolete and moreover uncritical; with his other works it is being evaluated in a series of papers by LAUENER and others (see **860–80**). The large collections made by European travellers and residents from the 1870s until World War II have been reported upon in widely scattered reports and papers, usually not specifically related to the province; the most significant are also accounted for at **860–80**. From the 1920s, a great deal of systematic field work was carried out by Chinese botanists in Yunnan; over the years, their collections have been covered in revisionary studies in *Acta Phytotaxonomica Sinica* and elsewhere. The first volume of a modern provincial flora appeared in 1977, incorporating as far as possible all previous botanical work, with a second volume in 1979.

[KUNMING INSTITUTE OF BOTANY, ACADEMIA SINICA], 1977–9. *Flora yunnanica* (ed. Wu Cheng-i). Vols. 1–2. Illus. Peking (Beijing): K'o hsüeh chu pan she (Academia Sinica Press).

Illustrated descriptive flora, presented as a series of provincial family revisions; each treatment includes keys to all taxa, synonymy, references and abbreviated citations, Chinese transliterations and local names, generalized indication of distribution, sometimes lengthy critical commentary, and notes on habitat, occurrence, etc.; tables of uses and complete indices at end. The first volume contains treatments of 28 families, in no predetermined sequence; volume 2 covers 22 families, again at random. Further volumes will appear as family treatments are ready, in the same manner as *A revised handbook to the flora of Ceylon* (**829**) or *Conspectus florae asiae mediae* (**750**). Volume 2 (1979) was the latest available at the time of writing (late 1980).

LÉVEILLÉ, H., L'abbé, 1915–17. *Catalogue des plantes du Yun-Nan avec renvoi aux diagnoses originales, observations et descriptions d'espèces nouvelles.* 299 pp.,

69 text-figs. Le Mans: The author. (Limited, holographed edition from author's MS.)

Systematic list of vascular plants reported from Yunnan, with descriptions of new taxa as well as references; no keys and no index. E. H. M. Cox considered this work to have been carelessly prepared. To a large extent, it accounts for collections in the province by French missionary-priests.

Partial work: Southern districts

WU CHENG-I and LI HSI-WEN (eds.), 1965. *Yün-nan jo ti ya jo tai chih wu ch'ü hsi yen chiu pao kao.* Fasc. 1. 146 pp., 38 pls. Peking: K'o hsüeh chu pan she (Academia Sinica Press).

Comprises a series of floristic contributions on the flora of tropical and subtropical southern Yunnan, with descriptions of new taxa, indication of new records, and taxonomic commentary. It represents a continuation of three papers by the senior author published in volumes 6 and 7 of *Acta Phytotaxonomica Sinica* (1959–60). No more parts have been published.

882

Kweichow (Guizhou)

No other flora is available as yet as a successor to the obsolete and uncritical enumeration by Léveillé (1914–15), originally published in an edition of only 20 copies. It is being evaluated, along with his other works, in a series of papers by LAUENER and others (see **860–80** under Partial works).

LÉVEILLÉ, H., L'abbé, 1914–15. *Flore de Kouy-Tchéou.* 532 pp., portrait. Le Mans: The author. (Limited, holographed edition from author's MS.)

Systematic enumeration of vascular plants, with keys to all taxa and brief descriptions of novelties; citations of *exsiccatae*, with localities; index to genera. Based largely on the collections of resident French missionary-priests active in the province in the late-nineteenth and early-twentieth centuries.

883

Kwangsi (Guangxi)

Since 1949 this province has been administered as an 'Autonomous Region'.

WANG CHÊN-JU [Wang Yen-chieh] *et al.*, 1940–2. [An enumeration of seed plants collected in Kwangsi]. Parts 1–9. *Kwangsi Agric.* **1**(1940): 68–77, 403–15; **2**(1941): 134–41, 223–9, 285–94, 371–84, 468–72; **3**(1942): 57–60, 121–4.

Systematic enumeration of seed plants, with synonymy, botanical names with Chinese equivalents, citations of *exsiccatae* with localities, and field data; no taxonomic commentary or indices. In part 1, the catalogue is preceded by a brief introduction. Evidently not completed; covers the Gymnospermae and families from Ranunculaceae through Rhizophoraceae (first Hutchinson system) in parts 1–8 (part 9 not seen by this reviewer). A revised (and presumably completed) version of this work, entitled *Catalogue of the plants of Kwangsi*, was apparently published in 1955 but no details have been available to this reviewer.

884

Kwangtung (Guangdong), Hong Kong, and Macao

In addition to Dunn and Tutcher's now-incomplete general flora (also still the most recent for Hong Kong), separate florulas have also been published for the Canton (Guangzhou) district and Hong Kong. For the minute Portuguese territory of Macao, see SÁ NOGUEIRA, A. C. DE, 1933. *Catálogo descritivó de 380 especies botânicas da Colóni de Macau.* 138 pp. Macao.

DUNN, S. T. and TUTCHER, W., 1912. *Flora of Kwangtung and Hong Kong.* 370 pp., map (Bull. Misc. Inform. (Kew), Add. Ser. 10). London.

Concise manual-key to vascular plants, with diagnoses, synonymy and references, fairly detailed general indication of local range, and notes on habitat, phenology, special features, etc.; gazetteer and index to family and generic names. The introductory section includes remarks on physical features, characteristics of the flora, history of botanical investigation, and desiderata. Does not include records for Hainan.

Local works: Canton (Guangzhou) district

HOW FOON-CHEW [Hou K'wan-chao] (ed.), 1956. *Kuangchou chih wu chih.* 8, 953 pp., 415 text-figs., 4 pls. Peking: K'o hsüeh chu pan she (Academia Sinica Press). (Reprinted 1959; unofficially reprinted in Hong Kong or Taiwan, n.d.)

Descriptive manual-flora of vascular plants of Canton and its environs, with keys, synonymy, Chinese transliterations, local names, local range (with some localities), taxonomic remarks, and notes on habitat, etc.; complete index. The work also includes chapters on the basics of taxonomy, collecting, etc. for students, lexica, and remarks on local floristics and vegetation. For a condensed version for field use, see How FOON-CHEW *et al.*, 1957. *Kuangchou chih wu chien so piao.* [i], 226 pp. Kuangchou: Academia Sinica Press. [Field manual-key to vascular plants in a pocket-sized format.]

Local works: Hong Kong and New Territories

Apart from the manual of Dunn and Tutcher, two works are available specifically for this British colony (although that by Bentham does not cover the New Territories). A number of attractive recent semi-popular illustrated works on different aspects of the local flora have also appeared in recent years, but their scope does not permit inclusion in this Guide. Work is in progress on a new descriptive flora (S. Thrower, personal communication).

BENTHAM, G., 1861. *Flora hong-kongiensis.* 20, li, 482 pp., map. London: Reeve.

Descriptive flora of vascular plants; with keys, limited synonymy, citation of *exsiccatae* with localities, taxonomic commentary, and miscellaneous other notes; glossary and index. Introductory chapters include remarks on botanical exploration, floristics and phytogeography as well as 'Outlines of Botany'. Forms part of the 'Kew Series' of British colonial floras initiated by W. J. Hooker. For additions, see HANCE, H. F., 1872. *Supplement to the* 'Flora hong-kongiensis'. 59 pp. London.

HONG KONG HERBARIUM, 1974 (1975). *Check list of Hong Kong plants.* 4th edn. x, 115 pp. (Hong Kong Agriculture and Fisheries Dept., Bull., 1). Hong Kong: Government Printer. (First edn. 1962; 3rd edn., 1966, mimeographed.)

Unannotated systematic checklist of vascular plants with Chinese and English vernacular names and limited synonymy; indices. A brief introductory section (bilingual) includes a list of references.

885

Fukien (Fujian)

The provincial flora begun by Metcalf in 1942 has never been continued.

METCALF, F. P., 1942. *Flora of Fukien and floristic notes on southeastern China.* Fasc. 1. xiv, [3], 82 pp., 2 maps. Canton: Lingnan University.

Descriptive flora of seed plants, with keys to all taxa, synonymy, references and citations, localities with indication of *exsiccatae*, general summary of extralimital distribution, taxonomic commentary, and miscellaneous other notes. An introductory section, with reference to southeastern China in general, treats floristics, vegetation formations, and the history of botanical investigation in the subregion (with particular reference to the work of the botanical group at Lingnan University). Covers only the Gymnospermae and Dicotyledoneae (Ranunculaceae through Fagaceae) on the standard Englerian system; no more published.

886

Taiwan (and the Pescadores)

The island of Taiwan, or Formosa, was under Japanese administration from 1895 to 1945. From 1949 it has been the seat of the Republic of China (Nationalist). A modern descriptive flora for the island (in English) has finally come to fruition in six large volumes (Li *et al.*, 1975–9), in fulfilment of a wish expressed by C. G. G. J. van Steenis in the late 1960s.[15] The older *Icones plantarum formosanarum* of Hayata (1911–21), with supplement by Yamamoto (1925–32), will, however, continue to remain useful for its illustrations. The woody plants are in addition separately covered in two substantial recent works.

HAYATA, B., 1911–21. *Icones plantarum formosanarum nec non et contributiones ad floram Formosanam.* 10 vols., supplement. Illus. Taihoku: Government of Formosa, Bureau of Productive Industry.

Comprises for the most part illustrated revisions of families and genera of vascular plants represented in Taiwan, with keys to species, synonymy and references, literature citations, indication of local and extralimital range, taxonomic commentary, and notes on habitat, special features, etc.; indices (vol. 10). A number of descriptions of new taxa from the Ryukyu Islands, the Bonin Islands, Hainan, and Fukien are also included. Volume 10 (pp. 97–233) contains a detailed exposition of the author's well-known 'dynamic' system of higher-plant phylogenetic classification. For additions, see YAMAMOTO, Y., 1925–32. *Supplementa iconum plantarum formosanarum.* Fasc. 1–5. Illus. Taihoku: Department of Forestry, Research Institute.

LI, HUI-LIN *et al.* (eds.), 1975–9. *Flora of Taiwan.* 6 vols. Illus. Taipei: Epoch.

Copiously illustrated descriptive manual-flora of

vascular plants, with keys to all taxa, synonymy, references, citations (especially to works with figures), Chinese transcriptions of names, local distribution (with citation of *exsiccatae*) and general summary of extralimital distribution, critical commentary of varying length, and miscellaneous notes; indices to Latin names in each volume. Arrangement of families is on the 1936 Engler-Diels system, but with monocots following dicots. An introductory section (in vol. 1) gives accounts of physical features, geology and soils, climate, features of the vegetation, and the plan of the flora, along with the keys to families. The style of the flora is relatively conservative; no karyotypes and little ecological or biological information are included and the emphasis, as in Walker's Okinawan flora (855) and other works of this 'school' of eastern Asiatic floristic writers, is largely on description, taxonomic references, geographical distribution and critical commentary. Volume 6, contains a lengthy comprehensive bibliography and a systematic enumeration of vascular plants (as a successor to Masamune's list of 1936) as well as addenda, statistical tables, and complete indices. [An edition of the whole *Flora* in Chinese is also projected.]

MASAMUNE, G., 1936. *Saishu Taiwan shokubutsu so-mokuroku* [Short flora of Formosa. An enumeration of the higher cryptogamic and phanerogamic plants hitherto known from the island of Formosa]. [vi], 410 pp., frontispiece. Taihoku: Editorial Department, 'Kudoa.'

Systematic list of native, naturalized, and commonly cultivated vascular plants, with concise indication of synonymy, references, and citations; Japanese vernacular and scientific names; concise indication of local range; bibliography and indices to generic and family names and their equivalents. The brief introductory section includes statistical tables of the flora.

Special groups – woody plants

LI, HUI-LIN, 1963. *Woody flora of Taiwan.* 992 pp., 371 text-figs., maps (end papers). Philadelphia: Morris Arboretum; Narberth: Livingston Publ. Co. (Also unofficially in Taiwan, no date.)

Briefly descriptive flora of woody plants, with keys to all taxa; synonymy and references; generalized indication of local range, with citation of *exsiccatae*; summary of extralimital distribution; excluded species; taxonomic commentary and other miscellaneous notes; list of references (post-1936) and new taxa; index to all botanical names. The introductory section includes remarks on physical features,

climate, soils, vegetation, forests, botanical exploration, floristics and phytogeography.

LIU TANG-SHUI, 1960–1. *Illustrations of native and introduced ligneous plants of Taiwan.* Vols. 1–2. 1388 pp., 1109 text-figs., 20 pls. (some in color, incl. maps). Taipei: National Taiwan University.

Atlas of woody plants, with descriptive text in Chinese (1 species per page); includes synonymy and references, Chinese and English vernacular names, and extensive notes on local range, habitat, special features, properties, uses, etc.; indices to all botanical and vernacular names at end of vol. 2. No keys are provided.

887

Hainan Tao (Hainan Dao)

This interesting tropical/subtropical island in the South China Sea, administratively part of Kwangtung Province but botanically always considered in its own right, shows a number of distinct floristic links with southeast Asia and Malesia within an essentially southern Chinese framework. A general flora in four volumes has recently been completed after an interruption due to the Cultural Revolution (Chun *et al.* 1964–5, 1974–7).

CHUN WOON-YOUNG *et al.*, 1964–77. *Flora hainanica.* 4 vols. 1272 text-figs., color map. Kwangchou: Academia Sinica Press.

Briefly descriptive flora of vascular plants, with keys to genera and species, synonymy, references and citations, Chinese transcriptions and local names, generalized indication of local and extralimital range, taxonomic commentary, and notes on habitat, occurrence, uses, etc.; complete indices in each volume. In volume 4 is an account of vegetation types (with map) as well as a key to families, addenda, and complete general indices of botanical and Chinese names. Represents the culmination of botanical work on the island begun by the late senior author and others in the 1920s.

MASAMUNE, G., 1943. *Kainan-to shokubutsu-shi.* [Flora kainantensis.] xv, 443 pp., 2 maps. Taihoku: Taiwan Sotokufu Gaijabu. (Reprinted.)

Systematic enumeration of vascular plants (largely compiled), without keys; includes relevant synonymy, references and citations, Japanese transliterations of names, indication of local range (with some localities), summary of extralimital distribution, and indices. An introductory section gives accounts of special features of the flora, phytogeography, vegetation formations, and the history of botanical investigation as well as a list of references. For additions, see MASAMUNE, G. and SYÔZI, Y., 1950–1. Florae novae kainantensis, I–II. *Acta Phytotax. Geobot.* **12**: 199–203; **14**: 87–90. [The main work has been reprinted in Japan during the middle 1970s but details have not been available to this reviewer.]

MERRILL, E. D., 1927. An enumeration of Hainan plants. *Lingnan Sci. J.* **5**: 1–186.

Systematic enumeration of vascular plants, with descriptions of new taxa; limited synonymy, with occasional references; citation of *exsiccatae*, with localities; general indication of extralimital range; taxonomic commentary and notes on habitat, etc.; no index. The introductory section includes remarks on phytogeography as well as a list of references. This early work provided a basis for the later elaboration of *Flora hainanica* (see above), and although by now much outdated is retained here as it is in English.

888

South China Sea Islands

Within this unit are collected three groups of islands in the South China Sea, all consisting of low islands, cays, sandbanks and reefs: the Pratas (Tung-sha) Islands, the Paracel Islands (Hsi-sha) and the Spratly (Nan-sha) Islands. Although shown on Chinese maps as wholly part of China, these islands have at different times been claimed by no less than six other countries; only Pratas is definitively Chinese.

I. Pratas (Dong-sha) Islands

Included here are the Vereker Banks. No modern checklist appears to be available for Pratas Island, the only land mass, but the scattered literature cited in Fosberg and Sachet, *Island bibliographies*, suggests that it is very poor in vascular plants although trees (*Pisonia* spp.) are present. Useful, though elderly, introductions include COLLINGWOOD, C., 1868. *Rambles of a naturalist on the shores and waters of the China Sea.* xiii, 445 pp., 3 pls., 7 figs. London. [Pratas Island, pp. 22–34.]; and HANCE, H. F., 1871. Note on *Portulaca psammotropha. J. Bot.* **9**: 201–2. [Discussion of Pratas with species list including among other plants the given *Portulaca*].

II. Paracel (Xi-sha) Islands

These low islands form a fairly compact group in the South China Sea south of Hainan and east of central Viet Nam. Included here is Macclesfield (Chung-sha) Bank.

CHANG HUNG-TA, 1948. The vegetation of the Paracel Islands. *Sunyatsenia*, **7**: 75–88, map.

This study includes an enumeration of 35 species of vascular plants, with indication of synonymy, references and citations, localities (with *exsiccatae*), and summary of extralimital range as well as occasional notes on habitat, frequency, biology, etc. The plant list is preceded by a description of the archipelago's physical features, geology, vegetation, and cultivated plants.

III. Spratly (Nan-sha) Islands

These widely scattered low islands and cays, with their numerous associated reefs, shoals and banks, are situated on an extensive marine rise west of Palawan and north of Borneo long considered to be a serious hazard for shipping but featuring potentially valuable marine and underground resources. References to vegetation and flora are few; the following appears to be the only more or less specific account available.

GAGNEPAIN, F., 1934. Quelques plantes des îlots de la Mer de Chine. *Bull. Mus. Hist. Nat. Paris*, II, **6**: 286.

Comprises a list of vascular plants collected from North Danger, Loaita, Itu-Aba, and Spratly (Storm) Islands as well as parts of the Paracels during the course of three fisheries surveys in 1930–3. Only 11 species were recorded.

Region (Superregion)

89

Southeastern Asia

Within this region as here circumscribed are the countries of Indochina (Vietnam, Laos, and Cambodia), Thailand, Burma, and the Andaman and Nicobar Islands. For the Paracels and Spratlys, see Region 88.

The progress of botanical exploration in Southeast Asia has been, and continues to be, fragmented due to geography, colonial history, and political and other developments. No single modern flora covers the area as a whole, and available national and subregional floras form a very heterogeneous assemblage. Furthermore, in botanical terms it has been overshadowed by Greater Malesia, India, and China.

Early botanical explorations were few, due to travel difficulties and restrictions and limited foreign contacts. Until well into the nineteenth century, the only significant contribution was *Flora cochinchinensis* (1790) by Loureiro, long court physician at Hué in what was then 'Cochinchina'. In the first half of the nineteenth century, in the wake of increased trade and colonial penetration, more detailed botanical exploration began in Martaban, Lower Burma (including Pegu), and Tennaserim as well as in the Andamans and Nicobars, chiefly through Griffith, Helfer, and Kurz. After the 1850s exploration extended to Upper Burma and, through Pierre, Harmand and Thorel, to southern Indochina and up the Mekong Basin; and in the last two decades of the century to Tonkin, Annam, the Shan States and Kachin, with much of the serious exploration being accomplished only after World War I. Indeed, the Chaine Annamitique and some parts of northern Burma were seriously explored botanically only after 1920, and one collector, Kingdon Ward, also made significant geographical observations. In Thailand, detailed botanical survey began after 1900 through the work of Kerr and other collectors (on behalf of the government) and Schmidt (in the southeast), since continued by Thai, Danish, and other botanists. Some contributions were made in southern Thailand by botanists from Malaya. By the 1960s modest botanical centers had been established in Rangoon, Bangkok, Saigon (Ho Chi Minh City) and Hanoi. Grave disruption was occasioned by World War II, and in Burma and Indochina, after a moderate revival following the war, collecting gradually wound down in the 1950s owing to political turbulence, guerrilla activities and finally war, except in northern Viet Nam where Vietnamese, Russian, and eastern European botanists were active. At the present time, the region remains unevenly explored, ranging from well known to poorly known, the latter being particularly true of northern and eastern Burma, Laos, and parts of the Chaine Annamitique. In recent years much new botanical work has been carried out in the Andamans and Nicobars, pursuant to the establishment in 1972 of a station of the Botanical Survey of India at Port Blair, South Andaman.

The first modern general floras for any part of the region are those of Kurz, including *Contributions towards a knowledge of the Burmese flora* (1874–7, not completed) and *Forest flora of British Burma* (1877). His early death was a serious loss to Burmese botany. Apart from coverage in *Flora of British India* (see **810–40**) and other Indian works, no general works have appeared save the government checklist (last revised in 1961) and some students' handbooks. More fortunate have been Thailand and the Indochina subregion, where the more or less provisional *Flora siamensis enumeratio* (1925–62) and *Flore générale de l'Indochine* (1907–51) are gradually being supplanted by, respectively, the more definitive *Flora of Thailand* (1970–), a Thai-Danish-Dutch project, and *Flore du Cambodge, du Laos, et du Viet-Nam* (1960–), like its predecessor a French project. The progress of these two current floras, while not rapid, is marked by mutual consultation, with many family treatments prepared by the same authorities for both works. Like the *Flora of Panama* (see **237**), the *Flore générale* has been extensively criticized for being too 'premature' and insufficiently critical. Nevertheless, in coordination the region lags behind Greater Malesia, except for indirect stimuli. Other important works include *Flore forestière de la Cochinchine* by Pierre (1881–1907, not completed), *Cây-cỏ miên-nam Viêt-Nam*, a students' manual by Pham Hoàng Hô and Nguyên Van Duong (1960; revised 1970–2), and *Forest flora of the Andaman Islands* (Parkinson, 1923). It may thus be seen that documentation is very uneven; here, as with the Mediterranean, the Himalaya, Mexico and Central America, and the West Indies, a general synoptic enumeration is much to be desired.

Progress

No overall review is available, though passing references occur in the essay of DE WIT, *Short history of the phytography of Malaysian vascular plants* (see Superregion 91–93, Progress). The current situation is sketchily surveyed in LEGRIS, P., 1974. *Vegetation and floristic composition of humid tropical continental Asia*. In *Natural resources of humid tropical Asia* (UNESCO), pp. 217–38 (Natural resources research, 12). Paris; other pertinent remarks appear in the general survey of tropical floristics by Prance (1978).

The history of Burmese phytography is covered in passing by BURKILL (1965), BISWAS (1943), and SANTAPAU (1958) among the reviews of South Asian work (see under Superregion 81–84, Progress). These same works may also be consulted for the Andamans and Nicobars; for recent specific reports, reference should be made to papers by K. Thothathri and N. P. Balakrishnan in BOTANICAL SURVEY OF INDIA, 1977. *All-India symposium on floristic studies in India: present status and future strategies* (*Howrah, 1977*). *Abstracts*, pp. 18–19. Howrah, West Bengal. For Thailand up to World War II, see JACOBS, M., 1962. Reliquiae kerrianae. *Blumea*, 11: 427–93. The countries of Indochina have been sketchily treated in TARDIEU-BLOT, M. L., 1957. L'oeuvre botanique de la France en Indochine. In *Proceedings of the Eighth Pacific Science Congress* (*Manila, 1953*) (Pacific Science Association), vol. 4, pp. 545–53. Manila; some historical material is also included in the Tome préliminaire (1944) of the *Flore générale de l'Indochine*.

Ongoing developments have also since 1947 been regularly reported in *Flora Malesiana Bulletin*.

Bibliographies. General bibliographies as for Division 8.

Regional bibliographies

ALLIED GEOGRAPHICAL SECTION, 1944. *An annotated bibliography of the Southwest Pacific and adjacent areas*. Vol. 3: *Malaya, Thailand, Indo-China*, [v], 256 pp., map. [Not restricted to botany or biology.]

DEPARTMENT OF EDUCATION, JAPAN, 1942. *Tōa kyō-ei-ken sigenkagaku bunken-mokuroku* [Bibliographic index for the study of the natural resources of the Greater East Asia Co-Prosperity Sphere]. Vol. 2: [French Indo-China and Thailand.] 253, 81 pp. Tokyo. [In Japanese, apart from references in other languages; not annotated.]

REED, C. F., 1969. *Bibliography to floras of Southeast Asia*. 191 pp., map. Baltimore: The author. [Evidently rather hastily compiled; not annotated.]

SMITHSONIAN INSTITUTION. 1969. *A bibliography of the botany of South East Asia*. 161 pp. Washington.

Indices. General indices as for Division 8.

Regional index

STEENIS, C. G. G. J. VAN and JACOBS, M. (eds.), 1948– . *Flora malesiana bulletin*, 1– . Leiden: Flora Malesiana Foundation (later Rijksherbarium). Mimeographed.) [Nowadays issued annually, this contains a substantial bibliographic section which rather thoroughly covers ongoing new literature on Southeast Asia. For fuller description, see Superregion 1–93.]

890

Region in general

No overall general treatments exist, but notice should be taken of the following: LAZARIDES, M., 1980. *The tropical grasses of Southeast Asia* (*excluding bamboos*). 225 pp. (Phanerogamarum monographiae, 12). Vaduz, Liechtenstein: Cramer (*apud* Gantner).

891

'Indo-China' (in general)

This term is used here in a geographical sense to account for works covering the whole of Viet Nam, Laos, Cambodia, and the lower Mekong Basin in general; it thus also encompasses the former territory of French Indo-China. The general flora begun under the direction of H. Lecomte, *Flore générale de l'Indochine*, was premature given the state of botanical exploration existing in the subregion in the early part of this century, and is now very incomplete; moreover, it has a lesser reputation than the *Flora of British India* or most of the other 'Kew floras' of the period. Its successor, *Flore du Cambodge, du Laos, et du Viêt-Nam*, is more critical and possesses a better basis but is progressing only relatively slowly. Other literature on the area is rather scattered and for this reference should be made to the standard bibliographies.[16]

Bibliographies

PÉTELOT, A., 1955. *Bibliographie botanique de l'Indochine*. 102 pp. (Arch. Rech. Agron. Cambodge Laos Viêtnam, 24). Saigon. [Classified listing, including itemization of monographs and revisions by families.]

VIDAL, J. E., 1972. *Bibliographie botanique indochinoise*. 100 pp. (Bull. Soc. Études Indochinoises, 47(4)). Paris. [Additions and amendments to Pételot's bibliography.]

AUBRÉVILLE, A. *et al.* (eds.), 1960– . *Flore du Cambodge, du Laos, et du Viêt-Nam*. Fasc. 1– . Illus. Paris: Laboratoire de Phanérogamie, Muséum National d'Histoire Naturelle.

Large-scale descriptive 'research' flora of vascular plants, published in fascicles (each with one or more families); family treatments, similar in format to the *Flore générale de l'Indo-Chine* (see below), include keys to genera and species, synonymy, references and citations, typification, localities with citations of *exsiccatae* as well as general indication of local and extralimital distribution, taxonomic commentary, and notes on habitat, phenology, ecology, uses, etc. (the latter to a rather greater extent than in the earlier work), and complete indices. Represents essentially a revised version of the *Flore générale* (to which further additions were discontinued in 1951); as of this writing (1980) 17 fascicles have been published, the latest in 1979.

LECOMTE, H. and HUMBERT, H. (eds.), 1907–51. *Flore générale de l'Indo-Chine*. 7 vols., plus Tome préliminaire and Supplément (in 9). Illus. Paris: Masson (later Laboratoire de Phanérogamie, Muséum National d'Histoire Naturelle).

Comprehensive descriptive flora of vascular plants, with keys to genera and species, full synonymy, references and citations, vernacular names, indication of *exsiccatae* with localities, generalized indication of local and extralimital distribution, taxonomic commentary, and notes on uses, special features, etc.; index to genera in each volume. In the *Tome préliminaire* (1944) are chapters on physical features, climate, geology, vegetation and forests, botanical exploration (with maps), and a cyclopedia of collectors and contributors to the *Flore* as well as a bibliography, general keys to families, and complete indices to genera and to vernacular names. The limits of the work encompass Viet Nam, Laos, Cambodia, eastern Thailand (Mekong Basin), and Hainan. The *Supplément* (1938–51) comprises in 10 parts additions and corrections to the original work (not, however, covering all families).

892

Viet Nam

The only separate works seen cover the southern part. An identification manual by Le Kha Khê, Vu Van Chuyên, and Thai Van Trung (Hanoi, 1961) exists for the northern part, but has not been available for examination.

Partial and special-group works

PHAM HOÀNG HÔ, 1970–2. *Cây-co miên-nam Viêt-Nam*. Revised edn. 2 vols. 6, 5272 + text-figs. Saigon: Bô Giáo-duc Trung-tâm Hoc-liêu. (Original edn., 1960, Saigon.)

Illustrated atlas-flora of the non-vascular (except Algae) and vascular plants recorded from southern Viet Nam,

with descriptive text and keys to families, genera and some groups of species; includes also very limited synonymy, vernacular names, and notes on local range, occurrence, phenology, etc.; addenda (vol. 2) and indices to family, generic, and vernacular names (in both volumes). An introductory section (vol. 1) incorporates an illustrated organography. The work is laid out along the lines of Coste's *Flore de France* (651) with up to three species on each page (small figures with accompanying text). Much of the work has been compiled from other sources such as *Flore générale de l'Indochine* and its successors.

PIERRE, J. B. L., 1881–99, 1907. *Flore forestière de la Cochinchine*. Vols. 1–5 (fasc. 1–26). 800 pp., 400 pls. Paris: Doin. (Reprinted in one volume, Lehre, Germany: Cramer.)

A sumptuous atlas-flora of woody plants, comprising superbly produced lithographic plates (by E. Delpy) with accompanying descriptive text; the latter includes extensive botanical descriptions, full synonymy with references, indication of *exsiccatae* with localities and general summary of distribution, and fairly detailed taxonomic commentaries. Complete general revisions are presented for some genera, notably *Garcinia*. The text and plates occupy fasc. 1–25; fasc. 26 (1907) includes a preface and a complete index to botanical names. [Owing partly to the author's death in 1905 the work was left incomplete; all the published and unpublished drawings are now filed in the phanerogamic herbarium of the Muséum National d'Histoire Naturelle, Paris.]

893

Cambodia (Kampuchea)

See also **891** (all works); **892** (PIERRE). One separate account of timber trees is also in existence, but evidently very rare.

Special group – forest trees

BÉJAUD, M. and CONRARD, M. L., 1932. *Essences forestières du Cambodge*. 4 vols. 830 pls. Phnom Penh: Service Forestier, Cambodge. (Possibly mimeographed.)

Vol. 1 comprises the text (484 pp.); vols. 2–4 the plates. [Not seen by this reviewer; bibliographic information partly supplied by J. Vidal. The work is most likely dendrological, with botanical, ecological and wood descriptions and notes on properties, uses and potential.]

894

Laos

See **891** (all works). No separate general floras are available, and considerable parts of the country remain poorly explored botanically. However, a useful introduction to the plant life of the area may be found in VIDAL, J., 1956–60. *La végétation du Laos.* 2 parts. 120, 462 pp., 37 pls., 6 maps (some in color) (Trav. Lab. Forest. Toulouse, 5(1), part 1). Toulouse.

895

Burma

See also **810–40** (BRANDIS; HOOKER). Considerable parts of the country remain botanically poorly explored (or not at all), and much of the collecting done in the decades up to the 1950s has not been reported upon or only haphazardly. Of all of tropical Asia, it 'has had the smallest proportion of its flora collected' (Prance). Current work is at a low level, with limited resources, and no new flora is in sight.

In addition to the works described below, note should also be taken of MERRILL, E. D., 1938–9. The Upper Burma plants collected by Capt. F. Kingdon-Ward on the Vernay–Cutting expedition. *Brittonia*, **4**: 20–188.

Keys to families

NATH NAIR, D. M., 1963. *The families of Burmese flowering plants.* 2 vols. Rangoon: Rangoon University Press. [Students' handbook, with keys to and descriptions of families; includes also a glossary and technical notes. Not seen; cited from Legris (see under Region 89, Progress).]

HUNDLEY, H. G. and U CHIT KO KO, 1961. *List of trees, shrubs, herbs and principal climbers, etc., recorded from Burma.* 3rd edn. xiv, 532, v pp. Rangoon: Superintendent of Government Printing and Stationery. (First edn. 1912.)

Systematically arranged list (in tabular form), with brief indication of habit and local range (relatively detailed); fairly extensive synonymy: vernacular names; lexica of vernacular names with botanical equivalents; index to generic and family names. The introductory section includes a glossary and a short bibliography. (Earlier editions of this work were limited to woody plants.)

KURZ, S., 1874–7. Contributions towards a knowledge of the Burmese flora. Parts 1–4. *J. Asiat. Soc. Bengal* **42**/2: 39–141; **44**/2: 128–90; **45**/2: 204–310; **46**/2: 49–258.

An enumeration of the seed plants of Burma, with non-dichotomous keys to genera and species together with family conspectuses; includes full synonymy (with references), general indication of range, with indication of some *exsiccatae* and localities, taxonomic commentary, and notes on habitat, phenology, etc.; no separate indices. Due to the death of the author the work was left incomplete; eight parts in all were planned, of which the four published cover families from Ranunculaceae to Apocynaceae (Bentham and Hooker system).

KURZ, S., 1877. *Forest flora of British Burma.* 2 vols. Calcutta: Government Printer. Reprinted 1974, Dehra Dun, Bishen Singh Mahendra Pal Singh.)

Briefly descriptive treatment (covering about 2000 species) of woody plants, with non-dichotomous keys to genera and species; limited synonymy, with references; vernacular names; generalized indication of local range; abbreviated notes on habitat, habit, phenology, timbers, etc.; indices to all botanical and vernacular names in vol. 2. The introductory section includes remarks on physical features, geology, forests, etc., together with a synopsis of families. The work is most useful in Tenasserim, Pegu, and Arakan; coverage of the Shan States, Upper Burma, and Kachin is relatively poor.

Partial work: Shan States

NATH NAIR, D. M., 1960. Botanical survey of the southern Shan States. In *Burma Research Society Fiftieth Anniversary Publications* (BURMA RESEARCH SOCIETY), vol. 1, pp. 161–418, figs. 1–12, 2 maps. [Rangoon]: distributed by Rangoon University Press. (Reprinted separately.)

Systematic enumeration of vascular plants, with limited synonymy, localities, citation of *exsiccatae*, vernacular names, and notes on habitat, uses, etc.; indices. An introductory section covers physical features, climate, floristics, botanical exploration, etc. Based largely on local field work carried out during the late 1950s.

896

Thailand (Siam)

Except for parts of the Mekong Basin and in Tenasserim, Thailand was botanically poorly known until after 1900. In the first four decades of this century, A. F. G. Kerr travelled extensively in the kingdom and with others built up the collections on which Craib's *Flora siamensis enumeratio* (1925–62) is based. Much additional collecting has been carried out since World War II and by the latter 1960s a sufficient basis had been laid for a descriptive *Flora of Thailand*, publication of which began in 1970 following much precursory work carried out primarily under Thai and Danish auspices. However, publication of this flora has been proceeding relatively slowly, with few large families covered to date.

Bibliography

WALKER, E. H., 1952. A contribution toward a bibliography of Thai botany. *Nat. Hist. Bull. Siam Soc.* **15**: 27–88.

HANSEN, B., 1973. Bibliography of Thai botany. *Ibid.*, **24**: 319–408. [Titles through 1969.]

CRAIB, W. G. and KERR, A. F. G. (eds.), 1925–62. *Flora siamensis enumeratio.* Vols. 1–2; 3, parts 1–3. Bangkok: The Siam Society.

Systematic enumeration of seed plants, without descriptions; synonymy, with references and citations; vernacular names; detailed summary of local range, with indication of *exsiccatae*; general indication of extralimital range; some taxonomic commentary; indices to generic and family names. The introductory section gives an outline of the general features of the flora, a discussion of local vernacular names, and lists of major and minor references. Not completed; covers families from Ranunculaceae to Gesneriaceae (Bentham and Hooker system).

SMITINAND, T. and LARSEN, K. (eds.), 1970– . *Flora of Thailand.* In several volumes. Illus. Bangkok: Applied Scientific Research Corporation of Thailand.

Briefly descriptive, well-annotated serial flora of vascular plants, with keys to genera and species, full synonymy, references and citations, typification, vernacular names, rather detailed indication of local range, generalized summary of extralimital distribution, and notes on habitat, ecology, phenology, special features, uses, etc.; representative illustrations; indices. The volumes are published in fascicles, with each containing one or more families at random. The interpretation of local distribution is aided by maps on the back covers. At this writing (late 1980) parts 1–3 of volume 2 and part 1 of volume 3 have been published; other parts are in press.

Partial work

Perhaps the most extensive and useful of these is the following:

LARSEN, K. *et al.*, 1961–9. Studies in the flora of Thailand. 59 parts. *Dansk Bot. Ark.* **20**: 1–275; **23**: 1–540; **27**(1): 1–107.

Comprises local revisions of a large number of families, based largely on collections by Thai-Danish botanical parties, with extensive annotations and descriptions of new taxa; complete index to genera in part 59. Forms a precursor to the new *Flora of Thailand* (see preceding entry).

897

Preparis and the Coco Islands

These islands, which lie between the Irrawaddy Delta (in Burma) and the Andaman Islands, are here separately recognized as they are under Burmese administration (the Andamans and Nicobars belong to India).

PRAIN, D., 1892. The vegetation of the Coco group. *J. Asiat. Soc. Bengal*, **60**/2: 283–406. (Reprinted 1894 in the author's *Memoirs and memoranda, chiefly botanical*).

Incorporates an annotated list of 358 species of which 307 are vascular plants, with indication of extralimital distribution, followed by a statistical study of distribution and the probable origin of the flora. An introductory section accounts for physical features and the general nature of the flora. [Not seen by this reviewer; information and reference from Blake and Atwood, 1942.]

898

Andaman Islands

See also 810–40 (BRANDIS; HOOKER); 911 (KING and GAMBLE). The Andamans are here considered als to include Narcondam Island, which lies some way to the east of North Andaman. For Preparis and the Coco Islands, see **897**. Existing floristic works for this group and the Nicobars are all more or less obsolete, the latest being Parkinson's forest flora of 1923. The known flora of both groups now stands at some 2200 species, of which 550 have been discovered since establishment of a 'circle' of the Botanical Survey of India at Port Blair in 1972.[17]

PARKINSON, C. E., 1923. *A forest flora of the Andaman Islands.* v, v, xiii, 325 pp., frontispiece, 6 pls. Simla: Superintendent, Government Central Press. (Reprinted 1972, Dehra Dun: Bishen Singh Mahendra Pal Singh.)

Descriptive field-manual of woody plants, with non-dichotomous, partly artificial keys to all taxa; literature citations; fairly detailed indication of local range; vernacular names; some taxonomic commentary; notes on habitat, special features, phenology, uses, etc.; index to all botanical and vernacular names. The introductory section includes remarks on physical features, climate, vegetation and forests, and botanical exploration as well as a list of major references and a synopsis of families; an appendix incorporates lists of plants in special categories.

ROGERS, C. G., 1903. *A preliminary list of the plants of the Andaman Islands.* ii, 51 pp. Port Blair, Andaman Islands: Chief Commissioner's Press.

Tabular systematic list of genera and species of seed plants arranged by families, with abbreviated notes on habitat and status. Compiled for the author by J. S. Gamble from the records in *Flora of British India* and other sources.

899

Nicobar Islands

See also **810–40** (BRANDIS; HOOKER); **911** (KING and GAMBLE). The group includes the islands from Car Nicobar south to Little and Great Nicobar. For further remarks, see **898** (Andaman Islands).

KURZ, S., 1876. A sketch of the vegetation of the Nicobar Islands. *J. Asiat. Soc. Bengal*, 45/2: 105–64, pls. 12–13.

Systematic enumeration of vascular plants, with descriptions of new taxa, limited synonymy, general indication of local range, citation of some *exsiccatae*

with localities, taxonomic commentary, and notes on habitat, frequency, etc.; no separate index. An introduction incorporates remarks on the geology, vegetation, and forests of the islands. Covers 624 species.

Notes

1 S. K. Jain in the first fascicle of *Flora of India* (1978); see *Flora Malesiana Bull.* **32**: 3203 (1979).

2 Much assistance in boundary delimitation has been gained from Schwartsberg, J. E. (ed.), 1978. *A historical atlas of South Asia.* Chicago: University of Chicago Press.

3 *Taxon*, **28**: 168 (1979).

4 G. L. Shah and A. R. Menon in *BSI Symposium Abstracts* (1977): 11.

5 B. A. Razi, *ibid.*, 16.

6 The cost was US$ 75000.

7 J. Joseph, *ibid.*, 15–16; a like comment was made respecting Andhra Pradesh. Similar points were also raised in the report by C. E. Ridsdale and A. J. G. H. Kostermans (*Flora Malesiana Bull.* **30**: 2759–66) on the Rijksherbarium field trip in Kerala and Tamil Nadu.

8 C. E. Ridsdale, personal communication.

9 The old Arab name for the island, from which the English word *serendipity* is derived (Goonetileke, 1970).

10 The history of the project is briefly discussed by Fosberg in the introductory part of the *Revised handbook*, volume 1, part 1 (1973). Further advice has been supplied by S. H. Sohmer.

11 A. S. Rao in *BSI Symposium Abstracts* (1977): 9.

12 C. L. Malhotra and P. K. Hajra, *ibid.*: 10.

13 M. M. J. van Balgooy in *Plant-geography of the Pacific*, 2nd edn. (1971; see **001**).

14 F. A. Stafleu, *Taxon*, **23**: 198–9 (1974) and personal communication.

15 C. G. G. J. van Steenis, The herb flora of Taiwan; in *Flora Malesiana Bull.* 22: 1562–7 (1968).

16 Recent statistics on the Indochina flora, reflecting changes brought about by modern revisionary work in comparison with family treatments in the *Flore générale*, are given in Vidal, J. E., 1964. Endémisme végétal et systématique en Indochine. *Comptes Rend. Soc. Biogéogr.* (Paris) 41: 153–59.

17 K. Thothathri and N. P. Balakrishnan in *BSI Symposium Abstracts* (1977) 18–19.

Division

9

Malesia and Oceania (tropical Pacific islands)

The great 'Horn of Plenty', the cornucopia of our Malaysian Flora, which was opened by van Rheede and by Rumphius shall still flow for a long time. I wish great joy to those that shall have the privilege to examine its contents.

L. G. M. Baas Becking,
Flora Malesiana, I, 4: iii (1948)

It is not merely new species and rarities about which we want to learn, but the real occurrence of widespread plants. It would be useful to draw up lists of at least 100 species...which botanical explorers should know in order to record their distribution.

E. J. H. Corner, *Pacific Science Information Bulletin*, 24(3–4): 19 (1972)

In the end, a practical taxonomy must rest on a limited number of characters which are observable without very elaborate equipment, which is one reason why uninformed academic botanists regard it as unscientific. I hope and believe that this Flora is producing new contributions to that end.

R. E. Holttum, *Flora Malesiana*, II, 1: (17) (1982)

This almost entirely insular division comprises peninsular Malaysia and all islands east and south of the Nicobar Islands, the Paracel and Spratly Islands, Botel Tobago (off the south coast of Taiwan), and the Bonin (Ogasawara) and Volcano (Kazan) Islands and north of the continent of Australia (with the Torres Strait and Ashmore and Cartier Islands), Lord Howe and Norfolk Islands, and the Kermadecs. The outer limits are marked by the Hawaiian chain and by Rapa Nui (Easter) and Sala-y-Gomez Islands. Australia, Tasmania and New Zealand are in Division 4, and the eastern Pacific islands from Guadelupe to Juan Fernandez are in Division 0 (as region 01). The division as a whole largely corresponds to the *northern half* (with the inclusion of New Caledonia) of the 'Pacific region' as delimited and subdivided by van Balgooy (1971).

Two superregions are recognized, based upon biogeographical, biohistorical and bibliographical criteria: Malesia (Superregion 91–93) and Oceania (Superregion 94–99). The dividing line lies east and north of Papuasia (including the Solomon Islands). The individual regions are arranged roughly in a west-to-east configuration, with delimitation on practical and biogeographical criteria. In Oceania, the richest regions come first, followed by those containing most of the low islands. The Hawaiian Islands, with their high endemism and greater percentage of American elements, are placed last.

The whole of the division is covered in the series

of maps published in *Pacific plant areas* (**001**, Steenis and Balgooy); accompanying these maps are descriptive notes and copious literature citations. The most thorough recent introduction to floristic plant geography in the division, with emphasis on Oceania, is BALGOOY, M. M. J. VAN, 1971. *Plant-geography of the Pacific*. 222 pp., maps, tables (Blumea, Suppl., 6). Leiden. [Map of subdivisions, p. 28.]

General bibliographies. Bay, 1910; Blake and Atwood, 1942; Frodin, 1964; Goodale, 1879; Holden and Wycoff, 1911–14; Jackson, 1881; Pritzel, 1871–7; Rehder, 1911; Sachet and Fosberg, 1955, 1971; USDA, 1958. See also under the superregions.

General indices. BA, 1926– ; BotA, 1918–26; BC, 1879–1944; BS, 1940– ; CSP, 1800–1900; EB, 1959– ; FB, 1931– ; IBBT, 1963–9; ICSL, 1901–14; JBJ, 1873–1939; KR, 1971– ; NN, 1879–1943; RŽ, 1954– . See also under the superregions.

Conspectus

903

Alpine and upper montane zones

Since the initiation of extensive studies by C. G. G. J. van Steenis in the 1930s, the mountains of Malesia have received a considerable amount of attention from botanists and some significant works have been published. Van Steenis' initial floristic analysis (1934–6) has in recent years been followed by two major works: *The Mountain flora of Java* (1972) and *Alpine flora of New Guinea* (1979–83). Only the latter is a proper flora in the sense of this *Guide*, but as the other two are of basic importance the occasion has been taken to include them.

STEENIS, C. G. G. J. VAN, 1934–6. On the origin of the Malaysian mountain flora, I–III. *Bull. Jard. Bot. Buitenz*. III, **13**: 135–262, 289–417; **14**: 56–72, 10 figs. (incl. maps), 2 folding maps, table.

Pages 155–260 in the first part comprise a checklist of bryophyte and vascular plant genera occurring above 1000 m in the mountains of Malesia, the Philippines and Papuasia and possessing temperate affinities, arranged within families; includes more or less detailed information on the species present, distribution, localities, and altitudinal range. The remainder of the work is phytogeographical. [Now rather in need of revision.]

Java

Partial work

HAMZAH, A., TOHA, M., and STEENIS, C. G. G. J. VAN, 1972. *The mountain flora of Java*. ix, 90 pp., frontispiece, 26 text-figs., 71 halftones, 57 color pls. Leiden: Brill.

Handsomely produced atlas of colored illustrations of a wide range of the more conspicuous members of the Javanese mountain flora, with accompanying descriptive text; the latter includes botanical details, local and extralimital range, and notes on ecology, biology, special features, related species, etc., as well as limited taxonomic commentary; index. The general part (80 pp.) covers the geography, environmental factors, and biota of the mountains as well as vegetation, plant formations, phytogeography, geobotanical history, the 'mountain mass elevation effect' question, flower biology, and introduced plants.

New Guinea

ROYEN, P. VAN, 1979–83. *The alpine flora of New Guinea*. 4 vols. lxviii, 3160 pp., *c*. 993 text-figs., 218 pls., frontispieces. Vaduz, Liechtenstein: Cramer (*apud* Gantner).

Volumes 2–4 (reproduced by offset lithography direct from the typescript) constitute a copiously illustrated documentary vascular flora of the high mountain areas of New Guinea above around 3000 m, with keys to all taxa, generally lengthy descriptions, synonymy with references and citations, vernacular names, typification, indication of internal and (where appropriate) extralimital range with citation of *exsiccatae*, taxonomic commentary, and brief notes on habitat, altitudinal zonation, phenology, etc.; indices in each volume (general index to appear at end of work). The first volume, which is typeset and fairly extensively illustrated, contains a mishmash of chapters (by a number of authors) on physical features, geology, climate, soils, general ecology, vegetation formations and communities, vegetation history, and the origin, affinities and distribution of the high-altitude flora along with chapters (by the senior author) in general geography, botanical exploration, and 'languages and native names'.

Volume 2 (1979) covers gymnosperms and monocotyledons, with a lengthy treatment of the Orchidaceae, while volumes 3–4 are to be devoted to the dicotyledons, save for an enumeration of high-altitude ferns (by J. Croft) at the end. Some of the seed plant treatments are (at least in part) by specialists. [Only vol. 1 seen by the writer at the time of this writing (mid-1980); description therefore based partly on publisher's advance leaflet and Koeltz Catalogue 262 (1980), pp. 204–5. Prior knowledge of this project, initiated in the late 1960s, and subsequent observation indicates that it is a bulky 'phytographic' flora. Volume 2 evidently employs a narrow concept of species and for orchids is unsatisfactory both in the field and in other respects (P. Cribb and T. Reeve, personal communication). Like the *Handbooks* series (see **930**), the *Alpine Flora* is a 'prestigious' – perhaps even pretentious – rather than practical work. The high posted price for the set (on a per-page basis corresponding to the *Handbooks*) will hardly render the work widely available in Papua New Guinea, save for a few institutional libraries and odd individuals. Moreover, the execution is inexcusably atrocious: muddy photographs and unevenly printed text in volume 1 and scores of printing errors in volume 2 (necessitating a seven-page errata leaflet).]

908

Wetlands

Comparatively little has been written upon the aquatic and marsh plants of Malesia and Oceania, although in recent years more attention has begun to be paid to them. The basic study is that of van Steenis (1932) on the collections of the German Limnological Sunda Expedition; added to this is a checklist of Philippine aquatics (1967).

MENDOZA, D. R. and DEL ROSARIO, R. M., 1967. *Philippine aquatic flowering plants and ferns*. 53 pp. (Museum Publication, 1). Manila: National Museum of the Philippines.

Annotated checklist, with synonymy, references, literature citations, and indication of distribution and habitat; index. Includes mangroves and sea-grasses as well as hydrophytes.

STEENIS, C. G. G. J. VAN, 1932. Die Pteridophyten und Phanerogamen der deutschen limnologischen Sunda-Expedition. *Archiv Hydrobiol.*, *Suppl.* XI, 3: 231–387, 36 pls., 8 figs., 4 tables.

Illustrated systematic account, with descriptions of novelties and extensive discussion of special features, distribution, ecology, and biology of plants accounted for; no keys. Encompasses hydrophytes and helophytes, similarly as in recent wetlands floras in the United States. Does not extend to New Guinea or the Philippines.

Superregion

91–93

Malesia

The fabled East Indies or Malay Archipelago, the greatest island empire in the world, was given the name Malesia in 1857 by Zollinger, a Swiss botanist-explorer in the southern part of the archipelago in the 1840s and 1850s who at the same time proposed botanical subdivisions within it.[1] Here it is considered to extend from the northern end of the Malay Peninsula (botanically part of this area rather than continental Asia) and the tip of Sumatra south and east in a flood of islands, large and small, all the way to San Cristobal in the southwestern Pacific. It includes among others **Bangka, Java, Borneo, the Anambas and Natunas, Sulawesi (Celebes), the Philippines, Maluku (the Moluccas or Spice Islands), New Guinea, the Bismarcks, and the Solomons.** This is similar to the delimitation adopted for *Flora Malesiana* by van Steenis, except that here the whole of Papuasia is included (the Louisiades and the Solomons being essentially biogeographic extensions of New Guinea) and that the Torres Strait Islands, save those nearest New Guinea, are as a group referred to eastern Australia (region 43).

All works relating generally to Indonesia (i.e., the former Netherlands East Indies), including *Flora Malesiana*, are here placed under **910–30**. Works on the constituent parts of Malaysia are placed respectively under **911** and **917**; those on the Philippines appear at **925**; while floras and other works for Papua New Guinea are given at **930** and those on the Solomon Islands at **938**. The system of units and regions adopted here for Malesia is based on geographical and biological criteria; political boundaries are largely arbitrary.

The long, illustrious history of botanical exploration and floristic writing in Malesia, in which hundreds of people and many institutions have been involved, has been well digested by de Wit in his *Short history of the phytography of Malaysian vascular plants*, published in volume 4 of *Flora Malesiana* in 1949. That essay is further supported by *Malaysian plant collectors and collections* by van Steenis-Kruseman (1950), the first volume of the *Flora*. Both contributions are thoroughly documented, and the latter was supplemented in 1958 and 1974. Progress and literature since 1947 have been chronicled in *Flora Malesiana Bulletin*. Only a summary of events as they relate to the production of the current standard floras can be given here.

The era of botanical 'discovery' in Malesia begins largely with the work of Rumphius in the latter part of the seventeenth century. His *Herbarium amboinense* was prepared at Ambon in southern Maluku, then the center of the spice trade, from about 1662 to 1697, with the *Auctuarium* (supplement) by 1702, but for security as well as other reasons publication was delayed until 1741–50 (*Auctuarium*, 1755) following editorial work by J. Burman. Its significance for the Malesian flora, however, was not really realized by Linnaeus, who 'bypassed' it, and (due to the almost complete lack of voucher specimens and involuntary deficencies in the plates) not adequately 'interpreted' in modern terms until 1917 although the earliest effort dates from 1754. It remains a 'standard' work for Maluku in the absence of any modern treatment for these islands, although containing many references to plants elsewhere in Malesia. Other early contributions were made by Bontius (Java), Kamel (Philippines), N. Burman (*Flora indica*, 1768), and (from India) Rheede, but of large-scale botanical exploration there was little until late in the eighteenth century.

In 1778, there was established in present-day Jakarta the *Bataviaasch Genootschap van Kunsten en Wetenschappen*, the first local learned society, with house, library, museum, and herbarium. Early contributors included Radermacher and von Wurmb, but after their deaths in the early 1780s the bottom dropped out of Dutch botanical work. From then until 1817 the major contributions were made by outsiders, in later years in Java once more partly through the Society. Even then, the island interiors, save Java, were scarcely visited, and most collecting was confined to a limited number of areas surrounding ports of call. The Spanish Malaspina expedition spent a considerable time in the Philippines in 1792, but most other contributions were by naturalists from the famous British and French exploring expeditions of the era, as well as by the American naturalist Horsfield.

It was Horsfield, following upon the Frenchmen Deschamps and Leschenault, who first effectively explored the interior of any part of Malesia. Although no *Flora javana* was realized as planned by him, the large collections of the Philadelphian, resident in Java for over 16 years and from 1811 attached to Raffles' 'court', were later partly worked up for *Plantae javanicae rariores* (1838–52) by Bennett and Brown and in the 1850s loaned to Miquel for his *Flora indiae batavae*. Penetration of the interiors of the Malay Peninsula and the other large islands only gradually followed, New Guinea being last (in the 1870s and 1880s).

The reforms of Raffles, which included plans for a central botanical garden, strongly influenced the new Dutch administration which in 1816 took control from the British. Under Governor-General van der Capellen, the present Botanical Garden at Bogor was founded in 1817 by Reinwardt, who had arrived with Capellen. In 1820 there was organized the Natural Sciences Commission, which conducted, with marked loss of life, extensive exploration and land surveys until 1850. Between them, these organizations assumed most of the scientific work of the *Bataviaasch Genootschap*. Very considerable botanical collections resulted, which are for the most part at Leiden and (from 1844) Bogor. Major contributors included Reinwardt, Blume, Teysmann, Hasskarl, Kuhl, van Hasselt, Kurz, Zipelius, Korthals, Junghuhn, and Zollinger. Outside Dutch territory, Cuming operated in the Philippines (and peninsular Malaysia, Singapore, and Sumatra) in 1836–40 and Griffith in Malacca in the early 1840s. There were also many collections from Penang, the first of the British 'Straits Settlements'. Many parts of the superregion were visited by the great voyages of discovery, exploration and mercantile diplomacy which were mounted in the decades following Waterloo.

The first major floristic work of the nineteenth century on the Indies was *Bijdragen tot de flora van Nederlandsch Indië* by Blume (1825–7). Although not a proper flora and mainly concerned with Javan plants, it has become fundamental to almost all later work on the superregion. Parallel to this effort were *Flora de Filipinas* by Blanco (1837; 2nd edn., 1845; 3rd edn., with colored plates, 1877–83), written in almost total isolation and originally published under royal decree, and *Descriptions of Malayan Plants* by Jack (1820–2), sadly limited by the author's early death and the loss of many collections shortly afterwards. Jack's work also encompassed southern Sumatra, as Bengkulu was under British control until 1824. No other unit floras were published until after 1859, except for the incomplete *Enumeratio plantarum Javae* (1827–8) and the 'show' work *Flora Javae* (1828–51, 1858) by Blume, and two florulas of Timor, respectively by Decaisne (1834) and Spanoghe/Schlechtendahl (1841).

Increasing interest in the Malesian flora and the continuing isolationism of Blume led others in the mid-nineteenth century to attempt various syntheses of the flora, among them the Dutch botanist Miquel. Experience with the preparation of *Plantae junghuhnianae* (1851–7) led Miquel to the conception of a major overall flora for Malesia, *Flora van Nederlandsch Indië* [Flora indiae batavae]. Initiated in 1854 with a grant-in-aid from the Ministry of Colonies, this work, inspired also by *Flora brasiliensis* and the Candollean *Prodromus*, appeared between 1855 and 1859. It covered all Malesia (save the Philippines and eastern New Guinea) and prompted Zollinger's phytogeographical proposals. A more or less critical compilation of what was then known and available to the author, but less refined than *Flora indica* and the future 'Kew floras', this *Flora* represents the major Dutch contribution to the great group of synthetic floras written or begun in Europe in the mid-nineteenth century. However, it was prepared without effective support from the Rijksherbarium at Leiden, direct access to the collections being yet refused by Blume although by goverment order loans had to be made. The *Flora* was thus deprived of its most important potential foundation, and, although the important collections of Horsfield, Junghuhn, and Zollinger, as well as those held by Miquel, Teysmann (at Bogor), and others were utilized freely, it perforce had to rely heavily on other published sources. It also lacks keys. The enthusiastic Miquel was as much concerned with speed and service as with painstaking accuracy. Never a visitor to the tropics, the author only gradually became aware of how incomplete and imperfect his work would be, a defect most noticeable in the first supplement to the *Flora*, the uncritical *Prodromus florae sumatranae* (1860–1).

However, with Miquel's accession to the directorship of the Rijksherbarium in 1862, the many unstudied collections uncovered there were critically analyzed and written up in a series of studies in the four-volume series *Annales musei botanici Lugduno-Batavi* (1863–9), which with the *Prodromus florae sumatranae* may be considered an extension of the *Flora*. Yet all these works probably covered less than half of the presently known vascular flora of the superregion (over 28 000 species). The task of general compilation had become too great for one man; even in the *Annales* Miquel was assisted by a number of others, including one of his two systematics students and only effective successor, R. Scheffer. Apart from Boerlage's *Handleiding*, no successor to Miquel's syntheses was to appear until *Flora Malesiana* in the mid-twentieth century.

Contemporaneously with the *Flora* there was published Zollinger's survey of the plant geography of Malesia (1857), in both Dutch and German versions; this included a botanical subdivision of the archipelago which is fundamentally sound but which has long been overshadowed by the more 'showy' zoological subdivisions of Huxley (1868) and especially Wallace (1860, 1876), whose 'Line' has acquired lasting popular fame although conceptually it is imperfect.[2] For plants, however, Zollinger – not of course very aware of the peculiarities of the Papuasian flora, then almost unknown – stressed the overall unity of the Malesian flora as distinct from that of Australia or mainland Asia north of the Isthmus of Kra. He had earlier been the first to propose the writing of a general Malesian flora, a project not, however, realized by himself.

After the conclusion of Miquel's works and, shortly afterwards, his death (1871), direct Dutch contributions diminished notably until well after 1900, evidently due to an onset of provincialism as well as a lack of students. Despite Miquel's final words, Malesian botany did not again become a major activity of the Rijksherbarium until the mid-twentieth century. Within Malesia, the period from 1850, the end of the Natural Sciences Commission and the publication of Junghuhn's definitive *Java* (1850–4), to 1880, when Treub was named director at Bogor, was comparably quiescent, with relatively little work on systematics until 1868 when Scheffer, the author of the first separate New Guinea checklist (1876), became director at Bogor and revived local phytographic studies. Nevertheless, much collecting and observation were accomplished in island interiors at this time and miscellaneous reports published; notable contributors were Teysmann and Beccari who, apart from visits to many other islands, inaugurated serious botanical exploration in New Guinea, the 'last frontier'. They were shortly afterwards followed by a large number of collectors in eastern Papuasia, most of them working at the behest of F. von Mueller in Melbourne. From

1876 local phytography in the Philippines was revived through the work of the forest botanist Vidal y Soler, but upon his death in 1889 again lapsed until after 1900. Continuous resident botanical work in peninsular Malaysia also had its beginnings in the 1870s.

The period from 1880 until the First World War was one of the most active in Malesian floristic exploration and phytography, associated as it was with markedly increased colonial development and economic activities, but to a large extent it was conducted not from metropolitan but from local (Bogor, Penang, Perak, Manila, Singapore) or peripheral (Calcutta, Brisbane, Melbourne) centers. Each major local center had its dominant personality: Treub at Bogor, Ridley at Singapore, and (from 1902) Merrill at Manila. Treub was not himself a floristic botanist like the other two but made good use of whomever became available, Herbarium Bogoriense playing a key role in his increasingly famous 'Institut Botanique de Buitenzorg'. As had earlier been the case at Calcutta, a small professional taxonomic staff gradually came into being there, followed by Manila after 1902 and Singapore in later years. On the other hand, nearly all descriptive work on northeastern New Guinea was accomplished in Berlin. Many organized biological survey and collecting expeditions, often part of general exploration and 'contact', were made throughout Malesia, aided by improving administrative control and communications, and by the end of the period a floristic picture much improved over that available to Zollinger had emerged, especially with regard to Borneo and Papuasia (several large expeditions were active in the latter between 1900 and 1915, with more to come between 1920 and 1926).

Most floristic work and publication in this period was area-centered, occasioned by the sheer size of the vascular flora in relation to the limited scientific manpower available. Gone were the days when one or two men could cover the whole flora of Malesia, but at the same time the modern tradition of international interdependence had not yet developed. The fashion for large, synthetic floras had also decayed by 1900, in part to be succeeded by 'contributions'. Rapid colonial development demanded 'results' and, coupled with the vertical national–imperial structure of activities then prevailing, evidently led to an acceptance of local work as expedient, with the regrettable consequence that, as in neighboring Australia, particularly after von Mueller's time, and southeast Asia, synonymy multiplied unnecessarily and species concepts and critical standards in published work varied widely. There also came to be a belief that the richness of the Malesian flora was 'inexhaustible' – leading Merrill later to claim that the total was around 45 000 species – and that every difference had to be 'optimistically' exploited in print. A large, fragmented 'literature' was thus created, ever-growing in complexity.

These constraints doubtless influenced Treub in advising against a replacement for Miquel's flora, and guiding taxonomic work at Bogor into more *ad hoc* activities: studies of economic groups, Javan flora, descriptions of novelties, etc. Almost the only overall work of the period on phanerogams was the generic flora, *Handleiding tot de kennis der flora van Nederlandsch Indië* (1890–1900), by Boerlage. Publication of this work, which included some lists of species but was largely based upon Bentham and Hooker's *Genera plantarum*, was aided by the Dutch colonial ministry but owing to the author's death in Ternate in North Maluku it was not completed. No successor has been published, current professional opinion among Malesian specialists being skeptical of the value of such works, and its usage is limited by having been written in Dutch. However, at the end of the period there appeared the first general treatises on pteridophytes, all by van Alderwerelt van Rosenburgh: *Malayan ferns* (1908), *Malayan fern allies* (1915), and a *Supplement* (1917). The works of both authors covered Malesia as a whole, but with the omission of eastern New Guinea in the *Handleiding*. On the other hand, the *ad hoc*, locally oriented activities of the various botanical centers, both within and without Malesia, resulted in a number of key area floras and related works, including some series of 'contributions'. Most are still more or less current, lacking effective successors although often of limited practical value, and are here described under their appropriate headings. Some of these were continued or did not appear until after World War I, but all were based upon work initiated before 1914.

The collections made by the various botanical centers and by locally resident botanists before 1914 were increasingly supplemented by large contributions from the developing forestry services, beginning in Java and the Philippines (and later in other parts of Indonesia, the British territories in the Malay Peninsula and Borneo, and eastern New Guinea), and by visiting phytographers, among them many Germans. One of the most travelled of the latter was Warburg in 1885–9 who, in underlining the overall floristic unity

of his 'Monsunia', demonstrated the essentially Malesian character of Papuasia.

World War I, while by and large not directly affecting Malesia, nevertheless caused marked disruptions which, accompanied by postwar economic recession and other developments, were to alter significantly the general patterns of Malesian botanical work, a process also influenced by conceptual and methodological changes in biology. In Indonesia, with stimuli from other agencies around 1917 and the appointment of Docters van Leeuwen as director at Bogor in 1918, increased emphasis was given to floristic synthesis and revisionary studies in Herbarium Bogoriense. This policy was continued until 1942, despite severe difficulties caused by the Depression of the 1930s. In addition to many family revisions, published as a series *Contributions à l'étude de la flore des Indes Néerlandaises* from 1923, new collecting (particularly by botanical field parties) was carried out in most islands, large and small. Collecting also continued in other parts of Malesia, notably New Guinea. To these contributions of botanical centers were added those of the forestry services and from many visiting and resident individuals, amateur and professional. Much of the inter-war effort in certain areas was accomplished in one way or another under American auspices, largely 'coordinated' by Merrill (firstly in Manila, later in the United States) as 'headquarters' botanist, with assistance from Bartlett on collections from Americans in Sumatra. Direct or indirect sponsorship also came from other 'metropolitan' countries. A notable feature of the period was an increase in international cooperation and division of labor, and communication through such media as the periodic Pacific Science Congresses, begun in 1920. By 1941 enough new work had been accomplished to make feasible a more definitive assessment of Malesian plant geography and floristics, and thoughts began to turn again towards an overall handbook-flora, either of Indonesia alone or of Malesia in general. These developments are well outlined by van Steenis (1938).

By contrast, the period between the World Wars was marked by a comparatively small number of new area floras. While external factors doubtless played their role, the wisdom of adding further to a stock of, at best, partially critical works in a superregion then generally sketchily known botanically, and with an inadequate critical foundation – *Flora indiae batavae* being inferior in this respect to *Flora australiensis* and *Flora of British India* – came to be questioned. Merrill in 1915 considered that only the environs of Singapore, Manila, Jakarta, and Bogor were by northern hemisphere standards sufficiently well known to enable passable local florulas to be written. Most local floristic works were essentially reports on places of botanical interest, with annotated checklists. The only larger local works based on *specific* initiatives were to some extent specialized: *Geïllustreerd Handboek der Javaansche Theeonkruiden* (on tea plantation weeds) by Backer and van Slooten (1924), *Onkruidflora der Javasche Suikerrietgronden* by Backer (1928–34; atlas completed in 1973), covering sugar plantation weeds, a *Mountain Flora of Java* by van Steenis (completed by 1945 but not published until 1972), and, finally, *Wayside trees of Malaya* by Corner (1940), a true 'field flora'.

It was in the difficult years of the 1930s that the idea of a new Malesian flora gradually took root in the minds of certain botanists at Bogor and elsewhere. One of them, H. J. Lam, became director at Leiden in 1933 and subsequently, in developing the Rijksherbarium into a significant systematics center, brought it back into the mainstream of Malesian botany. Others included Danser, later at Groningen, and van Steenis, at Bogor until 1946. Experience with the *Contributions*, which by 1941 extended to 34 numbers with detailed coverage of some 2000 species, had suggested that they were not suitable in the wider context envisaged, being laboriously conceived and aimed mainly at specialists. A more practical 'handbook', *Flora Malesiana*, first publicly advocated in 1938, was officially adopted as a project in 1940 under Baas Becking. However, the Japanese invasion postponed provisional launching of the *Flora* until 1948, giving time, though, for preparation of much background material. Formal launching took place in 1950 with the creation of the Flora Malesiana Foundation, headed by van Steenis as general editor. An accompanying newsletter, *Flora Malesiana Bulletin*, begin its career in 1947.

For the last three decades, *Flora Malesiana* has dominated progress in Malesian botany, and as of 1980 had covered some 15 per cent of the vascular flora: about 3900 species in Series I (spermatophytes) and 400 species in Series II (pteridophytes). Additional precursory studies extend these figures by five to ten per cent or so. At current rates of production, completion would take place in about the year 2100. The present 'state of the union' is described by Kalkman and Vink (1978, 1979) and van Steenis (1979). Its published historical, philosophical, and biblio-graphical foundation is without peer among tropical floras at the present time.

Some area works, mainly forest floras, have continued to be published, notably in Malaysia and Papua New Guinea. Among these may be mentioned *Flora of Malaya* (1953–), *Tree Flora of Malaya* (1972–), *Trees of Sabah* (1976–) and related works on East Malaysian Dipterocarpaceae, and *Handbooks of the Flora of Papua New Guinea* (1978–). For Indonesia, apart from the publication in the Netherlands of *Flora of Java* (1963–8), the school-flora of van Steenis, first published in Dutch in 1949, appeared in an Indonesian version in 1975. Another work for educational use is *Orders and families of Malayan seed plants* by Keng (1969; 2nd edn., 1978). However, good field floras for students remain few and far between in the superregion as a whole; most works, even those of recent vintage, are still written in terms of what the botanist thinks best despite the criticisms of Symington and Corner in the 1940s. For areas of special botanical interest and/or where regular teaching programmes are in progress, there is room for the elaboration of carefully prepared local florulas comparable to those recently published on Barro Colorado Island in Panama (**237**) and the Rio Palenque Science Center in Ecuador (**329**). Areas such as the Gedeh-Pangrango reserve in Java, Gunong Ulu Kali in peninsular Malaysia, and Mt Kaindi in Papua New Guinea are representative of possible sites. Checklists already exist for Pulau Tioman east of Johor, Mt Wilhelm in New Guinea, and Bougainville in the northern Solomon Islands.

At the present time, the superregion is on the average sketchily to moderately well explored, with a basis large enough for a meaningful critical general flora but less complete than Central America and well behind South Asia or much of tropical Africa. Only two units have been rather well explored: Java and peninsular Malaysia. The least-collected major areas are western New Guinea (Irian Jaya), Sulawesi, and Sumatra. It is believed, however, that on average 98 per cent of all vascular species have now been collected at least once.

Progress

For a detailed but somewhat fragmented historical survey, see WIT, H. C. D. DE, 1949. Short history of the phytography of Malaysian vascular plants. *Flora Malesiana*, I, 4: lxx–clxi, illus. Supplementary material has appeared annually in *Flora Malesiana Bulletin* since 1947. Further information on the role of the garden and institutes at Bogor (Buitenzorg) appears in HONIG, P. and VERDOORN, F., 1945. *Science and scientists in the Netherlands Indies.* New York. Current institutional organization is described in KALKMAN, C. and VINK, W., 1979. *Report on a visit to centres of systematic botany in Southeast Asia, September–October 1978.* 18 pp. Leiden. (Mimeographed.) The general review of tropical floristic botany by Prance (1978) should also be consulted.

The best overall introduction to the flora, vegetation, plant geography, patterns of field work, and the organization and recent development of Malesian botany is JACOBS, M., 1974. Botanical panorama of the Malesian archipelago (vascular plants). In *Natural resources of humid tropical Asia* (UNESCO), pp. 263–94, illus., maps, table (Natural resources research, 12). Paris.

Flora Malesiana. For an early conception of probable natural geographical limits, see ZOLLINGER, H., 1857. Ueber den Begriff und Umfang einer 'Flora Malesiana'. *Vierteljahrsschr. Naturf. Ges. Zürich,* 2: 317–49, map; also in Dutch as *idem,* 1857. Over het begrip en den omvang eener 'Flora Malesiana'. *Natuurk. Tijdschr. Ned. Ind.* 13: 293–322. On the genesis and development of the modern flora project, see STEENIS, C. G. G. J. VAN, 1938. Recent progress and prospects in the study of the Malaysian flora. *Chron. Bot.* 4: 392–7; *idem,* 1947. Doel, opzet en omvang der Flora Malesiana. *Chron. Nat.* 103: 67–70; *idem,* 1948. Introduction. *Flora Malesiana,* I, 4: v–xii; *idem,* 1949. De Flora Malesiana en haar beteekenis voor de Nederlandsche botanici. *Vakbl. Biol.* 29: 24–33; JACOBS, M., 1973. De Flora Malesiana en haar intellectuele bekoring. *Ibid,* 53: 287–9. map; STEENIS, C. G. G. J. VAN, 1979. The Rijksherbarium and its contribution to the knowledge of the tropical Asiatic flora. *Blumea,* 25: 57–77; and KALKMAN, C. and VINK, W., 1978. *General information on 'Flora Malesiana'.* 11 pp. [Leiden.] (Mimeographed.)

On the state of exploration and needs for collecting, see CORNER, E. J. H. (comp.), 1972. Urgent exploration needs: Pacific floras. *Pacific Sci. Inform. Bull.* 24(3/4): 17–27. A further review by M. M. J. van Balgooy, presented to the Thirteenth Pacific Science Congress, Vancouver in 1975 has yet to be published at this writing (1981).

General bibliographies. Bay, 1910; Blake and Atwood, 1942; Frodin, 1964; Goodale, 1879; Holden and Wycoff, 1911–14; Jackson, 1881; Pritzel, 1871–7; Rehder, 1911; Sachet and Fosberg, 1955, 1971; USDA, 1958.

Supraregional bibliographies

ALLIED GEOGRAPHICAL SECTION, SOUTHWEST PACIFIC AREA, 1944. *An annotated bibliography of the southwest Pacific and adjacent areas.* 3 vols. Maps. N.p. [Annotated lists, not primarily botanical. Volume 1 covers the present area of East Malaysia, Indonesia, and the Philippines; vol. 2, Papua New Guinea, the British Solomon Islands, the New Hebrides, and Micronesia; and vol. 3, Malaya, Thailand, Indo-China, and parts of Eastern Asia. Based on Australian library holdings.]

DEPARTMENT OF EDUCATION, JAPAN, 1942–4. *Tōa kyō-ei-ken sigenkagaku bunken mokuroku* [Bibliographic index for the study of the natural resources of the Greater East Asia Co-Prosperity Sphere]. Vols. 1, 3, 4, 6. Tokyo. [In Japanese, apart from references in other languages; titles arranged by subject areas, but not annotated. Volume 1 (1942) covers New Guinea (Botany, pp. 20–61); vol. 3 (1942), the Philippines (Botany, pp. 35–155); vol. 4 (1943), the Malay Peninsula (Botany, pp. 38–108); and vol. 6 (1944), the (Indonesian and Malaysian) East Indies.]

STEENIS, C. G. G. J. VAN, 1955. Annotated selected bibliography. In his *Flora Malesiana,* ser. I (*Spermatophyta*) (ed. C. G. G. J. van Steenis), vol. 5, pp. i–cxliv. Groningen: Noordhoff. (Reprinted separately.) [Provides detailed coverage, except for the Solomon Islands; briefly annotated. Arranged by areas as well as by families (and genera).]

General indices. BA, 1926– ; BotA, 1918–26; BC, 1879–1944; BS, 1940– ; CSP, 1800–1900; EB, 1959– ; FB, 1931– ; IBBT, 1963–9; ICSL, 1901–14; JBJ, 1873–1939; KR, 1971– ; NN, 1879–1943; RŽ, 1954–

Supraregional indices

STEENIS, C. G. G. J. VAN and JACOBS, M. (eds.), 1948– . *Flora malesiana bulletin,* 1– . Leiden: Flora Malesiana Foundation. (Mimeographed.) [Nowadays issued annually, though originally semiannual; contains a substantial bibliographic section classified by major plant groups with entries arranged by author and usually tersely annotated. Larger works may be separately reviewed. Covers very thoroughly a much wider area than Malesia, extending to the whole of Australasia, the Pacific, and monsoon Asia to the Indus and Japan.]

910–30

Superregion in general

The great modern *Flora Malesiana*, commenced in 1948, is the only modern descriptive treatment on the higher plants of the superregion. The two series, I (Spermatophyta) and II (Pteridophyta) have at this writing (late 1979) each covered some 15 per cent of the species in their respective great groups (25 000 spermatophytes; 3000 pteridophytes). Revisionary studies made and in a state of readiness for the work account for another five per cent, while other more or less recent revisions, or parts thereof, account for another 13 per cent, giving a rough total of one-third of the species having practical, more or less modern coverage.[3] The other general works on Malesia, *Flora indiae batavae* by Miquel and the *Handleiding* by Boerlage, are now principally of historico-documentary interest.

For the tree flora, only one overall work is available: the difficult *Geslachtstabellen* or generic keys by Endert (1928; 2nd edn., 1953), also available in English (1956). With regard to pteridophytes, the only nominally complete coverage prior to Series II of *Flora Malesiana* is the trio of works by van Alderwerelt van Rosenburgh (1908–17), headed by *Malayan ferns*.

Keng's *Orders and families of Malayan seed plants* is given under **911**.

Keys to genera – trees

ENDERT, F. H., 1928. *Geslachtstabellen voor Nederlandsch-Indische boomsoorten naar vegetatieve kenmerken*. 242 pp. (Meded. Bosbouwproefstat., 20). Buitenzorg. (Second edn. 1953, Bogor, revised F. H. Hildebrand.)

ENDERT, F. H., 1956. *Key to the tree genera in Indonesia*. 2nd edn., trans. R. D. Hoogland. [iii], 78, [9] pp. Canberra: Division of Land Research and Regional Survey, CSIRO, Australia. (Mimeographed.)

Comprises artificial keys, based almost entirely on bark and vegetative features, to the genera of Indonesian trees attaining at least 10 m in height, a bole of 2 m and a d.b.h. of 0.40 m.; index. Malaysia, New Guinea as a whole, and the Philippines are only imperfectly covered through lack of knowledge or exclusion. Effective use of these keys requires considerable practice.

STEENIS, C. G. G. J. VAN (ed.), 1948– . *Flora malesiana*. Series I: Spermatophyta. Vols. 1– . Illus.,

maps. Jakarta, Groningen (later Leiden, Alphen a/d Rijn): Noordhoff.

HOLTTUM, R. E. (ed.), 1959– . *Flora malesiana*. Series II: Pteridophyta. Vols. 1– . Illus., maps. Jakarta (etc., as above): Noordhoff.

Comprehensive descriptive floras respectively of seed plants (Series I) and pteridophytes (Series II), with keys to genera and species, rather detailed synonymy (with references and many citations), vernacular names (with provenance), generalized indication of internal and extralimital distribution (with a number of maps), taxonomic commentary, and extensive notes on habitat, phenology, special features, ecology, phytochemistry, etc. (the latter aspects with particular emphasis at family and generic levels, as in the *Pflanzenfamilien* and *Flora Neotropica*), and numerous drawings and photographs of representative species; index to all botanical names as well as addenda and corrigenda in each volume of revisions. The introductory sections in volumes 4 and 5 of Series I contain chapters on the scope and basis for the work, general botanical considerations affecting species delimitation, and the long essays *Specific and infraspecific delimitation* by van Steenis (vol. 5, clxvii–ccxxxiv) and *Short history of the phytography of Malaysian vascular plants* by H. C. D. de Wit (vol. 4, lxx–clxi) as well as the *Annotated selected bibliography* (see above). Volume 1 in Series I is largely given over to a detailed (and widely acclaimed) cyclopedia of collectors by Mme M. J. van Steenis-Kruseman. In the first volume of Series II may be found a bibliography of pteridological works on Malesia published since 1934 (i.e., subsequent to the third supplement of Christensen's *Index filicum*). In both series, one or more families are published in a fascicle, but no systematic order is followed (an index to those published appears in the cover of the latest fascicle of each series). In Series I, of the fifteen volumes originally projected, volumes 1, 4, 5, 6, 7, 8, and 9, part 1 have, at this writing (early 1980), been published. Volumes 2 and 3 have been projected to cover, respectively, Malesian phytogeography and vegetation. In Series II, of the three volumes projected, volume 1 (with five fascicles) was completed in 1982. [The conception and development of this flora project have been touched upon in the general introduction to **910–30**, where additional references are given.]

Phanerogams

For Series I of *Flora Malesiana*, see above.

BOERLAGE, J. G., 1890–1900. *Handleiding tot de kennis der flora van Nederlandsch-Indië*. Vols. 1–3, part 1. Leiden: Brill.

Descriptive generic flora of seed plants (in style resembling the *Genera plantarum* of Bentham and Hooker), including synoptic keys, summaries of sections within genera, and taxonomic commentary. In some instances species recorded for the region are listed at the end of the family, with brief indication of distribution. Not completed, owing to the author's untimely death in Ternate in 1900; covers the dicotyledons and gymnosperms.

MIQUEL, F. A. W., 1855–9. *Flora van Nederlandsch Indië* [Flora indiae batavae]. 3 vols. (in 4). 44 pls., 2 maps, tables. Amsterdam: van der Post.

Descriptive flora, with synoptic keys to species, of the seed plants of the former Netherlands East Indies, with inclusion of those from adjacent regions such as the Malay Peninsula and the Philippines (thus anticipating *Flora Malesiana* in its biogeographic rather than political delimitation); species entries include synonymy, with references and citations, vernacular names (where known), general indication of internal range, with some citations given for less frequently encountered species; taxonomic commentary and notes on habitat, uses, etc.; index to species at end of each volume. The appendices in volume 3 include statistics of the flora and a summary of the phytogeography of Java. The work, to a large extent compiled – speed being a characteristic of Miquel, as with Standley in his works on Middle America – is laid out in a format resembling that of the 'Kew Series' of British colonial floras; the botanical text is in Latin, the commentary in Dutch. Species found entirely outside the Netherlands East Indies are printed in smaller type. For the *Supplementum*, wholly concerned with Sumatra, see **913**. Numerous other additions are presented as family revisions and other notes in the author's *Annales musei botanici Lugduno-Batavi*. 4 vols. Leiden, 1863–9. [The genesis of, and unusual circumstances surrounding, this work are mentioned in the general introduction to **910–30** (see above).]

Pteridophytes

For Series II of *Flora Malesiana*, see above.

ALDERWERELT VAN ROSENBURGH, C. R. W. K. VAN, 1908. *Malayan ferns. Handbook to the determination of the ferns of the Malayan islands*. xl, 899, 11 pp. Batavia: Department of Agriculture, Netherlands India.

ALDERWERELT VAN ROSENBURGH, C. R. W. K. VAN, 1915. *Malayan fern allies. Handbook to the determination of the fern allies of the Malayan islands*. xvi, 261, 1 pp. Batavia: Department of Agriculture, Industry and Commerce, Netherlands India.

ALDERWERELT VAN ROSENBURGH, C. R. W. K. VAN, 1917. *Malayan ferns and fern allies*. Supplement 1. 577, 73 pp. Batavia: Department of Agriculture, Industry and Commerce, Netherlands India.

These three works together form a briefly descriptive comprehensive treatment, with partly dichotomous keys to species and genera, of the pteridophytes of Malesia; entries include synonymy (with references and citations), generalized indication of internal distribution, short taxonomic commentaries, and other notes. Indices to all botanical names appear at the end of each volume. The work, although by now much out of date, remains an important contribution, especially as series II of *Flora Malesiana* is still far from complete.

Region

91

Western and southern Malesia

This region comprises the Malay Peninsula, Sumatra, Palawan (with Calamian), and Borneo together with all associated island chains and groups, which together make up *Western Malesia* ('Sundaland' of many authors), and Java and the Lesser Sunda Islands as far as the Timor arc, which constitute *Southern Malesia*. Except for the inclusion of the last-named, necessary on phytogeographical grounds, the region is bounded on the east by the Asian continental shelf and its associated biogeographical line, Huxley's line (as modified by Simpson).[4]

While the region as a whole is served by *Flora Malesiana* and (to varying degrees) other works on the Malesian superregion, a goodly number of works is available for many of the individual units here recognized. Of these, the Malay Peninsula (with Singapore) and Java are the best provided for, owing to their long histories as centers of colonial development and scientific activity. Sumatra and Borneo each have one general work, neither very recent, and phytogeographic surveys exist for Nusa Tenggara (Lesser Sunda Islands) and the Anambas and Natuna Islands. A rather elderly florula is available for Timor, but in general the smaller groups in the region have very limited separate coverage.

The major available works on peninsular Malaysia and Singapore include, apart from the earlier *Materials for a flora of the Malay Peninsula* by King and Gamble (1889–1936) and the uncritical *Flora of the Malay Peninsula* by Ridley (1922–5), the 'revised' *Flora of Malaya* (1953–71), covering ferns, grasses and orchids; the *Tree Flora of Malaya* (1972–), of which three of the four volumes planned have now appeared; and the famous, unconventional *Wayside Trees of Malaya* by Corner (1940; 2nd edn., 1952; 3rd edn., in press). For student use, reference should also be made to *Orders and families of Malayan seed plants* by Keng (1969; 2nd edn., 1978). A number of more local florulas and checklists are also available, as well as a much-appreciated work on mangrove trees widely used outside as well as inside the Peninsula.

For Java, the major works are the notorious *Exkursionsflora von Java* by Koorders (1911–12; atlas, 1913–37) and the critical but mannered *Flora of Java* by Backer and Bakhuizen van den Brink, Jr. (1963–8). The woody plants are covered in the 13-part *Bijdragen tot de kennis der boomsoorten van Java* by Koorders and Valeton (1894–1914; atlas, 1913–18). An especially attractive album is *Mountain flora of Java* by Hamzah, Toha and van Steenis (1972). For school use, there is *Flora untuk sekolah di Indonesia* by van Steenis (1975; based on the Dutch version of 1949).

With regard to other units, Borneo features two general checklists, *A bibliographic enumeration of Bornean plants* by Merrill (1921) and the less critical *Enumeratio phanerogamarum bornearum* and its companion *Enumeratio pteridophytarum bornearum*, both by Masamune (1942, 1945), while at a more local level (mainly in the Malaysian states and Brunei) there are several works, notably *Forest trees of Sarawak and Brunei* by Browne (1955), *Trees of Sabah* by Cockburn *et al.* (1976–), and some major works on the tree family Dipterocarpaceae. In Sumatra coverage is much less satisfactory, with the only general work being

Prodromus florae sumatranae by Miquel (1860–1), which constituted a 'supplement' to his *Flora indiae batavae*. Timor is sketchily covered in *Prodromus florae timorensis*, a checklist by Britten *et al.* (1885), and by a paper on woody plants by Meijer Drees (1950) and a treatment of ferns (also including the remaining Lesser Sundas) by Posthumus (1944), but proper florulas of other islands or island groups are almost non-existent.

Attention should also be drawn to the tree lists which between them cover every part of the region (except Sarawak). The most complete of these is *Pocket check-list of timber trees*, originally compiled by Wyatt-Smith in the 1950s and now under revision for a third edition. Pocket lists are also available for Brunei and Sabah. In the Indonesian islands, uncritical mimeographed lists in A4 format have been available in a series originally begun by the Forest Research Institute, Bogor, in 1940; most lists, each in general corresponding to one province, are now in their third issues, mostly without significant revision. Sumatra (with the Western Islands, the Riau and Lingga groups, Banka and Billiton) is covered in nine such lists, Java in three, Kalimantan (Indonesian Borneo) in five, and Nusa Tenggara (with Timor) in two. The current versions are all published in *Laporan-laporanan Lembaga Penelitian Hutan* (Reports, Forest Research Institute). For many areas, they constitute the only recent floristic lists of any kind.

Many 'collection reports' for areas of greater or lesser extent as well as relatively detailed phytogeographic surveys are also available, but for the most part are not listed herein.

The level of collecting varies somewhat directly with the level of organized documentation. Peninsular Malaysia (with Singapore) and Java (with Madura) are well ahead, and rank very high among tropical countries in general. In the remaining units, the overall average is low although certain areas such as North Sumatra, Sabah, and Timor have received considerable attention.

Bibliographies
General and supraregional bibliographies as for Superregion 91–93. For a useful introduction to the main botanical literature on Region 91, see GAUDET, J. J. and STONE, B. C., 1968. Plant life of Malaysia. *Quart. Rev. Biol.* **43**: 306–10.

Indices. General and supraregional indices as for Superregion 91–93.

911

Peninsular Malaysia (Malaya, Tanah Melayu) and Singapore

The several available works together provide a relatively thorough coverage of a very rich flora, although treatment of much of the non-woody element is not particularly recent and in part is uncritical. For an introduction to area floristics and phytogeography, see KENG, H., 1970. Size and affinities of the flora of the Malay Peninsula. *J. Trop. Geog.* **31**: 43–56.

– Phanerogam flora: 8000–8500 species (Keng), 6766 species (Ridley).

Keys to families and genera

KENG, H., 1978. *Orders and families of Malayan seed plants.* 2nd edn. xl, 437 pp., 211 text-figs., frontispiece. Kuala Lumpur: University of Malaya Press. [Students' introductory manual, with analytical keys to orders, families, and some genera of seed plants in peninsular Malaysia and Singapore (and, by and large, for Western and Southern Malesia in general) accompanied by descriptive notes, illustrations, a glossary, and list of references; index. A most useful work.]

HOLTTUM, R. E. *et al.*, 1953–71. *A revised flora of Malaya.* Vols. 1–3. Illus. Singapore: Government Printer. (Third edn. of vol. 1, 1968; 2nd edn. of vol. 2, 1968.)

Well-illustrated descriptive systematic treatments, with keys to genera and species; synonymy, with references and principal citations; generalized indication of local and extralimital range; taxonomic commentary and notes on habitat, occurrences, affinities, special botanical features, etc.; indices to all botanical names in each volume. Each volume has an introductory section with chapters on general morphology, ecology, systematics, etc., of the group concerned. Of this series, only three volumes have appeared, viz.: vol. 1, *Orchids* (by Holttum); vol. 2, *Ferns* (by Holttum), and vol. 3, *Grasses* (mainly by H. B. Gilliland).

KING, G. and GAMBLE, J. S., 1889–1936. *Materials for a flora of the Malay Peninsula.* 5 vols. (in 26 parts). Calcutta. (Reprinted with separate pagination from various issues of *J. Asiat. Soc. Bengal*, **58**, sect. II–75.)

RIDLEY, H. N., 1907–8. *Materials for a flora of the Malay Peninsula: Monocotyledons.* 3 vols. Singapore: Government Printer.

These two works together comprise a systematic series, in flora form, of detailed revisions of angiosperm families occurring in peninsular Malaysia and Singapore, with keys to genera and species, full synonymy with references and citations, general indication of local and extralimital range (with citation of *exsiccatae* for less well-known species), and indices to all botanical names in each volume. The volumes on the dicotyledons (King and Gamble) also encompass the Andaman and Nicobar Islands (**898**, **899**); they were largely concluded by the First World War except for the Euphorbiaceae (by A. T. Gage, 1936). The work was originally recommended by Sir Hugh Low as an extension from the incomplete coverage provided by the *Flora of British India*, and its preparation and publication were subsidized by the Straits Settlements, Perak, and Selangor. Although formally admitted as a 'Kew flora', much of the work was carried out at Calcutta and Singapore. The two sets of *Materials* together were to form the prime basis for Ridley's *Flora of the Malay Peninsula* (1922–5; see below).

RIDLEY, H. N., 1922–5. *Flora of the Malay Peninsula.* 5 vols. Illus. London (later Ashford, Kent): Reeve. (Reprinted 1968, Amsterdam: Asher.)

Briefly descriptive flora of vascular plants, with non-dichotomous keys to species, synonymy with indication of some references, vernacular names, generalized summary of local and extralimital range, taxonomic commentary, and notes on habitat, dispersal, etc.; indices to botanical and vernacular names. The introductory section has an account of botanical exploration, while volume 5 incorporates a supplement. The work, based in large part upon the *Materials* (see preceding entry), shows signs of hasty compilation and is regarded to a large extent as uncritical. For field work, it is useless.[5]

WHITMORE, T. C. and NG, F. S. P. (eds.), 1972–8. *Tree flora of Malaya.* Vols. 1–3. Illus. (Malayan Forest Rec., 22). Kuala Lumpur: Longman.

Illustrated, briefly descriptive flora of native, naturalized, and commonly cultivated trees (except Dipterocarpaceae; see below), with keys to genera and species, limited synonymy and essential references, vernacular names, generalized indication of local and extralimital range, and notes on habitat, ecology, occurrence, frequency, etc.; complete indices at end of each volume. Under family and generic headings are given citations of significant taxonomic works as well as additional notes on classification, biology, timber, properties and uses, etc. An introductory section (vol. 1) includes a synopsis of tree families, a general description of a tree and its important features (for students), and easily recognized trees, but no general key to tree families has yet appeared. At this writing (mid-1979) three volumes (of a projected four) have appeared; as in *Flora Malesiana* no systematic sequence is followed. Vols. 1–2 were edited by T. C. Whitmore; vol. 3 by F. S. P. Ng. For dipterocarps, see SYMINGTON, C. F., 1943. *Foresters' manual of dipterocarps.* xliii, 244 pp., 114 text-figs. (Malayan Forest Rec., 16). Kuala Lumpur. (Reprinted 1976, Kuala Lumpur: University of Malaya Press, with added halftones.) [Well-illustrated systematic treatment, with keys, descriptions, and summary of local range. The photographs meant for the original publication of this forest-botanical classic were temporarily lost, only coming to light after World War II; they have been incorporated in the reprint.]

Special groups – common trees, wildflowers, etc.

For pteridophytes, orchids and grasses, see *Flora of Malaya* (ed. R. E. Holttum) above.

CORNER, E. J. H., 1940. *Wayside trees of Malaya.* 2 vols. vii, 772, v pp., 259 text-figs.; atlas of 228 pls. (halftones). Singapore: Government Printing Office. (2nd edn., 1952, with slight additions and alterations.)

Semi-popular illustrated descriptive treatment of the more commonly encountered trees in the Malay Peninsula and Singapore, with artificial keys to families (and some genera) and keys to genera and species, Malay and English vernacular names, generalized indication of local and extralimital range, with many notes on particular occurrences, taxonomic commentary, and numerous observations on habitat, phenology, special botanical and biological features, dispersal, uses, etc.; indices to vernacular names and to families and genera. In the introductory section are chapters on terminology, hints to identification, tree form and growth, phenological patterns, trees of local interest, vegetation formations, and a short list of references. The so-called second edition of this great tropical botanical classic differs only slightly from the first but a more extensive revision was carried out by its author in 1977 in the Peninsula and this version is now awaiting publication. For a related nontechnical work on shrubby, lianous and herbaceous plants, see HENDERSON, M. R., 1949–54. *Malayan wild flowers.* 2 vols. Illus. Kuala Lumpur: Malayan Nature Society. (Vol. 1 originally pub. in 3 parts in *Malayan Nature J.* 4(3–4), 1949; 6(1), 1950; 6(2), 1951; reprinted in 1 vol., 1951; 2nd reprint, 1959.) [Vol. 1 covers dicotyledons; vol. 2, monocotyledons.]

WYATT-SMITH, J., 1965. *Pocket check-list of timber trees.* Revised by K. M. Kochummen. vi, 428, 126 pp. illus. (incl. 46 halftones) (Malayan Forest Rec., 17). Kuala Lumpur.

Copiously illustrated manual for field identification of more important tree species, including both dipterocarps and non-dipterocarps, with extensive keys based almost entirely on vegetative attributes, vernacular and standardized names, and many illustrations; index. A new revision was in progress 1977–8 (K. M. Kochummen, personal communication); this has lately been published but no details are available at this writing (1981).

912

Western islands of Sumatra

See **913** (MIQUEL; SOEWANDA and TANTRA, 1973 (trees, Aceh), 1973 (trees, North Sumatra), 1973 (trees, Bengkulu), 1974 (trees, West Sumatra)). The Western Islands of Sumatra, which form a distinct chain running from Simeuluë in the north through the Banyak group, Nias, the Batu group, and the Mentawai Islands (including Siberut, Sipora and the Pagais) to the small, isolated Enggano in the south, have rarely been the subject of separate floristic works. Only the Mentawai group features any sort of a checklist, and that certainly incomplete, viz: RIDLEY, H. N., 1926. The flora of the Mentawi Islands. Pp. 57–94 in BODEN KLOSS, C. Spolia mentawiensia. *Bull. Misc. Inform.* (Kew), *1926*: 56–94. [Based on collections made in Siberut and Sipora in 1924 during a zoological field trip.] References to tree species in the various Western Islands appear also in the respective Bogor tree species lists for the different mainland provinces to which the islands are administratively attached.

913

Sumatra (Sumatera)

'Sumatra still wants its florist', wrote the naturalist and armchair traveler Pennant in the fourth volume of his *Outlines of the Globe* (1800). Collection checklists and other papers relating to the flora of Sumatra, together with the associated Western Islands (**912**) and the islands from the Riau group to Belitung (**914**), published since the completion in 1861 of

Miquel's *Prodromus* (see below) are widely scattered and with one exception – the series of tree species lists produced by the Forest Research Institute at Bogor, also described here – do not fall into distinctive series; reference for more specific details should perforce be made to standard bibliographies and indices. A useful introduction, however, may be found in the following: MERRILL, E. D. *et al.*, 1934. *An enumeration of plants collected in Sumatra by W. N. and C. M. Bangham.* 178 pp., 13 pls. (Contr. Arnold Arbor., no. 8). Jamaica Plain, Mass.

MIQUEL, F. A. W., 1860–1. *Florae indiae batavae, supplementum primum: Prodromus florae sumatranae.* [Flora von Nederlandsch Indië, eerste bijvoegsel. Sumatra, zijne plantenwereld en hare voortbrengselen.] xxiv, 656 pp., 4 pls. Amsterdam: van der Post; Leipzig: Fleischer. (Reissued 1862 as *Sumatra, zijne plantenwereld en hare voortbrengselen.*) German edn.: *idem*, 1862. *Sumatra, seine Pflanzenwelt und deren Erzeugnisse.* xxiv, 656 pp., 4 pls.

Pages 104–276 of this work comprise a systematic enumeration of the known seed plants of Sumatra (including Bangka and the Riau Archipelago), with synonymy, references and essential citations, brief indication of local range, and vernacular names. The latter part of the work consists of descriptions and notes on new and critical Sumatran taxa *not* included in the original *Florae indiae batavae* of 1855–9; this is followed by an additional supplement (pp. 618–26) and concluded by an index to all botanical names. The introductory section includes a list of references and a lengthy description of physical features, climate, vegetation, agriculture, etc. Text is respectively in Dutch or German except for botanical descriptions which are in Latin. Many additions may be found in MIQUEL, F. A. W., 1863–70. *Annales musei botanici Lugduno-Batavi.* Vols. 1–4. Leiden.

Special group – forest trees

Listed here are the several enumerations of forest tree species compiled in the botany section of the Forest Research Institute, Bogor (now known as Lembaga Penelitian Hutan) from 1940 onwards (save for the period during and immediately after World War II) for the different residencies (later provinces) of Sumatra, with a few of them for *afdelingen* or districts. They are largely based on collections and other observations made by forestry officers from 1917 onwards, with comparatively little control from other sources, and therefore must *not* be considered in botanical terms as critical or reliable. Until 1954 they were largely the work of

F. H. Hildebrand; subsequent reissues by R. Soewanda Among Prawira and others, though labeled as revisions, are largely unchanged save for the addition of generally good illustrations. [Limited staff and resources have militated against extensive revision, however much it may be desired (I. G. M. Tantra, personal communication).] All lists are similar in style, though varying in relative completeness, so that only a single annotation is given following the list of titles which are here arranged roughly from north to south, from Aceh to Lampung.

SOEWANDA AMONG PRAWIRA, R. and TANTRA, I. G. M. (comp.), 1973. *Daftar nama pohon-pohonan* (List of treespecies): *Aceh.* Revised edn. Illus. (Lap. Lemb. Penilit. Hutan, no. 179). Bogor. (Mimeographed. Originally pub. 1941, Buitenzorg, as HILDEBRAND, F. H. (comp.). *Lijst van boomnamen van Atjeh en onderhorigheden* (Bosbouwproefstation, Serie Boomnamenlijsten, no. 8); revised 1950, Bogor, as *idem, Daftar nama pohon-pohonan: Atjeh-Simalur.* [i], 70 pp. (Lap. Balai Penjelidik. Kehut., no. 32; also designated as Seri Daftar nama pohon-pohonan, no. 23).)

Idem, 1973. *Daftar nama pohon-pohonan: Sumatera Utara* [North Sumatra]. Revised edn. 123 pp., 57 pls. (Lap. Lemb. Penilit. Hutan, no. 171). Bogor. (Mimeographed. Originally pub. in two numbers as HILDEBRAND, F. H. (comp.), 1949. *Lijst van boomsoorten verzameld in Sumatera-Timur* [East coast of Sumatra]. [iv], 50 pp. (Rapp. Bosbouwproefsta., no. 9; also designated as Serie Boomnamenlijsten, no. 15). Buitenzorg, and *idem*, 1950. *Daftar nama pohon-pohonan: Tapanuli* [former Residency of Tapanuli]. [i], 43 pp. (Lap. Balai Penjelidik. Kehut., no. 29; also designated as Seri Daftar nama pohon-pohonan, no. 22). Bogor; latter revised 1954 as *idem, Daftar nama pohon-pohonan: Tapanuli.* [i], 50 pp. (Lap. Balai Penjelidik. Kehut., no. 67; also designated as Seri Daftar nama pohon-pohonan, no. 42).)

Idem, 1974. *Daftar nama pohon-pohonan: Sumatera Barat* [West Sumatra]. Revised edn. 64 pp., 46 pls. (Lap. Lemb. Penelit. Hutan, no. 187). Bogor. (Mimeographed. Originally pub. as HILDEBRAND, F. H. (comp.), 1950. *Daftar nama pohon-pohonan: Sumatera-Barat.* [i], 60 pp. (Lap. Balai Penjelidik. Kehut., no. 26; also designated as Seri Daftar nama pohon-pohonan, no. 21). Bogor; revised 1953 as *idem, Daftar nama pohon-pohonan: Sumatera-Barat.* [i], 67 pp. (Lap. Balai Penjelidik. Kehut., no. 64; also designated as Seri Daftar nama pohon-pohonan, no. 40).)

SOEWANDA AMONG PRAWIRA, R. (comp.), 1970. *Daftar nama pohon-pohonan: Riau.* Revised edn. Illus. (Lap. Lemb. Penelit. Hutan, no. 106). Bogor. (Mimeographed. Originally pub. as HILDEBRAND, F. H. (comp.), 1949. *Lijst van boomsoorten verzameld in Riau, Bengkalis en Indragiri.* [iv], 67 pp. (Rapp. Bosbouwproefsta., no. 11; also designated as Serie Boomnamenlijsten, no. 16). Buitenzorg; revised in two numbers, 1953, Bogor, as *idem, Daftar nama pohon-pohonan: Bengkalis* [northern Riau]. 26 pp. (Lap. Balai Penjelidik.

Kehut., no. 59; also designated as Seri Daftar nama pohon-pohonan, no. 36), and *idem, Daftar nama pohon-pohonan: Riau dan Indragiri* [southern Riau]. 67 pp. (Lap. Balai Penjelidik. Kehut., no. 61; also designated as Seri Daftar nama pohon-pohonan, no. 38). The 1970 revision represents a recombination of the 1953 numbers.)

SOEWANDA AMONG PRAWIRA, R. and TANTRA, I. G. M. (comp.), 1974. *Daftar nama pohon-pohonan: Jambi.* Revised edn. 26 pp., 29 pls. (Lap. Lemb. Penelit. Hutan, no. 185). Bogor. (Mimeographed. Originally pub. 1949, Buitenzorg, as [HILDEBRAND, F. H., comp.] *Lijst van boomsoorten verzameld in Djambi-Sumatra.* [iv], 14 pp. (Rapp. Bosbouwproefsta., no. 8; also designated as Serie Boomnamenlijsten, no. 14), with supplement (Rapp. Bosbouwproefsta., no. 8a).)

Idem, 1973. *Daftar nama pohon-pohonan: Bengkulu.* Revised edn. 60 pp., 42 pls. (Lap. Lemb. Penelit. Hutan, no. 159). Bogor. (Mimeographed. Originally pub. 1949, Buitenzorg, as HILDEBRAND, F. H. (comp.). *Lijst van boomsoorten verzameld in Benkuelen.* [iv], 47 pp. (Rapp. Bosbouwproefsta., no. 22; also designated as Serie Boomnamenlijsten, no. 20).)

Idem, 1972. *Daftar nama pohon-pohonan: Palembang* (*Sumatera Selatan*). Revised edn. 91, 4 pp., 60 pls. (Lap. Lemb. Penelit. Hutan, no. 141). Bogor. (Mimeographed. Originally pub. 1949, Buitenzorg, as HILDEBRAND, F. H. (comp.), *Lijst van boomsoorten verzameld in Palembang.* [iv], 78 pp. (Rapp. Bosbouwproefsta., no. 9; also designated as Serie Boomnamenlijsten, no. 15).)

Idem, 1972. *Daftar nama pohon-pohonan: Lampung.* Revised edn. 29, 2 pp., 40 pls. (Lap. Lemb. Penelit. Hutan, no. 143). Bogor. (Mimeographed. Originally pub. 1949, Buitenzorg, as HILDEBRAND, F. H. (comp.) *Lijst van boomsoorten verzameld in de Lampungse districten.* [iv], 18 pp. (Rapp. Bosbouwproefsta., no. 20; also designated as Serie Boomnamenlijsten, no. 19); revised 1954, Bogor, as *idem, Daftar nama pohon-pohonan: Lampung.* [i], 27 pp. (Lap. Balai Penjelidik. Kehut., no. 65; also designated as Seri Daftar nama pohon-pohonan, no. 41).)

Sparsely annotated, uncritical checklists of forest trees for each of the (at present) eight provinces of Sumatra, arranged alphabetically by families and genera and including local-vernacular and standardized trade names, indication of districts where collected, number of collections (but not *exsiccata*-numbers), and durability class (on a scale of five), all on a single line. Each of the systematic lists proper is preceded by a list with all records, fully annotated, arranged alphabetically by vernacular names; at the end of each treatment are lists of amended botanical names, a lexicon of Indonesian names, an alphabetical lexicon of genera with family equivalents, and an indexed album of full-page illustrations of key species depicting botanical features. No keys are included. Tree species covered are those known to attain a diameter (at breast height) of 40 cm or more (considered in theory to be the limits for effective

exploitation). [Moderately reliable as to genera but of limited value at species level; much modern revisionary work evidently not accounted for in current reissues. Records for the Western Islands (912) are given in the treatments for the respective mainland provinces to which administratively they are attached, as are also the islands from the Riau (Rhiow) group to Bilitung (Billiton) (914).]

914

Islands from the Riau group to Belitung

See also 913 (MIQUEL; SOEWANDA AND TANTRA 1972 (trees, Palembang); SOEWANDA, 1970 (trees, Riau)). For the Riau (Rhiow) and Lingga Archipelagoes, works on peninsular Malaysia and Singapore (911) may be found useful. Bangka (Banka) and Belitung (Billiton) and associated islets formerly comprised a separate residency (they currently are part of the province of Palembang or South Sumatra) and thus featured independent treatment in the Bogor Forest Research Institute tree species lists during their first 'cycle' of publication. Little else of significance exists except, perhaps, the following: KURZ, S., 1864. Korte schets der vegetatie van het eiland Bangka. *Natuurk. Tijdschr. Ned.-Indië,* **27**: 142–258. [Includes an annotated checklist of 959 species and a survey of the vegetation, based primarily upon a field trip by the author.]

Special group – forest trees

HILDEBRAND, F. H. (comp.), 1952. *Daftar nama pohon-pohonan: Bangka dan Billiton.* 46 pp. (Lap. Balai Penjelidik. Kehut., no. 57; also designated as Seri Daftar nama pohon-pohonan, no. 33). Bogor. (Mimeographed. Originally pub. 1949, Buitenzorg, as *idem, Lijst van boomsoorten verzameld in Bangka en Billiton.* [iv], 29 pp. (Rapp. Bosbouwproefsta., no. 12; also designated as Serie Boomnamenlijsten, no. 17).)

Uncritical checklist of forest tree species having a known diameter of 40 cm or more, with indication of occurrence and vernacular names; no keys or illustrations provided. [For background information and full description, see 913 under Forest trees.]

915

Anambas and Natuna Islands

See 913 (SOEWANDA, 1970 (trees, Riau)). Works for peninsular Malaysia and Singapore (911) will also be found useful, along with those on Borneo (917). The Anambas and Natuna Islands lie at the southern end of the South China Sea between peninsular Malaysia and Borneo and represent partly drowned mountains on a shallow continental shelf. Unfortunately, no full checklist is available for either group; the best source of information appears to be STEENIS, C. G. G. J. VAN, 1932. *Botanical results of a trip to the Anambas and Natoena Islands* (Bull. Jard. Bot. Buitenz., ser. III, vol. 12: 151–211, 11 figs.). Bogor. [Includes lists of collections preceded by some general remarks on the flora and its affinities.] Other coverage is provided in the Forest Research Institute tree species list for Riau Province (Sumatra), to which both groups are administratively attached, but this is very haphazard and unreliable.

916

Palawan (with Calamian)

For this long, narrow island (together with the more northerly Calamian and some associated smaller islands, including the Cagayan Sulu group), biogeographically intermediate between Western Malesia and the Philippines although rather nearer the former, no separate checklist is available and other floristic papers are few and scattered. However, records as far as known by the 1920s are indicated in *An Enumeration of Philippine flowering plants* by MERRILL (see 925). In addition, there is a checklist for Banguey Island, off the Sabah coast towards Palawan, viz.: MERRILL, E. D., 1926. The flora of Banguey Island. *Philipp. J. Sci.* 29: 341–427.

917

Borneo (Kalimantan)

Two general enumerations of the flora have been made respectively by Merrill (1921) and Masamune (1942, 1945), the latter a somewhat imperfect and uncritical work based to a large extent on herbarium labels; both works, however, are now rather out of date. Subsequent taxonomic literature pertaining to Bornean plants is very scattered, and reference should be made to standard bibliographies and indices. Since before World War II, the forest trees have received much attention, and several publications have resulted; the most complete originate from Sarawak and Sabah.

– Vascular flora: 10 000–11 000 species (Merrill); 5250 recorded as of 1926.

Bibliography

MERRILL, E. D., 1915. A contribution to the bibliography of the botany of Borneo. *Sarawak Mus. J.* 2: 99–136. [For additions to 1921, see the author's *Bibliographic enumeration*, pp. 2–6.]

MASAMUNE, G., 1942. *Enumeratio phanerogamarum bornearum*. 739 pp., map. Taihoku: Taiwan Sotukufu Gaijabu.

MASAMUNE, G., 1945. *Enumeratio pteridophytarum bornearum*. ii, 124 pp. Taihoku: Taiwan Sotukufu Gaijabu.

These two works comprise systematic enumerations of known Bornean seed plants and pteridophytes, with synonymy, references and principal citations, Japanese scientific names, indication of local range (with some localities), and indices to generic and family names. The introductory section of the 1942 volume gives an account of botanical exploration, notes on special features of the flora, statistics, and a short list of references, while the 1945 volume features a discussion of the fern flora. These compilations are somewhat uncritical, and should be used with caution. [Description prepared in part from material supplied by Prof. H. Hara, Tokyo.]

MERRILL, E. D., 1921. *A bibliographic enumeration of Bornean plants*. 637 pp. (J. Straits Branch Roy. Asiat. Soc. Special Number.) Singapore.

Systematic enumeration of seed plants, with synonymy, references, and literature citations; generalized indication of local and extralimital range, with localities and citations of *exsiccatae* for less well-known species; limited taxonomic commentary; index to all botanical names. The introductory section incorporates a summary of general features of the flora and vegetation as well as an account of botanical exploration in Borneo; there are also additions to the author's 1915 bibliography (see above). Much of Masamune's 1942 *Enumeratio* is based on this work. For supplements, see Additions...[I]–II. *Philipp. J. Sci.* **21** (1922): 515–34; **30** (1926): 79–87.

Special group – forest trees

Since 1940 a considerable literature has appeared on the tree flora of Borneo, but all of it is politically delimited according to state or province. The selection described below will be grouped under four subheadings: I. Sabah; II. Brunei; III. Sarawak; and IV. Indonesian Borneo.

I. Sabah

COCKBURN, P. F. *et al.*, 1976– . *Trees of Sabah*. Vol. 1. xv, 261 pp., 54 text-figs., 16 pls., maps (end papers) (Sabah For. Rec. 10). Kuching; Sarawak: Borneo Literature Bureau (for Jabatan Hutan, Sabah).

Illustrated dendrological treatment of 22 non-dipterocarp tree families, with keys, descriptions, synonymy, local vernacular and trade names, indication of local distribution, some critical remarks, and notes on ecology, special features, uses, etc.; complete index at end. Sources of information, etc., are given in the general part. No *exsiccatae* are cited. For a companion work on Dipterocarpaceae, see MEIJER, W. and WOOD, G. H. S., 1964. *Dipterocarps of Sabah*. 344 pp., 59 text-figs., 30 pls., map (Sabah For. Rec., 5). Kuching: Borneo Literature Bureau (for the Sabah Forest Department). [Copiously illustrated dendrological treatment, with keys, citation of *exsiccatae*, and notes on local distribution, special features, ecology, etc.] (Volume 2 of the main work appeared in 1980, covering 23 families.)

FOX, J. E. D., 1970. *Preferred check-list of Sabah trees*. 65 pp. (Sabah For. Rec. 7). Kuching: Borneo Literature Bureau (for the Sabah Forest Department).

Tabular checklist in pocket form, with notes on tree size, local range, and ecology; list of vernacular names, with botanical equivalents.

II. Brunei

Brunei is not part of Malaysia, but is a semi-independent sultanate in treaty relations with the United Kingdom.

PUKUL, HASAN BIN and ASHTON, P. S., 1966. *A checklist of Brunei trees*. 132 pp. Brunei: Government of Brunei (distributed by Borneo Literature Bureau, Kuching).

Tabular checklist of non-dipterocarp trees, arranged according to vernacular names (with botanical equivalents), and including notes on local range, ecology, etc.; list of botanical names, with vernacular equivalents. For the Dipterocarpaceae, see the junior author's *Manual of the dipterocarp trees of Brunei State* (under Sarawak).

III. Sarawak

The relatively early account of Browne (1955) is highly selective. P. S. Ashton and others now have an expanded version in preparation. No pocket checklist appears to be available. The Dipterocarpaceae have been treated fairly fully by Ashton (1964–8).

BROWNE, F. G., 1955. *Forest trees of Sarawak and Brunei*. 369 pp. Kuching: Sarawak Government Printer.

Illustrated dendrological treatment of the principal tree species of the two states, with descriptions, notes on distribution, special features, uses, etc.; index. The Dipterocarpaceae have been separately treated in greater depth by ASHTON, P. S., 1964. *A manual of the dipterocarp trees of Brunei State*. xii, 242 pp., 58 pls., 20 text-figs. Oxford: Oxford University Press; and *idem*, 1968. *A manual of the dipterocarp trees of Brunei State and Sarawak. Supplement*. viii, 129 pp., 23 pls., 15 text-figs. Kuching: Sarawak Forest Department.

IV. Indonesian Borneo

The whole of Indonesian Borneo, or Indonesian Kalimantan, is sketchily treated in several of the tree species lists published for the different parts of Indonesia by the Forest Research Institute (Lembaga Penelitian Hutan), Bogor, initiated in 1940 and subsequently revised or merely reissued. Five numbers are currently available for the four provinces (of which East Kalimantan, the largest, has two forest districts), although that for South Kalimantan (Banjarmasin and Hulu Sungai) evidently has had no reissue since 1953. They are not critical, and are included here mainly for the sake of completeness. [For further details, see under Sumatra (913).]

SOEWANDA AMONG PRAWIRA, R. (comp.), 1974. *Daftar nama pohon-pohonan: Bulungan dan Berau (Kalimantan Timur)*. Revised edn. [ii], 89 pp., 80 pls., map (Lap. Lemb. Penelit. Hutan, no. 196). Bogor. (Mimeographed. Originally pub. 1941, Buitenzorg, as HILDEBRAND, F. H. (comp.). *Boomnamenlijsten van Boeloengan en Beraoe (Oost-Borneo)*. (Bosbouwproefstation, Serie Boomnamenlijsten, no. 9); revised 1952, Bogor, as *idem*, *Daftar nama pohon-pohonan: Bulungan dan Berau (Kalimantan-Timur)*. 83 pp. (Lap. Balai Penjelidik. Kehut., no. 55; also designated as Seri Daftar nama pohon-pohonan, no. 32).) [District of Bulungan and Berau, comprising all of East Kalimantan Province north of the 'W' Range (north of the Kutai Basin); it lies east of G. Mulu in Sarawak and also just south of Sabah, both in East Malaysia.]

SOEWANDA AMONG PRAWIRA, R. and TANTRA, I. G. M. (comp.), 1971. *Daftar nama pohon-pohonan: Samarinda* (*Kalimantan Timur*). Revised edn. [ii], 118, 2 pp., 61 pls. (Lap. Lemb. Penelit. Hutan, no. 123). Bogor. (Mimeographed. Originally pub. 1949, Buitenzorg, as HILDEBRAND, F. H. (comp.). *Lijst van boomsoorten verzameld in Samarinda* (*Oost-Borneo*). [iv], 76 pp. (Rapp. Bosbouwproefsta., no. 2; also designated as Serie Boomnamenlijsten, no. 11); revised 1952, Bogor, as *idem, Daftar nama pohon-pohonan: Samarinda* (*Kalimantan-Timur*). 105 pp. (Lap. Balai Penjelidik. Kehut., no. 58; also designated as Seri Daftar nama pohon-pohonan, no. 34).) [District of Samarinda, comprising the southern part of East Kalimantan Province including the Kutai Basin, the Mahakam valley, and the oil region of Balikpapan, all south of the 'W' Range.]

HILDEBRAND, F. H. (comp.), 1953. *Daftar nama pohon-pohonan: Bandjarmasin dan Hulu Sungei* (*Kalimantan-Tenggara*). 62 pp. (Lap. Balai Penjelidik. Kehut., no. 63; also designated as Seri Daftar nama pohon-pohonan, no. 40). Bogor. (Mimeographed. Originally pub. 1949, Buitenzorg, as *idem. Lijst van boomsoorten verzameld in Bandjermasin en Hoeloe Sungei* (*Zuid-Oost-Borneo*). [iv], 48 pp. (Rapp. Bosbouwproefsta., no. 5; also designated as Serie Boomnamenlijsten, no. 13).) [Province of South Kalimantan, including the near-offshore island of Laut; formerly the districts of Banjarmasin and Hulu Sungai in the Residency of South and East Borneo.]

SOEWANDA AMONG PRAWIRA, R. (comp.), 1971. *Daftar nama pohon-pohonan: Kapuas-Barito* (*Kalimantan Tengah*). Revised edn. 90 pp., 90 pls. (Lap. Lemb. Penelit. Hutan, no. 121). Bogor. (Mimeographed. Coverage for part of area first pub. 1941, Buitenzorg, as HILDEBRAND, F. H. (comp.) *Boomnamenlijst van Sampit* (*Onderafdeling, Zuid-Borneo*). Bosbouwproefstation, Serie Boomnamenlijsten, no. 7); revised and expanded to cover whole area, 1949, Buitenzorg, as *idem. Lijst van boomsoorten verzameld in Kapoeas-Barito* (*Zuid-Borneo*). [iv], 61 pp. (Rapp. Bosbouwproefsta., no. 3; also designated as Serie Boomnamenlijsten, no. 12); 2nd revision, 1953, Bogor, as *idem. Daftar nama pohon-pohonan: Kapuas-Barito* (*Kalimantan-Selatan*). 77 pp. (Lap. Balai Penjelidik, Kehut., no. 62; also designated as Seri Daftar nama pohon-pohonan, no. 39).) [Province of Central Kalimantan; formerly the Kapuas-Barito district of the Residency of South and East Borneo.]

Idem, 1978. *Daftar nama pohon-pohonan: Kalimantan Barat*. 3rd revised edn. 107 pp., illus. (Lap. Lemb. Penelit. Hutan, no. 265). Bogor. (Mimeographed. Originally pub. 1941, Buitenzorg, as HILDEBRAND, F. H. (comp.) *Boomnamenlijst van Westerafdeling van Borneo* (Bosbouwproefstation, Serie Boomnamenlijsten, no. 10); first revision, 1952, Bogor, *idem. Daftar nama pohon-pohonan: Kalimantan-Barat*. 80

pp. (Lap. Balai Penjelidik. Kehut., no. 54; also designated as Seri Daftar nama pohon-pohonan, no. 31); second revision, 1970, as SOEWANDA AMONG PRAWIRA, R. *Daftar nama pohon-pohonan: Kalimantan Barat*. 99 pp. (Lap. Lemb. Penelit. Hutan, no. 107).) [Province, formerly Residency, of West Kalimantan (West Borneo).]

Sparsely annotated, uncritical checklists of forest trees known to attain a diameter (at breast height) of 40 cm or more, arranged alphabetically by families and genera; each entry (usually on one line) includes a vernacular name (Dayak or Malay), distribution (by subdistricts), number of collections (but no *exsiccata*-numbers), and durability class (on a scale of five). Each of the systematic lists proper is preceded by a list with all records, fully annotated, arranged alphabetically by vernacular names; at the end of each treatment are lists of amended botanical names, a lexicon of Indonesian names with more important local vernacular equivalents, an alphabetical lexicon of genera with family equivalents, and an indexed album (in most current issues) of full-page illustrations of key species depicting botanical features. No keys are included. [Much modern revisionary work has not been accounted for in current reissues, as noted in a review of the West Kalimantan list in *Flora Malesiana Bulletin*, p. 2129. However, these lists represent the sole organized coverage of Indonesian Borneo apart from the enumerations of Merrill and Masamune and general works on Malesia.]

918

Java (with Madura)

The associated islands of Krakatoa, Bawean and Kangean are included here. For Christmas Island, see **046**. Along with Maluku, Java was one of the earliest parts of Malesia to be botanized, but detailed exploration only began late in the eighteenth century and, despite a number of attempts, no complete and critical flora was to appear in print until the 1960s. Rivalries, false starts, a bitter controversy, varying institutional policies and support, and social and political changes marked the intervening period of botanical history in Java, nearly two centuries long. The flora – or what is left of it – is in general now well-documented, with a number of works of varying scope and style.

– Vascular flora: 4911 species (4598 native) (Backer and Bakhuizen).

BACKER, C. A. and BAKHUIZEN VAN DEN BRINK, R. C., JR., 1963–8. *Flora of Java*. 3 vols. 32 halftones, portrait, 2 maps. Groningen: Noordhoff (later Wolters-Noordhoff).

BACKER, C. A. and POSTHUMUS, O., 1939. *Varenflora voor Java*. 370 pp., 90 text-figs. Buitenzorg: s'Lands Plantentuin.

The first of this complementary pair of works is a concise, largely unillustrated descriptive manual of seed plants (6100 species, with over 4000 native), with species treatments in the form of keys; these treatments incorporate limited synonymy and abbreviated indication of local range, altitudinal zonation, habitat, and phenology, with taxonomic commentary (and some literature references) in footnotes. A complete index appears in vol. 3. In the introductory section (vol. 1) is a biography of Backer (with list of publications) and a brief account of the preparation and editing of the work, along with a somewhat forbidding key to families, while vol. 2 features an illustrated chapter on the vegetation and phytogeography of Java (by C. G. G. J. van Steenis and A. F. Schippers-Lammertse).

Like Koorders' *Exkursionsflora von Java* (see below), this work was patterned after the 'excursion-floras' of Europe, but is more technically descriptive and highly mannered, with less documentation and auxiliary information than in Koorders' work. A notable omission for a 'critical' work is a somewhat more complete synonymy. The absence of illustrations was deliberate, as the senior author believed them to have no place in a flora. The original Dutch version, by Backer alone, was revised and reproduced as *Beknopte flora van Java* (22 parts, 1942–61, Leiden: Rijksherbarium), an 'emergency' edition mimeographed as a precautionary measure.

The companion work in Dutch on the pteridophytes is similar in format to the main *Flora* but includes figures of representative species and is preceded by an introductory section with chapters on organography, fern communities, fern geography, a glossary, and a list of references, as well as general keys to all genera.

KOORDERS, S. H., 1911–12. *Exkursionsflora von Java, umfassend die Blütenpflanzen*. Vols. 1–3. xxiv, 2528 pp., 20 pls. (1 in color), 139 text-figs., 4 maps. Jena: Fischer.

KOORDERS, S. H., 1913–37. *Exkursionsflora von Java, umfassend die Blütenpflanzen*. Vol. 4, *Atlas*. pp. 1–688, 865–1020, pl. 1–976, 1166–1313. Jena: Fischer.

This notorious work comprises a sparsely illustrated manual-key to native, naturalized, and commonly cultivated seed plants (5067 species), with family and generic descriptions of variable length and

separate, more detailed accounting of montane species. Diagnoses in key leads include limited synonymy, references, localities and Javan range, altitudinal zonation, vernacular names, status, and terse notes on habit, habitat, biology, etc.; entries on montane species include additional synonymy, citations, indication of *exsiccatae*, localities, altitudes, etc., as well as extra-Javan distribution. Literature citations also appear in family and generic headings, and partial indices are given for each volume (addenda, corrigenda, and complete index in vol. 3). Arrangement follows the Englerian system, with use of Dalla Torre and Harms numbers. In the accompanying *Atlas* are large figures of most species in families Cycadaceae through Ranunculaceae as well as part of the Leguminosae (1110 pls. in all).

Designed after the fashion of European 'excursion-floras', this hastily compiled, uneven and rather uncritical work, for long severely criticized, has a few features not in its more critical successor: vernacular names, more detailed indication of local distribution, more synonymy and citations, and illustrations. Its origins as a work on the mountain flora are clearly evident. When published, this work sparked off a long and bitter controversy, but reasoned judgment has not on the whole been favorable.[6]

KOORDERS, S. H. and VALETON, T., 1894–1914. *Bijdragen tot de kennis der boomsoorten van (op) Java*. 13 parts (parts 1–10 in *Meded. Lands Plantentuin* 11, 14, 16, 17, 33, 40, 42, 59, 61, 68; parts 11–13 in *Meded. Dept. Landb. Ned.-Indië* 2, 10, 18.) Batavia, The Hague.

KOORDERS, S. H., 1913–18. *Atlas der Baumarten von Java* (im Anschluss an die 'Bijdragen tot de kennis der boomsoorten van Java'). 16 parts in 4 vols. 800 pls. Leiden.

The *Bijdragen* comprise a thoroughly documented, copiously descriptive flora of woody plants, comprising detailed family revisions with keys to genera and species with full synonymy (including references), literature citations, vernacular names, generalized indication of local and extralimital range (with some indication of *exsiccatae* and specific localities, notably concerning the Herbarium Koordersianum), taxonomic commentary, and notes on habitat, phenology and silviculture; general index to all botanical and vernacular names in part 13. Part 1 includes a short introduction to the work. Text in both Dutch and Latin. The *Atlas* comprises finely executed lithograph plates with short captions fully cross-referenced to the *Bijdragen*; vernacular names are included as well as brief synonymy, and the specimens used for the drawings have been cited. An outstanding work for its day, largely worked up at what is now Bogor with perforce limited reference to collections in Holland or elsewhere.

Partial work

STEENIS, C. G. G. J. VAN *et al.*, 1975. *Flora untuk sekolah di Indonesia.* Trans. at Universitas Gadja Mada, Yogyakarta. 495 pp., 46 text-figs. Jakarta-Pusat: Pradnya Paramita. (Originally pub. 1949 in Dutch as *Flora voor de scholen in Indonesië*, Jakarta; 2nd edn., 1951.)

Concise students' illustrated manual-key of native, naturalized and cultivated lowland plants commonly encountered, with notes on habit, distribution, phenology, special features, etc.; vernacular names; index. Covers about 400 species. The Indonesian translation is based on the 1949 version, and is in a slightly larger format than the original.

Special group – forest trees

The Forest Research Institute tree species list for Java, one of a series covering the whole of Indonesia (for background, see under **913**), first appeared in one number in 1950; the most recent reissue, however, is in three parts corresponding to the long-standing basic geographical and administrative divisions of Java and Madura.

SOEWANDA AMONG PRAWIRA, R. (comp.), 1976. *Daftar nama pohon-pohonan* [*di*] *Java-Madura*, I: Java Barat (West Java). 124 pp., 51 pls. (Lap. Lemb. Penelit. Hutan, no. 219). Bogor. (Mimeographed.)

Idem, 1977. *Daftar nama pohon-pohonan* [*di*] *Java-Madura*, II: Java Tengah (Central Java). 131 pp., 54 pls. (*ibid.*, no. 244).

Idem, 1977. *Daftar nama pohon-pohonan* [*di*] *Java-Madura*, III: Java Timur (East Java). 143 pp., 55 pls. (*ibid.*, no. 253).

Briefly annotated checklists of tree species (1280 in all) known to attain a height of 10 m and a diameter at breast height of 15 cm, arranged alphabetically by families and genera and including some synonymy, vernacular names, and notes on distribution, altitudinal range, wood properties, and (variously) notes on habitat, frequency and special features. Some introduced trees have also been included. Alphabetical lists of vernacular names (Indonesian, Sundanese, Javanese, Madurese, Dutch and English) with species reference numbers are also given. These lists relied heavily on published floras, including some of these described above, as comparatively less botanical material had been collected here by the Forestry Service in comparison with other parts of Indonesia. They are thus relatively more reliable than the lists produced for the 'Outer Provinces'. [However, the present issues are only slightly changed from the 1951 edition.] Earlier versions: HILDEBRAND, F. H. (comp.), 1950. *Daftar*

nama pohon-pohonan: Djawa-Madura. [i]. 171 pp. (Lap. Balai Penjelidik. Kehut., no. 35; also designated as Seri Daftar nama pohon-pohonan, no. 24). Bogor; revised and expanded 1951 as *idem*. [i], 183 pp. (Lap. Balai Penjelidik. Kehut., no. 50; also designated as Seri Daftar nama pohon-pohonan, no. 30). (Both issues mimeographed.)

919

Lesser Sunda Islands (Nusa Tenggara and Bali)

Comprises the island arc east of Java, from Bali to the Damar group, Timor, and Babar. Tree species lists are available for most areas, and a consolidated enumeration of the ferns was prepared by Posthumus (1944); other floristic lists are, however, few and scattered except for Timor. On the other hand, the vascular flora is essentially a depauperate extension of the Javanese flora and hence works on that area are useful. For an introduction to and analysis of the Lesser Sunda Islands flora, see KALKMAN, C., 1955. A plant-geographical analysis of the Lesser Sunda Islands. *Acta Bot. Neerl.* **4**: 200–25. A current bibliography appears in *Bot. J. Linn. Soc.* **79**: 175–6 (1979) as part of a review of East Malesian plant geography by C. G. G. J. van Steenis.

Partial works

BRITTEN, J., *et al.*, 1885. Prodromus florae timorensis; compiled in the Botanical Department of the British Museum. In *A naturalist's wanderings in the Eastern Archipelago* (ed. H. O. FORBES), App. VI, pp. 497–523. London: Sampson Low; New York: Harper.

Briefly annotated systematic list of vascular plants known from Timor, with descriptions of new taxa and brief notes on habitat, previous records, etc.; occasional taxonomic commentary; localities with some citations of *exsiccatae* (mostly of the author's collections); no separate index. A short introductory section includes a brief sketch of prior botanical exploration in the island (prepared by H. O. Forbes with assistance from H. N. Ridley). [The latest overall checklist for this botanically interesting island, climatically strongly influenced by Australia. Apart from the work of Meijer Drees listed below, little suitable floristic material exists. About the only Portuguese work of more recent date seen is SANTOS, P. E. C., 1934. *Aportamentos para o estudio da flora de Macau e de Timor.* 76 pp., 4 pp. addenda (loose insert), 27 halftones. [Lisbon: Jardim Colonial de Lisboa]; it is of little or no value and serves here merely as a measure of the very limited extent of Portuguese botanical work concerning Timor.]

MEIJER DREES, E., 1950. *Distribution, ecology and silvicultural possibilities of the trees and shrubs from the savanna-forest region in eastern Sumbawa and Timor.* 146 pp. (Lap. Balai Penjelidik. Kehut., no. 33). Bogor.

Included in this work are lists of trees and shrubs from Timor and eastern Sumbawa (both within the range of savanna and woodland species of *Eucalyptus*) with short descriptions and notes on distribution and ecology. [Not seen; reference and description from *Flora Malesiana*, ser. I, vol. 5: p. xxi (1955).]

Special groups – forest trees and pteridophytes

Included here are the only systematic works with overall coverage of the Lesser Sunda Islands and Timor. The first entry treats the area lists of forest trees produced in the series issued by the Forest Research Institute, Bogor; the second is a checklist of fern species by Posthumus (1944).

SOEWANDA AMONG PRAWIRA, R. and TANTRA, I. G. M. (comp.), 1972. *Daftar nama pohon-pohonan: Bali dan Lombok.* Revised edn. 24, 2 pp., 41 pls. (Lap. Lemb. Penelit. Hutan, no. 145). Bogor. (Mimeographed. Originally issued 1940, Buitenzorg, as HILDEBRAND, F. H. (comp.), *Boomnamenlijst van Bali en Lombok.* (Bosbouwproefstation, Serie Boomnamenlijsten, no. 2).)

HILDEBRAND, F. H. (comp.), 1953. *Daftar nama pohon-pohonan: Timor (Kepulauan Sunda Ketjil).* 51 pp. (Lap. Balai Penjelidik. Kehut., no. 60; also designated as Seri Daftar nama pohon-pohonan, no. 35). Bogor. (Mimeographed. Originally issued 1940, Buitenzorg, as *idem, Boomnamenlijst van Timor en onderhorigheden.* (Bosbouwproefstation, Serie Boomnamenlijsten, no. 3.) No further revision seen.)

Sparsely annotated, uncritical checklists of forest trees known to attain a minimum diameter, arranged alphabetically by families and genera; each entry (usually on one line) includes a vernacular name, distribution (by subdistricts), number of collections (but no *exsiccata*-numbers), and durability class (on a scale of five). Each of the systematic lists proper is preceded by a list with all records, fully annotated, arranged alphabetically by vernacular names; at the end of each treatment are lists of amended botanical names, a lexicon of Indonesian names with more important local vernacular equivalents, an alphabetical lexicon of genera with family equivalents, and an indexed album of full-page illustrations of key species depicting botanical features (only in the Bali-Lombok list). No keys are included. [The deficiencies of these lists are noted under **913** and **917**, with an example given under **927**.]

POSTHUMUS, O., [1944]. *The ferns of the Lesser Sunda Islands.* (Ann. Bot. Gard. Buitenzorg, *hors série*, pp. 35–113.) Buitenzorg.

Systematic enumeration of known ferns, with synonymy, references and citations, detailed indication of local range (with *exsiccatae* listed), taxonomic commentary, and notes on habitat; no separate index. The introductory section includes an account of previous botanical work in the area along with remarks on climate, etc., of representative localities.

Region

92

Central Malesia (including the Philippines)

In this region are included Celebes (Sulawesi), the Philippines (except for Palawan), and the Moluccas, as well as all associated islands. This region, along with the Lesser Sunda Islands (here included in Region 91 on botanical evidence), has been called 'Wallacea' by a number of authors and an 'unstable zone' by Simpson – a kind of 'no man's land' between the Oriental and Australasian zoogeographic regions – and is traversed by 'Weber's Line of faunal balance' as redefined by Mayr.[2] The western and eastern limits of this region are here largely conventionally defined by the respective continental shelves, with the exception indicated above. Both on botanical and zoological evidence, Region 92 is not uniform as an intermediate zone but is marked by various clines of biogeographic transition, a concept advanced by Simpson and subscribed to here. The limits or units in this region have therefore been rather finely drawn.

Although the region is as a whole served by *Flora Malesiana* and (to varying degrees) other works on the Greater Malesian superregion, separate documentation for the various units here recognized is very uneven, a situation to some extent matched by the current level of botanical exploration. Much of the available literature is now comparatively old.

The best-documented unit is certainly the Philippines, where *Enumeration of Philippine flowering plants* by Merrill (1923–6) has long been a standard work. With this may be associated *Fern flora of the Philippines* by Copeland (1958–60). A very good local flora is *Flora of Manila* by Merrill (1912; reprinted 1974). However, forest-botanical literature is scanty, apart from *Lexicon of Philippine trees* by Salvosa (1963); this reflects the lack of significant development of forest botany as a research field in the Philippines, despite the example of Vidal in the nineteenth century. The isolated Batanes Islands to the north were treated by Hatusima (1966) in his *An enumeration of the plants of Batan Islands, northern Philippines*. No separate accounts exist for the Sulu Archipelago.

Coverage of Sulawesi and Maluku is very imperfect. For the former, there is little else but the *Enumeratio specierum phanerogamarum Minahassae* by Koorders (1898; additions 1902–4, 1918–22) and *Die Farnflora von Celebes* by H. Christ (1898; additions 1904). The phytogeography has been surveyed by Lam (1942, 1950). For Maluku, herein divided into three units, the only major coverage remains the ancient classic *Herbarium amboinense* by Rumphius (1741–50; Auctuarium, 1755), subsequently interpreted by many authors, notably Merrill in 1917; the latest list of modern equivalents appeared in 1959. The islands of southeastern Maluku were sketchily covered by Hemsley (1885), while the Talaud group has had one collection report, relatively substantial, by Holthuis and Lam (1942). No separate accounts exist for the Sanghie Islands, northeast of Sulawesi, nor for the Northern Maluku islands.

As in western and southern Malesia, the Indonesian islands are all covered by mimeographed lists of forest tree species in A4 format, part of a series produced by the Forest Research Institute, Bogor, since 1940. Two lists cover Sulawesi and one deals with North and South Maluku. They are currently in their third version, but are in need of thorough critical revision like their counterparts further west.

Sulawesi as a whole is the least explored island in Greater Malesia, with the exception of Irian Jaya; only the northeastern Minahasa and the southwestern peninsula have had more than a few collectors. Maluku is also unevenly known although overall collecting is at a higher level; in particular, North Maluku is very poorly known and documented.

Bibliographies. General and supraregional bibliographies as for Superregion 91–93.

Indices. General and supraregional indices as for Superregion 91–93.

921

Celebes (Sulawesi)

With the main island are here included the adjacent islands of Salajar, the Butung group and the Banggai Archipelago. The Sula Islands have been grouped with the North Moluccas. It is regrettable that for an area of great phytogeographic interest, only scattered trip reports and checklists are available; easily the most substantial of these is Koorders' report on the northeastern Minahasa district (see below). Tree species lists are furthermore available for the whole area, and there is an old enumeration of pteridophytes by Christ (1898). A general introduction to Sulawesian floristics and plant geography (in Dutch) is LAM, H. J., 1950. Proeve eener plantengeografie van Celebes. *Tijdschr. Kon. Ned. Aardrijksk. Genootsch.* **67**: 566–604, 3 text-figs., 7 halftones; for an earlier version in English, see *idem*, 1942. Notes on the historical phytogeography of Celebes. *Blumea*, **5**: 600–40. [Lists of references included.]

Partial works: Minahassa

KOORDERS, S. H., 1898. *Verslag eener botanische dienstreis door de Minahasa, tevens eerste overzicht der Flora van N. O. Celebes*. xxvi, 716 pp., 17 pls. (incl. maps) (Meded. Lands Plantentuin, 19). Batavia, The Hague.

Part III (pp. 253–645) of this work comprises the author's *Enumeratio specierum phanerogamarum Minahassae*, an annotated systematic account of the seed plants known from northeastern Celebes; this includes brief descriptive notes, limited synonymy, vernacular names, detailed indication of local range, and remarks on habitats, uses, etc.; index to vernacular names. The introductory section to the whole work (pp. i–xxvi) includes chapters on physical features, geography, and botanical exploration; part I (pp. 1–110) describes Koorders's itinerary; while part II (pp. 111–252) includes an account of useful plants. For additions, see *idem*, 1902–4. [1.]–3. Nachtrag.... *Natuurk. Tijdschr. Ned.-Indië*, **61**: 250–61; **63**: 76–89, 90–99; also *idem*, 1918–22. *Supplementum op het eerste overzicht der Flora van N. O. Celebes*. Parts 1, 2/3. 50, 121 pp.; 13, 127 pls. Batavia: The author (part 2/3 issued by Mme A. Koorders-Schumacher). (Part 1(2) also pub. as *Bull. Jard. Bot. Buitenz.*, III, **2**: 242–60 (1920).) [The pteridophytes were included in *Die Farnflora von Celebes* by Christ (see below).]

Special groups – forest trees and pteridophytes

Included here are the only systematic works with overall coverage of Sulawesi. These include two tree species lists in the series produced by the Forest Research Institute, Bogor (for background, see under **913**), and a checklist (with supplement) of pteridophytes by Christ (1898, 1904).

SOEWANDA AMONG PRAWIRA, R. (comp.), 1972. *Daftar nama pohon-pohonan: Sulawesi Selatan, Tenggara dan sekitarnja* (South and Southeast Celebes and dependencies). Revised edn. 113, 4 pp., 55 pls. (Lap. Lemb. Penelit. Hutan, no. 151). Bogor. (Mimeographed. Originally issued 1940, Buitenzorg, as HILDEBRAND, F. H. (comp.). *Boomnamenlijst van Selebes en onderhorigheden.* (Bosbouwproefstation, Serie Boomnamenlijsten, no. 4); revised 1950, Bogor, as *idem, Daftar nama pohon-pohonan: Selebes.* [i], 105 pp. (Lap. Balai Penjelidik. Kehut., no. 43; also designated as Seri Daftar nama pohon-pohonan, no. 26).)

SOEWANDA AMONG PRAWIRA, R. (comp.), 1972. *Daftar nama pohon-pohonan: Sulawesi Tengah dan Utara* (Central and North Celebes). Revised edn. Illus. (Lap. Lemb. Penelit. Hutan, no. 156). Bogor. (Mimeographed. Originally issued 1940, Buitenzorg, as HILDEBRAND, F. H. (comp.). *Boomnamenlijst van Manado.* (Bosbouwproefstation, Serie Boomnamenlijsten, no. 1); revised 1951, Bogor, as *idem. Daftar nama pohon-pohonan: Manado.* [i], 55 pp. (Lap. Balai Penjelidik. Kehut., no. 44; also designated as Seri Daftar nama pohon-pohonan, no. 27).)

Sparsely annotated, uncritical checklists of forest trees known to attain a diameter (at breast height) of 40 cm or more, arranged alphabetically by families and genera; each entry (usually on one line) includes a vernacular name (Buginese, Malay, Tobelo, Macassarese, Toraja, Minahassan, etc.), distribution (by subdistricts), number of collections (but no *exsiccata*-numbers), and durability class (on a scale of five). Each of the systematic lists proper is preceded by a list with all records, fully annotated, arranged alphabetically by vernacular names; at the end of each of the two treatments are lists of amended botanical names, a lexicon of Indonesian names with more important local vernacular equivalents, an alphabetical lexicon of genera with family equivalents, and an indexed album (in current issues) of full-page illustrations of key species depicting botanical features. No keys are included. [For limitations of these lists, see discussion under 913 or 917.]

CHRIST, H., 1898. Die Farnflora von Celebes. *Ann. Jard. Bot. Buitenz.*, 15: 73–186, pls. 13–17 (incl. map).

CHRIST, H., 1904. Zur Farnflora von Celebes, II. *Ibid.*, 19: 33–44.

The main work comprises a systematic enumeration of pteridophytes, with descriptions of some novelties and inclusion of synonymy, localities with citations of *exsiccatae*, and notes on local and extralimital distribution, habitat, etc.; figures of selected species and map with collecting localities. [Both it and the supplement are based on collections by the Sarasin brothers, Koorders, and Warburg along with others of earlier date in the Bogor herbarium.]

922

Sangihe Islands

This chain of partly active volcanic islands extends from the end of the Minahassa Peninsula in northeastern Celebes north towards the Philippines. No separate checklists appear to be available.

923

Sulu Archipelago

No separate checklists are available for this island chain stretching from the southern Philippines and eastern Sabah, in which Jolo and Tawitawi are the largest islands. However, records as far as available are indicated in *Enumeration of Philippine flowering plants* by MERRILL (see 925). The area is, like Palawan, intermediate between Borneo and the Philippines in its biogeography but has more links with the latter.

924

Batanes Islands

This isolated small group, located between the Philippines and Taiwan but biogeographically most closely related to the former, has one recent checklist by Hatusima (see below).

HATUSIMA, S., 1966. An enumeration of the plants of Batan Islands, northern Philippines. *Mem. Fac. Agric. Kagoshima Univ.* 5: 13–70, 10 half tones, map.

Systematic enumeration, based on the results of a staff-student field trip but incorporating earlier published records; includes synonymy, references, localities with *exsiccatae*, and notes on occurrence, habitat, special features, etc.; discussion of floristics. Features some new species and many new records additional to those in Merrill's *Enumeration of Philippine Flowering Plants*.

925

The Philippines (Pilipinas)

The substantial amount of exploration and floristic and systematic work done in the Philippines, especially after 1900, was summarized by Merrill in his well-known *Enumeration* (1923–6) with respect to the seed plants, while the ferns were revised in a descriptive manual by Copeland (1958–60). There is also a useful local flora of the Manila district by Merrill (1912). However, little else of substantive value as a flora has been published. With regard to the forest flora, the lack of a strong tradition of forest botany comparable with that in Malaysia, Indonesia and New Guinea is the most likely reason why no significant manuals or checklists on the forest or tree flora comparable to those in western Malesia or Papuasia have ever been published.

– Vascular flora: about 9100 species (Merrill, 1926).

Bibliographies

MERRILL, E. D., 1926. Bibliography of Philippine botany. In his *Enumeration of Philippine flowering plants*, vol. 4, pp. 155–239. Manila.

NEMENZO, C. A., 1969. *The flora and fauna of the Philippines, 1851–1966: an annotated bibliography*. Part I: *Plants*. 307 pp., 1493 entries (Nat. Appl. Sci. Bull. Univ. Philipp., 21). Quezon City.

COPELAND, E. B., 1958–60 (1958–61). *Fern flora of the Philippines*. 3 parts. iv, 557, [iv, iv] pp. (Monogr. Philipp. Inst. Sci. Tech., 6). Manila: National Institute of Science and Technology, Philippines.

Briefly descriptive systematic treatment of true ferns, with keys to genera and species; full synonymy, with references and citations; taxonomic commentary and generalized indication of local and extralimital distribution, with citations of *exsiccatae* for less common species; index to all botanical names. The introductory section includes an account of fern geography as well as an explanation of terminology.

MERRILL, E. D., 1922(1923)–26. *An enumeration of Philippine flowering plants*. 4 parts. 3 text-figs., 6 maps (Publ. Bur. Sci. Philipp., 18). Manila: Department of Agriculture and Natural Resources, Philippine Islands. (Reprinted 1968, Amsterdam: Asher.)

Detailed systematic enumeration of seed plants, with full synonymy, references, and literature citations; generalized indication of local and extralimital distribution, with citations of *exsiccatae* for lesser-known species; taxonomic commentary and brief notes on habitat, etc., as recorded; vernacular names. Volume 4, apart from additions and corrections to the enumeration proper, is devoted to a general introduction, with chapters on physical features, climate, Philippine peoples and alphabets, origin of local vernacular names, botanical exploration and research, ecology, the biogeographical subdivisions of the archipelago, overall relationships of the flora and fauna, and geological correlations; a detailed bibliography (pp. 155–239) and complete indices to botanical and vernacular names are also provided. For additions, see QUISUMBING, E., 1930–44. New or interesting Philippine plants, I–II. *Philipp. J. Sci.* **41**: 315–71, 28 figs., 3 pls.; **76**: 37–56; also QUISUMBING, E. and MERRILL, E. D., 1928–53. New Philippine plants, I–II. *Ibid.* **37**: 133–213, 4 pls.; **82**: 323–39, 5 pls.

Partial work: Manila district

MERRILL, E. D., 1912. *A flora of Manila*. 490 pp. (Publ. Bur. Sci. Philipp., 5). Manila: Department of Agriculture and Natural Resources, Philippine Islands. (Reprinted 1968, Lehre, Germany: Cramer; 1974, Manila: Bookmark.)

Descriptive manual of wild and more commonly cultivated vascular plants of an area of about 100 km² around the city of Manila; includes analytical keys, essential synonymy, English and Tagalog vernacular names, indication of local and extralimital range, and notes on habitat, phenology, origin, cultivation, special features, etc.; indices to vernacular, generic, and family names. The introductory section includes an introduction to descriptive botany together with a glossary. Useful generally in inhabited lowland areas of the Philippines, as it covers adequately the widespread secondary and adventive flora. [Prepared in part as a students' manual following the author's part-time academic appointment in the new University of the Philippines in 1909. Although now sorely in need of revision, it is still in use for this purpose as the only such work in the Philippines.]

Special group – forest trees

As noted above, scarcely any suitable manuals on the woody flora and/or forest trees, even the principal species, have been produced in the Philippines in the twentieth century. The only works which may be noted here in passing are AHERN, G. P., 1901. *Compilation of notes on the most important timber tree species of the Philippine Islands*. 112 pp., 43 color pls.

Manila; and SALVOSA, F. M., 1963. *Lexicon of Philippine trees.* 136 pp. (Forest Products Res. Inst. Bull., 1). Los Baños, Laguna.

926

Talaud Islands

This small, non-volcanic group of islands lies between the Philippines and North Maluku. No general checklist is available, but the substantial report of a botanical survey of 1926 involving the group is accounted for below.

HOLTHUIS, L. B. and LAM, H. J., 1942. A first contribution to our knowledge of the flora of the Talaud Islands and Morotai. *Blumea*, 5: 93–256.

Systematic enumeration of collections made on a 1926 expedition, with descriptions of novelties, critical remarks, and notes on phytogeography, etc.; citations of *exsiccatae* included.

927

Maluku (The Moluccas): general

No floras or enumerations cover Maluku as a whole save for one (actually two under one cover) of the tree species lists issued as part of the Indonesia-wide series produced by the Forest Research Institute, Bogor (for background, see under 913). Key partial works are here given under 928.

Special group – forest trees

SOEWANDA AMONG PRAWIRA, R. (comp.), 1975. *Daftar nama pohon-pohonan: Maluku Utara dan Selatan* (North and South Moluccas). Revised edn. 130 pp., 61 pls. (Lap. Lemb. Penelit. Hutan, no. 210). Bogor. (Mimeographed. Originally issued 1941, Buitenzorg, as HILDEBRAND, F. H. (comp.) *Boomnamenlijst van Molukken.* (Bosbouwproefstation, Serie Boomnamenlijsten, no. 5); revised 1951, Bogor, in two parts as *idem, Daftar nama pohon-pohonan: Maluku-Utara.* [i], 53 pp. (Lap. Balai Penjelidik. Kehut., no. 45; also designated as Seri Daftar nama pohon-pohonan, no. 28). and *Daftar nama pohon-pohonan: Maluku-Selatan.* [i], 59 pp. (Lap. Balai Penjelidik. Kehut., no. 49; also designated as Seri Daftar nama pohon-pohonan, no. 29). In the current version, the two lists of 1951 remain as discrete entities, but share one album of illustrations.)

Sparsely annotated, uncritical checklists of forest trees respectively of Maluku Selatan and Maluku Utara known to attain a diameter (at breast height) of 40 cm or more, arranged alphabetically by families and genera; each entry (usually on one line) includes a vernacular name (Alfur, Malay, etc.), distribution (by subdistricts), number of collections (but not *exsiccata*-numbers), and durability class (on a scale of five). Each of the systematic lists proper is preceded by a list with all records, fully annotated, arranged alphabetically by vernacular names; at the end of each treatment are lists of amended botanical names, a lexicon of Indonesian names with more important local vernacular equivalents, and an alphabetical lexicon of genera with family equivalents. Following the separate lists is an indexed album of full-page illustrations of key species depicting botanical features. No keys are included. [Among the defects of these lists is the lack of consideration of much more recent taxonomic work. For example, under Maluku Selatan there is listed *Tristania* sp. 15 which in the list of name changes (pp. 58–9) was amended to *Metrosideros nigroviridis*; but nowhere is there an indication of the further change to *Lindsayomyrtus nigroviridis*, the name currently accepted. Similar instances could be quoted. Nevertheless, the two lists represent the sole organized floristic work covering all of Maluku.]

928

Maluku: subdivisions

Under this heading are listed works referring to one or another of the three provinces into which Maluku is currently divided: Maluku Utara (North Moluccas), Maluku Tengah (Central Moluccas), and Maluku Tenggara (Southeastern Moluccas). The last two together constituted the former division of Maluku Selatan (South Moluccas).

I. Maluku Utara

This province comprises the northern Moluccas from Morotai in the north through Halmahera, Bacan and Obi to the Sula group, along with associated smaller islands such as Gebe, Ternate and Tidore. As befitting a part of the 'Spice Islands', botanical contacts have been prolonged, but collecting has been sporadic and even now some of the islands are not well know although efforts continue. The only checklist available is the section on Maluku Utara (pp. 67–126) in the tree species list by SOEWANDA (927).

ii. Maluku Tengah

The main islands of the present province of Maluku Tengah – the heart of what was formerly Maluku Selatan (South Moluccas) – are Buru, Seram (Ceram) and Ambon; but also included are the Banda group and the southeastern islands stretching from the end of Seram out to Seram Laut. The only passably general work for this heart of the 'Spice Islands' is the classical *Herbarium amboinense* by G. E. Rumphius, the 'Plinius indicus' who lived on Ambon for nearly 50 years (1653–1702). The only other overall checklist available is the section on Maluku Selatan (pp. 2–66) in the tree species list by SOEWANDA (**927**).

RUMPHIUS, G. E., 1741–50. *Herbarium amboinense* [Het amboinsche kruid-boek]. (Trans. and ed. J. Burman.) 6 vols. 669 pls., portrait. Amsterdam: Uytwerf. (Reissued 1750.)

RUMPHIUS, G. E., 1755. *Herbarium amboinense auctuarium.* (Trans. and ed. J. Burman.) 74, [20] pp., 30 pls. Amsterdam.

Amply descriptive, copiously illustrated treatment (with parallel Latin and Dutch text) of vascular and some larger non-vascular plants (and corals, then thought to be partly plant-like), including indications of localities where observed or reported and copious notes on life-form, natural history, special features, cultivation, properties, uses, folklore, etc., as well as local vernacular names. The author's preface in volume 1, dated 1690, explains the arrangement of the work (in twelve 'books' or primary classes based on habit, practical criteria, local custom and other features) and its geographical coverage (not wholly restricted to Ambon), while the *Auctuarium* includes additions as well as a complete index (the latter with some Linnean nomenclatural equivalents). For modern nomenclatural interpretations of this fundamental work, see MERRILL, E. D., 1917. *An interpretation of Rumphius's* Herbarium amboinense. 595 pp., 2 maps (Publ. Bur. Sci. Philippines, no. 9). Manila; and WIT, H. C. D. DE, 1959. A checklist to Rumphius's *Herbarium amboinense.* In *Rumphius memorial volume* (ed. *idem*), pp. 339–460. Baarn: Hollandia. [The literature on *Herbarium amboinense* is considerable, but attention should be drawn to the following: PEETERS, A., 1979. Nomenclature and classification in Rumphius's *Herbarium amboinense.* In *Classifications in their social context* (ed. R. F. ELLEN and D. REASON), pp. 145–66. London: Academic Press.]

III. Maluku Tenggara

The Southeastern Moluccas, biogeographically more or less intermediate between the Lesser Sunda Islands, the central and northern Moluccas, and Papuasia, includes the Tenimber (Timor Laut) group, the Kai (Ké or Kei) Islands, and (for practical reasons) the Aru Islands. The province also encompasses the so-called Southwestern Islands (Kepulauan Barat Daya, covering the islands from Wetar to the Damar group and from Leti to Babar), which form the end of the Lesser Sundas island chain. Only two organized floristic lists cover this area: the section on Maluku Selatan (pp. 2–66) in the tree species list by Soewanda (see above under **927**) and Hemsley's enumeration, described below.

HEMSLEY, W. B., 1885. The south-eastern Moluccas. In *Reports on the scientific results of the voyage of 'HMS Challenger' during the years 1873–76* (ed. J. Murray), *Botany*, vol. 1, part 3, pp. 101–226, pls. 64–5. London: HMSO. (Reprinted 1965, New York: Johnson.)

Includes among other writings an annotated enumeration of non-vascular and vascular plants from the Kei, Tenimber, and Aru Islands as well as other parts of the southeastern Moluccas, with descriptions of novelties, entries include synonymy, references, brief indication of range, *exsiccatae*, critical remarks, and notes on occurrence, special features, cultivation, etc., as well as phytogeography. No separate index is provided, but a discussion of Beccari's species from the area is given in an appendix. [The plates actually appeared in 1884.]

Region

93

Papuasia (East Malesia)

Papuasia, a biogeographical term first introduced by O. Warburg in the 1890s, here includes the great island of New Guinea together with associated islands and extends eastwards to the end of the Solomon Islands group. Its western limit is defined by the continental shelf with which Lydekker's Line is associated. The bulk of the Torres Strait Islands, however, has been included with eastern Australia (region 43) as **439**. The region thus includes the Indonesian province of Irian Jaya, Papua New Guinea (i.e., the former Territory of Papua and the Trust Territory of New Guinea), and the greater part of the state of the Solomon Islands (formerly British Solomon Islands Protectorate).

Of the general works on Malesia given under **910–30**, only *Flora Malesiana* covers Papuasia at all adequately, and even then omits the Louisiades (**935**) and Solomons and the outer atolls (**938–39**). At the time the *Flora Malesiana* project was being developed, these groups, then very imperfectly known botanically, were loosely thought to be 'Pacific'. Subsequent work has shown that their affinity is Papuasian, although in both groups unusual elements exist.

No comprehensive flora for Papuasia is yet available, although a beginning has been made in this direction with publication of the first volume of the series *Handbooks of the flora of Papua New Guinea* (1978–), a project of the Division of Botany of the Papua New Guinea Office of Forests at Lae. A forerunner to this work, terminated far from completion, was *Manual of the forest trees of Papua and New Guinea* (1964–70), also prepared at the Division of Botany. While concentrating on Papua New Guinea, both works extend coverage to Irian Jaya (West New Guinea) and the Solomon Islands through utilization of literature records and collections available at Lae. However, as of 1980 neither series covers more than 11 families and, even given favorable conditions, completion of the *Handbooks* series will take many decades. It is also the wrong kind of flora project for Papua New Guinea given the present level of botanical knowledge and current national needs.

The literature as a whole comprises a bewildering variety of expedition checklists, series of 'contributions', partial floras and florulas, and works on special groups (including trees), as well as monographs and revisions, some not limited to Papuasia. Coming to grips with it is almost as formidable a task as that which faced Lam in the 1930s, although bibliographic aids in the interim have improved.[7] To a very large extent, this reflects the confused but compelling biohistory of the region: different political/administrative regimes in space and time, the lack of local botanical centers until the 1940s, imperial rivalries, availability of resources and motivation, the disruptions of war, occupation and other events, and above all the magnetic attraction of the 'Last Unknown' for natural historians from many countries.[8] Collections and associated materials are thus widely scattered, in contrast to, say, peninsular Malaysia or Indochina. Any synthesis, while sorely needed, would thus be a formidable undertaking, better handled through the *Flora Malesiana* project than as a separate regional series like the *Handbooks*. However, for regional use in the shorter term, production of a 'generic flora' may be the best progressive step, in association with a reproduced synthesis from the three or four nomenclatural/bibliographic card files known to exist.

The most comprehensive group of contributions, although not covering all families, is the great *Beiträge zur Flora von Papuasien* (1912–42), with 150 numbers in 26 parts successively edited by Lauterbach and Diels. Containing many keyed regional revisions, it was largely completed by 1943, the year of destruction of the greater part of the Berlin-Dahlem collections through bombing (F. Markgraf, personal communication). This, along with the two Division of Botany works already referred to, are treated here as major regional works. Other leading contribution-series are the botanical volumes of the original *Nova Guinea* (1909–36), edited by Pulle, and the 'botanical results' of the Archbold Expeditions, including *Plantae papuanae archboldianae* by Merrill and Perry (1939–53; the last three parts by Perry alone). However, as practical contributions these two latter series are of lesser consequence, being relatively miscellaneous in nature, although both are for the most part separately indexed. Some contributions in *Nova Guinea*, a work largely limited to present-day Irian Jaya, were written by authors also contributing to the *Beiträge*. Many other collection reports, some relatively substantial or significant like those by Gibbs, Ridley, Valeton, or Warburg, also exist but as they are scarcely floras or checklists cannot be considered here. the same limitation applies to most of the several area florulas and checklists which have appeared from the late-nineteenth century to the 1970s, as well as to the disorganized *Descriptive notes on Papuan plants* by von Mueller (1875–90), for the southeastern (British/Australian) zone a counterpart of the *Beiträge* and *Nova Guinea*.

Listings are here given, however, for the two parts of *Flora der deutschen Schutzgebiete in der Südsee* by Schumann and Lauterbach (1901–5), the first general account for former German New Guinea (as well as Micronesia and western Samoa in the first part), and *Guide to the forests of the British Solomon Islands* by Whitmore (1966), which contains the only recent checklist for the whole of the Solomon Islands. Also worthy of note are the accounts by Hemsley (1885) for the Aru Islands and the Admiralty group, Wagner and Grether's treatment of pteridophytes in the Admiralties (1948), Foreman's checklist of the flora of Bougainville (1972), and, finally, Peekel's remarkable *Illustrierte Flora des Bismarckarchipels für Naturfreunde*, a manuscript completed in 1947 and now available on microfiche. However, knowledge of the island groups associated with New Guinea is in general still very imperfectly organized, if now more extensive than 40 years ago, and good separate accounts are much needed apart from those on the Solomons.

The major forest trees of Irian Jaya, Papua New Guinea and the Solomons have each been covered in one or more processed or printed publications of varying quality. These have been grouped under the subheading Special group – forest trees under **930**, together with those on the Solomon Islands, given under **938**.

A rather imperfect set of keys to families and genera, with a bibliography of sources, was distributed by van Royen in 1959 and is accounted for below under **930**. New versions are reportedly in preparation.

Botanically, the region is at present on average about as well collected as Borneo but coverage is still quite uneven, Irian Jaya lagging considerably behind Papua New Guinea and the Solomons and the southern fall of the central cordillera somewhat neglected. Undercollected areas of more limited extent exist elsewhere, but much work in recent years has been carried out in the Louisiades, Bismarcks and Admiralties, making sketchy floristic checklists feasible.

For the major new flora of the alpine and subalpine belts by VAN ROYEN, see **903** (under New Guinea).

Bibliographies. General and supraregional bibliographies as for Superregion 91–93. Particular attention should be given to volume 1 of *Tōa kyō-ei-ken sigenkagaku bunken mokuroku*, which specifically covers New Guinea (botany, pp. 20–61).

Regional bibliography

LAM, H. J., 1934. Materials towards a study of the flora of the island of New Guinea. *Blumea*, 1:115–59, 2 maps. [This fine survey includes *inter alia* a selection of more important references on New Guinea, as well as detailed indices to the family revisions in *Beiträge*

zur Flora von Papuasien and *Nova Guinea, Botanique* as far as then available.[1]]

Indices. General and supraregional indices as for Superregion 91–93.

930

Region in general

The major works on the region, all of which have keys, are *Beiträge zur Flora von Papuasien* (1912–42), *Handbooks of the flora of Papua New Guinea* (1978–), and the incomplete *Manual of the forest trees of Papua and New Guinea* (1964–70). Key subsidiary works include *Flora der deutschen Schutzgebiete in der Südsee* (1901–5), *Nova Guinea: Botanique* (1909–36), and *Plantae Papuanae Archboldianae* (1939–53), here given under Partial works. These are followed by a selected group of specialized works on forest trees.

– Vascular flora: 13000–15000 species (Frodin, unpublished).

Keys to families and genera

ROYEN, P. VAN, [1959]. *Keys to the families and genera of higher plants in New Guinea*. 3 parts. Leiden. (Mimeographed.) [Largely compiled, artificial analytical keys to the genera and families of seed plants in mainland New Guinea, with family descriptions; index. An appendix gives a useful list of source references, arranged by families. Now rather incomplete and greatly in need of revision.]

LAUTERBACH, C. and DIELS, L. (eds.), 1912–42. Beiträge zur Flora von Papuasien, I–XXVI (in 150 parts). Illus. *Bot. Jahrb. Syst.* **49–72**, *passim*. (Reprinted separately.)

Comprises regional revisions and miscellaneous contributions, often with keys, on numerous families and genera, containing descriptions, synonymy (with references and citations), indication of localities with *exsiccatae*, and notes on classification, habitat, special features, etc.; detailed indices in Beiträge VI and X.[9] Many novelties have been described in this series, and much material also published in *Nova Guinea* incorporated. Some treatments are limited to former German New Guinea south of the equator, but others deal with Papuasia as a whole and even beyond. General introductions to many families were written from a floristic point of view by the senior editor, who also devoted a number of parts to a floristically based vegetation survey (1928–30). Up to the present, the

Beiträge collectively have remained the most important single contribution to Papuasian botany, with more lasting value than the contributions in *Nova Guinea* or in *Plantae papuanae archboldianae* (and its associated series), with some notable exceptions. It is thus a matter for regret that copyright problems have so far kept it from being reprinted. A number of family revisions were, however, published elsewhere, the most notable being SCHLECHTER, R., 1911–14. *Die Orchidaceen von Deutsch Neu-Guinea*. lxvi, 1079 pp. (Repert. Spec. Nov. Regni Veg. (Fedde), Beih., 1). Berlin: Fedde. (Reprinted 1974, Königstein/Ts., W. Germany: Koeltz.)

ROYEN, P. VAN *et al.*, 1964–70. *Manual of the forest trees to Papua and New Guinea*. 9 parts (in 6 fascicles). Illus., maps. Port Moresby, Papua New Guinea: Department of Forests, Territory of Papua and New Guinea (reprinted). (First edn. of part 1, 1964; 2nd edn. 1970.)

Descriptive treatments, in part compiled, of individual families including forest trees (with emphasis on the territories now comprising Papua New Guinea and on the more important species), with keys, partial synonymy, local distribution (with some citations of *exsiccatae*), and notes on taxonomy, habitat, life-form, wood, properties, uses, etc.; some vernacular as well as trade names included. The best treatment is the revised version of the Combretaceae (by M. Coode); others cover Anacardiaceae, Apocynaceae, Dipterocarpaceae, Eupomatiaceae, Himantandraceae, Magnoliaceae, Sapindaceae, and Sterculiaceae. The series has evidently been discontinued; some family treatments prepared for the work were not published.

WOMERSLEY, J. S. (ed.), 1978– . *Handbooks of the flora of Papua New Guinea*. Vol. 1, xvii, 278 pp., 115 text-figs., 2 maps (one folding). Melbourne: Melbourne University Press (for the Papua New Guinea Government).

Comprises a random group of well-illustrated regional family revisions by various authors; each treatment includes keys to genera and species, descriptions, some synonymy, generalized indication of internal range (by 'district', i.e., province, in Papua New Guinea; or regency, in Irian Jaya) and extralimital distribution, critical remarks, notes on ecology, special features, biology, dispersal, uses, etc., and family/generic literature sources; illustrated glossary, list of higher plant families in Papua New Guinea (209 in all), selected references (pp. 273–4), and botanical name index. The general part includes technical notes and a brief introduction to the vegetation. The first volume treats 164 species in 11 families (about 1.5 per cent of all species in the region); subsequent volumes will contain groups of additional families as they are ready (as in *Flora Malesiana*), without a systematic sequence. Emphasis in this work has been placed on Papua New Guinea, although it covers the whole of Papuasia as recognized here. [Volume 2, covering mostly small families but including Elaeocarpaceae and the Loranthaceae *sensu stricto*, was not released until well into 1982.]

Secondary and partial works

MERRILL, E. D. and PERRY, L. M. (eds.), 1939–49. Plantae papuanae archboldianae, I–XVIII. *J. Arnold Arbor.* **20–30**, *passim*.

PERRY, L. M., 1949–53. Plantae papuanae archboldianae, XIX–XXI. *Ibid.*, **30**: 139–65; **32**: 369–89; **34**: 191–257.

SMITH, A. C., 1941–4. Studies of Papuasian plants, I–VI. *Ibid.*, **22**: 60–80, 231–52, 343–74, 497–528; **23**: 417–43; **25**: 104–298.

These sets of contributions, which together make up the bulk of the 'Botanical results of the Archbold Expeditions' (others, mostly single entities, were published both in the *Journal of the Arnold Arboretum* and *Brittonia*), comprise treatments of a majority of flowering plant families in Papuasia. In the papers by Merrill and Perry, they mostly take the form of descriptions of novelties, critical notes, and new records with citations of *exsiccatae* and localities; while in the papers by Perry alone and by Smith they comprise full revisions of families and genera, sometimes (in Perry's treatments) with keys but in all cases including descriptions, localities with citations of *exsiccatae*, and critical notes. The last part of the Merrill and Perry series has an index to genera and families covered. [Based principally on the collections of L. J. Brass made on the first three Archbold Expeditions of 1933–4, 1936–7, and 1938–9. Many treatments limited to material available in the USA.]

PULLE, A. A. *et al.*, 1909–36. Botanique. In *Nova Guinea. Résultats de l'expédition scientifique néerlandaise à la Nouvelle Guinée* (ed. H. A. Lorentz *et al.*), vols. 8, 12, 14, 18. Illus. Leiden: Brill.

Comprises miscellaneous contributions in many families and genera, with few keys, based for the most part on botanical collections from West New Guinea (Irian Jaya) made between 1907 and 1921. A great many treatments consist principally of descriptions of new taxa, miscellaneous records of known species and critical notes, and are not truly revisionary; of these, easily the most significant is the series of papers on Orchidaceae by J. J. Smith, accompanied by numerous delicately executed plates. Some papers are complementary to their counterparts by the same authors in *Beiträge zur Flora von Papuasien* (see above). No detailed index is, however, available.[10] Additional botanical contributions on these and later collections are found in *Nova Guinea*, N.S. 1–10 (1937–59) and *Nova Guinea, Botany*, parts 1–24 (1960–6).

SCHUMANN, K. and LAUTERBACH, C., 1901 (1900). *Die Flora der deutschen Schutzgebiete in der Südsee*. xvi, 613 pp., 22 pls., map. Leipzig: Borntraeger; and *idem*, 1905. *Nachträge zur Flora der deutschen Schutzgebiete in der Südsee* (*mit Ausschluss Samoa's und der Karolinen*). 446 pp., 14 pls., portrait. Leipzig. (Reprinted in 1 vol., 1976, Vaduz, Liechtenstein: Cramer (*apud* Gantner).

The main work comprises a systematic enumeration of the then-known non-vascular and vascular plants of then-German New Guinea, Samoa, and Micronesia, with descriptions of new taxa, detailed synonymy (with references and citations), indication of *exsiccatae* with localities, summary of overall range both within and outside the area, miscellaneous notes, and vernacular names (without provenance); list of references; index to genera. The *Nachträge* similarly documents collections made between 1899 and 1904, particularly those gathered in 1901–2 by R. Schlechter. The introductory sections include chapters on the history of botanical collecting as well as short biographies of important collectors, while a complete general index to the whole work appears at the end. Encompasses 2208 species (1560 vascular), perhaps 15 per cent of the total flora of Papuasia (exclusive of the Solomon Islands) but including many secondary and weed species. [The reprint is in a considerably reduced and in some respects more convenient format than the original volumes.]

Special groups – pteridophytes and monocotyledons

Under this subheading are listed two recently initiated series of students' manuals, which despite their titles provide coverage for all of Papuasia. They may be considered as 'gap-filling' but not critical nor providing full and consistent coverage to species level. They may also be viewed as precursors to a much-needed (and feasible) illustrated generic flora.

JOHNS, R. J. and BELLAMY, A., 1979 (1980). *The ferns and fern allies of Papua New Guinea*. Parts 1–5 (in one fascicle). Illus., maps. [Bulolo, PNG: Forestry College.]

JOHNS, R. J., 1981. *The ferns and fern-allies of Papua New Guinea*. Parts 6–12 (in one fascicle). Illus., maps (Papua New Guinea University of Technology Research Reports, no. 48–81). Lae, PNG.

Students' manual, with keys to genera and species, short descriptions, synonymy, citations of *exsiccatae* (without localities), generalized indication of distribution and ecology, and (sometimes) taxonomic commentary; illustrations (figures and/or monochrome prints) of most taxa covered as well as dot distribution maps; literature references. Part 1 comprises a brief introductory section and a synopsis of genera based on the classification of Crabbe *et al.* (*Fern Gaz.* 11: 141–62, 1975). Parts 2–12 cover Parkeriaceae, Matoniaceae, Cheiropleuraceae, Dipteridaceae, Ophioglossaceae, Marattiaceae (in part), Osmundaceae, Psilotaceae, Marsileaceae, Salviniaceae, and Azollaceae. [The layout of this work is wasteful; judicious editing and printing reduction would have saved considerable space without any real loss in content. Nevertheless, it will serve as a useful supplement to the outdated treatments by Brause (ferns), Herter (Lycopodiaceae) and Hieronymus (Selaginellaceae) in the *Beiträge* of Lauterbach and Diels (see above), Wagner and Grether for the Admiralty Islands (see below), and VAN ALDERWERELT VAN ROSENBURGH (see **910–30**), as well as to Series II of *Flora Malesiana*.]

JOHNS, R. J. and HAY, A. (eds.), [1981]. *A students' guide to the monocotyledons of Papua New Guinea*. Part 1. iii, 90 pp., 36 pls. (Training manual for the Forestry College, vol. 13). Bulolo.

Students' manual, with keys to genera and species, generic and family descriptions, lists of species, and notes on distribution, identification (with *exsiccatae* sometimes cited), ecology, and other features; many line drawings; references (only those in English) at end of each family treatment. An introduction and a page 'How to recognize a monocotyledon' precede the work, but no index is provided. Families treated, all by Hay, cover Stemonaceae, Dioscoreaceae, Taccaceae, Philydraceae, Philesiaceae, Pontederiaceae, Araceae and Hypoxidaceae. [The deficiencies of this work are much as in the foregoing; moreover, only 300 copies have been printed. It will, however, usefully complement the *Handbooks of the flora of Papua New Guinea* (see above) whose volumes so far cover only dicotyledons.]

Special group – forest trees

Papuasia in general, along with each of its three major polities (Irian Jaya, Papua New Guinea, and the Solomon Islands), are covered by a heterogeneous collection of treatments of the principal forest trees, all but one comparatively limited in scope. The sole exception is the set of manuals by van Royen (and others) of 1964–70 which have therefore been treated above as a principal general work. Whitmore's treatment of the 'big trees' of the Solomons, in his *Guide to the forests of the British Solomon Islands*, is listed under **938**. The remaining works are accounted for below under two subheadings: Irian Jaya and Papua New Guinea.

I. Irian Jaya (Irian Barat)

The Irian Jaya Province of Indonesia, earlier known as Irian Barat (West Irian), has its most extensive (but least critical) coverage of forest trees in one of the tree species lists produced by the Forest Research Institute, Bogor (for background details, see under **913**). The other treatments listed are rather more detailed but cover fewer species.

HILDEBRAND, F. H., 1953. *Daftar nama pohon-pohonan: Irian*. 46 pp. (Lap. Balai Penjelidik. Kehut., *hors série*; also designated as Seri Daftar nama pohon-pohonan, no. 35). Bogor. (Originally pub. 1941, Buitenzorg, as *idem*, *Boomnamenlijsten van Nieuw-Guinea*. (Bosbouwproefstation, Serie Boomnamenlijsten, no. 6.) 1953 version reissued 1963 as *Daftar nama pohon-pohonan: Irian Barat*, comprising no. 91 of 'Laporan-laporanan Balai Penjelidikan Kehutanan', but full details have been unavailable.)

Sparsely annotated, uncritical checklist of forest trees known to attain a diameter (at breast height) of 40 cm or more, arranged alphabetically by families and genera; each entry (usually on one line) includes a vernacular name, distribution (by subdistricts), number of collections (but not *exsiccata*-numbers), and durability class (on a scale of five). The systematic list is preceded by a list with all records, fully annotated, arranged alphabetically by vernacular names, and other lexica appear at the end. No keys are included. [The deficiencies of these lists are further discussed under **913** and **917**, with an example under **927**.]

VERSTEEGH, C., 1971. *Key to the most important native timber trees of Irian Barat (Indonesia) based on field characters*. 63 pp., 7 pls. (Meded. Landbouwhoogesch. Wageningen, 71–19). Wageningen.

Largely unillustrated dendrological treatment of 108 leading tree species recorded from the northern lowlands of Irian Jaya, with a glossary, key for identification and standardized descriptive notes stressing habit, trunk and bank characters arranged under subheadings. A related work, with numerous coloured illustrations of habit, bark and wood, is DIREKTORAT JENDERAL KEHUTANAN, INDONESIA, 1976. *Mengenal beberapa jenis kayu Irian Jaya*. Vol. (Jilid) 1. ix, 114 pp., color illus., 2 maps. Jakarta. [Indonesian-language dendrological guide to 70 species (or species-groups) of forest trees, with colored illustrations of boles (with crowns) and woods and descriptive text; includes many vernacular names, but no keys.]

II. Papua New Guinea

JOHNS, R. J., 1975–77. *Common forest trees of Papua New Guinea*. 13 parts. Illus. (Training manual for the Forestry College, vol. 8). Bulolo, Papua New Guinea.

Copiously illustrated, largely compiled descriptive students' guide to more common forest tree species, but with many remarks on other trees and other forms of plants found in the forest; includes many keys, ecological notes, etc.; short lists of references in each part; index (part 13). The work is regrettably rather uneven. See also HAVEL, J. J., 1975 (1976).

Forest botany. 2 parts. Illus. (*Ibid.*, vol. 3). [Part 1 covers terminology; part 2 comprises an illustrated dendrological treatment, without keys, of principal timber species. The publication of part 2 was delayed for 10 years, and its contents have been to a large extent incorporated in Johns's work.]

WHITE, C. T., 1961. *Forest botany lectures*. Revised by K. J. White and E. E. Henty. 89 pp. [Port Moresby: Department of Forests, Territory of Papua and New Guinea.] (Mimeographed.)

The closest approach in Papua New Guinea to a forest flora, this is an introductory treatment in forest botany with families grouped into ten 'lectures' or sections; partial keys, descriptions (with emphasis on principal tree species), and miscellaneous biological and other notes are included, with an index to families and genera at end. Originally written for use during World War II, but not revised since 1961 and now somewhat outdated; more recent works, however, lack its scope and readability.

931

Aru Islands

See **928** (HEMSLEY). No separate checklists are available, though sporadic collections have been made over a long period. Although biologically Papuasian, the islands are administratively part of South Maluku (**928**).

932

Western Islands

These comprise Misool, Salawati, Batanta, Waigeo, and several smaller islands. No checklists are available except for parts of Waigeo; see ROYEN, P. VAN, 1960. The vegetation of some parts of Waigeo Island. *Nova Guinea, Bot.* **3**: 25–62, illus.

933

Islands in Cendrawasih (Geelvink) Bay

The main islands here are Japen and Biak. No separate checklists are available.

934

Trobriand and Woodlark Islands

This unit covers the chain of islands from the Trobriands through Woodlark (Murua) to the Laughlans. Extensive collections have been made only since World War II, and no separate checklists are yet available. [The large, high d'Entrecasteaux Islands are here considered part of the mainland.]

935

Louisiade Archipelago

This comprises the large islands of Misima, Tagula (Sudest) and Rossel along with many associated islets. As with the preceding unit, extensive collecting has only been carried out since World War II and no separate checklist is available at present. The islands are of considerable botanical interest, with certain links with the flora of New Caledonia, and here the important timber tree family Dipterocarpaceae reaches its eastern limit. However, they are *not* included in *Flora Malesiana* and it would be highly desirable for a separate checklist to be prepared.

936

Admiralty Islands

Included here are Manus, several associated smaller islands, and the long chain running west to Wuvulu. Two references, one very old, refer to the area but with much new collecting in recent years both are now very incomplete. The fern flora, however, has very useful keys for identification.

HEMSLEY, W. B., 1885. The Admiralty Islands. In *Reports on the scientific results of the voyage of 'HMS Challenger' during the years 1873–76* (ed. J. Murray), *Botany*, vol. 1, part 3, pp. 227–75. London: HMSO. (Reprinted 1965, New York: Johnson.)

Includes an enumeration of plants known from the north-west of Manus (Nares Harbour) and other areas, based to a large extent on the collections of the expedition's naturalist, H. N. Moseley, with descriptions of novelties and incorporating synonymy, references, distribution (with citations of *exsiccatae*), and notes on occurrence, dispersal, etc.; no separate index.

WAGNER, W. H., Jr. and GRETHER, D. F., 1948. *The pteridophytes of the Admiralty Islands*. (Univ. Calif. Publ. Bot. vol. 23, part 2, pp. 17–110, pls. 5–25). Berkeley, Calif.

Keyed, annotated enumeration of all known ferns and fern-allies of (mainly) Manus Island, with descriptions of new taxa; includes essential synonymy (with references), localities (with citations of *exsiccatae*), general indication of overall range, and notes on distinguishing features, taxonomy, habitat, frequency, life-form, special features, etc.; photographs of new or interesting species; no separate index. The introductory section gives a general description of Manus and its fern flora; species concepts are also discussed. Useful well beyond its nominal coverage (J. Croft, personal communication).

937

Bismarck Archipelago

This is here considered to include the islands of New Britain, New Ireland, Lavongai (New Hanover) and many associated smaller islands. Apart from the scattered citations provided by general works on Papuasia, especially Schumann and Lauterbach's *Flora* and the *Beiträge*, no useful coverage is available except for the works described below. That by Peekel was never actually published, but facsimiles of various kinds are now available in a number of botanical libraries. Both works are now very incomplete, as recent exploration has added greatly to the known flora, especially in the mountains.

PEEKEL, G., Fr. [1947.] Illustrierte Flora des Bismarck-Archipels für Naturfreunde, I–XII, Supplements I, II. folios 1–40, 15–2016, 1–129, 1–12. Illus. (MS.; reproduced as Inter-Documentation Company (IDC) microfiche set B-8096.)

Copiously illustrated, unpublished, descriptive account of the flora of the Bismarck Archipelago, chiefly lowland New Ireland and the Gazelle Peninsula of New Britain, indexed but without keys; includes native, naturalized, and commonly cultivated plants with vernacular names in up to four local languages and notes on special features, cultivation, uses, etc., appendices and index to vernacular and botanical names. Local range is somewhat sketchily indicated, and there are no *exsiccatae* cited. Most of Peekel's collections were determined by overseas (mainly German) specialists, and the novelties and more interesting records written up in numerous articles in botanical journals. [The work, although unpublished, is of considerable value for this area; the original manuscript is now housed in the library of the *Provinzialät* of the missionary *Orden des Heiligen Jesu* in Hiltrup, Rhein-Westphalia, West Germany, but facsimiles of various kinds are now fairly well spread through larger botanical libraries, including a moderately priced commercial microfilm version. An English translation is now (1981) in progress at the Division of Botany, Lae, Papua New Guinea with a view to future publication in book form.]

Partial work: Gazelle Peninsula

SCHUMANN, K., *et al.*, 1898. *Die Flora von Neu-Pommern.* pp. 59–158, map (Notizbl. Bot. Gart. Berlin, vol. 2). Berlin.

Annotated systematic enumeration of known non-vascular and vascular plants, with descriptions of new taxa, synonymy, citations of literature and *exsiccatae*, localities, and commentary; also includes an introduction covering botanical exploration, physiography, and vegetational/floristic formations along with an index. [Coverage limited to the Gazelle Peninsula in northeastern New Britain, centering on Simpson Harbor and treating particularly the large collections of F. Dahl, in 1896 resident in the area under the patronage of 'Queen' Emma Forsayth. All records are included by SCHUMANN and LAUTERBACH (**930**, Partial works).]

938

Solomon Islands

In this archipelago are the islands from Bougainville (with Buka) to San Cristobal, along with Sikaiana, Rennell, and Bellona. All of them except the first two now form part of the state of the Solomon Islands; Bougainville and Buka are part of Papua New Guinea. For the outer atolls, see **939**; for the Santa Cruz Islands, see **951**. One general but not entirely critical enumeration combined with a more detailed treatment of larger forest trees (Whitmore, 1966) is available for this area, partly supplemented by a checklist for Bougainville and Buka along similar lines. The

Solomon Islands, however, are *not* included within the limits of *Flora Malesiana*, which terminate east of New Ireland.

WHITMORE, T. C., 1966. *Guide to the forests of the British Solomon Islands.* 226 pp., 7 text-figs., maps (in end papers). London: Oxford University Press.

The latter part of this work comprises an alphabetical list of all species of vascular plants recorded from the Solomon Islands (including Bougainville and Santa Cruz), with Kwara'ae vernacular names (where known) and references for those species originally described from the area. The remainder of the book is devoted mainly to an illustrated descriptive treatment of the more common large forest trees; the introductory section includes notes on vegetation formations and methods of tree description, with lists of special characteristics (milky sap, deciduousness, etc.). A list of references is found at the end. For additional information on Bougainville and Buka, see FOREMAN, D. B. 1972. *A check list of the vascular plants of Bougainville, with descriptions of some common forest trees.* 194 pp., illus., map. (Bot. Bull. (Lae), 5). Lae: Division of Botany, Department of Forests, Papua New Guinea. [Includes *inter alia* a list of species, arranged by families, with references and citations of *exsiccatae*.]

939

Outer northeastern atolls

Included here are the atolls of Nuguria, Kilinailau, Tauu, Nukumanu and Ontong Java, all in Papua New Guinea except for the last-named (part of Solomon Islands). Scattered collections have been made, but no checklists appear to be available.

Superregion

94–99

Oceania

This vast superregion – the real 'South Seas' – comprises all Pacific islands and island groups from western Micronesia, the Santa Cruz Islands, and New Caledonia (with the Coral Sea Islands) across to Rapa Nui (Easter Island) and Sala-y-Gomez and north to the Hawaiian Islands. It thus corresponds largely to the primary limits of coverage of Merrill's *Botanical bibliography of the islands of the Pacific* (1947) and from a biogeographical point of view to the Polynesian and Neocaledonian subregions of the Oriental Region of Thorne (1963).[11] Botanically the affinities are with Malesia, although in the Hawaiian Islands there is an appreciable tropical American element (Balgooy, 1971).[12]

Oceania has attracted considerable attention from botanists of many countries ever since the latter part of the eighteenth century, when the classical contributions of Commerson, Banks, Solander, and the Forsters were made, and at the present time is relatively well, if unevenly, known. Bibliographically it is exceptionally well covered through *A botanical bibliography of the islands of the Pacific* by Merrill (1947; subject index by Walker), *Flora Malesiana Bulletin*, and several local cumulations including *Island bibliographies* (1955, 1971).

This exceptional degree of attention, however, has also led to marked fragmentation in botanical progress and floristic publication, with attempts at synthesis comparatively few and then more monographic than floristic. Broadly speaking, though, four stages in the development of the present level of botanical knowledge of Oceania can be recognized.

Until the 1840s, botanical exploration in the superregion was largely the province of the omnibus voyages of discovery and exploration, ranging from Bougainville and Cook to Du Petit-Thouars and Wilkes. The majority carried naturalists, these often also acting as medical officers, and from their observations and collections a rough overall picture of the flora was pieced together. There were, however, large critical gaps resulting from uneven geographical coverage, local hostility, and the shore-bound nature of

much local exploration. Thus the period of primary floristic investigation could not be said to have closed by 1842. The 'high' islands were very imperfectly known, and 'low' islands and atolls were largely ignored as their 'monotonous' flora was thought to be just that – a fallacy not realized until the present century.

The second stage, comprising the years from the 1840s until about 1920, was characterized largely by single island or island-group exploration, and by and large was related to colonial penetration and annexation, a process largely complete by the 1900s. Contributions were also made by detailed coastal surveying and charting expeditions. A large proportion of collecting was accomplished by resident or visiting non-specialists, amateur or semi-professional, with their finds being sent to public and private herbaria in several metropolitan countries, along with Australia and New Zealand. Official biological expeditions were few, largely taking the form of resource surveys. Among the most important of the latter was the British Mission to Fiji of 1860–1, to which Seemann was attached as botanist; one of its results was *Flora vitiensis* (1865–73), a key work also incorporating other collections, not only from Fiji but other South Pacific islands as well, and only now (1980) being supplanted by a new standard work on that group. Other major contributions of the period were *Flora of the Hawaiian Islands* by Hillebrand (1888), a synthesis based on the author's long experience in those islands and the first true manual for any part of the superregion; *Flore de la Polynésie française* by Drake (1893), *Die Flora von Samoa-Inseln* by Reinecke (1896–8), later incorporated in *Die Flora der deutschen Schutzgebiete in der Südsee* by Schumann and Lauterbach (1901; see **930**) which also consolidated what little was then known of Micronesia; and *Catalogue des plantes phanérogames de la Nouvelle-Calédonie et dépendances* by Guillaumin (1911). All remain standard works today, none ever having been fully supplanted except for Schumann and Lauterbach's work in Micronesia.

The work which may be considered to have ended the primary phase of exploration in the eastern half of the superregion, as far west as Fiji, is *Illustrationes florae insularum maris Pacifici* by Drake (1886–92), produced in the tradition of Hooker's southern zone works of the mid-century. In the western half, this level was not reached until well into the present century, with the extended primary exploration of rich New Caledonia, the explorations of Morrison and others in the New Hebrides and of Volkens, Krämer, and Ledermann in Micronesia up to 1914. Attention also began to be drawn to the low islands and atolls through the work of Guppy (centered on Fiji), the Australian Museum in Tuvalu, the Bishop Museum in Marcus, Midway, and Palmyra (the latter leading to the fine illustrated *Palmyra Island* (1916) by Rock), and (after World War I) Setchell (in central and southeastern Polynesia). However, since Drake's *Illustrationes* no supraregional synthetic floras or enumerations have been published.

The third stage recognized here comprises the period between the world wars, when Pacific biological exploration, developing on a larger and more independent scale with the aid of private wealth, was dominated by large American museums, notably the Bishop Museum. Established in 1889, this museum, the first major institution of its kind in Oceania, at first largely limited its activities to the Hawaiian Islands, but with a change of direction after 1919 work expanded into nearly all other parts of Oceania, covering anthropology, botany, and zoology. A rapid expansion of botanical and other collections ensued through the efforts of several expeditions and many resident and visiting individuals, either under institutional auspices or self-sponsored (the latter sometimes with the aid of small grants-in-aid). Collaborative work with scientists from Japan, the only other metropolitan power then significantly active in Oceania (mainly in Micronesia), was carried out in the 1930s. Development of the Museum botanical programme, which included the publication of island floras and other contributions, was from 1920 until the early 1950s guided by E. D. Merrill as consultant botanist. By 1941 the basis had been laid for what is now 'without doubt the largest and most valuable collection of Pacific plant materials in the world',[13] with further substantial contributions arriving in the decennium after 1945 from the extensive American surveys of Micronesia and other sources.

A goodly number of current standard floras and related works were produced in the 25 years from 1920 through 1945, and the basis laid for others which were not to appear until after World War II or have still to be fully realized.[14] The greater part of these was the work of botanical staff or 'associates' of the Bishop Museum, and published in that institution's *Bulletin* or *Occasional Papers* series. Among them were floras of the northern Line, Howland and Baker Islands (1927) and of Samoa (1935–8) by Christophersen; the Hawaiian

Leeward Islands (1931) by Christophersen and Caum; Rarotonga (1931) and Makatea (1934) by Wilder; a less than critical treatment of parts of southeastern Polynesia (1931–5) by Brown and Brown; some fern floras (1929–38) by Copeland; and floras of Niue (1943) and the Manua Islands in Samoa (1945) by Yuncker. Under other auspices there appeared Skottsberg's flora of Easter Island (1920–2, with ferns by Christensen who also treated Samoan pteridophytes for the Bishop Museum in 1943); Kanehira's illustrated one-volume manual (1933) and enumeration (1935) of the flora of the Japanese mandated islands in Micronesia; Setchell's contributions on Tutuila and Rose Atoll in Samoa (1924) and on Tahiti (1926); Guillaumin's many contributions on New Caledonia and the New Hebrides throughout the period; and, finally, Degener's inimitable *Flora hawaiiensis* (1932–). Most of these works were, however, documentary in nature, lacking keys and sometimes descriptions, and some were little more than vehicles for ethnobotanical information. Perhaps the most practical work was Kanehira's *Flora micronesica*, written in Japanese but well-illustrated and with keys later made available in English (1956). Coverage of the superregion was thus improved but still remained very patchy.

In 1928 Merrill (1929) surveyed the status of exploration in Oceania, noting that the various islands were still very unevenly known, with overall knowledge insufficient to warrant production of an annotated enumeration along the lines of those on Borneo (**917**) and the Philippines (**925**). A call was made for an organized Pacific botanical survey, with emphasis on the 'high' islands, especially in the south-central sector (Region 95) which then was considered very poorly known and phytogeographically crucial. Nevertheless, as an extension of his Polynesian botanical bibliography, the first version of which appeared in 1924 (with new editions in 1937 and 1947), Merrill began compilation of a nomenclatural card file as a contribution towards a future general enumeration, adding to this until the end of the 1940s when retirement stopped work. The enumeration project has evidently not since been resumed, but the bibliographical materials were later utilized for *Island bibliographies*.

This unrealized attempt at synthesis is symptomatic of the more fragmented pattern of floristic work which has prevailed since World War II, and particularly since the early 1950s – the fourth of the developmental stages recognized here. Institutionally, this period has been characterized by a marked decline in the level of botanical contributions from the Bishop Museum with respect to Oceania (due to financial stringency and changes in priorities) and the concomitant rise of the Washington-based atoll research programme (Office of Naval Research; later National Research Council and the Smithsonian Institution) and the Pacific floristic mapping project at the Rijksherbarium at Leiden, the latter originally conceived by Lam in 1939 and later prosecuted by van Steenis and van Balgooy with their collaborators. The last floras published in the Bishop Museum *Bulletin* were *Flora of Ponape* (1952) by Glassman and *Plants of Tonga* (1959) by Yuncker, while work on some of the floristic projects initiated under the inter-war programme largely lapsed after the early 1950s. The new *Atoll Research Bulletin*, organ of the atoll research programme, began publication in the early 1950s, and the first volume of *Pacific Plant Areas* appeared in 1963 (followed by the second in 1966 and the third in 1975).

The 'Pacific' botanical unit at the Smithsonian Institution, until 1978 headed by Fosberg but since partly disbanded, has been responsible for many florulas of low islands and atolls in the Pacific (and other oceans), produced either separately or as parts of larger studies and for the most part published in the *Atoll Research Bulletin*. That the plants of these low islands exhibit phytogeographical patterns paralleling those of the 'high' islands has been a significant result of the atoll program.[15] Other work of the unit related to Pacific floristic botany has included the preparation of *Flora of Micronesia* (1975–) and a flora of the Marquesas (1975–) as well as the only completed portion of Grant's flora of the Society Islands (1974) – the absence of which had been deprecated by Merrill in 1951 – and an enumeration for the northern Marianas (1975), all in *Smithsonian Contributions to Botany*, and the tripartite *Island bibliographies* (1955; supplement 1971), covering Micronesian botany, Pacific Islands vegetation, and low islands and atolls in all oceans. In Hawaii, there were published a new checklist of Hawaiian native and introduced flowering plants (1973) by St John, a revised reprint of Rock's fine *Indigenous trees of Hawaii* (1974), and the first volume of *Flora vitiensis nova* by A. C. Smith (1979), all through the Pacific Tropical Botanical Garden. From Guam there appeared *Flora of Guam* by Stone (1970). A number of these floras and other works, however, represent projects begun in the inter-war

period; more recent American taxonomic research in the Pacific has, apart from many small contributions in whole or in part by St John, tended to be synthetic – also in contination of a pre-war tradition exemplified by the work of Sherff, Fosberg, and others in their revisionary studies.

A renewed awareness of the Pacific Islands on the part of European metropolitan powers as well as New Zealand also took place, which in botanical terms has manifested itself in much new field work and publication, notably in New Caledonia, Fiji, the New Hebrides, and the Santa Cruz group as well as the adjacent Solomon Islands (938). Some of this work was conducted from already existing (Suva) or newly established (Honiara, Guam, Noumea) herbaria. Collections made from the early 1950s onwards have greatly improved botanical knowledge in New Caledonia and the south central Pacific Islands (the 'Fijian region' of the many regional revisions of Smith and collaborators), and a number of interesting discoveries have been made. Notable contributions to date have included *Flore de la Nouvelle-Calédonie et dépendances* (1966–), a large-scale documentary work, and *Plants of the Fiji Islands* by Parham (2nd edn., 1972), an annotated checklist. From New Zealand, Brownlie has revised the pteridophytes of several islands, notably New Caledonia and Fiji, and Sykes has made substantial additions to Yuncker's Niue flora in *Contributions to the flora of Niue* (1970). A number of revisions of Pacific genera have resulted from work on *Flora Malesiana* at Leiden, and these and others have been used for *Pacific Plant Areas*. Guillaumin's earlier work was encapsulated in his ill-documented, partly compiled *Flore analytique et synoptique de la Nouvelle-Calédonie* (1948), now quite incomplete but one of the limited number of Oceanic identification manuals.

The present state of knowledge has been surveyed by Corner (1972) and Prance (1978), the latter based on an unpublished report by Fosberg written for the Thirteenth Pacific Science Congress in 1975. These represent the latest reports prepared under the aegis of the Standing Committee on Pacific Botany of the Pacific Science Association, the main coordinating committee for the Pacific (and Malesian) superregions. Several islands and groups are still considered inadequately explored and/or documented, and almost throughout there are marked threats from increasing human pressure and 'development'. Many raised limestone flat-topped islands have been comparatively neglected and require urgent attention; among these are Makatea and Nauru.[16] Nevertheless, the level of exploration as a whole compares favorably with the West Indies or the Mediterranean.

The considerable fragmentation of floristic work and collections, both in the past and more recently as was more or less inevitable with such widely dispersed islands, will, however, make preparation of a new general enumeration a most difficult and challenging task, although the strictures of Merrill are less applicable now than in 1928. To this end, attempts should be made to strengthen botanical work at the Bishop Museum where the most extensive resources are to be found, and obtain collaboration elsewhere. The creation of an 'information center' should also be encouraged. At a more local level, a new manual of the Hawaiian flora on 'biosystematic' principles should be written, while in other areas a step forward would be simple florulas for educational purposes, like Gowers' illustrated, locally produced treatment of common New Hebrides trees (1976).

Progress

No overall historical review for the superregion has been seen. The state of knowledge in 1928 and some suggestions for advancement were given in MERRILL, E. D., 1929. Pacific botanical survey. In *Annual report of the Director for 1928* (comp. H. E. Gregory), pp. 53–7 (*Bernice P. Bishop Mus. Bull.*, 65). Honolulu. Modern reviews include CORNER, E. J. H. (comp.), 1972. Urgent exploration needs: Pacific floras. *Pacific Sci. Inform. Bull.* 24(3/4): 17–27, and an extract from an unpublished paper by Fosberg in the general survey by Prance (1978). Additional details may be found in Bishop Museum annual reports and the various Pacific Science Congress proceedings.

General bibliographies. Bay, 1910; Blake and Atwood, 1942; Frodin, 1964; Goodale, 1879; Holden and Wycoff, 1911–14; Jackson, 1881; Pritzel, 1871–7; Rehder, 1911; Sachet and Fosberg, 1955, 1971; USDA, 1958.

Supraregional bibliographies

LEESON, I., 1954. *A bibliography of bibliographies of the South Pacific*. vii, 61 pp. London: Oxford University Press.

MERRILL, E. D. and WALKER, E. H., 1947. A botanical bibliography of the islands of the Pacific; and A subject index to Elmer D. Merrill's 'A botanical bibliography of the islands of the Pacific'. *Contr. US Natl. Herb.* 30(1): 1–404. [Concisely annotated bibliography by E. D. Merrill; detailed subject index by E. H. Walker. In addition to all the islands in Superregion 93–97, full coverage is extended to Juan Fernandez, the Kermadec group, and Norfolk and Lord Howe Islands. Contains 3850 entries.]

STEENIS, C. G. G. J. VAN, 1955. Annotated selected bibliography. In his *Flora Malesiana*, ser. I, 5: i–cxliv. Groningen: Noordhoff. (Reprinted separately.) [Provides a limited selection of more important floras.]

General indices. BA, 1926– ; BotA, 1918–26; BC, 1879–1944; BS, 1940– ; CSP, 1800–1900; EB, 1959– ; FB, 1931– ; IBBT, 1963–9; ICSL, 1901–14; JBJ, 1873–1939; KR, 1971– ; NN, 1879–1943; RŽ, 1954– .

Supraregional indices

STEENIS, C. G. G. J. VAN and JACOBS, M. (eds.), 1948– . *Flora malesiana bulletin*, 1– . Leiden: Flora Malesiana Foundation (mimeographed). [Issued nowadays annually, this contains a substantial bibliographic section also very thoroughly covering the Pacific/ Oceanic literature. A fuller description appears under Superregion 91–93.]

940–90

Superregion in general

Only one general floristic work for the superregion, and then only dealing with its eastern half, has even been compiled: *Illustrationes florae insularum maris Pacifici*, a reference flora now almost a century old and intended partly for the connoisseur. A projected enumeration by Merrill was initiated and a card file prepared but nothing was ever published apart from the bibliography (see above).

DRAKE DEL CASTILLO, E., 1886–92. *Illustrationes florae insularum maris Pacifici*. 458 pp., 50 pls. Paris: Masson. (Reprinted 1977, Vaduz, Liechtenstein: Cramer.)

Pages 103–408 of this work contain a systematic enumeration of the known vascular plants of Polynesia (2189 species), with synonymy, references and citations, indication of localities with *exsiccatae*, general summary of internal and extralimital distribution, and taxonomic commentary on individual species, genera, and families; index to all botanical names. The lengthy introductory section (pp. 1–100) encompasses chapters on the physical characteristics of the islands, general features of the flora, floristics and phytogeography, and the progress of botanical exploration. At the end of the work is an atlas of 50 plates of representative species, each with accompanying descriptive text. Coverage by this work extends to the Fiji Islands, Tonga, Samoa, the Marquesas, the Societies, the Tuamotu Archipelago, and the Hawaiian Islands.

Region

94

New Caledonia and dependencies

This region, nearly all under French administration, encompasses the main island of New Caledonia ('La Grande Terre', including Kunié and Îles Belep), the Loyalty Islands, the islands from Walpole through Matthew and Hunter to Conway Reef, and the Coral Sea groups of Chesterfield and the (Australian) Coral Sea Islands Territory. 'La Grande Terre' is distinguished by one of the most peculiar floras in the Pacific (and indeed the world), with high endemism, many archaic taxonomic and morphological 'relicts', and very mixed floristic affinities compressed into a relatively small area. For practical and biohistorical reasons, though, it has been placed, together with its 'dependencies', with tropical Oceania in Division 9, although its biogeographic position vis-à-vis the Oriental and Australian floristic kingdoms is a moot point.[17]

A useful introduction to the floristics and plant geography of New Caledonia may be found in THORNE, R. F., 1965. Floristic relationships of New Caledonia. *Univ. Iowa Stud. Nat. Hist.* **20**(7): 1–14.

For very many years, New Caledonia has been served largely by the numerous, scattered contributions of A. Guillaumin, for the most part published in two series: *Matériaux*, containing family and generic revisions precursory to a general flora, and *Contributions*, comprising more or less extensive lists of new collections, florulas of small islands or local areas, and miscellaneous matters of record. Only the *Matériaux* have been indexed, and then only through 1945, just prior to publication of the author's *Flore analytique et synoptique de la Nouvelle-Calédonie* (1948). This work, largely compiled, part uncritical, and without bibliographic aids or commentary, is now in need of thorough

revision or replacement. The latter function is being effected through the new 'research' work, *Flora de la Nouvelle-Calédonie et dépendances*, begun in 1967 and of which eight parts have now (1979) been published. Although accounting for the great amount of new material and information which has become available since World War II, full realization of this large-scale flora will, however, as with *Flora Malesiana* and the new *Flora of Micronesia*, take several decades and even then be a reference rather than a practical work. There is thus scope for preparation and publication of a concise 'interim' work along the lines of Merxmüller's *Prodromus einer Flora von Südwestafrika* (**521**).

Of the separate island units here recognized, no properly published coverage exists for the under-collected Loyalties and only sketchy documentation is available for the eastern islands and for the Huon, Chesterfield, and Coral Sea groups. The eastern islands are small, 'high' volcanic entities, while those in the Coral Sea are widely scattered low cays and atolls, some on the drowned 'Queensland Rise'. The Coral Sea group is Australian territory, while all other islands are administered from New Caledonia.

Bibliographies. General and supraregional bibliographies as for Superregion 94–99.

Indices. General and supraregional indices as for Superregion 94–99.

940

Region in general

The only nominally complete works on the flora are the two outdated compilations of Guillaumin: the *Catalogue* (1911) and the *Flore analytique et synoptique* (1948), both now very incomplete. The many separate contributions to the flora by the same author are largely grouped in two key series: *Matériaux pour la flore de la Nouvelle-Calédonie*, part I– (1914)– , which consists very largely of family revisions, and *Contributions à la flore de la Nouvelle-Calédonie*, Nos. 1 (1911)–130 (1973), based mostly on miscellaneous collections by various botanists and others. An index to the *Matériaux* up to part LXXXV was provided by the author in *Bull. Soc. Bot. France*, **92**: 76–7 (1946); for further details, see the bibliographies of Blake and Atwood (1942) and Merrill (1947) and the annual cumulations in *Flora Malesiana Bulletin* (1947–). Of the new *Flore de la Nouvelle-Calédonie et dépendances*, only a few angiosperm families in addition to the pteridophytes and gymnosperms have to date been published. Many 'collection reports' also exist. Major forest trees have been treated in a somewhat uncritical work by Sarlin (1954), now in need of revision and amplification.

AUBRÉVILLE, A. *et al.* (eds), 1967– . *Flore de la Nouvelle-Calédonie et dépendances.* Fasc. 1– . illus., maps. Paris: Muséum National d'Histoire Naturelle, Laboratoire de Phanérogamie.

Extensively descriptive, copiously illustrated flora in the form of regional monographs of vascular plant families, with keys to genera and species, full synonymy, localities with citations and general statement of overall range (where applicable); distribution maps; notes on habitat, distribution, frequency, biology, juvenile forms (bizarre blastogenic forms are a characteristic feature in the flora) and special features of individual species; each fascicle separately indexed. At this writing (late 1980), nine fascicles have appeared, of which fasc. 3 constitutes a detailed revision of the pteridophytes (by G. Brownlie) and fasc. 8 the Orchidaceae (by N. Hallé).

GUILLAUMIN, A., 1911. Catalogue des plantes phanérogames de la Nouvelle-Calédonie et dépendances (Îles des Pins et Loyalty). *Ann. Inst. Bot.-Géol. Colon. Marseille* **19**: 77–290, map. (Reprinted separately in 3 parts.)

Systematic enumeration of then-known seed plants, with synonymy, vernacular names, citation of localities together with *exsiccatae*, and indication of useful plants; index to family and vernacular names. The introductory chapter includes a fairly extensive section on the history of local botanical exploration and an annotated list of collectors. Provides a useful summary of earlier work on the flora and serves as a starting-point for the author's long series of *Matériaux* and *Contributions* (see above).

GUILLAUMIN, A., 1948. *Flore analytique et synoptique de la Nouvelle-Calédonie* (Phanérogames). 369 pp. Paris: Office de la Recherche Scientifique Coloniale.

Unannotated manual-key to the seed plants of New Caledonia and its dependencies, with keys to families, genera, and species; brief descriptions of the characteristics of each family; index to generic and family names. The work is in effect a distillation from the author's series of *Matériaux* and *Contributions*; it is now in need of thorough revision (M. Schmid, personal communication).

Partial works

The two most extensive precursory series are Guillaumin's *Matériaux* and *Contributions*, already noted above (the latter was continued virtually up to the time of his death in 1974). However, they are not described here because of their scattered mode of publication and consequent bibliographic complications; moreover, those published before 1948 have been accounted for in his *Flore analytique et synoptique* (see above). However, some significant unit contributions have been made, both by Guillaumin and by non-French botanists (notably Schlechter, Rendle *et al.*, Däniker, and Thorne). The most substantial, by Däniker (1932–43) and Guillaumin (1957–74), are separately accounted for below.

DÄNIKER, A. U., 1932–43. *Ergebnisse der Reise von Dr A. U. Däniker nach Neu-Caledonien und den Loyaltäts-Inseln, 4: Katalog der Pteridophyta und Embryophyta siphonogama.* 5 parts. 507 pp. (Vierteljahrsschr. Naturf. Ges. Zürich, 77–8, Beibl. no. 19). Zurich. (Also pub. as *Mitt. Bot. Mus. Univ. Zürich*, 142.)

Critical systematic enumeration of vascular plants in the Däniker 1924–5 collection, with descriptions of new taxa; includes full synonymy, references and citations, vernacular names, localities with *exsiccatae*, taxonomic commentary, and extensive notes on habitat, biology, etc.; index to generic and family names (part 5). The introductory section includes a general description of the several islands as well as Däniker's own itinerary and an account of past botanical exploration.

GUILLAUMIN, A., 1957–74. *Résultats scientifiques de la mission franco-suisse de botanique en Nouvelle-Calédonie, 1950–2,* I–V. 5 parts (Mém. Mus. Natl. Hist. Nat., II/B, Bot. 8(1), 8(3), 15(1), 15(2), 23). Paris.

Comprises annotated lists of collections, arranged systematically in each part, with indication of localities and *exsiccatae*, descriptions of novelties, references, and some taxonomic commentary; keys given for a few groups (e.g., *Araucaria* in part 1; *Dysoxylum* in part 4 – mainly intended as supplementation for the 1948 *Flore analytique* (see above); no index. Parts 4 and 5 contain extensive addenda and corrigenda to treatments in parts 1–3, along with new material. Many treatments are by specialists – part 5 represents the author's last contribution to New Caledonian botany, and appeared posthumously.

Special group – forest trees

SARLIN, P., 1954. *Bois et forêts de la Nouvelle-Calédonie.* 303 pp., 131 pls. (incl. halftones), 3 folding maps (Publ. CTFT, 6). Nogent-sur-Marne, France: Centre Technique Forestier Tropical.

In Part III of this work (pp. 81–212) there is a dendrological treatment of principal forest trees of New Caledonia, with botanical descriptions, trade and vernacular names, occurrence, habitat, properties, structure of wood and its description, and provenances, with cross-referenced illustrations at the end of the work. The introduction (Part I) deals with background (lack of previous forest-botanical studies and collections, including wood samples), topography, climate, soils, biotic factors, land use and forest exploitation, while in Part II is found a descriptive treatment of vegetation and the forests. Part IV gives conclusions, with a classification of actual and potential uses. Most useful for its plates but otherwise uncritical, having been hastily prepared by a non-specialist without adequate assistance (M. Schmid, personal communication). See also CENTRE TECHNIQUE FORESTIER TROPICAL, 1975. *Inventaire des ressources forestières de la Nouvelle-Calédonie.* 227 pp. Nogent-sur-Marne. (Mimeographed.)

945

Loyalty Islands

These comprise the more or less raised atolls of Ouvéa, Lifou, and Maré and their associated islets. The flora (550–600 native and naturalized vascular species) has generally been accounted for in overall works on New Caledonia, but it is still considered undercollected. Of the known flora, 360 species are considered truly indigenous, with ten per cent endemism at the most. The only known recent survey is SCHMID, M., 1967. *Note sur la végétation des îles Loyauté.* 70 pp. Noumea: ORSTOM, Centre Nouméa. (Mimeographed.)

946

Eastern dependencies

These comprise Walpole, Matthew and Hunter Islands, with an extension to Conway Reef. No botanical collections had been made before 1972; this last of Guillaumin's *Contributions* is thus the first report for the area. However, only Walpole and Matthew are covered; no collections are yet available for Hunter.
GUILLAUMIN, A., 1973. Contributions à la flore de la Nouvelle-Calédonie, 130: Plantes des îles Walpole et Matthew. *Bull. Mus. Natl. Hist. Nat.* (Paris), sér. III, no. 192 (Bot., no. 12): 180–183.
Includes unannotated lists of plants collected from Walpole (45 species) and Matthew (10 species).

947

Huon Islands (with Petrie Reef)

These low islands lie to the northwest of 'La Grande Terre', beyond Îles Belep.
GUILLAUMIN, A. and VEILLON, J. M., 1969. Plantes des archipels Huon et Chesterfield. *Bull. Mus. Natl. Hist. Nat.* (Paris), sér. II, 41: 606–7.
Unannotated list of 10 species.

948

Chesterfield Islands

These isolated low islands lie on the Coral Sea Rise between New Caledonia and the Great Barrier Reef of Australia.
COHIC, F., 1959. *Report on a visit to the Chesterfield Islands, September 1957.* 11 pp., maps (Atoll Research Bull. no. 63). Washington.
Pages 3–4 contain an unannotated list of vascular plants collected by the author (20 species). The remainder of the report describes previous visits to the islands, the vegetation, and the terrestrial fauna; a list of references is appended. For additions, see GUILLAUMIN and VEILLON (1969; under **947**).

949

Coral Sea Islands Territory

These scattered low islands, partly associated with the Coral Sea Rise, lie northeast of the Great Barrier Reef and generally west of the Chesterfields. They have in recent years been constituted as an Australian external territory, with a focus at Willis Island with its weather station. This was described, with inclusion of a plant list, by Davis (1923).
DAVIS, J. K., 1923. *Willis Island, a storm warning station in the Coral Sea.* [v], 119 pp., illus. Melbourne: Critchley Parker.
Appendix B (pp. 111–12), entitled 'Some plants from Willis Island', includes a list of seven vascular plants collected by the author, with indication of overall

distribution (prepared with assistance from C. T. White). Most of this book deals with the geography and climatology of the region.

Region

95

South central Pacific islands

Within this region are included the 'high' islands of eastern Melanesia and southwestern Polynesia, as well as a number of low islands and atolls. It is sometimes known as the 'Fijian Region'. The area is bounded by the Santa Cruz Islands to the northwest to Samoa and Niue in the east, and extends south from a line between the Santa Cruz group and Samoa to Aneityum (New Hebrides) and the Minerva Reefs south of Fiji. Conway Reef is included in region 94.

The region is more or less covered by a heterogeneous assemblage of floras and encumerations, of which only *Flora vitiensis nova* by A. C. Smith (1979–) can, by modern standards, be considered adequate. The Fiji Islands also, fortunately, have a comparatively recent checklist, *Plants of the Fiji Islands*, by Parham (2nd edn., 1972). Niue is also relatively thoroughly covered through the flora of Yuncker (1943) and the supplementary contribution of Sykes (1970), but neither work has keys. Tonga has also been covered by Yuncker (1959) but this likewise lacks keys; moreover, it does not effectively account for the group as a whole, in which some islands are under-collected. Divided Samoa is covered by a variety of accounts, none of them complete and the latest nearly 40 years old; parts of the group are also considered inadequately collected. The same also applies to the Wallis and Futuna Islands, although a checklist by St John and Smith appeared in 1971.

Coverage of the western subregion (Rotuma, the Reef Islands, Santa Cruz, and the New Hebrides) is even more fragmented and uneven, both in literature and in available collections, although much new field work has been carried out in Santa Cruz and the New Hebrides since 1960. The former (along with the Reef Islands) are part of the state of the Solomon Islands (**938**) and as such have been sketchily accounted for in Whitmore's *Guide to the forests of the British Solomon*

Islands (1966) with its scarcely annotated checklist. For the New Hebrides, only the scattered contributions of Guillaumin and (more recently) Schmid are available, apart from a nominally complete enumeration by the former, the *Compendium* (1948), now virtually useless. The important trees have been treated in an illustrated atlas by Gowers (1976; corrected reprint, 1978). Considerable collecting has been carried out in Rotuma but only the pteridophytes have had a recent treatment (St John, 1954). However, several families occurring in this western subregion have been accounted for in Fiji-centered revisions by A. C. Smith and his associates.

Bibliographies. General and supraregional bibliographies as for Superregion 94–99.

Indices. General and supraregional indices as for Superregion 94–99.

950

Region in general

No general works are available for the region as a whole, due in large part to differing levels of botanical knowledge and, as a corollary to politico-administrative fragmentation, a relatively localized and sporadic approach to botanical work. However, many family revisions relating to the region as a whole can be found in A. C. Smith's series of precursory papers for a new flora of Fiji, *Studies of Pacific Island plants*, in progress since 1941 and now with more than 30 parts published (noted under **955**). The eastern island groups (Fiji, Samoa, and Tonga) are also covered by Drake del Castillo's *Illustrationes florae insularum maris Pacifici* (1886–92; see **940–90**), but this work is now largely of historical interest.

951

Santa Cruz Islands

Included herein are the Taumako (Duff) Islands as well as the Reef Islands, Ndeni (Santa Cruz), Utupua and Vanikoro, the last-named once noted for its stands of kauri (*Agathis obtusa*). For Tikopia, Anuta, and Fatutaka (Mitre), see **952**. All known records up to 1966 are, in theory, accounted for by WHITMORE in his Solomons checklist (938), as the group is administratively part of the Solomon Islands. No adequate separate checklists are available and, despite considerable collecting since the early 1960s, the group is still considered undercollected.

952

Tikopia, Anuta, and Fatutaka (Mitre)

These three comparatively small islands form a loose group lying east of the Santa Cruz Islands, with which they are administratively associated. Any records have in theory been accounted for by Whitmore (938). No separate checklists appear to be available. They are evidently as undercollected as the Santa Cruz group itself.

953

Vanuatu (New Hebrides)

This group, an Anglo–French condominium (colloquially referred to as the 'Pandemonium') from 1906 to 1980, extends from the Torres Islands in the north through the Banks Group, Santo, and Efate to Aneityum. For Matthew and Hunter Islands, see 946. No satisfactory modern flora or checklist is available apart from the sketchy *Compendium* by Guillaumin (1948), now virtually useless, which summarized several earlier reports by the same author (given below). Some individual island florulas, however, have become available and are also accounted for here, while forest trees were treated in an atlas by Gowers (1976). Much new collecting has been accomplished, beginning in the 1960s, by both anglophone and francophone workers but the group is still considered unevenly known. Available collections are scattered through a number of herbaria so that preparation of a new consolidated account will require a certain amount of international collaboration.

A useful introduction to the biota of the New Hebrides is CORNER, E. J. H. and LEE, K. E. (coord.), 1975. Discussion on the results of the 1971 Royal

Society expedition to the New Hebrides. *Philos. Trans. Roy. Soc.* (*London*), B, **272**: 267–486. [The flora is considered in papers by W. L. Chew, E. J. H. Corner, A. Gillison, and M. Schmid.]

GUILLAUMIN, A., 1947–8 (1948). Compendium de la flore phanérogamique des Nouvelles-Hébrides. *Ann. Inst. Bot.-Géol. Colon. Marseille*, **55/56**: 5–58.

Briefly annotated systematic enumeration of seed plants, with limited synonymy; includes indication of local distribution according to islands, with summary of extralimital distribution (if applicable), and occasional notes; index to families. A goodly number of entries are merely referred to as 'sp.'. The brief preface gives a list of the principal collectors in the islands up to World War II. The work is based largely on several precursory papers by the same author, viz: Contribution à la flore des Nouvelles-Hébrides, I–III. *Bull. Soc. Bot. France*, **66**(1919):267–77; **74**(1927):693–712; **76**(1929):298–303; also Contribution to the flora of the New Hebrides. *J. Arnold Arbor.* **12**(1931): 221–64, figs. 1–3 (incl. map); **13**(1932): 1–29, 81–124, figs. 1–3, pl. 43; **14**(1933): 53–61; also Contribution à la flore des Nouvelles-Hébrides. *Bull. Soc. Bot. France*, **82**(1935): 316–54, map; and *Bull. Mus. Natl. Hist. Nat.* (Paris), II, **9**(1937): 283–306, 1 fig.

Partial works

A number of florulas now exist for individual islands but should not be regarded as critical nor, in the case of those by Schmid, formally published. The florula of Santo by Guillaumin is now very incomplete. Apart from those listed below, a florula of Tanna by Schmid was also distributed but no copy has been available for examination.

GUILLAUMIN, A., 1938. A florula of the island of Espiritu Santo, one of the New Hebrides. *J. Linn. Soc. Bot.* **51**: 547–66. [Checklist, with introductory matter, based mainly on the results of the Baker expedition of 1933–34.]

SCHMID, M., 1970. *Florule d'Anatom.* 53 pp. Nouméa: Centre ORSTOM, Nouméa (mimeographed). [Briefly annotated checklist.]

SCHMID, M., 1973. *Espèces de végétaux superieurs observés à Vaté.* 42 pp. Nouméa (mimeographed). [Briefly annotated checklist; excludes Orchidaceae.]

SCHMID, M., 1974. *Florule de Erromango.* 52 pp. Nouméa (mimeographed). [Briefly annotated checklist; excludes Orchidaceae.]

SCHMID, M., 1974. *Florule de Pentecôte.* 25 pp. Nouméa (mimeographed). [Briefly annotated checklist.]

Special group – forest trees

GOWERS, S., 1976. *Some common trees of the New Hebrides and their vernacular names.* 189 pp., illus. Port Vila, New Hebrides: Forestry Branch, Department of Agriculture, New Hebrides Condominium Government (mimeographed). (Reprinted with corrections, 1978.)

Illustrated dendrological treatment of the more common (about 60) species of trees, including a few naturalized or commonly cultivated, with large figures and accompanying botanical and timber descriptions, local vernacular and Bislama (New Hebrides pidgin) names, and notes on uses, special features, etc. An introductory section includes notes on the limits of the work, remarks on vernacular names, and a general key (pp. 10–14), while at the end are an illustrated glossary and indices to botanical and vernacular names. Arrangement of tree species is alphabetical by genus and species.

954

Rotuma

See also **955** (PARHAM); the island is administratively part of Fiji. While a 'manual' for the ferns has been produced by St John (1954), no separate work of recent date for the seed plants is available for this most easterly floristic outpost of 'Melanesia', although substantial collections have been made by St John.

ST JOHN, H., 1954. Ferns of Rotuma Island, a descriptive manual. *Occas. Pap. Bernice P. Bishop Mus.* 21(9): 161–208, 11 figs. (incl. map).

Briefly descriptive fern flora, with keys to all taxa, with full synonymy, references and citations; critical taxonomic remarks; notes on frequency, habitat, uses, etc.; localities given, with listing of *exsiccatae* or sight records; list of references; index to genera. The introduction treats extensively of geography, climate, geology, deforestation, the economy, features of the fern flora and its relationships, and general attributes of vernacular names; a short history of exploration is also included.

955

Fiji Islands

This comprises the islands of the Fijian Archipelago as well as the limits of the state of Fiji (except Rotuma; see **954**). At this writing (August, 1979), the first volume of the long-awaited new flora by A. C. Smith has made its appearance, and is described below. It represents the cumulation of some 140 years of more or less detailed botanical exploration of the islands, beginning with the United States Exploring Expedition. Apart from the classic flora of Seemann and the useful checklist of Parham (2nd edn., 1972), described below, other important precursory works include *New plants from Fiji*, I–III (1930–32) by J. W. Gillespie as well as *Fijian plant studies*, I–II (1936–42) and *Studies of Pacific Island plants*, I– (1941–) by A. C. Smith; for details, see the bibliography in Parham's checklist or other standard sources.

Bibliography

A detailed chronological bibliography appears in *Plants of the Fiji Islands* by Parham (see below).

PARHAM, J. W., 1972. *Plants of the Fiji Islands.* Revised edn. xv, 462, xxix pp., 104 figs., 1 color pl. Suva: Government Printer. (First edn. 1964.)

Annotated systematic enumeration of vascular plants (including those naturalized, adventive, or commonly cultivated), with essential synonymy, vernacular names, generalized indication of local range (with more details for less common species); numerous literature citations; extensive chronologically arranged bibliography; glossary and general index to family, generic, and vernacular names. The introductory section includes a brief history of botanical work in the islands as well as general notes on their physical features, flora, and vegetation.

SEEMANN, B., 1865–73. *Flora vitiensis: a description of the plants of the Viti or Fiji Islands with an account of their history, uses and properties.* xxxiii, 453 pp., 100 pls. (mostly in color; incl. map). London: Reeve. (Reprinted 1977, Vaduz, Liechtenstein: Cramer.)

Comprehensive descriptive flora of vascular and non-vascular plants, with principal synonymy as well as reference and literature citations; vernacular names; indication of localities, with *exsiccatae*, and general summary of extralimital range; taxonomic and phytogeographical notes, often referring to plants outside the Fiji archipelago; extensive commentary on local uses; index to all botanical names. The introductory section includes surveys of previous exploration as well as of the physical features, flora, and vegetation of the islands. Although now much outdated, Seemann's classic remains the fundamental reference text on the flora of Fiji. The superb hand-colored plates are by W. H. Fitch.

SMITH, A. C., 1979. *Flora vitiensis nova: a new flora of Fiji*. Vol. 1. 501 pp., 101 figs., color. pls. Lawai, Kauai, Hawaii: Pacific Tropical Botanical Garden.

Copiously annotated, briefly descriptive illustrated flora of seed plants, with keys to all taxa, full synonymy with references and citations, vernacular names, general indication of local range with citation of representative *exsiccatae*, extralimital distribution where applicable, and notes on habitat, frequency, occurrence, special features, properties, local uses, etc.; typification with historical remarks; critical commentary. The introductory section (in volume 1), with several scenic colored plates, gives the plan and genesis of the work (for which preliminary studies began in 1933 as part of the Pacific botanical programme of the Bishop Museum, in continuation of work initiated by J. W. Gillespie in the late 1920s), physical features, climate, floristics, vegetation, and phytogeography of the islands, a discussion of the delimitation of taxa, and an account of botanical exploration. Covers gymnosperms and monocotyledons (except Orchidaceae), encompassing native, naturalized, adventive, and commonly cultivated species. Dicotyledons will appear in volumes 2 and 3, with a general index at the end of the work (no index in volume 1). Within the major flowering plant groups, the linear version of the Takhtajan system (1969) is being followed. [An interesting feature of the introduction is the author's discussion of the question of 'splitting' or 'lumping' of taxa in the context of Pacific botany. The often sweeping 'lumping' of the van Steenis 'school' as exemplified in *Flora Malesiana* is strongly criticized.]

Special group – pteridophytes

BROWNLIE, G., 1977. The pteridophyte flora of Fiji. 397 pp., 44 pls., 3 maps (Beih. Nova Hedwigia, 55). Vaduz, Liechtenstein: Cramer. (*apud* Gantner).

Descriptive treatment of ferns and fern-allies, with keys, synonymy (with references), general indication of local and extralimital distribution, with citation of *exsiccatae*, critical commentary, distinguishing features, and notes on habitat, etc.; typification; illustrations of representative species; complete botanical index. An introduction includes an account of exploration, sources of information, remarks on introduced and excluded species, technical notes, and a synoptic list of species; major references, pp. 7–8. Supersedes COPELAND, E. B., 1929. *Ferns of Fiji*. 105 pp., 5 pls. (Bernice P. Bishop Mus. Bull., 59). Honolulu.

956

Tonga

Comprises the island chain, and the Kingdom of Tonga, from Niuafo'ou and Tafahi in the north to the Minerva Reefs in the south. Despite the relatively modern flora by Yuncker, the islands are unevenly explored botanically.

YUNCKER, T. G., 1959. *Plants of Tonga*. 283 pp., 16 text-figs. (Bernice P. Bishop Mus. Bull., 220). Honolulu.

Systematic enumeration of known vascular plants, bryophytes, and fungi, with very brief descriptions of all vascular species; includes essential synonymy, vernacular names, localities with citations of *exsiccatae* and relevant earlier literature, general indication of external range, and notes on habitat and frequency as well as status (native naturalized, etc.); indices to all taxa and to vernacular names. The introduction includes a general description of the islands and their climate as well as an account of previous botanical works.

957

Niue

Here is also included the Beveridge Reef. The floras of Yuncker (1943) and Sykes (1970) are complementary. The island is a self-governing state associated with New Zealand.

SYKES, W. R., 1970. *Contributions to the flora of Niue*. 321 pp., 45 halftones (New Zealand Div. Sci. Indust. Res. Bull., 200). Wellington.

Systematic enumeration of vascular plants and Bryophyta, with descriptions given for many species; limited synonymy; vernacular names; indication of localities with *exsiccatae*; extensive taxonomic commentary and notes on habitat, special features, biology, etc.; brief glossary, list of references, and indices to all botanical and vernacular names. The introductory section gives general remarks on climate and vegetation, botanical exploration, and the results of the 1965 survey by the author. Intended as a supplement to Yuncker's *Flora* (see below).

YUNCKER, T. G., 1943. *The flora of Niue Island*.

ii, 126 pp., 3 text-figs. (incl. map), 4 pls. (Bernice P. Bishop Mus. Bull., 178). Honolulu.

Briefly descriptive flora of vascular plants and bryophytes, with critical remarks and notes on habitat uses and status (native, naturalized, etc.); vernacular names (Niuean and English); general indication of range, with citation of some *exsiccatae*; index to all taxa and to vernacular names. The introductory section deals with the geography, soils, climate, history, and economy of the island as well as the general features of the flora and vegetation. No keys are provided.

958

Samoa

Samoa as here delimited includes the islands from Savai'i in the west to Rose Atoll in the east. For Swains Island, see **974**. The group is politically divided, with Savai'i and Upolu comprising the independent state of Western Samoa, while the other islands make up American Samoa. Although botanical exploration has been extensive, no fully retrospective modern flora with keys is available. The general works of Christophersen and Reinecke are more or less complementary, and are supplemented by two partial works dealing with American Samoa.

CHRISTOPHERSEN, E., 1935–8. *Flowering plants of Samoa*, [I–II]. 2 parts, 53 text-figs., 3 pls. (Bernice P. Bishop Mus. Bull., 128, 154). Honolulu.

Systematic enumeration of angiosperms, with descriptions of new taxa; synonymy, with references and citations; vernacular names; indication of localities with *exsiccatae*; critical remarks and notes on habitat, local uses, etc.; list of references (in both parts); indices to all botanical and vernacular names. Based largely on collections at the Bishop Museum made after 1920, with relatively little reference to earlier German work. For a companion treatment of the Pteridophyta, see CHRISTENSEN, C., 1943. A revision of the Pteridophyta of Samoa. 138 pp., 4 pls. (Bernice P. Bishop Mus. Bull., 177). Honolulu. [Critical enumeration, with a few keys to species.]

REINECKE, F., 1896–8. Die Flora der Samoa-Inseln, [I–II.] *Bot. Jahrb. Syst.* **23**: 237–368, pls. 4–5; **25**: 578–708, pls. 8–13.

Systematic enumeration of the then-known non-vascular and vascular plants of Samoa, with descriptions of new taxa; limited synonymy; vernacular names; localities, with citations of *exsiccatae*; brief mention of overall distribution (if applicable); sometimes extensive taxonomic notes, along with general observations on individual species; general index to all taxa in both parts at the end of part II. The introduction includes sections on physical features, geology, physiography, and climate of the islands, as well as the general features of the flora. For additions, see LAUTERBACH, C., 1908. Beiträge zur Flora von Samoa-Inseln. *Bot. Jahrb. Syst.* **41**: 215–38.

Partial works: American Samoa

SETCHELL, W. A., 1924. *American Samoa*. vi, 275 pp., 57 text-figs., 37 pls. (Publ. Carnegie Inst. Wash., 341). Washington.

Includes an enumeration (pp. 41–129) of non-vascular and vascular plants of Tutuila including all records up to 1924, with descriptions of new taxa; notes on habitats, biology, uses, special features, etc.; vernacular names; much critical taxonomic discussion; localities, with citations of *exsiccatae*; list of references and indices. The introduction deals with geography, physiography, geology, soils, climate, vegetation associations, and origin of the flora, as well as the history of botanical exploration. Part III of the work gives an account of the plants known from Rose Atoll.

YUNCKER, T. G., 1945. *Plants of the Manua Islands*. 73 pp., map (Bernice P. Bishop Mus. Bull., 184). Honolulu.

Systematic enumeration of the known vascular plants and mosses, with descriptions of new or little-known taxa; indication of habitat; vernacular names; some taxonomic remarks; localities given, with citation of *exsiccatae*; general indication of overall range; index to all botanical and vernacular names. The introduction includes a general description of the island group. For supplement, see *idem*, 1946. Additions to the flora of the Manua Islands. *Occas. Pap. Bernice P. Bishop Mus.* **18**(14): 207–9.

959

Wallis and Horne (Futuna) Islands

Included here are the Wallis and Horne (Futuna) Islands, a group under French administration. The islands are considered insufficiently known botanically, and the recent list by St John and Smith (1971) is to some extent provisional.

ST JOHN, H. and SMITH, A. C., 1971. The

vascular plants of the Horne and Wallis Islands. *Pacific Sci.* 25(3): 313–48, 2 figs.

Systematic enumeration of known vascular plants, with descriptions of some new taxa; very limited synonymy; indication of localities, with *exsiccatae* and literature citations; summary of extralimital range; critical remarks and notes on habitat, uses, special features, etc.; list of references; no separate index. The introductory section includes notes on geography, previous botanical work in the islands, and a brief analysis of the vascular flora.

Region

96

Micronesia

Included in this region, which covers a very large expanse of the western Pacific, are all the islands of the now gradually dissolving Trust Territory of the Pacific Islands as well as Guam, Marcus, Wake, Nauru, Banaba (Ocean) and Kiribati (the Gilberts).

Current knowledge of the region has been summarized by Fosberg and Sachet (1975). Only two general compendia, both relatively preliminary in nature, have for long been available: the illustrated *Flora micronesica* (1933), in Japanese but later provided with English translations of the keys, and the *Enumeration of Micronesian plants* (1935), both by Kanehira and emphasizing the islands of the Trust Territory. These works, however, served to consolidate most earlier botanical knowledge of these islands, including the work of the exploring expeditions before 1840 (mainly in the Marianas) and the German and (after 1915) Japanese botanical surveys. The present *Flora of Micronesia*, publication of which began in 1975, is a definitive critical work, unfortunately lacking illustrations, which has been based on the results of all earlier field work in the Trust Territory and other Micronesian islands including the extensive American surveys of the post-World War II period and afterwards together with a manuscript enumeration of Hosokawa, extensive critical studies by the authors and others, and relevant published floras, monographs and revisions. However, it has been admitted that realization of this flora will be slow and recently (1979) a preliminary name list of the dicotyledons has been

published in *Micronesica*. This is of course more up-to-date than Kanehira's list but is less detailed, lacking citations of *exsiccatae* and commentary.

Coverage of individual units within the region is patchy. Guam has had two descriptive floras, by Safford (1905) as well as Stone (1971); only the latter is included here. For the remaining islands, more or less extensively annotated enumerations have been published on Palau (Otobed, 1971), Yap (Volkens, 1901), Truk (Hosokawa, 1937), Ponape (Glassman, 1952), the northern Marianas (Fosberg *et al.*, 1975; Hosokawa, 1934), Marcus (Sakagami, 1961), Wake (Fosberg, 1959; Fosberg and Sachet, 1969), the northern Marshalls (Fosberg, 1955), Kiribati (Volkens, 1903, also covering the Marshalls), and a number of the low Carolines, including Kapingamarangi. No florulas are available for Nauru and Banaba or for Kosrae (Kusaie), all considered more or less undercollected. Some other less visited islands, particularly in the western Carolines, are also considered undercollected.

Progress

FOSBERG, F. R. and SACHET, M.-H., 1975. Micronesia: status of floristic knowledge. In *Proceedings of the Thirteenth Pacific Science Congress* (Pacific Science Association), vol. 1, p. 98. Vancouver. [Abstract only.]

Bibliographies. General and supraregional bibliographies as for Superregion 94–99. See also Division 1 (LAWYER *et al.*, in press.).

Regional bibliographies

SACHET, M.-H. and FOSBERG, F. R., 1955. Annotated bibliography of Micronesian botany. In *Island bibliographies* (ed. M.-H. Sachet and F. R. Fosberg), pp. 1–132 (Natl. Acad. Sci. – Natl. Res. Council Publ., 335). Washington; also *idem*, 1971. Supplement to Annotated bibliography of Micronesian botany. In *Island bibliographies: Supplement* (ed. M.-H. Sachet and F. R. Fosberg), pp. 3–75 (*Ibid.*, 335, Supplement). Washington. [These complementary works comprise briefly annotated lists, with subject, geographical, and systematic indices.]

UTINOMI, H., 1944. Bibliographia micronesica scientiae naturalis et cultis. In *Tōa kyō-ei-ken sigenkagaku bunken mokuroku* [Bibliographic index for the study of the natural resources of the Greater East Asia Co-Prosperity Sphere] (Department of Education, Japan), vol. 5 (*Micronesia*). 208 pp. Tokyo. English edn.: UTINOMI, H., 1952. *Bibliography of Micronesia*. Trans. and ed. O. A. Bushnell *et al.* 157 pp. Honolulu: University of Hawaii Press. [Unannotated lists; for botany, see pp. 1–21; pp. 3–16 in English edn.]

Indices. General and supraregional indices as for Superregion 94–99.

960

Region in general

As noted above, the only nominally complete overall works are *Flora micronesica* and the *Enumeration of Micronesian plants*, both by Kanehira. Although the former is illustrated, it is in Japanese. Both deal largely with the Trust Territory (thus only sketchily accounting for Nauru, Banaba, Kiribati, Marcus, and Wake). These islands are also nominally covered by Schumann and Lauterbach in their German Pacific colonial flora (see **930**). These will eventually all be superseded by *Flora of Micronesia*, begun in 1975, which covers the whole region with the addition of Mapia north of Biak (**933**); however, up to this writing (1979) its progress has been slow. A new name list of dicotyledons appeared in 1979.

No introductory section has yet appeared in *Flora of Micronesia*, and a *Botanical report on Micronesia* by F. R. Fosberg (350 pp., 445 figs., 1946) prepared for the US Commercial Company was never published. The best modern general and topographical work on the region from a biological point of view, which includes a gazetteer, is GRESSITT, J. L., 1954. *Insects of Micronesia: introduction*. ix, 257 pp., 70 text-figs., map (B. P. Bishop Museum: Insects of Micronesia, 1). Honolulu.

FOSBERG, F. R. and SACHET, M.-H., 1975– . *Flora of Micronesia*. Part 1– (Smithsonian Contr. Bot., 20, etc.). Washington.

Critical descriptive 'research' flora of native, naturalized, and common introduced seed plants, with keys to genera and species, full synonymy including references and citations, detailed indication of regional distribution (with citation of *exsiccatae*) and extralimital range, vernacular names, critical commentary, and notes on habitat, biology, ethnobotany (the last often extensive), etc.; references and index in each fascicle. While more than one family may appear in a fascicle, and an ultimate arrangement on the traditional Englerian sequence is envisaged, family revisions are not appearing in any systematic order; to date (mid-1979), three fascicles have been published. [It is regrettable that a separate institutional series was not established for this work, as was done for *Flora of New South Wales* and *Flora of Ecuador*.]

FOSBERG, F. R., SACHET, M.-H., and OLIVER, R., 1979. A geographical checklist of the Micronesian Dicotyledoneae. *Micronesica*, **15**: 41–295, map.

Systematic name list, with synonymy and indication of islands of occurrence. In the introductory section it is indicated that the checklist was prepared in advance of the flora, which by its nature would be slow in realization. Covers 1342 species, of which 662 are considered native. No indication is given of whether a corresponding checklist for the rest of the seed plants has been, or is to be, published. A list of references and a family index are given at the end of the work.

KANEHIRA, R., 1933. *Flora micronesica*. 8, 468, 37 pp., 211 text-figs., 21 pls., map. Tokyo: South Seas Bureau.

Illustrated descriptive treatment of trees and shrubs, with keys, together with an enumeration of all known vascular plants; synonymy, with references; indication of local and extralimital range; vernacular names; index to botanical names. The introductory section includes a general sketch of the climate, vegetation, and general features of the flora; an account of botanical exploration; statistics of the flora; an analysis of the vegetation on different islands. Wholly in Japanese; for a free English translation of the keys, see ST JOHN, H., 1956. A translation of the keys in 'Flora Micronesica' of Ryôzo Kanehira (1933). *Pacific Sci.* **10**: 96–102.

KANEHIRA, R., 1935. Enumeration of Micronesian plants. *J. Dept. Agric. Kyushu Imp. Univ.* **4**: 237–464, pl. 2 (map).

Systematic enumeration of vascular plants, with synonymy and references; vernacular names; localities with citations of *exsiccatae*; index to all botanical names. The introductory section includes an account of botanical exploration. Both this and the preceding work cover the Marianas, Palau, Caroline, and Marshall Islands, i.e., the then-Japanese Mandate (together with Guam).

961

Palau (Belau) Islands

The island chain of Palau (Belau), sometimes treated as geographically part of the Carolines, extends from Ngaiangl (Kayangel) in the north through Babeldaob (Babelthuap) and Koror to Tobi (and Helen Reef) in the south. [With the current dismemberment of the US Trust Territory these islands are scheduled for separate political status.]

FOSBERG, F. R. *et al.*, 1980. *Vascular plants of Palau with vernacular names.* ii, 43 pp. Washington, DC: Department of Botany, Smithsonian Institution.

Systematically arranged but unannotated checklist, with equivalent English and Palauan vernacular names (where known) and indication of status (if exotic). An introductory section gives an indication of the scope and method of development of the work. [Partly succeeds the checklist of Otobed, see below.]

OTOBED, D. O., 1977. *Guide list of plants of the Palau Islands.* Revised edn. vii, 52 pp. Koror, Palau: Biology Laboratory, Trust Territory of the Pacific Islands. (Mimeographed; first issued 1961 with subsequent versions in 1967, 1971 and 1972.)

Part 1 of this unannotated work comprises a systematically arranged checklist of scientific names, with English and Palauan equivalents where known. Part 2 is a lexicon of Palauan names with scientific equivalents. On p. i there is a brief introduction and on pp. ii–vii an index to families; at the end of the work is a rather brief list of references. Covers vascular plants and some mosses and algae. [Based partly on a local herbarium now at the Conservation Service in Koror, and prepared with assistance from F. R. Fosberg and others. I thank Demei Otobed for sending a copy of this work for annotation.]

962

Caroline Islands

This immense chain, stretching through 25 degrees of latitude, is here considered to comprise the islands from Ngulu and Yap in the west to Kosrae (Kusaie) in the east, with a southward extension to the isolated atoll of Kapingamarangi. Both 'high' and 'low' islands and several types of atolls are represented. Botanical collecting has been uneven, with more remote places seldom visited; until after World War II, little work was done on most atolls and low islands.[18] Floristic writing has, furthermore, been rather localized, and for convenience entries are grouped under seven subheads. [With the current dissolution of the Trust Territory, the islands will receive separate status as a federation with four states, administered from Kolonia (Ponape).]

I. Yap district

Includes Ngulu, Yap, Ulithi, Fais, and Sorol.

FOSBERG, F. R. and EVANS, M., 1969. *A collection of plants from Fais, Caroline Islands.* 15 pp. (Atoll Res. Bull., 133). Washington.

Enumeration of plants collected by the authors in 1965 during a 2½ hour visit from a 'field trip ship', with citation of *exsiccatae*, indication of status, local names, and ecological and biological notes. The list is preceded by general remarks on the vegetation. Covers 120 species, considered only a 'fraction' of the total. [Fais is a 19 m high raised atoll of 2.8 km² with some phosphate deposits; compare Makatea in the Tuamotus.]

VOLKENS, G., 1901. Die Vegetation der Karolinen, mit besonderer Berücksichtigung der von Yap. *Bot. Jahrb. Syst.* **31**: 412–77; pls. 11–14.

Includes a briefly annotated list of the vascular and some non-vascular plants known from Yap, with vernacular names; based on a field trip of 1899. The remainder of this work is taken up with a consideration of physical features, climate, geology, vegetation formations, etc., on this and other islands in the Carolines.

II. Truk group

HOSOKAWA, T., 1937. [A preliminary account of the phytogeographical study on Truk, Caroline.] *Bull. Biogeogr. Soc. Japan*, 7: 171–255, figs. 1–51.

Includes a systematic enumeration of vascular plants, with vernacular names and indication of localities (with *exsiccatae*), earlier records, general occurrence, habitat, life-form, and special features. The introductory section gives accounts of physical features, climate, and botanical exploration, while at the end are accounts of vegetation, floristics, and phytogeography as well as lists of plants by life-forms. A summary (in English) and list of references are also given. Text in Japanese, apart from the enumeration proper, which is in English.

III. Ponape

GLASSMAN, S. F., 1952. *The flora of Ponape*. iii, 152 pp., 19 pls. (Bernice P. Bishop Mus. Bull., 209). Honolulu.

Systematic enumeration of vascular plants, with descriptions of new taxa; synonymy, with references and citations; indication of *exsiccatae*, with localities; generalized summary of extralimital range; taxonomic commentary and notes on uses; list of references and full index. The introductory section includes chapters on physical features, general attributes of the flora and vegetation, botanical exploration, and agriculture and economic botany.

IV. Kosrae (Kusaie)

No separate flora or checklist is available. [The island is now a separate administrative district within the recently organized Federated States of Micronesia.]

V. Kapingamarangi

This atoll, the most isolated of the Caroline Islands, lies due north of Nuguria (**939**) at about 1 °N latitude.

NIERING, W. A., 1956. *Bioecology of Kapingamarangi Atoll, Caroline Islands: terrestrial aspects*. iv, 32 pp., 33 text-figs. (incl. maps) (Atoll Res. Bull., 49). Washington.

Primarily ecological, but includes a table of species of vascular plants, divided into trees, shrubs, and herbs, with each islet being separately accounted for; notes on localities, frequency and mode of origin of each species. See also the author's Terrestrial ecology of Kapingamarangi Atoll, Caroline Islands. *Ecol. Monogr.* 33: 131–60.

VI. Eastern low islands and atolls

Includes the islands from Pingelap to Nukuoro, associated with Ponape and Kosrae. Florulas are available for Pingelap, Ant and Mokil.

GLASSMAN, S. F., 1953. New plant records from the eastern Caroline Islands, with a comparative study of the plant names. *Pacific Sci.* 7: 291–311.

Includes reports on collections from Mokil, Ant and Pingelap, based on one-day visits, with annotated checklists

of the higher (and lower) plants found in each island unit featuring local names, habitat, frequency of occurrence, and *exsiccatae*; linguistic comparison and list of references at end. For Pingelap, see also ST JOHN, H., 1948. Report on the flora of Pingelap Atoll, Caroline Islands, and observations on the vocabulary of the native inhabitants. *Ibid.*, **2**: 96–113.

VII. Central and western low islands and atolls

Included here are all islands from Oroluk west to Eauripik, passing through, among others, Namoluk (SE of the Truk group), Satawal (W of the Truk group), and Namonuito and the Hall Islands (N and NW of the Truk group), for which separate florulas are available.

FOSBERG, F. R., 1969. *Plants of Satawal Island, Caroline Islands*. 13 pp. (Atoll Res. Bull., 132). Washington.

Enumeration of plants collected in 1965 on a two-day visit, with citation of *exsiccatae*, indication of status, local names, and notes on biology, ecology, etc.; preceded by a general description of the vegetation.

MARSHALL, M., 1975. *The natural history of Namoluk Atoll, eastern Caroline Islands*. ii, 53 pp., 13 pls. (Atoll Res. Bull., 189). Washington.

Comprises a general natural history and ethnobiological survey, with an account of the vascular flora including a list of 113 species; local names, biology, uses, and voucher specimens are indicated. 12 photographs of vegetation are appended. [The atoll is subject to severe typhoons with accompanying destruction of vegetation.]

STONE, B. C., 1959. Flora of Namonuito and the Hall Islands. *Pacific Sci.* **13**: 88–104.

Annotated enumeration of higher plants, with keys to species, indication of localities (with citation of some *exsiccatae*), and notes on habitat, special features, cultivation, etc.

963

Marianas Islands (including Guam)

These comprise the chain of islands from Guam in the south through Rota, Saipan and Pagan to Farallon de Pajaros in the north. [Guam is United States territory, and the northern Marianas have recently been constituted as a self-governing common-wealth in association with the United States.]

I. Guam

STONE, B. C., 1970. *The flora of Guam.* vi, 629 pp., 97 text-figs., 16 pls. (3 in color), 4 maps (Micronesica, 6). Agaña.

Briefly descriptive flora of native, naturalized, and commonly cultivated vascular plants, with keys to all taxa; synonymy, references, and citations of pertinent literature; vernacular names; generalized indication of local, Marianan, and extralimital range, with citation of some *exsiccatae*; short notes on habitat, special botanical features, uses, etc.; indices to English and Chamorro vernacular names and to all botanical names. The introductory section includes chapters on the history of Guam, floristics and vegetation formations, phytogeography, forest and other plant resources, natural reserves, agriculture and economic botany, botanical exploration, and suggestions for future work.

II. Northern Marianas

FOSBERG, F. R., FALANRUW, M. V. C., and SACHET, M.-H., 1975. Vascular flora of the northern Marianas Islands. *Smithsonian Contr. Bot.* 22: 1–45, 2 maps.

Annotated enumeration of vascular plants, with citation of *exsiccatae* and localities and notes on habitat, special features, variability, wider distribution, etc.; list of references; index to families. The introduction *inter alia* gives an account of previous work on the flora. No keys or synonymy are included. For additions, see *idem*, 1977. Additional records of vascular plants from the Northern Marianas Islands. *Micronesica* 13: 27–31. [N.B. This work does *not* account for Rota, Tinian or Saipan; the southernmost island included is Farallon de Medinilla!]

HOSOKAWA, T., 1934. [Preliminary account of the vegetation of the Marianne Islands group]. *Bull. Biogeogr. Soc. Japan,* 5: 124–172, figs. 1–9, pls. 10–14 (incl. map).

Includes (pp. 129–51) a systematic list of vascular plants, with indication of islands on which present; tabular list of local distribution, with summary of extralimital range (if applicable). The remainder of the work consists of sections on physical features, climate, botanical collectors, vegetation, and phytogeography; a summary and list of references is appended. Text in Japanese, except for the main list and tables.

964

Parece Vela (Okino-tori Shima)

This islet, which is associated with a reef, lies in the Philippine Sea at 20 ° 25 ′N. by 136 ° 05 ′E., well west of Farallon de Pajaros in the Marianas and southwest of Kazan Retto (Volcano Islands). No separate accounts of the vascular plants, if any exist there, appear to be available.

965

Marcus Island

The vascular flora of this isolated atoll is very limited; the following paper (although primarily zoological) appears to provide the most recent information.

SAKAGAMI, S. F., 1961. An ecological perspective of Marcus Island, with special reference to land animals. *Pacific Sci.* 15: 82–104, illus., map.

Includes a list of plants and description of the vegetation, with map showing distribution of the plant formations and accompanying photographs. A historical sketch is also provided. Supplants (at least in part) BRYAN, W. A., 1903. A monograph of Marcus Island. *Occas. Pap. Bernice P. Bishop Mus.* 2: 77–124.

966

Wake Island

See also **998** (CHRISTOPHERSEN).

FOSBERG, F. R., 1959. *Vegetation and flora of Wake Island.* 20 pp. (Atoll Res. Bull., 67). Washington.

Enumeration of the native, naturalized, and commonly cultivated vascular plants and algae known from the atoll, with notes on the occurrence, frequency, habitats, and general features of each species; some taxonomic commentary; local range, with citation of *exsiccatae* and literature records; list of references. The introductory section covers the climate, soils, and main vegetation features. For further data, see FOSBERG, F. R. and SACHET, M.-H., 1969. *Wake Island vegetation and flora, 1961–1963.* 15 pp. (Atoll Res. Bull., 123). Washington.

967

Marshall Islands

The only purportedly general work, now long out of date, is that by Volkens described below. The several modern florulas of individual atolls or groups thereof are grouped under two subheads: one for the Radak (northern) chain and the other for the Ralik (southern) chain. [With the current dismemberment of the Trust Territory these islands are scheduled for separate political status.]

VOLKENS, G., 1903. Die Flora der Marshallinseln. *Notizbl. Königl. Bot. Gart. Berlin*, **4**: 83–91.

Briefly annotated list of the known vascular plants and fungi of the Marshall (and Gilbert) Islands, with some descriptive remarks; vernacular names; citations of collections, but local range not given; notes on uses. The introductory section includes some general remarks on the vegetation.

I. Radak chain

Comprises the northern islands from Bikini and Pokak through Majuro to Mili and Narik.

FOSBERG, F. R., 1955. *Northern Marshalls expedition 1951–1952: Land biota; vascular plants*. 22 pp. (Atoll Res. Bull., 39). Washington.

Systematic enumeration of vascular plants, with indication of presence by atoll or islet (including citation of *exsiccatae*) and notes on habitat, occurrence, frequency, special features, etc.; vernacular names; critical remarks. Based mainly on military expeditions of 1951–2 with other records from field trips in 1946 and 1950. For additions, see *idem*, 1959. *Additional records of phanerogams from the northern Marshall Islands*. 9 pp. (Atoll Res. Bull., 68). Washington.

Individual atolls

STONE, B. C. and ST JOHN, H., 1960. *A brief field guide to the plants of Majuro, Marshall Islands*. Majuro, Marshall Is.: Marshall Islands Intermediate School.

Provides a key to species, descriptive text, and many illustrations. [Not seen by this reviewer; cited from Lawyer, J. *et al. A guide to selected current literature on vascular plant floristics for the contiguous United States*... (see under Division 1).]

TAYLOR, W. R., 1950. *Plants of Bikini and other northern Marshall Islands*. 227 pp., 79 pls. (incl. map), frontispiece. Ann Arbor: University of Michigan Press.

Copiously illustrated, briefly descriptive flora of non-vascular and vascular plants, without keys; limited synonymy; citation of *exsiccatae*; notes on ecology, special features, biology, cultivation, etc.; list of references. Especially useful for marine algae.

II. Ralik chain

Comprises the southern islands from isolated Ujelang and Eniwetok in the northwest through Kwajalein and Jaluit to Ebon in the south. In contrast to the Radak chain, no general florula is available.

Individual atolls

FOSBERG, F. R. and SACHET, M.-H., 1962. *Vascular plants recorded from Jaluit Atoll*. 39 pp. (Atoll Res. Bull., 92). Washington.

Annotated enumeration of vascular plants, with descriptive remarks, vernacular names, citations, and indication of *exsiccatae*; historical notes (Jaluit was German administrative headquarters for the Marshalls); list of references.

ST JOHN, H., 1960. Flora of Eniwetok Atoll. *Pacific Sci.* **14**: 313–36.

Amply descriptive flora of vascular plants (42 species), with keys, synonymy, local range (with citation of *exsiccatae*), taxonomic commentary, and notes on distribution and status; list of references.

968

Kiribati (Gilbert Islands)

See also **967** (Volkens). Although the new nation of Kiribati, granted independence in 1979, extends to include Banaba (in **969**) and many Central Pacific islands (region 97), only the Gilberts proper, from Makin Meang through Tarawa, Tabiteuea and Onotoa to Arorai (north to south) are considered here. In addition to the florulas accounted for below, an unannotated plant list also appears in WOODFORD, C. M., 1895. The Gilbert Islands. *Geogr. J.* (London) **6**: 325–50, map. No proper general flora or enumeration is available.

Individual atolls

LUOMALA, K., 1953. *Ethnobotany of the Gilbert Islands*. v, 129 pp., map (Bernice P. Bishop Mus. Bull., 213). Honolulu.

Includes a plant checklist for Tabiteuea Atoll, together with notes on earlier records for other atolls and islands in the group. Other parts of the paper cover physical features, soil, vegetation, etc.

Moul, E. T., 1957. *Preliminary report on the flora of Onotua Atoll, Gilbert Islands.* ii, 48 pp., map (Atoll Res. Bull., 57). Washington.

A description of the vegetation and an annotated enumeration of non-vascular and vascular plants, with notes on habitat, occurrence, biology, vernacular names, uses, etc.; localities given, with citation of *exsiccatae*.

969

Banaba (Ocean Island) and Nauru

These famous phosphate islands are regrettably amongst the most poorly known botanically in Micronesia, the more so because of great disturbance to their vegetation in the wake of extensive mining operations. With access in the past being difficult through isolation and company policy, only casual collections have been made and botanical references are few and scattered, with no florulas available. A topographical account of the islands, with references to the vegetation, appears in Ellis, A. F., 1936. *Ocean Island and Nauru.* 319 pp. Sydney: Angus & Robertson; and notes on the vegetation of Nauru appear also in Burges, N. A., 1934. Nauru. *Sci. J.* (Sydney), **13**: 30–5.

Region

97

Central Pacific islands

This spread-out region includes all the low islands and atolls extending from Tuvalu (Ellice Islands) and Howland and Baker in the west towards the Line Islands in the east and the northern Cook Islands and the Tokelau group (with Swains) in the south.

The comparative poverty and monotony of the vascular flora, the great diffusion of the various islands and groups, and administrative fragmentation and disinterest have doubtless contributed to the lack of any general flora or enumeration. In addition, the level of botanical survey is variable: from very well-known

(Phoenix Islands, which include the one-time air stopover of Canton Island) to poorly collected (Tuvalu, apart from the somewhat better known Funafuti).

However, most of the seven units designated here have nominally complete checklists or florulas, or have one or more individual islands with such a work, here deemed 'representative'. Units with overall coverage include the Howland and Baker Islands together with the northern Line Islands (Christophersen, 1927) and the Tokelau Islands (Parham, 1971). The Ellice Islands are 'represented' by Funafuti (Maiden, 1904); the Phoenix Islands by Canton (Degener and Gillaspy 1955); the northern Cook Islands by Manihiki (Cranwell, 1933), and the southern Line Islands by Caroline (Clapp and Sibley, 1971). Vostok (Fosberg, 1937; Clapp and Sibley, 1971) and Flint (St John and Fosberg, 1937). In the northern Line Islands, supplementary coverage is available for Christmas Island (Chock and Hamilton, 1962) and Fanning (St John, 1974), as well as that classic of atoll florulas, *Palmyra Island* (Rock, 1916). Other information is very scattered, and standard bibliographies should be consulted.

Bibliographies. General and supraregional bibliographies as for Superregion 94–99.

Indices. General and supraregional bibliographies as for Superregion 94–99.

971

Tuvalu (Ellice Islands)

No flora or enumeration covering the whole of Tuvalu is available. Of the nine atolls in the group, only Funafuti, with the largest land area, has a florula – and that now rather elderly. Background information appears in Hedley, C., 1896. *General account of the atoll of Funafuti.* 71 pp. (Austral. Mus. Mem., 3). Sydney.

Individual florula

Maiden, J. H., 1904. The botany of Funafuti, Ellice group. *Proc. Linn. Soc. New South Wales*, **29**: 539–56.

Comprises an enumeration of vascular plants, bryophytes and lichens, with limited synonymy and generalized indication of distribution; extensive notes on habitat, form and uses, as well as vernacular names, included. A brief introduction is attached, while at the end are remarks on mechanisms and methods of dispersal. Based on collections made during the Australian Museum expedition of the 1890s.

972

Howland and Baker Islands

For the flora of these two islands, see **976** (Christophersen). No more recent reports are available.

973

Phoenix Islands

These comprise a relatively compact group of eight islands, of which the largest is Canton, a former air base and way station for commercial flights between the United States and Australasia and the South Pacific. It is also the only island with a published florula, although surveys have been made elsewhere in the group. No general checklist is available. They are now part of Kiribati (**968**).

Individual florula

Degener, O. and Gillaspy, E., 1955. *Canton Island, South Pacific.* ii, 51 pp. (Atoll Res. Bull., 41). Washington.

Pages 16–31 comprise a discursive enumeration of the seed plants recorded from the atoll, with general notes on habitat, biology, occurrence, and frequency, and including citations of *exsiccatae*. An introductory section contains accounts of geography, climate, topography, and history of human contacts, while at the end are found notes on the vegetation and on the fauna and aquatic flora. Additions may be found in Degener, O. and I., 1959. *Canton Island, South Pacific* (*resurvey of 1958*). 24 pp. (Atoll Res. Bull., 64). Washington.

974

Tokelau Islands

In addition to the three atolls of Atafu, Nukunonu, and Fakaofo, which constitute the political entity of the Tokelau Islands, this unit also includes Swains Island (administratively part of Samoa). A set of checklists has recently become available for the Tokelaus (Parham, 1971).

Parham, B. E. V., 1971. The vegetation of the Tokelau Islands with special reference to the plants of

Nukunonu Atoll. *New Zealand J. Bot.* 9: 576–609, 11 text-figs. (incl. maps), table.

Table 1 comprises an enumeration of known vascular plants of the Tokelau Islands (both indigenous and adventive), with indication of life-form and their occurrence elsewhere in the South Pacific. Separate lists are given for each of the three atolls (Nukunonu, Fakaofo, and Atafu), with local vernacular names and remarks on biogeography. The remainder of the paper deals with physical features, vegetation, and human influence; a list of references is appended.

975

Northern Cook Islands

These include the Danger Islands (Pukapuka), Nassau, Suvorov (Suwarrow), Manihiki (with Rakahanga), and Tongareva (Penrhyn). No general enumeration is available for these scattered entities, and the only island florula is that by Cranwell (1933).

Individual florula

Cranwell, L., 1933. Flora of Manihiki, Cook group. *Rec. Auckland Inst. Mus.* 1: 169–71.

Briefly annotated list of vascular and non-vascular plants, with notes on habitats and life-forms; includes limited synonymy and vernacular names. The introductory section incorporates general remarks on Manihiki, and a list of references is attached.

976

Northern Line Islands

Here are included Jarvis, Christmas, Fanning, Washington and Palmyra Islands, together with Kingman Reef. Christophersen's overall treatment of this group remains an important reference point for later floristic studies, now available for Fanning and Christmas. The classic early florula by Rock (1916) for Palmyra has also been included.

Christophersen, E., 1927. *Vegetation of the Pacific equatorial islands.* 79 pp., 13 text-figs., 7 pls. (Bernice P. Bishop Mus. Bull., 44). Honolulu.

Includes separate enumerations of the thenknown vascular plants of Jarvis, Christmas, Fanning, Washington, and Palmyra Islands as well as those of

Howland and Baker Islands; each list contains notes on life-form, habitat, occurrence, and biology for the species covered, as well as some taxonomic remarks and citations of *exsiccatae*. Each island list is also accompanied by general remarks on climate, soils, vegetation, and land use. At the end are notes on dispersal methods of the plants, introduced species, and a list of references. Pages 73–6 furthermore give a tabular summary of geographical distribution of species within this group of islands.

Individual florulas

CHOCK, A. K. and HAMILTON, D. C., Jr. 1962. *Plants of Christmas Island*. 7 pp. (Atoll Res. Bull., 90). Washington.

Briefly annotated list of vascular plants (41 species, of which some 35 spontaneous) and some fungi, with occasional taxonomic notes and including citations of *exsiccatae* as well as references to earlier records; list of references at end. The introductory section includes an account of previous collecting on this island.

ROCK, J. F., 1916. *Palmyra Island: with a description of its flora*. 53 pp., 20 halftones, map (Coll. Hawaii Publ., Bull. 4). Honolulu.

Annotated, copiously illustrated enumeration of non-vascular and vascular plants, with notes on their habitat, life-form, and biology; includes synonymy, with references, and citations of *exsiccatae* with generalized indication of extralimital range. The introductory section gives a general description of the atoll, with remarks on the climate, vegetation, animal life, and history of human contact. The work is one of the earliest yet most outstanding examples of an atoll florula, with superb photographs by the author.

ST JOHN, H., 1974. Vascular flora of Fanning Island, Line Islands, Pacific Ocean. *Pacific Sci.* 28: 339–55.

Systematic enumeration of all reported vascular plants irrespective of status (indigenous species in bold face), with citations of *exsiccatae*. 102 taxa are accounted for, with 20 indigenous and 28 adventive. The remainder of the paper comprises descriptions of new taxa of *Pandanus*.

977

Southern Line Islands

Includes the scattered islands of Caroline, Vostok, Flint, Starbuck, and Malden. No overall flora is available, although florulas exist for the first three named. The range of spontaneous vascular plants on some is very small: on Vostok, only two species are so far known.

Individual florulas

CLAPP, R. B. and SIBLEY, F. C., 1971. Notes on the vascular flora and terrestrial vertebrates of Caroline Atoll, southern Line Islands. ii, 16 pp., 5 text-figs. (Atoll Res. Bull., 145). Washington.

Includes an enumeration of known vascular plants (35 species, of which 25–6 are native or accidentally introduced), with some general observations for each species; localities, with citations of *exsiccatae*, and older references included. An introductory section includes a description of the atoll, summaries of the vegetation, and notes on the history of human contact.

FOSBERG, F. R., 1937. Vegetation of Vostok Island, central Pacific. *Proc. Hawaiian Acad. Sci.* 11(1935–36): 19 (also appearing as *Bernice P. Bishop Mus. Spec. Publ.*, 30: 19).

Includes a 'list' of one non-vascular and two vascular plants (*Boerhaavia diffusa* and *Pisonia grandis*). A fuller account of the vegetation, based on a new survey in 1965, is given in CLAPP, R. B. and SIBLEY, F. C., 1971. *The vascular flora and terrestrial vertebrates of Vostok Island, South-Central Pacific*. 10 pp., illus., map (Atoll Res. Bull., no. 144). Washington. [Reports the presence of the same two vascular plants in 1965 as in 1935.]

ST JOHN, H. and FOSBERG, F. R., 1937. *Vegetation of Flint Island, central Pacific*, pp. 1–4 (Occas. Pap. Bernice P. Bishop Mus., vol. 12, no. 24).

Includes a list of native and more significant introduced vascular plants, with notes on their status (and method of introduction) and citations of *exsiccatae*. A brief prefatory section features a description of the island and its extant vegetation.

Region

98

Islands of southeastern Polynesia

This region covers all the islands from the southern Cooks eastwards and southeastwards to Rapa Nui (Easter) and Sala-y-Gomez and northeastwards to the Marquesas, thus also encompassing the Societies, the Austral Islands and Rapa Iti (Rapa), the Tuamotu (Low) Archipelago (including Makatea), and the Mangareva (Gambier) and Pitcairn groups.

Because of the scattered and diverse nature of the area, its varied political affinities, and a history of fragmented and episodic botanical work under a number of 'metropolitan' auspices which has resulted in widely scattered collections, no overall flora, apart from Drake's more general *Illustrationes* (**940–90**), has ever been published. The level of botanical exploration also varies widely, although in general it is lower than in Hawaii, with many islands still considered poorly collected. Documentation is, with some exceptions, largely unsatisfactory by modern standards.

The two general floras, *Flore de la Polynésie française* (Drake, 1893) and *Flora of Southeastern Polynesia* (Brown and Brown, 1931–5) neither encompass the entire region (for political or other reasons) nor can be considered definitive by current standards. Drake's work, doubtless largely derived from his *Illustrationes*, represents sources available in France, while that of the Browns was evidently based mostly on American collections made from 1920 onwards and housed in the Bishop Museum. The latter was severely criticized by Merrill in his *The Botany of Cook's Voyages* (1954) as having been overambitious in conception and prepared without proper guidance, which resulted in 'erratic writing and non-existent editing', according to Sachet in the introduction to her *Flora of the Marquesas* (**989**).

While a new, critical regional flora would be a very worthwhile addition to the literature, it would require extensive international collaboration and would have to be based on extensive new revisionary studies, such as that on *Cyrtandra* by the late G. W. Gillett and on the Rubiaceae by S. P. Darwin and others. However, a useful preliminary step might be the preparation of a prodromus in the form of a keyed enumeration, thus enabling taxonomic problems to be pinpointed.

Of the several floras, florulas, and enumerations for the nine units designated here (or for parts thereof), mostly dating from after 1920 when intensive botanical exploration expanded from a few readily accessible islands, the most critical are Sachet's Marquesian flora as well as the fragment of a Societies flora by Grant *et al.* (1974). Both represent complete reviews of available collections, literature, and unpublished data. It is most regrettable that Grant's work was never completed, as the latest general coverage of the Societies dates from 1893. Because it, along with the projected Raiatean flora of J. W. Moore, originated as a Bishop Museum project parallel to that of the Browns (the latter originally intended to cover only the Marquesas), the Societies (and, for similar reasons, the Cooks) were not

dealt with in *Flora of Southeastern Polynesia*, although this would be expected on phytogeographic grounds. Another group of critical works are the papers on Rapa Nui (Easter Island) by Skottsberg (1922, 1951) and Christensen (1920).

Among other 'standard' contributions, the southern Cooks are covered by 'representative' florulas for Rarotonga (Wilder, 1931) and Aitutaki (Fosberg, 1975) as well as a general survey of pteridophytes (Brownlie and Philipson, 1971), while the Societies have coverage for Raiatea (Moore, 1933, 1963 – neither proper florulas), Tahiti (Nadéaud, 1873, Setchell, 1926 – the latter not a florula), with the pteridophytes as a whole treated by Copeland (1932). In the Tuamotus, separate coverage exists for Makatea (Wilder, 1934), Raroia (Doty *et al.*, 1954) and Rangiroa (Stoddart and Sachet, 1969). The Gambiers (Mangareva) are treated in an unsatisfactory paper by Huguenin (1974), while fragmented coverage is available for Pitcairn Island (Brownlie, 1961; Maiden, 1901 – the former only on the pteridophytes) and Oeno and Henderson (St John and Philipson, 1960, 1962). No separate coverage is available for the Australs and Rapa Iti, save for the Hallé checklist (1980).

The pteridophytes for the region as a whole were last revised in 1931 by the Browns and in 1938 by Copeland. More recent treatments are available only for some units, either separately or as parts of vascular or general florulas.

Bibliographies. General and supraregional bibliographies as for Division 9 and Superregion 93–97.

Regional bibliographies

O'REILLY, P. and REITMAN, E., 1967. *Bibliographie de Tahiti et de la Polynésie française*. xvi, 1046 pp. (Publ. Soc. Océan., 14). Paris: Société des Océanistes. [Botany, pp. 248–88.]

Indices. General and supraregional indices as for Division 9 and Superregion 93–97.

980

Region in general

Neither of the two general floras accounted for below covers the region as a whole, and that by the Browns is in part nominal. Drake's flora (1893) covers those islands under French control (thus excluding the southern Cooks and the Pitcairn group), while that by the Browns omits the southern Cooks and the Societies and gives only sketchy coverage outside its originally projected limits, the Marquesas. As previously noted, a new, critical general flora is much needed. [One other work has not been seen: ROBERTSON, R., 1952. Catalogue des plantes vasculaires de la Polynésie française. *Bull. Soc. Études Océanien.* (Papeete) 8: 371–406; this covers 254 species recorded after 1900.)]

BROWN, F. B. H. and BROWN, E. W., 1931–5. *Flora of southeastern Polynesia*, I–III. 3 vols. Illus. (Bernice P. Bishop Mus. Bull., 84, 89, 130). Honolulu.

Descriptive flora of native, naturalized, and 'village' plants, with keys to all taxa; synonymy occasionally indicated, without references; some citations of standard floras and other works in the text; citation of *exsiccatae* if present in the Bishop Museum; indication of localities and summary of overall range; sometimes extensive taxonomic commentary and notes on sources of introduced species, local varieties of cultivated plants, special features, etc.; vernacular names. The introductory sections in each part deal with physical features, vegetation (rather sketchily), floristics, native agricultural systems, etc., in the Marquesas (other areas virtually neglected), as well as the itinerary of the Bayard Dominick expedition (which yielded most of the materials on which this flora is based). Lists of references and complete indices conclude each part. The work is most complete for the Marquesas but also includes records from the Tuamotus, the Austral Islands and Rapa Iti, and (sketchily) the Mangareva and Pitcairn groups, but *omits* the Societies and southern Cook Islands and does not extend to Rapa Nui. For an important supplement on the ferns, accounting for later expeditions and field work and extending to the Societies, see COPELAND, E. B., 1938. *Ferns of southeastern Polynesia.* pp. 45–101, 25 text-figs. (Occas. Pap. Bernice P. Bishop Mus., vol. 14, no. 5). Honolulu.

DRAKE DEL CASTILLO, É., 1893. *Flore de la Polynésie française.* xxiv, 352 pp., color map. Paris: Masson.

Briefly descriptive flora of native, naturalized and 'village' vascular plants, with keys to all taxa; includes synonymy (with references and literature citations), localities with citations of *exsiccatae* and summary of external distribution or origin, critical remarks, brief habitat and other notes, and vernacular names; index to all botanical names. An introduction includes notes on physical features, floristics, etc. of the French territories in southeastern Polynesia (as well as the Wallis Islands), a brief history of botanical exploration, statistics of the flora, and a list of references. Much of this work was, presumably, based on the author's *Illustrationes florae insularum maris Pacifici* (1886–92; see **940–90**). It remains the most recent nominally overall work with respect to the Societies (except for the treatments by Copeland and Grant *et al.* on parts of the vascular flora), and elsewhere within the limits of its coverage is in a way complementary to the Browns' flora.

981

Society Islands

Drake's obsolete *Flore de la Polynésie française* of 1893 (see **980** above) is still the only nominally comprehensive work and is also to date the last French flora on the Society Islands. Of the considerable botanical results of American expeditions and local field programmes (the latter principally by M. L. Grant and J. W. Moore) in the 1920s and 1930s, only a regrettably small proportion has ever appeared in the form of parts of a general vascular flora (Copeland, 1932; Grant *et al.*, 1974 (but largely written by 1936)) and no continuation is evidently in sight. Miscellaneous contributions have been published since 1920 on individual islands, but florulas projected for Tahiti and Raiatea were never realized.[19] For vegetation and phytogeography, see PAPY, H. R. (1951–55) 1954–5. *Tahiti et les îles voisines. La végétation des îles de la Société et de Makatéa (Océanie française).* 2 parts, 386 pp., illus., map (Trav. Lab. Forest. Toulouse, Sér. V, vol. 2, part 1, no. 3). Toulouse.

COPELAND, E. B., 1932. *Pteridophytes of the Society Islands.* ii, 86 pp., 16 pls. (Bernice P. Bishop Mus. Bull., 93). Honolulu.

Briefly descriptive treatment of lower vascular plants, with keys to all taxa; essential synonymy; localities, with citations of *exsiccatae*; general summary of extralimital range; more or less extensive critical commentary; index. Based mainly on the collections of Grant in 1930–1 and other American sources, without full revision of earlier French collections covered by Drake. Some additions are included in the author's *Ferns of southeastern Polynesia* (see **980**, BROWN and BROWN).

GRANT, M. L., FOSBERG, F. R., and SMITH, H. M., 1974. *Partial flora of the Society Islands: Ericaceae to Apocynaceae.* vii, 85 pp. (Smithsonian Contr. Bot., 17). Washington.

Rambling systematic treatment of certain 'sympetalous' families, with long or moderately long description and analytical keys to genera, species, and lower taxa, full synonymy (with references), vernacular names, localities (with indication of *exsiccatae*), extralimital range (if applicable), critical remarks, and extensive (largely compiled) notes on ethnobotany as well as on habitats and associates; index to botanical names. The introductory section includes a biographical sketch of the senior author as well as chapters on geography, climate, ethnobotany, and sources for the work, while at the end is an account of botanical work in the area (by H. M. Smith) and a bibliography. No continuation of this work is in sight, as far as is known.

I. Tahiti

The only florula, an annotated checklist, is NADÉAUD, J., 1873. *Énumération des plantes indigènes de l'île de Tahiti.* v, 86 pp. Paris; all records have been incorporated into Drake's general flora (980). Many critical notes and additions appear in SETCHELL, W. A., 1926. Tahitian spermatophytes collected by W. A. Setchell, C. B. Setchell and H. E. Parks. *Univ. Calif. Publ. Bot.* 12: 143–213, pls. 23–6.

II. Other islands

No florulas are available for any of the 'leeward' islands in the Societies. J. W. Moore's projected florula on Raiatea has never been published although two precursory papers appeared, viz: MOORE, J. W., 1933. *New and critical plants from Raiatea.* 53 pp. (Bernice P. Bishop Mus. Bull., 102); and *idem*, 1963. *Notes on Raiatean flowering plants, with descriptions of new species and varieties.* 26 pp., illus. (*ibid.*, no. 226).

982

Southern Cook Islands

Included here are the 'high' islands of Rarotonga and Mangaia, the 'almost-atoll' of Aitutaki, and associated atolls (including, for geographical reasons, Palmerston which botanically is closer to the Northern Cooks (975)). No general flora is available; the only group treated for the whole of the southern Cooks to date are the pteridophytes (Brownlie and Philipson, 1971). Of individual islands, significant florulas exist only for Aitutaki (Fosberg, 1975) and Rarotonga (Cheeseman, 1903; Wilder, 1931). Considerable new collecting was, however, undertaken in the islands in 1974 by Sykes and work is in progress towards a general flora (W. R. Sykes, personal communication).

Bibliography

KINLOCH, D. I. (ed.) 1980. *Bibliography of research on the Cook Islands.* 164 pp. (New Zealand MAB Report, no. 4). Lower Hutt, NZ: Soil Bureau, Department of Scientific and Industrial Research, New Zealand (for New Zealand National Commission for UNESCO). [Entries with detailed annotations; botany, pp. 9–67, by W. R. Sykes with, on pp. 11–17, a state-of-knowledge review.]

BROWNLIE, G. and PHILIPSON, W. R., 1971. Pteridophyta of the southern Cook Group. *Pacific Sci.* 25: 502–11.

Annotated systematic enumeration covering 80 taxa, with synonymy, references, citations of *exsiccatae*, critical commentary, and notes on frequency, habitat, etc. An introductory section includes plant-geographical considerations. Includes collecting and observations made during the Cook Bicentenary expedition of 1969 and supersedes for these plants the works of Cheeseman and Wilder described below.

I. Rarotonga and Mangaia

CHEESEMAN, T. F., 1903. *The flora of Rarotonga, the chief island of the Cook group.* (*Trans. Linn. Soc. London*, ser. II, Bot., vol. 6, pp. 261–313, pls. 31–5, map.) London.

Annotated enumeration of vascular plants, with Latin descriptions of novelties (these partly by W. B. Hemsley) and commentaries on habitats, biology, special features, uses, etc., with summaries of external distribution (where appropriate); not separately indexed. The introductory section gives a general account of the island of Rarotonga, with descriptions of the vegetation, noteworthy plants, crops, and introduced species as well as of geographical features, climate, geology, etc. Based largely upon a three-months' visit in 1899, this work accounts for 334 species (of which 233 are considered indigenous). [Not effectively superseded by the later florula by Wilder, whose additions were largely among cultivated, adventive and naturalized species which by the end of the 1920s had become more numerous.]

WILDER, G. P., 1931. *Flora of Rarotonga.* 113 pp., 3 text-figs., 8 pls. (Bernice P. Bishop Mus. Bull. no. 86). Honolulu.

Systematic enumeration of native and introduced

(including many cultivated) vascular plants, without citation of *exsiccatae* or references but giving brief diagnostic notes together with remarks on habitat, status, frequency, phenology, uses, etc., vernacular names, and indication of local range (with for some less common species the mention of particular localities). The introductory section gives a general description and historical sketch of the island along with notes on general features of the flora, while the index includes cross references to Cheeseman's flora (see above). [The work has a strong ethnobotanical slant and is moreover by current standards only semi-critical.] Recent contributions usefully supplementing the two aging florulas are PHILIPSON, W. R., 1971. Floristics of Rarotonga. *Bull. Roy. Soc. New Zealand*, 8: 49–54; and STODDART, D. R. and FOSBERG, F. R., 1972. *Reef islands of Rarotonga, with list of vascular flora.* 14 pp. (Atoll Res. Bull., no. 160). Washington.

II. Other islands

FOSBERG, F. R., 1975. Vascular plants of Aitutaki. In STODDART, D. R. and GIBBS, P. E. (eds.), *Almost-atoll of Aitutaki* (*Cook Islands*), pp. 73–84 (Atoll Res. Bull., no. 190). Washington.

Briefly annotated systematic list, with citations of *exsiccatae* and localities along with indication of status, habitat, biology, etc. [Constitutes the first florula for the island. The vegetation is described by D. R. Stoddart elsewhere in the main work.]

983

Tubuai (Austral) Islands

These include Maria, Rimatara, Rurutu, Tubuai and Raivavae. Until 1980, the only general coverage of this group had nominally been provided by the regional flora of BROWN and BROWN (980), which contained very few pertinent records. This gap has now been filled, after a fashion, in the paper by N. Hallé here described; this at least accounts for the numerous collections made after 1930 (i.e., by the 1934 Mangarevan Expedition and later field trips) which hitherto had been partly treated in many separate papers.

HALLÉ, N., 1980. Les orchidées de Tubuaï (archipel des Australes, Sud Polynésie): suivies d'un catalogue des plantes à fleurs et fougères des îles Australes. *Cah. Indo-Pacifique*, 2(3–4): 69–130, Pls. 1–12.

Pages 85–129 comprise an alphabetically arranged checklist of 627 species of vascular plants of

the Austral Islands and Rapa Iti, with citations of *exsiccatae* arranged by island; bibliography, pp. 129–30. See also FOSBERG, F. R., and ST JOHN, H., 1952. Végétation et flore de l'atoll Maria, îles Australes. *Rev. Sci. Bourbonn.*, (1951): 1–7.

984

Rapa Iti and Bass Islets (Morotiri)

This comprises the extra-tropical reefless 'high' island of Rapa Iti (Rapa) and the tiny Bass Islets to the southeast. Although the large island is of considerable botanical interest, until 1980 the only general coverage was nominally in regional floras (see 980). The group has now been covered by Hallé (see preceding entry).

985

Tuamotu Archipelago (excluding Mangareva)

Almost all islands in this large archipelago are low, with a small, monotonous vascular flora. Only Makatea, a raised, phosphate-bearing atoll, has a relatively rich, diverse flora, monographed by Wilder (1934) but considered to be still in need of more collecting. For convenience, the works are divided into two groups: one for Makatea and the other for all the low islands, for which few florulas have been published. The latter, however, are nominally covered in the regional flora of BROWN and BROWN (see 980), and in the Raroia florula by Doty *et al.* (see below) there is provided a key to the species commonly expected in the low islands as a whole.

I. Makatea

WILDER, G. P., 1934. *Flora of Makatea.* 49 pp., 5 pls., map. (Bernice P. Bishop Mus. Bull., 120). Honolulu.

Systematic enumeration of native, naturalized and commonly cultivated vascular plants, without synonymy or references and but few citations of *exsiccatae*; includes brief diagnostic notes together with sometimes extensive remarks on habitat, biology, cultivation, uses, etc.; vernacular names; generalized indication of local range; index only to genera and species. An introductory section includes notes on physical features, the fauna and vegetation and features of the flora as well as on the now-defunct phosphate extraction industry. The flora is reportedly incomplete and by modern standards is only semi-critical.

II. Low Islands

Separate florulas appear to be available only for Rangiroa in the western islands (near Makatea) and for Raroia (of *Kon-Tiki* fame) in the central islands.

DOTY, M. S. *et al.*, 1954. *Floristics and plant ecology of Raroia atoll, Tuamotus.* 58 pp. (Atoll Res. Bull., 33). Washington.

Includes an enumeration (pp. 24–35) of native and aboriginally introduced seed plants (by M. Doty) with an appendix (pp. 36–41) on 'village' plants; includes citations of *exsiccatae* with localities and notes on occurrence, habitat, biology, etc.; vernacular names. Pteridophytes (by K. Wilson) appear on pp. 57–8, and treatments of some non-vascular plants are also given. Pages 14–23 feature a 'key to commonly expected Tuamotuan vascular plants' originally prepared by F. R. Fosberg. [The total of native and aboriginally introduced vascular plants numbers some 51 species.]

STODDART, D. R. and SACHET, M.-H., 1969. *Reconnaissance geomorphology of Rangiroa Atoll, Tuamotu Archipelago, with list of vascular flora of Rangiroa.* 44 pp. (Atoll Res. Bull., 125). Washington.

Pages 33–44 feature an annotated checklist (by M.-H. Sachet) based on a 1963 field trip, with localities, citations of *exsiccatae*, and notes on habitat, occurrence, etc., along with introductory remarks. There is a total of 121 species recorded, but only some 40 are considered indigenous.

986

Mangareva group (Gambier Islands)

This is here considered to comprise the Mangareva archipelago proper – three 'high' islands surrounded by a reef – and the atoll of Timoe to the southeast. Coverage is nominally provided by the regional floras of BROWN and BROWN and of DRAKE (see **980**) but no critical florula is available for other records including those of the 1934 American expedition. However, an unannotated checklist (Huguenin, 1974) has recently appeared in a French monograph on the group, based on a new field survey. Mangareva is noted for megalithic churches but little 'bush' and an indigenous vascular flora of only 41 species.

HUGUENIN, B., 1974. La végétation des îles Gambier. Relevé botanique des espèces introduites. *Cah. Pacifique*, **18**(2): 459–71.

Sparsely annotated checklist of native and introduced vascular plants, with indication of status, origin, and local occurrence; no references. An introductory note gives the actual indigenous vascular flora as having 41 species (12 pteridophytes), with 27 per cent endemism; there is also a large range of more recent introductions, of which some 200 were recorded by the 1966 field party.

987

Pitcairn Islands

Here are included Pitcairn Island (of *Bounty* fame) as well as Oeno, Henderson (Elizabeth) and Ducie, a structurally varied lot but united politically as the last of the British colonies in the Pacific. Florulas are available for all islands except impoverished Ducie, although the phanerogams of Pitcairn need revision. Patchy coverage is also provided in the general flora of BROWN and BROWN (see **980**).

I. Pitcairn

BROWNLIE, G., 1961. The pteridophyte flora of Pitcairn Island (Studies in Pacific ferns, IV). *Pacific Sci.* **15**: 297–300.

Enumeration of 20 species (based partly on new collections), with synonymy, references, citation of *exsiccatae*, critical remarks, and notes on habitat, frequency, etc.; floristic affinities are with Malesia (including Papuasia).

MAIDEN, J. H., 1901. Notes on the botany of Pitcairn Island. *Rep. Meetings Australasian Assoc. Adv. Sci.* **8**: 262–71.

Briefly annotated list of vascular plants, based mainly on collections by Miss R. A. Young; includes vernacular names and notes on local uses. An introduction includes remarks on botanical work in the island and a short list of references. [Now superseded for pteridophytes by Brownlie (see above).]

II. Oeno, Henderson, and Ducie

No separate account appears to exist for Ducie, whose vascular flora is reportedly extremely poor (one species). Henderson, a raised atoll, has a much richer flora than Oeno, which is low and small.

St John, H. and Philipson, W. R., 1960. List of the flora of Oeno Atoll. *Trans. Roy. Soc. New Zealand*, 88(2): 401–3.

Systematic enumeration of 17 vascular plants and one alga, based largely on collections from the 1934 Mangarevan expedition; includes synonymy, citation of *exsiccatae*, and notes on habitats and local distribution.

St John, H. and Philipson, W. R., 1962. An account of the flora of Henderson Island, south Pacific Ocean. *Trans. Roy. Soc. New Zealand, Bot.* 1(14): 175–94, 4 text-figs., 7 halftones.

Annotated, illustrated systematic enumeration of vascular and non-vascular plants of this interesting atoll, with descriptions of new taxa; synonymy, with references and citations (the latter mainly from the Browns' *Flora of southeastern Polynesia*), indication of *exsiccatae*, critical remarks, and notes on habitat, occurrence, etc.; list of references. The introductory section includes descriptive remarks on the island and its vegetation as well as an account of previous botanical work.

988

Rapa Nui (Easter) and Sala-y-Gomez Islands

These comprise the very isolated, extratropical islands of Rapa Nui (Easter Island) and Sala-y-Gomez. Rapa Nui, renowned for its 'mysterious' megalithic statues, has been visited by many scientists, including botanists, and a good critical flora has been produced by Skottsberg (1922; additions 1951), with an accompanying treatment of pteridophytes by Christensen (1920). The island is now virtually treeless and the distinctive *toro-miro* tree (*Sophora toromiro*) survives only in cultivation in Göteborg, Sweden. For casual notes on the vascular plants (two or three species) of the more easterly Sala-y-Gomez, an island group consisting of little more than rocks, see Norris, R. M., 1960. Sala-y-Gomez – lonely landfall. *Pacific Disc.* 13(6): 20–5.

Skottsberg, C. J. F., 1922. The phanerogams of Easter Island. In *Natural history of Juan Fernandez and Easter Island* (ed. C. J. F. Skottsberg), part 2 (Botany), pp. 61–84, 2 text-figs., pls. 6–9. Uppsala: Almqvist & Wiksell.

Christensen, C. and Skottsberg, C. J. F., 1920. The ferns of Easter Island. *Ibid.*, pp. 49–53, 3 text-figs. Uppsala.

Skottsberg, C. J. F., 1951. A supplement to the pteridophytes and phanerogams of Juan Fernandez and Easter Island. *Ibid.*, pp. 763–92, 1 text-fig., pls. 55–7.

The original contributions respectively include critical systematic enumerations of flowering plants and of ferns, with descriptions of novelties, synonymy with references, citation of *exsiccatae* with localities, summary of extralimital distribution (where applicable), and extensive notes on habitat, frequency, etc., as well as critical commentary. The introductory section of the flowering plants account includes an account of botanical work on the island by the author and his predecessors. Additional material is given in the 1951 supplement, mixed with that from Juan Fernández.

989

Marquesas

A new, critical flora (Sachet, 1975–) of this floristically distinctive group of 'high' islands to the north of southeastern Polynesia has begun to make its appearance; it represents a considerable improvement over the general flora of Brown and Brown, hitherto the only relatively recent work on the group, as well as the much earlier one of Drake (see **980**).

Sachet, M.-H., 1975– . *Flora of the Marquesas*, 1: *Ericaceae-Convolvulaceae*. iv, 34 pp., map (Smithsonian Contr. Bot., 23). Washington.

Critical, detailed 'research' flora (including native, naturalized, and 'village' plants), with keys, long descriptions, detailed synonymy with references and citations, indication of *exsiccatae* from all sources with localities, general summary of distribution, taxonomic commentary, and notes on habitat, biology, occurrence, frequency, etc.; vernacular names and local uses; no index (in first fascicle). An introductory section covers the history of botanical collecting, background to the work, and the logistical and other problems of preparing such a work on a Pacific island group. The faults of the Browns' flora are also reviewed. To date (mid-1979), only this one fascicle has appeared; arrangement follows the classical Englerian system.

Region

99

Hawaiian Islands

This comprises the Hawaiian Islands chain from the 'big island' of Hawaii northwest through the other main islands (to Niihau) and the smaller Leeward Islands (in part extratropical) to Midway and Kure. Johnston Island is also included here.

The indigenous vascular flora of the 'high' islands, about 2000 species and varieties (including 272 pteridophytes) is for its size about the most peculiar and isolated in the world, with over 94 per cent endemism, and at the same time one of the most fragile, with a long official list of extinct, endangered and threatened vascular plant taxa. However, although the literature on the vascular flora is large, no good modern manual is available nor is there a good recent discrete botanical bibliography. All present major general works have shortcomings and it is often necessary to utilize critical revisions and other papers which are very scattered. On the other hand, recent floristic reviews are available for Johnston and the various Leeward Islands (as part of island monographs) which supplement the enumerations of Christophersen produced between the world wars.

The only single-volume manual, which covers all vascular plants then known, is still Hillebrand's *Flora of the Hawaiian Islands* (1888), written following a twenty-year residence in the then-kingdom but published posthumously. In style, it is very much in the classic Anglo–American tradition which could well serve as a model for a new field manual, which is very much needed. An extensive supplement to the Hillebrand flora was published by Heller (1897), based to a large extent upon his own extensive collections in the islands. Of more recent general works, the ponderous and uniquely loose-leaf illustrated *Flora hawaiiensis* of Degener (1932–), now in its seventh volume, seems no nearer completion after nearly half a century. St John's 1973 checklist is little more than a name list based partly upon work done before 1930 by E. H. Bryan, Jr., and is padded out with scores of garden and other exotics not even part of the large adventive and naturalized element in the modern Hawaiian flora.

Of partial works, the best is without doubt Rock's *Indigenous trees of Hawaii* (1913), whose usefulness has been enhanced for current users through the revisions by Herbst for the 1974 reprint. For the exotic (and some indigenous) flora, reference should also be made to Neal's *In gardens of Hawaii* (1965). A excellent general introduction to Hawaiian biota remains the first volume of Zimmermann's *Insects of Hawaii* (1948; floristics by Fosberg).

Bibliographies. General and supraregional bibliographies as for Superregion 94–99. See also Division 1 (LAWYER *et al.*, in press).

Indices. General and supraregional indices as for Superregion 94–99.

990

Region in general

As already noted, no good modern manual for the vascular flora is available. All available 'standard' works have limitations or shortcomings, including the three general treatises (respectively by Degener, Hillebrand (with supplement by Heller), and St John) and the two major special works (respectively by Rock and Neal, the latter not separately listed). However, the recent reprint of Rock's book includes revised nomenclature and new commentary by D. E. Herbst.

The exotic (and some indigenous) vascular flora is, in addition to its treatment in the works of Degener and St John, covered in the following illustrated, keyed manual: NEAL, M. C., 1965. *In gardens of Hawaii*. xix, 924 pp., illus. (Bernice P. Bishop Mus. Spec. Publ., 50). Honolulu. (Reprinted; originally pub. 1928 as *In Honolulu gardens*.)

The best general introduction to the Hawaiian environment and biota remains the following: ZIMMERMANN, E. C., 1948. *Insects of Hawaii*, 1: Introduction. xx, 206 pp., 52 text-figs. (incl. maps). Honolulu: University of Hawaii Press. [Includes a chapter on the flora by F. R. Fosberg, with statistical table by families from which the overall figures used above have been taken. St John in his 1973 checklist credits the indigenous seed-plant flora with 2668 species and varieties (1442 species), while Degener has used a fanciful figure of 50 000 taxa.]

Keys to families and genera

ST JOHN, H. and FOSBERG, F. R., 1938–40. *Identification of Hawaiian plants*. I–II. 53, 47 pp. (Occas. Pap. Univ. Hawaii 36, 41). Honolulu. [Keys to and descriptions of families, with keys to genera; indices. Dicotyledoneae in part 1; Monocotyledoneae and Gymnospermae in part 2.]

DEGENER, O., 1932– . *Flora hawaiiensis; or the new illustrated flora of the Hawaiian Islands*. Vols. 1– . Illus. Honolulu (later Waialua, Oahu), Hawaii: The author. (Vols. 1–4 reprinted under one cover, 1946, Honolulu.)

This, the only 'loose-leaf flora' in the world, is a copiously descriptive, profusely illustrated work covering spontaneous and commonly cultivated vascular plants, with keys to all taxa; includes full synonymy (with references and literature citations), general indication of local and extralimital distribution, type localities, and more or less extensive taxonomic commentary and notes on status, ecology, occurrence, uses, special features, etc.; vernacular names (Hawaiian and English). The introductory leaves include historical matter as well as a glossary and index to families; successive 'temporary' indices to all botanical and vernacular names have also been provided. At this writing (mid-1979) leaves to the extent of six volumes (by 1963) and a substantial part of vol. 7 have been issued, but no taxonomic progression in publication has been observed and the 'volumes' merely serve as receptacles. However, each sheet has been numbered according to the Englerian sequence for purposes of sorting. The sheets are published with text on the front and (usually) the illustration on the reverse side for each taxon. The offset reprint of 1946, combining vols. 1–4 but retaining the loose-leaf format, was made following destruction of stock in a tidal wave.[20]

HILLEBRAND, W., 1888. *Flora of the Hawaiian Islands*. xcvi, 673 pp., frontispiece, 4 maps (at end). Heidelberg, Germany: Winter; London: Williams & Norgate; New York: Westermann. (Reprinted 1965, New York: Hafner.)

Briefly descriptive manual-flora of native and then-naturalized vascular plants (999 species), with keys to genera and species, concise synonymy (with references and literature citations), vernacular names, general indication of local range, with localities given for less common taxa, taxonomic commentary, and brief notes on habitat, uses, and special features; index to all botanical names. The introductory section includes chapters on physical features of the islands, general aspects of the flora (with statistics), and the geography and vegetation of the different islands. Also included is a reprint of Bentham's 'Outline of Botany' which had originally appeared in the colonial 'Kew floras'. Written in the last years of the Hawaiian monarchy and based on much field experience in the 1850s and 1860s, this classic work in the Benthamian tradition has yet to be wholly superseded. Due to the author's death in 1886 publication of this work was cared for by W. F. Hillebrand, but the introduction is only fragmentary, not having been completed. The following important work is regarded by Blake and Atwood as a supplement: HELLER, A. A., 1897. Observations on the ferns and flowering plants of the Hawaiian Islands. *Bull. Geol. & Nat. Hist. Surv. Minnesota* (Bot. Ser. 2) 9: 760–922 (Minnesota Botanical Studies, 1).

ST JOHN, H., 1973. *List and summary of the flowering plants in the Hawaiian Islands*. 519 pp. (Mem. Pacific Tropical Bot. Gard., 1). Lawai, Kauai, Hawaii.

Briefly annotated systematic list of the accepted botanical names, basionyms, and more important synonyms of all native, naturalized, adventive, and cultivated flowering plants in the Hawaiian Islands; entries include year of publication, citations of major works on the Hawaiian flora in which the plant is mentioned, indication of status (and country of origin, if applicable), and vernacular names. An addendum, a list of new names and combinations, and a complete general index are also provided. The introductory section gives an explanation of the work as well as statistics of the flora (1442 indigenous species). The list is based on a manuscript compilation made before 1930 by E. H. Bryan, Jr. Related works on pteridophytes include CHRISTENSEN, C., 1925. Revised list of Hawaiian Pteridophyta. *Bernice P. Bishop Mus. Bull.*, **25**: 1–30; and WAGNER, W. H., Jr., 1950. Ferns naturalized in Hawaii. (Bernice P. Bishop Mus. Occas. Pap. 20: 95–121).

Special groups – trees

ROCK, J. F., 1913. *The indigenous trees of the Hawaiian Islands*. [viii], 518 pp., 215 pls. (halftones). Honolulu: The author (under patronage). (Reprinted 1974 as *Indigenous trees of Hawaii*. xx, 548 pp. Rutland, Vt., and Tokyo: Tuttle (in association with Pacific Tropical Botanical Garden).)

Amply descriptive account of Hawaiian trees, with non-dichotomous keys to genera and species and a general key to families; includes full synonymy (with references and literature citations), Hawaiian and English vernacular names, rather detailed indication of local and extralimital distribution, critical remarks, and extensive notes on habitat, field attributes, ecology, timber properties, and uses; complete indices to botanical and vernacular names. The introductory section includes an illustrated account of vegetation and floristic regions in the islands, as well as of forest zones. The excellent photographs by the author are all from life or from freshly collected material. This 'tree book' has no rivals in the Pacific and but few elsewhere in the tropics. The 1974 reprint includes a new introduction by S. Carlquist and a detailed table of nomenclatural changes by D. Herbst.

995

Hawaiian Leeward Islands

These include all islands and atolls west of the main islands from Nihoa west through Laysan and Midway to Kure. The only work covering the chain as a whole remains the 1931 enumeration of Christophersen and Caum, although revised lists for all islands (see below) have appeared in more recent years.

CHRISTOPHERSEN, E. and CAUM, E. L., 1931. *Vascular plants of the Leeward Islands, Hawaii*. 41 pp., 16 pls., 3 maps (Bernice P. Bishop Mus. Bull., 81). Honolulu.

Enumeration of known vascular plants of the islands from Nihoa in the east to Midway and Kure in the west, with synonymy and references; description of new or critical taxa; localities with citations of *exsiccatae* and general indication of overall range; remarks on ecology and taxonomy; list of references at end. The introductory section includes chapters on the physical features, flora, and vegetation of each island or island group.

Individual islands

Within the last two decades, detailed biological and habitat surveys have been carried out on all the Leeward Islands (except Midway) under the general direction of R. B. Clapp. The comprehensive reports, all published in *Atoll Research Bulletin*, have included new floristic accounts with checklists. As these are all similar in style and content, providing accepted names, locality and collection records, references, and extensive ecological notes and records of changes since

earlier reports, only minimum references are given here. It is to be hoped that a consolidated botanical report can be published so as to provide a new reference point.

Nihoa: Bull. 207 (Clapp and Kridler, 1977).
Necker: Bull. 206 (Clapp, Kridler, and Fleet, 1977).
French Frigate Shoals: Bull. 150 (Amerson and Clapp, 1971).
Gardner Pinnacles: Bull. 163 (Clapp, 1972).
Laysan: Bull. 171 (Ely and Clapp, 1973).
Lisianski: Bull. 186 (Clapp and Wintz, 1975).
Pearl and Hermes Reef: Bull. 174 (Amerson, Clapp, and Wintz, 1974).
Kure Atoll: Bull. 164 (Woodward, 1972).

For Midway, see NEFF, J. A. and DUMONT, P. A., 1955. *A partial list of the plants of the Midway Islands*. 11 pp. (*Atoll Res. Bull.*, 45). Washington. [Based on a resurvey of 1954. Extensive changes to the flora had occurred as a result of World War II and other human activities.]

998

Johnston Atoll

As a result of human activities from World War II onwards, this is about the most altered atoll in the Pacific, physically and floristically. In 1923 it was still more or less in its natural state. By the 1970s all natural vegetation had disappeared and the shapes of the islets on the atoll had been completely altered, with a great increase in areas through dredging. Christophersen in 1931 recorded three vascular plant species as found on the 1923 survey; Amerson and Shelton have recorded 124. All the additions have been accidentally or intentionally introduced, as for some four decades the atoll has been a United States military base and storage dump for obsolete chemical and biological agents.

AMERSON, A. B., Jr. and SHELTON, P. C., 1976. *The natural history of Johnston Atoll, Central Pacific Ocean*. xx, 479 pp., illus., maps (Atoll Res. Bull., 192). Washington.

Pages 47–65 of this exhaustive topographic account deal with plant life, with a tabular check list on pp. 52–60 and bibliography. Appendices 2–6 (pp. 387–442) give detailed records of plant arrivals for each of the five islets.

CHRISTOPHERSEN, E., 1931. *Vascular plants of Johnston and Wake Islands*. 20 pp., 5 pls., 3 maps (Occas. Pap. Bernice P. Bishop Mus. vol. 9, no. 13). Honolulu.

Separate enumerations of the vascular plants for

each of Johnston and Wake Islands, with citations of *exsiccatae* and literature records; extensive notes on habitat, frequency, biology, and taxonomy. The introduction includes remarks on the general features of the islands, their flora, and their vegetation.

999

Luna

No vascular or non-vascular plants have yet been found for study by exobiologists and others, despite costly explorations from 1969 to 1973.

Notes

1 Zollinger, H., 1857. Ueber den Begriff und Umfang einer 'Flora Malesiana'. *Vierteljahrsschr. Naturf. Ges. Zürich* 2: 317–49, map; reviewed in Lam, H. J., 1937. On a forgotten floristic map of Malaysia (H. Zollinger 1857). *Blumea*, Suppl. 1: 176–82, map. For the modern delimitation of Malesia upon which *Flora Malesiana* is based, see Steenis, C. G. G. J. van, 1948. Hoofdlijnen van de plantengeographie van de Indische Archipel op grond van de verspreiding der Phanerogamen geslachten. *Tijdscher. Kon. Ned. Aardrijksk. Genootsch.* 65: 193–208.

2 The latest reviews of the famous biogeographical transition from west to east within Malesia are (for zoology) Simpson, G. G., 1977. Too many lines; the limits of the Oriental and Australian zoogeographic regions. *Proc. Amer. Philos. Soc.* 121(2): 107–20, and (for botany) Steenis, C. G. G. J. van, 1979. Plant geography of east Malesia. *Bot. J. Linn. Soc.* 79(2): 97–178.

3 Details are given in the papers by VAN STEENIS (1979) and KALKMAN and VINK (1979) referred to under Progress at the end of the supraregional introduction (p. 477).

4 Simpson (1977), p. 117.

5 Ridley's *Flora* was reviewed critically by Corner in *J. Malay Br. Roy. As. Soc.* (1933), 42 and by Symington in *Bull. Misc. Inform.* (Kew), *1937*: The work was compiled within a traditional structure and philosophy and is useful largely as a checklist, being valueless for field work.

6 See Wit, H. C. D. de, in *Flora Malesiana*, I, vol. 4, pp. cxxvi–cxxvii (1949) who labelled the work 'unfortunate'. Koorders was originally contracted to write a mountain flora of Java, but on his own account enlarged his brief. With the short time available, quality was sacrificed in favor of shortcuts and compilation. Nevertheless, he had bested Backer who had entertained the ambition to write the complete critical flora of Java (and to that effect had published some trial works, including one volume of a *Schoolflora voor Java* (1911)). The result was that Backer never forgave Koorders for the rest of his life (to 1963); but were it not for strong representations by Lam, Backer's own flora might never have been completely realized, even in its so-called 'emergency' edition. Two polemical tracts, with sometimes colorful language, appeared shortly after publication of the text of the *Exkursionsflora*, viz: BACKER, C. A., 1913. *Kritiek*

op de 'Exkursionsflora von Java' (bearbeitet von Dr S. H. Koorders). i, 67 pp. Weltevreden, Java: Visser; and KOORDERS, S. H., 1914. *Opmerkingen over eene Buitenzorgsche kritiek op mijne 'Exkursionsflora von Java'*. vii, 201 pp. Batavia: Kolff. The controversy, the most bitter and long-lasting in the whole history of floristic botany, highlights an almost insoluble equation in flora-writing: scholarship versus expediency, with time (and finance) as governing factors.

7 Lam, H. J. in *Blumea*, 1: 115–59 (1934). The present author is currently preparing a new annotated systematic bibliography.

8 The best overall review of exploration in English is Souter, G., 1964. *New Guinea: the last unknown*. Sydney.

9 Families in Beiträge I–XX have been indexed by Lam (1934).

10 Families in volumes 8, 12, and 14 have been indexed by Lam (1934).

11 Thorne, R. F., 1963. Biotic distribution patterns in the tropical Pacific. In *Pacific basin biogeography* (ed. J. L. Gressitt), pp. 311–50. Honolulu.

12 Balgooy, M. M. J. van, 1971. *Plant geography of the Pacific as based on a census of phanerogam genera.* (*Blumea*, Suppl. 6.) Leiden.

13 Force, R. W., 1977. *Annual reports of the Director, Bernice P. Bishop Museum, 1971–5*, p. 6. Honolulu.

14 On the prewar contributions, see Blake and Atwood in *Geographical Guide*, 1 (1942), 11–12, and St John in *Bernice P. Bishop Mus. Bull.* 133 (1935), 56.

15 Fosberg, F. R., 1976. Phytogeography of atolls and other coral islands. In *Proceedings of the Second International Coral Reef Symposium* (Brisbane, 1974), vol. 1: pp. 389–96. Brisbane.

16 *Ibid*.

17 By Thorne (in Gressitt, *Pacific Basin Biogeography* (Honolulu, 1963) 326) treated as a subregion of his Oriental Region, but by van Balgooy (*Plant geography of the Pacific* (Leiden, 1971); cf. also *Blumea* 25 (1979) 81) placed in his Australian kingdom, 'albeit in a high hierarchical rank.'

18 One of the most comprehensive general surveys in the Carolines remains that made by the German research vessel 'Peiho' in 1908–10, which visited many atolls and low islands; however, little attention was paid to the terrestrial flora and only incidental collections were made by the ethnologist Krämer. A good opportunity for the collection of meaningful floristic data was thus lost, or perhaps not realized at the time.

19 Referred to as 'in preparation' in *Bernice P. Bishop Mus. Bull.* 175: 15 (1942).

20 In 1979 as in 1939 the method of publication of this work is 'unique among floras at present', but Blake and Atwood's hope of 'forever' has now inverted itself in this botanical Spandau.

Appendix A

Major general bibliographies, indices, and library catalogues covering world floristic literature

Introduction

The following remarks are intended to serve as a background to an annotated enumeration of the principal general retrospective bibliographies, indices and abstracting journals, and library catalogues likely to be consulted in a search for floristic literature references, particularly in areas not adequately covered by regional source bibliographies and indices. Not included here are general taxonomic works, as these have been dealt with in Division 0 and moreover are more or less well described in standard textbooks and other introductory accounts.

As discussed in Chapter 1 of the General Introduction, current floristic literature, including new monographic floras and enumerations, is in my view now less efficiently reported on and indexed than before World War II, in spite of the growth of new abstracting and indexing services and other means of communication. One cause may be the waste of effort resulting from the existence of too many overlapping services, as has been noted by some commentators (although even before World War I this was also regarded as a problem); but more significantly it appears that a number of the major services covering biological sciences presently concentrate efforts where demand is greatest (and financial security most probable) and/or on material of more general interest, thus neglecting to cover, or at best covering unevenly, much which seemingly is mainly regional in scope. A third cause is structural: material is emanating from many more sources and in far greater volume than 40 years ago when the first volume of the *Geographical Guide* by Blake and Atwood appeared. Finally, the disruptions caused by the two World Wars resulted in the extinction of a number of key services, with lasting effects not entirely overcome by their often differently organized successors.

The most useful general indexing journals of the present time specifically related to systematic botany are *Excerpta Botanica* (from 1959), a partial successor to the prewar *Botanisches Centralblatt* (1880–1945), and *Kew Record of Taxonomic Literature* (from 1971). These are supported by a number of regional indices such as *Index to American Botanical Literature* (from 1886), *AETFAT Index* (from 1952), and *Flora Malesiana Bulletin* (from 1948); only the last has commentaries, abstracts, or full-length reviews. The regional indices as a whole have slightly more currency than their general counterparts, which range from 1–3 years behind. Those wishing more timely coverage must plunge into *Biological Abstracts* (and *Biological Abstracts/RRM*) – since 1959 without geographical indices as such – or *Referativnyj Žurnal* or *Bulletin Signalétique*, or perhaps have a search profile established through a computerized database service such as AGRICOLA or BIOSIS Previews. The botanical profession may be in modern terms too small to support a special quick-index of the *Current Contents* type, as was evidently demonstrated by the short life (1974–5) of *Asher's Guide to Botanical Periodicals*. It may be noted in general that a rather large proportion of contributions classifiable as floristic is published in more or less obscure channels (including many journals of local interest), causing it not to be reported in the general indices or but tardily so. In addition, many journals publishing such contributions are irregular, which results in negative discrimination by at least one major service. It is here that regional indices are of great value, particularly with regard to literature from areas off the beaten track'.

At the present time, one of the sources first consulted comprises the useful annual reviews of progress in systematics as well as floristic geobotany which have for many years appeared regularly in *Progress in Botany* [Fortschritte der Botanik]; these contain fairly extensive lists of new taxonomic and floristic literature. Particularly is it useful for monographic floras, enumerations and related works, as

since the termination of *Naturae novitates* in 1944 these have lacked a dedicated periodical index and current reporting in standard general abstracting and indexing journals, as well as professional outlets carrying announcements such as *Taxon*, is at best haphazard (save for the more tardy *Excerpta Botanica* and *Kew Record*). Of greater value, however, are the serial annotated catalogues and announcement lists of two specialist book firms in Central Europe – who thus carry on the tradition of the old Berlin firm of Junk – as well as other, less highly organized trade channels (e.g., publishers' lists, leaflets, and advertisements as well as booksellers' catalogues). Supplementing these sources are, as already noted, special sections in *Taxon* and other standard botanical and biological journals, but none of them represents a systematic overall coverage of new literature.

The several primary sources herein considered in detail, both monographic and periodical, are variously treated in an extensive range of secondary literature of greater or lesser scope, much of it arguably better known to the librarian or bibliographer than to the practising botanist. Among large, general reference works for libraries may be mentioned *A world bibliography of bibliographies* by T. Besterman (4th edn., 5 vols., 1965–6, Lausanne), a massive empirical compilation in which full bibliographic details as well as (where necessary) collations, along with the number of entries accounted for, are out of its enormous range given for most of the source works described here, and *Les sources du travail bibliographique* compiled under the direction of L.-N. Malclès (3 vols., 1952–8, Geneva; reprinted 1965), which features an annotated selection of monographs, key serials, and other works as well as bibliographies thought to characterize a given discipline, analytically arranged so as to facilitate an overview. [For the benefit of biologists, all the biological entries in *A world bibliography of bibliographies* have been separately reprinted in a handy small format as Besterman, T. 1971. *Biological sciences: a bibliography of bibliographies.* 471 pp. Totowa, N. J.: Rowman & Littlefield.]

Several introductory texts covering the use of biological and botanical literature are also available, ranging in style and texture from the empirical to the analytical and aimed at students of biology or library science (and bibliography) or both. They are mentioned here as all of them give consideration to floristic literature along with related works on geographical botany and various kinds of bibliographies. Broadest in scope, and serving to place the biological sciences at a glance in relation to the whole range of human knowledge, is the *Manuel de bibliographie* by L.-N. Malclès (1963, Paris; 2nd edn., 1969); in this work, chapters 27 and 29 respectively cover general biology and botany. For biology as a whole, the significant works are *The use of biological literature* by R. T. Bottle and H. V. Wyatt (1966, London: 2nd edn., 1971); *Éléments d'un guide bibliographique du naturaliste* by F. Bourlière (1940–1, Mâcon, Paris), and *Führer zur biologischen Fachliteratur* by G. Ewald (1973, Stuttgart). In botany, key modern guides are *Botanical bibliographies* by L. H. Swift (1970, Minneapolis, Minn.; reprinted 1974, Königstein/Ts., W. Germany) and *Esbôco de un Guia da literatura botânica* by C. T. Rizzini (1957, Rio de Janeiro). [These last two are comparable in scope, and serve the same function as, the baroque *Bibliotheca botanica* of Linnaeus (1736; revised 1751).] Supplementing these standard texts are two select lists: *Basic books for the library* by G. P. de Wolf (1970, in *Arnoldia*, 30(3): 107–13) and *Some botanical reference works* by P. I. Edwards (1971, *Biol. J. Linn. Soc.* 3(3): 269–75). The lists of key reference works given by S. F. Blake in the introductions of the two volumes of the *Geographical Guide* should also be consulted.[1]

Well-founded systematics textbooks containing more or less extensive lists of literature, with indication of general taxonomic works and varying selections of floras, include *Taxonomy of vascular plants* by G. H. M. Lawrence (1951, New York), *Taxonomy of flowering plants* by C. L. Porter (1967, San Francisco), and *Vascular plant systematics* by A. E. Radford *et al.* (1974, New York). Texts of more restricted scope with lists are *An introduction to plant taxonomy* by C. Jeffrey (1968, London), *A guide to the practice of herbarium taxonomy* by P. Leenhouts (1968, Utrecht), *Pflanzensystematik* by F. Weberling and H. Schwantes (2nd edn., 1975, Stuttgart), and *Plant systematics* by S. B. Jones, Jr., and A. E. Luchsinger (1979, New York).

The only overall review of the structure of the information network catering for systematic biology is a collection of articles of varying scope entitled *Storage and retrieval of biological information*, edited by P. I. Edwards (1971, in *Biol. J. Linn. Soc.* 3(3): 165–299) from a series of lectures given at the Linnean Society of London in 1970. Also serving as reviews are the textbooks by Bottle and Wyatt, Bourlière, and Ewald earlier referred to. Rather more succinct but

serving to place biological bibliography in relation to the art as a whole are the interpretative treatments of L. E. Bamber (for biology) and J. R. Blanchard (for agriculture) in *Bibliography: current state and future prospects* (ed. R. B. Downs and F. B. Jenkins) (1967, Urbana, Ill.) and L.-N. Malclès in her already-mentioned *Manuel de bibliographie*. A classified list covering a large range of recent bibliographies, including many national and regional works, has been presented in *A checklist of natural history bibliographies and bibliographical scholarship, 1966–1970*, by G. Bridson and A. P. Harvey (1971, in *J. Soc. Bibliog. Nat. Hist.* 5: 428–67). Finally, there must be mentioned the significant recent theoretical study of the whole biological information system, *Die Bibliographie der Biologie*, by H.-R. Simon (1977, Stuttgart) to which reference has been made in the introductory chapters of this book.

In concluding these remarks, it is perhaps worthy of note that few, if any, analytico–critical surveys of the systematics information network have been written – an indication, perhaps, of a certain continuing mindlessness in the profession. The above-mentioned collection of Edwards was based as has been seen on a series of lectures and naturally is somewhat haphazard, while the work of Simon – as he himself noted – is only a beginning to serious research (and further development) on the biological information system, both in itself and so that its effectiveness can be improved. [It is understood, early 1981, that the Linnean lecture series is being used as a basis for a more definitive work on biological information sources, to be published by Butterworths (P. I. Edwards, personal communication).] However, the initiative represented by the 1970 lecture series did not, so far as has been ascertained lead to a continuing 'study group'; and Stafleu's review of the state of botanical bibliography and indexing at the 1975 International Botanical Congress in Leningrad suggested to the present writer that, while activity was considerable (with even a mention of the *Guide* project itself!) and the importance of biological information work was in some quarters becoming increasingly recognized in its own right, little coordination among projects existed, with a consequent waste of effort in some cases. Put more simply, taxonomic botanists haven't collectively taken a stand on what they want in an information network.

Developments since 1975 have remained fairly conventional with highlights few apart from the commencement of the monumental second edition of *Taxonomic literature* (for the editing of which a diskette-based word-processing system was brought into use in 1980) and the publication, at long last, of the complete *Index nominum genericorum* (*plantarum*) after 25 years of effort. An apparently low level of professional interest in problems of botanical information retrieval has furthermore manifested itself in the lack of symposia devoted to botanical bibliography and/or the application of information systems technology at the Thirteenth International Botanical Congress in Sydney in 1981, despite past attention at such major meetings as well as rapid technological advances since 1975, including the advent of microcomputers with ample storage capacity. The ultimate goal of a world systematic and floristic information system, as first conceived by Leopold Nicoltra of Messina in 1884 and strongly pushed by J. P. Lotsy and others after 1900, thus seems still far from realization despite considerable advances (and setbacks) in the present century and may be yet not possible (A. L. Takhtajan in Presidential Address, *Proceedings of the Twelfth International Botanical Congress*, pp. 61–2 (1979, Leningrad)). No resource for systematic, floristic, and geographical botany comparable, for instance, to the Human Relations Area Files at New Haven in the field of anthropology exists, although suitable nuclei are present at centers such as the Hunt Institute for Botanical Documentation and the International Bureau for Plant Taxonomy and Nomenclature (the latter housing the large files for *Taxonomic Literature*) and major botanical libraries are also 'information resource centers' albeit in a less structured fashion than implied here. The announced development of the European Floristic and Taxonomic Information System (as recorded in the annual reports of the European Science Foundation) will therefore be watched with interest, not least its location and organization.[2]

It is all but a certainty, though, that future progress will be dependent upon some form of computer and related information-systems technology, whose relative availability, simplicity and cost-effectiveness is yearly increasing. At one level, which is not without consequences for botanical bibliography at the more personal level, progress has been such that recently there was made an, admittedly jocular, comment 'never trust a computer you can't lift' in a suggestion that small computers might soon be 'the only game in town' (editorial in *Byte*, 6(7): 10 (1981)).

On a larger but certainly current dimension, large network organizations such as the On-Line Computer Library Center (OCLC), the Research Libraries Network (RLIN) and the Washington Library Network (WLN) in the United States are creating large databases of varying capabilities encompassing part or all of the holdings of many libraries; the first-named in particular has processed those of the Missouri Botanical Garden and the New York Botanical Garden, both major resources for floristic literature (cf. p. xii). Together with biological databases such as AGRICOLA (from 1970) and BIOSIS Previews (from 1969), the means are thus being created for the eventual automated production of various kinds of specialized source bibliographies, in this way contributing to a need recently expressed editorially in *Nature* (**293**: 341, 1 October 1981), as well as for remote accessioning of bibliographic information – with profound consequences for many current forms of botanical documentation. Considerable problems remain, however, in progressing towards computerized botanical bibliography at any but a relatively small scale, as has been shown by the compilers of what is claimed to be the first largely computer-processed and typeset botanical bibliography, the Strandell Catalogue of Linnaeana (G. H. M. Lawrence and R. W. Kiger in *Svenska Linnésällskap. Årsskr.* (1978): 276–95, 1979); among these are a paucity of suitable specialized computer programs, differing data standards, older material not covered by any data bases, non-accessibility of some data bases by information classes, and financial and other human limitations. The best watchwords seem therefore to be 'hasten slowly'.[3]

The works described in this appendix are given in three sections under the following headings:

General bibliographies

General indices (and abstracting journals)

Major library catalogues

In the first two of these, each title is preceded by an abbreviated designation of (for general bibliographies) author(s) and date(s) of publication or (for general indices) the initials of the title and year(s) of coverage. These abbreviations correspond to those given under the subheadings 'General bibliographies' and 'General indices' at the beginning of each division and superregion in the *Guide* itself. Annotations have been based upon personal knowledge and examination (regrettably for some too hurried) as well as on the previously mentioned works of Besterman, Ewald, Malclès, Simon, Stafleu and Cowan, and others, *Bibliographies of botany* by a former head of the John Crerar Library (Chicago), J. Christian Bay (described below), *Bibliographies: subject and national* by R. L. Collison (1952, London; 3rd edn., 1968), and other sources (including articles directly or indirectly referring to particular works).[4]

General bibliographies[5]

BAY, 1909.   Bay, J. C., 1909. *Bibliographies of botany*. (Proressus rei botanicae, vol. 3, part 2, pp. 331–456.) Jena. [Written as a precursor to a projected bibliography of botany for 1870 through 1899, this compilation, subtitled *A contribution toward a bibliotheca bibliographica*, includes annotated lists of 'general and comprehensive bibliographies' (heading 3) and 'National (regional) bibliographies' (heading 3a). Very good coverage of nineteenth-century (and earlier) source bibliographies, with full collations of some bibliographic periodicals up to 1908 also given (heading 2). The work was compiled successively at the Missouri Botanical Garden, the Library of Congress, and the John Crerar Library in Chicago (of which last Bay was later director/librarian); regrettably, his larger work was never published. (The year 1909 is the correct date of publication, not 1910 as given in some references.)]

BLAKE and ATWOOD, 1942.   Blake, S. F. and Atwood, A. C., 1942. *Geographical guide to floras of the world*, 1: *Africa, Australia, North America, South America, and islands of the Atlantic, Pacific and Indian Ocean*. 336 pp. (USDA Misc. Publ., 401). Washington. (Reprinted 1963, 1967, New York: Hafner; 1974, Königstein, W. Germany: Koeltz.) [Comprehensive, briefly annotated critical bibliography of general, regional, and local floras and enumerations, covering all parts of the world except Eurasia. Continued by Blake, 1961 (see below). Primary sources were the USDA botany subject catalogue (see below under USDA, 1958), Bay (see above), Holden and Wycoff, and Rehder (see below), the Arnold Arboretum Library catalogue (see under Major library catalogues), and searches on shelves and in 'certain' periodicals, mainly in Washington, DC. Ewald, p. 129; total entries 3025 (Ewald).]

BLAKE, 1961.   Blake, S. F., 1961. *Geographical guide to floras of the world*, 2: *Western Europe*. 742 pp. (USDA Misc. Publ., 797). Washington. (Reprinted 1973, Königstein, W. Germany: Koeltz.) [A continua-

tion of the preceding; covers Western Europe from Scandinavia and Iceland to Italy but omits Western Germany and Austria. Based primarily upon the same sources as part 1 but with the addition of a number of European libraries, bibliographies and indices. No further continuation published or even contemplated. Ewald, p. 129; 3757 primary and 3085 subsidiary entries (Blake).]

FRODIN, 1964.   Frodin, D. G., 1964. *Guide to the standard floras of the world*. iv. 59 pp. Knoxville, Tenn.: Department of Botany, University of Tennessee. (Mimeographed.) [The predecessor of the present work, this classified list is limited to publications providing more or less complete coverage of vascular (or seed) plants for a given area. Based primarily on the botanical library of the Field Museum of Natural History, Chicago, with additions from several other sources. 900 entries.]

GOODALE, 1879.   Goodale, G. L., 1879. *The floras of different countries*. 12 pp. (Bibl. Contr. Libr. Harvard Univ., 9). Cambridge, Mass. [Briefly annotated, geographically arranged list of the more important independently published floristic works of the period. Few smaller local floras are included. Locations (if held) within Harvard libraries are given. Largely compiled from Pritzel (see below) and Harvard sources. 400 entries (Besterman).]

HOLDEN and WYCOFF, 1911–14.   Holden, W. and Wycoff, E., 1911–14. *Bibliography relating to the floras*. 513 pp. (Bibliographic contributions from the Lloyd Library, Cincinnati, Ohio, vol. 1, nos. 2–13.) Cincinnati, Ohio. [Each number of this work comprises a self-contained, unannotated list, arranged by author, of the floristic literature of a given geographical area, either large or small, as known to the compilers (who were senior Lloyd librarians) with an indication of those actually held in the Lloyd Library. The geographical classification regrettably is less detailed than in most other works of the same genre. In the main, only independently published works are accounted for and, apart from Lloyd Library holdings, compilation was at second hand – as also with the Goodale bibliography (see above). 6750 entries (Besterman).]

HULTÉN, 1958.   Hultén, E., 1958. [Bibliography.] In *The amphi-Atlantic plants*, pp. 298–330 (Kungl. Svensk. Vetenskapsakad. Handl., IV, vol. 7, part 1.) Stockholm. (Whole work reprinted 1973, Königstein, W. Germany: Koeltz.) [Geographically arranged, unannotated list of floristic literature for almost the entire Holarctic region. An exceptional feature is that the areal coverage of nearly every work is shown on an accompanying map. Compiled in Stockholm from sources also used for *Index Holmensis*.]

JACKSON, 1881. Jackson, B. Daydon, 1881. *Guide to the literature of botany*. xl, 626 pp. (Index Society Publications, no. 8). London: Longmans. (Reprinted 1964, New York: Hafner; 1974, Königstein/Taunus, W. Germany: Koeltz.) [Compiled as an avocation by a then-'City man' of London, with a practically-oriented credo which aimed at being 'suggestive' rather than 'exhaustive', this subject-arranged short-title bibliography, which is largely unannotated, contains in its sections 72 through 103 (pp. 221–405) a geographically arranged listing of independently published general, regional and local floristic works. Many titles not in Pritzel (see below) are included. Coverage runs through 1880. Ewald, pp. 133–4; 9000 titles in all according to Jackson himself but Besterman gives a figure of 10000 (of which the writer estimates about one-third to be floristic).]

PRITZEL, 1871–7. Pritzel, G. A., 1871–7. *Thesaurus literaturae botanicae*. iv, 577 pp. Leipzig: Brockhaus. (Reprinted 1950, Milan: Görlich; 1972, Königstein, W. Germany: Koeltz 1st edn. 1847–51.) [The main body of this classic work has entries arranged by author, but its appendix comprises a short-title classified list arranged under subject and geographical headings. Under the latter (numbers 20–36) are found independently published regional, national and local floras and other floristic works, all accounted for more fully in the alphabetical division. This bibliography is essential for historical regional floristic literature, largely omitted from the present *Guide*. (For a commemorative retrospective, see STAFLEU, F. A., 1973. Pritzel and his *Thesaurus*. *Taxon*, 22: 119–30.) Ewald, pp. 135–6. 10871 numbered entries covering some 15000 works (Pritzel).]

REHDER, 1911. Rehder, A., 1911. *The Bradley bibliography*. Volume 1, Dendrology, part 1 (Publications of the Arnold Arboretum, no. 3). Cambridge, Mass.: Riverside Press. (Reprinted 1976, Königstein/Taunus, W. Germany: Koeltz.) [This volume, the first of five published in the years 1911–18, includes among other topics geographically arranged listings of regional and local works on trees and on woody plants in general published before 1900, the closing date for the whole work. Volume 2 is an author and general subject index, with full details given for each entry (making cross-checking optional). Prepared largely at the library of the Arnold Arboretum in Jamaica Plain and based in the first instance on it and other Boston-area libraries, this very comprehensive work, the fruit of a large private benefaction, covers in all 145000 entries (including the fully detailed author and subject indices) on almost 3900 pages, with 75000 titles alone on dendrology in volumes 1 and 2 (Besterman).]

SACHET and FOSBERG, 1955, 1971. Sachet, M. H. and Fosberg, F. R., 1955, 1971. *Island bibliographies*. v, 577 pp.; and *idem*. *Island bibliographies: supplement*. ix, 427 pp. (National Academy of Sciences/National Research Council Publ., 335). Washington. [Each of the two parts of this work comprises *three* annotated bibliographies, each separately indexed. The first in both parts deals with Micronesian botany; the second with geography, physical features, etc., of coral atolls and other low islands in *all* oceans; and the third with vegetation on tropical Pacific islands of all kinds. A number of relatively obscure florulas and floristic lists are herein accounted for. 7500 entries in original work, of which 3500 comprise the section on Micronesian botany (Besterman).]

USDA, 1958. United States Department of Agriculture, National Agricultural Library, 1958. *Botany subject index*. 15 vols. Boston: Microphotography Co. [G. K. Hall]. [A reproduction in book form of one of the two greatest classified card catalogues covering systematic botany, the Botany Subject Catalogue in the USDA central library in Washington (now the National Agricultural Library at Beltsville, Md.). Begun in 1906 and continued until July 1952, this catalogue has a world-wide scope with countless headings, including among others geographical units in great detail. It is considered to be the most complete guide to the botanical literature of the first half of the twentieth century, a period otherwise imperfectly served, especially between 1914 and 1926 (at least for English-language materials). It was the fundamental source for the *Geographical Guide* by Blake and Atwood (see above) and its discontinuance through lack of funds can only be regarded as short-sighted. The book version contains in each volume a table of contents. (For an early description, see ATWOOD, A. C., 1911. *Description of the comprehensive catalogue of botanical literature in the libraries of Washington* (USDA Bureau of Plant Industry Circular, 87). Washington.) 315000 entries (Simon, p. 82).]

General indices[6]

BA, 1926– . *Biological abstracts*. Vol. 1– . Baltimore (later Philadelphia), 1926– . [Large-scale classified abstracting journal, now published semi-monthly, with at present five comprehensive indices: author, subject (KWIC-type), 'CROSS' (*concept relation of subject specialities*, now simply termed a concept index), 'biosystematic' (for groups of organisms), and (from 1974) generic-specific (for specific taxa), with cumulation semi-annual (and every five years on microform). Since 1965 accompanied by *BioResearch Index* (now *Biological Abstracts/RRM*) which extracts titles from books, book contents, annual reports, conference materials and reports, and review journals as well as reporting bibliographies and developments in nomenclature; its contents are *additional* to those in *Biological Abstracts* proper. This latter appears monthly, and is indexed similarly to the parent work (but without long-term cumulations). As of 1980, these leading stablemates of BIOSIS (for *Biosciences Information Service*), the largest such service in biology, cover 290 000 items per year (165 000 in *BA*; 125 000 in *BA/RRM*). From 1969 all titles have also been made available on magnetic tape (as BIOSIS Previews), with retrospective searches possible, and a related tape service offers the abstract texts; under their 'CLASS' system, special individual or group subject 'profiles' can also be established to be searched regularly on a subscription basis. Despite its extent, however, BIOSIS does not cover all biological literature; for floristic material it should be regarded largely as a secondary source unless used with a custom search profile. The discontinuance in 1959 of geographical and conventional subject indices in favor of the more quickly produced contextual KWIC index has also been disadvantageous in relation to floristic botany. Supplementation from other indexing and abstracting journals is therefore required.][7]

BotA, 1918–26. *Botanical Abstracts*. Vols. 1–15. Baltimore, 1918–26. [Monthly abstracting journal, with (usually) two volumes per year, with in each issue the abstracts arranged under broad subject headings; semi-annual cumulative author indices. Coverage restricted to periodical and serial articles. Initiated by B. E. Livingston of Johns Hopkins University and others as a reaction to the suspension in transmission to the United States in 1917 of *Botanisches Centralblatt*, but nonetheless continued

after the close of World War I; in 1926 absorbed into *Biological Abstracts* (see above). Coverage, originally small and largely restricted to American serials, rose to 9929 items per year by 1926 (Besterman), but, like its successor, it was considered by Blake and Atwood as a source of only secondary importance. For other descriptions, see Ewald, p. 111, and Collison, p. 61. (Some account of the genesis of this journal, and its relationship with *Botanisches Centralblatt*, can be found in Verdoorn, F., 1945. Farlow's interest in an international abstracting journal. *Farlowia*, 2(1): 71–82.)]

BC, 1879–1944. *Botanisches Centralblatt* (from 1938 *Botanisches Zentralblatt*). Jahrgang 1–40 (Bände 1–142); NS, Jg. 1–37 (Bde. 143–79, no. 8). Cassel, 1880–1901; Leiden, 1902–5; Jena, 1906–45. (Suspended 1920–1.) [The first 22 years of this work, a key resource for botany particularly before and during the World War I, comprised, in quarterly volumes, a miscellany of material – not unlike the present-day journal *Taxon* – including reports of new books and periodical literature together with reviews. Growing professional pressure for a more properly organized abstracting/indexing journal gradually manifested itself on both sides of the Atlantic and led to a change of ownership and editorial reorganization in 1902. From then until 1919, publication was weekly, with three volumes per year (two containing notices or reviews, and the third – issued in parallel – devoted to 'Neue Literatur', a classified index in which only very short abstracts appeared with the titles). Close links between Holland and Germany in the years before 1914 led to a change of publishers in 1906 which was to have unfortunate consequences in the difficult aftermath of the World War I, as well as during it when severe restrictions upon its entry into the USA were imposed; it was the latter which finally unleashed latent interest in an American abstracting journal and in 1918 *Botanical Abstracts* came into being. In 1919 the Association Internationale des Botanistes, sponsors of the *Centralblatt*, collapsed and not until 1922 did it reappear, again under new auspices – this time the Deutsche Botanische Gesellschaft. However, the *Centralblatt* of the inter-war years was comparatively less international in scope, with a lower level of coverage than in the 1900s. The last issue appeared in 1945; a partial successor is *Excerpta Botanica* (see below), also published by Fischer. The *Centralblatt* was viewed by Blake and Atwood as a key secondary resource for their *Geographical Guide*. Some 330 000 items were accounted for, averaging 2500–3000 per year (Besterman) with its highest level attained during 1902–10. It covered monographs as well as serial and periodical articles, and some cumulative indices were published, particularly in the New Series. (Some historical information on the *Centralblatt* is given in Verdoorn, F., 1945. The plant scientist in the world's turmoils. In *Plants and plant science in Latin America* (ed. F. Verdoorn), pp. xv–xxii. Waltham, Mass., as well as in the same author's article in *Farlowia* cited above under *Botanical abstracts*.)][8]

BS, 1940– . *Bulletin signalétique* (through 1955 *Bulletin analytique*). Vol. 1– . Paris, 1940– . [Produced by the Centre National de la Recherche Scientifique, this comprises a monthly index serial for all fields of science with, in more recent years, separate series for different subject areas (division 370 (formerly 17), *Biologie et physiologie végétales*, as from 1969 accounted for floristic botany). Contents of each division are subdivided in some detail, but as each entry is accompanied only by a brief *précis* it is less useful as an information source than *Biological Abstracts* or *Referativnyj Žurnal, Biologija*. Only periodical articles are covered. With respect to floristic botany it appears to have weaknesses similar to *Biological Abstracts*, but counterbalancing this is one advantage: the botanical division is under separate covers, something not available in *Biological Abstracts* since 1959. As a 'newsletter', *Bulletin Signalétique* has some value but must be used with other sources for effective coverage of floristic botany. Also described by Ewald, p. 112.]

CSP, 1800–1900. *Catalogue of scientific papers*. 22 vols. London (later Cambridge), 1867–1925. [A comprehensive index, published under the auspices of the Royal Society of London, covering periodical articles published in all scientific fields, from 1800 through 1900, in four series and one supplementary volume (19 volumes in all), with a much-criticized continuous arrangement by author in each chronological series (1800–63; 1864–73; 1874–83; 1884–1900). Considered as of inestimable value for the historian of science, and useful for tracing papers by nineteenth-century authors, particularly as the contemporary standard bibliographies all deal with independently published works; indeed, the existence of the *Catalogue* was specifically acknowledged by Pritzel in the introduction to the second edition of his *Thesaurus*. In all, nearly 2.3 million titles are accounted for. Three

further volumes of cumulative indices were published, but regrettably these cover only pure mathematics, mechanics, and physics. The work was initiated by Act of Parliament in 1864, and continued until 1925 when available resources could not cope with a continuation of coverage beyond 1900. An additional description appears in Ewald, pp. 130–1, who defines it as a monographic bibliography and notes the peculiarities of its system of title abbreviations. For annual coverage of the years 1901–14, see its successor ICSL, 1901–14 below.]

EB, 1959– . *Excerpta botanica*, section A: *Taxonomia et chorologia*. Vol. 1– . Stuttgart, 1959– . [This abstracting journal, begun through an agreement between Fischer, publishers of the defunct *Botanisches Centralblatt*, and the International Association for Plant Taxonomy, contains briefly annotated listings of new books and periodical articles in the field of systematics, including floristics, classified in the case of the latter under only seven geographical headings. Book reviews are also included. Published bimonthly (later monthly), with an additional cumulative annual (now semi-annual) index appearing later; seven parts comprise a volume. Each subject area and geographical unit, however, is covered only once in any volume. Up through 1963 accounted for 1000–2000 titles per year (Besterman), with perhaps 3000 at present; appears to be most thorough for Eurasia but generally requires to be used with other sources. Multilingual (German, English, French). For another description see Ewald, p. 116.]

FB, 1931– . *Fortschritte der Botanik* (now also entitled *Progress in Botany*). Vol. 1– . Berlin, 1932– . [Concise annual review of all fields of botany, organized under several major headings and prepared by specialists. Systematics and floristics of higher plants usually comprise one major heading, which, at least in recent years, has normally included lists of the more important new floristic works brought to notice in the year under review. Useful as a secondary information source, and corroborative in terms of assessment of relative importance, a key factor in determining its inclusion or not in a selective work such as this *Guide*. Now bilingual (German and English); coverage rising from 1000 at the outset to 2000 in 1962 (Besterman). For another description, see Ewald, p. 116.]

IBBT, 1963–9. *Index bibliographique de botanique tropicale*. 9 vols. Paris, 1964–71. [Trimestrial annual compilation of literature relating to tropical botany, including tropical floristics. Each annual volume arranged by author, with a final part devoted to a cumulative subject index. Published under the auspices of the French Office de la Recherche Scientifique et Technique d'Outre-Mer but suspended in 1971 due to unspecified technical difficulties.]

ICSL, 1901–14. *International catalogue of scientific literature*, section M: *Botany*. 14 vols. London, 1902–19. [A continuation of the *Catalogue of Scientific Periodicals* (see above), this journal, one of a stable of 17 covering all fields of science, comprised an annual index to new botanical literature in general, including that on floristics. Covers 82 582 items in all, with annual coverage ranging from 4728 to 7355 (Besterman), but although covering both monographic and periodical literature it reportedly did not provide comprehensive coverage. Utilized for the *Geographical Guide* by Blake and Atwood together with other secondary sources. For another description, see Ewald, p. 119–20.][9]

JBJ, 1873–1939. *Just's botanischer Jahresbericht*. Jahrgang 1–67. Berlin, 1874–1943. [Subtitled *Sytematisch geordnetes Repertorium der botanischen Literatur aller Länder*, this comprised an annual analytical classified guide, in two major divisions and a number of subdivisions, to botanical literature in all fields; one subdivision (in Part II) dealt with floristic (geographical) botany. This floristic subdivision comprised two sections (α) for titles, arranged by author, and (β) for reviews in abstract form, arranged geographically. Until World War I coverage was anywhere from one to three years behind, but afterwards, with more difficult circumstances and a rapidly rising volume of literature, compilation lagged still further behind and some of the last volumes were never completed. Much use was made of the Botanical Museum and other special libraries in Berlin by its compilers and from Jg. 11 (1885), when it received its permanent title upon transfer of the editorship from Leopold Just to Emil Koehne, its direction was linked with the Museum itself. The *Jahresbericht* was thus forced into suspension upon the destruction of the Museum library in March 1943. Total coverage had by then reached over 1.1 million items; the annual increment ranged up to 19 500 (for 1911: Besterman does not provide figures for later years). It was, though, in its time – the six decades from the 1870s into the early 1930s – the most complete guide to botanical literature available, and was furthermore well indexed by author and subject. It was early in the present century rated by W. G. Farlow at Harvard University (see Verdoorn in *Farlowia*, cited above under *Botanical Abstracts*) as being of better value than (although not as timely as) *Botanisches Centralblatt*, and later served as one of the secondary sources for Blake and Atwood. (From 1906, when Friederich Fedde became editor, there was introduced as an *Anhang* to the *Jahresberichte* the journal *Repertorium specierum novarum regni vegetabilis*, now known as *Feddes Repertorium*, for the publication (or republication) of taxonomic novelties, much as Walpers had attempted in the nineteenth century by way of supplementing the Candollean *Prodromus*. A predecessor of the *Jahresbericht* was *Repertorium der periodischen botanischen Literatur*, produced as *Beiblätter* to the journal *Flora* (Regensburg) during 1864–73.) For another description, see Ewald, pp. 121–2.][10]

KR, 1971– . *Kew record of taxonomic literature*. Vol. 1– . London, 1974– . [Elaborately classified annual index to new botanical names and new literature in systematics, floristics, and related fields, arranged by subject and region. The geographical classification based upon the system used in the Kew Herbarium, is exceptionally detailed. An index to authors appears at the end. With 6200–6400 entries per year (some, inevitably, replicated) through 1974, it is currently the most comprehensive indexing source in systematic and floristic botany, forming a pendant to the long-established *Zoological Record*. Its sources comprise the libraries of Kew Gardens, the British Museum (Natural History), the Commonwealth Forestry Institute (at Oxford) and the Commonwealth Mycological Institute. At the time of writing (1981), *Kew Record* has regrettably shown a serious lag (three to four years or more) and cannot be used as a 'news service' although serious efforts are in hand to reduce this gap. The quality of the geographical classification has also declined consequent to a production change made starting with the 1976 number.][11]

NN, 1879–1943. *Nature novitates*. Vols. 1–65. Berlin, 1879–1943. [Subtitled *Bibliographie neuer Erscheinungen aller Länder auf dem Gebiete der Naturgeschichte und der exacten Wissenschaften*, this bimonthly trade guide reported the appearance of new books and other key contributions in biology and other sciences; each volume was separately indexed. Coverage through 1912 ranged from 5000 to 9492 items per year, but after World War I it dropped to an average of 3500 until 1938; during World War II there was a further fall to less than half this figure (Besterman). It was moreover of much reduced importance to botany after World War I. For another description, see Ewald, p. 123.]

Mention may here also be made of an earlier repertorium of books and monographs, *Bibliotheca historico-naturalis, physiochemica et mathematica* (37 vols., 1851–87, Göttingen; published semiannually).

RŽ, 1954– . *Referativnyj Žurnal: (04) Biologija*. Vol. [1] (1953)– . Moscow, 1954– . [One of 61 different series published by the Soviet All-Union Institute for Scientific and Technical Information (VINITI), Moscow, *RŽ Biologija* comprises a monthly abstracting journal, classified into 15 divisions, of which one (04Б) deals with botany. Coverage is worldwide, with some emphasis on Soviet-bloc literature. Within each division, material is further subdivided, sometimes to a higher degree than in *Biological Abstracts*. Designation of abstracts is alphanumeric and coordinated with month of issue and principal subdivisions; each abstract is moreover provided with a UDC number. In 1972, 106000 items were accounted for (Simon). Subject and author indices are provided in an annual cumulation. Generally speaking, *RŽ Biologija* is less comprehensive than *Biological Abstracts* but at the same time less markedly interdisciplinary or biomedically oriented. For another description, see Ewald, p. 125. For new Soviet botanical books, *Novye Knigi SSSR*, available in English translation from 1964 for the biological sciences, may be consulted.]

Major library catalogues

Only those published in the present century and listing a substantial number of floras in relation to the totality of entries are given here. Nineteenth-century catalogues are well treated by Bay in his *Bibliographies of Botany* (see above). For a complete accounting of published catalogues, see Collison, R., 1973. *Published library catalogues*. London: Mansell. Of exceptional value for bibliographical control, particularly in the final stages of preparation and correction of copy, has been the *National Union Catalog: Pre-1956 Imprints* (1968–80, London: Mansell) and its supplements, although these are not here described. Unfortunately, there remain a number of major botanical libraries without, as yet, modern published catalogues, at least since the end of the nineteenth century.

Royal Botanic Gardens, Kew, 1974. *Author catalogue of the Royal Botanic Gardens Library, Kew, England*. 5 vols. Boston: Hall.

Idem, 1974. *Classified catalogue of the Royal Botanic Gardens Library, Kew, England*. 4 vols. Boston.

This catalogue of one of the largest of botanical libraries, photo-reduced from cards, includes some 100000 items in its author division and 64680 items in its classified division. Supersedes the printed catalogue of 1899–1919, which covered author entries through 1915.

Tucker, E. M., 1914–33. *Catalogue of the library of the Arnold Arboretum of Harvard University*. 3 vols. (Publications of the Arnold Arboretum, 6). Cambridge, Mass.

A printed catalogue with author and subject divisions. Volumes 1–2 cover holdings up through 1917. Volume 3 (authors only) comprises a supplement for 1917–33. A projected fourth volume to contain the complementary subject index never materialized.

United States Department of Agriculture, 1967–70. *Dictionary catalog of the National Agricultural Library, 1862–1965*. 73 vols. New York: Rowman & Littlefield.

Idem. 1972–3. *Dictionary catalog of the National Agricultural Library, 1966–70*. Totowa, New Jersey: Rowman & Littlefield.

The main work comprises a photoreproduced card catalogue, with authors and subject entries in one sequence; in vol. 73 are translations of articles. The first supplement has authors in vols. 1–8 and subjects and translations in vols. 9–12. Since 1966 the *Dictionary Catalog* has also been updated by Rowman and Littlefield on a monthly basis, with one volume per year and cumulations every five years. Automation of cataloguing from 1970 and the concomitant creation of the AGRICOLA database have enabled supplements from vol. 6 (1971) onwards to be arranged on a broad subject basis, with inclusion of indices. [This library was the primary resource for the compilation of the *Geographical Guide* – its junior author, Alice Cary Atwood, was on the library staff – and is at this writing (1981) serving as a principal resource for compilation of the expanded second edition of *Taxonomic Literature*. It contains the largest single botanical collection in the USA.]

Woodward, B. B. and Townsend, A. C. (eds.), 1903–40. *British Museum (Natural History): catalogue of the books, manuscripts, maps and drawings*. 8 vols. London: Trustees of the British Museum. (Reprinted

1964, Weinheim, W. Germany: Cramer (in association with Wheldon & Wesley, Codicote near Hitchin, Herts., England).)

Printed catalogue, arranged solely by author (save some special categories) and published in two series (volumes 1–5, 1903–15 and 6–8, 1922–40). Coverage thus runs up to anywhere from 1920–38 depending upon the letter of the alphabet. Botany is not separately treated.

Notes

1 Two other important biological literature guides, which came to the writer's notice too late for inclusion in the text, are *Smith's guide to the literature of the life sciences*, by R. C. Smith, W. M. Reid and A. E. Luchsinger (9th edn., 1980, Minneapolis; 1st edn., 1942), and *Putevoditel′ dlja biologov po bibliografičeskim izdanijam* by V. L. Levin, V. G. Levina and D. V. Lebedev (1978, Leningrad). [The latter contains illuminating graphical tables showing the chronological ranges of many of the sources described in this Appendix.]

2 See p. 23, note 32.

3 The subject of on-line information retrieval methodology and how it relates to botanical bibliography deserves fuller treatment than has been possible here. This includes also a consideration of the different utilities, networks and databases which supply and handle bibliographic (and other) information. Since its effective emergence in the early 1970s this field has developed rapidly, and is continuing to advance fast; but it is only in quite recent years that these facilities have also become accessible by individuals having personal computers with 'dial-up' capabilities, as reported in a number of magazines during 1981–2, e.g., Online information retrieval by S. K. Roberts (*Byte* 6(12): 452–61, *passim* (December 1981)). As mentioned in note 32 on p. 23 of this book, useful reviews of the 'state-of-the-art' in videotex at the time of final proof are the several articles in *Microcomputing* 5(10) (October 1981) and *Which Computer?* (May 1982). In addition to the problems mentioned in the text, the use of videotex on-line in botanical bibliography also has the drawback of the relatively high cost of 'connect time' to the networks (including DIALOG, ORBIT and BRS) necessitating a well-thought-out search strategy (more effectively, it has been argued, by intermediaries). Some key networks, such as OCLC's On-line Union Catalog, require dedicated, purpose-designed terminals; moreover, this catalog lacks text and has hitherto not been searchable by subject. Thus, as one botanical librarian has noted (I. MacPhail, Information resources for botanical gardens; in *AABGA Bulletin* (July 1981): 90–5, it may be some little time, if ever, before the systematic botanical information system dispenses entirely with 'traditional' resources and means of communication, despite hyperbolical forecasts by some writers of a 'paperless' future. [Suitable general introductions to on-line information retrieval include HENRY, W. M., *et al.*, 1980. *Online searching: an introduction*. London; and LANCASTER, F. W., 1979. *Information retrieval systems: characteristics, testing and evaluation*. 2nd edn. New York. A comparison of the three large *agricultural* databases – AGRICOLA, AGRIS and

CAB – is provided by LONGO, R. M. J. and MACHADO, U. D., 1981. Characterization of databases on the agricultural sciences. *Journal of the American Society for Information Science*, 32(2): 83–91. The entry of microcomputers as tools for personal information systems and for on-line searching is considered by LUNDEEN, G., 1981. Microcomputers in personal information systems. *Special Libraries*, 72(4): 127–37. Other references may be found in the Wright and Hawkins bibliography cited on p. 23 of this book.]

4 For an early-twentieth century review of botanical bibliography, including comparisons of some abstracting and indexing services of the day, see BAY, J. C., 1906. Contributions to the theory and history of botanical bibliography. *Bibliographical Society of America, Proceedings and Papers*, 1(1) (1904–5): 75–83. [The botanical section of ICSL was viewed as 'not decidedly satisfactory' and stood a poor third to BC and JBJ. It was also seen as 'absurdly' expensive.]

5 Descriptions of many of these are also given in Levin *et al.*, *Putevoditel′* (see note 1).

6 Descriptions of most of these are also given in Levin *et al.*, *Putevoditel′* (see note 1).

7 For zoological systematic literature it is claimed (E. O. Wiley, *Phylogenetic systematics*, 1981, New York) that BA is 'much less complete' than *Zoological Record*. Additionally, BA is, in the experience of the present writer, weak in its coverage of independently published works like floras and monographs.

8 The *Centralblatt* was by Bay (note 4) considered to be the most timely, with its *Neue Literatur* section being of the greatest value. Its nearest effective modern equivalent (for taxonomic literature) is the mimeographed *Current Awareness List* issued by the Library of the Royal Botanic Gardens, Kew, on a monthly basis since 1973. Editorial policies and style of the *Centralblatt* were reviewed in memorials by E. Bornet and J. P. Lotsy, *ibid.*, 89(1) [Jg. 23]: 1–7 (1902).

9 The background and organization for the ICSL, and its editorial and bureaux system, are given as prefaces to each annual volume, along with the classification of contents (which followed a code peculiar to the work). B. D. Jackson was disciplinary referee in botany for the ICSL board of control, based in London.

10 Some later annual reports in JBJ were classified lists of literature resembling the modern *Kew Record* (KR); an example is the 1935 review of literature on higher-plant morphology and systematics by K. Krause, the last of its kind ever to appear (during 1943–4). Over the years there was a steady reduction in the length and discursiveness of reviews, as well as some changes in organization of contents. This, and the increasing 'lag' as noted by SIMON, may well have been inducements for von Wettstein to establish *Progress in Botany* (originally *Fortschritte der Botanik*) (FB).

11 The backlog has, since 1981, been gradually cleared; the 1979 and 1980 volumes are to appear during 1983 (S. Fitzgerald, personal communication).

Appendix B

Abbreviations of serials cited

Introductory remarks

The following lexicon lists abbreviations of serials and periodicals cited in the bibliographic entries along with the introductory digests in Part II of this work.

The standard for abbreviations largely follows that in *Botanico-periodicum-huntianum* (*B-P-H*), edited by G. H. M. Lawrence *et al.* (1968, Pittsburgh: Hunt Library), the most comprehensive currently available reference specifically covering the field. However, it is not without errors and moreover in terms of the present work distinctly incomplete, even accounting for the lapse of time since publication. *B-P-H*, though, was conceived in terms of the retrospective *Bibliographia huntiana* project (a definitive critical bibliography of botanical literature from 1730 through 1840) and also, to some extent, Stafleu's *Taxonomic literature* (1967, Zug) and its successor, *TL-2* (1976– , Utrecht).

To complete the listings of serials given below, other sources were therefore consulted. These include: *British union-catalogue of periodicals* (and its supplements); *New serial titles 1950–70, 1971–5,* and *1976* (and *Subject guide to New serial titles 1950–70*); *Scientific serials in Australian libraries*; *Serial publications of foreign governments 1815–1931* (Gregory); *Union list of serials in the libraries of the United States and Canada,* 3rd edn.; *Union list of serials in New Zealand libraries* (and its supplement), and *World list of scientific periodicals 1900–60* (and its supplements). Use has also been made of some specialized periodical listings including *An index of state geological survey publications issued in series*, by J. B. Corbin (1965, New York); *List of periodical publications in the Library, Royal Botanic Gardens, Kew* (1978, London); *Serial publications in the British Museum* (*Natural History*) *Library*, third edition (1980, London), and *Serial sources for the BIOSIS Data Base* (1979 and 1980 editions), as well as the lexica in Blake and Atwood's *Geographical guide to floras of the world*, Merrill and Walker's *A bibliography of Eastern Asiatic botany* and its *Supplement*, and Rehder's *Bibliography of cultivated trees and shrubs hardy in the cooler regions of the Northern Hemisphere*.

Only indefinite serials (and periodicals) as conventionally defined are accounted for here.[1] Expedition reports are not accounted for, nor are multi-volumed symposium and congress proceedings; these are spelt out in full in the text. Also omitted here are flora-series themselves, although these are sometimes treated as (and a few indeed are) indefinite serials.

The dates of commencement (and termination) have been given as far as available resources reasonably allow. In those few instances for which no data have been available, the volume and year of publication concerned are given instead. Commentary has been given where deemed necessary.

This lexicon is not meant to be a scholarly work in itself, and the user is at all times advised to refer also to the various aforementioned standard works. It may be noted here, however, that a modern work giving collations and 'phylogenies' for a large range of botanical and 'parabotanical' serials remains a desideratum, despite J. Christian Bay's remarks in closing the introduction to his *Bibliographies of botany* (here described in Appendix A) and the example of Jonas Dryander in his *Banksian Library catalogue* now nearly two centuries old. Even in the massive second edition of *Taxonomic literature* by Stafleu and Cowan only a few serials are fully collated.

Reference

1 Huff, W. H., 1967. Periodicals. In *Bibliography: current state and future trends* (ed. R. B. Downs and F. B. Jenkins), pp. 62–83. Urbana: University of Illinois Press.

Lexicon

AIBS Education Review: American institute of biological sciences education review. Arlington, Va., 1972– .

Abh. Auslandsk. (Hamburg), Reihe C, Naturwiss.: Abhandlungen aus dem Gebiet der Auslandskunde. Reihe C: Naturwissenschaften. [Hamburg University.] Hamburg, 1920– .

Abh. Königl. Akad. Wiss. Berlin: Abhandlungen der königlichen Akademie der Wissenschaften in Berlin, [Physikalischen-Mathematischen Klasse]. Berlin, 1815–1900.

Abh. Königl. Ges. Wiss. Göttingen: Abhandlungen der Königlichen Gesellschaft der Wissenschaften zu Göttingen. Göttingen, 1843–92.

Acad. Ci. Cuba, Ser. Biol.: Academia de Ciencias de Cuba, seria Biologia. Havana, 1967– .

Acta Amazonica: Acta amazonica. Manaus, Brazil, 1972– .

Acta Boreali, A, Sci.: Acta Borealia, A: Scientia. Tromsø, Norway, 1951– .

Acta Bot. Neerl.: Acta botanica neerlandica. Amsterdam, 1952– .

Acta Bot. Venez.: Acta botanica venezuelica. Caracas, 1965– .

Acta Horti Gothob.: Acta horti gothoburgensis. Goteborg, Sweden, 1924–66.

Acta Horti Petrop.: Acta horti petropolitani. *See* Trudy Imperatorskago S.-Peterburgskago botaničeskago sada.

Acta Phytotax. Barcinon.: Acta phytotaxonomica barcinonensia. Barcelona, 1968– .

Acta Phytotax. Geobot.: Acta phytotaxonomica et geobotanica. Kyoto, Japan, 1932– .

Acta Phytotax. Sin.: Acta phytotaxonomica sinica. Peking, 1951– .

Acta Soc. Fauna Fl. Fenn.: Acta societatis pro fauna et flora fennica. Helsinki, 1875– .

Acta Univ. Lund.: Acta universitatis lundensis [Lunds universitets årsskrift]. Lund, 1864–1904; N.S., 1905– .

Adansonia, sér. 2: Adansonia, série 2. Paris, 1961– .

Agric. Handb. USDA: Agriculture Handbooks, United States Department of Agriculture. Washington, 1950– .

Akad. Wiss. Wien, Math.-Naturwiss. Kl., Denkschr.: Akademie der Wissenschaften in Wien. Mathematisch-Naturwissenschaftliche Klasse. Denkschriften. Vienna, 1919–51.

Aliso: (El) Aliso. Claremont, California, 1948– .

Amer. Midl. Naturalist: American midland naturalist. South Bend, Ind., 1909– .

Anais Junta Invest. Colon.: Anais do Junta de Investigações Colonais. *See* next entry.

Anais Junta Invest. Ultram.: Anais do Junta de Investigações do Ultramar. Lisbon, 1946– .

Anales Jard. Bot. Madrid: Anales del jardin botánico de Madrid. Madrid, 1941–50.

Anales Mus. Nac. Hist. Nat. Buenos Aires: Anales del museo nacional de historia natural de Buenos Aires. Buenos Aires, 1864–1931.

Anales Mus. Nac. Montevideo: Anales del museo nacional de Montevideo. Montevideo, 1894–1909; ser. 2, 1925– .

Anales Univ. Chile: Anales de la universidad de Chile. Santiago, 1843– .

Ann. Arid Zone: Annals of arid zone. Jodhpur, Rajasthan, India, 1962– .

Ann. Bot. (London): Annals of botany. London, 1887– .

Ann. Bot. Gard. Buitenzorg: Annals of the botanic gardens, Buitenzorg. Buitenzorg (Bogor), Indonesia, 1941–52.

Ann. Carnegie Mus.: Annals of the Carnegie Museum. Lancaster, Pennsylvania, 1901– .

Ann. Inst. Bot.-Géol. Colon. [Mus. Colon.] Marseille: Annales de l'institut botanico-géologique colonial de Marseille. Paris, etc., 1893– . [Several successive series.]

Ann. Jard. Bot. Buitenzorg: Annales du jardin botanique de Buitenzorg. Batavia (Jakarta), Leiden, 1876–1940.

Ann. Kentucky (Soc.) Nat. Hist.: Annals of (the) Kentucky (society of) natural history. Louisville, Ky., 1941– (vol. 2, 1968).

Ann. Missouri Bot. Gard.: Annals of the Missouri botanical garden. St Louis, 1914– .

Ann. Mus. Goulandris: Annales musei Goulandris (Epeteris mouseiou Goulandre). Kifissia, Greece, 1973– .

Ann. Mus. Roy. Afrique Centr. (Tervueren), Série 8vo, Sci. Econ.: Annales de la musée royale d'Afrique centrale, série-en-octavo: sciences économiques. Tervueren, Belgium, 1947– (originally as Annales...série-en-octavo: sciences historiques et économiques).

Ann. New York Acad. Sci.: Annals of the New York academy of sciences. New York, 1879– .

Ann. Roy. Bot. Gard. (Peradeniya): Annals of the royal botanic gardens. Peradeniya, Ceylon, 1901– .

Ann. Sci. Nat. Bot.: Annales des sciences naturelles; botanique. Paris, 1834– . (Several successive series.)

Ann. Soc. Sci. Nat. Charente-Infèr.: Annales de la société de sciences naturelles de Charente-Infèrieur. La Rochelle, France, 1854– .

Ann. Transvaal Mus.: Annals of the Transvaal museum. Pretoria, 1908– .

Annual Rep. Div. Plant Industry, CSIRO, Australia: Annual reports, Division of plant industry, Commonwealth scientific and industrial research organization of Australia. Canberra, 1960/61– .

Annual Rep. Geol. Surv. Arkansas: Annual report of the Geological Survey of Arkansas. Little Rock, Arkansas, 1887–92.

Annual Rep. Geol. & Nat. Hist. Surv. Minnesota: Annual reports of the geological and natural history survey of Minnesota. St Paul, Minn., 1873–99.

Annual Rep. Michigan Acad. Sci.: Annual report of the Michigan academy of science, arts and letters. Lansing, Mich., 1894– .

Annual Rep. Missouri Bot. Gard.: Annual report of the Missouri botanical garden. St Louis, 1889–1912.

Annual Rep. New Jersey State Mus.: Annual reports of the New Jersey state museum. Trenton. Report for 1910, 1911.

Annuario Reale Ist. Bot. Roma: Annuario del Reale istituto botanico di Roma. Rome, 1884–1907.

Arch. Bot. Biogeogr. Ital. (Arch. Bot. (Forlì)): Archivio botanico e biogeografico italiano. Forlì, Emilia-Romagna, Italy, 1925– . (*Originally as* Archivio botanico per la sistematica, fitogeografia e genetica...; *from 1935–55 as* Archivio botanico.)

Arch. Hydrobiol. Suppl.: Archiv für Hydrobiologie: Supplementbände. Stuttgart, 1911– .

Arch. Nauk Biol. Towarz. Nauk. Warszawsk.: Archiwum nauk biologicznych towarzystwa naukowego warszawskiego. Warsaw, 1921– .

Arch. Rech. Agron. Cambodge Laos Viêtnam: Archives des recherches agronomiques au Cambodge, au Laos et au Viêtnam. Saigon, 1951– .

Ark. Bot.: Arkiv för botanik. Stockholm, 1903–49; N.S. 1952– .

Arnoldia: Arnoldia. Jamaica Plain, Mass., 1941– .

Arq. Jard. Bot. Rio de Janeiro: Arquivos do jardim botanico do Rio de Janeiro. Rio de Janeiro, 1915– .

Arq. Mus. Nac. Rio de Janeiro: Arquivos do museu nacional do Rio de Janeiro. Rio de Janeiro, 1876– .

Atoll Res. Bull.: Atoll research bulletin. Washington, 1951– .

Atti Ist. Bot. 'Giovanni Briosi' (R. Univ. Pavia): Atti dell'istituto botanico 'Giovanni Briosi' e laboratorio crittogamico italiano della Reale università di Pavia. *See* next entry.

Atti Ist. Bot. Lab. Crittogam. Univ. Pavia: Atti dell'istituto botanico della (Reale) università di Pavia, (Reale) laboratorio crittogamico. Ser. I *et seq.*; Suppl. A–J. Milan, etc., 1874–8, 1888– (Suppl. A–J to ser. V, Forlì, 1944–55). (Ser. I as Archivio triennale del laboratorio di botanica crittogamica; Ser. II–VI as above with slight variations, notably as in the preceding entry.)

Atti Soc. Acclim. Agric. Sicilia: (Giornale ed) Atti della società di acclimazione ed agricoltura in Sicilia. Palermo, 1860–91.

Austral. Conservation Found., Spec. Publ.: Australian conservation foundation; special publications. Canberra, 1968– .

Austral. J. Sci.: Australian journal of science. Sydney, 1938–70. [*Succeeded by* Search.]

Austral. Mus. Mem.: Australian museum memoirs. Sydney, 1851– .

Austral. Syst. Bot. Soc. Newsletter: Australian systematic botany society newsletter. Brisbane, etc., 1973– .

BSBI Conf. Rep.: Botanical society of the British Isles: conference reports. London, etc., 1949– .

Beih. Bot. Centralbl.: Beihefte zum Botanischen Centralblatt. Kassel, 1891–1944.

Beih. Nova Hedwigia: Beihefte zur Nova Hedwigia. Weinheim, Germany, etc., 1962– .

Ber. Deutsch. Bot. Ges.: Berichte der Deutschen Botanischen Gesellschaft. Berlin, 1883– .

Bergens Mus. Skr.: Bergens museums skrifter. Bergen, 1878–1943.

Bernice P. Bishop Mus. Bull.: Bernice P. Bishop museum bulletin. Honolulu, Hawaii, 1922– .

Bibliogr. Bull. USDA: Bibliographical bulletin, United States department of agriculture. Washington, 1943–54.

Biblioth. Bot.: Bibliotheca botanica. Kassel, 1886– .

Biol. Conservation: Biological conservation. Barking, Essex, 1968– .

Biol Sci. Ser. Western Illinois Univ.: Biological sciences series, Western Illinois University. Macomb, Illinois. Vol. 10, 1972.

Biologia (Lahore): Biologia. Lahore, Pakistan, 1955– .

BioScience: BioScience. Washington, 1964– .

Biotropica: Biotropica. Pullman, Washington State, 1969– .

Blumea: Blumea. Leiden, Holland, 1934– .

Bol. Bot. Latinoamer.: Boletín botánico latinoamericano. Bogotá, 1978– .

Bol. Inst. Centr. de Biociências (Porto Alegre) (Bol. Inst. Ci. Nat. Univ. Rio Grande do Sul): Boletim instituto central de biociências. Porto Alegre, Rio Grande do Sul, 1955– (originally as Boletim do instituto de ciências naturais da universidade do Rio Grande do Sul).

Bol. Minist. Agric. [Brazil]: Boletim do ministerio de agricultura. Rio de Janeiro, 1912– .

Bol. Minist. Agric., Serv. Florest. [Brazil]: Boletim do ministerio de agricultura, serviço florestal. Rio de Janeiro, 1929–31, 1956– .

Bol. Mus. Munic. Funchal: Boletim do museu municipal do Funchal. Funchal, Madeira, 1945– .

Bol. Mus. Nac. Hist. Nat. (Santiago): Boletín del museo nacional de historia natural. Santiago, 1937– .

Bol. Mus. Nac. Rio de Janeiro: Boletim do museu nacional de Rio de Janeiro. Rio de Janeiro, 1923–41.

Bol. Soc. Brot.: Boletim da Sociedade broteriana. Coimbra, Portugal, 1880–1920; sér. II, 1922– .

Bot. Arch.: Botanisches Archiv. Königsberg, Berlin, 1922–44.

Bot. Bull. (Lae): Botany bulletin, Department of Forests, Papua New Guinea. Lae, Papua New Guinea, 1969– .

Bot. J. Linn. Soc.: Botanical journal of the Linnean society of London. London, 1969– . [*Title changed from* Journal of the Linnean society, Botany.]

Bot. Jahrb. Syst.: Botanische Jahrbücher für Systematik, Pflanzengeschichte und Pflanzengeographie. Leipzig, 1881– .

Bot. Mag. (Tokyo): Botanical magazine. Tokyo, 1887– .

Bot. Mater. Gerb. Bot. Inst. Komarova Akad. Nauk SSSR: Botaničeskie materialy Gerbarija Botaničeskogo instituti imeni V. L. Komarova Akademii nauk SSSR. (Notulae systematicae ex herbario instituti botanici nomine V. L. Komarovi academiae scientiarum URSS.) Leningrad, 1919–26; 1931–63.

Bot. Notis.: Botaniska notiser. Lund, Sweden, 1841–1980.

Bot. Soc. Brit. Isles, Proc.: Botanical society of the British Isles. Proceedings. Arbroath, Scotland, 1954– .

Bot. Zurn. SSSR: Botaničeskij žurnal SSSR. Moscow & Leningrad, 1916– (originally as Zurnal Russkago Botaničeskago Obščestva pri Imperatorskoj Akademii Nauk, Petrograd).

Brittonia: Brittonia. New York, 1931– .

Brotéria (*later* Brotéria, Sér. Bot.): Brotéria. Lisbon, 1902–31.

Bul. Grăd. Muz. Bot. Univ. Cluj: Buletinul grădinii botanice şi al muzeului botanic de la universitatea din Cluj. Cluj, Romania, 1926–48.

Bull. Acad. Int. Géogr. Bot.: Bulletin de l'académie internationale de géographie botanique. Le Mans, France, 1899–1919.

Bull. Agric. Fish. Dept. [Hong Kong]: Bulletin, Agriculture and Fisheries Department. Hong Kong, 1974– .

Bull. Biogeogr. Soc. Japan: Bulletin of the biogeographical society of Japan. Tokyo, 1929– .

Bull. Biol. Sci. Washington: Bulletin of the biological society of Washington. Washington, DC, 1918– (vol. 2, 1972).

Bull. Bot. Soc. Bengal: Bulletin of the botanical society of Bengal. Calcutta, 1947– .

Bull. Bot. Surv. India: Bulletin of the botanical survey of India. Calcutta, 1959– .

Bull. Brit. Antarctic Surv.: Bulletin of the British antarctic survey. London, 1963– .

Bull. Brit. Mus. [Nat. Hist.], Bot.: Bulletin of the British museum [natural history]. Botany. London, 1951– .

Bull. Conservation Dept. Univ. Nebraska: Bulletin of the conservation department, University of Nebraska. Lincoln, Nebraska, 1928– .

Bull. Dept. Medicinal Pl., Nepal: Bulletin of the department of medicinal plants, Nepal. Kathmandu, ?1969– .

Bull. Fac. Sci. (Cairo Univ.): Bulletin of the faculty of science, Cairo University. Cairo, 1934– .

Bull. Forests Dept. Western Australia: Bulletin of the Forests Department, Western Australia. Perth, Western Australia, 1919– .

Bull. Geol. & Nat. Hist. Surv. Minnesota: Bulletin of the geological and natural history survey of Minnesota. St Paul, Minn. 1885– .

Bull. Herb. Boissier: Bulletin de l'herbier Boissier. Geneva, 1893–1908.

Bull. Inst. Jamaica, Sci. Ser.: Bulletin of the Institute of Jamaica, Science series. Kingston, Jamaica, 1940– .

Bull. Jard. Bot. Buitenzorg: Bulletin du jardin botanique de Buitenzorg. Buitenzorg (Bogor), Indonesia, 1911–50.

Bull. Jard. Bot. Etat: Bulletin du jardin botanique de l'état (now Bulletin du jardin national de Belgique). Brussels, 1902– .

Bull. Josselyn Bot. Soc. Maine: Bulletin of the Josselyn botanical society of Maine. Portland, Maine, 1907– .

Bull. Mauritius Inst.: Bulletin of the Mauritius institute. Port Louis, Mauritius, 1937– .

Bull. Misc. Inform. (Kew): Bulletin of miscellaneous information. Kew, England, 1887–1942.

Bull. Misc. Inform. (Kew), Addit. Ser.: Bulletin of miscellaneous information, Kew: additional series. London, 1898–1936.

Bull. Mus. Hist. Nat. (Paris): Bulletin du muséum d'histoire naturelle. Paris, 1895– .

Bull. Nat. Hist. Soc. New Brunswick: Bulletin of the natural history society of New Brunswick. St John, New Brunswick, 1882–1914.

Bull. Natl. Mus. Canada: Bulletin of the national museum of Canada. Ottawa, 1913–68. (Includes several series, among them Biological series and Contributions to botany.)

Bull. Natl. Sci. Mus. (Tokyo): Bulletin of the National science museum, Tokyo, Japan. Tokyo, 1939–54; N.S., 1954–74. [From 1974 further new series initiated in separate subject areas.]

Bull. New York Bot. Gard.: Bulletin of the New York botanical garden. Lancaster, Pa., etc., 1896–1932.

Bull. Ohio Biol. Surv.: Bulletin of the Ohio biological survey. Columbus, O., 1913– .

Bull. Oklahoma Agric. Mech. Coll.: Bulletin of the Oklahoma agricultural and mechanical college (*later* Oklahoma State University). Stillwater, Okla., ?1915– .

Bull. Peking Soc. Nat. Hist.: Bulletin of the Peking society of natural history. Peking, 1926–30.

Bull. Raffles Mus. Singapore: Bulletin of the Raffles Museum, Singapore. Singapore, 1928–70.

Bull. Roy. Soc. New Zealand: Bulletin of the royal society of New Zealand. Wellington, 1910–30, 1953– .

Bull. Soc. Bot. France: Bulletin de la société botanique de France. Paris, 1854– .

Bull. Soc. Etudes Indochin.: Bulletin de la société d'études indochinoises (de Saïgon). Saigon, etc., 1883– (N.S., 1926–).

Bull. Soc. Études Océanien. (Papeete): Bulletin de la société des études océaniennes. Papeete, Tahiti, 1917.

Bull. Soc. Roy. Bot. Belgique: Bulletin de la société royale de botanique de Belgique. Brussels, 1862– .

Bull. Soc. Sci. Nat. (Phys.) Maroc: Bulletin de la société des sciences naturelles (et physiques) du Maroc. Rabat, Morocco, 1921– .

Bull. Techn. Sci. Serv., Minist. Agric. [Egypt]: Bulletin of the technical and scientific service, Ministry of agriculture, Egypt. Cairo, 1916– .

Bull. Torrey Bot. Club: Bulletin of the Torrey botanical club. Lancaster, Pennsylvania, 1870– .

Bull. Univ. Tokyo Mus.: Bulletin of the University of Tokyo museum (Sogo kenkyu shiryokan). Tokyo, 1970– .

Bull. Wellington Bot. Soc.: Bulletin of the Wellington botanical society. Wellington, NZ, 1941– .

Cah. Indo-Pacifique: Cahiers de l'Indo-Pacifique. Paris, 1979–80. [*Successor to* Cahiers du Pacifique.]

Cah. Pacifique: Cahiers du Pacifique. Paris, 1958–78.

Candollea: Candollea. Geneva, 1922– .

Caribbean Forest.: Caribbean forester. Rio Pedras, Puerto Rico, 1939–63.

Castanea: Castanea. Morgantown, West Virginia, 1936– .

Ceiba: Ceiba. Tegucigalpa, Honduras, 1950– .

Ceylon J. Sci., Biol. Sci.: Ceylon journal of science. Biological sciences. Colombo, Sri Lanka, 1957– .

Chron. Bot.: Chronica botanica. Leiden (later Waltham, Mass.), 1935–53/4.

Chron. Nat.: Chronica naturae. Batavia (Jakarta), 1947–50.

Ciencia y naturaleza (Quito): Ciencia y naturaleza. Quito, 1957– .

Coll. Hawaii Publ. Bull.: College of Hawaii publications. Bulletins. Honolulu, Hawaii, 1911–16.

Collecc. Ci. INTA: Collección científica, Instituto national de técnologia agropecuaria. Buenos Aires, 1959– .

Colloq. Int. CNRS: Colloques internationaux du Centre national de la recherche scientifique, France. Paris, 1946– .

Connecticut State Geol. Surv. Bull.: Connecticut state geological and natural history survey bulletin. Hartford, Connecticut, 1903– .

Contr. Arnold Arbor.: Contributions from the Arnold Arboretum, Harvard University. Jamaica Plain, Mass., 1932–8.

Contr. Biol. Lab. Chin. Assoc. Advancem. Sci. Sect. Bot.: Contributions from the biological laboratory of the Chinese association for the advancement of science. Section botany. Nanking, 1930–39.

Contr. Gray Herb.: Contributions of the Gray herbarium of Harvard university. New series. Cambridge, Mass., 1891– .

Contr. Inst. Bot. Univ. Montréal: Contributions de l'institut botanique de l'université de Montréal. Montreal, 1938– .

Contr. Inst. Ecuat. Ci. Nat.: Contribuciones del Instituto ecuatoriano de ciencias naturales. Quito, < 1950– .

Contr. New South Wales Natl. Herb.: Contributions from the New South Wales national herbarium. Sydney, 1939–74.

Contr. NSW Natl. Herb., Flora Ser.: Contributions from the New South Wales national herbarium, flora series. Sydney, 1961–74. [Now monographic.]

Contr. Ocas. Mus. Hist. Nat. Colegio 'La Salle': Contribuciones ocasionales del museo de historia natural del colegio 'de La Salle'. Havana, 1944–60.

Contr. US Natl. Herb.: Contributions from the United States national herbarium. Smithsonian Institution. Washington, 1890– .

Contr. Univ. Michigan Herb.: Contributions from the University of Michigan herbarium. Ann Arbor, Michigan, 1939– .

Cornell Univ. Agric. Exp. Sta. Mem.: Cornell university agricultural experiment station memoirs. Ithaca, New York, 1913– .

Dansk Bot. Ark.: Dansk botanisk arkiv udgivet af Dansk botanisk forening. Copenhagen, 1913–80.

Dir. Gen. Agric. [Iraq], Bull.: Directorate-general of agriculture, Iraq. Bulletin. Baghdad, < 1943–?60.

Dir. Gen. Agric. Res. [Iraq], Tech. Bull.: Directorate-general of agricultural research, Iraq (Mudiryat al-Buhuth Wa-al-Mashari al-Zira'iyah al-Ammah). Technical bulletin. Baghdad, 1960– .

Ecol. Monogr.: Ecological monographs. Durham, North Carolina, 1931– .

Ecos: Ecos: CSIRO environmental research. Melbourne, 1974– .

Estudos, Ensaios e Documentos, Junta Invest. Ultram.: Estudos, ensaios e documentos, Junta de Investigações do Ultramar, Portugal. Lisbon, 1950– .

Etudes Bot. IEMVPT: Etudes botaniques, Institut d'élevage et de médecine vétérinaire des pays tropicaux. Maisons-Alfort, Val-de-Marne, France, 1972– .

FNA Rep.: Flora North America reports. *Abbreviated as* Fl. N. Amer. Rep.

Fern Gaz.: (British) fern gazette. Kendal, etc., 1909– .

Field Studies: Field studies. London, 1959– .

Fieldiana, Bot.: Fieldiana: botany. Chicago natural history museum (*later* Field museum of natural history). Chicago, 1946– .

Fl. Males. Bull.: Flora malesiana bulletin. Leiden, Holland, 1947– .

Fl. N. Amer. Rep.: Flora North America reports. Washington, DC, 1967–78. [84 nos.]

Forest Bull. (Georgetown): Forest bulletin. Georgetown, Guyana, 1948– .

Forest Bull. India: Forest bulletin, India. Calcutta, 1906–7, 1911– .

Forest Prod. Res. Inst. Bull. [Philippines]: Forest products research institute bulletin. Los Baños, Laguna, Philippines, 1963– .

Fortschr. Bot.: Fortschritte der Botanik. *See* Progress in botany.

Garcia de Orta: Garcia de Orta. Lisbon, 1953– . (Comprises several series, including a ser. Botânica.)

Geogr. J. (London): Geographical journal. London, 1893– .

Geogr. Rev. (New York): Geographical review. New York, 1916– .

Glasn. Zemaljsk. Muz. Bosni Hercegovini: Glasnik zemaljskog muzeja u Bosni i Hercegovini. Sarajevo, Yugoslavia, 1889–1937.

Göteborgs Kungl. Vetensk. Vitterh. Samhälles Handl.: Göteborgs Kungliga Vetenskaps- och Vitterhets-Samhälles handlingar. Goteborg, 1778–1966. [Series V, 1928–40, and Series VI, 1941–66, each divided into Sect. A, Humanistiska skrifter, and Sect. B, Matematiska och naturvetenskapliga skrifter.]

Gulf Res. Rep.: Gulf research reports. Ocean Springs, Mississippi, 1961– .

Handb. Peking Soc. Nat. Hist.: Handbooks of the Peking society of natural history. Peking, 1926– . (Originally as Educational series).

Health Bulletin (India): Health bulletin, Medical department, (Government of) India. Calcutta, etc., 1916– .

Hooker's Icon. Pl.: Hooker's icones plantarum. London, 1867– .

Illinois State Mus. Sci. Pap. Ser.: Illinois state museum: scientific papers series. Springfield, Illinois, 1940– .

Indian Forester: Indian forester. Allahabad, India, 1875– .

Indian For. Rec.: Indian forest records. Calcutta, etc., 1907–34; N.S., Botany, 1937– (one of several concurrent series).

Iowa State Coll. J. Sci.: Iowa state college journal of science. Ames, Iowa, 1926– .

Izv. Bot. Inst. (Sofia): Izvestija na botaničeskaja institut. Sofia, Bulgaria, 1950– .

J. Agric. Trop. Bot. Appl.: Journal d'agriculture tropicale et de botanique appliqué. Paris, 1954– .

J. Appl. Ecol.: Journal of applied ecology. Oxford, 1964– .

J. Arnold Arbor.: Journal of the Arnold arboretum of Harvard University. Jamaica Plain (later Cambridge), Mass., 1920– .

J. Asiat. Soc. Bengal: Journal of the Asiatic society of Bengal. Calcutta, 1834–1936. [After 1864 divided into parts, with natural history in Part II.]

J. Bombay Nat. Hist. Soc.: Journal of the Bombay natural history society. Bombay, 1886– .

J. Bot.: Journal of botany, British and foreign. London, 1863–1942.

J. Coll. Sci. Imp. Univ. Tokyo: Journal of the college of science, Imperial university of Tokyo. Tokyo, 1887–1925.

J. Coll. Sci. Sun Yat-sen Univ.: Journal of the college of science, Sun Yat-sen University. Canton (Guangzhou), 1928– .

J. Dept. Agric. Kyushu Imp. Univ.: Journal of the department of agriculture of the Kyushu imperial university, Fukuoka, Japan, 1923–46.

J. Fac. Agric. Hokkaido Univ.: Journal of the faculty of agriculture of the Hokkaido university. Sapporo, Japan, 1902– .

J. Forest. (Tsinan): Journal of forestry [Chinese forestry associations.] Tsinan, China, 1921.

J. Gen. Biol. (USSR): Journal of general biology, USSR. *See* Žurnal obščej biologii (SSSR).

J. Indian Bot. Soc.: Journal of the Indian botanical society. Madras, 1921– . [First 2 vols. entitled Journal of Indian botany.]

J. Linn. Soc., Bot.: Journal of the Linnean society. Botany. London, 1865–1968.

J. Mysore Univ., B.: Journal of the Mysore university, series B. Mysore, India, 1940– .

J. & Proc. Roy. Soc. Western Australia: Journal and proceedings of the royal society of Western Australia. Perth, Australia, 1914– .

J. S. African Bot.: Journal of South African botany. Kirstenbosch, Cape Province, 1935– .

J. S. African Bot., Suppl.: Journal of South African botany, supplementary volumes. Kirstenbosch, Cape Province, 1944– .

J. Soc. Bibliogr. Nat. Hist.: Journal of the society for the bibliography of natural history. London, 1936– .

J. Straits Branch Roy. Asiat. Soc.: Journal of the Straits branch of the Royal Asiatic Society. Singapore, 1878–1922. [Includes a Special Number, 1921.]

J. Tennessee Acad. Sci.: Journal of the Tennessee academy of science. Nashville, Tennessee, 1926– .

J. Trans. Victoria Inst. (London): Journal of the transactions of the Victoria Institute. London, 1866– .

J. Trop. Geog.: Journal of tropical geography. Singapore, 1953– (originally as Malayan journal of tropical geography).

Kashmir Sci.: Kashmir science. Srinigar, Kashmir, 1964– .

Kew Bull.: Kew bulletin. London, 1946– .

Kew Bull. Addit. Ser.: Kew bulletin, additional series. London, 1958– .

Kirkia: Kirkia. Harare (Salisbury), Zimbabwe, 1960– .

Kongel. Danske Vidensk. Selsk. (Biol. Skr.): Kongelige dansk videnskabernes selskab. Biologiske skrifter. Copenhagen, 1939– .

Kongl. Svenska Fysiograf. Sällskap. Handl.: Kongl. Svenska fysiografiska sällskapets handlingar. *See* Acta universitstis lundensis.

Kongl. Svenska Vetenskapsakad. Handl.: Kongl. svenska vetenskapsakademiens handlingar. Stockholm, 1855– .

Kwangsi Agric.: Kwangsi agriculture. Liuzhou, China, 1940– .

Lap. Balai Penjelidik. Kehut.: Laporan-laporanan Balai penjelidikan kehutanan. *See* Laporan-laporanan Lembaga penelitian hutan.

Lap. Lembaga Penelit. Hutan: Laporan-laporanan Lembaga penelitian hutan. Bogor (Buitenzorg), 1948– (*originally* Rapports van het bosbouwproefstation, *then* Laporan-laporanan Balai penjelidikan kehutanan).

Lav. Ist. Bot. Reale Univ. Modena: Lavori dell'istituto botanico della Reale università di Modena. Modena, Italy. Vol. 3, 1932.

Leafl. W. Bot.: Leaflets of western botany. San Francisco, 1932–66.

Lilloa: Lilloa. Tucumán, Argentina, 1937– .

Lingnan Sci. J.: Lingnan science journal. Canton, 1927–50.

Lloydia: Lloydia. Cincinnati, Ohio, 1938– .

Louisiana Forest. Commiss. Bull.: Louisiana forestry commission bulletin. Baton Rouge, Louisiana, 1945– .

Madroño: Madroño. Berkeley, Calif., 1916– .

Magyar Bot. Lapok: Magyar botanikai lapok. Budapest, 1902–34.

Malayan Forest Rec.: Malayan forest records. Singapore & Calcutta, 1922– .

Malayan Nat. J.: Malayan nature journal. Kuala Lumpur, 1940– .

Malpighia: Malpighia. Messina, Italy, 1887–1937.

Maria Moors Cabot Foundation, Spec. Publ.: Maria Moors Cabot Foundation, special publications. Petersham, Mass., 1947– .

Maryland Weather Serv., Spec. Publ.: Maryland weather service, special publications. Baltimore, 1899–1910.

Mater. Pozn. Fauny Fl. SSSR, Otd. Bot.: Materialy k poznaniju fauny i flory SSSR, izdavaemye Moskovskim obščestvom ispytatelej prirody. Otdel botaničeskij. Moscow, 1940– .

Meded. Bosbouwproefstat.: Mededeelingen van het bosbouwproefstation. Buitenzorg (Bogor), Indonesia, 1915–50.

Meded. Bot. Mus. Herb. Rijks Univ. Utrecht: Mededeelingen van het botanisch museum en herbarium van de rijks universiteit te Utrecht. Utrecht, 1932– .

Meded. Dept. Landb. Ned.-Indië: Mededeelingen uitgeven van het departement van landbouw in Nederlandsch-Indië. Batavia (Jakarta), Indonesia, 1905–14.

Meded. Landbouwhogeschool (Wageningen): Mededeelingen van de landbouwhogeschool. Wageningen, 1908– .

Meded. Lands Plantentuin: Mededeelingen uit 's lands plantentuin. Batavia (Jakarta), Indonesia, 1884–1904.

Meded. Rijks-Herb. Leiden: Mededeelingen van 's Rijks-Herbarium. Leiden, 1910–33.

Mem. Boston Soc. Nat. Hist.: Memoirs of the Boston society of natural history. Boston, Mass., 1862–1936.

Mem. Bot. Surv. S. Africa: Memoirs of the botanical survey of South Africa. Pretoria, 1919– .

Mem. Brooklyn Bot. Gard.: Memoirs of the Brooklyn botanical garden. Brooklyn, New York, 1918–36.

Mem. Fac. Agric. Kagoshima Univ.: Memoirs of the faculty of agriculture, Kagoshima University. Kagoshima, Japan, 1952– .

Mém. Inst. Égypte: Mémoires de l'institute d'Égypte. Cairo, 1862– .

Mém. Inst. Études Centrafr.: Mémoires de l'institut d'études centrafricaines. Brazzaville, Congo Republic, 1948– .

Mém. Inst. Franç. Afrique Noire: Mémoires de l'institut français d'Afrique noire. Dakar, Senegal, 1939– .

Mem. Inst. Invest. Agron. Moçambique: Memórias do instituto de investigação agronomia de Moçambique, Centro de Documentação Agrario. Maputo, Mozambique, 1966– .

Mém. Inst. Rech. Saharien. Univ. Alger: Mémoires de l'institut des recherches sahariennes de l'Université d'Alger. Algeria, 1954– .

Mem. Junta Invest. Ultram., II. sér.: Memórias do Junta de Investigaçoes do Ultramar, séria 2. Lisbon, 1958– .

Mém. Mus. Natl. Hist. Nat., sér. II/A, Zool.: Mémoires du muséum national d'histoire naturelle. Série II/A, Zoologie. Paris, 1950– .

Mém. Mus. Natl. Hist. Nat., sér, II/B, Bot.: Mémoires du muséum national d'histoire naturelle. Série II/B, Botanique. Paris, 1950– .

Mem. New York Bot. Gard.: Memoirs of the New York botanical garden. New York, 1900– .

Mem. Pacific Tropical Bot. Gard.: Memoirs of the Pacific Tropical botanical garden. Lawai, Kauai, Hawaii, 1973– .

Mem. Real Acad. Ci. Barcelona: Memorias de la real academia de ciencias y artes de Barcelona. Barcelona, 1892– .

Mem. Sci. Soc. China: Memoirs of the science society of China. Shanghai, 1924–32.

Mem. Soc. Brot.: Memórias da Sociedade broteriana. Coimbra, Portugal, 1930–1, 1943– .

Mém. Soc. Linn. Normandie: Mémoires de la société linnéenne de Normandie. Caen, Paris, 1824–8, 1835–1924.

Mém. Soc. Sci. Nat. Maroc: Mémoires de la société des sciences naturelles du Maroc. Rabat, Morocco, 1921–52.

Mem. Torrey Bot. Club: Memoirs of the Torrey botanical club. New York, 1889– .

Michigan Bot.: Michigan botanist. Ann Arbor, Michigan, 1962– .

Micronesica: Micronesica. Agana, Guam, 1964– .

Milwaukee Publ. Mus. Publ. Bot.: Milwaukee public museum, publications in botany. Milwaukee, 1955– .

Minnesota Geol. Surv. Rep., Bot. Ser.: Minnesota geological survey reports: botanical series. Minneapolis, Minnesota, 1892–1912.

Misc. Publ. USDA: Miscellaneous publications, United States department of agriculture. Washington, DC, 1927– .

Misc. Rep. Res. Inst. Nat. Resources [Japan]: Miscellaneous reports of the research institute for natural resources. Tokyo, 1943– .

Mississippi State Geol. Surv. Bull. Mississippi state geological survey bulletin. Jackson, Mississippi, 1907– .

Missouri Bot. Gard. Monogr. Syst. Bot.: Missouri botanical garden monographs in systematic botany. St Louis, 1978– .

Mitt. Bot. Staatssamml. München: Mitteilungen (aus) der botanischen Staatssammlung, München. Munich, 1950– .

Mitt. Inst. Allg. Bot. Hamburg: Mitteilungen aus dem Institut für allgemeine Botanik in Hamburg. Hamburg, 1916–39.

Monde Pl.: Le monde des plantes. Le Mans (later Toulouse), 1891– .

Monogr. Alabama Geol. Surv.: Monographs of the Alabama geological survey. University, Alabama, 1883–1945.

Monogr. Amer. Midl. Naturalist: Monographs of the American midland naturalist. South Bend, Indiana, 1944– .

Monogr. Canad. Dept. Agric. Res. Br.: Monographs of the Canadian department of agriculture, research branch. Ottawa, < 1966– .

Monogr. Inst. Estud. Pirenaicos: Monografias del Instituto

de estudios pirenaicos [del Consejo superior de Investigaciones Cientificas, Spain]. Zaragoza, Spain, 1948– . (Comprises several series, including a ser. Botánica.)

Monogr. Philipp. Inst. Sci. Tech.: Monographs of the Philippine (national) institute of science and technology. Manila, 1951–63. [*Successor to* Publications of the Bureau of Science, Philippines.]

Muelleria: Muelleria. Melbourne, 1955– .

Mus. Nac. Costa Rica, Ser. Bot.: Museo nacional de Costa Rica, serie Botánica. San José, Costa Rica, 1937– .

Nat. Appl. Sci. Bull. Univ. Philipp.: Natural and applied bulletin; university of the Philippines. Manila, 1930– .

Nat. Hist. Bull. Siam Soc.: Natural history bulletin of the Siam society. Bangkok, 1947– .

Nat. Res.: Nature and resources. Paris, 1965– .

Natl. Acad. Sci. – Natl. Res. Council Publ.: National Academy of Sciences – National Research Council publications. Washington, DC, 1951– .

Natl. Mus. Nat. Sci. Canada, Publ. Bot.: National museum of natural sciences of Canada, publications in botany. Ottawa, 1969– . [Successor in part to Bulletin of the National Museum of Canada.]

Natura Mosana: Natura mosana. Liége, Belgium, 1948– .

Natural resources research: Natural resources research, [UNESCO]. Paris, 1963– .

Naturaliste Canad.: Le naturaliste canadien. Quebec, 1869– .

Naturaliste Malg.: Le naturaliste malgache. Tananarive, 1949–62.

Nature: Nature. London, 1869– .

Natuurk. Tijdschr. Ned.-Indië: Natuurkundig tijdschrift voor Nederlandsch-Indië. Batavia (Jakarta), Indonesia, 1850–1940.

New York State Mus. Bull.: New York state museum bulletin. Albany, New York, 1887– .

New Zealand Div. Sci. Indust. Res. Bull.: New Zealand division of scientific and industrial research bulletins. Wellington, New Zealand, 1927– . [Including also Cape Expedition Series bulletins.]

New Zealand J. Bot.: New Zealand journal of botany. Wellington, 1963– .

New Zealand J. Sci. (Wellington): New Zealand journal of science. Wellington, 1958– .

Northw. Sci.: Northwest science. Cheney, Washington State, 1927– .

Notes Roy. Bot. Gard. Edinburgh: Notes from the royal botanic garden, Edinburgh. Edinburgh, 1900– .

Notizbl. (Königl.) Bot. Gart. Berlin (-Dahlem): Notizblatt des (Königlichen) botanischen Gartens und Museums zu Berlin (-Dahlem). Berlin, 1895–1955.

Notul. Syst. (Paris): Notulae systematicae. Paris, 1909–60.

Nouv. Arch. Mus. Hist. Nat.: Nouvelles archives du muséum d'histoire naturelle. Paris, 1865–1914.

Nov. Sist. Vyss. Rast.: Novosti sistematiki vyssikh rastenij. Leningrad, 1964– .

Nova Acta Regiae Soc. Sci. Upsal.: Nova acta regiae societatis scientiarum upsaliensis. Uppsala, 1773– (ser. IV, 1905–).

Nova Guinea: Nova Guinea. Leiden, 1909–36.

Nova Guinea, Bot.: Nova Guinea, botany. Leiden, 1960–66.

OPTIMA Newsletter: Organization for the Phyto-taxonomic Investigation of the Mediterranean Area: newsletter. Geneva, etc., 1975– .

Occas. Pap. Bernice P. Bishop Mus.: Occasional papers of the Bernice P. Bishop museum of Polynesian ethnology and natural history. Honolulu, Hawaii, 1898– .

Occas. Pap. Univ. Hawaii: Occasional papers of the university of Hawaii. Honolulu, Hawaii, 1923– .

Ohio J. Sci.: Ohio journal of science. Columbus, O., 1901– .

Ohio State Univ. Bull.: Ohio state university bulletin. Columbus, 1896–1906, 1909– .

Oklahoma Agric. Exp. Sta. Bull.: Oklahoma agricultural experiment station bulletin. Stillwater, Oklahoma, 1892– .

Oklahoma Agric. Mech. Coll. Bull.: Oklahoma agricultural and mechanical college, bulletin [biological (science) series]. Stillwater, Oklahoma, 1950– .

Opera Bot.: Opera botanica a societate botanica lundensi. Lund, Sweden, 1953– .

Opera Bot., B: Opera botanica, series B [Flora of Ecuador]. Lund, 1973– .

Opera Lilloana: Opera lilloana. Tucumán, Argentina, 1957– .

Pacific Disc.: Pacific discovery. San Francisco, 1948– .

Pacific Insects, Monogr.: Pacific insects, monographs. Honolulu, Hawaii, 1961– .

Pacific Sci.: Pacific science. Honolulu, Hawaii, 1947– .

Pacific Sci. (Assoc.) Inform. Bull.: Pacific science association information bulletin. Honolulu, Hawaii, 1947– .

Pakistan J. Forest.: Pakistan journal of forestry. Upper Topa, Pakistan, 1951– .

Pakistan Syst.: Pakistan systematics. Rawalpindi, 1977– .

Philipp. J. Sci.: Philippine journal of science. Manila, 1906– .

Phil. Trans.: Philosophical transactions of the royal society of London. London, 1665–1886. [Succeeded by Philosophical transactions of the royal society of London. Series B.]

Phil. Trans., Ser. B.: Philosophical transactions of the royal society of London. Series B. London, 1887– .

Phytologia: Phytologia. New York, 1933– .

Phytologia, Mem.: Phytologia; memoirs. Plainfield, NJ, 1980– .

Phyton (Horn): Phyton. Horn, Austria, 1948– .

Pl. Sci. Bull.: Plant science bulletin. Urbana, Ill., etc., 1955– .

Polar Rec.: Polar record. Cambridge, England, 1931– .

Popular Sci. Monthly: Popular science monthly. New York, 1872– .

Posebna Izd. Biol. Inst. & Zemaljsk. Muz. u Sarajevu: Posebna izdanja biologii institut i zemaljskog museja (Bosni Hercegovini) u Sarajevu. Sarajevo, 1950– .

Posebna Izd. Srpska (Kral.) Akad.: Posebna izdanja Srpska (Kral'evsk) Akademia. Belgrade, 1890– .

Posebni Izd. Filoz. Fak. Univ. Skoplje: Posebni izdanija, Filozofski fakultet na Universitet Skoplje. Skoplje, Macedonia, Yugoslavia, 1950– .

Prace Monogr. Komis. Fizjogr.: Prace monograficzne komisji fizjograficznej. Cracow, Poland, 1925–31.

Preslia: Preslia. Prague, 1914– .

Proc. Amer. Philos. Soc.: Proceedings of the American philosophical society. Philadelphia, Pa., 1838– .

Proc. Boston Soc. Nat. Hist.: Proceedings of the Boston society of natural history. Boston, Mass., 1841–1942.

Proc. Calif. Acad. Sci.: Proceedings of the California academy of sciences. San Francisco, 1854– .

Proc. Hawaiian Acad. Sci.: Proceedings of the Hawaiian academy of science. Honolulu, Hawaii, 1926– .

Proc. Iowa Acad. Sci.: Proceedings of the Iowa academy of science. Des Moines, Ia., 1887– .

Proc. Linn. Soc. New South Wales: Proceedings of the Linnean society of New South Wales. Sydney, 1875– .

Proc. Natl. Acad. Sci. India, sect. B., Biol. Sci.: Proceedings of the national academy of sciences of India. Section B. Biological sciences. Allahabad, India, 1936– .

Proc. Nova Scotian Inst. Sci.: Proceedings of the Nova Scotian institute of science. Halifax, Nova Scotia, 1863– .

Proc. Roy. Irish Acad.: Proceedings of the royal Irish academy. Dublin, 1841– .

Proc. Roy. Soc. Arts Sci. Mauritius: Proceedings of the royal society of arts and science of Mauritius. [New series.] Port Louis, 1950– .

Proc. Roy. Soc. Queensland: Proceedings of the Royal society of Queensland. Brisbane, 1883– .

Proc. & Trans. Roy. Soc. Canada: Proceedings and transactions of the Royal society of Canada. Montreal, 1882– (three successive series).

Prog. Bot.: Progress in botany (*also* Fortschritte der Botanik). Berlin, 1932– .

Provancheria: Provancheria. Quebec, 1966– .

Publ. Arnold Arbor.: Publications of the Arnold Arboretum. Cambridge, Mass., 1891–1921 (–1933).

Publ. Biol. (Jeddah): Publications in biology, King Abdulaziz University. Jeddah, 1978– .

Publ. Bur. Sci. Philipp.: Publications (later Monographs) of the Bureau of Science, Philippines. Manila, 1908–34. [Succeeded by Monographs of the Philippine institute of science and technology.]

Publ. CTFT: Publications du Centre technique de forestier tropical. Nogent-sur-Marne, 1950– .

Publ. Cairo Univ. Herb.: Publications of the Cairo University herbarium. Cairo, 1968– .

Publ. Canada Dept. Agric., Res. Br.: Publications of the Canada department of agriculture (later Agriculture Canada), research branch. Ottawa, 1951– (as unified series; no. 846 onwards).

Publ. Carnegie Inst. Wash.: Publications of the Carnegie institution of Washington. Washington, 1902– .

Publ. Comitié Etud. Hist. Sci. AOF: Publications du Comitié d'Etudes historiques et scientifiques de l'Afrique occidentale française. Dakar/Gorée, Sénégal, 1934–9 (series A, 1934–9; series B, 1935–9).

Publ. Field Columbian Mus., Bot. Ser.: Publications of the Field Columbian museum. Botanical series. Chicago, 1895–1932.

Publ. Field Mus. Nat. Hist., Bot. Ser.: Publications of the Field museum of natural history. Botanical series. Chicago, 1930–44.

Publ. Inst. Nac. Hielo Continental Patagonica: Publicaciones, Instituto nacional de hielo continental patagonica. Buenos Aires. Vol. 8, 1965.

Publ. Inst. Natl. Rech. Sci. Rwanda: Publications de l'institut national de recherches scientifiques du Rwanda. Butare, Rwanda, < 1971– .

Publ. Natl. Sci. Res. Council, Iran: Publications of the national science research council, Iran. Teheran. Vol. 21, 1978.

Publ. Soc. Océan.: Publications de la société des océanistes. Paris, 1951– .

Publ. Syst. Assoc.: Publications of the systematics association. London, 1940, 1953– .

Publ. Univ. Pretoria: Publications of the university of Pretoria. Pretoria, Transvaal, 1921– (Natural science, 1936– ; new series, 1956–).

Quart. J. Taiwan Mus.: Quarterly journal of the Taiwan museum. Taipei, 1948– .

Quart. Rev. Biol.: Quarterly review of biology. Baltimore, Maryland, 1926.

Queensland Naturalist: Queensland naturalist. Brisbane, 1908– .

Rapp. Bosbouwproefsta.: Rapporten van het bosbouwproefstation. *See* Laporan-laporanan Lembaga penelitian hutan.

Razpr. Slovensk. Akad. Znan.: Razprave Slovenskie akademija znanosti in umjetnosti v Ljubljani.

Ljubljana, 1950– . (Comprises a number of concurrent series, including ser. IV, 1951– .)

Rec. Auckland Inst. Mus.: Records of the Auckland institute and museum. Auckland, 1930– .

Rec. Bot. Surv. India: Records of the botanical survey of India. Calcutta, 1893– .

Rec. Domin. Mus.: Records of the Dominion museum. Wellington, New Zealand, 1942– . (Includes also a Miscellaneous Series.)

Recueil Trav. Bot. Néerl.: Recueil des travaux botaniques néerlandais. Nijmegen, Netherlands, 1904–50.

Regnum Veg.: Regnum vegetable. Utrecht, 1953– .

Rendiconti Seminarii Fac. Sci. Reale Univ. Cagliari: Rendiconti seminarii (Rendiconti del seminario) della facoltà di scienze di reale università di Cagliari, Cagliari, Sardinia, Italy, 1931– .

Rep. Austral. Natl. Antarctic Res. Exped.: Reports of the Australian national antarctic research expeditions. Melbourne, 1948– .

Rep. Inst. Sci. Res. Manchukuo: Report of the institute of scientific research, Manchukuo. Hsinking, Manchuria, (= Changchun, China), 1936–40.

Rep. Meetings Australas. Assoc. Advancem. Sci.: Reports of meetings of the Australasian association for the advancement of science. Sydney, 1888–1954.

Rep. State Biol. Surv. Kansas: Reports of the state biological survey of Kansas. [Lawrence], Kansas, < 1976– .

Repert. Spec. Nov. Regni Veg. [Fedde]: Repertorium specierum novarum regni vegetabilis, ed. F. Fedde. Berlin, 1905–42.

Repert. Spec. Nov. Regni Veg. [Fedde], Beih.: Repertorium specierum novarum regni vegetabilis, ed. F. Fedde. Beihefte. Berlin, 1911–44.

Rev. Int. Bot. Appl. Agric. Trop.: Revue international de botanique appliquée et d'agriculture tropicale. Paris, 1921–53.

Revista Acad. Col. Ci. Exact.: Revista de la academia colombiana de ciencias exactas, fisicas y naturales correspondiente de la española. Bogotá, 1936– .

Revista Agron. (Rio Grande do Sul): Revista agronómica. Porto Alegre, Brazil, 1937– .

Revista Fac. Agron. (Maracay): Revista de la facultad de agronomía. [Universidad central de Venezuela.] Maracay, Venezuela, 1952– .

Revista Fac. Ci. Univ. Lisboa, sér. 2, C, Ci. Nat.: Revista da faculdads de ciências, universidade de Lisboa. Séria 2, C, ciências naturais. Lisbon, 1950– .

Revista Fac. Nac. Agron., Medellín: Revista. Facultad nacional de agronomía, universidad de Antioquia, Medellín. Medellín, 1939– .

Revista Mus. La Plata, Secc. Bot.: Revista del museo de La Plata. Sección botánica. La Plata, Argentina, 1936– .

Revista Mus. Nac. (Lima): Revista del museo nacional. Lima, Peru, 1932– .

Revista Mus. Prov. Ci. Nat. Córdoba: Revista del museo provincial de ciencias naturales de Córdoba. Córdoba, Argentina. Vol. 3, 1938.

Revista Soc. Colomb. Ci. Nat.: Revista (*originally* Boletín) de la sociedad colombiana de ciencias naturales. Bogotá, 1913–31.

Revista Sudamer. Bot.: Revista sudamericana de botánica. Montevideo, 1934–56.

Revista Univ. Nac. Córdoba: Revista de la Universidad nacional de Córdoba. Córdoba, Argentina, 1914– .

Revista Univ. (Santiago): Revista universitaria. Santiago, Chile, 1915– .

Rhodesian Agric. J.: Rhodesian agricultural journal. Salisbury, Zimbabwe, 1902– .

Rhodora: Rhodora. Lancaster, Pennsylvania, etc., 1899– .

Rodriguésia: Rodriguésia. Rio de Janeiro, 1935– .

S. African Biol. Soc., Pamphl.: South African biological society, pamphlets. Pretoria, 1931–59.

S. African J. Sci.: South African journal of science. Cape Town, 1903– (*originally* Reports of the annual meetings of the South African association for the advancement of science).

Sabah Forest Rec.: Sabah forest records. Sandakan, Sabah (later also Kuching, Sarawak), 1938– (originally as North Borneo forest records.)

Sarawak Mus. J.: Sarawak museum journal. Kuching, Sarawak, 1911– .

Sargentia: Sargentia. Jamaica Plain, Mass., 1942–9.

Sarracenia: Sarracenia. Montreal, 1959– .

Saudi Biol. Soc. Publ.: Saudi biological society: publications. N.p., < 1979– .

Sci. Bull. Brigham Young Univ., Biol. Ser.: Science bulletin of the Brigham Young University, biological series. Provo, Utah, 1955– .

Sci. J. (Sydney): Science journal. [University of Sydney science association.] Sydney, 1917–38.

Sci. Rep. Kanazawa Univ., Biol.: Science reports of the Kanazawa university; biology. Kanazawa, Japan, 1951– .

Scient. Rep. Brit. Antarctic Surv.: Scientific reports of the British antarctic survey. London, etc., 1953– .

Search: Search. Sydney, 1971– .

Selbyana: Selbyana. Sarasota, Florida, 1975– .

Sellowia: Sellowia. Itajaí, Brazil, 1949– (*originally* Anais botânicos).

Sida: Sida; contributions to botany. Dallas, Texas, 1962– .

Skr. Norsk Polarinst.: Skrifter Norsk polarinstitutt. Oslo, 1948– .

Skr. Svalbard Ishavet: Skrifter om Svalbard og Ishavet. Oslo, 1927–40. (See also Skrifter Norsk polarinstitutt.)

Slovenská Vlastiv.: Slovenská vlastiveda. [Slovenské akadémia vied a umení.] Bratislava, Czechoslovakia, 1943–8.

Smithsonian Contr. Bot.: Smithsonian contributions to botany. Washington, 1969– .

South Australian Woods & Forests Dept. Bull.: South Australian woods and forests department, bulletin. Adelaide, 1928– .

Southwestern Louisiana J.: Southwestern Louisiana journal. Lafayette, Louisiana, 1951– .

Special Number Educ. Misc. Hunan: Special number of the educational miscellany of Hunan. [Changsha], Hunan. Vol. 4, 1924.

Special Publ. Brit. Columbia Forest Serv.: Special publications of the British Columbia forest service [B-series]. Victoria, BC., < 1949– .

Special Publ. Brit. Columbia Prov. Mus. Nat. Hist.: Special publications of the British Columbia provincial museum of natural history and anthropology. Vancouver, 1947– .

State Biol. Surv. Kansas, Techn. Publ.: State biological survey of Kansas: technical publications. [Lawrence], Kansas, < 1976– .

Stud. Nat. Hist. Iowa Univ.: Studies in natural history, Iowa university. Iowa City, Iowa, 1888– (*originally* as Bulletin of Iowa university's laboratories of natural history).

Sudan Forests Bull.: Sudan forests bulletins. Khartoum, 1958– .

Sudan Notes & Rec.: Sudan notes and records. Khartoum, 1918– .

Sunyatsenia: Sunyatsenia. Canton, China, 1930–48.

Svensk Bot. Tidskr.: Svensk botanisk tidskrift. Stockholm, 1907–80.

Symb. Bot. Upsal.: Symbolae botanicae upsalienses. Uppsala, Sweden, 1932– .

Syst. Bot.: Systematic botany. Tallahassee, Florida, 1976– .

TC Ankara Üniv. Fen Fakult., Yayinlari: Türkiye cumhuriyet Ankara üniversitesi, fen fakültesi. Yayinlari. Ankara, 1948– . (Botanik, 1952– .)

Taxodium: Taxodium. 1943.

Taxon: Taxon. Utrecht, 1951– .

Techn. Bull. Lafayette Nat. Hist. Mus.: Technical bulletin of the Lafayette natural history museum. Lafayette, Louisiana, 1969– .

Techn. Bull. Univ. Brit. Columbia Bot. Gard.: Technical bulletin of the University of British Columbia botanical garden. Vancouver, 1972– .

Techn. Rep. Meterol. Climatol. Arid Regions, Inst. Atmos. Phys., Univ. Arizona: Technical reports, meteorology and climatology of arid regions, institute of atmospheric physics of the University of Arizona. Tucson, 1956– .

Texas Agric. Expt. Sta. Publ.: Texas agricultural experiment station. Publications. College Station, Texas.

Tijdschr. Kon. Ned. Aardrijksk. Genootsch.: Tijdschrift van het koninklijk nederlandsch aardrijkskundig genootschap. Amsterdam, 1874– .

Trab. Soc. Ci. Nat. 'La Salle': Trabajos de la sociedad de ciencias naturales 'de La Salle'. Caracas. Vol. 20, 1960.

Trans. Acad. Sci. St Louis: Transactions of the academy of science of St Louis. St Louis, 1860– .

Trans. Kentucky Acad. Sci.: Transactions of the Kentucky academy of science. Lexington, Kentucky, 1914– .

Trans. Linn. Soc. London, II, Bot.: Transactions of the Linnean society of London. Ser. 2, Botany. London, 1875–1922.

Trans. Linn. Soc. London. II, Zool.: Transactions of the Linnean society of London. Ser. 2, Zoology. London, 1875–1936.

Trans. Nat. Hist. Soc. Taiwan: Transactions of the natural history society of Taiwan. Taihoku (Taipei), Taiwan, 1914– .

Trans. Oceanogr. Inst. Moscow: Transactions of the oceanographic institute of Moscow. See Trudy Gosudarstvennogo okeanografičeskogo instituta.

Trans. & Proc. Bot. Soc. Edinburgh: Transactions and proceedings of the botanical society of Edinburgh. Edinburgh, 1876–84; 1893– .

Trans. & Proc. New Zealand Inst.: Transactions and proceedings of the New Zealand institute. Wellington, New Zealand, 1868–1933.

Trans. & Proc. Roy. Soc. New Zealand: Transactions and proceedings of the royal society of New Zealand. Wellington, New Zealand, 1934–52.

Trans. & Proc. Roy. Soc. South Australia: Transactions and proceedings of the royal society of South Australia. Adelaide, 1877– .

Trans. Roy. Soc. Edinburgh: Transactions of the royal society of Edinburgh. Edinburgh, 1788– .

Trans. Roy. Soc. New Zealand: Transactions of the royal society of New Zealand. Wellington, New Zealand, 1953–60. [From 1970 divided into series; see text entry.]

Trans. Roy. Soc. New Zealand, Bot.: Transactions of the royal society of New Zealand. Botany. Wellington, New Zealand, 1960–71.

Trans. Roy. Soc. South Africa: Transactions of the royal society of South Africa. Cape Town, 1908– .

Trans. Wisconsin Acad. Sci.: Transactions of the Wisconsin academy of sciences, arts and letters. Madison, Wisconsin, 1870– .

Trav. Inst. Sci. Chérifien, Sér. Bot.: Travaux de l'institut scientifique chérifien. Série botanique. Tangier, etc., 1952– .

Trav. Lab. Forest. Toulouse: Travaux du laboratoire forestier de Toulouse. Toulouse, 1928– . (Comprises several concurrent series.)

Tree species lists, [Buitenzorg]: Tree species lists of the Forest Research Institute, Buitenzorg. Buitenzorg (Bogor), Indonesia, 1940–1.

Trop. Woods: Tropical woods. New Haven, Connecticut, 1925–60.

Trudy Bot. Inst. Akad. Nauk Tadžiksk. SSR: Trudy. Botaniceskij institut. Akademija nauk Tadžikakej SSR. Dushanbe, Moscow/Leningrad, 1962– .

Trudy Bot. Inst. AN SSSR: Trudy Botaničeskego institut Akademii nauk SSSR. Moscow/Leningrad, 1933– . (Comprises several concurrent series.)

Trudy Bot. Muz. Imp. Akad. Nauk: Trudy Botaničeskago muzeja Imperatorskej Akademii Nauk. St Petersburg (Leningrad), 1902–32. (After 1916 'Imperatorskej' was replaced by 'Rossijskej'.)

Trudy Dal'nevostočnogo Fil. 'Komarova', ser. Bot.: Trudy Dal'nevostočnogo filial imeni V. L. Komarova. Serija botaničeskaja. Moscow/Leningrad, 1956– .

Trudy Glavn. Bot. Sada SSSR: Trudy Glavnogo botaničeskogo sada SSSR (*earlier* Trudy Imperatorskago S-Peterburgskago botaničeskago sada, etc.; *also titled* Acta horti petropolitani). *See under earlier title.*

Trudy Gosud. Nikitsk. Bot. Sada: Trudy Gosudarstvennogo Nikitskogo botaničeskogo sada. Yalta, Ukranian SSR, 1934– . (Slight change of title after 1953.)

Trudy Gosud. Okeanogr. Inst. (Moscow): Trudy Gosudarstvennogo okeanografičeskogo instituta. Moscow, 1931–34, 1947– .

Trudy Imp. S.-Peterburgsk. Bot. Sada: Trudy Imperatorskago S.-Peterburgskago botaničeskago Sada. Acta horti petropolitani. St Petersburg, 1871–1930. (From 1915 title changed to Trudy Glavnago Botaničeskogo Sada.)

Trudy Inst. Biol. Akad. Nauk SSSR Ural'sk Fil.: Trudy Instituta biologii, Akademija nauk SSSR, Ural'skij filial. Moscow/Leningrad, 1948.

Trudy Kazakhstansk. Fil. Akad. Nauk SSSR: Trudy Kazakhstanskij filial, Akademiha nauk SSSR. Moscow/Leningrad, 1937–41.

Trudy Mongol'sk. Kemiss.: Trudy Mongol'skoj komissii. Moscow/Leningrad, 1928– .

Trudy Poljarn. Komiss.: Trudy Poljarnoj komissii. Moscow/Leningrad, 1930–37.

Trudy Rostovsk. Obl. Biol. Obšč.: Trudy Rostovskogo oblastnogo biologičeskogo obščestva. Rostov, Russian SFSR, 1937–1940.

Trudy Severn. Bazy AN SSSR: Trudy Severnoj bazy. Moscow/Leningrad. Vol. 2, 1937.

Trudy Tiflissk. Bot. Sada: Trudy Tiflisskago (*later* Trudy Tbilsskogo) botaničeskago sada. Tbilisi (Tiflis), 1895–1917, 1920–34, 1938–49.

Tuatara: Tuatara. Wellington, New Zealand, 1947– .

TÜRDOK, Bibliogr. Ser.: TÜRDOK, Bibliografiya seriya. Ankara. Vol. 5, 1972.

US Forest Surv. Res. Pap.: United States forest service: research papers. Washington, DC, etc., 1964– . (Includes several concurrent series issued by regional centers, each with separate numeration.)

Uitgaven Natuurw. Studiekring Suriname Ned. Antillen: Uitgaven van de natuurwetenschapplijke studiekring voor Suriname en de Nederlandse Antillen. The Hague (later Utrecht), 1946– . (*Originally as* Uitgaven...voor Suriname en Curaçao.)

Uitgaven Natuurw. Werkgroep Ned. Antillen: Uitgaven van de 'Natuurwetenschappelijke werkgroep Nederlandse Antillen'. Willemstad, Curaçao, 1951–71.

Univ. Calif. Publ. Bot.: University of California publications in botany. Berkeley, etc., California, 1902– .

Univ. Florida Agric. Exp. Sta. Bull.: University of Florida agricultural experiment station bulletin. Gainesville, Florida, 1910– .

Vakbl. Biol.: Vakblad voor biologen. Amsterdam, 1919– .

Verh. Kon. Ned. Akad. Wetensch., II. reeks, Afd. Natuurk.: Verhandelingen der Koninklijke Nederlandse akademie van wetenschappen, tweede reeks (*until 1951* Tweede sectie), afdeling Natuurkunde. Amsterdam, 1892– .

Vermont Agric. Exp. Sta. Bull.: Vermont state agricultural college, agricultural experiment station. Bulletin. Burlington, Vermont, 1887– .

Victorian Naturalist: Victorian naturalist. Melbourne, 1884– .

Vierteljahrsschr. Naturf. Ges. Zürich: Vierteljahrsschrift der Naturforschenden Gesellschaft in Zürich. Zürich, 1856– .

Vierteljahrsschr. Naturf. Ges. Zürich, Beibl.: Vierteljahrsschrift der Naturforschenden Gesellschaft in Zürich. Beiblätter. Zürich, 1923– .

Virginia Agric. Exp. Sta. Tech. Bull.: Virginia agricultural experiment station. Technical bulletin. Blacksburg, Virginia, 1915– .

W. Virginia Univ. Bull.: West Virginia University bulletins. Morganstown, West Virginia, 1888– .

Webbia: Webbia. Florence, 1905– .

Willdenowia: Willdenowia. Berlin, 1953– .

Wiss. Mitt. Bosnien-Herzegowina: Wissenschaftliche Mittheilungen aus Bosnien und der Herzegowina. Vienna, 1893–1916.

Wyoming Agric. Exp. Sta. Bull.: Wyoming agricultural experiment station. Bulletin. Laramie, Wyoming, 1891– .

Wyoming Agric. Exp. Sta. Res. J.: Wyoming agricultural
    experiment station. Research journal. Laramie,
    Wyoming, 1966– .

Zambia Forest Res. Bull.: Zambia forest research bulletin.
    Ndola, 1960– (*originally* Research bulletin, Forest
    department, Northern Rhodesia).

Zap. Imp. Akad. Nauk, Fiz.-Mat. Otd.: Zapiski
    Imperatorskoj akademii nauk po
    Fiziko-matematičeskomu otdeleniju. Ser. 8. St
    Petersburg, 1894–1916.

Zap. Južno-Ussurijsk. Otd. Gosud. Russk. Geogr. Obšč.:
    Zapiski Južno-Ussurijskogo otdela
    Gosudarstvennogo russkogo geografičeskogo
    obščestva. Vladivostok, Russian SFSR, 1927–9.

Žurn. Obščej Biol. (SSSR): Žurnal obščej biologii (Journal
    of general biology, USSR). Moscow, 1940– .

Geographical index

The reference numbers in the following index stand for the geographical unit headings (**000–999**) as well as the designations used for major divisions (**D0–D9**), superregions (numbers prefixed by S) and regions (**R01–R99**). These all appear in the text in bold sans-serif along with the name of the polity or other unit in question, and should enable a desired heading to be found more quickly than otherwise.

Each number basically consists of three digits: the first, or third-order digit, represents the appropriate major geographical division; the second, or second-order digit, represents the geographical region or, if it be **0**, denotes that the category is of very broad scope or is physiographically or synusially founded; and the third, or first-order digit, represents the basic geographical unit or, if it be **0**, that the said category is of regional scope (and is used for regional floras, etc.). Some numbers are connected by a dash; these generally represent superregions or superregional floras and are inclusive from the first to the last given region. A few regions are linked by a slash (/); these are 'double' regions wherein a significant number of units has to be recognized without there being a basis for the designation of normal regions. A fuller account of the geographical system appears in Part I, Chapter 2.

An attempt has been made to account for a number of synonymous geographical names, with the guidance in particular of the geographical scheme given in the entomological informatic system proposal of Travis *et al.* (1962)[1]. However, 'colonial' and other geographical names superseded 10–15 or more years ago have, for the most part, been omitted. Use of a good world or historical atlas is recommended in cases of omission or doubt.

Reference

1 Travis, B. V., Caswell, H. H., Jr., Rowan, W. B., Starcke, H. and Ross, C. H., 1962. *Classification and coding system for compilations from the world literature on insects and other arthropods that affect the health and comfort of man.* 59 pp., illus., map (United States Army, Quartermaster Research and Engineering Center, Technical Report ES-4). Natick, Mass.

Author index

The following index accounts for all authors cited in Part II of this book, save for those only casually mentioned in heading and other commentaries as well as those responsible for such reference items as general bibliographies, general indices and major library catalogues. The latter items are fully covered in Appendix A, wherein each class is alphabetically arranged (the bibliographies and catalogues by author(s), the indices (including abstracting works) by title).

No attempt has been made to index Part I, as it is basically an extended introduction to the *Guide* proper, which constitutes Part II.

The year following the author's name is the first (or only) date of publication. Numerals in **bold** type refer to geographical headings (see geographical index for full explanation). Page numbers, in *italic*, are given with either an 'a' (left half of page) or a 'b' (right half of page).

A few entries are given with a '—' in place of a year of (first) publication. This indicates that the work in question had not, so far as was known by January 1982, been published and/or become generally available.

The format of the index is modeled on that used by Blake and Atwood for the first volume of their *Geographical Guide* (1942) as well as by Blake alone for the second volume of that work (1961). Vital statistics of authors, however, have here been omitted for practical reasons and because there is now a greater range of taxonomic reference works containing this kind of information than was the case in 1939.

For authors whose works were (first) published before about 1941, users are in particular advised to consult the greatly enlarged second edition of *Taxonomic Literature*, by F. A. Stafleu and R. S. Cowan (1976– ,Utrecht); as of 1984 this mammoth work has reached 'Sak'. The great majority of floras and their kin published or begun before World War II are there treated in bibliographical detail and with recensional and other pertinent information (but without a précis and commentary on the contents as is the case in the present work). It should be noted, however, that *Taxonomic Literature* (or *TL-2* for short) treats only works which can for at least some copies be demonstrated as having been independently published and/or distributed.

More 'recent' authors (who, as before 1941, may have also been, or acted only as, editors) are not so singularly treated, and a considerable variety of sources, too numerous to detail here, must therefore be consulted for further information. The introductory parts of each volume of *Taxonomic Literature*, though, collectively account for much the greatest part of this range, thus providing a starting point for further studies on any given author and his works.

> And in such indexes, although small pricks
> To their subsequent volumes, there is seen
> The baby figure of the giant mass
> Of things to come at large.

Shakespeare, *Troilus and Cressida*, Act 1, Scene 3.

Wyk, P. van
 1972, Kruger National Park (South Africa)   **517**
  265a

Yamamoto, Y.
 1925, Taiwan (and the Pescadores)   **886**   *459b*
Yanagihara, M.
 1941, Daito Islands   **854**   *436b*
Yepez T., G.
 1960, Territorio Colón (Las Aves)   **283**   *177b*
Young, S. B.
 1971, Bering Sea Islands (St Lawrence)   **071**   *85a*
Yü, Te-tsun
 1979, China   **860–80**   *446a*
Yuncker, T. G.
 1938, Honduras   **234**   *161a*
 1940, Honduras   **234**   *161b*
 1943, Niue   **957**   *513b*
 1945, Samoa (Manua Islands)   **958**   *514b*
 1946, Samoa (Manua Islands)   **958**   *514b*
 1959, Tonga   **956**   *513a*
Yunnan, Institute of Botany; *see* China, Kunming
   Institute of Botany, Academia Sinica
Yurtsev, B. A.
 1968, eastern Siberia (Arctalpine zone)   **067**   *84a*
 1978, North Polar Regions   **S05–07**   *79a*

Žamsran, C.
 1972, Mongolia (Ulan Bator district)   **761**   *387b*
Zangheri, P.
 1976, Italian peninsula and surrounding areas
   620   *315a*
Zardini, E. M.
 1978, Buenos Aires (metropolitan region)   **386**   *213b*
Zerov, D. K. *et al.*
 1965, Ukrainskaja SSR (in general)   **694**   *355b*
Zeybek, N.
 1972, Turkey   **771**   *392b*
Zimmermann, E. C.
 1948, Hawaiian Islands   **990**   *530b*
Zohary, M.
 1935, Sinai Peninsula (Egypt)   **776**   *395a*
 1949, Israel   **775**   *395a*
 1950, Iraq   **778**   *395a*
 1966, southwestern Asia   **S77–79**   *390b*
 1966, Palestine (*sensu lato*)   **775**   *394b*
 1976, Palestine (*sensu lato*)   **775**   *394b*
 1980, southwestern Asia   **770–90**   *391a*
Zollinger, H.
 1857, Malesia (Dutch version)   **S91–93**   *477a*
 1857, Malesia (German version)   **S91–93**   *477a*
Zotov, V. D.
 1965, islands south of New Zealand   **415**   *223b*
Zuccoli, T.
 1973, North Polar Regions   **050–70**   *79b*
Zuloaga, G.
 1955, Aves Island   **279**   *176a*
Zzyzz, C. L., Ultimo
 DOOM, Luna   **999**   *533a*